FLUID
MECHANICS

FLUID MECHANICS

FOUNDATIONS & APPLICATIONS

Problem Solving Using Mathematica®

SERGEY P. KISELEV
EVGENII V. VOROZHTSOV
VASILY M. FOMIN

JAICO PUBLISHING HOUSE

Ahmedabad Bangalore Bhopal Chennai
Delhi Hyderabad Kolkata Lucknow Mumbai

Published by Jaico Publishing House
A-2 Jash Chambers, 7-A Sir Phirozshah Mehta Road
Fort, Mumbai - 400 001
jaicopub@jaicobooks.com
www.jaicobooks.com

Published in arrangement with
Birkhäuser Boston
c/o Springer-Verlag New York, Inc.
175 Fifth Avenue
New York, NY 10010, USA

FLUID MECHANICS
ISBN 978-81-7992-918-6

First Jaico Impression: 2008

Printed by
Pashupati Printers (P) Ltd., Delhi-95

Contents

Preface

Fluid mechanics (FM) is a branch of science dealing with the investigation of flows of continua under the action of external forces. The fundamentals of FM were laid in the works of the famous scientists, such as L. Euler, M.V. Lomonosov, D. Bernoulli, J.L. Lagrange, A. Cauchy, L. Navier, S.D. Poisson, and other classics of science. Fluid mechanics underwent a rapid development during the past two centuries, and it now includes, along with the above branches, aerodynamics, hydrodynamics, rarefied gas dynamics, mechanics of multiphase and reactive media, etc.

The FM application domains were expanded, and new investigation methods were developed. Certain concepts introduced by the classics of science, however, are still of primary importance and will apparently be of importance in the future. The Lagrangian and Eulerian descriptions of a continuum, tensors of strains and stresses, conservation laws for mass, momentum, moment of momentum, and energy are the examples of such concepts and results. This list should be augmented by the first and second laws of thermodynamics, which determine the character and direction of processes at a given point of a continuum. The availability of the conservation laws is conditioned by the homogeneity and isotropicity properties of the Euclidean space, and the form of these laws is related to the Newton's laws. The laws of thermodynamics have their foundation in the statistical physics. These concepts and laws are insufficient for the description of continuum motion, however, they should be augmented by the equations of state or, as one says sometimes, by the closure relationships, which relate the tensors of strains and stresses to their derivatives. Each continuum possesses its own equations of state, which are found by processing the experimental data or by solving the corresponding problems from the statistical physics.

The concept of continuum does not contain in itself the information about the discrete character of substance at the molecular level, all this information is laid in thermodynamics laws and closure relationships (equations of state). It follows from here that the FM is no closed field of knowledge; it has a tight interaction with other branches of physics and mathematics.

The teaching of FM has its remarkable traditions laid in the lecture courses of L.D. Landau and E.M. Lifschitz, L.I. Sedov, N.I. Kochin, L. Prandtl, G. Batchelor, W. Prager, P. Germain, and other authors. New interesting results have appeared in FM after the publication of these lecture courses, however, and there is now a need in presenting them in the educational literature, so that students can be rapidly introduced into the scope of present-day FM problems and methods.

The development of computer algebra and of a powerful universal software system *Mathematica* has led to the fact that the task of the FM presentation with the use of the *Mathematica* system has become topical. The present authors have undertaken an attempt in this book at solving this task. A large number of programs are presented that have enabled us to perform in the process of material presentation both analytical and numeric computations with the aid of a personal computer. All *Mathematica 3.0* programs are stored on the Birkhäuser server. The url address of which is as follows (see Appendix B for further details of our programs):

`http://www.birkhauser.com/book/isbn/0-8176-3995-0`

In addition, we aimed at taking into account the international character of science. Therefore, we considered it necessary to familiarize Western readers in more detail with the achievements of the Russian scientists in the field of fluid mechanics.

The present book has been written on the basis of the lecture courses, which were presented by the authors during the past few years at the Novosibirsk State Technical University and at the Novosibirsk State University. The lectures were intended for graduate and postgraduate students that have already attended the introductory lecture courses in FM. At the same time, this book gives all of the necessary FM concepts and the derivations of all formulas are presented. There are also a large number of problems with the solutions. All of these features enable one to use this book both for an initial and a deeper study of FM.

Note that, although we present FM as a theoretical discipline, its development is closely related to the experiment and the practical needs of industry. The role of experiment in the formation of the basic concepts and closing relationships of FM is very significant. After these concepts have been established, however, the deductive methods of mathematics take on deciding significance.

The book consists of seven chapters. Chapter 1 presents the basic concepts of continua: the Lagrangian and Eulerian description, tensors of strains and stresses, equations of continuity, momentum, and energy. We use throughout the chapter a tensor invariant form, which does not depend on the choice of coordinate system. Therefore, the presentation of the basic FM concepts is preceded by a brief introduction in tensor analysis. The basic definitions of the tensor analysis, which are used in the following, are briefly introduced. From the very beginning, the strains are not assumed to be small; therefore, various tensors of strains and stresses are introduced that are related to the initial and current configurations at the Lagrangian description and in a fixed Eulerian frame.

In Chapter 2, we present the derivation of the differential equations for continuity, energy, and motion. The concept of local thermodynamic equilibrium as well as the the thermodynamics laws enable us to indicate the general form of the closure relationships for the FM governing equations. We formulate the Hamilton–Ostrogradsky variational principle, which enables us, on the one hand, to find the FM motion equations and, on the other hand, to establish a relation between the integral conservation laws for energy and momentum and the isotropicity and homogeneity properties of space and time.

In Chapter 3, the fundamentals of the similarity and dimension theory as well as the mathematical methods for studying the weak discontinuities (the characteristics) and strong discontinuities in fluid mechanics are presented.

Chapter 4 is devoted to the fundamentals of the theory of dynamics of ideal incompressible fluid. For the ideal fluid, we derive the Bernoulli and Thomson integrals and consider the planar and axisymmetric irrotational flows. We also study the general properties of the vortex flows. We further consider in detail the methods for the solution of problems on the ideal fluid flow around planar and axisymmetric bodies.

Chapter 5 deals with the viscous fluid flows. The Navier–Stokes equations are derived, and a number of the solutions of these equations are obtained at small Reynolds numbers. The basics of the Prandtl's boundary layer theory are presented. Some approaches to the description of turbulent fluid flows are considered.

Chapter 6 is devoted to the gas dynamics of ideal and viscous compressible gases. We present the theory of the Laval nozzle, shock waves, and Riemann waves. We also give the solution of a problem on shock wave structure in a viscous gas. The Chaplygin's theory for the transformation of gas dynamics equations to the hodograph plane is presented in sufficient detail.

Chapter 7 is devoted to a new FM branch, the mechanics of multiphase heterogeneous media. This branch of mechanics has appeared during the past 20–30 years, and it now enjoys a period of intense development. One can speak today about the fact that the general approaches have been developed, which are applicable to the description of an arbitrary multiphase medium. Various methods of averaging belong to them, which enable one to go over to an averaged description, as well as the idea of interpenetrating continua, each of which refers to a corresponding phase.

A mathematical model of the gas–particle flow was historically the first model describing multiphase flows. This was related to the practical applications in the field of two-phase gas dynamics of Laval nozzles and the flow of dusty gas around the flying vehicles. The developed methods were then used in the models of bubbly fluid flows, porous materials, and gas mixtures. For this reason, sufficient attention is paid here to the model of gas–particle flow. It should be noted that there are at present in the literature a number of monographs devoted to the mechanics of multiphase media. These monographs are cited in the list of References. Our presentation is nevertheless different from them, since a number of the results presented in our book belongs to the present authors. This refers to the theory of thin discontinuities in the gas–particle mixtures, which carry a finite surface mass; the continual/discrete model, which enables one to model the flows with the intersection of particles trajectories; the theory of caustics in the pseudogas of particles, which gives a limitation for the total number of particles lying on a caustic as well as the condition for its formation; the theory of shock wave interaction with a particle's cloud within the framework of which an explanation is given for the formation of a collective shock wave upstream of the particle's cloud at a volume concentration of particles of the order of several percents.

A triple numbering of formulas is used in the book. The first number indicates the chapter number, the second number is the section number, and the third number is the formula number in the section.

The authors hope that this book will be useful for students of universities and higher technical colleges as well as for specialists working in the field of FM.

The authors express their deep gratitude to professional colleagues whose discussions contributed to the elucidation of many complex questions of FM.

1

Definitions of Continuum Mechanics

The purpose of the present chapter is to provide a systematic introduction of the basic concepts and definitions of the tensors of strains and stress. The presentation begins with a section, in which we briefly present the elements of tensor analysis. Tensor analysis enables one to present in a simple and elegant form the fundamantals of continuum mechanics, and it is used systematically subsequently throughout the book. We then introduce the definition of the tensors of strains and stress, which characterize a continuum, in a reference frame and in an actual frame without any assumptions on the smallness of strains.

1.1 Vectors and Tensors

Tensor analysis plays an important role in formulating the basic notions of continuum mechanics, such as the tensors of strains and stresses, and in obtaining the partial differential equations for the functions sought for in different coordinate systems. Therefore, the presentation of the fundamentals of continuum mechanics is preceded by a brief insight into tensor analysis. The formulas presented here will be used throughout our book. Note that a thorough presentation of tensor analysis may be found in[1–4].

In order to identify the position of some point in space, it is necessary to define a coordinate system, which is specified with the aid of coordinate lines x^1, x^2, x^3 and basis vectors $\vec{e}_1, \vec{e}_2, \vec{e}_3$ (see Fig. 1.1). Along the coordinate line x^1, the quantities x^2 and x^3 are constant and the variation of x^1 corresponds to a shift along the coordinate line x^1. The same is true for the coordinate lines x^2 and x^3. Let us define a certain small vector $d\vec{r}$ with the aid of the equation

$$d\vec{r} = dx^i \vec{e}_i, \tag{1.1.1}$$

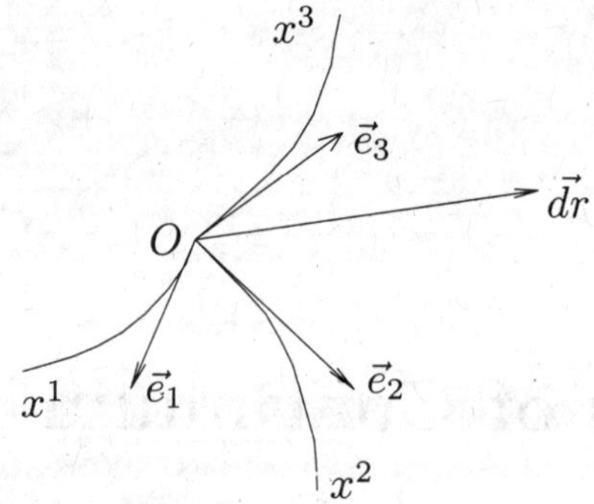

Figure 1.1: The curvilinear coordinate system.

where the summation is assumed over the repeating indices. This is a commonly accepted rule, the exceptions of which will be mentioned specifically. By representing the left-hand side of (1.1.1) in the form $d\vec{r} = \frac{\partial \vec{r}}{\partial x^i} dx^i$ and by comparing with the right-hand side, we find the expression for the basis vectors, which are tangent to the coordinate lines x^i

$$\vec{e}_i = \frac{\partial \vec{r}}{\partial x^i}. \tag{1.1.2}$$

The coordinate system should be chosen in such a way that the vectors $\vec{e}_i$ do not lie in the same plane; therefore, it is assumed that the condition $\vec{e}_i \cdot (\vec{e}_j \times \vec{e}_k) \neq 0$ is always valid, and one can introduce the mutually inverse basis

$$\vec{e}^{\,i} = \frac{\vec{e}_j \times \vec{e}_k}{\vec{e}_i \cdot (\vec{e}_j \times \vec{e}_k)}. \tag{1.1.3}$$

It follows from the vector product definition that

$$\vec{e}^{\,i} \cdot \vec{e}_j = \delta^i_j = \begin{cases} 0, & i \neq j \\ 1, & i = j, \end{cases} \tag{1.1.4}$$

where δ^i_j is the Kronecker symbol and the dot denotes the scalar product.

Let us consider an arbitrary coordinate transform $y^i = y^i(x^1, x^2, x^3)$ with the Jacobian $|\frac{\partial y^i}{\partial x^j}| \neq 0$. The quantity $d\vec{r}$ does not depend on the choice of the coordinate system; therefore, along with (1.1.1) the equation

$$d\vec{r} = dy^i \vec{e}^{\,\prime}_i \tag{1.1.5}$$

is valid, from which it follows that

$$\vec{e}^{\,\prime}_i = \frac{\partial \vec{r}}{\partial y^i} = \frac{\partial \vec{r}}{\partial x^j}\frac{\partial x^j}{\partial y^i} = \frac{\partial x^j}{\partial y^i}\vec{e}_j. \tag{1.1.6}$$

Differentiating $y^i = y^i(x^j)$, we obtain that dx^i and $\vec{e}_i$ are transformed reciprocally as

$$dy^i = \frac{\partial y^i}{\partial x^j}dx^j, \quad \vec{e}\,'_i = \frac{\partial x^j}{\partial y^i}\vec{e}_j. \tag{1.1.7}$$

By analogy with (1.1.7), let us define the covariant a_i and contravariant b^j quantities

$$a'_i(y^k) = \frac{\partial x^j}{\partial y^i}a_j(x^k(y^l)), \quad b'^i(y^k) = \frac{\partial y^i}{\partial x^j}b^j(x^k(y^l)), \tag{1.1.8}$$

where the primes denote the transformed components at a given coordinate transform $y^i = y^i(x^j)$.

The quantity $\vec{C} = C^i\vec{e}_i$, which is invariant (invariable) under the coordinate transform $y^i = y^i(x^j)$, is called the vector $\vec{C}$:

$$\vec{C} = C^i\vec{e}_i = C'^i\vec{e}\,'_i. \tag{1.1.9}$$

The quantity $\hat{T} = T^{ij\ldots s}\vec{e}_i\vec{e}_j\ldots\vec{e}_s$, which is invariant under the coordinate transform $y^i = y^i(x^j)$, is called a *tensor* $\hat{T}$:

$$\hat{T} = T^{ij\ldots s}\vec{e}_i\vec{e}_j\ldots\vec{e}_s = T'^{ij\ldots s}\vec{e}\,'_i\vec{e}\,'_j\ldots\vec{e}\,'_s. \tag{1.1.10}$$

The quantity of indices $i, j, \ldots, s$ is called the *tensor rank.* The products $\vec{e}_i\vec{e}_j\ldots\vec{e}_s$ are called the *polyadic products* and form a base with respect to which the tensor $\hat{T}$ is expanded. A permutation of the vectors in the polyadic product (the polyad) leads to a new polyad. For example, the permutation of the vectors $\vec{e}_i$ and $\vec{e}_j$ gives a new polyad $\vec{e}_j\vec{e}_i\ldots\vec{e}_s$. When passing to a new coordinate system the polyadic product is transformed in accordance with the law

$$\vec{e}\,'_i\vec{e}\,'_j\ldots\vec{e}\,'_s = \frac{\partial x^k}{\partial y^i}\frac{\partial x^l}{\partial y^j}\cdots\frac{\partial x^n}{\partial y^s}\vec{e}_k\vec{e}_l\ldots\vec{e}_n.$$

In the particular case of two vectors the polyadic products $\vec{e}_i\vec{e}_j$ are called the *dyads.* There exist nine linearly independent dyads $\vec{e}_1\,\vec{e}_1$, $\vec{e}_1\,\vec{e}_2, \ldots,$ $\vec{e}_3\,\vec{e}_3$ in the base of which the second-rank tensor will have the form

$$\hat{T} = T^{ij}\vec{e}_i\vec{e}_j, \quad T'^{ij}(y^n) = \frac{\partial y^i}{\partial x^k}\frac{\partial y^j}{\partial x^l}T^{kl}(x^n(y^s)). \tag{1.1.11}$$

The second equation in (1.1.11) follows from the invariance of the tensor $\hat{T}$ under the coordinate transformation $y^i = y^i(x^j)$ and the law for the transformation of the base dyads

$$\vec{e}\,'_i\vec{e}\,'_j = \frac{\partial x^k}{\partial y^i}\frac{\partial x^l}{\partial y^j}\vec{e}_k\vec{e}_l.$$

The components of the second-rank tensor T^{ij} form a matrix in which the element T^{ij} stands at the intersection of the ith row and jth column. For the purpose of brevity, we will omit in the following the word "components" and call the components T^{ij} not quite strictly the tensor. This convention is often encountered in the literature and should not mislead the reader. Each time the T^{ij} are encountered the components of the tensor $\hat{T} = T^{ij}\vec{e}_i\vec{e}_j$ are meant.

The tensor T^{ij} can be decomposed into a sum of a symmetric S^{ij} and antisymmetric A^{ij} tensor in accordance with the rule

$$T^{ij} = \frac{1}{2}(T^{ij} + T^{ji}) + \frac{1}{2}(T^{ij} - T^{ji}) = S^{ij} + A^{ij}, \qquad (1.1.12)$$
$$S^{ij} = S^{ji}, \quad A^{ij} = -A^{ji}.$$

The metric tensor g_{ij} plays an important role; its specification determines the coordinate system. Let $d\vec{r} = dx^i\vec{e}_i$. Then the squared distance is

$$ds^2 = d\vec{r} \cdot d\vec{r} = \vec{e}_i \cdot \vec{e}_j dx^i dx^j = g_{ij} dx^i dx^j.$$

From here, the formulas

$$ds^2 = g_{ij} dx^i dx^j, \qquad g_{ij} = \vec{e}_i \cdot \vec{e}_j \qquad (1.1.13)$$

follow. Using the contravariant basis, we can determine in a similar way that

$$g^{ij} = \vec{e}^{\,i} \cdot \vec{e}^{\,j}. \qquad (1.1.14)$$

Let us expand $\vec{e}^{\,i}$ with respect to the basis $\vec{e}_j$, so that the equation $\vec{e}^{\,i} = b^{ij}\vec{e}_j$ is valid. We now multiply both sides of this expression by $\vec{e}^{\,k}$ as a scalar product, and with regard for (1.1.3) and (1.1.14) we find

$$\vec{e}^{\,i} \cdot \vec{e}^{\,k} = g^{ik} = b^{ij}\vec{e}_j \cdot \vec{e}^{\,k} = b^{ij}\delta_j^k = b^{ik},$$

consequently, the equalities

$$\vec{e}^{\,i} = g^{ij}\vec{e}_j, \quad \vec{e}_i = g_{ij}\vec{e}^{\,j}. \qquad (1.1.15)$$

With the aid of g_{ij} and g^{ij}, one can lower and raise the indices

$$\hat{T} = T_{ij}\vec{e}^{\,i}\vec{e}^{\,j} = T_{ij}g^{ik}g^{jl}\vec{e}_k\vec{e}_l = T^{kl}\vec{e}_k\vec{e}_l,$$

from where we obtain the rule:

$$T^{kl} = g^{ik}g^{jl}T_{ij}, \quad T_{kl} = g_{ik}g_{jl}T^{ij}. \qquad (1.1.16)$$

If the tensor g_{ij} is given, then g^{ij} is determined by the formula

$$\| g^{ij} \| = \| g_{ij} \|^{-1} = \frac{1}{g} \| A(g_{ij}) \|, \quad g = \det \| g_{ij} \|, \qquad (1.1.17)$$

where $\| A(g_{ij}) \|$ is a matrix constructed from the cofactors (minors) of the matrix g_{ij}. To prove formula (1.1.17), let us consider the scalar product $\vec{e}_i \cdot \vec{e}^{\,j}$:

$$\delta_i^j = \vec{e}_i \cdot \vec{e}^{\,j} = g_{ik}\vec{e}^{\,k} g^{js}\vec{e}_s = g_{ik}g^{js} \underbrace{\vec{e}^{\,k} \cdot \vec{e}_s}_{\delta_s^k} = g_{ik}g^{jk}.$$

It follows from the definition (1.1.13) that g_{ij} is a symmetric tensor whose components can be written as

$$g_{ij} = \vec{c}_i \cdot \vec{c}_j = |\vec{e}_i| \cdot |\vec{e}_j| \cdot \cos\varphi_{ij}. \tag{1.1.18}$$

For $i = j$ we have $g_{ii} = \vec{e}_i \cdot \vec{e}_i = |\vec{e}_i|^2$, from where we find with regard for (1.1.18)

$$|\vec{e}_i| = \sqrt{g_{ii}}, \quad \cos\varphi_{ij} = \frac{g_{ij}}{\sqrt{g_{ii}}\sqrt{g_{jj}}} \quad \text{for } i \neq j, \tag{1.1.19}$$

where the summation is absent over the repeating indices. In the orthogonal coordinate system $g_{ij} = 0$ at $i \neq j$,

$$\| g_{ij} \| = \left\| \begin{matrix} g_{11} & & 0 \\ & g_{22} & \\ 0 & & g_{33} \end{matrix} \right\|. \tag{1.1.20}$$

By using (1.1.13), we find:

$$ds^2 = g_{11}(dx^1)^2 + g_{22}(dx^2)^2 + g_{33}(dx^3)^2 = ds_1^2 + ds_2^2 + ds_3^2,$$

from where we obtain the expressions for the physical coordinates ds_i in terms of the coordinates dx^i and g_{ii}:

$$ds_1 = \sqrt{g_{11}}dx^1, \quad ds_2 = \sqrt{g_{22}}dx^2, \quad ds_3 = \sqrt{g_{33}}dx^3.$$

In the case of the Cartesian coordinate system the metric tensor has the simplest form:

$$g_{ij} = g^{ij} = \delta_{ij} = \delta^{ij} = \begin{cases} 0, & i \neq j \\ 1, & i = j, \end{cases}$$

where δ_{ij}, δ^{ij} are the Kronecker symbols.

1.1.1 Covariant Differentiation

Let the vector $\vec{B}(x^i) = b^i\vec{e}_i$ be given in the curvilinear coordinate system $x^i, \vec{e}_i$ whose derivative is

$$\frac{\partial \vec{B}}{\partial x^i} = \frac{\partial b^j}{\partial x^i}\vec{e}_j + b^j\frac{\partial \vec{e}_j}{\partial x^i}.$$

Let us expand the vector $\vec{A} = \frac{\partial \vec{e}_j}{\partial x^i}$ with respect to the basis $\vec{e}_i$:

$$\frac{\partial \vec{e}_j}{\partial x^i} = \Gamma^k_{ji}\vec{e}_k, \tag{1.1.21}$$

where $\Gamma^k_{ji}(x^i)$ are the Christoffel symbols (Γ^k_{ji} is not a tensor). Substituting (1.1.21) into the foregoing formula, we obtain:

$$\frac{\partial \vec{B}}{\partial x^i} = \left(\frac{\partial b^k}{\partial x^i} + b^j\Gamma^k_{ji}\right)\vec{e}_k = \nabla_i b^k \vec{e}_k, \tag{1.1.22}$$

where we have introduced the covariant derivative of b^k

$$\nabla_i b^k = \frac{\partial b^k}{\partial x^i} + b^j\Gamma^k_{ji}. \tag{1.1.23}$$

It follows from the relation $\vec{e}^{\,i} \cdot \vec{e}_j = \delta^i_j$ and from (1.1.21) that

$$\frac{\partial \vec{e}^{\,j}}{\partial x^i} = -\Gamma^j_{ki}\vec{e}^{\,k}; \tag{1.1.24}$$

therefore, the derivative of the vector $\vec{B} = b_k\vec{e}^{\,k}$ is equal to

$$\frac{\partial \vec{B}}{\partial x^i} = \left(\frac{\partial b_k}{\partial x^i} - \Gamma^j_{ki}b_j\right)\vec{e}^{\,k} = \nabla_i b_k \vec{e}^{\,k},$$

from where the formula follows for the covariant derivative of b_k:

$$\nabla_i b_k = \frac{\partial b_k}{\partial x^i} - b_j\Gamma^j_{ki}. \tag{1.1.25}$$

The derivative of the tensor $\hat{T} = T^{jk}\vec{e}_j\vec{e}_k$ is computed by the formula

$$\begin{aligned}\frac{\partial \hat{T}}{\partial x^i} &= \frac{\partial T^{jk}}{\partial x^i}\vec{e}_j\vec{e}_k + T^{jk}\frac{\partial \vec{e}_j}{\partial x^i}\vec{e}_k + T^{jk}\vec{e}_j\frac{\partial \vec{e}_k}{\partial x^i} \\ &= \frac{\partial T^{jk}}{\partial x^i}\vec{e}_j\vec{e}_k + T^{jk}\Gamma^l_{ji}\vec{e}_l\vec{e}_k + T^{jk}\Gamma^l_{ki}\vec{e}_j\vec{e}_l.\end{aligned}$$

Interchanging in the second item the indices $j \leftrightarrow l$, and in the third item $k \leftrightarrow l$, we obtain:

$$\frac{\partial \hat{T}}{\partial x^i} = \nabla_i T^{jk}\vec{e}_j\vec{e}_k, \tag{1.1.26}$$

where the covariant derivative $\nabla_i T^{jk}$ is determined by the formula

$$\nabla_i T^{jk} = \frac{\partial T^{jk}}{\partial x^i} + T^{lk}\Gamma^j_{li} + T^{jl}\Gamma^k_{li}. \tag{1.1.27}$$

One can show that the covariant derivative of the metric tensor is equal to zero: $\nabla_k g_{ij} = 0$. Consider the tensor $T_{ij} = \nabla_i b_j$, which can be presented in the form $\nabla_i b_j = g_{jk}\nabla_i b^k$. On the other hand, $b_j = g_{jk}b^k$; therefore, $\nabla_i(g_{jk}b^k) = g_{jk}\nabla_i b^k$, from where it follows that

$$\nabla_i g_{jk} = 0. \tag{1.1.28}$$

The covariant derivative of a scalar is

$$\nabla_i \varphi = \frac{\partial \varphi}{\partial x^i}.$$

In the Euclidean space

$$\Gamma^k_{ij} = \Gamma^k_{ji}. \tag{1.1.29}$$

To prove this formula, let us differentiate the basis vector $\vec{e}_j$, and we obtain with regard for (1.1.1):

$$\frac{\partial \vec{e}_j}{\partial x^k} = \frac{\partial^2 \vec{r}}{\partial x^k \partial x^j} = \frac{\partial^2 \vec{r}}{\partial x^j \partial x^k} = \frac{\partial \vec{e}_k}{\partial x^j},$$

from where

$$\Gamma^s_{jk}\vec{e}_s = \Gamma^s_{kj}\vec{e}_s.$$

Let us find the expression for Γ^k_{ij} in terms of the derivatives of g_{ij}:

$$\begin{aligned} \frac{\partial g_{ij}}{\partial x^k} &= \frac{\partial \vec{e}_i}{\partial x^k}\cdot\vec{e}_j + \frac{\partial \vec{e}_j}{\partial x^k}\cdot\vec{e}_i = \Gamma^s_{ik}\vec{e}_s\cdot\vec{e}_j + \Gamma^s_{jk}\vec{e}_s\cdot\vec{e}_i \\ &= \Gamma^s_{ik}g_{sj} + \Gamma^s_{jk}g_{is} = \Gamma_{ik.j} + \Gamma_{jk.i}. \end{aligned} \tag{1.1.30}$$

Interchanging the indices $i \leftrightarrow k$, $j \leftrightarrow k$, we find:

$$\frac{\partial g_{ik}}{\partial x^j} = \Gamma_{ij.k} + \Gamma_{kj.i}, \quad \frac{\partial g_{jk}}{\partial x^i} = \Gamma_{ki.j} + \Gamma_{ji.k}. \tag{1.1.31}$$

Adding two equations (1.1.31), subtracting (1.1.30), and taking into account (1.1.29), we obtain:

$$\begin{aligned} \Gamma_{ij.k} &= \frac{1}{2}\left(\frac{\partial g_{ik}}{\partial x^j} + \frac{\partial g_{jk}}{\partial x^i} - \frac{\partial g_{ij}}{\partial x^k}\right), \\ \Gamma^k_{ij} &= \frac{1}{2}g^{ks}\left(\frac{\partial g_{is}}{\partial x^j} + \frac{\partial g_{js}}{\partial x^i} - \frac{\partial g_{ij}}{\partial x^s}\right). \end{aligned} \tag{1.1.32}$$

1.1.2 The Levi–Civita Tensor

The absolutely antisymmetric Levi–Civita tensor is equal to

$$\begin{aligned} \varepsilon &= \varepsilon^{ijk}\vec{e}_i\vec{e}_j\vec{e}_k, \quad \varepsilon^{ijk} = \vec{e}^{\,i}\cdot(\vec{e}^{\,j}\times\vec{e}^{\,k}), \\ \varepsilon &= \varepsilon_{ijk}\vec{e}^{\,i}\vec{e}^{\,j}\vec{e}^{\,k}, \quad \varepsilon_{ijk} = \vec{e}_i\cdot(\vec{e}_j\times\vec{e}_k). \end{aligned} \tag{1.1.33}$$

It follows from the definition (1.1.33) that, at a permutation of any two indices, the Levi–Civita tensor changes its sign:

$$\varepsilon_{123} = -\varepsilon_{213} = \varepsilon_{231} = -\varepsilon_{321} = \dots. \tag{1.1.34}$$

If two or three indices coincide, then the Levi–Civita tensor is equal to zero:

$$\varepsilon_{112} = \dots = \varepsilon_{333} = 0. \tag{1.1.35}$$

Writing the mixed product (1.1.33) in Cartesian coordinates with the basis $\vec{E}_i$, we can find ε_{123}.

$$\begin{aligned}
(\vec{e}_1 \cdot (\vec{e}_2 \times \vec{e}_3))^2 &= \begin{vmatrix} e_1^1 & e_1^2 & e_1^3 \\ e_2^1 & e_2^2 & e_2^3 \\ e_3^1 & e_3^2 & e_3^3 \end{vmatrix}^2 \\
&= \left| \begin{pmatrix} e_1^1 & e_1^2 & e_1^3 \\ e_2^1 & e_2^2 & e_2^3 \\ e_3^1 & e_3^2 & e_3^3 \end{pmatrix} \begin{pmatrix} e_1^1 & e_2^1 & e_3^1 \\ e_1^2 & e_2^2 & e_3^2 \\ e_1^3 & e_2^3 & e_3^3 \end{pmatrix} \right| \\
&= \begin{vmatrix} \vec{e}_1 \cdot \vec{e}_1 & \vec{e}_1 \cdot \vec{e}_2 & \vec{e}_1 \cdot \vec{e}_3 \\ \vec{e}_2 \cdot \vec{e}_1 & \vec{e}_2 \cdot \vec{e}_2 & \vec{e}_2 \cdot \vec{e}_3 \\ \vec{e}_3 \cdot \vec{e}_1 & \vec{e}_3 \cdot \vec{e}_2 & \vec{e}_3 \cdot \vec{e}_3 \end{vmatrix} = \det \parallel g_{ij} \parallel = g,
\end{aligned}$$

where e_i^j is the jth Cartesian component of the vector

$$\vec{e}_i = e_i^k \vec{E}_k; \quad \vec{E}_i \cdot \vec{E}_j = \delta_{ij}.$$

It follows from here that

$$\varepsilon_{123} = \sqrt{g}, \quad \varepsilon^{123} = \frac{1}{\sqrt{g}}. \tag{1.1.36}$$

We determine from the definitions

$$\vec{e}^{\,i} = \frac{\vec{e}_j \times \vec{e}_k}{\vec{e}_i \cdot (\vec{e}_j \times \vec{e}_k)}, \quad \varepsilon_{ijk} = \vec{e}_i \cdot (\vec{e}_j \times \vec{e}_k)$$

that $\vec{e}_j \times \vec{e}_k = \varepsilon_{ijk}\vec{e}^{\,i} = \varepsilon_{jki}\vec{e}^{\,i}$, from where

$$\vec{e}_i \times \vec{e}_j = \varepsilon_{ijk}\vec{e}^{\,k}, \quad \vec{e}^{\,i} \times \vec{e}^{\,j} = \varepsilon^{ijk}\vec{e}_k. \tag{1.1.37}$$

The formulas (1.1.37) enable us to write the vector product in the form

$$\vec{a} \times \vec{b} = a^i b^j \vec{e}_i \times \vec{e}_j = \varepsilon_{ijk} a^i b^j \vec{e}^{\,k} = \varepsilon^{ijk} a_i b_j \vec{e}_k. \tag{1.1.38}$$

The value of an infinitesimal volume dV constructed on the vectors $\vec{e}_1 dx^1, \vec{e}_2 dx^2, \vec{e}_3 dx^3$ is equal to

$$\begin{aligned}
dV &= \vec{e}_1 dx^1 \cdot (\vec{e}_2\, dx^2 \times \vec{e}_3\, dx^3) = \vec{e}_1 \cdot (\vec{e}_2 \times \vec{e}_3)\, dx^1 dx^2 dx^3 \\
&= \varepsilon_{123}\, dx^1\, dx^2\, dx^3 = \sqrt{g}\, dx^1\, dx^2\, dx^3.
\end{aligned} \tag{1.1.39}$$

1.1.3 Differential Operations

Let us define the basic operations grad, div, rot on the scalar φ, the vector $\vec{B} = b^j\vec{e}_j$, and the tensor $\hat{T} = T^{ij}\vec{e}_i\vec{e}_j$. From the definition $\nabla = \vec{e}^{\,i}\partial/\partial x^i$, we have

the operation grad:

$$\nabla\varphi = \nabla_i\varphi\vec{e}^{\,i}, \quad \nabla\vec{B} = \nabla_i b^j\vec{e}^{\,i}\vec{e}_j, \quad \nabla\hat{T} = \nabla_i T^{jk}\vec{e}^{\,i}\vec{e}_j\vec{e}_k; \tag{1.1.40}$$

the operation div:

$$\operatorname{div}\vec{B} = \nabla\cdot\vec{B} = \nabla_i b^j \underbrace{\vec{e}^{\,i}\cdot\vec{e}_j}_{\delta^i_j} = \nabla_j b^j, \tag{1.1.41}$$

$$\operatorname{div}\hat{T} = \nabla\cdot\hat{T} = \nabla_i T^{jk} \underbrace{\vec{e}^{\,i}\cdot\vec{e}_j}_{\delta^i_j}\vec{e}_k = \nabla_i T^{ik}\vec{e}_k;$$

the operation rot:

$$\operatorname{rot}\vec{B} = \nabla\times\vec{B} = \vec{e}^{\,i}\nabla_i\times b^j\vec{e}^{\,j} = \nabla_i b^j(\vec{e}^{\,i}\times\vec{e}^{\,j}) = \nabla_i b_j\varepsilon^{ijk}\vec{e}_k. \tag{1.1.42}$$

1.1.4 Physical Components of Vectors and Tensors

The physical components b^i of the vector $\vec{B} = b^i\vec{e}_i$ depend not only on the physical processes, but also on the chosen coordinate system g_{ij}. Let us determine for the vector $\vec{B}$ its physical components $\tilde{b}^i$ by relating them to the unit vectors $\vec{E}_i$ of an orthogonal coordinate system $g_{ij} = 0$ at $i \neq j$ [if the coordinate system is nonorthogonal, then one can go over to an orthogonal coordinate system with the aid of some nonsingular transformation $y^i = y^i(x^k)$]. We have in the orthogonal coordinate system that

$$\vec{B} = b^i\vec{e}_i = |\vec{e}_i|b^i\frac{\vec{e}_i}{|\vec{e}_i|} = \sqrt{g_{ii}}b^i\vec{E}_i = \frac{b^i}{\sqrt{g^{ii}}}\vec{E}_i = \tilde{b}^i\vec{E}_i,$$

from where it follows that

$$\tilde{b}^i = \sqrt{g_{ii}}b^i = \frac{b^i}{\sqrt{g^{ii}}}, \tag{1.1.43}$$

where $\vec{E}_i = \vec{e}_i/|\vec{e}_i|$ is a unit vector, which is parallel with the basis vector $\vec{e}_i$. For the contravariant components of the tensor, we can obtain in a similar way the formulas

$$\tilde{T}^{ij} = \sqrt{g_{ii}g_{jj}}T^{ij} = T^{ij}/\sqrt{g^{ii}g^{jj}}. \tag{1.1.44}$$

For the covariant components, we have the formulas

$$\tilde{T}_{ij} = \sqrt{g^{ii}g^{jj}}T^{ij} = T_{ij}/\sqrt{g_{ii}g_{jj}}, \tag{1.1.45}$$

and for the mixed components, we have

$$\tilde{T}^i_j = \sqrt{g_{ii}g^{jj}}T^i_j = T^i_j/\sqrt{g^{ii}g_{jj}}, \tag{1.1.46}$$

where the summation is not carried out over the repeating indices.

1.1.5 Eigenvalues and Eigenvectors of a Symmetric Tensor

Let us multiply the tensor $\hat{T} = T^{ij}\vec{e}_i\vec{e}_j$ in a scalar way by a vector $\vec{b} = b^k\vec{e}_k$:

$$\hat{T} \cdot \vec{b} = T^{ij}\vec{e}_i\vec{e}_j \cdot \vec{e}_k b^k = T^{ij}g_{jk}b^k\vec{e}_i = T^i_k b^k \vec{e}_i = \vec{c}. \tag{1.1.47}$$

Thus, the tensor $\hat{T}$ is a linear operator whose action on the vector $\vec{b}$ yields the vector $\vec{c} = T^i_k b^k \vec{e}_i$. There exist among the vectors $\vec{b}$ the *eigenvectors*, which satisfy the condition

$$\hat{T} \cdot \vec{b} = \lambda\vec{b}, \tag{1.1.48}$$

where λ is some number, which is called the *eigenvalue*, or the *principal value.* With regard for (1.1.47), one can rewrite equation (1.1.48) in the form

$$(T^i_k - \lambda\delta^i_k)b^k = 0. \tag{1.1.49}$$

It is well known from the algebra that a system of linear equations with the zero right-hand side has a nonzero solution if and only if

$$|T^i_k - \lambda\delta^i_k| = 0. \tag{1.1.50}$$

Computing the third-order determinant of (1.1.49), we obtain the equation

$$\lambda^3 - J_1\lambda^2 + J_2\lambda - J_3 = 0, \tag{1.1.51}$$

where J_1, J_2, J_3 are the first, second, and third invariants of the tensor $\hat{T}$, respectively:

$$J_1 = T^i_i, \quad J_2 = \frac{1}{2}(T^i_i T^j_j - T^j_i T^i_j), \quad J_3 = \det \parallel T^j_i \parallel. \tag{1.1.52}$$

In the case of a *symmetric* tensor, all eigenvalues are *real.* Let λ and λ^* be the complex conjugate values to which the eigenvectors $\vec{b}$ and $\vec{b}^*$ correspond. Then

$$\hat{T} \cdot \vec{b} = \lambda\vec{b}, \quad \hat{T} \cdot \vec{b}^* = \lambda^*\vec{b}^*, \quad \vec{b} \cdot \vec{b}^* = |\vec{b}|^2 \neq 0.$$

Multiplying the first equation by $\vec{b}^*$ and the second equation by $\vec{b}$ and subtracting one equation from another, we obtain:

$$(\hat{T}\cdot\vec{b})\cdot\vec{b}^* - (\hat{T}\cdot\vec{b}^*)\cdot\vec{b} = (\lambda - \lambda^*)(\vec{b}\cdot\vec{b}^*) = 2i\,\mathrm{Im}\,\lambda|\vec{b}|^2.$$

On the other hand, by virtue of the equality $T^{ij} = T^{ji}$, we have

$$(\hat{T}\cdot\vec{b})\cdot\vec{b}^* - (\hat{T}\cdot\vec{b}^*)\cdot\vec{b} = T^{ij}b_jb_i^* - T^{ij}b_j^*b_i = (T^{ij} - T^{ji})b_jb_i^* = 0,$$

from where it follows that $\mathrm{Im}\,\lambda = 0$; i.e., λ is real.

If all eigenvalues are different, i.e., $\lambda_\alpha \neq \lambda_\beta$, *then the eigenvectors* $\vec{b}_\alpha$ *are orthogonal.* To prove this property, let us multiply both sides of the equation $\hat{T}\cdot\vec{b}_\alpha = \lambda_\alpha\vec{b}_\alpha$ by $\vec{b}_\beta$ and the equation $\hat{T}\cdot\vec{b}_\beta = \lambda_\beta\vec{b}_\beta$ by $\vec{b}_\alpha$. Subtracting one equation from another, we obtain:

$$\begin{aligned}(\hat{T}\cdot\vec{b}_\alpha)\cdot\vec{b}_\beta - (\hat{T}\cdot\vec{b}_\beta)\cdot\vec{b}_\alpha &= (T^{ij} - T^{ji})(b_\alpha)_j(b_\beta)_i = 0\\ &= (\lambda_\alpha - \lambda_\beta)(\vec{b}_\alpha\cdot\vec{b}_\beta),\end{aligned}$$

from where it follows that

$$(\vec{b}_\alpha\cdot\vec{b}_\beta) = 0. \tag{1.1.53}$$

The eigenvectors $\vec{b}_\alpha$ are determined from the system (1.1.49) at $\lambda = \lambda_\alpha$. Since the determinant of the system is equal to zero, one of the equations is linearly dependent, and we add to (1.1.49) an additional normalization condition $|\vec{b}_\alpha|^2 = 1$, which can be rewritten with regard for (1.1.53) as

$$\vec{b}_\alpha\cdot\vec{b}_\beta = \delta_\alpha^\beta. \tag{1.1.54}$$

Let us choose a coordinate system whose basis vectors coincide with the eigenvectors $\vec{e}_i = \vec{b}_i$; therefore, $\hat{T} = T^{ij}\vec{b}_i\vec{b}_j$. Multiplying this relation by $\vec{b}_k$ and $\vec{b}_l$, we obtain:

$$\vec{b}_k\cdot\hat{T}\cdot\vec{b}_l = T^{ij}(\vec{b}_i\cdot\vec{b}_k)(\vec{b}_j\cdot\vec{b}_l) = T^{ij}\delta_{ik}\delta_{jl} = T^{kl}. \tag{1.1.55}$$

It follows from the definition of eigenvectors (1.1.48) that

$$\vec{b}_k\cdot\hat{T}\cdot\vec{b}_l = \lambda_l(\vec{b}_k\cdot\vec{b}_l) = \lambda_l\delta_{kl}. \tag{1.1.56}$$

Comparing (1.1.55) and (1.1.56), we obtain that, in the basis $\vec{b}_i$, the tensor $\hat{T}$ is diagonal and has the form

$$T^{ij} = \lambda_i\delta^{ij}, \tag{1.1.57}$$

where there is no summation with respect to the index i. In the basis $\vec{b}_i$, the invariants $\hat{T} = \lambda_i\delta^{ij}\vec{b}_i\vec{b}_j$ have an especially simple form

$$J_1 = \lambda_1 + \lambda_2 + \lambda_3, \quad J_2 = \lambda_1\lambda_2 + \lambda_1\lambda_3 + \lambda_2\lambda_3, \quad J_3 = \lambda_1\lambda_2\lambda_3. \tag{1.1.58}$$

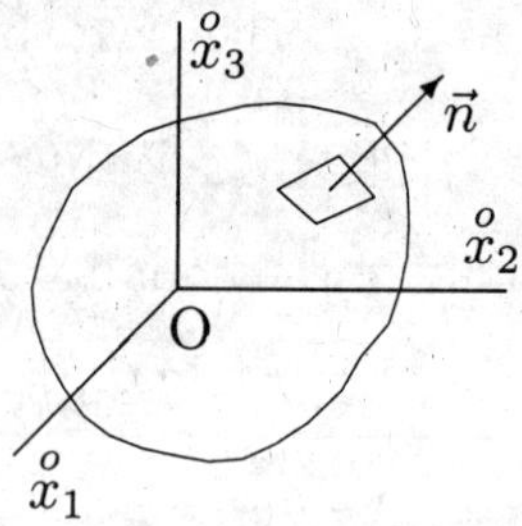

Figure 1.2: To the proof of the Ostrogradsky–Gauss theorem.

We have in the case of multiple roots that

1) at $\lambda_2 = \lambda_3 \neq \lambda_1$, all vectors that are perpendicular to $\vec{b}_1$ are the eigenvectors.

2) at $\lambda_2 = \lambda_3 = \lambda_1 = \lambda$, any vector is an eigenvector.

In the basis $\vec{b}_i$, we have $\hat{T} = \lambda \vec{b}_i \vec{b}_j$. If $\vec{c}$ is an arbitrary vector, then

$$\hat{T} \cdot \vec{c} = \lambda \vec{b}_i (\vec{b}_i \cdot \vec{c}) = \lambda c^i \vec{b}_i = \lambda \vec{c}.$$

A symmetric tensor whose principal values are equal is called a *spherical tensor*. In an arbitrary basis $\vec{e}_i$, the spherical tensor has the form

$$\hat{T} = \lambda g^{ij} \vec{e}_i \vec{e}_j. \tag{1.1.59}$$

Multiplying (1.1.59) by an arbitrary vector $\vec{c} = c^k \vec{e}_k$, we obtain $\hat{T} \cdot \vec{c} = \lambda g^{ij} (\vec{e}_j \cdot \vec{e}_k) c^k \vec{e}_i = \lambda \vec{c}$, which implies that any vector $\vec{c}$ is an eigenvector of tensor (1.1.59).

1.1.6 The Ostrogradsky–Gauss Theorem

In mathematical analysis courses, the Ostrogradsky–Gauss theorem is proved, which states that, for any three continuously differentiable functions Q_1, Q_2, Q_3 in Cartesian coordinates $\overset{o}{x}_i$, the following formula is valid:

$$\int_S (Q_1 \cos(n, \overset{o}{x}_1) + Q_2 \cos(n, \overset{o}{x}_2) + Q_3 \cos(n, \overset{o}{x}_3))\, dS$$
$$= \int_V \left(\frac{\partial Q_1}{\partial \overset{o}{x}_1} + \frac{\partial Q_2}{\partial \overset{o}{x}_2} + \frac{\partial Q_3}{\partial \overset{o}{x}_3} \right) dV, \tag{1.1.60}$$

where $dV = d\overset{o}{x}_1\, d\overset{o}{x}_2\, d\overset{o}{x}_3$ is an infinitesimal element of the volume V, S is a surface bounding the volume V, $\vec{n}$ is a normal to the surface S (see Fig. 1.2), and $\cos(n, \overset{o}{x}_i)$ are the cosines of the angles between the normal

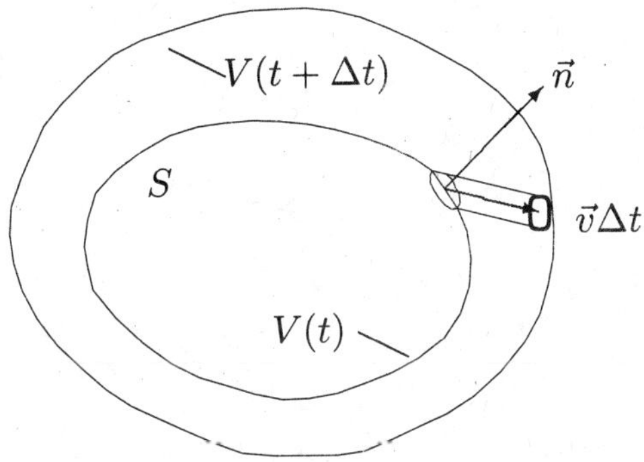

Figure 1.3: To the derivation of the formula (1.1.66) for the differentiation of an integral over the moving volume $V(t)$.

$\vec{n}$ and the coordinate axes $\overset{o}{x}_i$. One can consider the functions Q_1, Q_2, Q_3 as the components of the vector $\vec{B}$, and one can rewrite equation (1.1.60) in the form

$$\int_S \vec{B}\cdot\vec{n}\,dS = \int_V \operatorname{div}\vec{B}\,dV. \tag{1.1.61}$$

The integrands in (1.1.61) are the scalar quantities, which do not depend on the choice of the coordinate system. By choosing an arbitrary coordinate system x^i with the basis vectors $\vec{e}_i$ and taking into account the relationships

$$\vec{B} = b^k\vec{e}_k,\ \vec{n} = n_j\vec{e}^{\,j},\ \vec{B}\cdot\vec{n} = b^k n_k,\ d\overset{o}{x}_1\, d\overset{o}{x}_2\, d\overset{o}{x}_3 = \sqrt{g}\,dx^1dx^2dx^3, \tag{1.1.62}$$

$$\operatorname{div}\vec{B} = \nabla_k b^k = \frac{\partial b^k}{\partial x^k} + b^i\Gamma^i_{ki},\quad d\tau = dx^1dx^2dx^3,\ dV = \sqrt{g}\,d\tau,$$

let us write the Ostrogradsky–Gauss theorem in an arbitrary coordinate system:

$$\int_S b^k n_k dS = \int_V \nabla_k b^k\,dV, \tag{1.1.63}$$

where the expression for the area element dS is obtained below [see (1.1.76)]. Let us find with the aid of the Gauss–Ostrogradsky theorem the formula for the differentiation of an integral taken over a moving volume $V(t)$:

$$\frac{d}{dt}\int_{V(t)} f(x^i,t)\,dV$$
$$= \lim_{\Delta t\to 0}\frac{1}{\Delta t}\left(\int_{V(t+\Delta t)} f(x^i,t+\Delta t)\,dV - \int_{V(t)} f(x^i,t)\,dV\right)$$

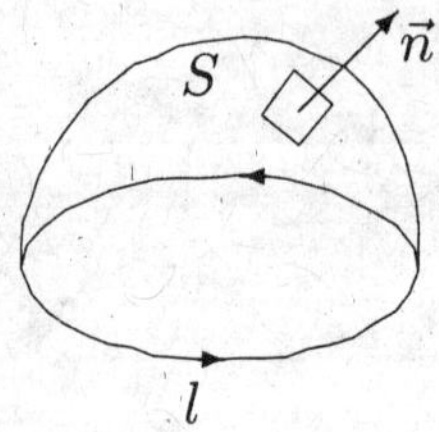

Figure 1.4: To the proof of the Stokes theorem.

$$= \lim_{\Delta t \to 0} \frac{1}{\Delta t} \left(\int_{V(t)} f(x^i, t + \Delta t)\, dV - \int_{V(t)} f(x^i, t)\, dV \right.$$
$$\left. + \int_{\Delta V} f(x^i, t)\, dV \right), \tag{1.1.64}$$

where $\Delta V = V(t + \Delta t) - V(t)$. Let us specify the volume V at the moment of time t. This volume will go over into the volume $V(t + \Delta t)$ at $t + \Delta t$ (see Fig. 1.3). It follows from Fig. 1.3 that the volume alteration ΔV is equal to

$$\Delta V = \int_S \vec{v} \cdot \vec{n} \Delta t\, dS.$$

The last integral in (1.1.64) can be written with regard for this formula and the Ostrogradsky–Gauss theorem (1.1.63) as

$$\int_{\Delta V} f(x^i, t)\, dV = \int_S f v^i n_i \Delta t\, dS = \int_V \nabla_i (f v^i) \Delta t\, dV. \tag{1.1.65}$$

Substituting (1.1.65) into (1.1.64), we obtain the desired formula:

$$\frac{d}{dt} \int_{V(t)} f(x^i, t)\, dV = \int_{V(t)} \left(\frac{\partial f}{\partial t} + \nabla_i (f v^i) \right) dV. \tag{1.1.66}$$

1.1.7 The Stokes Theorem

Let a closed contour l be given on which a smooth surface S is spanned (see Fig. 1.4). It is assumed that the region is simply connected; therefore, the contour l can be contracted into a point by a continuous deformation. If a continuously differentiable vector field $\vec{B}(x^i)$ is given in the region and on the surface, then the *Stokes theorem* is valid:

$$\oint_l \vec{B} \cdot d\vec{r} = \int_S \vec{n} \cdot \operatorname{rot} \vec{B}\, dS, \tag{1.1.67}$$

or in a componentwise form $\vec{B} = b_j\vec{e}^j$

$$\oint_l b_j dx^j = \int_S \varepsilon^{ijk}\nabla_i b_j n_k \, dS, \tag{1.1.68}$$

where the contour l and the surface S are shown in Fig. 1.4.

1.1.8 The Weyl Formula

Consider a convolution (the summation over repeating indices) of Christoffel symbols (1.1.32):

$$\Gamma^i_{ij} = \frac{1}{2}g^{is}\Big(\frac{\partial g_{is}}{\partial x^j} + \frac{\partial g_{js}}{\partial x^i} - \frac{\partial g_{ij}}{\partial x^s}\Big). \tag{1.1.69}$$

By changing the notations of the dummy indices $i \leftrightarrow s$ in the second item

$$g^{is}\frac{\partial g_{js}}{\partial x^i} = g^{si}\frac{\partial g_{ji}}{\partial x^s} = g^{is}\frac{\partial g_{ij}}{\partial x^s},$$

we obtain the formula:

$$\Gamma^i_{ij} = \frac{1}{2}g^{is}\frac{\partial g_{is}}{\partial x^j}. \tag{1.1.70}$$

The determinant differential is

$$dg = dg_{is}A_{is}, \tag{1.1.71}$$

where A_{is} is the cofactor of the element g_{is}. On the other hand, the elements of the inverse matrix are equal to (1.1.17):

$$g^{is} = \frac{1}{g}A_{is},$$

from where we find with regard for (1.1.71):

$$dg = gg^{is}dg_{is}. \tag{1.1.72}$$

Substituting (1.1.72) into (1.1.70), we obtain the *Weyl formula*:

$$\Gamma^i_{ij} = \frac{1}{2g}\frac{\partial g}{\partial x^j} = \frac{\partial}{\partial x^j}(\ln\sqrt{g}). \tag{1.1.73}$$

By using the Weyl formula let us prove the Ostrogradsky-Gauss theorem in an arbitrary coordinate system. According to (1.1.41) and (1.1.73), the divergence of the vector $\vec{B}$ is equal to

$$\begin{aligned}\nabla_i b^i &= \frac{\partial b^i}{\partial x^i} + b^i\Gamma^j_{ij} = \frac{\partial b^i}{\partial x^i} + \frac{b^i}{2g}\frac{\partial g}{\partial x^i} \\ &= \frac{1}{\sqrt{g}}\Big(\sqrt{g}\frac{\partial b^i}{\partial x^i} + b^i\frac{\partial\sqrt{g}}{\partial x^i}\Big) = \frac{1}{\sqrt{g}}\frac{\partial}{\partial x^i}(\sqrt{g}b^i).\end{aligned} \tag{1.1.74}$$

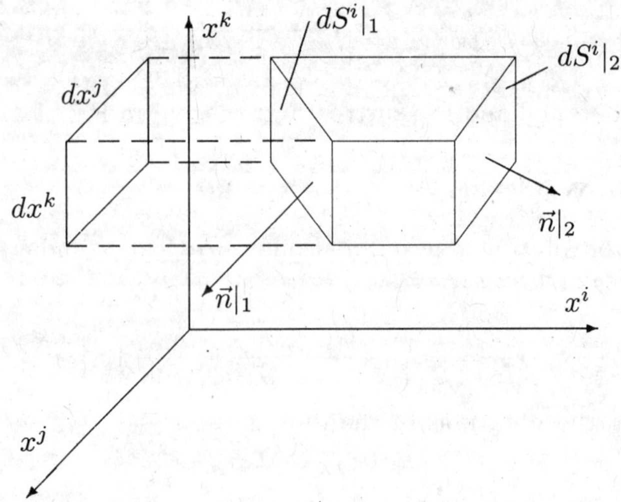

Figure 1.5: To the proof of the Ostrogradsky–Gauss theorem in the arbitrary coordinate system.

Let us choose a convex prism as an integration volume (see Fig. 1.5), for which we determine with regard for (1.1.39) that

$$\begin{aligned}\int_V \nabla_i b^i dV &= \int_V \frac{\partial}{\partial x^i}(\sqrt{g}b^i)d\tau = \int_V \frac{\partial}{\partial x^i}(\sqrt{g}b^i)dx^i dx^j dx^k \\ &= \sum_i \left(\int_{S_2^i} \sqrt{g}b^i \Big|_2 dx^j\, dx^k - \int_{S_1^i} \sqrt{g}b^i \Big|_1 dx^j\, dx^k \right), \quad (1.1.75)\end{aligned}$$

where the summation is carried out over all lateral facets i, j, k. Let us identify two opposite lateral facets with the normals $\vec{n}|_1$ and $\vec{n}|_2$. As will be shown below [see (1.3.3) and (1.3.6)] the area of a lateral prism facet is determined by formula

$$dS\, n_i = \sqrt{g}\, dx^j\, dx^k, \qquad \vec{n} = n_i \vec{e}^{\,i}. \tag{1.1.76}$$

Since the projections of the normals to the opposite prism facets onto the x^i axis have the opposite signs (see Fig. 1.5), we have with regard for (1.1.76) that

$$dx^j\, dx^k = -\frac{1}{\sqrt{g}} n_i dS^i \Big|_1 = \frac{1}{\sqrt{g}} n_i dS^i \Big|_2, \tag{1.1.77}$$

where

$$n_i \Big|_1 = \vec{n} \cdot \vec{e}_i \Big|_1, \quad n_i \Big|_2 = \vec{n} \cdot \vec{e}_i \Big|_2.$$

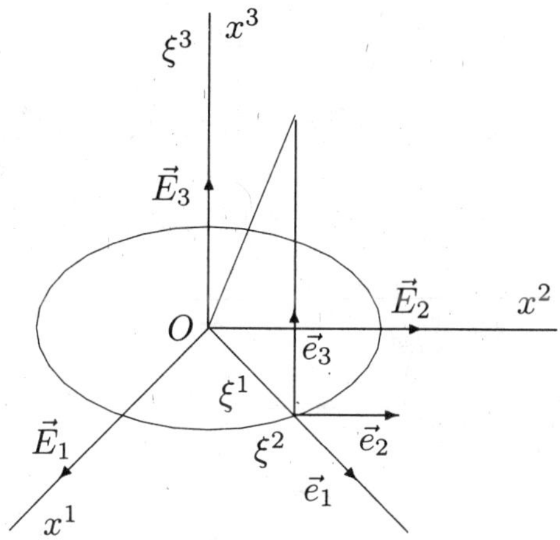

Figure 1.6: The cylindrical coordinate system.

Substituting (1.1.77) into (1.1.75), we obtain the Ostrogradsky–Gauss formula (1.1.63)

$$\int_V \nabla_i b^i dV = \sum_i \left(\int_{S_2^i} b^i n_i dS^i \Big|_2 + \int_{S_1^i} b^i n_i dS^i \Big|_1 \right) = \int_S b^i n_i dS.$$

Since an arbitrary volume can be partitioned into a finite number of convex prisms, this formula can be obtained for an arbitrary volume by a summation of this expression written for each convex prism. The integrals over all internal facets will cancel (since the outer normals to the neighboring prisms have the opposite signs and are equal in their modulus) and there will remain only the integral over the external surface of an arbitrary volume.

Problem 1.1. Find the metric tensor components g_{ij} in the cylindrical coordinate system. Write the expression for the length element ds. Determine the components g^{ij} and the basis vectors $\vec{e}_i$.

Solution: Let us introduce the cylindrical coordinate system ξ^l shown in Fig. 1.6.

It follows from (1.1.2) and (1.1.13) that

$$g_{ij} = \frac{\partial \vec{r}}{\partial \xi^i} \frac{\partial \vec{r}}{\partial \xi^j} = \frac{\partial x^k}{\partial \xi^i} \frac{\partial x^l}{\partial \xi^j} \vec{E}_k \cdot \vec{E}_l = \frac{\partial x^l}{\partial \xi^i} \frac{\partial x^l}{\partial \xi^j},$$

since the equation $\vec{E}_k \cdot \vec{E}_j = \delta_{kl}$ is valid in the Cartesian coordinate system x^i. The relation between the Cartesian and cylindrical coordinates is given by the formulas

$$\begin{aligned}
x^1 &= \xi^1 \cos\xi^2; \quad x^2 = \xi^1 \sin\xi^2; \quad x^3 = \xi^3;\\
\frac{\partial x^1}{\partial \xi^1} &= \cos\xi^2; \quad \frac{\partial x^1}{\partial \xi^2} = -\xi^1 \sin\xi^2; \quad \frac{\partial x^1}{\partial \xi^3} = 0;\\
\frac{\partial x^2}{\partial \xi^1} &= \sin\xi^2; \quad \frac{\partial x^2}{\partial \xi^2} = \xi^1 \cos\xi^2; \quad \frac{\partial x^2}{\partial \xi^3} = 0;\\
\frac{\partial x^3}{\partial \xi^1} &= 0; \quad \frac{\partial x^3}{\partial \xi^2} = 0; \quad \frac{\partial x^3}{\partial \xi^3} = 1.
\end{aligned}$$

We have from here that

$$\begin{aligned}
g_{11} &= \frac{\partial x^i}{\partial \xi^1}\frac{\partial x^i}{\partial \xi^1} = \left(\frac{\partial x^1}{\partial \xi^1}\right)^2 + \left(\frac{\partial x^2}{\partial \xi^1}\right)^2 + \left(\frac{\partial x^3}{\partial \xi^1}\right)^2\\
&= (\cos\xi^2)^2 + (\sin\xi^2)^2 = 1;\\
g_{22} &= \frac{\partial x^i}{\partial \xi^2}\frac{\partial x^i}{\partial \xi^2} = (\xi^1 \sin\xi^2)^2 + (\xi^1 \cos\xi^2)^2 = (\xi^1)^2;\\
g_{33} &= \frac{\partial x^i}{\partial \xi^3}\frac{\partial x^i}{\partial \xi^3} = 1;\\
g_{12} &= \frac{\partial x^i}{\partial \xi^1}\frac{\partial x^i}{\partial \xi^2} = -\xi^1 \cos\xi^2 \sin\xi^2 + \xi^1 \cos\xi^2 \sin\xi^2 = 0.
\end{aligned}$$

One can show in a similar way that $g_{ij} = 0$ at $i \neq j$. According to (1.1.13), $g_{ij} = \vec{e}_i \cdot \vec{e}_j$; therefore, the basis vectors $\vec{e}_i$ of the cylindrical coordinate system are orthogonal to each other (see Fig. 1.6). The coordinate system in which $g_{ij} = 0$ at $i \neq j$ is called orthogonal. For the cylindrical coordinate system, the ξ^i are usually denoted by the letters $\xi^1 = r$, $\xi^2 = \varphi$, $\xi^3 = z$; therefore, $g_{11} = 1$, $g_{22} = r^2$, $g_{33} = 1$:

$$ds^2 = dr^2 + r^2 d\varphi^2 + dz^2.$$

It follows from equation (1.1.17) that the matrix $g^{ij} = (g_{ij})^{-1}$; therefore,

$$g^{11} = 1, \quad g^{22} = \frac{1}{r^2}, \quad g^{33} = 1, \quad g^{ij} = 0 \text{ at } i \neq j.$$

The basis vectors $\vec{e}_i$ are directed along the tangents to the corresponding coordinate lines ξ^i (see Fig. 1.6). The modulus $|\vec{e}_i|$ is determined by the formula $g_{ii} = |\vec{e}_i|^2$ (there is no summation over i) and is equal to

$$|\vec{e}_1| = 1; \quad |\vec{e}_2| = r; \quad |\vec{e}_3| = 1.$$

The relation between the $\vec{e}_i$ and the basis vectors of the Cartesian coordinate system is determined from equation (1.1.2) and the formulas for

$\frac{\partial x^i}{\partial \xi^j}$:

$$\begin{aligned}
\vec{e}_1 &= \frac{\partial \vec{r}}{\partial \xi^1} = \frac{\partial x^k}{\partial \xi^1}\vec{E}_k = \frac{\partial x^1}{\partial \xi^1}\vec{E}_1 + \frac{\partial x^2}{\partial \xi^1}\vec{E}_2 + \frac{\partial x^3}{\partial \xi^1}\vec{E}_3 \\
&= \cos\xi^2\vec{E}_1 + \sin\xi^2\vec{E}_2 = \cos\varphi\vec{E}_1 + \sin\varphi\vec{E}_2; \\
\vec{e}_2 &= \frac{\partial \vec{r}}{\partial \xi^2} = -\xi^1\sin\xi^2\vec{E}_1 + \xi^1\cos\xi^2\vec{E}_2 \\
&= -r\sin\varphi\vec{E}_1 + r\cos\varphi\vec{E}_2; \\
\vec{e}_3 &= \frac{\partial \vec{r}}{\partial \xi^3} = \vec{E}_3.
\end{aligned}$$

The above task can easily be solved with the aid of the software system *Mathematica 3.0* (see our notebook `prog1-1.nb`). The syntax of all *Mathematica* functions, which we use in our book, is explained in Appendix A. In what follows, we present the output of the *Mathematica* notebook `prog1-1.nb`.

The Expressions for Metric Tensor Components

$$g = \begin{pmatrix} 1 & 0 & 0 \\ 0 & \xi_1^2 & 0 \\ 0 & 0 & 1 \end{pmatrix}$$

The Squared Length Element

$$ds\hat{}2 = dr^2 + dz^2 + d\varphi^2 r^2$$

The Expressions for the Basis Vectors

$$e(1) = \cos(\xi_2)B^{\rightharpoonup}{}_1 + \sin(\xi_2)B^{\rightharpoonup}{}_2$$

$$e(2) = \cos(\xi_2)\xi_1 B^{\rightharpoonup}{}_2 - \sin(\xi_2)\xi_1 B^{\rightharpoonup}{}_1$$

$$e(3) = B^{\rightharpoonup}{}_3$$

Note that the letter `E` is reserved by the *Mathematica* system as a base of the natural logarithms (`E`= 2.71828...). In this connection, we have used the notation $\vec{B}_1, \vec{B}_2, and \vec{B}_3$ instead of $\vec{E}_1$, $\vec{E}_2$, and $\vec{E}_3$, respectively. The system *Mathematica 3.0* enables the user to produce the output of the symbolic computations in conventional mathematical notation using the Greek letters, if needed. In the above notebook, we have used the

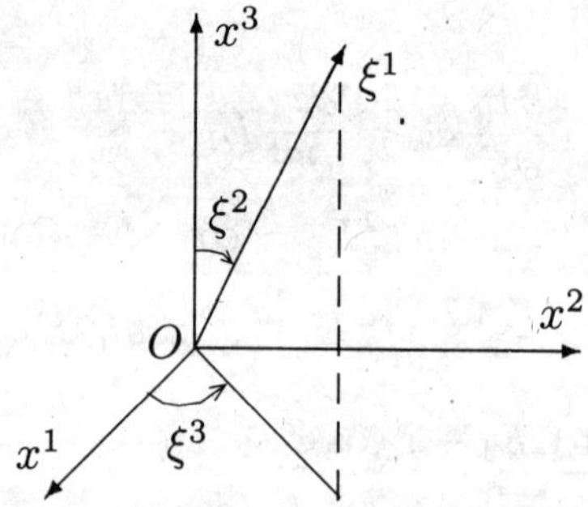

Figure 1.7: The spherical coordinate system.

same notations as in the above solution, which we have found by hand. It is easy to see that our calculations coincide with the output of the *Mathematica* program.

Problem 1.2. Find the components of a metric tensor in the spherical coordinate system. Write the expression for the element of the length dS.

Solution: Introduce the spherical coordinate system ξ^i shown in Fig. 1.7. As has been shown in the foregoing problem, the metric tensor g_{ij} in an arbitrary coordinate system is determined by the formula

$$g_{ij} = \frac{\partial x^k}{\partial \xi^j}\frac{\partial x^k}{\partial \xi^j},$$

where $x^i = x^i(\xi^j)$ determines the relation of arbitrary coordinates ξ^i to the Cartesian coordinates x^i. The expressions for the Cartesian coordinates x^i in terms of the spherical coordinates ξ^i (see Fig. 1.7) have the form

$$x^1 = \xi^1 \sin\xi^2 \cos\xi^3, \quad x^2 = \xi^1 \sin\xi^2 \sin\xi^3, \quad x^3 = \xi^1 \cos\xi^2.$$

Computing the derivatives $\frac{\partial x^i}{\partial \xi^j}$, we can find the g_{ij}:

$$\begin{aligned} g_{11} &= \frac{\partial x^k}{\partial \xi^1}\frac{\partial x^k}{\partial \xi^1} = 1, \quad g_{22} = \frac{\partial x^k}{\partial \xi^2}\frac{\partial x^k}{\partial \xi^2} = (\xi^1)^2, \\ g_{33} &= \frac{\partial x^k}{\partial \xi^3}\frac{\partial x^k}{\partial \xi^3} = (\xi^1)^2 \sin^2\xi^2, \quad g_{ij} = 0 \text{ at } i \neq j. \end{aligned}$$

The notations $\xi^1 = r$, $\xi^2 = \theta$, and $\xi^3 = \varphi$ are usually employed for spherical coordinates, in which the metric tensor components have the form

$$g_{11} = 1, \quad g_{22} = r^2, \quad g_{33} = r^2 \sin^2 \theta, \tag{1.1.78}$$

and the length element is equal to

$$ds^2 = dr^2 + r^2 \, d\theta^2 + r^2 \sin^2 \theta \, d\varphi^2. \tag{1.1.79}$$

The spherical coordinate system is orthogonal; therefore,

$$g^{11} = 1, \quad g^{22} = \frac{1}{r^2}, \quad g^{33} = \frac{1}{r^2 \sin^2 \theta}. \tag{1.1.80}$$

We have made the *Mathematica* notebook `prog1-2.nb`, which solves the same problem. We present in what follows the results of symbolic computations by this notebook.

The Expressions for Metric Tensor Components

$$g = \begin{pmatrix} 1 & 0 & 0 \\ 0 & r^2 & 0 \\ 0 & 0 & r^2 \sin^2(\theta) \end{pmatrix}$$

The Squared Length Element

$$ds\hat{}2 = dr^2 + d\theta^2 r^2 + d\varphi^2 r^2 \sin^2(\theta)$$

It is easy to see that the results obtained by the above notebook coincide with the results that we have presented above and have obtained by hand, i.e., without using a computer.

Problem 1.3. Find the metric tensor g_{ij} for the oblique-angled coordinate system shown in Fig. 1.8. Determine the covariant coordinate system ξ_i, $\vec{e}^i$, g^{ij}.

Solution: Let a Cartesian coordinate system x^1, x^2 be given together with the basis $\vec{E}_1$, $\vec{E}_2$, $|\vec{E}_1| = 1$, $|\vec{E}_2| = 1$. Construct an oblique-angled coordinate system ξ^1, ξ^2 by a clockwise rotation of the x^2-axis by the angle $\frac{\pi}{2} - \varphi$ (see Fig. 1.8). It follows from the construction that the basis vectors of the oblique-angled coordinate system will satisfy the condition $|\vec{e}_1| = 1$, $|\vec{e}_2| = 1$. Let M be a point with the radius vector $\vec{R}$ in the plane, which can be written as

$$\vec{R} = x^1 \vec{E}_1 + x^2 \vec{E}_2; \quad \vec{R} = \xi^1 \vec{e}_1 + \xi^2 \vec{e}_2,$$

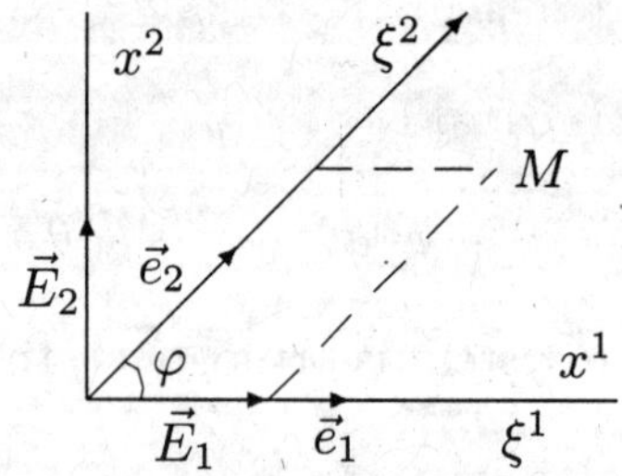

Figure 1.8: The oblique-angled coordinate system.

from where it follows that $x^1\vec{E}_1 + x^2\vec{E}_2 = \xi^1\vec{e}_1 + \xi^2\vec{e}_2$. Multiplying this equation by $\vec{E}_1$ and $\vec{E}_2$, we obtain:

$$x^1 = \xi^1(\vec{e}_1 \cdot \vec{E}_1) + \xi^2(\vec{e}_2 \cdot \vec{E}_1); \quad x^2 = \xi^1(\vec{e}_1 \cdot \vec{E}_2) + \xi^2(\vec{e}_2 \cdot \vec{E}_2).$$

Taking into account the equalities

$$\vec{e}_1 \cdot \vec{E}_1 = 1, \quad \vec{e}_2 \cdot \vec{E}_1 = \cos\varphi, \quad \vec{e}_1 \cdot \vec{E}_2 = 0, \quad \vec{e}_2 \cdot \vec{E}_2 = \sin\varphi$$

we find:

$$x^1 = \xi^1 + \xi^2 \cos\varphi, \quad x^2 = \xi^2 \sin\varphi,$$

from where it follows that

$$\frac{\partial x^1}{\partial \xi^1} = 1; \quad \frac{\partial x^1}{\partial \xi^2} = \cos\varphi; \quad \frac{\partial x^2}{\partial \xi^1} = 0; \quad \frac{\partial x^2}{\partial \xi^2} = \sin\varphi.$$

It follows from equation $g_{ij} = \frac{\partial x^k}{\partial \xi^i}\frac{\partial x^k}{\partial \xi^j}$ that

$$\begin{aligned} g_{11} &= \left(\frac{\partial x^1}{\partial \xi^1}\right)^2 + \left(\frac{\partial x^2}{\partial \xi^1}\right)^2 = 1; \quad g_{22} = \left(\frac{\partial x^1}{\partial \xi^2}\right)^2 + \left(\frac{\partial x^2}{\partial \xi^2}\right)^2 = 1; \\ g_{12} &= \frac{\partial x^1}{\partial \xi^1}\frac{\partial x^1}{\partial \xi^2} + \frac{\partial x^2}{\partial \xi^1}\frac{\partial x^2}{\partial \xi^2} = \cos\varphi. \end{aligned}$$

Since $g^{ij} = (g_{ij})^{-1}$, then

$$g_{ij} = \begin{pmatrix} 1 & \cos\varphi \\ \cos\varphi & 1 \end{pmatrix}; \quad g^{ij} = \frac{1}{\sin^2\varphi}\begin{pmatrix} 1 & -\cos\varphi \\ -\cos\varphi & 1 \end{pmatrix}.$$

The covariant coordinate system is determined by the equations

$$\xi_i = g_{ij}\xi^j; \quad \vec{e}^{\,i} = g^{ij}\vec{e}_j.$$

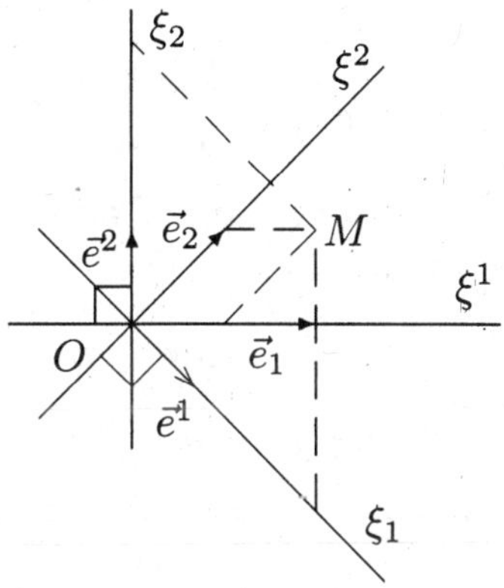

Figure 1.9: The covariant ξ_i and contravariant ξ^i coordinates of the point M in the oblique-angled coordinate system.

Taking into account the explicit expressions for g_{ij} and g^{ij}, we obtain:

$$\begin{aligned} \xi_1 &= g_{11}\xi^1 + g_{12}\xi^2 = \xi^1 + \cos\varphi\,\xi^2; \\ \xi_2 &= g_{21}\xi^1 + g_{22}\xi^2 = \xi^2 + \cos\varphi\,\xi^1; \end{aligned}$$

$$\vec{e}^{\,1} = \frac{1}{\sin^2\varphi}(\vec{e}_1 - \cos\varphi\,\vec{e}_2); \quad \vec{e}^{\,2} = \frac{1}{\sin^2\varphi}(\vec{e}_2 - \cos\varphi\,\vec{e}_1); \qquad (1.1.81)$$

$$|\vec{e}^{\,1}| = \sqrt{g^{11}} = 1/\sin\varphi; \quad |\vec{e}^{\,2}| = \sqrt{g^{22}} = \frac{1}{\sin\varphi}.$$

The scalar products

$$\vec{e}^{\,1}\cdot\vec{e}_2 = \frac{1}{\sin^2\varphi}(\vec{e}_1\cdot\vec{e}_2 - \cos\varphi(\vec{e}_2)^2) = 0, \quad \vec{e}^{\,2}\cdot\vec{e}_1 = 0,$$

because $\vec{e}_1\cdot\vec{e}_2 = \cos\varphi$, $(\vec{e}_1)^2 = (\vec{e}_2)^2 = 1$. Thus, the vector $\vec{e}_1$ is perpendicular to $\vec{e}_2$, and $\vec{e}^{\,2}$ is perependicular to $\vec{e}_1$. We show in Fig. 1.9 the covariant coordinate system ξ_1, ξ_2.

This task can also be solved with the aid of *Mathematica 3.0*. See the following output of our *Mathematica* program `prog1-3.nb`.

The Expressions for Metric Tensor Components

$$g = \begin{pmatrix} 1 & \cos(\varphi) \\ \cos(\varphi) & 1 \end{pmatrix}$$

The Covariant Components $g^{ij}, i, j = 1, 2$

$$\begin{pmatrix} \frac{1}{1-\cos^2(\varphi)} & -\frac{\cos(\varphi)}{1-\cos^2(\varphi)} \\ -\frac{\cos(\varphi)}{1-\cos^2(\varphi)} & \frac{1}{1-\cos^2(\varphi)} \end{pmatrix}$$

The Expressions for the Covariant Basis Vectors $\vec{e}^1$ and $\vec{e}^2$ in Terms of the Contravariant Basis Vectors $\vec{e}_1$ and $\vec{e}_2$

$$e\hat{\ }1 = \csc^2(\varphi)e\rightharpoonup_1 - \cot(\varphi)\csc(\varphi)e\rightharpoonup_2$$

$$e\hat{\ }2 = \csc^2(\varphi)e\rightharpoonup_2 - \cot(\varphi)\csc(\varphi)e\rightharpoonup_1$$

The *Mathematica 3.0* system has presented the expressions for the basis vectors $\vec{e}^1$ and $\vec{e}^2$ in a very compact form in terms of the trigonometric functions $\cot\varphi$ and $\csc\varphi$. By definition, $\cot\varphi = \cos\varphi/\sin\varphi$ and $\csc\varphi = 1/\sin\varphi$. It is easy to see that the substitution of these definitions into the expressions for $\vec{e}^1$ and $\vec{e}^2$ yields the familiar formulas (1.1.81).

1.2 Eulerian and Lagrangian Description of a Continuum: Strain Tensor

1.2.1 Lagrangian and Eulerian Description of a Continuum

Continuum mechanics (CM) serve for the description of the motion of gases, deformable solids, plasma, and other media consisting of a large number of atoms, molecules, ions, electrons, and other particles. An accurate computation of the motion of all particles represents an unrealistic task; therefore, the following extraordinarily efficient technique is used for the description. Certain infinitesimal material volume is chosen, which nevertheless contains a very large number of particles. Then an averaging is performed over this volume. As a result of this, the density, velocity, temperature, and other parameters are determined at each point. We will denote their set by $\varphi^{i...s}$, which characterize the medium state at a given point. Thus, the continuum, or as the physicists say, the field, is defined at each point, where a tensor is defined as a function $\varphi^{i...s} = \varphi^{i...s}(x^i, t)$. The CM problem consists of the formulation of equations for $\varphi^{i...s}$ and their solution at given initial and boundary conditions. To solve this problem, it is necessary to introduce a fixed coordinate system of the observer x^i, which is termed the *Eulerian* coordinate system in CM. In addition, one should identify each point of a continuum. Therefore, the coordinates ξ^i are assigned to each point at $t = 0$, which do not change subsequently, i.e., $\xi^i = \text{const}$, and are termed the *Lagrangian coordinates.* It is usually assumed that at the initial moment of time $t = 0$ the Eulerian and Lagrangian coordinates

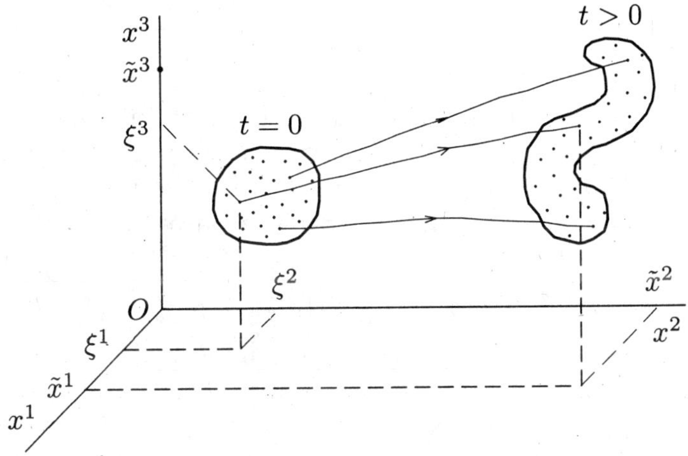

Figure 1.10: The relation between the Eulerian and Lagrangian coordinates of points of a continuum.

coincide: $x^i|_{t=0} = \xi^i$. Then the continuum motion is determined by the equation

$$x^i = x^i(\xi^j, t), \tag{1.2.1}$$

which determines under the condition

$$\det\left|\left|\frac{\partial x^i}{\partial \xi^j}\right|\right| \neq 0, \tag{1.2.2}$$

a one-to-one correspondence between the Eulerian x^i and Lagrangian ξ^i coordinates. Equation (1.2.1) is sometimes called the motion law or the particle trajectory since, for a given particle with ξ^i = const, it determines the coordinate of this particle at any moment of time t. [Note that the condition $x^i|_{t=0} = \xi^i$ is not a necessary condition for the existence of the trajectory (1.2.1) because one can always determine some coordinate transform $x^i|_{t=0} = f(\xi^j)$ such that $x^i = x^i(f(\xi^j), t) = \tilde{x}^i(\xi^j, t)$ (see Fig. 1.10)]. To determine the invariant objects (vector, tensor), one must define a coordinate system, which is specified with the aid of the basis vectors $\vec{e}_i$. For example, by specifying an orthonormal basis of a Cartesian coordinate system

$$\vec{e}_i \cdot \vec{e}_j = \delta_{ij} = \begin{cases} 0, & i \neq j \\ 1, & i = j, \end{cases} \tag{1.2.3}$$

we can define in the neighborhood of the coordinate origin at the moment of time $t = 0$ a particle by the vector

$$\vec{r}_0 = \xi^i \vec{e}_i, \tag{1.2.4}$$

which at the moment of time t will have the vector

$$\vec{r} = x^i \vec{e}_i. \tag{1.2.5}$$

Along with a fixed coordinate system, the *accompanying coordinate system* is often used in CM, in which the particle coordinates ξ^i remain constant at all times, the basis vectors $\hat{\vec{e}}_i$ depend on time, and the particle position at the moment of time t is determined by the vector

$$\vec{r} = \xi^i \vec{e}_i(t). \tag{1.2.6}$$

In this case, the coordinate system is frozen in the continuum and it deforms together with it and is generally curvilinear. Consequently, equation (1.2.6) is local and determines the radius vector of particles located in a small neighborhood of a given particle whose position is taken as a coordinate origin $\xi^i = 0$. While moving from one point to another, however, one can determine the radius vector of an arbitrary point of a continuum with respect to a given point with $\xi^i = 0$.

Thus, there exist two techniques for the description of a continuum. In the Lagrangian description technique, the Lagrangian coordinates ξ^i and the time t are the independent variables. The Lagrangian coordinates are generally curvilinear and deform in an accompanying coordinate system together with a continuum. In the Lagrangian approach, a complete description of the motion of a given particle with the coordinate ξ^i is given, which is determined by equation (1.2.1). Let (1.2.1) be given in a Cartesian basis of an Eulerian coordinate system (1.2.3). Then the velocity $\vec{v}$ and the acceleration $\vec{a}$ of a particle ξ^i will be equal to

$$\vec{v} = v^i \vec{e}_i = \left(\frac{\partial x^i}{\partial t}\right)_{\xi^i} \vec{e}_i, \quad \vec{a} = a^i \vec{e}_i = \left(\frac{\partial v^i}{\partial t}\right)_{\xi^i} \vec{e}_i. \tag{1.2.7}$$

And, inversely, the integration of the relationship $\frac{\partial x^i}{\partial t} = v^i(\xi^j, t)$ at initial conditions $x^i|_{t=0} = \xi^i$ determines (1.2.1).

In the Eulerian description technique, the independent coordinates are the coordinates of points of the space x^i and the time t. The functions $\varphi^{i\ldots s}(x^i, t)$ determine at fixed x^i the temporal variation of the function $\varphi^{i\ldots s}$ at a given point of space through which different points of a continuum pass. The passage from the Eulerian description technique to the Lagrangian one and a reverse passage are carried out with the aid of the motion law (1.2.1) and consist in a passage from the independent coordinates x^i to the independent coordinates ξ^i. Let a function $\varphi^{i\ldots s} = \varphi^{i\ldots s}(x^i, t)$ be given in the Eulerian coordinates. Then, in the Lagrangian coordinates, we obtain:

$$\varphi^{i\ldots s} = \varphi^{i\ldots s}(x^i(\xi^j, t), t) = \tilde{\varphi}^{i\ldots s}(\xi^j, t), \tag{1.2.8}$$

where the motion law (1.2.1) is found by integrating the system of equations $\frac{dx^i}{dt} = v^i(x^j, t)$ under the initial conditions $x|_{t=0} = \xi^i$. If the function $\varphi^{i\dots s} = \varphi^{i\dots s}(\xi^j, t)$ is given conversely in the Lagrangian coordinates, then, in the Eulerian coordinates, we will have

$$\varphi^{i\dots s} = \varphi^{i\dots s}(\xi^i(x^j, t), t) = \tilde{\tilde{\varphi}}^{i\dots s}(x^j, t), \tag{1.2.9}$$

where $\xi^i = \xi^i(x^j, t)$ is the inverse transformation from the Eulerian coordinates to the Lagrangian coordinates, which by virtue of (1.2.2) always exists and is determined from the solution of equations (1.2.1). Equation (1.2.9) enables us to determine in the Eulerian coordinates the *substantive derivative* with respect to the time of the function $\varphi^{i\dots s}(x^k, t)$ for a fixed particle ξ^i:

$$\left(\frac{\partial \varphi^{i\dots s}}{\partial t}\right)_{\xi^i} = \left(\frac{\partial \varphi^{i\dots s}}{\partial t}\right)_{x^i} + \left(\frac{\partial \varphi^{i\dots s}}{\partial x^j}\right)_t \left(\frac{\partial x^j}{\partial t}\right),$$

which can be written with regard for (1.2.7) in the form

$$\left(\frac{\partial \varphi^{i\dots s}}{\partial t}\right)_{\xi^i} = \frac{\partial \varphi^{i\dots s}}{\partial t} + v^j \frac{\partial \varphi^{i\dots s}}{\partial x^j}. \tag{1.2.10}$$

Choosing, in particular, the velocity components $v^i(x^k, t)$ as $\varphi^{i\dots s}(x^k, t)$, we obtain from (1.2.10) an expression for acceleration in the Cartesian coordinate system:

$$\vec{a} = a^i \vec{e}_i, \quad a^i = \frac{\partial v^i}{\partial t} + v^j \frac{\partial v^i}{\partial x^j}. \tag{1.2.11}$$

Let us present a formula determining the substantive derivative in Lagrangian coordinates:

$$\left(\frac{\partial \varphi^{i\dots s}}{\partial t}\right)_{\xi^j} = \frac{\partial}{\partial t}\varphi^{i\dots s}(\xi^j, t). \tag{1.2.12}$$

The stationary (steady) processes are of great importance in mechanics. For these processes, the motion characteristics in the Eulerian description depend only on the coordinates x^i and not explicitly on the time t:

$$\frac{\partial}{\partial t}\varphi^{i\dots s}(x^k) = 0. \tag{1.2.13}$$

Note that the same motion can be stationary in one reference frame and nonstationary in another reference frame. It is convenient to represent the flow pattern at a stationary flow with the aid of streamlines, which are determined in such a way that the tangent to a streamline at each point is directed along or in a counterdirection of the velocity vector.

Consequently, the vector of an infinitesimal increment $d\vec{r} = dx^i \vec{e}_i$ along the streamline is parallel with the velocity vector $\vec{v} = v^i \vec{e}_i$; therefore, $d\vec{r} = d\lambda\, \vec{v}$, from where the equation for the streamlines follows:

$$\frac{dx^1}{v^1} = \frac{dx^2}{v^2} = \frac{dx^3}{v^3} = d\lambda. \tag{1.2.14}$$

Equations (1.2.14) differ from the equation of trajectories

$$\frac{dx^i}{dt} = v^i(x^i, t), \tag{1.2.15}$$

since the time t enters the right-hand side of equation (1.2.15). It is seen from the comparison of (1.2.14) and (1.2.15) that the streamlines and the trajectories coincide in the stationary case. If we choose some line l, which is no streamline, and if we draw a streamline through each point of l, then we obtain in the result a stream surface, the tangent to which is parallel to the velocity vector $\vec{v}$. If l is a closed line, then the stream surface forms a stream pipe.

It follows from the foregoing that the Lagrangian and Eulerian descriptions are in the mechanical respect complete and equivalent to each other. The experience shows, however, that in the problems of the mechanics of deformable solids (the elasticity and plasticity theory) a preference is given to the Lagrangian description, and in the aerohydrodynamics problems (the flows of fluid and gas) the Eulerian description technique is usually employed.

1.2.2 Strain Tensor

Let us specify a Eulerian coordinate system $x^i, \vec{e}_i$ in which we choose an infinitesimal particle and study its deformation between the moments of time t_0 and t. During the time $\Delta t = t - t_0$, the particle shifts as a solid body by a vector $\vec{u}$, making a rotation $\vec{\omega} = \frac{1}{2}\text{rot}\, \vec{u}$, and undergoes some deformation consisting of a change in particle form. To determine the deformation, let us consider two states of the particle. We will assume that, at the moment of time t_0 (the *reference frame*), the coordinates of particles $x_0^i = x_0^i(\xi^1, \xi^2, \xi^3, t_0)$, where ξ^j are the Lagrangian coordinates (see Fig. 1.11). Let us define in the particle the coordinate system ξ^i with the origin at point A and the basis $\overset{o}{\vec{e}}_i = \frac{\partial \vec{r}_0}{\partial \xi^i}$. Then the distance from point A to an infinitesimally close point B at $t = t_0$ is equal to

$$ds_0^2 = \overset{o}{g}_{ij}\, d\xi^i d\xi^j, \quad \overset{o}{g}_{ij} = \overset{o}{\vec{e}}_i \cdot \overset{o}{\vec{e}}_j\,. \tag{1.2.16}$$

At the moment of time t (the actual configuration), the distance between the points A and B is

$$ds^2 = \hat{g}_{ij} d\xi^i d\xi^j, \quad \hat{g}_{ij} = \hat{\vec{e}}_i \cdot \hat{\vec{e}}_j, \tag{1.2.17}$$

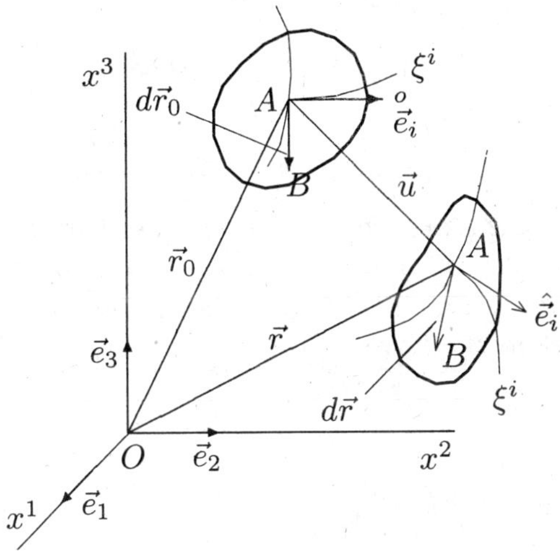

Figure 1.11: To the derivation of the strain tensor.

where $\hat{\vec{e}}_i = \frac{\partial \vec{r}}{\partial \xi^i}$ is the basis vector of the accompanying coordinate system at the time t. In the accompanying coordinate system, the basis vectors $\hat{\vec{e}}_i$ are "frozen" in the continuum and vary at the medium deformation $\hat{\vec{e}}_i = \hat{\vec{e}}_i(t)$. Thus, in the Lagrangian coordinates, the *accompanying coordinate system* is determined by the basis $\hat{\vec{e}}_i$, and a fixed coordinate system is determined by the basis $\overset{o}{\vec{e}}_i$. The alteration of the squared distance between the points A and B is equal to

$$ds^2 - ds_0^2 = (\hat{g}_{ij} - \overset{o}{g}_{ij})\, d\xi^i\, d\xi^j,$$

which enables us to introduce the strain tensor $ds^2 - ds_0^2 = 2\varepsilon_{ij} d\xi^i d\xi^j$ with the components[1,2]

$$\varepsilon_{ij} = \frac{1}{2}(\hat{g}_{ij} - \overset{o}{g}_{ij}). \tag{1.2.18}$$

The components of the *strain tensor* (1.2.18) relate the state in two configurations; therefore, one can introduce two strain tensors. Let us determine the strain tensor in the reference frame basis, which is called in the literature the *Green's strain tensor*

$$\overset{o}{E} = \varepsilon_{ij}\, \overset{o}{\vec{e}}{}^i\, \overset{o}{\vec{e}}{}^j. \tag{1.2.19}$$

The contravariant components of the strain tensor in the reference frame basis are

$$\overset{o}{\varepsilon}{}^{ij} = \overset{o}{g}{}^{ik}\, \overset{o}{g}{}^{jl} \varepsilon_{kl}. \tag{1.2.20}$$

In the accompanying coordinate system, the Almansi strain tensor is determined, for which we have

$$\hat{E} = \varepsilon_{ij}\hat{\vec{e}}^{\,i}\hat{\vec{e}}^{\,j}, \quad \hat{\varepsilon}^{ij} = \hat{g}^{\,ik}\hat{g}^{\,jl}\varepsilon_{kl}. \tag{1.2.21}$$

If the motion law of a continuum $x^i = x^i(\xi^j, t)$, $x^i\Big|_{t_0} = x_0^i$ is given, then one can express ε_{ij} in terms of the first derivatives of x^i. Consider the actual configuration. Since ds^2 is a scalar, its magnitude is the same in the accompanying and Eulerian coordinate systems, $\hat{g}_{ij}d\xi^i d\xi^j = g_{kl}dx^k dx^l$, from where we have

$$\hat{g}_{ij} = g_{kl}\frac{\partial x^k}{\partial \xi^i}\frac{\partial x^l}{\partial \xi^j}. \tag{1.2.22}$$

Substituting (1.2.22) into (1.2.18), we obtain:

$$\hat{\varepsilon}_{ij} = \frac{1}{2}\Big(g_{kl}\frac{\partial x^k}{\partial \xi^i}\frac{\partial x^l}{\partial \xi^j} - \overset{o}{g}_{ij}\Big). \tag{1.2.23}$$

In the reference frame, we have for ds_0 that

$$\overset{o}{g}_{ij}\, d\xi^i d\xi^j = g_{kl}dx_0^k dx_0^l,$$

from where

$$\overset{o}{g}_{ij} = g_{kl}\frac{\partial x_0^k}{\partial \xi^i}\frac{\partial x_0^l}{\partial \xi^j}, \tag{1.2.24}$$

where $\frac{\partial x_0^k}{\partial \xi^i}$ is determined from the equation $x_0^k = x_0^k(\xi^i, t_0)$. Substituting (1.2.24) into (1.2.18), we obtain the formula:

$$\overset{o}{\varepsilon}_{ij} = \frac{1}{2}\Big(\hat{g}_{ij} - g_{kl}\frac{\partial x_0^k}{\partial \xi^i}\frac{\partial x_0^l}{\partial \xi^j}\Big). \tag{1.2.25}$$

Employing the formula for the transformation of the strain tensor components from the accompanying coordinate system to the Eulerian coordinate system

$$\varepsilon_{ij} = \frac{\partial \xi^k}{\partial x^i}\frac{\partial \xi^l}{\partial x^j}\hat{\varepsilon}_{kl} = \frac{1}{2}\Big(\frac{\partial \xi^k}{\partial x^i}\frac{\partial \xi^l}{\partial x^j}\hat{g}_{kl} - \overset{o}{g}_{kl}\frac{\partial \xi^k}{\partial x^i}\frac{\partial \xi^l}{\partial x^j}\Big)$$

and the invariance $ds^2 = \hat{g}_{kl}d\xi^k d\xi^l = g_{ij}dx^i dx^j$, we obtain the strain tensor in the Eulerian coordinate system:

$$\varepsilon_{ij} = \frac{1}{2}\Big(g_{ij} - \overset{o}{g}_{kl}\frac{\partial \xi^k}{\partial x^i}\frac{\partial \xi^l}{\partial x^j}\Big). \tag{1.2.26}$$

Let us find the expressions for $\overset{o}{\varepsilon}_{ij}$ and $\hat{\varepsilon}_{ij}$ in terms of the displacement vector $\vec{u}$. As follows from Fig. 1.11, $\vec{r} = \vec{r}_0 + \vec{u}$, from where

$$\frac{\partial \vec{u}}{\partial \xi^i} = \frac{\partial \vec{r}}{\partial \xi^i} - \frac{\partial \vec{r}_0}{\partial \xi^i} = \hat{\vec{e}}_i - \overset{o}{\vec{e}}_i .$$

Expressing $\hat{\vec{e}}_i = \overset{o}{\vec{e}}_i + \frac{\partial \vec{u}}{\partial \xi^i}$, we obtain the formula:

$$\hat{g}_{ij} = \hat{\vec{e}}_i \cdot \hat{\vec{e}}_j = \overset{o}{\vec{e}}_i \cdot \overset{o}{\vec{e}}_j + \overset{o}{\vec{e}}_i \cdot \frac{\partial \vec{u}}{\partial \xi^j} + \overset{o}{\vec{e}}_j \cdot \frac{\partial \vec{u}}{\partial \xi^i} + \frac{\partial \vec{u}}{\partial \xi^i} \cdot \frac{\partial \vec{u}}{\partial \xi^j}.$$

Substituting this expression into (1.2.18), we find the Green's strain tensor as

$$\overset{o}{\varepsilon}_{ij} = \frac{1}{2}\Big(\overset{o}{\vec{e}}_i \cdot \frac{\partial \vec{u}}{\partial \xi^j} + \overset{o}{\vec{e}}_j \cdot \frac{\partial \vec{u}}{\partial \xi^i} + \frac{\partial \vec{u}}{\partial \xi^i} \cdot \frac{\partial \vec{u}}{\partial \xi^j}\Big). \qquad (1.2.27)$$

Expressing $\overset{o}{\vec{e}}_i = \hat{\vec{e}}_i - \frac{\partial \vec{u}}{\partial \xi^i}$ and

$$\overset{o}{g}_{ij} = \hat{\vec{e}}_i \cdot \hat{\vec{e}}_j - \hat{\vec{e}}_i \cdot \frac{\partial \vec{u}}{\partial \xi^j} - \hat{\vec{e}}_j \cdot \frac{\partial \vec{u}}{\partial \xi^i} + \frac{\partial \vec{u}}{\partial \xi^i} \cdot \frac{\partial \vec{u}}{\partial \xi^j},$$

we find with regard for (1.2.18) the Almansi strain tensor

$$\hat{\varepsilon}_{ij} = \frac{1}{2}\Big(\hat{\vec{e}}_i \cdot \frac{\partial \vec{u}}{\partial \xi^j} + \hat{\vec{e}}_j \cdot \frac{\partial \vec{u}}{\partial \xi^i} - \frac{\partial \vec{u}}{\partial \xi^i} \cdot \frac{\partial \vec{u}}{\partial \xi^j}\Big). \qquad (1.2.28)$$

Assuming $\vec{u} = \overset{o}{u}{}^i \overset{o}{\vec{e}}_i = \hat{u}^i \hat{\vec{e}}_i$ and using the definition of a covariant derivative $\frac{\partial \vec{u}}{\partial \xi^i} = \overset{o}{\nabla}_i \overset{o}{u}{}^k \overset{o}{\vec{e}}_k$, $\frac{\partial \vec{u}}{\partial \xi^j} = \hat{\nabla}_j \hat{u}^k \hat{\vec{e}}_k$, we find the scalar products:

$$\begin{aligned}
\overset{o}{\vec{e}}_i \cdot (\overset{o}{\nabla}_j \overset{o}{u}{}^k) \overset{o}{\vec{e}}_k &= \overset{o}{\nabla}_j \overset{o}{u}_i, \qquad \hat{\vec{e}}_i \cdot (\hat{\nabla}_j \hat{u}^k) \hat{\vec{e}}_k = \hat{\nabla}_j \hat{u}_i, \\
\frac{\partial \vec{u}}{\partial \xi^i} \cdot \frac{\partial \vec{u}}{\partial \xi^j} &= \overset{o}{\nabla}_i \overset{o}{u}{}^k \overset{o}{\nabla}_j \overset{o}{u}{}^l \overset{o}{\vec{e}}_k \cdot \overset{o}{\vec{e}}_l = \overset{o}{g}_{kl} \overset{o}{\nabla}_i \overset{o}{u}{}^k \overset{o}{\nabla}_j \overset{o}{u}{}^l = \overset{o}{\nabla}_i \overset{o}{u}_k \overset{o}{\nabla}_j \overset{o}{u}{}^k, \\
\frac{\partial \vec{u}}{\partial \xi^i} \cdot \frac{\partial \vec{u}}{\partial \xi^j} &= \hat{\nabla}_i \hat{u}_k \hat{\nabla}_j \hat{u}^k.
\end{aligned}$$

Substituting these products in (1.2.27) and (1.2.28), we obtain the *Green's strain tensor in a reference frame*:

$$\overset{o}{\varepsilon}_{ij} = \frac{1}{2}(\overset{o}{\nabla}_i \overset{o}{u}_j + \overset{o}{\nabla}_j \overset{o}{u}_i + \overset{o}{\nabla}_i \overset{o}{u}_k \overset{o}{\nabla}_j \overset{o}{u}{}^k) \qquad (1.2.29)$$

and the *Almansi tensor in the actual configuration*

$$\hat{\varepsilon}_{ij} = \frac{1}{2}(\hat{\nabla}_i \hat{u}_j + \hat{\nabla}_j \hat{u}_i - \hat{\nabla}_i \hat{u}_k \hat{\nabla}_j \hat{u}^k). \qquad (1.2.30)$$

In the case of small strains $\overset{o}{\nabla}_i \overset{o}{u}_j \ll 1$, $\hat{\nabla}_i \hat{u}_j \ll 1$, the difference between the reference frame and the actual configuration disappears: $\hat{g}_{ij} \approx \overset{o}{g}_{ij}$; therefore, the strain tensors of Green and Almansi coincide:

$$\overset{o}{\varepsilon}_{ij} = \frac{1}{2}(\overset{o}{\nabla}_i \overset{o}{u}_j + \overset{o}{\nabla}_j \overset{o}{u}_i) \approx \frac{1}{2}(\hat{\nabla}_i \hat{u}_i + \hat{\nabla}_j \hat{u}^i) \approx \hat{\varepsilon}_{ij}. \tag{1.2.31}$$

Let us find the expressions for the strain deformation ε_{ij} in the Eulerian coordinates in terms of the displacement vector $\vec{u}$. Since $\vec{r} = \vec{r}_0 + \vec{u}$, the variation of an infinitesimal vector $d\vec{r}$ will be given by the formula $d\vec{r} - d\vec{r}_0 = d\vec{u}$. Since $d\vec{r} = dx^i \vec{e}_i$, $d\vec{u} = \frac{\partial \vec{u}}{\partial x^i} dx^i$, therefore, we have that $d\vec{r}_0 = dx^i(\vec{e}_i - \frac{\partial \vec{u}}{\partial x^i})$. With regard for these relations, we can find the variation of a squared distance between the points A and B (see Fig. 1.11):

$$\begin{aligned} ds^2 - ds_0^2 &= d\vec{r} \cdot d\vec{r} - d\vec{r}_0 \cdot d\vec{r}_0 \\ &= dx^i dx^j \vec{e}_i \cdot \vec{e}_j - dx^i dx^j \left(\vec{e}_i - \frac{\partial \vec{u}}{\partial x^i}\right) \cdot \left(\vec{e}_j - \frac{\partial \vec{u}}{\partial x^j}\right) \\ &= \left(\frac{\partial \vec{u}}{\partial x^i} \cdot \vec{e}_j + \frac{\partial \vec{u}}{\partial x^j} \cdot \vec{e}_i - \frac{\partial \vec{u}}{\partial x^i} \cdot \frac{\partial \vec{u}}{\partial x^j}\right) dx^i dx^j. \end{aligned} \tag{1.2.32}$$

Using the definition of the strain tensor

$$ds^2 - ds_0^2 = 2\varepsilon_{ij} dx^i dx^j \tag{1.2.33}$$

and the definition of the covariant derivative $\frac{\partial \vec{u}}{\partial x^i} = \nabla_i u^k \vec{e}_k$, we obtain with regard for equations:

$$\begin{aligned} \frac{\partial \vec{u}}{\partial x^i} \cdot \vec{e}_j &= \nabla_i u^k \vec{e}_k \cdot \vec{e}_j = \nabla_i u^k g_{kj} = \nabla_i u_j, \\ \frac{\partial \vec{u}}{\partial x^i} \cdot \frac{\partial \vec{u}}{\partial x^j} &= \nabla_i u^k \nabla_j u^l \vec{e}_k \cdot \vec{e}_l \\ &= g_{kl} \nabla_i u^k \nabla_j u^l = \nabla_i u^k \nabla_j u_k, \end{aligned}$$

the strain tensor in the Eulerian coordinates

$$\varepsilon_{ij} = \frac{1}{2}(\nabla_i u_j + \nabla_j u_i - \nabla_i u^k \nabla_j u_k). \tag{1.2.34}$$

It is seen from the comparison of (1.2.30) and (1.2.34) that the Eulerian strain tensor coincides with the Almansi strain tensor written in the Eulerian coordinate system. Expression (1.2.34) can be obtained directly from (1.2.30) if one chooses in the reference frame such a system of accompanying coordinates $\overset{o}{g}_{ij}$ that it coincides in the actual configuration with the Eulerian coordinate system $\hat{g}_{ij} = g_{ij}$. In this case, one can perform in (1.2.30) a substitution $\hat{\nabla}_i \to \nabla_i$, $\hat{u}_i \to u_i$, $\hat{u}^k \to u^k$, and then (1.2.30) will coincide with (1.2.34).

Let us give a geometric interpretation of the components ε_{ij} following[1]. Let the coordinate system

$$\overset{o}{g}_{ij}=\overset{o}{\vec{e}}_i \cdot \overset{o}{\vec{e}}_j=|\overset{o}{\vec{e}}_i||\overset{o}{\vec{e}}_j|\cos\varphi^0_{ij} \tag{1.2.35}$$

be specified in the reference frame. As a result of deformation, this system will go over into an actual configuration with the basis vectors $\hat{\vec{e}}_i$ and the metric tensor

$$\hat{g}_{ij} = \hat{\vec{e}}_i \cdot \hat{\vec{e}}_j = |\hat{\vec{e}}_i||\hat{\vec{e}}_j|\cos\hat{\varphi}_{ij}. \tag{1.2.36}$$

By using the definition of the basis vectors, we can find the relative variation of their moduli

$$\frac{|\hat{\vec{e}}_i|}{|\overset{o}{\vec{e}}_i|} = \frac{|\partial\vec{r}/\partial\xi^i|}{|\partial\vec{r}_0/\partial\xi^i|} = \frac{|d\vec{r}_i|}{|d\vec{r}_{0i}|} = \frac{ds_i}{ds_{0i}} = l_i + 1, \tag{1.2.37}$$

where ds_i and ds_{0i} are the infinitesimal elements of the length along the coordinate lines ξ^i and l_i is the coefficient of a relative lengthening along ξ^i. Substituting (1.2.35)–(1.2.37) into (1.2.18), we obtain:

$$2\varepsilon_{ij} = ((1+l_i)(1+l_j)\cos\hat{\varphi}_{ij} - \cos\overset{o}{\varphi}_{ij})|\overset{o}{\vec{e}}_i||\overset{o}{\vec{e}}_j|. \tag{1.2.38}$$

Let us choose in the reference frame a Cartesian coordinate system, in which we find from (1.2.38) the expressions for the diagonal components:

$$2\varepsilon_{ii} = (1+l_i)^2 - 1. \tag{1.2.39}$$

In the case of small deformations, when $\varepsilon_{ii} \ll 1$, it follows from (1.2.39) that the diagonal components of the strain tensor in Cartesian coordinates are equal to the relative lengthenings along the corresponding axes; i.e., $\varepsilon_{ii} \approx l_i$. For the off-diagonal components $i \neq j$, the angle between the vectors $\overset{o}{\vec{e}}_i$ and $\overset{o}{\vec{e}}_j$ prior to the deformation is equal to $\overset{o}{\varphi}_{ij}=\frac{\pi}{2}$, and after the deformation between the vectors $\hat{\vec{e}}_i$ and $\hat{\vec{e}}_j$ is equal to $\hat{\varphi}_{ij} = \frac{\pi}{2} - \hat{\theta}_{ij}$. Substituting these relationships into (1.2.38), we find that $2\varepsilon_{ij} = |\hat{\vec{e}}_i||\hat{\vec{e}}_j|\sin\hat{\theta}_{ij}$, from where it follows that

$$\sin\hat{\theta}_{ij} = \frac{2\varepsilon_{ij}}{\sqrt{\hat{g}_{ii}}\sqrt{\hat{g}_{jj}}}, \tag{1.2.40}$$

where $i \neq j$ and there is no summation over the repeating indices. For small deformations with $\varepsilon_{ij} \ll 1$, we determine from (1.2.40) that $\varepsilon_{ij} = \theta_{ij}/2$. Thus, if we choose an infinitesimal, rectangular parallelepiped in the reference frame, it will go over after the deformation in an oblique parallelepiped, the side lengthenings of which are determined by (1.2.39), and the skewness angles $\hat{\theta}_{ij}$ are determined by formula (1.2.40).

As it follows from (1.2.18), the strain tensor is a symmetric tensor, which can be reduced to a diagonal form (1.1.57) by a choice of the coordinate system. The eigenvectors $\vec{b}_\alpha$ form an orthogonal coordinate system (1.1.57), which can shift and rotate as an absolutely rigid body in the process of deformation. The axes of a Cartesian coordinate system, which are directed along the principal axes, are called the *principal axes* of the strain tensor. The strain tensor components in the coordinate system related to the principal axes are called the principal components or the principal values of the strain tensor ε_i. The principal values and the eigenvectors are determined from the system of linear equations (1.1.49) of the form $(\varepsilon^i_j - \varepsilon\delta^i_j)b^j = 0$, which has the solution if the determinant $|\varepsilon^i_j - \varepsilon\delta^i_j| = 0$. Calculating the determinant, we obtain a third-order equation $\varepsilon^3 - J_1\varepsilon^2 + J_2\varepsilon - J_3 = 0$, where J_1, J_2, and J_3 are the first, second, and third invariants of the strain tensor (1.1.52), respectively:

$$J_1 = \varepsilon^i_i, \quad J_2 = \frac{1}{2}((\varepsilon^i_i)^2 - \varepsilon^i_j\varepsilon^j_i), \quad J_3 = |\varepsilon^j_i|. \tag{1.2.41}$$

If one chooses a coordinate system coinciding with principal axes, then the strain tensor will be diagonal, $\varepsilon_{ij} = \varepsilon_i\delta_{ij}$ (there is no summation over i), and the invariants will have a simple form:

$$J_1 = \varepsilon_1 + \varepsilon_2 + \varepsilon_3, \quad J_2 = \varepsilon_1\varepsilon_2 + \varepsilon_1\varepsilon_3 + \varepsilon_2\varepsilon_3, \quad J_3 = \varepsilon_1\varepsilon_2\varepsilon_3. \tag{1.2.42}$$

In the case of infinitesimal deformations, the first invariant determines the relative change of an infinitesimal volume:

$$\frac{dV - dV_0}{dV_0} \approx J_1 = \varepsilon^i_i. \tag{1.2.43}$$

To prove this fact, let us choose in the reference frame the coordinate system coinciding with the principal axes of the strain tensor: $\overset{o}{\vec{e}}_\alpha = \vec{b}_\alpha$ and $\overset{o}{\varphi}_{\alpha\beta} = \pi/2$. It follows from (1.2.40) that these basis vectors will go over again to orthogonal basis vectors $\hat{\vec{e}}_\alpha$ with $\hat{\varphi}_{\alpha\beta} = \pi/2$ because $\sin\hat{\theta}_{\alpha\beta} \sim \varepsilon_{\alpha\beta} = 0$. If we construct at the moment of time t_0 on the basis vectors $\overset{o}{\vec{e}}_\alpha = \vec{b}_\alpha$ an infinitesimal rectangular parallelepiped with the volume $dV_0 = ds_{01}ds_{02}ds_{03}$, then after deformation it will go over at time t to a rectangular parallelepiped with the volume $dV = ds_1ds_2ds_3$, where ds_{0i} and ds_i are the length elements along the ith axis at times t_0 and t. (The principal axes of the tensors $\overset{o}{\varepsilon}_{ij}$ and $\hat{\varepsilon}_{ij}$ pass through the same points of the medium in the corresponding spaces $\overset{o}{g}_{ij}$ and $\hat{g}_{ij}$; however, their components are generally different: $\overset{o}{\varepsilon}_\alpha \neq \hat{\varepsilon}_\alpha$.) The elements ds_{0i} along the ith axes change and go over to ds_i, where $ds_i^2 = \hat{g}_{ii}(d\xi^i)^2$ and $ds_{0i}^2 = \overset{o}{g}_{ii}(d\xi^i)^2$. Using the definition of the strain tensor (1.2.18) in the actual orthogonal coordinate system, we obtain:

$$\begin{aligned} ds_i^2 - ds_{0i}^2 &= (\hat{g}_{ii} - \overset{o}{g}_{ii})(d\xi^i)^2 = 2ds_i^2\varepsilon_{ii}/\hat{g}_{ii} \\ &= 2\hat{g}^{ii}\varepsilon_{ii}ds_i^2 = 2\hat{\varepsilon}_i^i ds_i^2 = 2\hat{\varepsilon}_i ds_i^2, \end{aligned}$$

from where we can find the variation of the length of a parallelepiped edge along the ith axis:

$$\frac{ds_i}{ds_{0i}} = \frac{1}{\sqrt{1-2\hat{\varepsilon}_i}}. \tag{1.2.44}$$

In the case of infinitesimal deformations $\hat{\varepsilon}_i \ll 1$, $\hat{\varepsilon}_i \approx \varepsilon_i$, we obtain from (1.2.44):

$$ds_i/ds_{0i} \approx 1 + \varepsilon_i, \quad dV/dV_0 \approx 1 + \varepsilon_1 + \varepsilon_2 + \varepsilon_3, \tag{1.2.45}$$

from where we obtain (1.2.43).

1.2.3 A Condition for Compatibility of Deformations

The problem in determining ε_{ij} at a given motion law $x^i = x^i(\xi^j, t)$ was solved above. One can formulate an inverse problem in finding x^i (or $\vec{r} = x^i\vec{e}_i$, which is the same) by using a given strain tensor ε_{ij} in a simply connected continuum. This problem can be solved under certain limitations for the components ε_{ij}, which are called the compatibility conditions[1,5]. As a matter of fact, the strain tensor ε_{ij} has six different components, on the basis of which one must find only three components of the vector $\vec{r}$, which is not possible for arbitrary ε_{ij}.

Let a vector $\vec{r}_1$ exist, which determines a point where the basis $\hat{\vec{e}}_i(\xi_1^i)$ is specified. It is required to determine the vector $\vec{r}$ on the basis of the given $\hat{\varepsilon}_{ij}(\xi^k)$ in some neighborhood of $\vec{r}_i$. We obtain from the definition of $\hat{\vec{e}}_i = \frac{\partial \vec{r}}{\partial \xi^i}$ that

$$\vec{r} = \vec{r}_1 + \int_{\xi_1^i}^{\xi^i} \hat{\vec{e}}_i d\xi^i. \tag{1.2.46}$$

The vector $\vec{r}$ is uniquely determined from (1.2.46) if the integrand is a total differential, i.e.,

$$\frac{\partial \hat{\vec{e}}_i}{\partial \xi^j} = \frac{\partial \hat{\vec{e}}_j}{\partial \xi^i}. \tag{1.2.47}$$

As was pointed out in Section 1.1 [see (1.1.21)], the derivative of the basis vector is expressed in terms of the Christoffel symbols

$$\frac{\partial \hat{\vec{e}}_i}{\partial \xi^j} = \Gamma^k_{ij} \hat{\vec{e}}_k; \tag{1.2.48}$$

therefore, substituting (1.2.48) into (1.2.47), we will find the integrability condition for (1.2.46):

$$\Gamma^k_{ij} = \Gamma^k_{ji}, \tag{1.2.49}$$

which is satisfied by virtue of the definition of the Christoffel symbols

$$\Gamma^k_{ij} = \frac{1}{2} \hat{g}^{ks} \Big(\frac{\partial \hat{g}_{is}}{\partial \xi^j} + \frac{\partial \hat{g}_{js}}{\partial \xi^i} - \frac{\partial \hat{g}_{ij}}{\partial \xi^s} \Big), \quad \text{where} \quad \hat{g}_{ij} = \overset{o}{g}_{ij} + 2\hat{\varepsilon}_{ij}.$$

The vector $\hat{\vec{e}}_i$ enters equation (1.2.46), which is determined from equation (1.2.48):

$$\hat{\vec{e}}_i(\xi^i) = \hat{\vec{e}}_i(\xi^i_1) + \int_{\xi^i_1}^{\xi^i} \Gamma^k_{ij} \hat{\vec{e}}_k d\xi^j. \tag{1.2.50}$$

The integrability condition for (1.2.50) has the form

$$\frac{\partial}{\partial \xi^l} (\Gamma^k_{ij} \hat{\vec{e}}_k) = \frac{\partial}{\partial \xi^j} (\Gamma^k_{il} \hat{\vec{e}}_k). \tag{1.2.51}$$

Using the relationships

$$\Gamma^k_{ij} \frac{\partial \hat{\vec{e}}_k}{\partial \xi^l} = \Gamma^k_{ij} \Gamma^s_{kl} \hat{\vec{e}}_s = \Gamma^s_{ij} \Gamma^k_{sl} \hat{\vec{e}}_k, \quad \Gamma^k_{il} \frac{\partial \hat{\vec{e}}_k}{\partial \xi^j} = \Gamma^k_{il} \Gamma^s_{kj} \hat{\vec{e}}_s = \Gamma^s_{il} \Gamma^k_{sj} \hat{\vec{e}}_k,$$

let us rewrite (1.2.51) in the form

$$R^k_{jli} = \frac{\partial \Gamma^k_{ij}}{\partial \xi^l} - \frac{\partial \Gamma^k_{il}}{\partial \xi^j} + \Gamma^s_{ij} \Gamma^k_{sl} - \Gamma^s_{il} \Gamma^k_{sj} = 0, \tag{1.2.52}$$

where we have denoted by R^k_{jli} the Riemann–Christoffel curvature tensor. By a passage to the covariant components

$$R_{jlik} = \hat{g}_{sk} R^s_{jli}, \tag{1.2.53}$$

let us rewrite the integrability condition as

$$R_{jlik} = 0. \tag{1.2.54}$$

Thus, the conditions (1.2.49) and (1.2.54) should be imposed on $\hat{\varepsilon}_{ij}$ for the determination of x^i by the given $\hat{\varepsilon}_{ij}$ (or $\hat{g}_{ij}$); i.e., *the space determined by $\hat{g}_{ij}$ should be Euclidean.* [The space is called *Euclidean* if at each point the curvature tensor (1.2.53) is equal to zero; therefore, one can introduce for the overall space a Cartesian coordinate system in which $ds^2 = d(x^1)^2 + d(x^2)^2 + d(x^3)^2$.] It follows from equations (1.2.52) and (1.2.54) that

$$R_{jlik} = -R_{ljik} = -R_{jlki} = R_{ikjl}; \tag{1.2.55}$$

therefore, there are in (1.2.54) only six independent equations for

$$R_{1212},\ R_{1313},\ R_{2323},\ R_{1223},\ R_{2123},\ R_{3132}.$$

At small strains, $\varepsilon_{ij} = \frac{1}{2}(\hat{g}_{ij} - \delta_{ij}) \ll 1$ and the product of the Christoffel symbols will be a small quantity of the second order with respect to ε_{ij}. Therefore, neglecting their products in (1.2.52), we have that

$$\frac{\partial \Gamma_{ij.k}}{\partial \xi^l} = \frac{\partial \Gamma_{il.k}}{\partial \xi^j}. \tag{1.2.56}$$

Substituting into (1.2.56) the formula

$$\Gamma_{ij.k} = \frac{1}{2}\left(\frac{\partial \hat{g}_{ik}}{\partial \xi^j} + \frac{\partial \hat{g}_{jk}}{\partial \xi^i} - \frac{\partial \hat{g}_{ij}}{\partial \xi^k}\right) = \frac{\partial \varepsilon_{ik}}{\partial \xi^j} + \frac{\partial \varepsilon_{jk}}{\partial \xi^i} - \frac{\partial \varepsilon_{ij}}{\partial \xi^k},$$

we obtain the *Saint–Venant conditions*:

$$\frac{\partial^2 \varepsilon_{jk}}{\partial \xi^i \partial \xi^l} + \frac{\partial^2 \varepsilon_{il}}{\partial \xi^k \partial \xi^j} - \frac{\partial^2 \varepsilon_{ij}}{\partial \xi^k \partial \xi^l} - \frac{\partial^2 \varepsilon_{lk}}{\partial \xi^i \partial \xi^j} = 0, \tag{1.2.57}$$

two of which have the form

$$\begin{gathered} \frac{\partial^2 \varepsilon_{11}}{\partial (\xi^2)^2} - 2\frac{\partial^2 \varepsilon_{12}}{\partial \xi^1 \partial \xi^2} + \frac{\partial^2 \varepsilon_{22}}{\partial (\xi^1)^2} = 0, \\ \frac{\partial^2 \varepsilon_{12}}{\partial (\xi^3)^2} - \frac{\partial^2 \varepsilon_{13}}{\partial \xi^3 \partial \xi^2} - \frac{\partial^2 \varepsilon_{32}}{\partial \xi^1 \partial \xi^3} + \frac{\partial^2 \varepsilon_{33}}{\partial \xi^1 \partial \xi^2} = 0, \end{gathered} \tag{1.2.58}$$

and the remaining four conditions are obtained by a cyclic permutation of the indices 1, 2, 3. Note that, in the process of the derivation of the compatibility conditions (1.2.54) and (1.2.57), it was implicitly assumed that the integration domain filled with a continuum was simply connected. In the case of a multiply connected domain, the compatibility conditions (1.2.49) and (1.2.54) can be violated; therefore, the vectors $\vec{r}$ and $\hat{\vec{e}}_i$ will already not be single-valued functions of coordinates. Such singular solutions indeed take place and are called *dislocations* and *disclinations*.

1.2.4 Rate-of-Strain Tensor: Cauchy-Helmholtz Theorem

Consider the deformation of an infinitesimal particle (see Fig. 1.11) during an infinitesimal time $dt = t - t_0$. Let a relative vector between the points B and A be equal to $d\vec{r}_0$ at $t = t_0$. After the deformation at time t, it will be equal to

$$d\vec{r} = d\vec{r}_0 + (\vec{v}_B - \vec{v}_A)dt. \tag{1.2.59}$$

Assuming the velocity field to be continuous, we will have with the accuracy up to the terms of the order $d\vec{r}_0$ that

$$\vec{v}_B = \vec{v}_A + \left(\frac{\partial \vec{v}}{\partial \xi^i}\right)_A d\xi^i. \tag{1.2.60}$$

By using the definition of the covariant derivative, let us rewrite (1.2.60) in the form

$$\vec{v}_B = \vec{v}_A + \frac{1}{2}(\nabla_i v_j + \nabla_j v_i)d\xi^i \overset{o}{\vec{e}}{}^j + \frac{1}{2}(\nabla_i v_j - \nabla_j v_i)d\xi^i \overset{o}{\vec{e}}{}^j. \tag{1.2.61}$$

Since the strains are infinitesimal during the time dt, no difference is made in (1.2.61) between the bases of the reference frame and the actual configuration $\overset{o}{\vec{e}}_i \approx \hat{\vec{e}}_i$. The Eulerian coordinate system is assumed to be Cartesian; therefore, $g_{ij} = \delta_{ij}$. Let us define the rate-of-strain tensor

$$\dot{\varepsilon}_{ij} = \frac{1}{2}(\nabla_i v_j + \nabla_j v_i) \tag{1.2.62}$$

and the tensor of rotation velocities

$$\omega_{ij} = \frac{1}{2}(\nabla_i v_j - \nabla_j v_i). \tag{1.2.63}$$

It can be seen from (1.2.62) and (1.2.63) that the tensor $\dot{\varepsilon}_{ij}$ is symmetric, $\dot{\varepsilon}_{ij} = \dot{\varepsilon}_{ji}$, and the tensor ω_{ij} is antisymmetric, $\omega_{ij} = -\omega_{ji}$. Since $\omega_{ii} = 0$, the ω_{ij} has three different components, for which one can define the corresponding vector (more exactly, a pseudovector) $\vec{\omega} = \omega^i \vec{e}_i$ of the angular velocity in accordance with

$$\omega^1 = \omega_{32}/\sqrt{\overset{o}{g}}, \quad \omega^2 = \omega_{13}/\sqrt{\overset{o}{g}}, \quad \omega^3 = \omega_{21}/\sqrt{\overset{o}{g}}. \tag{1.2.64}$$

(A pseudovector does not change its sign at a reflection of the coordinate axes, i.e., $\vec{\omega}' = \vec{\omega}$ at $\vec{r}' = -\vec{r}$. A conventional vector $\vec{b}$ changes its sign at a reflection of the coordinate axes, i.e., $\vec{b}' = -\vec{b}$ at $\vec{r}' = -\vec{r}$.) Substituting (1.2.64) into (1.2.61), let us rewrite (1.2.61) as follows:

$$\vec{v}_B = \vec{v}_A + \dot{\varepsilon}_{ij} d\xi^i \overset{o}{\vec{e}}{}^j + \vec{\omega} \times d\vec{\xi}, \tag{1.2.65}$$

where we have introduced in accordance with (1.2.64) the angular velocity vector

$$\vec{\omega} \;=\; \frac{1}{2}\mathrm{rot}\,\vec{v} = \frac{1}{2}\overset{o}{\varepsilon}{}^{ijk}\nabla_j v_k \overset{o}{\vec{e}}_i$$

$$= \frac{1}{2\sqrt{\overset{o}{g}}}(\nabla_j v_k - \nabla_k v_j)\ \overset{o}{\vec{e}}_i = \frac{1}{2\sqrt{\overset{o}{g}}} \begin{vmatrix} \overset{o}{\vec{e}}_i & \overset{o}{\vec{e}}_j & \overset{o}{\vec{e}}_k \\ \nabla_i & \nabla_j & \nabla_k \\ v_i & v_j & v_k \end{vmatrix}. \quad (1.2.66)$$

Let us prove that the equation (1.2.65) coincides identically with (1.2.61). Representing the vector product in terms of the Levi–Civita tensor, we obtain with regard for (1.2.66):

$$\vec{\omega} \times d\vec{\xi} = \overset{o}{\varepsilon}_{ijk}\, \omega^i d\xi^j\ \overset{o}{\vec{e}}^{\,k} = \frac{1}{2}\ \overset{o}{\varepsilon}_{ijk} \overset{o}{\varepsilon}^{\,ils} \nabla_l v_s d\xi^j\ \overset{o}{\vec{e}}^{\,k}, \quad (1.2.67)$$

where

$$\overset{o}{\varepsilon}_{ijk} = \overset{o}{\vec{e}}_i \cdot (\overset{o}{\vec{e}}_j \times \overset{o}{\vec{e}}_k), \quad \overset{o}{\varepsilon}^{\,ijk} = \overset{o}{\vec{e}}^{\,i} \cdot (\vec{e}^{\,j} \times \vec{e}^{\,k}).$$

In order to find the $\overset{o}{\varepsilon}_{ijk} \overset{o}{\varepsilon}^{\,ils}$, let us define the direct product and the convolution of two tensors. According to[1–4,6], the direct product of a tensor of rank n $\hat{A} = A^{i\ldots j}\vec{e}_i \ldots \vec{e}_j$ by a tensor $\hat{B} = B^{k\ldots s}\vec{e}_k \ldots \vec{e}_s$ of rank m is defined as a tensor of rank $n+m$ of the form $\hat{C} = C^{i\ldots jk\ldots s}\vec{e}_i \ldots \vec{e}_j \vec{e}_k \ldots \vec{e}_s$. The components $C^{i\ldots jk\ldots s}$ are obtained by a simple multiplication of the components of the tensors (matrices) $\hat{A}$ and $\hat{B}$

$$C^{i\ldots jk\ldots s} = A^{i\ldots j} B^{k\ldots s}. \quad (1.2.68)$$

At a convolution of a tensor $\hat{C}$, for example, over the indices j and k, it is necessary to make a substitution of the direct product of two basis vectors $\vec{e}_j$ and $\vec{e}_k$ by a scalar product, as a result of which the tensor rank will be reduced by two:

$$C^{i\ldots jk\ldots s}\vec{e}_i \ldots (\vec{e}_j \cdot \vec{e}_k) \ldots \vec{e}_s = g_{jk} C^{i\ldots jk\ldots s} \vec{e}_i \ldots \vec{e}_{j-1} \vec{e}_{k+1} \ldots \vec{e}_s;$$

therefore, the components of the convolved tensor of rank $n+m-2$ will be equal to

$$D^{i\ldots(j-1)(k+1)\ldots s} = g_{jk} C^{i\ldots jk\ldots s}. \quad (1.2.69)$$

Thus, the convolution of the tensor is performed by lowering (raising) the index, which is made dummy, and the summation is carried out over the corresponding dummy indices.

Let us introduce a Cartesian coordinate system with an orthonormal basis $\vec{i}$, $\vec{j}$, $\vec{k}$, in which

$$\overset{o}{\vec{e}}_i = \overset{o}{e}_{ix}\, \vec{i} + \overset{o}{e}_{iy}\, \vec{j} + \overset{o}{e}_{iz}\, \vec{k}, \quad \overset{o}{\vec{e}}^{\,i} = \overset{o}{e}{}^i_x \vec{i} + \overset{o}{e}{}^i_y \vec{j} + \overset{o}{e}{}^i_z \vec{k}.$$

Substituting these expansions into the definitions of the Levi–Civita tensors [the last two equations in (1.2.67)], we obtain:

$$\overset{o}{\varepsilon}_{ijk} = \begin{vmatrix} \overset{o}{e}_{ix} & \overset{o}{e}_{iy} & \overset{o}{e}_{iz} \\ \overset{o}{e}_{jx} & \overset{o}{e}_{jy} & \overset{o}{e}_{jz} \\ \overset{o}{e}_{kx} & \overset{o}{e}_{ky} & \overset{o}{e}_{kz} \end{vmatrix}, \quad \overset{o}{\varepsilon}^{\,ijk} = \begin{vmatrix} \overset{o}{e}{}^i_x & \overset{o}{e}{}^i_y & \overset{o}{e}{}^i_z \\ \overset{o}{e}{}^j_x & \overset{o}{e}{}^j_y & \overset{o}{e}{}^j_z \\ \overset{o}{e}{}^k_x & \overset{o}{e}{}^k_y & \overset{o}{e}{}^k_z \end{vmatrix}.$$

While determining the direct product of these tensors, let us make use of a well-known theorem of linear algebra[7] that states that the determinant of the product of two matrices is equal to the product of the determinants of these matrices:

$$|a_{ij}|\,|b_{ij}| = |a_{is}b_{sj}| = |a_{is}b_{js}|,$$

where the last equality is a consequence of the fact that the determinant value does not change if the rows are replaced with columns (the transposition). With regard for this theorem, the components of the tensor of rank six formed by a direct product of ε_{ijk} and ε^{nls} will be equal to

$$\begin{aligned}\varepsilon_{ijk}\,\varepsilon^{nls} &= \left|\begin{pmatrix} e_{ix} & e_{iy} & e_{iz} \\ e_{jx} & e_{jy} & e_{jz} \\ e_{kx} & e_{ky} & e_{kz}\end{pmatrix}\begin{pmatrix} e^n_x & e^l_x & e^s_x \\ e^n_y & e^l_y & e^s_y \\ e^n_z & e^l_z & e^s_z\end{pmatrix}\right| \\ &= \begin{vmatrix} \vec{e}_i\cdot\vec{e}^{\,n} & \vec{e}_i\cdot\vec{e}^{\,l} & \vec{e}_i\cdot\vec{e}^{\,s} \\ \vec{e}_j\cdot\vec{e}^{\,n} & \vec{e}_j\cdot\vec{e}^{\,l} & \vec{e}_j\cdot\vec{e}^{\,s} \\ \vec{e}_k\cdot\vec{e}^{\,n} & \vec{e}_k\cdot\vec{e}^{\,l} & \vec{e}_k\cdot\vec{e}^{\,s}\end{vmatrix} = \begin{vmatrix} \delta^n_i & \delta^l_i & \delta^s_i \\ \delta^n_j & \delta^l_j & \delta^s_j \\ \delta^n_k & \delta^l_k & \delta^s_k\end{vmatrix},\end{aligned} \tag{1.2.70}$$

where we have omitted for brevity of notation the index zero over the vector components. Calculating the determinant (1.2.70), we obtain the following compact expression:

$$\varepsilon_{ijk}\varepsilon^{nls} = \delta^n_i(\delta^l_j\delta^s_k - \delta^s_j\delta^l_k) - \delta^l_i(\delta^n_j\delta^s_k - \delta^s_j\delta^n_k) + \delta^s_i(\delta^n_j\delta^l_k - \delta^l_j\delta^n_k).$$

A convolution of this tensor over the indices k and n (1.2.69) determines a tensor of rank four with the components

$$\varepsilon_{ijk}\varepsilon^{kls} = \delta^l_i\delta^s_j - \delta^l_j\delta^s_i. \tag{1.2.71}$$

Note that formulas (1.2.70) and (1.2.71) for the product of the Levi–Civita tensors are valid in an arbitrary basis.

Substituting (1.2.71) into (1.2.67), we obtain:

$$\vec{\omega}\times d\vec{\xi} = \frac{1}{2}(\delta^l_j\delta^s_k - \delta^s_j\delta^l_k)\nabla_l v_s d\xi^j\,\overset{o}{\vec{e}}{}^{\,k} = \frac{1}{2}(\nabla_j v_k - \nabla_k v_j)d\xi^j\,\overset{o}{\vec{e}}{}^{\,k},$$

from where it follows with regard for (1.2.62) that (1.2.65) coincides identically with (1.2.61). Thus, the proof of (1.2.65) is completed.

The relationship (1.2.65) expresses the contents of the *Cauchy–Helmholtz theorem*: the velocity of any point of an infinitesimal particle is composed of the velocity $\vec{v}_A$ of the translational movement of the particle center, the rotation velocity $\vec{\omega}\times d\vec{\xi}$ of a particle as an absolutely rigid body with respect to the point A, and the velocity $\dot{\varepsilon}_{ij}d\xi^i\,\overset{o}{\vec{e}}{}^{\,j}$ related to the particle deformation.

Note that in the case of a finite deformation of an infinitesimal particle, its motion can also be presented in the form of a rotation as a solid and a deformation[8]. The main point in the proof of this fact is the Kelly's theorem, which states that any matrix, including the distortion matrix $B = \| \frac{\partial x^i}{\partial \xi^j} \|$, can be presented in the form of a product of a symmetric matrix S and an orthogonal matrix O:

$$C = SO, \quad S = S^T, \quad OO^T = E, \quad E = \begin{pmatrix} 1 & 0 & 0 \\ 0 & 1 & 0 \\ 0 & 0 & 1 \end{pmatrix}, \tag{1.2.72}$$

where the superscript T corresponds to the transposition of a matrix. Following[8], let us present S in terms of the orthogonal matrix U and the diagonal matrix K:

$$S = UKU^T, \quad UU^T = E, \quad K = \begin{pmatrix} k_1 & 0 & 0 \\ 0 & k_2 & 0 \\ 0 & 0 & k_3 \end{pmatrix}. \tag{1.2.73}$$

Substituting (1.2.73) into (1.2.72) and denoting $V = U^T O$, we obtain:

$$C = UKV, \tag{1.2.74}$$

where $UU^T = VV^T = E$, $\det U = \det V = 1$, $k_i > 0$, from where it follows that the arbitrary motion of an infinitesimal particle can be presented as a rotation V, as a result of which the coordinate axes will coincide with the principal axes of S, the tension (compression) along the principal ith axes by a factor of k_i, and one more rotation moving the coordinate axes to some other position. As a result of this, the particle undergoes the rotation as an absolutely rigid body as well as deformation.

Following (1.2.66), let us determine the angular velocity vector in the Eulerian coordinate system

$$\begin{aligned} \vec{\omega} &= \frac{1}{2}\operatorname{rot}\vec{v} = \frac{1}{2}\varepsilon^{ijk}\nabla_j v_k \vec{e}_i \\ &= \frac{1}{2\sqrt{g}}(\nabla_j v_k - \nabla_k v_j)\vec{e}_i = \frac{1}{2\sqrt{g}}\begin{vmatrix} \vec{e}_1 & \vec{e}_2 & \vec{e}_3 \\ \nabla_1 & \nabla_2 & \nabla_3 \\ v_1 & v_2 & v_3 \end{vmatrix} = \omega^i \vec{e}_i. \end{aligned} \tag{1.2.75}$$

In the case of a Cartesian coordinate system, $g_{ij} = \delta_{ij}$ and the formula (1.2.75) simplifies to

$$\vec{\omega} = \begin{vmatrix} \vec{e}_1 & \vec{e}_2 & \vec{e}_3 \\ \frac{\partial}{\partial x_1} & \frac{\partial}{\partial x_2} & \frac{\partial}{\partial x_3} \\ v_1 & v_2 & v_3 \end{vmatrix} = \omega^i \vec{e}_i. \tag{1.2.76}$$

The knowledge of the angular velocity vector $\vec{\omega} = \omega^i \vec{e}_i$ at each point $\vec{x} = x^i \vec{e}_i$ enables one to introduce similarly to (1.2.14) the vortex lines, the tangent to which $d\vec{r}$ is parallel with the angular velocity vector $\vec{\omega}$.

$$\frac{dx^1}{\omega^1} = \frac{dx^2}{\omega^2} = \frac{dx^3}{\omega^3}. \tag{1.2.77}$$

Choosing the line l that does not coincide with the vortex line and drawing the vortex lines through each point of l, we obtain a vortex surface. If l is closed, a *vortex tube* emerges.

Problem 1.4. The velocity field in Cartesian coordinates has the components[9]

$$v_1 = \frac{x_1}{1+t}, \quad v_2 = \frac{2x_2}{1+t}, \quad v_3 = \frac{3x_3}{1+t}.$$

Find 1) the motion trajectories, 2) the components of velocity and acceleration in the Eulerian and Lagrangian coordinates, and 3) the streamlines.
(*Remark:* in the above formulas, we have taken into account the fact that the covariant and contravariant components coincide in Cartesian coordinates.)

Solution:

1) The motion trajectories are determined from the equations (1.2.15)

$$\frac{dx_1}{dt} = \frac{x_1}{1+t}; \quad \frac{dx_2}{dt} = \frac{2x_2}{1+t}; \quad \frac{dx_3}{dt} = \frac{3x_3}{1+t}$$

under the initial conditions $x_i|_{t=0} = \xi_i$. For the x_1 coordinate, we have the following equations:

$$\frac{dx_1}{dt} = \frac{x_1}{1+t}; \ \frac{dx_1}{x_1} = \frac{dt}{1+t}; \ \log x_1 = \log(1+t) + \log \xi_1, \quad x_1 = \xi_1(1+t).$$

We can determine in a similar way that

$$x_2 = \xi_2\,(1+t)^2; \quad x_3 = \xi_3\,(1+t)^3.$$

2) The obtained trajectories $x_i = x_i(\xi_i, t)$ determine at a fixed moment of time t the relation between the Eulerian (E) and Lagrangian (L) coordinates. Taking these formulas into account, we can find the velocities $v_i(\xi_i, t)$ in the Lagrangian coordinates (LC):

$$\begin{aligned} v_1 &= \frac{x_1}{1+t} = \xi_1; \\ v_2 &= \frac{2x_2}{1+t} = \frac{2\xi_2(1+t)^2}{1+t} = 2\xi_2(1+t); \\ v_3 &= \frac{3x_3}{1+t} = \frac{3\xi_3(1+t)^3}{1+t} = 3\xi_3(1+t)^2. \end{aligned}$$

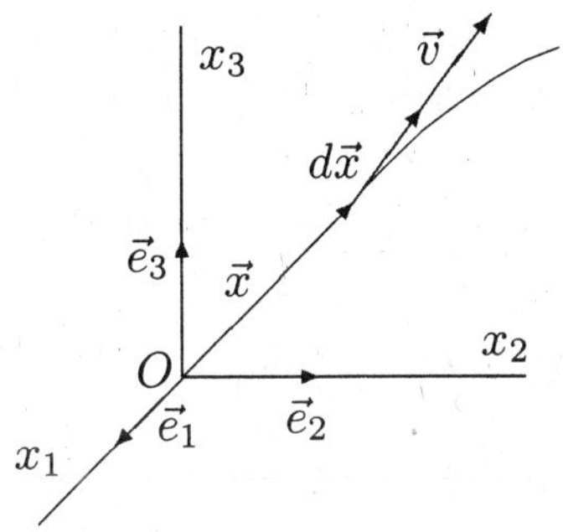

Figure 1.12: The trajectory of particle motion.

The acceleration in LC is determined by the second formula in (1.2.7) $a_i^L = \frac{\partial}{\partial t} v_i(\xi_i, t)$, from where we obtain with regard for $v_i = v_i(\xi_i, t)$:

$$a_1^L = 0, \quad a_2^L = 2\,\xi_2, \quad a_3^L = 6\,\xi_3(1+t).$$

The acceleration in Eulerian coordinates is determined from (1.2.11) as

$$a_i^E = \frac{\partial v_i}{\partial t} + v_j \frac{\partial v_i}{\partial x_j} = \frac{\partial v_i}{\partial t} + v_1 \frac{\partial v_i}{\partial x_1} + v_2 \frac{\partial v_i}{\partial x_2} + v_3 \frac{\partial v_i}{\partial x_3},$$

where $v_i = v_i(x_i, t)$. Therefore,

$$\begin{aligned} a_1^E &= \frac{\partial v_1}{\partial t} + v_1 \frac{\partial v_1}{\partial x_1} = -\frac{x_1}{(1+t)^2} + \frac{x_1}{(1+t)^2} = 0, \\ a_2^E &= \frac{\partial v_2}{\partial t} + v_2 \frac{\partial v_2}{\partial x_2} = -\frac{2x_2}{(1+t)^2} + \frac{4x_2}{(1+t)^2} = \frac{2x_2}{(1+t)^2}, \\ a_3^E &= \frac{\partial v_3}{\partial t} + v_3 \frac{\partial v_3}{\partial x_3} = \frac{6x_3}{(1+t)^2}. \end{aligned}$$

It is easy to see that, if the relation $x_i = x_i(\xi_i, t)$ is substituted into equations for a_i^E, then the equations are obtained, which determine a_i^L.

3) It follows from Fig. 1.12 that the vector tangent to the streamline (SL) $d\vec{x}$ is directed along the velocity $\vec{v}$; therefore, $\vec{v} \times d\vec{x} = 0$, or

$$\begin{vmatrix} \vec{e}_1 & \vec{e}_2 & \vec{e}_3 \\ v_1 & v_2 & v_3 \\ dx_1 & dx_2 & dx_3 \end{vmatrix} = 0.$$

Expanding the determinant into the first row, we obtain:

$$\vec{e}_1(v_2 dx_3 - v_3 dx_3) - \vec{e}_2(v_1 dx_3 - v_3 dx_1) - \vec{e}_3(v_1 dx_2 - v_2 dx_1) = 0;$$
$$v_2 dx_3 - v_3 dx_2 = 0, \quad v_1 dx_3 - v_3 dx_1 = 0, \quad v_1 dx_2 - v_2 dx_1 = 0,$$

from where we obtain equations (1.2.14):

$$\frac{dx_3}{v_3} = \frac{dx_2}{v_2}, \quad \frac{dx_1}{v_1} = \frac{dx_3}{v_3}, \quad \frac{dx_1}{v_1} = \frac{dx_2}{v_2}.$$

Substituting into these equations the dependence $v_i = v_i(x_i, t)$, we find:

$$\frac{dx_1}{x_1} = \frac{dx_2}{2x_2}, \quad \frac{dx_1}{x_1} = \frac{dx_3}{3x_3}, \quad \frac{dx_2}{2x_2} = \frac{dx_3}{3x_3}.$$

Integrating the first equation, we obtain that $\log x_1 = \log x_2^{1/2} + C$. At the time $t = 0$, we have

$$\log \xi_1 + \log \xi_2^{-1/2} = C, \quad C = \log \frac{\xi_1}{\xi_2^{1/2}};$$

therefore,

$$\log \frac{x_1}{\xi_1} = \log\Big(\frac{x_2}{\xi_2}\Big)^{1/2}, \quad \frac{x_2}{\xi_2} = \Big(\frac{x_1}{\xi_1}\Big)^2.$$

We can find in a similar way from the second and third equations that

$$\frac{x_3}{\xi_3} = \Big(\frac{x_1}{\xi_1}\Big)^3; \quad \Big(\frac{x_2}{\xi_2}\Big)^3 = \Big(\frac{x_3}{\xi_3}\Big)^2.$$

It is interesting to note that in the case under consideration the streamlines and the trajectories coincide (in a general case, they are different). The motion trajectories

$$x_1 = \xi_1(1+t); \quad x_2 = \xi_2(1+t)^2; \quad x_3 = \xi_3(1+t)^3$$

can indeed be written in the form

$$\frac{x_2}{\xi_2} = \Big(\frac{x_1}{\xi_1}\Big)^2; \quad \frac{x_3}{\xi_3} = \Big(\frac{x_1}{\xi_1}\Big)^3; \quad \Big(\frac{x_2}{\xi_2}\Big)^3 = \Big(\frac{x_3}{\xi_3}\Big)^2.$$

The above task can easily be solved with the aid of the software system *Mathematica 3.0.* We present below the output of the corresponding program `prog1-4.nb`.

The Velocity Field

$$\frac{x_1}{1+t}$$

$$\frac{2x_2}{1+t}$$

$$\frac{3x_3}{1+t}$$

Determination of the Motion Trajectories

$$\{\{x_1[t] \to (1+t)\xi_1\}\}$$

$$(1+t)\xi_1$$

$$\{\{x_2[t] \to (1+t)^2\xi_2\}\}$$

$$(1+t)^2\xi_2$$

$$\{\{x_3[t] \to (1+t)^3\xi_3\}\}$$

$$(1+t)^3\xi_3$$

The Velocities $v_i(\xi_i , t)$, i = 1,2,3, in the Lagrangian Coordinates

$$\xi_1$$

$$2(1+t)\xi_2$$

$$3(1+t)^2\xi_3$$

The Accelerations in Lagrangian Coordinates

$$0$$

$$2\xi_2$$

$$6(1+t)\xi_3$$

The Accelerations in Eulerian Coordinates

$$0$$

$$\frac{2x_2}{(1+t)^2}$$

$$\frac{6x_3}{(1+t)^2}$$

The Streamlines

$$\frac{x}{1+t}$$

$$\frac{2x_2[x]}{1+t}$$

$$\frac{3x_3[x]}{1+t}$$

$$\{\{x_2[x] \to \frac{x^2\xi_2}{\xi_1^2}\}\}$$

$$\{\{x_3[x] \to \frac{x^3\xi_3}{\xi_1^3}\}\}$$

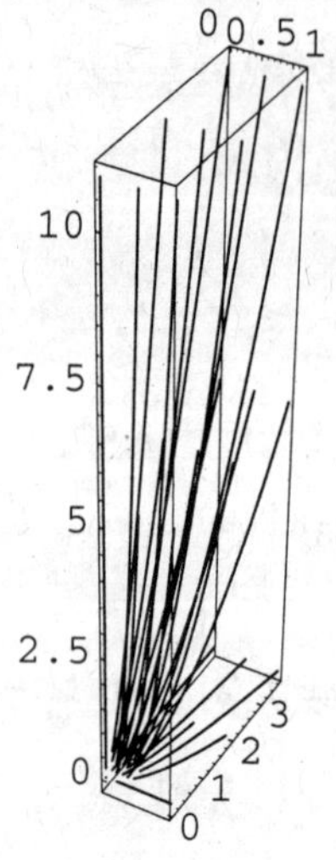

Figure 1.13: The trajectories of particles of a continuum.

It is easy to see that the symbolic computation results by the above Notebook `prog1-4.nb` coincide with the results presented above by us, which we have obtained by hand. The powerful built-in computer graphics enable the user of *Mathematica 3.0* a rapid visualization of the results. We show in Fig. 1.13 the trajectories of 27 particles of a continuum, which we have chosen in our Notebook in a sufficiently arbitrary way.

Problem 1.5. The displacement field is given in the Cartesian basis $\overset{o}{\vec{e}_i} = \vec{e}_i = \vec{E}_i$ in the form

$$\vec{u} = (\xi_2 - \xi_3)\vec{E}_1 + (\xi_2 + \xi_3)\vec{E}_2 - \xi_1\vec{E}_3.$$

Determine the Lagrangian $\overset{o}{\varepsilon}_{ij}$ and Eulerian ε_{ij} strain tensors.

Solution: From the definition of the displacement vector $\vec{x} = \vec{\xi} + \vec{u}$, we have that

$$x_i = \xi_i + u_i, \quad u_1 = \xi_2 - \xi_3, \quad u_2 = \xi_2 + \xi_3, \quad u_3 = -\xi_1.$$

From here, the formulas for the derivatives follow:

$$\begin{aligned} \frac{\partial u_1}{\partial \xi_1} &= 0; \ \frac{\partial u_1}{\partial \xi_2} = 1; \ \frac{\partial u_1}{\partial \xi_3} = -1; \ \frac{\partial u_2}{\partial \xi_1} = 0; \ \frac{\partial u_2}{\partial \xi_2} = 1; \\ \frac{\partial u_2}{\partial \xi_3} &= 1; \ \frac{\partial u_3}{\partial \xi_1} = -1; \ \frac{\partial u_3}{\partial \xi_2} = 0; \ \frac{\partial u_3}{\partial \xi_3} = 0. \end{aligned}$$

Substituting the values of $\frac{\partial u_i}{\partial \xi_j}$ into the Green's formula (1.2.29), we obtain:

$$\overset{o}{\varepsilon}_{11} = \frac{1}{2}\Big[2\frac{\partial u_1}{\partial \xi_1} + \Big(\frac{\partial u_1}{\partial \xi_1}\Big)^2 + \Big(\frac{\partial u_3}{\partial \xi_1}\Big)^2\Big] = \frac{1}{2};$$

$$\overset{o}{\varepsilon}_{12} = \frac{1}{2}\Big[\frac{\partial u_1}{\partial \xi_2} + \frac{\partial u_2}{\partial \xi_1} + \frac{\partial u_1}{\partial \xi_1}\frac{\partial u_1}{\partial \xi_2} + \frac{\partial u_2}{\partial \xi_1}\frac{\partial u_2}{\partial \xi_2} + \frac{\partial u_3}{\partial \xi_1}\frac{\partial u_3}{\partial \xi_2}\Big] = \frac{1}{2}.$$

We can further calculate in a similar way that

$$\overset{o}{\varepsilon}_{13}= -1, \quad \overset{o}{\varepsilon}_{22}= 2, \quad \overset{o}{\varepsilon}_{23}= \frac{1}{2}, \quad \overset{o}{\varepsilon}_{33}= 1.$$

Since the Eulerian coordinate system coincides with the coordinate system of the reference frame, we determine from the equation $x_i = \xi_i + u_i$ that

$$x_1 = \xi_1 + \xi_2 - \xi_3, \quad x_2 = 2\xi_2 + \xi_3, \quad x_3 = \xi_3 - \xi_1.$$

Using the elimination method, we find:

$$\xi_1 = x_2 - 2x_1 - 3x_3, \quad \xi_2 = x_1 + x_3, \quad \xi_3 = x_2 - 2x_1 - 2x_3.$$

It follows from here that

$$u_1 = 3x_1 - x_2 + 3x_3, \quad u_2 = x_2 - x_1 - x_3, \quad u_3 = 2x_1 - x_2 + 3x_3.$$

Differentiating the u_i with respect to x_j, we obtain the formulas:

$$\frac{\partial u_1}{\partial x_1} = 3; \quad \frac{\partial u_1}{\partial x_2} = -1; \quad \frac{\partial u_1}{\partial x_3} = -1; \quad \frac{\partial u_2}{\partial x_1} = -1;$$

$$\frac{\partial u_2}{\partial x_2} = 1; \quad \frac{\partial u_2}{\partial x_3} = -1; \quad \frac{\partial u_3}{\partial x_1} = 2; \quad \frac{\partial u_3}{\partial x_2} = -1; \quad \frac{\partial u_3}{\partial x_3} = 3.$$

Substituting the $\frac{\partial u_i}{\partial x_j}$ into the second Almansi formula (1.2.34), we find the ε_{ij}:

$$\varepsilon_{11} = \frac{1}{2}\Big[2\frac{\partial u_1}{\partial x_1} - \Big(\frac{\partial u_1}{\partial x_1}\Big)^2 - \Big(\frac{\partial u_2}{\partial x_1}\Big)^2 - \Big(\frac{\partial u_3}{\partial x_1}\Big)^2\Big] = -4;$$

$$\varepsilon_{12} = \frac{1}{2}\Big(\frac{\partial u_1}{\partial x_2} + \frac{\partial u_2}{\partial x_1} - \frac{\partial u_1}{\partial x_1}\frac{\partial u_1}{\partial x_2} - \frac{\partial u_2}{\partial x_1}\frac{\partial u_2}{\partial x_2} - \frac{\partial u_3}{\partial x_2}\Big) = 2.$$

We obtain in a similar way the following expressions for the remaining components:

$$\varepsilon_{13} = -\frac{11}{2}; \quad \varepsilon_{22} = -\frac{1}{2}; \quad \varepsilon_{23} = \frac{5}{2}; \quad \varepsilon_{33} = -\frac{13}{2}.$$

Below we present the solution of the same problem with the aid of *Mathematica 3.0*. The presented output of the *Mathematica* Notebook `prog1-5.nb` also contains the results of intermediate symbolic computations, so that the interested reader can compare them with the same expressions obtained above by us "by hand".

Calculation of the Components $(\epsilon^o)_{ij}$, $i, j = 1, 2, 3$

$$\epsilon\hat{}o = \begin{pmatrix} \frac{1}{2} & \frac{1}{2} & -1 \\ \frac{1}{2} & 2 & \frac{1}{2} \\ -1 & \frac{1}{2} & 1 \end{pmatrix}$$

Calculation of the Components ϵ_{ij} by Almansi Formula

$$-2x[1] + x[2] - 3x[3]$$

$$x[1] + x[3]$$

$$-2x[1] + x[2] - 2x[3]$$

The Calculation of Displacements u_i in Eulerian Variables

$$3x[1] - x[2] + 3x[3]$$

$$-x[1] + x[2] - x[3]$$

$$2x[1] - x[2] + 3x[3]$$

The Almansi Formulas

$$\epsilon = \begin{pmatrix} -4 & 2 & -\frac{11}{2} \\ 2 & -\frac{1}{2} & \frac{5}{2} \\ -\frac{11}{2} & \frac{5}{2} & -\frac{13}{2} \end{pmatrix}$$

Problem 1.6. Find the tensor of small strains ε_{ij}, the rotation tensor ω_{ij}, the principal strains ε_i, and the principal axes $\vec{n}_i$ of the strain tensor for the displacement field

$$\vec{u} = (4x_1 - x_2 + 3x_3)\vec{E}_1 + (x_1 + 7x_2)\vec{E}_2 + (4x_2 + 4x_3 - 3x_1)\vec{E}_3,$$

given in the Cartesian basis $\vec{E}_i$.

Solution: The components of the displacement vector are

$$u_1 = 4x_1 - x_2 + 3x_3,\ u_2 = x_1 + 7x_2,\ u_3 = -3x_1 + 4x_2 + 4x_3.$$

Differentiating the u_i with respect to x_j, we obtain the formulas:

$$\frac{\partial u_1}{\partial x_1} = 4;\ \frac{\partial u_1}{\partial x_2} = -1;\ \frac{\partial u_1}{\partial x_3} = 3;\ \frac{\partial u_2}{\partial x_1} = 1;\ \frac{\partial u_2}{\partial x_2} = 7;$$

$$\frac{\partial u_2}{\partial x_3} = 0;\ \frac{\partial u_3}{\partial x_1} = -3;\ \frac{\partial u_3}{\partial x_2} = 4;\ \frac{\partial u_3}{\partial x_3} = 4.$$

The strain tensor ε_{ij} and the rotation tensor are determined by formulas (1.2.62) and (1.2.63)

$$\varepsilon_{ij} = \frac{1}{2}\Big(\frac{\partial u_i}{\partial x_j} + \frac{\partial u_j}{\partial x_i}\Big); \quad \omega_{ij} = \frac{1}{2}\Big(\frac{\partial u_i}{\partial x_j} - \frac{\partial u_j}{\partial x_i}\Big).$$

Substituting here the expressions for $\frac{\partial u_i}{\partial x_j}$, we find:

$$\begin{aligned}
\varepsilon_{11} &= \frac{\partial u_1}{\partial x_1} = 3; \ \varepsilon_{12} = \frac{1}{2}\Big(\frac{\partial u_1}{\partial x_2} + \frac{\partial u_2}{\partial x_1}\Big) = \frac{1}{2}(-1+1) = 0;\\
\varepsilon_{13} &= 0; \quad \varepsilon_{22} = 7; \quad \varepsilon_{23} = 2; \quad \varepsilon_{33} = 4;\\
\omega_{12} &= \frac{1}{2}\Big(\frac{\partial u_1}{\partial x_2} - \frac{\partial u_2}{\partial x_1}\Big) = \frac{1}{2}(-1-1) = -1; \quad \omega_{13} = 3, \quad \omega_{23} = -2.
\end{aligned}$$

Note that the components ε_{ij} should satisfy the inequality $\varepsilon_{ij} \ll 1$; however, the components ε_{ij} have been chosen for the convenience of computations to be of the order of unity. The tensors ε_{ij} and ω_{ij} are written down in the form of the matrices:

$$\varepsilon_{ij} = \begin{pmatrix} 4 & 0 & 0 \\ 0 & 7 & 2 \\ 0 & 2 & 4 \end{pmatrix}; \quad \omega_{ij} = \begin{pmatrix} 0 & -1 & 3 \\ 1 & 0 & -2 \\ -3 & 2 & 0 \end{pmatrix}$$

The tensor ε_{ij} is symmetric: $\varepsilon_{ij} = \varepsilon_{ji}$; therefore, it generally has six different components. The tensor ω_{ij} is antisymmetric: $\omega_{ij} = -\omega_{ji}$; therefore, it has 3 different components, to which the components ω_i of the rotation vector correspond (more exactly, this a pseudovector, since it does not change its sign at a reflection $\vec{r} \to -\vec{r}$) in accordance with the law (1.2.64)

$$\omega_1 = \omega_{23}, \quad \omega_2 = \omega_{13}, \quad \omega_3 = \omega_{12}.$$

The principal values ε_i and the vectors $\vec{n}_i$ are determined from a system of three equations $(\varepsilon_{kl} - \varepsilon\delta_{kl})n^l = 0$ (the summation is carried out over the index l, but not over the k). The system of homogeneous linear equations has a nonzero solution provided that $\det(\varepsilon_{kl} - \varepsilon\delta_{kl}) = 0$. Substituting the values of ε_{ij} into this equation, we obtain:

$$\begin{vmatrix} (4-\varepsilon) & 0 & 0 \\ 0 & (7-\varepsilon) & 2 \\ 0 & 2 & (4-\varepsilon) \end{vmatrix} = 0, \ (4-\varepsilon)((7-\varepsilon)(4-\varepsilon)-4) = 0.$$

The cubic equation has three roots $\varepsilon_1 = 8, \ \varepsilon_2 = 4, and \ \varepsilon_3 = 3$. Let us write the system of equations $(\varepsilon_{kl} - \varepsilon_i\delta_{kl})n_i^l = 0$ in the matrix form:

$$\begin{pmatrix} (4-\varepsilon_i) & 0 & 0 \\ 0 & (7-\varepsilon_i) & 2 \\ 0 & 2 & (4-\varepsilon_i) \end{pmatrix}\begin{pmatrix} n_i^1 \\ n_i^2 \\ n_i^3 \end{pmatrix} = 0, \quad i = 1, 2, 3.$$

Since the determinant of this system is equal to zero, there are two equations for the determination of three components n_i^l. Therefore, it is necessary to augment them by the normalization condition $(n_i^1)^2 + (n_i^2)^2 + (n_i^3)^2 = 1$. Substituting $\varepsilon_1 = 8$ into the system of equations, we obtain two equations $4n_1^1 = 0$, $2n_1^3 - n_1^2 = 0$, from where $n_1^1 = 0$, $n_1^2 = 2n_1^3$, and from the normalization condition $4(n_1^3)^2 + (n_1^3)^2 = 1$, we have that $5(n_1^3)^2 = 1$. We finally obtain for $\varepsilon_1 = 8$: $n_1^l = (0, +\frac{2}{\sqrt{5}}, \frac{1}{\sqrt{5}})$. We can find in a similar way for $\varepsilon_2 = 4$ that $n_2^l = (1, 0, 0)$ and at $\varepsilon_3 = 3$, $n_3^l = (0, \frac{1}{\sqrt{5}}, -\frac{2}{\sqrt{5}})$. One can verify by a direct substitution that $\vec{n}_i \cdot \vec{n}_j = \delta_{ij}$; therefore, the vectors $\vec{n}_i$ can be taken as a new orthonormal basis in which the strain tensor will have the diagonal form

$$\varepsilon'_{ij} = \begin{pmatrix} 8 & 0 & 0 \\ 0 & 4 & 0 \\ 0 & 0 & 3 \end{pmatrix}$$

It follows from the definition of the strain tensor $\varepsilon_i = \Delta x_i / x_i$ and the form of ε'_{ij} that along the direction $\vec{n}_1$ the substance has extended by a factor of eight, along the direction $\vec{n}_2$ it has extended by a factor of four, and along the direction $\vec{n}_3$ by a factor of three.

In what follows, we present the output of the *Mathematica* Notebook `prog1-6.nb`, which solves this problem.

The Specification of the Displacement Vector Components u_1, u_1 , u_3 in the Cartesian Basis

$$4x_1 - x_2 + 3x_3$$

$$x_1 + 7x_2$$

$$-3x_1 + 4x_2 + 4x_3$$

Calculation of the Strain Tensor ϵ_{ij}

$$\epsilon = \begin{pmatrix} 4 & 0 & 0 \\ 0 & 7 & 2 \\ 0 & 2 & 4 \end{pmatrix}$$

Calculation of the Rotation Tensor ω_{ij}

$$\omega = \begin{pmatrix} 0 & -1 & 3 \\ 1 & 0 & -2 \\ -3 & 2 & 0 \end{pmatrix}$$

The Principal Strains ϵ_i and the Eigenvectors n_i

$$\{3, 4, 8\}$$

$$\{\{0, -1, 2\}, \{1, 0, 0\}, \{0, 2, 1\}\}$$

The Normalized Eigenvectors

$$\{0, -\frac{1}{\sqrt{5}}, \frac{2}{\sqrt{5}}\}$$

$$\{1, 0, 0\}$$

$$\{0, \frac{2}{\sqrt{5}}, \frac{1}{\sqrt{5}}\}$$

Problem 1.7. The velocity field is given by a formula

$$\vec{v} = (Ax_3 - Bx_2)\vec{e}_1 + (Bx_1 - Cx_3)\vec{e}_2 + (Cx_2 - Ax_1)\vec{e}_3,$$

where $A = \text{const}$, $B = \text{const}$, and $C = \text{const}$. Find the vortex lines.

Solution: Let us rewrite equations (1.2.77) for the vortex line in the form

$$\frac{dx_1}{\omega_1} = \frac{dx_2}{\omega_2}; \quad \frac{dx_1}{\omega_1} = \frac{dx_3}{\omega_3}; \quad \frac{dx_2}{\omega_2} = \frac{dx_3}{\omega_3}.$$

The components of the velocity field v_i are

$$v_1 = Ax_3 - Bx_2, \quad v_2 = Bx_1 - Cx_3, \quad v_3 = Cx_2 - Ax_1.$$

Substituting v_i into the determinant (1.2.76), we find: $\omega_1 = C$, $\omega_2 = A$, $\omega_3 = B$, from where we have:

$$\frac{dx_1}{C} = \frac{dx_2}{A}, \quad \frac{dx_1}{C} = \frac{dx_3}{B},$$

$$x_2 = \frac{A}{C}x_1 + k_1, \quad x_3 = \frac{B}{C}x_1 + k_2,$$

where k_1 and k_2 are the constants. The obtained equations show that the vortex lines are the straight lines in the case of the rotation of an absolutely rigid body.

In what follows, we present the output of the *Mathematica* Notebook `prog1-7.nb`, which solves the given task.

The Specification of the Velocity Field

$$-Bx_2 + Ax_3$$

$$Bx_1 - Cx_3$$

$$-Ax_1 + Cx_2$$

Calculation of the Components of the Angular Velocity $\omega^{\rightharpoonup}$

$$\omega_1 = C$$
$$\omega_2 = A$$
$$\omega_3 = B$$

$$x\_2 = \mathrm{C}[1] + \frac{\mathrm{Ax_1}}{\mathrm{C}}$$

$$x\_3 = \mathrm{C}[2] + \frac{\mathrm{Bx_1}}{\mathrm{C}}$$

We have used the built-in *Mathematica* function `DSolve[_]` to solve the ordinary differential equations (ODEs)

$$\frac{dx_2}{dx_1} = \frac{\omega_2}{\omega_1}, \quad \frac{dx_3}{dx_1} = \frac{\omega_3}{\omega_1}.$$

The function `DSolve[_]` denotes the integration constants, which appear in the general solution, by `C[1]`, `C[2]`,.... *Mathematica* considers each of the above two ODEs as an independent differential equation; therefore, it has denoted the integration constant appearing in $x_3(x_1)$ by `C[1]`. But this notation has already appeared as a result of the integration of the ODE for x_2. It is clear that the integration constants in the solutions of two independent ODEs are arbitrary, and consequently, they do not generally coincide. To express this fact explicitly, we have used the transformation rule `C[1] -> C[2]` while preparing the output of the solution $x_3(x_1)$.

Problem 1.8. For the velocity field

$$\text{a)} \quad \vec{v} = Cx_2\vec{e}_1; \quad \text{b)} \quad \vec{v} = \frac{Cx_2}{x_1^2 + x_2^2}\vec{e}_1 + \frac{Cx_1}{x_1^2 + x_2^2}\vec{e}_2,$$

find the streamlines and the vortex lines.

Solution: In case a) $\vec{v} = Cx_2\vec{e}_1$ we have that $v_1 = Cx_2$, $v_2 = v_3 = 0$; therefore, we find from (1.2.76):

$$\vec{\omega} = \frac{1}{2}\begin{vmatrix} \vec{e}_1 & \vec{e}_2 & \vec{e}_3 \\ \frac{\partial}{\partial x_1} & \frac{\partial}{\partial x_2} & \frac{\partial}{\partial x_3} \\ Cx_2 & 0 & 0 \end{vmatrix} = -\frac{C}{2}\vec{e}_3, \ \omega_3 = -\frac{C}{2}, \ \omega_1 = \omega_2 = 0.$$

Equations (1.2.14) for the streamlines $\frac{dx_1}{v_1} = \frac{dx_2}{v_2} = \frac{dx_3}{v_3}$ yield $\frac{dx_1}{Cx_1} = \frac{dx_2}{0} = \frac{dx_3}{0}$. From here we have that $dx_2 = dx_3 = 0$, $x_2 = k_2$, $x_3 = k_3$,

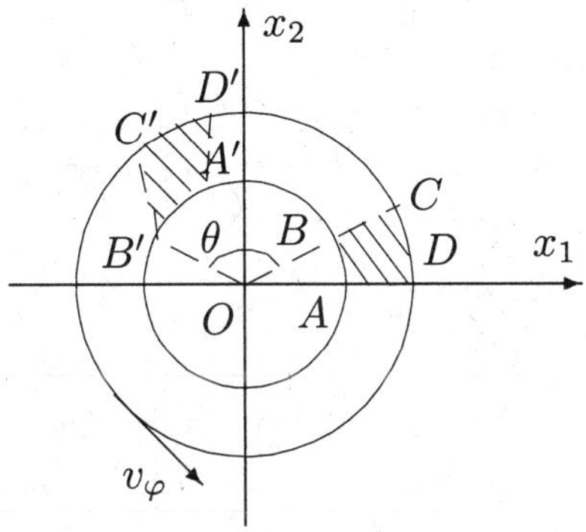

Figure 1.14: The positions of a liquid particle at two moments of time.

where k_i = const; i.e., the streamlines are the lines parallel with the x_1 axis.

We have in a similar way for the vortex lines (1.2.77) that

$$\frac{dx_1}{\omega_1} = \frac{dx_2}{\omega_2} = \frac{dx_3}{\omega_3}, \quad \frac{dx_1}{0} = \frac{dx_2}{0} = -\frac{2dx_3}{C},$$

from where it follows that the vortex lines are the lines parallel with the x_3 axis ($x_1 = k_4,\ x_2 = k_5,\ k_i = \text{const}$).

In case b), where

$$v_1 = -\frac{Cx_2}{x_1^2 + x_2^2}, \quad v_2 = \frac{Cx_1}{x_1^2 + x_2^2}, \quad v_3 = 0,$$

we have from (1.2.76) that

$$\vec{\omega} = \frac{1}{2} \begin{vmatrix} \vec{e}_1 & \vec{e}_2 & \vec{e}_3 \\ \frac{\partial}{\partial x_1} & \frac{\partial}{\partial x_2} & \frac{\partial}{\partial x_3} \\ -\frac{Cx_2}{x_1^2+x_2^2} & \frac{Cx_1}{x_1^2+x_2^2} & 0 \end{vmatrix}$$

$$= -\frac{\vec{e}_3}{2}\left(\frac{C}{x_1^2 + x_2^2} - \frac{2Cx_1^2}{(x_1^2 + x_2^2)^2} + \frac{C}{x_1^2 + x_2^2} - \frac{2Cx_2^2}{(x_1^2 + x_2^2)^2}\right) = 0,$$

from where it follows that $\omega_1 = \omega_2 = \omega_3 = 0,\ x_i \neq 0$.

Equations for the streamlines (1.2.14) have the form

$$\frac{(x_1^2 + x_2^2)\,dx_1}{-Cx_2} = \frac{(x_1^2 + x_2^2)\,dx_2}{Cx_1} = \frac{dx_3}{0},$$

from where we have that

$$dx_3 = 0,\ x_3 = \text{const},\ -\frac{dx_1}{Cx_2} = \frac{dx_2}{Cx_1},\ x_1 dx_1 + x_2 dx_2 = 0,\ x_1^2 + x_2^2 = C.$$

Thus, the streamlines are the circles (Fig. 1.14). A paradox arises here: the liquid particles rotate along a circle, however, the angular velocity $\vec{\omega} = 0$. If it were a solid body, then by virtue of $v_\varphi = \omega r$ the quantity $\omega \neq 0$ if $v_\varphi \neq 0$. There arises in the liquid an additional rotation of a liquid particle at the expense of the velocity gradient (see Fig. 1.14). We have for the given velocity field that

$$v_1^2 + v_2^2 = \frac{C^2}{r^2}, \quad v_\varphi^2 = v_1^2 + v_2^2, \quad v_\varphi = \frac{C}{r}.$$

Thus, the farther from the coordinate origin the particles are located, the slower their velocities. Let us choose at the moment of time $t = 0$ a liquid particle ABCD (see Fig. 1.14). By virtue of the dependence $v_\varphi = C/r$ it will revert at a later moment of time $t > 0$ into a particle $A'B'C'D'$, which has undergone a rotation as a solid by the angle θ and the reverse rotation by the same angle because of the velocity gradient. As a result, the total rotation angle of a liquid particle will be equal to zero. Note that $\omega = 0$ everywhere with the exception of the point $\vec{x} = 0$, where a vortex filament is located, which is perpendicular to the plane $x_1 x_2$. The dependence $\omega_3(x_1, x_2)$ in the case of a vortex filament is determined by the Dirac delta function:

$$\omega_3 = \Gamma \delta(x_1)\, \delta(x_2), \quad \Gamma = \int \omega_1\, dx_1\, dx_2 = \int_0^{2\pi} v_\varphi\, r\, d\varphi = 2\pi C,$$

hence, $\Gamma = 2\pi C$. The quantity Γ is called the *circulation.*

In what follows, we present the output of our *Mathematica* Notebook `prog1-8.nb`, which shows how the given task can be solved on a computer in case a) of the velocity field specification.

The Specification of the Velocity Field

$$\begin{matrix} C x_2 \\ 0 \\ 0 \end{matrix}$$

Calculation of the Components of the Angular Velocity $\omega^{\rightharpoonup}$

$$\omega\_1 = 0$$

$$\omega\_2 = 0$$

$$\omega\_3 = -\frac{C}{2}$$

$$x\_2 = k_2$$

$$x\_3 = k_3$$

1.3 Stress Tensor

There are two kinds of forces in nature: long-range forces and short-range forces. The gravitational and electromagnetic forces that are related first of all to the interaction of charges and ions belong to long-range forces. The long-range forces affect all particles of a continuum located within some volume V; for this reason, they are termed the body or mass forces. The body forces will be denoted by the letter $\vec{F}$ in the following, and the magnitude of this force depends on the coordinate and time: $\vec{F} = \vec{F}(\vec{x}, t)$.

Another type of force is the short-range force the example of which is the interaction of molecules in gas or in solid. The total charge of molecules is equal to zero; therefore, the interaction between them begins to manifest itself only when they draw closer at the distances of the order of several angstroms (10^{-8} cm.). (There exist in nature, besides the electric forces, the nuclear forces and weak interactions having a lesser interaction radius $r < 10^{-13}$ cm. These forces do not affect the motion of molecules and atoms in the classical mechanics, however; therefore, they will not be considered in the following.)

A natural idealization of the short-range forces is the notion of surface forces $\vec{f}$ introduced in mechanics by the great French mathematician O. Cauchy. The surface forces act only at a contact of two bodies $d\vec{f} = \vec{f}_n dS$ and satisfy the third Newton's law $\vec{f}_n = -\vec{f}_{-n}$, where $\vec{n}$ is a normal to the surface and dS is the contact area.

1.3.1 The Cauchy Stress Tensor in the Accompanying Coordinate System

Consider a tetrahedron $OA_1A_2A_3$ constructed on the vectors $\hat{\vec{e}}_1 d\xi^1$, $\hat{\vec{e}}_2 d\xi^2$, and $\hat{\vec{e}}_3 d\xi^3$, where $\hat{\vec{e}}_i$ are the basis vectors of an accompanying coordinate system ξ^i (see Fig. 1.15). The forces $\vec{f}^{-1} dS_1$, $\vec{f}^{-2} dS_2$, and $\vec{f}^{-3} dS_3$ act on the lateral facets of the tetrahedron, and the force $\vec{f}_n dS_n$ acts on the area with the normal $\vec{n}$. According to the second Newton's law, we have that

$$\vec{f}^{-i} dS_i + \vec{f}_n dS_n = \rho\, dV \,\frac{d\vec{v}}{dt}.$$

The volume $dV \approx \frac{1}{3} d\xi^1 d\xi^2 d\xi^3$, and the area $dS_i \approx d\xi^j d\xi^k$; therefore, taking $d\xi^i \to 0$, we can see that the right-hand side is a quantity of third-order smallness and the left-hand side is the quantity of the second-order smallness. Therefore, we can neglect the right-hand side and obtain the equation:

$$\vec{f}^{-i} dS_i + \vec{f}_n dS_n = 0. \tag{1.3.1}$$

With regard for the third Newton's law $\vec{f}^{-i} = -\vec{f}^{i}$, we can rewrite

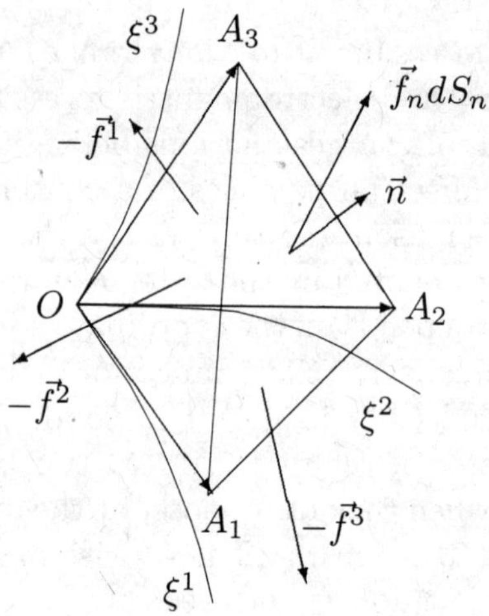

Figure 1.15: To the derivation of the Cauchy formula (1.3.10).

equation (1.3.1) in the form $\vec{f}^i dS_i = \vec{f}_n dS_n$, from where

$$\vec{f}_n = \frac{1}{dS_n}\vec{f}^i dS_i, \tag{1.3.2}$$

where dS_i is the ith facet area, and dS_n is the area of a facet with the normal $\vec{n}$. We have taken into account the fact that the vectors of external normal to the areas dS_i are directed into an opposite side with respect to the vectors $\hat{\vec{e}}_i$. It follows from Fig. 1.15 that

$$\vec{A_1A_2} = \hat{\vec{e}}_2 d\xi^2 - \hat{\vec{e}}_1\, d\xi^1, \quad \vec{A_1A_3} = \hat{\vec{e}}_3 d\xi^3 - \hat{\vec{e}}_1\, d\xi^1;$$

therefore,

$$dS_n \cdot \vec{n} = \frac{1}{2}(\vec{A_1A_2} \times \vec{A_1A_3}) = \frac{1}{2}((\hat{\vec{e}}_2 \times \hat{\vec{e}}_3)d\xi^2 d\xi^3 + (\hat{\vec{e}}_3 \times \hat{\vec{e}}_1)d\xi^1 d\xi^3$$
$$+(\hat{\vec{e}}_1 \times \hat{\vec{e}}_2)d\xi^1 d\xi^2) = \frac{1}{2}(\varepsilon_{231}\hat{\vec{e}}^{\,1} d\xi^2\, d\xi^3 + \varepsilon_{312}\hat{\vec{e}}^{\,2} d\xi^1\, d\xi^3 + \varepsilon_{123}\hat{\vec{e}}^{\,3} d\xi^1\, d\xi^2).$$

It follows from here that the area of a triangle constructed on the vectors $\hat{\vec{e}}_j d\xi^j$ and $\hat{\vec{e}}_k d\xi^k$ is equal to

$$dS_i = \frac{1}{2}|\hat{\vec{e}}_j \times \hat{\vec{e}}_k| d\xi^j d\xi^k = \frac{1}{2}|\varepsilon_{jki}\hat{\vec{e}}^{\,i}| d\xi^j d\xi^k = \frac{1}{2}\sqrt{\hat{g}}\sqrt{\hat{g}^{ii}} d\xi^j d\xi^k. \tag{1.3.3}$$

The area of a parallelogram constructed on the vectors $\hat{\vec{e}}_j d\xi^j$ and $\hat{\vec{e}}_k d\xi^k$ will be by a factor of two larger, respectively:

$$d\tilde{S}_i = \sqrt{\hat{g}\hat{g}^{\,ii}}\, d\xi^j\, d\xi^k, \tag{1.3.4}$$

where there is no summation over the repeating indices. Combining formula (1.3.3) with the foregoing formula, we obtain:

$$dS_n \cdot \vec{n} = \frac{dS_1}{\sqrt{\hat{g}^{11}}}\hat{\vec{e}}^{\,1} + \frac{dS_2}{\sqrt{\hat{g}^{22}}}\hat{\vec{e}}^{\,2} + \frac{dS_3}{\sqrt{\hat{g}^{33}}}\hat{\vec{e}}^{\,3}. \qquad (1.3.5)$$

Expanding $\vec{n}$ into the basis $\hat{\vec{e}}^{\,i}$, $\vec{n} = n_i \hat{\vec{e}}^{\,i}$, we rewrite (1.3.5) in the form

$$dS_n \cdot n_i = \frac{dS_i}{\sqrt{\hat{g}^{ii}}}, \qquad (1.3.6)$$

where there is no summation over the repeating indices. Substituting (1.3.6) into (1.3.2) and replacing i with j, we find[5]:

$$\vec{f}_n = \sqrt{g^{ii}}\,\vec{f}^{j} n_j = \hat{\vec{\sigma}}^{\,j} n_j, \qquad (1.3.7)$$

where we have introduced the notation

$$\hat{\vec{\sigma}}^{\,j} = \sqrt{\hat{g}^{jj}}\,\vec{f}^{j}. \qquad (1.3.8)$$

The quantity $\hat{\vec{\sigma}}^{\,j}$ is called the principal contravariant stress vector

$$\begin{cases} \hat{\vec{\sigma}}^{\,1} = \sqrt{\hat{g}^{11}}\,\vec{f}^{\,1}, \\ \hat{\vec{\sigma}}^{\,2} = \sqrt{\hat{g}^{22}}\,\vec{f}^{\,2}, \\ \hat{\vec{\sigma}}^{\,3} = \sqrt{\hat{g}^{33}}\,\vec{f}^{\,3}. \end{cases} \qquad (1.3.9)$$

Employing the expansion into the basis $\vec{f}_n = f_n^i \hat{\vec{e}}_i$, $\hat{\vec{\sigma}}^{\,j} = \hat{\sigma}^{ij}\hat{\vec{e}}_i$, let us rewrite (1.3.7) as

$$f_n^i = \hat{\sigma}^{ij} n_j. \qquad (1.3.10)$$

The obtained relation (1.3.10) is called the *Cauchy formula*, which enables one to find the force acting on an area with a normal $\vec{n}$, if the Cauchy contravariant stress tensor $\hat{\sigma}^{ji}$ is known in the basis $\hat{\vec{e}}_i$. The force acting along a perpendicular to the area with the normal $\vec{n}$ is equal to

$$N = \vec{f}_n \cdot \vec{n} = f_n^j \hat{\vec{e}}_j \cdot n_i \hat{\vec{e}}^{\,i} = \hat{\sigma}^{jk} n_k n_i \hat{\vec{e}}_j \cdot \hat{\vec{e}}^{\,i} = \hat{\sigma}^{jk} n_j n_k. \qquad (1.3.11)$$

The force tangent to the area is

$$T = \sqrt{(\vec{f}_n)^2 - N^2} = \sqrt{g_{ik}\hat{\sigma}^{ij}\hat{\sigma}^{kl} n_j n_l - (\hat{\sigma}^{jk} n_j n_k)^2}. \qquad (1.3.12)$$

If a continuum does not contain the internal moments, then the Cauchy stress tensor is symmetric, i.e., $\hat{\sigma}^{ij} = \hat{\sigma}^{ji}$. To prove this fact, let us choose an arbitrary infinitesimal volume V bounded by the surface S. Neglecting the terms proportional to the volume V, let us write the equilibrium equations for the forces and moments as

$$\int_S \vec{f}_n dS = 0, \quad \int_S (\vec{r} \times \vec{f}_n) dS = 0. \tag{1.3.13}$$

Using the Ostrogradsky–Gauss theorem (1.1.63) we obtain with regard for (1.3.7):

$$\int_S \vec{f}_n dS = \int_S \hat{\vec{\sigma}}^{\,j} n_j dS = \int_V \nabla_j \hat{\vec{\sigma}}^{\,j} dV = 0,$$

from where we obtain by virtue of the arbitrariness of V:

$$\nabla_j \hat{\vec{\sigma}}^{\,j} = 0, \tag{1.3.14}$$

where the covariant derivative (1.1.23) is equal to

$$\nabla_j \hat{\vec{\sigma}}^{\,j} = \frac{\partial \hat{\vec{\sigma}}^{\,j}}{\partial \xi^j} + \hat{\vec{\sigma}}^{\,j} \Gamma^i_{ji}. \tag{1.3.15}$$

Expanding $\hat{\vec{\sigma}}^{\,j}$ into the basis $\hat{\vec{\sigma}}^{\,j} = \hat{\sigma}^{ij} \hat{\vec{e}}_i$, let us rewrite (1.3.14) with regard for (1.1.27) in the form

$$\nabla_j \hat{\vec{\sigma}}^{\,j} = \nabla_j (\hat{\sigma}^{ij} \hat{\vec{e}}_i) = \nabla_j (\hat{\sigma}^{ij}) \hat{\vec{e}}_i = 0,$$

from where it follows that

$$\nabla_j \hat{\sigma}^{ij} = 0, \tag{1.3.16}$$

where

$$\nabla_j \hat{\sigma}^{ij} = \frac{\partial \hat{\sigma}^{ij}}{\partial \xi^j} + \hat{\sigma}^{ik} \Gamma^j_{kj} + \hat{\sigma}^{kj} \Gamma^i_{kj}. \tag{1.3.17}$$

For the second equation in (1.3.13), we determine in a similar way that

$$\int_S (\vec{r} \times \vec{f}_n)\, dS = \int_S (\vec{r} \times \hat{\vec{\sigma}}^{\,j} n_j)\, dS = \int_S (\vec{r} \times \hat{\vec{\sigma}}^{\,j}) n_j dS = \int_V \nabla_j (\vec{r} \times \hat{\vec{\sigma}}^{\,j})\, dV = 0,$$

from where by virtue of the arbitrariness of the V, we obtain:

$$\nabla_j (\vec{r} \times \hat{\vec{\sigma}}^{\,j}) = 0. \tag{1.3.18}$$

Since V is small, $\vec{r} = \xi^k \hat{\vec{e}}_k$, and with regard for $\hat{\vec{\sigma}}^{\,j} = \hat{\sigma}^{ij} \hat{\vec{e}}_i$ and the definition of the vector product (1.1.37), we find:

$$\nabla_j (\xi^k \hat{\sigma}^{ij} (\hat{\vec{e}}_k \times \hat{\vec{e}}_i)) = \nabla_j (\xi^k \hat{\sigma}^{ij} \hat{\varepsilon}_{kis} \hat{\vec{e}}^{\,s}) = \nabla_j (\xi^k \hat{\sigma}^{ij}) \hat{\varepsilon}_{kis} \hat{\vec{e}}^{\,s},$$

from where it follows that

$$\nabla_j(\xi^k\hat{\sigma}^{ij})\hat{\varepsilon}_{kis} = 0. \tag{1.3.19}$$

In the derivation of (1.3.19), we have taken into account the expression for the components of the Levi–Civita tensor $|\hat{\varepsilon}_{kis}| = (\det \parallel \hat{g}_{ij} \parallel)^{1/2}$; therefore, we have from the condition $\nabla_j\hat{g}_{kl} = 0$ that $\nabla_j\hat{\varepsilon}_{kis} = 0$. We can find from (1.3.19) with regard for (1.3.16) that

$$\nabla_j(\xi^k\hat{\sigma}^{ij})\hat{\varepsilon}_{kis} = \delta^k_j\hat{\sigma}^{ij}\hat{\varepsilon}_{kis} + \xi^k\nabla_j\hat{\sigma}^{ij}\hat{\varepsilon}_{kis} = \hat{\varepsilon}_{kis}\hat{\sigma}^{ik} = 0.$$

Assuming

$$\hat{\varepsilon}_{kis}\hat{\sigma}^{ik} = \sum_{k<i}\hat{\varepsilon}_{kis}\hat{\sigma}^{ik} + \sum_{k>i}\hat{\varepsilon}_{kis}\hat{\sigma}^{ik}$$

and interchanging in the second item the dummy indices k and i, we determine with regard for $\hat{\varepsilon}_{iks} = -\hat{\varepsilon}_{kis}$ that

$$\sum_{k<i}\hat{\varepsilon}_{kis}(\hat{\sigma}^{ik} - \hat{\sigma}^{ki}) = 0,$$

from where it follows that

$$\hat{\sigma}^{ik} = \hat{\sigma}^{ki}. \tag{1.3.20}$$

1.3.2 Piola–Kirchhoff Stress Tensors in the Reference Frame and in the Eulerian Coordinates

Let a Cartesian coordinate system $\overset{o}{g}_{ij} = \delta_{ij}$ be given at the moment of time $t = 0$ in the reference frame. Let us choose in the reference frame a tetrahedron constructed on the vectors $\overset{o}{\vec{e}}_i\ d\xi^i$, which at the time t will in the result of deformation go over to a tetrahedron formed by the vectors $\hat{\vec{e}}_i d\xi^i$ (see Fig. 1.16). The areas of the triangles $dS^0_i = \frac{1}{2}d\xi^j\, d\xi^k$ go over after the deformation to the areas $dS_i = \frac{1}{2}\sqrt{\hat{g}}\sqrt{\hat{g}^{ii}}\, d\xi^j\, d\xi^k$; therefore,

$$\frac{dS_i}{dS^0_i} = \sqrt{\hat{g}\hat{g}^{ii}}. \tag{1.3.21}$$

According to (1.3.6), the equations $dS_i = \sqrt{\hat{g}^{ii}}n_i\, dS_n$, $dS^0_i = n^0_i\, dS^0_n$ are valid. Substituting them into (1.3.21), we obtain:

$$n_i\, dS_n = \sqrt{\hat{g}}\, n^0_i\, dS^0_n. \tag{1.3.22}$$

Let us determine the force vector $\vec{P}_n$ related to an area dS^0_n in the refernce frame with the aid of the equation

$$\vec{P}_n\, dS^0_n = \vec{f}_n\, dS_n,$$

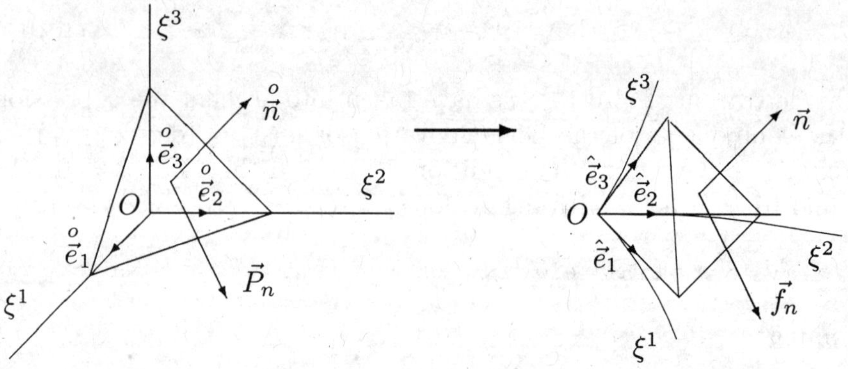

Figure 1.16: To the derivation of the formula for the Piola–Kirchhoff tensor.

from where we have with regard for (1.3.22) that

$$\vec{P}_n\, dS_n^0 = \hat{\vec{\sigma}}^{\,i} n_i\, dS_n = \sqrt{\hat{g}}\, \hat{\vec{\sigma}}^{\,i} n_i^0\, dS_n^0. \tag{1.3.23}$$

Applying formula (1.3.7) to the reference frame $\vec{P}_n = \overset{o}{\vec{\sigma}}{}^{i}\, n_i^0$, we find from (1.3.23) the contravariant stress vector in the reference frame basis[5]:

$$\overset{o}{\vec{\sigma}}{}^{i} = \sqrt{\hat{g}}\, \hat{\vec{\sigma}}^{\,i}. \tag{1.3.24}$$

Expanding $\overset{o}{\vec{\sigma}}{}^{i}$ into the basis $\overset{o}{\vec{e}}_i$, we can determine the *Piola–Kirchhoff tensor* in the reference frame basis $\overset{o}{\vec{\sigma}}{}^{i} = \overset{o}{\sigma}{}^{ji}\, \overset{o}{\vec{e}}_j$. Using (1.3.24), we determine the relation between $\overset{o}{\sigma}{}^{ij}$ and $\hat{\sigma}^{ij}$. Let the law of continuum deformation $x^i = x^i(\xi^j, t)$ be given at which the vector $d\, \overset{o}{\vec{r}}$ goes over to $d\vec{r}$ (the time t is a parameter). The expression for $d\vec{r}$ is the same in the accompanying Lagrangian and Eulerian Cartesian coordinate systems, i.e., $d\xi^j\, \hat{\vec{e}}_j = dx^i\, \overset{o}{\vec{e}}_i$, from where we have that

$$\overset{o}{\vec{e}}_j = \frac{\partial x^i}{\partial \xi^j}\, \overset{o}{\vec{e}}_i\, . \tag{1.3.25}$$

Substituting (1.3.25) into the relation $\hat{\vec{\sigma}}^{\,i} = \hat{\sigma}^{ji} \hat{\vec{e}}_j$, we find with regard for equation $\overset{o}{\vec{\sigma}}{}^{i} = \overset{o}{\sigma}{}^{ji}\, \overset{o}{\vec{e}}_j$ and equation (1.3.24) the expression for the Piola–Kirchhoff tensor in terms of the Cauchy tensor

$$\overset{o}{\sigma}{}^{ki} = \sqrt{\hat{g}}\, \frac{\partial x^k}{\partial \xi^j}\, \hat{\sigma}^{ji}. \tag{1.3.26}$$

It can be seen from (1.3.26) that the Piola–Kirchhoff tensor is no symmetric tensor. The Cauchy formula in the reference frame basis has the form $P_n^i = \overset{o}{\sigma}{}^{ij} n_j^0$.

Let us find the stress tensor in Eulerian coordinates. We will assume the Eulerian coordinate system to be for simplicity a Cartesian system. Let us at first determine a passage from the accompanying Lagrangian coordinate system to the Eulerian one for a given motion law $x^i = x^i(\xi^j, t)$. In accordance with the rule for the transformation of the tensor contravariant components, we obtain:

$$\sigma^{ij} = \frac{\partial x^i}{\partial \xi^k} \frac{\partial x^j}{\partial \xi^l} \hat{\sigma}^{kl}. \tag{1.3.27}$$

It follows from the relationships (1.3.27) that the stress tensor σ^{ij} is symmetric. Along with (1.3.27), we present the expression for $\hat{\sigma}^{ij}$ in terms of σ^{ij}

$$\hat{\sigma}^{ij} = \frac{\partial \xi^i}{\partial x^k} \frac{\partial \xi^j}{\partial x^l} \sigma^{kl}. \tag{1.3.28}$$

The stress tensor $\hat{\sigma}^{ij}$ can be expanded into the spherical and deviator components

$$\hat{\sigma}^{ij} = -\hat{P}\hat{g}^{ij} + \hat{S}^{ij}, \tag{1.3.29}$$

$$\hat{S}_i^i = 0, \quad \hat{P} = -\frac{1}{3}\hat{\sigma}_i^i,$$

where $\hat{S}^{ij}$ is called the *tensor of the stress deviator* and $\hat{P}$ is the *pressure.* In the Cartesian coordinate system, equation (1.3.29) simplifies to

$$\sigma_{ij} = -P\delta_{ij} + S_{ij}, \tag{1.3.30}$$

$$S_{ii} = 0, \quad P = -\frac{1}{3}\sigma_{ii},$$

where S_{ij} is the tensor of the stress deviator and P is the pressure.

1.3.3 Principal Values and Invariants of the Stress Tensor

The principal values of the symmetric stress tensor σ^{ij} and its invariants are determined by equations (1.1.49)–(1.1.52), in which one must make a substitution $T \to \sigma$. One can consider the same results, however, from the viewpoint of a tensor surface, which can be put into correspondence with the stress tensor. Let us introduce a Eulerian Cartesian coordinate system $g_{ij} = \delta_{ij}$ and consider at some point O a unit area with the normal $\vec{n}$. Then, in accordance with (1.3.11), the normal force acting on this area will be equal to

$$N = \sigma^{ij} n_i n_j, \tag{1.3.31}$$

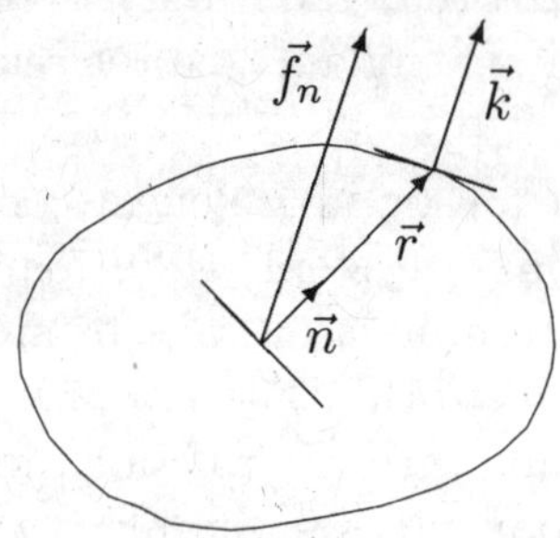

Figure 1.17: The construction of the normal $\vec{k}$ to the Cauchy stress surface.

where σ^{ij} is the stress tensor at point O. Let $\vec{r}$ be a vector emanating from the point O. Then

$$n_i = \cos(nx_i) = x_i/r. \tag{1.3.32}$$

Let us determine the Cauchy stress surface from the condition $Nr^2 =$ const, and then we obtain the equation for the *Cauchy stress surface* by substituting (1.3.32) into (1.3.31):

$$\sigma^{ij}x_ix_j = 2\Phi_\sigma(x_i) = \text{const}. \tag{1.3.33}$$

The tensor surface (1.3.33) represents a second-order surface and can be introduced in the same way for the strain tensor if we replace σ^{ij} with ε_{ij}. The differentiation of (1.3.33) with respect to x_j yields

$$f_n^j = \sigma^{ij}n_i = \sigma^{ij}\frac{x_i}{r} = \frac{1}{r}\frac{\partial\Phi_\sigma}{\partial x_j}, \tag{1.3.34}$$

from where it follows that the normal $\vec{k}$ to the Cauchy stress surface is parallel with a vector of the force acting on the area with the normal $\vec{n} = \frac{x_i}{r}\vec{e}_i$, where $\vec{r} = x_i\vec{e}_i$ is the vector $\vec{r}$ emanating from the coordinate origin O into a given point of the Cauchy stress surface (see Fig. 1.17).

It is known from the analytic geometry that the second-order surfaces have at least three principal directions for which the vector $\vec{k}$ is parallel with $\vec{r}$. (If the tensor surface is spherical, then there are infinitely many such directions, and the corresponding tensor $\sigma^{ij} = -P\delta^{ij}$ is called the spherical tensor.) With regard for (1.3.34), the condition for the parallelism of $\vec{k}$ and $\vec{r}$ may be written as

$$\sigma^{ij}n_j = \sigma\delta^{ij}n_j, \tag{1.3.35}$$

where σ is a proportionality coefficient. It follows from (1.3.35) that

$$(\sigma^{ij} - \sigma\delta^{ij})n_j = 0, \tag{1.3.36}$$

which coincides with the accuracy up to notation with (1.1.49). Note that the system (1.1.49) is valid in any coordinate system. The principal values are found from the condition that the determinant (1.1.50) be equal to zero, which yields the equation

$$\sigma^3 - I_1\sigma^2 + I_2\sigma - I_3 = 0, \tag{1.3.37}$$

where I_1, I_2, and I_3 are the first, second, and third invariants of the stress tensor, respectively:

$$I_1 = \sigma^i_i, \quad I_2 = \frac{1}{2}((\sigma^i_i)^2 - \sigma^i_j\sigma^j_i), \quad I_3 = \det \| \sigma^j_i \| . \tag{1.3.38}$$

The formulas (1.3.38) are valid in an arbitrary coordinate system, and the values of I_i do not depend on the choice of a coordinate system. Substituting the found eigenvalues σ_α into (1.3.36) and (1.1.49), we obtain with regard for equation

$$(n^\alpha_1)^2 + (n^\alpha_2)^2 + (n^\alpha_3)^2 = 1 \tag{1.3.39}$$

the corresponding eigenvector $\vec{n}^\alpha = n^\alpha_i \vec{e}_i$. According to (1.1.53) and (1.3.39), the eigenvectors form an orthonormal basis

$$\vec{n}^\alpha \cdot \vec{n}^\beta = \delta^{\alpha\beta}, \tag{1.3.40}$$

in which the stress tensor is diagonal

$$\sigma^{ij} = \begin{pmatrix} \sigma_1 & 0 & 0 \\ 0 & \sigma_2 & 0 \\ 0 & 0 & \sigma_3 \end{pmatrix}, \tag{1.3.41}$$

and the invariants I_i are equal to

$$I_1 = \sigma_1 + \sigma_2 + \sigma_3, \quad I_2 = \sigma_1\sigma_2 + \sigma_1\sigma_3 + \sigma_2\sigma_3, \quad I_3 = \sigma_1\sigma_2\sigma_3. \tag{1.3.42}$$

1.3.4 Differentiation of the Stress Tensor with Respect to Time

The rate of change of a stress tensor is an important characteristic of the continuum behavior. Following[10], one can show that the substantive derivative

$$\frac{d\sigma_{ij}}{dt} = \frac{\partial\sigma_{ij}}{\partial t} + v^k\frac{\partial\sigma_{ij}}{\partial x^k}$$

does not determine the rate of change of the stress tensor. Let us consider some parallelepiped, which rotates as an absolutely rigid body together with the forces applied to its boundaries. In this case, the components of the stress tensor in the Eulerian coordinate system $\sigma^{ij} = O_{ik}O_{jl}\hat{\sigma}^{kl}$

will change with time: $\frac{d\sigma_{ij}}{dt} \neq 0$, where O_{ik} is the rotation matrix. This change, however, is not related to the physical processes occurring in a chosen material volume because, in the accompanying coordinate system of the parallelepiped, $\frac{d\hat{\sigma}^{kj}}{dt} = 0$. It is seen from here that the quantity $\frac{d\sigma_{ij}}{dt}$ is no tensor, since, by virtue of the tensor definition (1.1.11), the equality $T_{ik} = O_{ik}O_{jl}\hat{T}^{kl}$ should be valid, and if $\hat{T}^{kl} = 0$, then $T_{ij} = 0$.

Let us determine the derivative of the tensor in such a way that we again obtain a tensor as a result of the differentiation of a tensor with respect to time. This can be done in different ways[5,6,10–13], however, we will follow[5], since it uses substantially the invariance property of a tensor Cauchy surface. Let the stress tensor $\hat{\sigma}^{kl}(\xi^k, t)$ be determined in some neighborhood of the point $\xi^k = \text{const}$. According to formula (1.3.31), we can find the magnitude of a normal force acting on a unit area with the normal $\vec{n}$, both in the Eulerian Cartesian coordinate system for which $g_{ij} = \delta_{ij}$ and in the accompanying curvilinear coordinate system

$$N = \sigma_{ij} n_i n_j = \hat{\sigma}^{kl} \hat{n}_k \hat{n}_l, \quad \hat{n}_k = \frac{\partial x^i}{\partial \xi^k} n_i. \tag{1.3.43}$$

Specifying some infinitesimal vector $d\vec{r}$, let us rewrite (1.3.32) in the form $n_i = dx^i/dr$. Then, by multiplying (1.3.43) with $(dr)^2 = \delta_{ij} dx^i dx^j = \hat{g}_{ij} d\xi^i d\xi^j$, we find the equation for the Cauchy stress surface

$$\Phi_\sigma = r^2 N = \sigma_{ij} dx^i dx^j = \hat{\sigma}_{ij} d\xi^i d\xi^j, \tag{1.3.44}$$

where we have used in the derivation of (1.3.44) the relationships

$$\hat{\sigma}^{kl} \hat{n}_k \hat{n}_l = \hat{\sigma}_{kl} \hat{n}^k \hat{n}^l, \quad \hat{n}^k = \frac{\partial \xi^k}{\partial x^i} n^i = \frac{\partial \xi^k}{\partial x^i} \frac{dx^i}{dr} = \frac{d\xi^k}{dr}.$$

Thus, it follows from (1.3.44) that $\sigma_{ij} dx^i dx^j$ forms an invariant quadratic form, which does not depend on the choice of a coordinate system. One can determine in a similar way a tensor surface for an arbitrary symmetric tensor $\hat{A}$ in the neighborhood of some point $\xi^i = \text{const}$.

$$\Phi(t) = a_{ij} dx^i dx^j = A_{ij} d\xi^i d\xi^j, \tag{1.3.45}$$

where $A_{ij}(\xi^k, t)$ are the covariant components of $\hat{A}$ in the accompanying Lagrangian coordinate system and $a_{ij}(x^k, t) = a^{ij}(x^k, t)$ are the components of $\hat{A}$ in the Eulerian Cartesian coordinate system. According to (1.3.45), the relation between the components $a_{ij}, A_{ij}, A_{k.}^{\cdot j}, A^{i}_{\cdot j}$ is given by the formulas

$$a_{ij} \frac{\partial x^i}{\partial \xi^k} \frac{\partial x^j}{\partial \xi^l} = A_{kl} = \hat{g}_{ki} \hat{g}_{lj} A^{ij} = \hat{g}_{lj} A_{k.}^{\cdot j} = \hat{g}_{ki} A^{i}_{\cdot l}. \tag{1.3.46}$$

Differentiating (1.3.45) with respect to time and equalling the first and third terms, we obtain:

$$\dot{\Phi}(t) = \frac{\partial A_{ij}}{\partial t} d\xi^j d\xi^k. \tag{1.3.47}$$

A scalar quantity is on the left-hand side of (1.3.47), which does not change under the coordinate transformations. The quantity $d\xi^j d\xi^k$ transforms as the contravariant tensor components; therefore, the

$$A'_{ij} = \frac{\partial A_{ij}}{\partial t} \tag{1.3.48}$$

are the covariant tensor components and determine the desired derivative in an accompanying coordinate system. Since equation (1.3.45) is valid for any symmetric tensor, let us determine with the aid of (1.3.45) the derivative of the tensor a'_{ij} in the Eulerian coordinates:

$$a'_{ij} dx^i dx^j = A'_{ij} d\xi^i d\xi^j. \tag{1.3.49}$$

To determine an explicit expression for a'_{ij}, we again differentiate (1.3.45) with respect to time:

$$\dot{\Phi}(t) = \frac{da_{ij}}{dt} dx^i dx^j + a_{ij} \frac{d(dx^i)}{dt} dx^j + a_{ij} \frac{d(dx^j)}{dt} dx^i, \tag{1.3.50}$$

where $a_{ij} = a_{ij}(x^k, t)$. Using the relations

$$\begin{aligned} dx^i &= \frac{\partial x^i}{\partial \xi^k} d\xi^k, \quad v^k = \left.\frac{\partial x^k}{\partial t}\right|_{\xi^i}, \quad \frac{da_{ij}}{dt} = \frac{\partial a_{ij}}{\partial t} + v^k \frac{\partial a_{ij}}{\partial x^k}, \\ \frac{d}{dt}(dx^i) &= \frac{\partial v^i}{\partial \xi^k} d\xi^k = \frac{\partial v^i}{\partial \xi^k} \frac{\partial \xi^k}{\partial x^s} dx^s = \frac{\partial v^i}{\partial x^s} dx^s, \end{aligned}$$

we obtain from (1.3.50):

$$\dot{\Phi}(t) = \left(\frac{da_{ij}}{dt} + a_{sj} \frac{\partial v^s}{\partial x^i} + a_{is} \frac{\partial v^s}{\partial x^j} \right) dx^i dx^j. \tag{1.3.51}$$

Comparing (1.3.49) and (1.3.51), we now find the derivative of the tensor in the Eulerian coordinates $(a_{ij})'$, which corresponds to the derivative with respect to time in the accompanying coordinates of A'_{ij}

$$a'_{ij} = \frac{\partial a_{ij}}{\partial t} + v_k \frac{\partial a_{ij}}{\partial x^k} + a_{sj} \frac{\partial v^s}{\partial x^i} + a_{is} \frac{\partial v^s}{\partial x^j}. \tag{1.3.52}$$

Choosing as $\hat{A}$ the stress tensor $\Sigma = \sigma_{ij} \vec{e}_i \vec{e}_j = \hat{\sigma}_{ij} \hat{\vec{e}}^i \hat{\vec{e}}^j$, we determine from (1.3.46), (1.3.48), (1.3.45), and (1.3.52) that

$$(\hat{\sigma}_{ij})' = (\hat{\sigma}_{ij})^{\cdot} = \frac{\partial}{\partial t} \hat{\sigma}_{ij}(\xi^k, t) \tag{1.3.53}$$

$$\begin{aligned}\sigma'_{ij} &= \frac{\partial\sigma_{ij}}{\partial t} + v^k\frac{\partial\sigma_{ij}}{\partial x^k} + \sigma_{is}\frac{\partial v^s}{\partial x^j} + \sigma_{sj}\frac{\partial v^s}{\partial x^i},\\ \dot{\Phi}_\sigma &= (\hat{\sigma}_{ij})'d\xi^i d\xi^j = \sigma'_{ij}dx^i dx^j,\\ \sigma'_{ij}\frac{\partial x^i}{\partial \xi^k}\frac{\partial x^j}{\partial \xi^l} &= (\hat{\sigma}_{kl})' = \hat{g}_{ki}\hat{g}_{lj}(\hat{\sigma}^{ij})' = \hat{g}_{lj}(\sigma^{\cdot j}_{k\cdot})' = \hat{g}_{ki}(\sigma^{i\cdot}_{\cdot l})'.\end{aligned}$$

Note that the mixed and contravariant components of the derivatives of tensors are obtained by raising one or two indices.

Let us find the time derivative of the strain tensor. Let at time $t = 0$ an infinitesimal length element $ds_0^2 = \delta_{ij}\, d\xi^i\, d\xi^j = \text{const}$. After the deformation, it will be equal to $ds^2 = \hat{g}_{ij}d\xi^i d\xi^j = \delta_{ij}dx^i dx^j$. The scalar invariant Φ_ε determining the strain tensor has the form

$$\Phi_\varepsilon(t) = \frac{1}{2}(ds^2 - ds_0^2) = \hat{\varepsilon}_{ij}d\xi^i d\xi^j = \varepsilon_{ij}dx^i dx^j.$$

Differentiating this formula with respect to time, we introduce the derivatives of the strain tensor:

$$\dot{\Phi}_\varepsilon = \frac{1}{2}\frac{d}{dt}(ds^2) = (\dot{\hat{\varepsilon}}_{ij})\, d\xi^i d\xi^j = (\varepsilon_{ij})'dx^i dx^j.$$

Comparing this formula with (1.3.45) and (1.3.49), we see that in our case

$$\Phi(t) = \frac{1}{2}ds^2, \quad A_{ij} = \frac{1}{2}\hat{g}_{ij}, \quad a_{ij} = \frac{1}{2}\delta_{ij}.$$

Substituting these formulas in (1.3.48) and (1.3.52), we obtain:

$$\begin{aligned}(\hat{g}_{ij})' &= (\hat{g}_{ij})^{\cdot} = \frac{\partial}{\partial t}\hat{g}_{ij}(\xi^k, t) = 2\frac{\partial}{\partial t}\hat{\varepsilon}_{ij} = 2\dot{\hat{\varepsilon}}_{ij}(\xi^k, t),\\ 2\,(\varepsilon_{ij})' &= (\delta_{ij})' = \frac{\partial v^i}{\partial x^j} + \frac{\partial v^j}{\partial x^i} = 2\dot{\varepsilon}_{ij}(x^k, t), \qquad (1.3.54)\\ \dot{\Phi}_\varepsilon &= \frac{1}{2}(\hat{g}_{ij})'d\xi^i d\xi^j = \dot{\varepsilon}_{ij}dx^i\, dx^j,\\ \dot{\varepsilon}_{ij}\frac{\partial x^i}{\partial \xi^k}\frac{\partial x^j}{\partial \xi^l} &= \dot{\hat{\varepsilon}}_{kl} = \hat{g}_{ki}\hat{g}_{lj}\dot{\hat{\varepsilon}}^{ij} = \hat{g}_{lj}\dot{\hat{\varepsilon}}^{\cdot j}_{k\cdot} = \hat{g}_{ki}\dot{\hat{\varepsilon}}^{i\cdot}_{\cdot l}.\end{aligned}$$

Differentiating with respect to time the relationship

$$\hat{g}^{is}(\xi^k, t)\hat{g}_{sj}(\xi^k, t) = \delta^i_j$$

we obtain:

$$(\hat{g}^{ik})^{\cdot} = -\hat{g}^{il}\hat{g}^{jk}(\hat{g}_{lj})^{\cdot} = -2\dot{\hat{\varepsilon}}^{ik}. \qquad (1.3.55)$$

Using the formulas (1.3.53)–(1.3.55), one can show that the $(\hat{\sigma}_{ij})^{\cdot}$, $(\hat{\sigma}^{ij})^{\cdot}$, $(\hat{\sigma}^{i\cdot}_{\cdot j})^{\cdot}$, $(\hat{\sigma}^{\cdot i}_{j\cdot})^{\cdot}$ form in the accompanying coordinate system different tensors for the rate of change of the stresses $\hat{\Sigma}$, to which different tensors of the stress rate in Eulerian coordinates correspond. Referring to Problem 1.4 of the present section, we now give the corresponding formulas for the stress rate:

1) the contravariant stress tensor $\hat{\sigma}^{ij}$

$$\begin{aligned}(\hat{\sigma}^{ij})^{\Delta} &= (\hat{\sigma}^{ij})^{\cdot} = \frac{\partial}{\partial t}\hat{\sigma}^{ij}(\xi^k, t), \\ (\sigma_{ij})^{\Delta} &= \frac{\partial \sigma_{ij}}{\partial t} + v^k \frac{\partial \sigma_{ij}}{\partial x^k} - \sigma_{ik}\frac{\partial v^j}{\partial x^k} - \sigma_{kj}\frac{\partial v^i}{\partial x^k};\end{aligned} \tag{1.3.56}$$

2) the mixed stress tensor $\hat{\sigma}^{i\cdot}_{\cdot j}$

$$\begin{aligned}(\hat{\sigma}^{i\cdot}_{\cdot j})^{-} &= (\hat{\sigma}^{i\cdot}_{\cdot j})^{\cdot} = \frac{\partial}{\partial t}\sigma^{i\cdot}_{\cdot j}(\xi^k, t), \\ (\sigma_{ij})^{-} &= \frac{\partial \sigma_{ij}}{\partial t} + v^k \frac{\partial \sigma_{ij}}{\partial x^k} - \sigma_{ik}\frac{\partial v^j}{\partial x^k} + \sigma_{jk}\frac{\partial v^k}{\partial x^i};\end{aligned} \tag{1.3.57}$$

3) the mixed stress tensor $\hat{\sigma}^{\cdot i}_{j\cdot}$

$$\begin{aligned}(\hat{\sigma}^{\cdot i}_{j\cdot})^{+} &= (\hat{\sigma}^{\cdot i}_{j\cdot})^{\cdot} = \frac{\partial}{\partial t}\sigma^{\cdot i}_{j\cdot}(\xi^k, t), \\ (\sigma_{ij})^{+} &= \frac{\partial \sigma_{ij}}{\partial t} + v^k \frac{\partial \sigma_{ij}}{\partial x^k} + \sigma_{ik}\frac{\partial v^k}{\partial x^j} - \sigma_{jk}\frac{\partial v^i}{\partial x^k}.\end{aligned} \tag{1.3.58}$$

The above tensors of the stress rates (1.3.53) and (1.3.56)–(1.3.58) are linearly dependent between each other:

$$(\sigma_{ij})' + (\sigma_{ij})^{\Delta} = (\sigma_{ij})^{+} + (\sigma_{ij})^{-} = 2\,(\sigma_{ij})^{\nabla},$$

and as a linearly independent tensors, one can take $(\sigma_{ij})'$, $(\sigma_{ij})^{\Delta}$, σ_{ij}^{∇}, where the tensor $(\sigma_{ij})^{\nabla} = \frac{1}{2}((\sigma_{ij})' + (\sigma_{ij})^{\nabla})$ is called the *Jaumann derivative* and has the components

$$(\sigma_{ij})^{\nabla} = \frac{\partial \sigma_{ij}}{\partial t} + v^k \frac{\partial \sigma_{ij}}{\partial x^k} - \sigma_{ik}\omega_{jk} - \sigma_{jk}\omega_{ik}, \tag{1.3.59}$$

where

$$\omega_{ij} = \frac{1}{2}\left(\frac{\partial v^i}{\partial x^j} - \frac{\partial v^j}{\partial x^i}\right)$$

is the tensor of the rotation rate. The Jaumann derivative is equal to the difference of a substantive derivative and the rate of change of the stress tensor at the expense of a rotation of an infinitesimal particle as an absolutely rigid body. It determines the rate of change of the stress tensor σ_{ij} in a coordinate system moving at a particle velocity $\vec{v}$ and rotating in space at a velocity $\vec{\omega}$. Since all the definitions of the stress tensor rate (1.3.53) and (1.3.56)–(1.3.59) are equivalent, one can use any of these definitions depending on the problem.

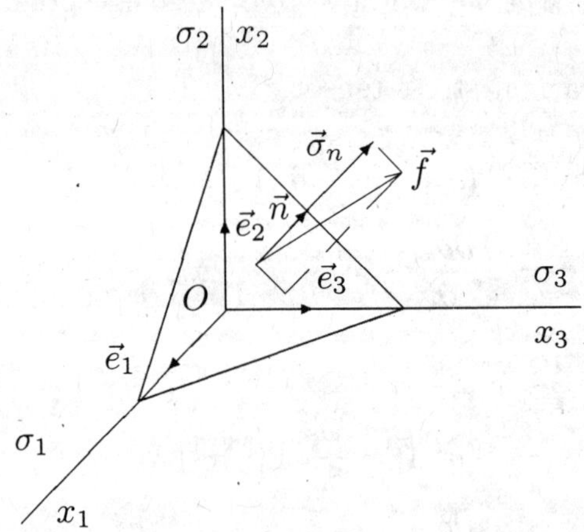

Figure 1.18: The octahedron area with the forces given on it.

Problem 1.9. Calculate the shear and normal stresses acting on an octahedron area (the octahedron area makes equal angles with the principal axes of the stress tensor).

Solution: Let us choose a coordinate system coinciding with the principal axes of the stress tensor σ_{ij} (Fig. 1.18). Then the stress tensor has the form

$$\sigma_{ij} = \begin{pmatrix} \sigma_1 & 0 & 0 \\ 0 & \sigma_2 & 0 \\ 0 & 0 & \sigma_3 \end{pmatrix}.$$

In the given coordinate system, the octahedron area is specified by the normal $\vec{n} = \frac{1}{\sqrt{3}}(\vec{e}_1 + \vec{e}_2 + \vec{e}_3)$ with the components $n_1 = n_2 = n_3 = \frac{1}{\sqrt{3}}$. The force acting on the area is determined from (1.3.10):

$$f_i = \sigma_{ij} n_j = \begin{pmatrix} \sigma_1 & 0 & 0 \\ 0 & \sigma_2 & 0 \\ 0 & 0 & \sigma_3 \end{pmatrix} \cdot \begin{pmatrix} \frac{1}{\sqrt{3}} \\ \frac{1}{\sqrt{3}} \\ \frac{1}{\sqrt{3}} \end{pmatrix} = \begin{pmatrix} \sigma_1/\sqrt{3} \\ \sigma_2/\sqrt{3} \\ \sigma_3/\sqrt{3} \end{pmatrix}.$$

The modulus of the normal force $\vec{\sigma}_n$ is determined by the projection of $\vec{f}$ onto the normal $\vec{n}$:

$$\sigma_n = \sigma_{ij} n_i n_j = f_i n_i = \left(\frac{1}{\sqrt{3}}, \frac{1}{\sqrt{3}}, \frac{1}{\sqrt{3}}\right) \begin{pmatrix} \sigma_1/\sqrt{3} \\ \sigma_2/\sqrt{3} \\ \sigma_3/\sqrt{3} \end{pmatrix} = \frac{1}{3}(\sigma_1 + \sigma_2 + \sigma_3).$$

It follows from here that σ_n is equal to the pressure p with the opposite sign: $\sigma_n = \frac{1}{3}\sigma_{ii} = -p$. We now find the modulus of the shear force $\vec{\sigma}_t$

from the Pythagor's theorem (see Fig. 1.18):

$$\begin{aligned}
\sigma_t^2 &= f_i f_i - \sigma_n^2, \; f_i f_i = f_1^2 + f_2^2 + f_3^2 = \frac{1}{3}(\sigma_1^2 + \sigma_2^2 + \sigma_3^2);\\
\sigma_t^2 &= \frac{1}{3}(\sigma_1^2 + \sigma_2^2 + \sigma_3^2) - \frac{1}{9}(\sigma_1 + \sigma_2 + \sigma_3)^2\\
&= \frac{1}{9}((\sigma_1^2 - 2\sigma_1\sigma_2 + \sigma_2^2) + (\sigma_1^2 - 2\sigma_1\sigma_3 + \sigma_3^2) + (\sigma_2^2 - 2\sigma_2\sigma_3 + \sigma_3^2))\\
&= \frac{1}{9}((\sigma_1 - \sigma_2)^2 + (\sigma_1 - \sigma_3)^2 + (\sigma_2 - \sigma_3)^2).
\end{aligned}$$

The found quantity is called the *octahedral shear stress*

$$\sigma_{\text{oct}} = \frac{1}{3}\sqrt{(\sigma_1 - \sigma_2)^2 + (\sigma_1 - \sigma_3)^2 + (\sigma_2 - \sigma_3)^2}. \tag{1.3.60}$$

Problem 1.10. Express σ_{oct} in terms of the principal values of the stress deviator S_i.

Solution: The stress deviator S_{ij} is determined by formula (1.3.30): $\sigma_{ij} = -p\delta_{ij} + S_{ij}$. It follows from here that the principal values S_i satisfy the equation $\sigma_i = -p + S_i$. Thus, we obtain:

$$\begin{aligned}
(\sigma_1 - \sigma_2)^2 + (\sigma_1 - \sigma_3)^2 + (\sigma_2 - \sigma_3)^2 = (S_1 - S_2)^2 + (S_1 - S_3)^2\\
+(S_2 - S_3)^2 = 2\,(S_1^2 + S_2^2 + S_3^2) - 2\,(S_1S_2 + S_1S_3 + S_2S_3).
\end{aligned}$$

It follows from the definition of S_i that $S_1 + S_2 + S_3 = 0$. Squaring this equation, we obtain:

$$S_1^2 + S_2^2 + S_3^2 + 2\,(S_1S_2 + S_1S_3 + S_2S_3) = 0.$$

Substituting this expression into the preceding formula, we find:

$$(\sigma_1 - \sigma_2)^2 + (\sigma_1 - \sigma_3)^2 + (\sigma_2 - \sigma_3)^2 = 3\,(S_1^2 + S_2^2 + S_3^2).$$

Consequently, the σ_{oct} has the form

$$\sigma_{\text{oct}} = \frac{1}{\sqrt{3}}\sqrt{S_1^2 + S_2^2 + S_3^2}. \tag{1.3.61}$$

Problem 1.11. Find the extremal values of the shear stresses τ_i.

Solution: Let us choose the coordinate system coinciding with the principal axes of the stress tensor. Then

$$\sigma_{ij} = \begin{pmatrix} \sigma_1 & 0 & 0 \\ 0 & \sigma_2 & 0 \\ 0 & 0 & \sigma_3 \end{pmatrix}.$$

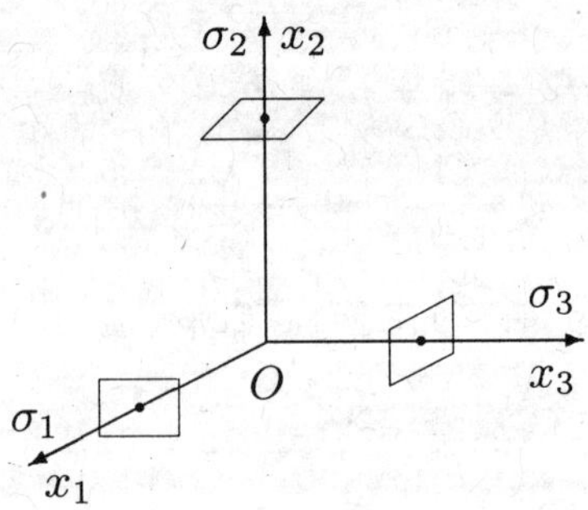

Figure 1.19: The principal areas of the stress tensor.

Choosing some area with the normal $\vec{n}$, we can determine the components f_i, σ_n, σ_t (see Problem 1.9).

$$\begin{aligned}
f_i &= \sigma_{ij}n_j = \begin{pmatrix} \sigma_1 & 0 & 0 \\ 0 & \sigma_2 & 0 \\ 0 & 0 & \sigma_3 \end{pmatrix} \begin{pmatrix} n_1 \\ n_2 \\ n_3 \end{pmatrix} = \begin{pmatrix} \sigma_1 n_1 \\ \sigma_2 n_2 \\ \sigma_3 n_3 \end{pmatrix} \\
\sigma_n &= f_i n_i = \sigma_1 n_1^2 + \sigma_2 n_2^2 + \sigma_3 n_3^2, \\
\sigma_t^2 &= f_i f_i - \sigma_n^2 = \sigma_1^2 n_1^2 + \sigma_2^2 n_2^2 + \sigma_3^2 n_3^2 - (\sigma_1 n_1^2 + \sigma_2 n_2^2 + \sigma_3 n_3^2)^2.
\end{aligned}$$

The Lagrange function $\Phi = \sigma_t^2 - \lambda n_i n_i$ satisfies the following extremum equations:

$$\frac{\partial \Phi}{\partial n_i} = 0, \quad n_i n_i = 1, \quad \sigma_1 > \sigma_2 > \sigma_3.$$

Assuming $i = 1, 2, 3$, we obtain:

$$\begin{aligned}
n_1(\sigma_1^2 - 2\sigma_1(\sigma_1 n_1^2 + \sigma_2 n_2^2 + \sigma_3 n_3^2) - \lambda) &= 0; \\
n_2(\sigma_2^2 - 2\sigma_2(\sigma_1 n_1^2 + \sigma_2 n_2^2 + \sigma_3 n_3^2) - \lambda) &= 0; \\
n_3(\sigma_3^2 - 2\sigma_3(\sigma_1 n_1^2 + \sigma_2 n_2^2 + \sigma_3 n_3^2) - \lambda) &= 0; \\
n_1^2 + n_2^2 + n_3^2 - 1 &= 0.
\end{aligned}$$

This system has the solutions

$$\begin{aligned}
n_1 &= \pm 1;\ n_2 = 0;\ n_3 = 0;\ \lambda = -\sigma_1^2;\ \tau_1 = 0; \\
n_1 &= 0;\ n_2 = \pm 1;\ n_3 = 0;\ \lambda = -\sigma_2^2;\ \tau_2 = 0; \\
n_1 &= 0;\ n_2 = 0;\ n_3 = \pm 1;\ \lambda = -\sigma_3^2;\ \tau_3 = 0,
\end{aligned}$$

where $\tau_i = (\sigma_t)_{\text{extrem}}$. It follows from here that the corresponding areas, where $\sigma_t = 0$, are perpendicular to the principal axes (see Fig. 1.19).

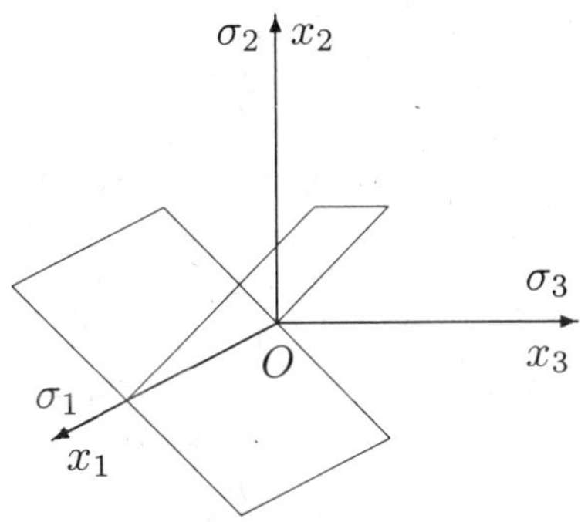

Figure 1.20: The areas of the maximum shear stress τ_1.

Another solution of the system is given by the formulas

$$\begin{aligned} n_1 &= 0,\ n_2 = \pm\frac{1}{\sqrt{2}},\ n_3 = \frac{\pm 1}{\sqrt{2}},\ \lambda = -\sigma_2\sigma_3,\ \tau_1 = \frac{\sigma_2 - \sigma_3}{2}, \\ n_1 &= \pm\frac{1}{\sqrt{2}},\ n_2 = 0,\ n_3 = \frac{\pm 1}{\sqrt{2}},\ \lambda = -\sigma_1\sigma_3,\ \tau_2 = \frac{\sigma_1 - \sigma_3}{2}, \qquad (1.3.62) \\ n_1 &= \pm\frac{1}{\sqrt{2}},\ n_2 = \frac{\pm 1}{\sqrt{2}},\ n_3 = 0,\ \lambda = -\sigma_1\sigma_2,\ \tau_3 = \frac{\sigma_1 - \sigma_2}{2}. \end{aligned}$$

It follows from here that the extremal values of τ_2 lie on the areas, which halve the right angles (see Fig. 1.20), where we show two such areas.

Problem 1.12. Find the rate of change of the contravariant A^{ij} and mixed $A^{i\cdot}_{\cdot j}$ components of the stress tensor $\hat{A}$ in the accompanying and Eulerian Cartesian coordinate systems:

$$\hat{A} = A^{ij}\hat{\vec{e}}_i\hat{\vec{e}}_j = A^{i\cdot}_{\cdot j}\hat{\vec{e}}_i\hat{\vec{e}}^j = a_{ij}\vec{e}_i\vec{e}_j.$$

Solution: Let us multiply equation (1.3.46) by the matrix $\frac{\partial \xi^k}{\partial x^n}\frac{\partial \xi^l}{\partial x^s}$. As a result, we obtain:

$$a_{ns} = \frac{\partial \xi^k}{\partial x^n}\frac{\partial \xi^l}{\partial x^s}\hat{g}_{ki}\hat{g}_{lj}A^{ij}. \qquad (1.3.63)$$

Employing the relations

$$\begin{aligned} \hat{g}_{ki} &= \hat{\vec{e}}_k \cdot \hat{\vec{e}}_i = \frac{\partial \vec{r}}{\partial \xi^k} \cdot \frac{\partial \vec{r}}{\partial \xi^i} = \frac{\partial x^p}{\partial \xi^k}\frac{\partial x^p}{\partial \xi^i}, \\ \hat{g}_{lj} &= \frac{\partial x^m}{\partial \xi^l}\frac{\partial x^m}{\partial \xi^j}, \quad \vec{e}_i \cdot \vec{e}_j = \delta_{ij}, \qquad (1.3.64) \end{aligned}$$

let us rewrite (1.3.63) as

$$a_{ns} = \frac{\partial x^s}{\partial \xi^j} \frac{\partial x^n}{\partial \xi^i} A^{ij}. \tag{1.3.65}$$

Differentiating (1.3.65) with respect to time, we obtain:

$$\frac{\partial a_{ns}}{\partial t} + \frac{\partial a_{ns}}{\partial x^k} \frac{dx^k}{dt} = \frac{\partial x^s}{\partial \xi^j} \frac{\partial x^n}{\partial \xi^i} \dot{A}^{ij} + A^{ij} \frac{\partial v^s}{\partial \xi^j} \frac{\partial x^n}{\partial \xi^i} + A^{ij} \frac{\partial x^s}{\partial \xi^j} \frac{\partial v^n}{\partial \xi^i}.$$

With regard for (1.3.65) and the relationships

$$\frac{dx^k}{dt} = v^k, \quad A^{ij} \frac{\partial x^n}{\partial \xi^i} \frac{\partial v^s}{\partial \xi^j} = A^{ij} \frac{\partial x^n}{\partial \xi^i} \frac{\partial x^k}{\partial \xi^j} \left(\frac{\partial v^s}{\partial \xi^j} \frac{\partial \xi^j}{\partial x^k} \right) = a_{nk} \frac{\partial v^s}{\partial x^k},$$
$$A^{ij} \frac{\partial x^s}{\partial \xi^j} \frac{\partial v^n}{\partial \xi^i} = A^{ij} \frac{\partial x^s}{\partial \xi^j} \frac{\partial x^k}{\partial \xi^i} \left(\frac{\partial v^n}{\partial \xi^i} \frac{\partial \xi^i}{\partial x^k} \right) = a_{ks} \frac{\partial v^n}{\partial x^k},$$

we obtain the equation

$$\frac{\partial a_{ns}}{\partial t} + v^k \frac{\partial a_{ns}}{\partial x^k} = (a_{ns})^{\Delta} + a_{nk} \frac{\partial v^s}{\partial x^k} + a_{ks} \frac{\partial v^n}{\partial x^k}, \tag{1.3.66}$$

where we have introduced the notation

$$(a_{ns})^{\Delta} = \frac{\partial x^s}{\partial \xi^j} \frac{\partial x^n}{\partial \xi^i} \dot{A}^{ij} \tag{1.3.67}$$

for the components of the rate of change of stresses in the Eulerian coordinate system. Expressing the quantity $(a_{ns})^{\Delta}$ in (1.3.66), we obtain:

$$(a_{ns})^{\Delta} = \frac{\partial a_{ns}}{\partial t} + v^k \frac{\partial a_{ns}}{\partial x^k} - a_{nk} \frac{\partial v^s}{\partial x^k} - a_{ks} \frac{\partial v^n}{\partial x^k}, \tag{1.3.68}$$
$$(A^{ij})^{\Delta} = \dot{A}^{ij} = \hat{g}^{ki} \hat{g}^{lj} (A_{kl})^{\Delta}.$$

Using (1.3.46) again, let us write the expression for the mixed components:

$$a_{ij} \frac{\partial x^i}{\partial \xi^l} \frac{\partial x^j}{\partial \xi^k} = \hat{g}_{li} A^{i\cdot}_{\cdot k}.$$

Multiplying this equation by $\frac{\partial \xi^k}{\partial x^n} \frac{\partial \xi^l}{\partial x^s}$, we find with regard for (1.3.64):

$$a_{sn} = \frac{\partial x^s}{\partial \xi^i} \frac{\partial \xi^k}{\partial x^n} A^{i\cdot}_{\cdot k}. \tag{1.3.69}$$

Differentiating this equation with respect to time, we obtain:

$$\frac{\partial a_{sn}}{\partial t} + \frac{\partial a_{sn}}{\partial x^k} \frac{dx^k}{dt} = \frac{\partial x^s}{\partial \xi^i} \frac{\partial \xi^k}{\partial x^n} \dot{A}^{i\cdot}_{\cdot k} + \frac{\partial v^s}{\partial \xi^i} \frac{\partial \xi^k}{\partial x^n} A^{i\cdot}_{\cdot k} + \frac{\partial x^s}{\partial \xi^i} \left(\frac{\partial \xi^k}{\partial x^n} \right)^{\cdot} A^{i\cdot}_{\cdot k}. \tag{1.3.70}$$

From the condition $\frac{\partial x^n}{\partial \xi^k}\frac{\partial \xi^k}{\partial x^s} = \delta^n_s$, we find the derivative

$$\left(\frac{\partial \xi^k}{\partial x^s}\right)^{\cdot} \frac{\partial x^n}{\partial \xi^k} = -\frac{\partial v^n}{\partial \xi^k}\frac{\partial \xi^k}{\partial x^s} = -\frac{\partial v^n}{\partial x^s},$$

where $v^n = v^n(x^k, t)$. Using the relations $\frac{dx^k}{dt} = v^k$,

$$\begin{aligned}
\frac{\partial x^s}{\partial \xi^i}\left(\frac{\partial \xi^k}{\partial x^n}\right)^{\cdot} A^{i\cdot}_{\cdot k} &= \left(\frac{\partial x^m}{\partial \xi^k}\left(\frac{\partial \xi^k}{\partial x^n}\right)^{\cdot}\right)\frac{\partial x^s}{\partial \xi^i}\frac{\partial \xi^k}{\partial x^m} A^{i\cdot}_{\cdot k} = -\frac{\partial v^m}{\partial x^n} a_{sm}, \\
\frac{\partial v^s}{\partial \xi^i}\frac{\partial \xi^k}{\partial x^n} A^{i\cdot}_{\cdot k} &= \frac{\partial x^m}{\partial \xi^i}\frac{\partial \xi^k}{\partial x^n}\frac{\partial v^s}{\partial \xi^i}\frac{\partial \xi^i}{\partial x^m} A^{i\cdot}_{\cdot k} = a_{mn}\frac{\partial v^s}{\partial x^m},
\end{aligned}$$

we find from (1.3.70):

$$\begin{aligned}
(a_{sn})^{-} &= \frac{\partial a_{sn}}{\partial t} + v^k \frac{\partial a_{sn}}{\partial x^k} - a_{mn}\frac{\partial v^s}{\partial x^m} + a_{sm}\frac{\partial v^m}{\partial x^n}, \qquad (1.3.71) \\
(A^{i\cdot}_{\cdot k})^{-} &= \dot{A}^{i\cdot}_{\cdot k} = \hat{g}^{ij}(A_{jk})^{\Delta}.
\end{aligned}$$

References

1. **Sedov, L.I.,** *Continuum Mechanics*, Vol. 1 (in Russian), Fifth Edition, Nauka, Moscow, 1994.
2. **Sokolnikov, I.,** *Tensor Analysis (Theory and Applications in Geometry and in Continuum Mechanics* (in Russian), Nauka, Moscow, 1971.
3. **McConnell, A.J.,** *Application of Tensor Analysis*, Dover Publications Inc., New York, 1957.
4. **Rashevskii, P.K.,** *The Riemann Geometry and Tensor Analysis* (in Russian), Nauka, Moscow, 1967.
5. **Ilyushin, A.A.,** *Continuum Mechanics* (in Russian), Moscow State University, Moscow, 1990.
6. **Germain, P.,** *Cours de Mécanique des Milieux Continus. Tome I. Théorie Générale*, Masson et Cie, Éditeurs, Paris, 1973.
7. **Smirnov, V.I.,** *A Course of Higher Mathematics* (in Russian), Vol. 3, Pt. 1, Nauka, Moscow, 1967.
8. **Godunov, S.K.,** *Elements of Continuum Mechanics* (in Russian), Nauka, Moscow, 1978.

9. **Mase, G.E.,** *Theory and Problems of Continuum Mechanics*, McGraw-Hill Book Company, New York, 1970.
10. **Prager, B.,** *Einführung in die Kontinuumsmechanik*, Birkhäuser Verlag, Basel und Stuttgart, 1961.
11. **Truesdell, C. and Noll, W.,** The non-linear field theories of mechanics, in *Encyclopedia of Physics, III/3*, Springer-Verlag, Berlin, 1965.
12. **Oldroyd, J.G.,** On the formulation of rheological equations of state, in *Proc. Roy. Soc. Lond. A*, 200:523, 1950.
13. **Cotter, B.A. and Rivlin, R.S.,** Tensors associated with time-dependent stress. *Quart. Appl. Math.*, 13(2):177, 1955.

2

Fundamental Principles and Laws of Continuum Mechanics

We derive in this chapter the governing differential equations of continuum mechanics with the aid of the above definitions and the conservation laws for the mass, momentum, energy, and momentum moment written for finite volumes of a continuum. The differential equations of continuum mechanics (equations of continuity, momentum, and energy) represent the partial differential equations written in the Lagrangian and Eulerian coordinates. They are applicable for the description of any continua. The specification of a continuum is achieved by specifying the equation of state. We discuss in the present chapter the general principles of the construction of the equations of state and their form in the simplest case of an ideal and viscous, heat-conducting gas.

The Hamilton–Ostrogradsky variational principle plays an important role in the continuum mechanics. On the one hand, the motion equations and the boundary conditions for them can be obtained with its aid. On the other hand, it enables one to elucidate the relation between the conservation laws underlying continuum mechanics and the symmetry properties of space and time. Both of these aspects are discussed in detail in the present chapter.

2.1 Equations of Continuity, Motion, and Energy for a Continuum

As pointed out in Section 1.2, the continuum description is based on the determination at some point $\vec{r}$ of an infinitesimal volume $dV \approx (dl)^3$, which nevertheless contains a very large number $dN \gg 1$ of molecules or other particles. After the execution of an averaging procedure over this volume, one can introduce the density ρ, mean velocity $\vec{v}$, and temperature T determining the velocity of the thermal (chaotic) molecules motion. (The mean velocity of the thermal motion of velocities is equal

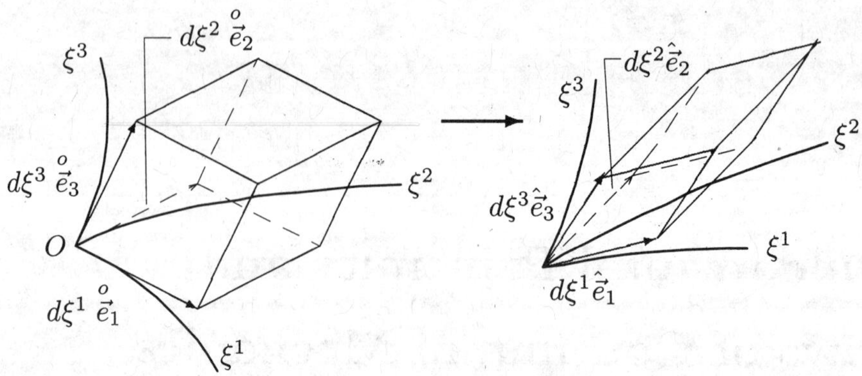

Figure 2.1: To the derivation of the continuity equation in the Lagrangian coordinates.

to zero.) If the characteristic size of the variation of the mean parameters L is much larger than the linear size dl of the material volume, then one can neglect the size dl as compared to L and assume that the parameters $\rho, \vec{v}$, and T are determined at point $\vec{r}$.

Note that we consider here a continuum, which does not contain the internal variables,and is often called the formless medium in the literature. For the media with an internal structure, in addition to $\rho(\vec{r}), \vec{v}(\vec{r})$, and $T(\vec{r})$, appear the new variables $\mu_i(\vec{r})$ such as, for example, the internal moment of momentum (the spin), the porosity, etc. The examples of such media will be considered in Chapter 7.

Let us determine at some initial moment of time $t = 0$ the physical fields $\rho(\vec{r}), \vec{v}(\vec{r})$, and $T(\vec{r})$. If the differential equations are known, which describe the variation of these fields in time, then their solution under given initial and boundary conditions will enable one to find the density $\rho(\vec{r}, t)$, velocity $\vec{v}(\vec{r}, t)$, and temperature $T(\vec{r}, t)$ of a continuum at each moment of time t at any space point. Thus, it is necessary to formulate the differential equations for the density ρ, velocity $\vec{v}$, and temperature T. The corresponding equations follow from the integral conservation laws for the mass, momentum, and energy, which are conditioned by the isotropicity and homogeneity properties of the Euclidean space and time, as will be shown below in Section 2.3.

2.1.1 Continuity Equation

Let us at first derive the *continuity equation in Lagrangian variables* in the accompanying coordinate system. For this purpose, we introduce the notion of an *individual volume* by which we shall mean a volume consisting of the same particles at any time. Let us choose at the time $t = 0$ in a reference frame the volume in the form of a parallelepiped constructed on the vectors $d\xi^1 \overset{o}{\vec{e}}_1$, $d\xi^2 \overset{o}{\vec{e}}_2$, $d\xi^3 \overset{o}{\vec{e}}_3$ (see Fig. 2.1). After the deformation, it will go over to a parallelepiped spanned on the vectors $d\xi^2\hat{\vec{e}}_1$, $d\xi^2\hat{\vec{e}}_2$, $d\xi^3\hat{\vec{e}}_3$, where the coordinates $d\xi^i$ remain constant and the basis vectors change in time, which leads to the alteration of the parallelepiped volume dV. Since the coordinates $d\xi^i$ refer at all times to the same particles (they are frozen in the continuum), the planes $d\xi^i\hat{\vec{e}}_i \times d\xi^j\hat{\vec{e}}_j$ also consist at any time of the same particles. This means that the particles do not penetrate the walls of the parallelepiped and do not leave it during the deformation process. Consequently, the given parallelepiped is an individual volume in which the number of particles is preserved: $dN = n\,dV$ as well as the particles mass: $dm = \rho\,dV$. Since the mass of a substance comprised within the parallelepiped is conserved, we obtain:

$$\rho_0 dV_0 = \rho\,dV, \tag{2.1.1}$$

where the subscript zero corresponds to the initial moment of time. For the volumes of the parallelepiped dV_0 and dV, the formulas

$$dV_0 = \overset{o}{\vec{e}}_1 \cdot (\overset{o}{\vec{e}}_2 \times \overset{o}{\vec{e}}_3)\, d\xi^1 d\xi^2 d\xi^3, \quad dV = \hat{\vec{e}}_1 \cdot (\hat{\vec{e}}_2 \times \hat{\vec{e}}_3)\, d\xi^1\, d\xi^2\, d\xi^3$$

are valid, which upon substitution into (2.1.1) yield, with regard for (1.1.33), the formula

$$\rho = \rho_0 \frac{\overset{o}{\vec{e}}_1 \cdot (\overset{o}{\vec{e}}_2 \times \overset{o}{\vec{e}}_3)}{\hat{\vec{e}}_1 \cdot (\hat{\vec{e}}_2 \times \hat{\vec{e}}_3)} = \rho_0 \frac{\varepsilon^0_{123}}{\hat{\varepsilon}_{123}}. \tag{2.1.2}$$

Since (1.1.36) holds, we have that $\varepsilon^0_{123} = \sqrt{g^0}$, $\hat{\varepsilon}_{123} = \sqrt{\hat{g}}$. Therefore, the continuity equation in Lagrangian coordinates will have the form

$$\rho = \rho_0 \frac{\sqrt{g^0}}{\sqrt{\hat{g}}}, \tag{2.1.3}$$

where $g^0 = \det \parallel g^0_{ij} \parallel$, $\hat{g} = \det \parallel \hat{g}_{ij} \parallel$.

Let us consider the Eulerian coordinates x^i with the metric tensor $g_{ij} = \vec{e}_i \cdot \vec{e}_j$ and find the *continuity equation in the Eulerian coordinates.* Let us identify in a continuum some individual volume $V(t)$ (see Fig. 2.2) within which the substance mass will remain constant; therefore, the following equation is valid:

$$\frac{dm}{dt} = 0, \quad m = \int_{V(t)} \rho\,dV. \tag{2.1.4}$$

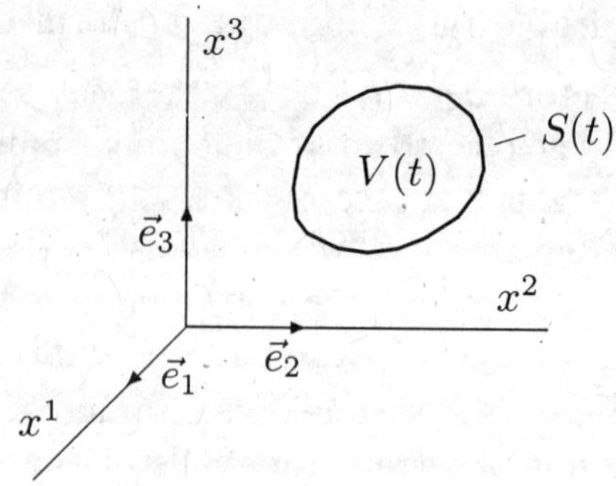

Figure 2.2: To the derivation of the continuity equation.

The individual volume surface $S(t)$ consists of the same particles; therefore, it changes in time, and the points of the volume $V(t)$ boundary are described by the trajectories equations

$$\frac{dx^i}{dt} = v^i, \quad x^i\Big|_{t=0} = \xi^i. \tag{2.1.5}$$

Applying the rule for the differentiation of an integral with respect to a moving volume (1.1.66) and taking (2.1.5) into account, let us rewrite the first equation in (2.1.4) as

$$\int_{V(t)} \left(\frac{\partial \rho}{\partial t} + \nabla_i \rho v^i \right) dV = 0. \tag{2.1.6}$$

Since (2.1.6) is valid for any individual volume $V(t)$, we obtain from (2.1.6) the continuity equation in the Eulerian coordinates:

$$\frac{\partial \rho}{\partial t} + \nabla_i \rho v^i = 0, \tag{2.1.7}$$

where

$$\nabla_i \rho v^i = \frac{\partial \rho v^i}{\partial x^i} + \rho v^j \Gamma^i_{ji}.$$

Using the Weyl formula (1.1.73), we can rewrite the continuity equation in the form

$$\frac{\partial \rho}{\partial t} + \frac{1}{\sqrt{g}} \frac{\partial}{\partial x^i} (\sqrt{g} \rho v^i) = 0. \tag{2.1.8}$$

2.1.2 Equations of Motion and of Momentum Moment

The motion equations follow from the conservation law for momentum, written for some arbitrary volume V of a continuum in the form

$$\int_V \rho \vec{a}\, dV = \int_V \rho \vec{F}\, dV + \int_S \vec{f}_n\, dS, \tag{2.1.9}$$

where $\vec{a}$ is the acceleration, $\vec{F}$ is the body force, and $\vec{f}_n$ is a surface force applied to the surface S bounding the volume V. Equation (2.1.9) points to the fact that the variation of the momentum $\int \rho\vec{a}\, dV$ of a continuum located within the volume V is equal to the sum of the body forces $\int \rho\vec{F}\, dV$ and the surface forces $\int \vec{f}_n dS$.

Consider at first equation (2.1.9) in the accompanying Lagrangian coordinates ξ^i, with the basis vectors $\hat{\vec{e}}_i$ and the metric tensor $\hat{g}_{ij} = \hat{\vec{e}}_i \cdot \hat{\vec{e}}_j$. Using equations (1.1.63) and (1.3.7), let us rewrite the expression for the surface force as follows:

$$\int_S \vec{f}_n\, dS = \int_S \hat{\vec{\sigma}}^{\,j} n_j\, dS = \int_V \nabla_j \hat{\vec{\sigma}}^{\,j}\, dV. \tag{2.1.10}$$

Substituting (2.1.10) into (2.1.9), we rewrite (2.1.9) in the form

$$\int_V (\rho\vec{a} - \rho\vec{F} - \nabla_j \hat{\vec{\sigma}}^{\,j})\, dV = 0, \tag{2.1.11}$$

where

$$\nabla_j \hat{\vec{\sigma}}^{\,j} = \frac{\partial \hat{\vec{\sigma}}^{\,j}}{\partial \xi^j} + \hat{\vec{\sigma}}^{\,j} \hat{\Gamma}^i_{ji}, \quad \hat{\Gamma}^i_{ji} = \Gamma^i_{ji}(\hat{g}_{kl}).$$

By virtue of the arbitrariness of the integration volume V (in the accompanying coordinates, V coincides with the individual volume), we obtain from (2.1.11) with regard for the Weyl formula (1.1.73) the equation:

$$\vec{a} = \frac{1}{\rho\sqrt{\hat{g}}} \frac{\partial}{\partial \xi^j}(\sqrt{\hat{g}}\, \hat{\vec{\sigma}}^{\,j}) + \vec{F}. \tag{2.1.12}$$

Expanding all vectors in (2.1.12) with respect to the basis $\hat{\vec{e}}_k$, we will have the relationships

$$\vec{a} = \hat{a}^k \hat{\vec{e}}_k, \quad \vec{F} = \hat{F}^k \hat{\vec{e}}_k, \quad \hat{\vec{\sigma}}^{\,j} = \hat{\sigma}^{kj} \hat{\vec{e}}_k,$$

upon substitution of which into (2.1.12), we will have the *motion equation*

$$\hat{a}^k = \frac{1}{\rho} \nabla_j \hat{\sigma}^{kj} + F^k, \tag{2.1.13}$$

where

$$\nabla_j \hat{\sigma}^{kj} = \frac{\partial \hat{\sigma}^{kj}}{\partial \xi^j} + \hat{\sigma}^{ki}\, \hat{\Gamma}^j_{ij} + \hat{\sigma}^{ij} \hat{\Gamma}^k_{ij}.$$

To determine $\hat{a}^k$, let us specify a Eulerian coordinate system x^i with the basis vectors $\vec{e}_i$ and the metric tensor $g_{ij} = \vec{e}_i \cdot \vec{e}_j$. Since the acceleration $\vec{a}$ in the Eulerian and accompanying coordinate systems is the same, we have:

$$\frac{\partial^2 x^j}{\partial t^2} \vec{e}_j = \hat{a}^i \hat{\vec{e}}_i, \tag{2.1.14}$$

where the second derivative $\frac{\partial^2 x^i}{\partial t^2}$ is taken for a fixed particle $\xi^j = \text{const}$ and the basis vectors are related by equations (1.3.25) $\hat{\vec{e}}_l = \frac{\partial x^j}{\partial \xi^l}\vec{e}_j$. Multiplying equation (2.1.14) by $\hat{\vec{e}}^k$, we obtain with regard for relations

$$\hat{\vec{e}}_i \cdot \hat{\vec{e}}^k = \delta_i^k, \quad \hat{\vec{e}}^k = \hat{g}^{kl}\hat{\vec{e}}_l, \quad g_{ij} = \vec{e}_i \cdot \vec{e}_j$$

the following expression for the acceleration:

$$\hat{a}^k = g_{ij}\hat{g}^{kl}\frac{\partial x^j}{\partial \xi^l}\frac{\partial^2 x^i}{\partial t^2}. \tag{2.1.15}$$

The body force components $\hat{F}^k$ in the accompanying coordinates can be expressed in a similar way in terms of the components F^i in the Eulerian coordinates in the form

$$\hat{F}^k = g_{ij}\hat{g}^{kl}\frac{\partial x^j}{\partial \xi^l}F^i, \tag{2.1.16}$$

where $F^i = \vec{F} \cdot \vec{e}^{\,i}$. Substituting (2.1.15) and (2.1.16) into (2.1.13), we can write the *motion equation* in the form

$$g_{ij}\hat{g}^{kl}\frac{\partial x^j}{\partial \xi^l}\left(\frac{\partial^2 x^i}{\partial t^2} - F^i\right) = \frac{1}{\rho}\nabla_j\hat{\sigma}^{kj}. \tag{2.1.17}$$

Equation (2.1.17) simplifies in the case of the Eulerian Cartesian coordinates $g_{ij} = \delta_{ij}$. Multiplying the left- and right-hand sides of (2.1.17) by $\hat{g}_{sk}$ under the condition $g_{ij} = \delta_{ij}$, we obtain:

$$\left(\frac{\partial^2 x^i}{\partial t^2} - F^i\right)\frac{\partial x^i}{\partial \xi^s} = \frac{1}{\rho}\nabla_j(\hat{g}_{sk}\hat{\sigma}^{kj}). \tag{2.1.18}$$

Equations (2.1.17) and (2.1.18) enable one to determine the motion law $x^i = x^i(\xi^j, t)$ at a given stress tensor $\hat{\sigma}^{kj}(\xi^i)$.

Let us find the motion equations in the Lagrangian coordinates in the reference frame basis $\overset{o}{\vec{e}}_i$. Let the reference frame basis $\overset{o}{\vec{e}}_i$ coincide with the Eulerian basis $\vec{e}_i$. Then we have the relationships

$$\overset{o}{g}_{ij} = g_{ij}, \quad \rho\sqrt{\hat{g}} = \rho_0\sqrt{g}, \quad \vec{a} = \frac{\partial^2 x^k}{\partial t^2}\overset{o}{\vec{e}}_k, \quad \vec{F} = F^k\,\overset{o}{\vec{e}}_k\,. \tag{2.1.19}$$

Employing equation (1.3.24) $\sqrt{\hat{g}}\hat{\vec{\sigma}}^{\,j} = \overset{o}{\vec{\sigma}}{}^{\,j} = \overset{o}{\sigma}{}^{kj}\,\overset{o}{\vec{e}}_k$ as well as the relations (2.1.19), let us rewrite the motion equation (2.1.12) in the basis $\overset{o}{\vec{e}}_i$ of the reference frame

$$\sqrt{g}\rho_0\frac{\partial^2 x^k}{\partial t^2} = \frac{\partial\,\overset{o}{\sigma}{}^{kj}}{\partial \xi^j} + F^k. \tag{2.1.20}$$

In particular, in the Cartesian basis of the reference frame $\overset{o}{g}_{ij} = \delta_{ij}$, the motion equation will have the form

$$\rho_0 \frac{\partial^2 x^k}{\partial t^2} = \frac{\partial \overset{o}{\sigma}{}^{kj}}{\partial \xi^j} + F^k. \tag{2.1.21}$$

If we express the Piola–Kirchhoff tensor $\overset{o}{\sigma}{}^{kj}$ with the aid of (1.3.26) in terms of the Cauchy tensor $\hat{\sigma}^{ij}$, then equation (2.1.21) may be rewritten as follows:

$$\rho_0 \frac{\partial^2 x^k}{\partial t^2} = \frac{\partial}{\partial \xi^j}\Big(\sqrt{\hat{g}}\frac{\partial x^k}{\partial \xi^i}\hat{\sigma}^{ij}\Big) + F^k. \tag{2.1.22}$$

Let us derive the motion equations in the *Eulerian coordinates* in the basis $\vec{e}_i$. For this purpose, we note that equation (2.1.12) is written in the vector form and it will be valid in the Eulerian coordinate system if we make the substitutions $\hat{g} \to g$, $\hat{\vec{\sigma}}{}^j \to \vec{\sigma}{}^j$, $\xi^j \to x^j$. (These substitutions can be formally substantiated as follows. At the time $t = 0$, one can choose such a basis in the reference frame that it coincides with the Eulerian basis in the actual configuration. Then $\hat{g}_{ij} = g_{ij}$, $\xi^i = x^i$, $\hat{\vec{\sigma}}{}^j = \vec{\sigma}{}^j$.) Using the expansion $\vec{\sigma}{}^j = \sigma^{kj}\vec{e}_k$ and the definition of the substantive derivative in the Eulerian coordinates (1.2.11), we obtain the *motion equations in the Eulerian coordinates*

$$\rho\Big(\frac{\partial v^k}{\partial t} + v^j \nabla_j v^k\Big) = \frac{1}{\sqrt{g}}\frac{\partial}{\partial x^j}(\sqrt{g}\,\sigma^{kj}) + \rho F^k. \tag{2.1.23}$$

Using the Weyl formula (1.1.73), we can write the motion equations in the Eulerian coordinates in a more conventional form

$$\rho\Big(\frac{\partial v^k}{\partial t} + v^j \nabla_j v^k\Big) = \nabla_j \sigma^{kj} + \rho F^k, \tag{2.1.24}$$

where

$$\nabla_j \sigma^{kj} = \frac{\partial \sigma^{kj}}{\partial x^j} + \sigma^{kl}\Gamma^j_{lj} + \sigma^{lj}\Gamma^k_{lj}.$$

Besides the conservation law for the momentum (2.1.9), the independent conservation law for the momentum moment takes place. In the case of a formless medium, when there is no internal moment of the momentum and the surface and body forces, the conservation law for the momentum moment, written for some arbitrary volume V of a continuum, has the form

$$\frac{d}{dt}\int_V (\vec{r}\times\vec{v})\,\rho\, dV = \int_V (\vec{r}\times\vec{F})\,\rho\, dV + \int_S (\vec{r}\times\vec{f}_n)\, dS. \tag{2.1.25}$$

Equation (2.1.25) implies that the rate of change of the momentum moment of the volume V is equal to the sum of the moments of the body force and the surface force acting on this volume V.

Consider for definiteness a Lagrangian, accompanying coordinate system in which V is the individual volume. To calculate the derivative with respect to time, let us derive the rule for the differentiation of an integral over a moving volume in the Lagrangian coordinates $\frac{d}{dt}\int_V \varphi\rho\,dV$. Since the substance mass remains constant within the individual volume V:

$$M = \int_{V(t)} \rho\,dV = \int_M dm = \text{const},$$

the following relationship is valid:

$$\frac{d}{dt}\int_{V(t)} \varphi\rho\,dV = \frac{d}{dt}\int_M \varphi\,dm = \int_M \frac{d\varphi}{dt}dm = \int_{V(t)} \frac{d\varphi}{dt}\rho\,dV. \quad (2.1.26)$$

Note that, if (2.1.26) is applied to the left-hand side of equation (2.1.9), then we obtain the relation:

$$\int_V \rho\vec{a}\,dV = \int_V \frac{d\vec{v}}{dt}\rho\,dV = \frac{d}{dt}\int_V \rho\vec{v}\,dV,$$

from which it follows that (2.1.9) represents the second Newton's law, written for the individual volume. Applying (2.1.26) to (2.1.25), we obtain with regard for the relations $\frac{d\vec{r}}{dt} = \vec{v}$, $\vec{f}_n = \hat{\sigma}^{ij}\hat{\vec{e}}_i n_j$, the equation:

$$\int_V \left(\vec{r}\times\frac{d\vec{v}}{dt}\right)\rho\,dV = \int_V (\vec{r}\times\rho\vec{F})\,dV + \int_S (\vec{r}\times\hat{\vec{e}}_i\hat{\sigma}^{ij})n_j\,dS. \quad (2.1.27)$$

Transforming the last integral to the integral over the volume, we find:

$$\begin{aligned}\int_S (\vec{r}\times\hat{\vec{e}}_i\hat{\sigma}^{ij})n_j\,dS &= \int_V \nabla_j(\vec{r}\times\hat{\vec{e}}_i\hat{\sigma}^{ij})\,dV = \int_V (\hat{\vec{e}}_j\times\hat{\vec{e}}_i)\hat{\sigma}^{ij}\,dV \\ &+ \int_V (\vec{r}\times\nabla_j\sigma^{ij}\hat{\vec{e}}_i)\,dV,\end{aligned}$$

where we have used the definition $\hat{\vec{e}}_i = \nabla_j\vec{r}$. Using this expression, let us rewrite (2.1.27) as

$$\int_V (\vec{r}\times(a^i - \rho\hat{F}^i - \nabla_j\hat{\sigma}^{ij})\hat{\vec{e}}_i)\,\rho\,dV = \int_V (\hat{\vec{e}}_j\times\hat{\vec{e}}_i)\hat{\sigma}^{ij}\,dV. \quad (2.1.28)$$

Substituting (2.1.13) into (2.1.28), we obtain:

$$(\hat{\vec{e}}_j\times\hat{\vec{e}}_i)\hat{\sigma}^{ij} = \hat{\varepsilon}_{jil}\hat{\sigma}^{ij}\hat{\vec{e}}^{\,l} = 0.$$

Since the basis vectors $\hat{\vec{e}}^{\,l}$ are linearly independent, we have for each component that $\hat{\varepsilon}_{jil}\hat{\sigma}^{ij} = 0$, from where the condition for the symmetry of the Cauchy stress tensor (1.3.20) follows:

$$\hat{\sigma}^{ij} = \hat{\sigma}^{ji}. \quad (2.1.29)$$

This formula coincides with formula (1.3.20), obtained for an infinitesimal volume. We have obtained a more general result here, which is valid also for a finite volume and consists of the following: *the condition for the symmetry of the Cauchy stress tensor follows from the conservation law for the momentum moment for a formless medium.*

Using the rule for the differentiation of an integral over a moving volume in the Eulerian coordinates (1.2.11), one can obtain the formula (2.1.27), in which one must make the substitutions

$$\frac{d\vec{v}}{dt} \to \left(\frac{\partial v^k}{\partial t} + v^j \nabla_j v^k\right)\vec{e}_k, \quad \hat{\vec{e}}_i \to \vec{e}_i, \quad \hat{\sigma}^{ij} \to \sigma^{ij},$$

where $\vec{e}_i$ are the basis vectors of the Eulerian coordinate system x^i. Similar transformations lead thereafter to the condition for the symmetry of the Cauchy tensor in the Eulerian coordinates

$$\sigma^{ij} = \sigma^{ji}. \tag{2.1.30}$$

In recent years, the models of micropolar media are widely used in the investigations of the media with microstructure[1–4], in which the material point (as a matter of fact, an infinitesimal volume is considered, which contains a large number of particles) possesses, along with the mean velocity $\vec{v}$, density ρ, and temperature, the internal momentum moment $\vec{l}$, surface moments $\vec{m}_n$, and body moments $\vec{M}$ of the pairs of forces. In this case, the conservation law for the moment has for the volume V the form

$$\frac{d}{dt}\left(\int_V \vec{l}\rho\, dV + \int_V (\vec{r}\times\vec{v})\rho\, dV\right) = \int_V (\vec{r}\times\vec{F})\,\rho\, dV$$
$$+ \int_S (\vec{r}\times\vec{f}_n)\, dS + \int_V \vec{M}\rho\, dV + \int_S \vec{m}_n dS. \tag{2.1.31}$$

The moments of the surface pairs of forces $\vec{m}_k$ act on the surface S, bounding the volume V; therefore, we can write in the accompanying basis $\hat{\vec{e}}_i$, similarly to (1.3.7), that

$$\vec{m}_n = \hat{\vec{m}}^j n_j = m^{ij}\hat{\vec{e}}_i n_j. \tag{2.1.32}$$

Then one can rewrite (2.1.31) with regard for (2.1.26), (2.1.28), and the expression

$$\int_S \vec{m}_n dS = \int_S \hat{\vec{m}}^j n_j\, dS = \int_V \nabla_j \hat{\vec{m}}^j\, dV$$

in the form

$$\rho\frac{d\vec{l}}{dt} = \rho\vec{M} + \nabla_j\hat{\vec{m}}^j + (\hat{\vec{e}}_j \times \hat{\vec{e}}_i)\hat{\sigma}^{ij}. \tag{2.1.33}$$

It is easy to see that, in this case, the Cauchy tensor $\hat{\sigma}^{ij}$ is not symmetric. Performing similar calculations in the Eulerian coordinate system, we obtain that the Cauchy tensor σ^{ij} is also not symmetric.

2.1.3 The Energy Conservation Law: The First and Second Laws of Thermodynamics

As was noted above, the state of an infinitesimal volume dV is characterized by the parameters $\rho(\vec{r},t)$, $\vec{v}(\vec{r},t)$, and $T(\vec{r},t)$. The temperature $T(\vec{r},t)$ determines the velocity of the thermal (chaotic) motion of molecules (particles) located in the volume dV. The temperature T is defined strictly for the thermodynamically equilibrium systems in which all thermodynamic processes proceed at an infinitely slow rate. In this case, at each moment of time t, a thermodynamically equilibrium state takes place. The thermodynamically equilibrium state is such a state in which all thermodynamic characteristics remain constant as long as desired under the invariable external conditions. An actually moving medium is a nonequilibrium system; therefore, the introduction of the temperature T as a parameter and the use of the thermodynamic description for finding the dependence $T = T(\vec{r},t)$ is some approximation of the physical processes. If the characteristic linear size of an individual volume dl satisfies the inequalities $\bar{\lambda} \ll dl \ll L$, then such approximation will be sufficiently good and one can then assume that a thermodynamically equilibrium state is realized inside the volume dV, where $\bar{\lambda}$ is the molecular-free path or the distance between the molecules and L is the characteristic size of the variation of the mean parameters of the medium. At the motion of an individual volume dV, the equilibrium state within it will change during the characteristic time $\Delta t \sim L/v$, where v is the velocity of the medium motion. The assumption on the realization of a state that is close to the equilibrium state will be justified if Δt is much larger than the characteristic time for achieving the equilibrium state (relaxation) $\tau_s \sim \bar{\lambda}/c$, where c is the velocity of the thermal motion of molecules. It is easy to see that, if the velocity of the medium motion v is less or of the order of the velocity of the thermal motion of molecules c, then, from the condition $\bar{\lambda} \ll L$, the inequality $\tau_s \ll \Delta t$ follows. If this inequality is violated, then it is necessary to use for the description the kinetic equation. This approach is not discussed, and the interested reader is referred to the monographs[5,6]. It will be assumed in the following that the inequalities $\bar{\lambda} \ll L$ or $\tau_s \ll \Delta t$ are satisfied.

Along with the temperature T, the state of a thermodynamic system is characterized by the entropy S and by some set of the parameters μ_i. The independent parameters μ_s form the basis of a thermodynamic space of a system, and all remaining thermodynamic parameters can be expressed in terms of these parameters: $\mu_l = f(\mu_s)$. It is clear that some other set of independent parameters μ'_s can be used as the basis

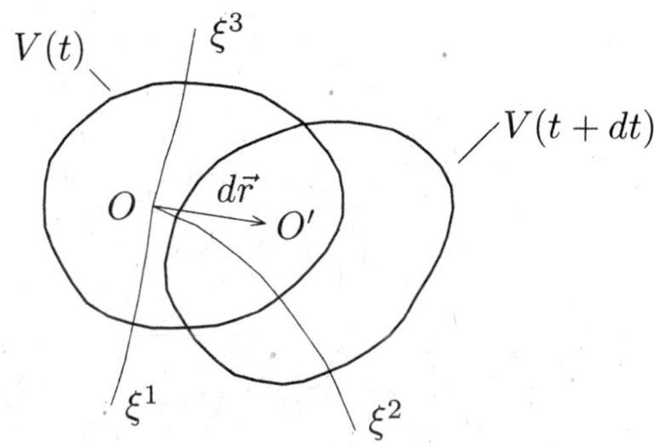

Figure 2.3: To the derivation of the energy equation.

parameters; only their overall number will be the same. The choice of the independent parameters is determined first of all by the problem formulation.

The most important characteristic of a thermodynamic system is the specific *internal energy* E, which is a function of the independent parameters S, μ_i, and the total energy of an individual volume dV with the mass $dm = \rho\, dV$ is found by formula

$$\mathcal{E}\, dm = \rho\Big(\frac{v^2}{2} + E(S, \mu_i)\Big)\, dV, \tag{2.1.34}$$

where $v^2/2$ is the specific kinetic energy.

Let us identify an individual volume $V(t)$ in a continuum and find the total energy variation $\mathcal{E}_M = \int_{V(t)} \mathcal{E}\, dm$ of a particle with the mass $M = \int_{V(t)} \rho\, dV$ during the time dt (see Fig. 2.3). The *energy conservation law* asserts that the total energy variation $d\mathcal{E}_M$ occurs at the expense of the work of the external body forces $dA_M^{(V)}$, external surface forces $dA_M^{(S)}$, fluxes of the body heat $dQ_M^{(V)}$, and surface heat $dQ_M^{(S)}$

$$d\mathcal{E}_M = dA_M^{(V)} + dA_M^{(S)} + dQ_M^{(V)} + dQ_M^{(S)}. \tag{2.1.35}$$

The variation of the total energy $d\mathcal{E}_M$ is composed of the kinetic energy variation $dK_M = d\int_V \rho\frac{v^2}{2} dV$ and the internal energy variation $dE_M = d\int_V E\rho\, dV$:

$$d\mathcal{E}_M = dK_M + dE_M. \tag{2.1.36}$$

Following[7], let us find the kinetic energy variation dK_M during the time dt. Consider the accompanying coordinate system $\xi^i, \hat{\vec{e}}_i$ in which the motion equation has the form (2.1.13)

$$\rho\frac{d\vec{v}}{dt} = \hat{\nabla}_j \hat{\sigma}^{ij} \hat{\vec{e}}_i + \rho\vec{F}. \tag{2.1.37}$$

Let us multiply the left- and right-hand sides of equation (2.1.37) by the displacement vector $d\vec{r} = \vec{v}\,dt = \hat{v}_k \hat{\vec{e}}^k\,dt$ and integrate over the volume $V(t)$ (see Fig. 2.3). As a result, we obtain the equation:

$$\int_{V(t)} \rho\vec{v}\frac{d\vec{v}}{dt} dV\,dt = \int_{V(t)} \hat{\nabla}_j \hat{\sigma}^{ij}\hat{v}_i dV\,dt + \int_{V(t)} \rho\vec{F}\cdot d\vec{r}\,dV. \tag{2.1.38}$$

Using (2.1.26), let us present the left-hand side of (2.1.38) in the form

$$\int_{V(t)} \rho\vec{v}\frac{d\vec{v}}{dt} dV\,dt = \left(\int_M \frac{d}{dt}\frac{v^2}{2}dm\right)dt = \frac{d}{dt}\left(\int_{V(t)} \rho\frac{v^2}{2}dV\right)dt = dK_M, \tag{2.1.39}$$

where $v^2 = \hat{g}_{ij}\hat{v}^i\hat{v}^j = \hat{v}_i\hat{v}^i$ is a scalar quantity. Let us transform the right-hand side of (2.1.38):

$$\int_{V(t)} \hat{\nabla}_j\hat{\sigma}^{ij}\hat{v}_i\,dV\,dt = \int_{V(t)} \hat{\nabla}_j(\hat{\sigma}^{ij}\hat{v}_i)\,dV\,dt - \int_{V(t)} \hat{\sigma}^{ij}\hat{\nabla}_j\hat{v}_i\,dV\,dt. \tag{2.1.40}$$

Employing the expansion

$$\hat{\nabla}_j\hat{v}_i = \frac{1}{2}(\hat{\nabla}_j\hat{v}_i + \hat{\nabla}_i\hat{v}_j) + \frac{1}{2}(\hat{\nabla}_j\hat{v}_i - \hat{\nabla}_i\hat{v}_j) = \dot{\hat{\varepsilon}}_{ij} + \hat{\omega}_{ij}$$

as well as the symmetry properties $\hat{\sigma}^{ij} = \hat{\sigma}^{ji}$, $\dot{\hat{\varepsilon}}_{ij} = \dot{\hat{\varepsilon}}_{ji}$, $\hat{\omega}_{ij} = -\hat{\omega}_{ji}$, we obtain:

$$\hat{\sigma}^{ij}\hat{\nabla}_j\hat{v}_i = \hat{\sigma}^{ij}\dot{\hat{\varepsilon}}_{ij}. \tag{2.1.41}$$

Let us transform the first integral in the right-hand side of (2.1.40) to the integral over the surface

$$\int_{V(t)} \hat{\nabla}_j(\hat{\sigma}^{ij}\hat{v}_i)\,dV\,dt = \int_{S(t)} \hat{\sigma}^{ij}n_j\hat{v}_i\,dS\,dt = \int_{S(t)} \hat{\vec{\sigma}}_n\cdot d\vec{r}\,dS = dA_M^{(S)}. \tag{2.1.42}$$

Substituting (2.1.41) and (2.1.42) into (2.1.40), we find:

$$\int_V \hat{\nabla}_j\hat{\sigma}^{ij}\hat{v}_i\,dV\,dt = dA_M^{(S)} - \int_V \hat{\sigma}^{ij}\dot{\hat{\varepsilon}}_{ij}dV\,dt. \tag{2.1.43}$$

Denoting $dA_M^{(V)} = \int_{V(t)} \rho\vec{F}\cdot d\vec{r}\,dV$, let us rewrite (2.1.38), with regard for (2.1.39) and (2.1.43), in the form

$$dK_M = dA_M^{(V)} + dA_M^{(S)} - \int_{V(t)} \hat{\sigma}^{ij}\dot{\hat{\varepsilon}}_{ij}\,dV\,dt. \tag{2.1.44}$$

Substituting (2.1.36) into (2.1.35), we obtain with regard for (2.1.44):

$$dE_M = \int_{V(t)} \hat{\sigma}^{ij}\dot{\hat{\varepsilon}}_{ij}\,dV\,dt + dQ_M^{(V)} + dQ_M^{(S)}. \tag{2.1.45}$$

Introducing the specific velocity (per unit mass) of the body heat supply $\dot{q}_V$ and the surface velocity of the heat supply $\dot{q}_n$, we will have for the volume $V(t)$ that

$$\begin{aligned} dQ_M^{(V)} &= -\int_V \dot{q}_V \rho \, dV \, dt, \qquad (2.1.46) \\ dQ_M^{(S)} &= -\int_S \dot{q}_n \, dS \, dt = -\int_S \dot{\hat{q}}^i n_i \, dS \, dt = -\int_V \hat{\nabla}_i \dot{\hat{q}}^i \, dV \, dt, \end{aligned}$$

where the minus sign is related to the fact that the positive heat fluxes $\dot{q}_V > 0$, $\dot{q}_n > 0$ lead to the energy reduction in the volume $V(t)$. Going over to the integration variable $dm = \rho \, dV$, we can write:

$$\begin{aligned} dE_M &= d\int_{V(t)} E\rho \, dV = d\int_M E \, dm = \int_M dE \, dm, \\ \int_{V(t)} \hat{\sigma}^{ij} \dot{\hat{\varepsilon}}_{ij} \, dV \, dt &= \int_M \frac{1}{\rho} \hat{\sigma}^{ij} \dot{\hat{\varepsilon}}_{ij} \, dm \, dt, \\ dQ_M^{(V)} &= -\int_{V(t)} \dot{q}_V \rho \, dV \, dt = -\int_M \dot{q}_V \, dm \, dt, \\ dQ_M^{(S)} &= -\int_{V(t)} \hat{\nabla}_i \dot{\hat{q}}^i dV \, dt = -\int_M \frac{1}{\rho} \hat{\nabla}_i \dot{\hat{q}}^i dm \, dt. \end{aligned}$$

Substituting these formulas into (2.1.45), we obtain:

$$\int_M \left(\frac{dE}{dt} - \frac{1}{\rho} \hat{\sigma}^{ij} \dot{\hat{\varepsilon}}_{ij} + \dot{q}_V + \frac{1}{\rho} \hat{\nabla}_i \dot{\hat{q}}^i \right) dt \, dm = 0.$$

Since this equation is valid for an arbitrary integration region M, the expression within the brackets should be equal to zero. As a result, the equation for internal energy will have the form

$$\frac{dE}{dt} = \frac{1}{\rho} \hat{\sigma}^{ij} \dot{\hat{\varepsilon}}_{ij} - \dot{q}_V - \frac{1}{\rho} \hat{\nabla}_i \dot{\hat{q}}^i. \qquad (2.1.47)$$

If we introduce the notation for the total heat flux to the unit mass per the unit of time

$$\frac{dq_e}{dt} = -\dot{q}_V - \frac{1}{\rho} \hat{\nabla}_i \dot{\hat{q}}^i, \qquad (2.1.48)$$

then equation (2.1.47) may be rewritten as

$$dE = \frac{1}{\rho} \hat{\sigma}^{ij} d\hat{\varepsilon}_{ij} + dq_e, \qquad (2.1.49)$$

where we have used the first relationship in (1.3.54) of the form $d\hat{\varepsilon}_{ij} = \dot{\hat{\varepsilon}}_{ij} dt$. Equation (2.1.49) [or (2.1.47)] is called the *energy equation*, or the *first law of thermodynamics.* Since the scalar quantities stand on the left- and right-hand sides of (2.1.47) and (2.1.48), these equations are valid in an arbitrary coordinate system. In particular, in the Eulerian coordinate system $g_{ij} = \vec{e}_i \cdot \vec{e}_j$, we will have:

$$\frac{dE}{dt} = \frac{1}{\rho}\sigma^{ij}\dot{\varepsilon}_{ij} + \frac{dq_e}{dt}, \tag{2.1.50}$$

or going over to the increments

$$dE = \frac{1}{\rho}\sigma^{ij}\dot{\varepsilon}_{ij}dt + dq_e, \tag{2.1.51}$$

where

$$\dot{\varepsilon}_{ij} = \frac{1}{2}(\nabla_i v_j + \nabla_j v_i), \ \vec{v} = v^i\vec{e}_i = v_i\vec{e}^{\,i}.$$

Note that in the Eulerian coordinates the $\dot{\varepsilon}_{ij}dt$ are not differentials of the components of the strain tensor ε_{ij}. It follows from (2.1.49) and (2.1.51) that the variation of the specific internal energy occurs at the expense of the work of the internal stresses $\frac{1}{\rho}\hat{\sigma}^{ij}\,d\hat{\varepsilon}_{ij}$ and the heat supply (removal) dq_e. The increment of the heat quantity dq_e is proportional to the temperature increment

$$dq_e = C\,dT, \tag{2.1.52}$$

where C is the specific heat. It follows from the energy equation (2.1.49) and equation (2.1.52) that one can choose the T and $\hat{\varepsilon}_{ij}$ as independent thermodynamic variables. Then $E = E(T, \hat{\varepsilon}_{ij})$. The specific heat C is a scalar quantity, and therefore it can only depend on the temperature T and the invariants of the strain tensor J_i. The experience shows that the specific heat depends on temperature and the first invariant $J_1 = \hat{\varepsilon}_{kk}$; therefore,

$$C = C(T, \hat{\varepsilon}_{kk}). \tag{2.1.53}$$

The stress $\hat{\sigma}^{ij}$ depends in the simplest case on the strain tensor $\hat{\varepsilon}^{\,ij}$ and the temperature T

$$\hat{\sigma}^{ij} = \hat{\sigma}^{ij}(T, \hat{\varepsilon}_{ij}). \tag{2.1.54}$$

Substituting (2.1.53) and (2.1.54) into equation (2.1.49), we can find the internal energy increment under the variation of the thermodynamic parameters from $T^{(1)}$, $\hat{\varepsilon}_{ij}^{(1)}$ to $T^{(2)}$, $\hat{\varepsilon}_{ij}^{(2)}$:

$$E(T^{(2)}, \hat{\varepsilon}_{ij}^{(2)}) - E(T^{(1)}, \hat{\varepsilon}_{ij}^{(1)}) = \int_{\hat{\varepsilon}_{ij}^{(1)}}^{\hat{\varepsilon}_{ij}^{(2)}} \frac{1}{\rho}\hat{\sigma}^{ij}(T, \hat{\varepsilon}_{ij})\,d\hat{\varepsilon}_{ij} + \int_{T^{(1)}}^{T^{(2)}} C(T, \hat{\varepsilon}_{kk})\,dT. \tag{2.1.55}$$

It follows from (2.1.49) that the quantity dE is a total differential; therefore, according to (2.1.55), the energy E is uniquely determined at each point of the thermodynamic space of the system T, $\hat{\varepsilon}_{ij}$. The energy variation ΔE is the difference of the energies taken at the corresponding points. At the same time, the work and the quantity of heat obtained by a system under the variation of parameters are the functions of the process and depend on the integration path. The work performed over the system is

$$\delta A = \int_{\hat{\varepsilon}_{ij}^{(1)}}^{\hat{\varepsilon}_{ij}^{(2)}} \frac{1}{\rho} \hat{\sigma}^{ij}(T, \hat{\varepsilon}_{ij})\, d\hat{\varepsilon}_{ij} \tag{2.1.56}$$

and it depends not only on the initial strain value $\hat{\varepsilon}_{ij}^{(1)}$ and the final strain value $\hat{\varepsilon}_{ij}^{(2)}$, but also on those values of the temperature T, which were taken during the overall thermodynamic process. Exactly the same heat quantity

$$\delta q_e = \int_{T^{(1)}}^{T^{(2)}} C(T, \hat{\varepsilon}_{kk})\, dT \tag{2.1.57}$$

depends on $T^{(1)}, T^{(2)}$ and on the values of $\hat{\varepsilon}_{kk}$, i.e., on the integration path, the equation of which is $\varphi(\hat{\varepsilon}_{ij}, T) = 0$, where φ is some given function.

Equation (2.1.49) is valid for the formless media. In the case of the availability of the internal structure characterized by the independent parameters μ_s, it is necessary to add to the right-hand side of (2.1.49) the work dA' on the increments of the parameters μ_s

$$dA' = G^s d\mu_s, \tag{2.1.58}$$

where $G^s = G^s(T, \hat{\varepsilon}_{ij}, \mu_s)$ is the corresponding thermodynamic force. The specification of energy as a function of the system state $E(T, \hat{\varepsilon}_{ij}, \mu_s)$ enables one to determine all of the remaining functions

$$\hat{\sigma}^{ij} = \rho \frac{\partial E}{\partial \hat{\varepsilon}_{ij}}, \quad C = \frac{\partial E}{\partial T}, \quad G^s = \frac{\partial E}{\partial \mu_s}. \tag{2.1.59}$$

The system (2.1.59) is usually called the *equations of state* of a system.

There exists, along with the internal energy, another thermodynamic function of state, which is called the *entropy S*. Before we give its strict definition, we introduce the notion of the reversible and irreversible processes. By the *reversible process* in the space of states, we mean such a process, which can proceed both in the forward and in the reverse direction, and the work δA and the heat flux δq [see (2.1.56), (2.1.57)] in the forward and reverse process differ only by their signs. All remaining processes are called the *irreversible* processes. The reversible processes proceed infinitely slowly, and the system trajectory passes consequently through a sequence of equilibrium states. We define among all possible processes the *adiabatic processes* at which $\delta q_e = 0$, i.e., the processes without the heat supply, and the *isothermal processes* at $T = \text{const}$, i.e., at a constant temperature. Both of these processes are the idealization of the actual processes. An isothermal process is only possible at an infinite specific heat $C \to \infty$, and the adiabatic process is only possible in the presence of ideal non-heat-conducting walls bounding the identified continuum volume.

Let us at first introduce the entropy at the reversible processes, for which the total differential of the specific entropy S exists, which is equal to

$$dS = \frac{1}{T}\, dq_e, \tag{2.1.60}$$

where the temperature T is an integrating factor, and dq_e is an infinitesimal increment of the heat quantity obtained by a unit mass. It follows from (2.1.60) that the entropy S is a single-valued function in the space of states $S = S(T, \hat{\varepsilon}_{ij})$; therefore, the integral over a closed contour in this space is equal to zero:

$$\oint dS = \oint \frac{1}{T}\, dq_e = 0. \tag{2.1.61}$$

In the case of irreversible processes, the formula

$$T\, dS = dq_e + dq', \quad dq' \geq 0, \tag{2.1.62}$$

takes place instead of (2.1.60), where dq' is called the noncompensated heat, and it determines the energy quantity dissipated into the heat (for the reversible processes $dq' = 0$). The relations (2.1.62) represent the contents of the *second law of thermodynamics*. The knowledge of the specific entropy enables one to find the entropy of the specific volume $V(t)$, having the mass $M = \int \rho\, dV$, by the formula

$$S_M = \int_{V(t)} S \rho\, dV = \int_M S\, dm. \tag{2.1.63}$$

For the determination of dq', the Gibbs formula is postulated, which has the following form for two-parametric gases:

$$dE = T\, dS - P \frac{d}{dt}\left(\frac{1}{\rho}\right). \tag{2.1.64}$$

The system of equations of continuity, motion, and energy is universal and valid for any continua. It is based on the conservation laws following from the Euclidean property of space, and on the thermodynamics laws.

This system of equations, however, is no closed system since the equations of state (2.1.59) are unknown. The equations of state (2.1.59) have a specific form for each substance and continuum. They are determined either with the aid of the basic experiments or with the aid of statistical physics methods. Their determination represents one of the most important problems of continuum mechanics. Continuum mechanics is closely related here to other branches of physics.

The first law of thermodynamics does not require further comment. The situation is different with the second law of thermodynamics. The nature of this law becomes understandable if one turns to the Boltzmann formula, which defines the entropy S in terms of the probability W of the realization of a system state at a given energy E:

$$S = k \ln W, \tag{2.1.65}$$

where k is the Boltzmann constant. The probability W is proportional to the number of possible states of a system in the phase space of the momenta and coordinates of all particles at a given total energy E. The process develops in time in such a way that, at each moment of time, the most probable state is realized; therefore, the inequalities $dW/dt > 0$, $dS/dt > 0$ are valid. Consider two states of a closed thermodynamic system with the same energy, which represents an infinitesimal individual volume of a continuum. We will assume in the first state that the total energy is equal to the kinetic energy, and there is only one technique to distribute it between the molecules: to assign the same energy to all molecules. In the second state, it is assumed that a part of the energy exists in the form of the thermal energy related to the chaotic motion of particles; therefore, there are a large number of techniques to distribute this energy between the molecules, which is proportional to the number of permutations in the phase space (the calculation details are presented in statistical physics courses). Consequently, the second state will realize with a larger probability and the kinetic energy of our closed system will go over to the thermal energy, but there will be no passage of the thermal energy to the kinetic energy. Thus, the process of a passage of the kinetic energy to the thermal energy will be irreversible, what is fixed by the second law of thermodynamics. Note that formula (2.1.65) is applicable both to reversible and irreversible processes, and the equilibrium is achieved when a state is achieved, which has the maximum probability; therefore, the entropy in the equilibrium state is maximum.

In connection with the Boltzmann formula, one more question about the validity of a probabilistic description of a system of molecules arises. Each molecule is described by deterministic equations of the classical mechanics, which enable one to determine, in principle, the position and velocity of a particle at any moment of time on the basis of a given initial state. The classical mechanics equations are reversible in time (they are invariant under the substitution of t for $-t$); therefore, one can say on the basis of a given molecule state what happened to a particular molecule at any previous moment of time. The answer to this question is that there are in nature unstable systems, the behavior of which differs fundamentally from the motion of a molecule, as discussed above. The infinitesimal disturbances, which always exist in nature, lead unstable dynamic systems to considerable uncontrollable deviations of their trajectories; therefore, the behavior of such systems obeys the probabilistic laws. It is easy to see that the reversibility property is absent for such systems. If we try at some moment of time t to return our system to its initial state by inverting the velocities of all molecules to the opposite ones ($\vec{v} \to -\vec{v}$), then an infinitesimal disturbance will always be found, which will lead to a significant deviation of the trajectory from the original one, and at the time $2t$, the system will be in a state, which differs considerably from the initial one.

Let us present an example of a dynamically unstable system consisting of gas and absolutely elastic balls. Let us assume that the free path of the balls between their collisions l is much larger than the ball diameter d. Consider the collision of two balls with the target distance b. Then it follows from the momentum conservation law that the angle of the deviation of the balls after the collision is $\varphi = \pi - 2\theta$, where $\sin\theta = b/d$. An infinitesimal disturbance δb of the target distance leads to an infinitesimal disturbance of the deviation angle of the balls $\delta\theta \approx \delta b/d$. Until the next collision, the ball passes the distance l; therefore, the disturbance of the target distance before the next collision is equal to $\delta b_1 = \delta\theta l \approx (l/d)\delta b$. After $N+1$ collisions, the target distance disturbance is

$$\delta b_n \approx (l/d)^N \delta b. \tag{2.1.66}$$

It follows from formula (2.1.66) that at $l \gg d$ the disturbances will quickly grow, and after several collisions the state of the system will differ significantly from the state, which was not subject to an infinitesimal disturbance.

The above-discussed questions on the relation between the second law of thermodynamics and the behavior of thermodynamically unstable systems are discussed in detail in the literature[8,9].

2.1.4 Equation of State (General Relations)

We will formulate in this section some general principles underlying the equation of state of formless media. We will consider for definiteness the motion of an individual volume $V(t)$ in the accompanying coordinates ξ^i with the basis $\hat{\vec{e}}_i$. The equation of state should satisfy the *causality principle*, which states that the stress at point ξ^i at the time t should be determined completely by the motion of the system up to this time. Thus, the stress should be represented by the functional[10]

$$\hat{\sigma}(\xi^i, t) = \hat{\mathcal{F}}_t(\xi^i, t, x^i(\bar{\xi}^j, \tau, T(\bar{\xi}^j, \tau)), \tag{2.1.67}$$

where $\tau < t$, $x^j = x^j(\xi^i, t)$ is the motion law, $\hat{\sigma} = \hat{\sigma}^{ij}\vec{e}_i\vec{e}_j$, $\mathcal{F}_t = \mathcal{F}_t^{ij}\hat{\vec{e}}_i\hat{\vec{e}}_j$. The conditions of the *homogeneity of space and time* lead to the fact that the functional $\hat{\mathcal{F}}_t$ should not depend explicitly on the coordinates ξ^i and the time t; therefore, equation (2.1.67) should have the form

$$\hat{\sigma}(\xi^i, t) = \hat{\mathcal{F}}_t(x^i(\bar{\xi}^j, t - \tau), T(\bar{\xi}^j, \tau)). \tag{2.1.68}$$

The next important principle of the locality, which is valid for the formless media, states that the functional $\hat{\mathcal{F}}_t$ depends only on those $\bar{\xi}^i$, which lie in an infinitesimal neighborhood of the ξ^i point. This is expressed practically by the fact that the functional $\hat{\mathcal{F}}_t$ depends only on the functions $x^i(\xi^j, t-\tau)$, $T(\xi^j, t-\tau)$ and the derivatives $\frac{\partial x^i}{\partial \xi^j}(\xi^j, t-\tau)$, i.e., on the strains $\hat{\varepsilon} = \hat{\varepsilon}_{ij}\hat{\vec{e}}_i\hat{\vec{e}}_j$. This conclusion follows in an obvious way from the fact that the stress state of an individual volume will change only as a result of its deformation and the variation of its thermodynamic parameters (the temperature T). The dependence of the coordinates x^i should be absent because of the medium *homogeneity*. In addition, the medium should be *isotropic*, and its properties should not change under the rotation of the coordinate system. This means that the functional $\hat{\mathcal{F}}_t$ may depend, besides the strain tensor, only on the identity tensor $\hat{\delta}$. With regard for the foregoing, the *equation of state of a homogeneous isotropic formless continuum* can be written down in the following general form:

$$\hat{\sigma}(\xi^i, t) = \hat{\mathcal{F}}_t(\hat{\varepsilon}((t-\tau), \xi^i), T((t-\tau), \xi^i)), \tag{2.1.69}$$

or in the form of an integral in time $t_0 < \tau < t$

$$\hat{\sigma}(\xi^i, t) = \int_{t_0}^{t} \hat{\Phi}(\hat{\varepsilon}((t-\tau), \xi^i), T((t-\tau), \xi^i))\, d\tau. \tag{2.1.70}$$

Let us require that $\hat{\Phi}$ be an analytic function of $\hat{\varepsilon}$

$$\hat{\Phi} = \sum_k \alpha_k(T)\hat{\varepsilon}^k, \tag{2.1.71}$$

where

$$\hat{\varepsilon}^k = \hat{\varepsilon}_{ik_1}\hat{\varepsilon}_{k_1k_2}\ldots\hat{\varepsilon}_{k-1,j}\hat{\vec{e}}_i\hat{\vec{e}}_j, \tag{2.1.72}$$

and

$$T = T((\tau - t), \xi^i), \quad \hat{\varepsilon}_{ij} = \hat{\varepsilon}_{ij}((\tau - t), \xi^k).$$

Denoting the principal values of the strain tensor $\hat{\varepsilon}_{ij}$ by the letters $\lambda_1, \lambda_2, \lambda_3$, let us write the equation for the λ_i (1.1.50), (1.1.51) in the form

$$|\hat{\varepsilon}_{ij} - \lambda\delta_{ij}| = -\lambda^3 + a\lambda^2 - b\lambda + c = 0, \tag{2.1.73}$$

where

$$a = \sum \lambda_i, \quad b = \sum \lambda_i\lambda_j, \quad c = \lambda_1\lambda_2\lambda_3.$$

Since equation (2.1.73) is invariant under the coordinate transformations, it can be rewritten in the arbitrary coordinates as

$$\hat{\varepsilon}^3 = a\hat{\varepsilon}^2 - b\hat{\varepsilon} + c\delta. \tag{2.1.74}$$

Multiplying this equation by $\hat{\varepsilon}^{n-3}$, we obtain the Hamilton–Cayley formula:

$$\hat{\varepsilon}^n = a\hat{\varepsilon}^{n-1} - b\hat{\varepsilon}^{n-2} + c\hat{\varepsilon}^{n-3}, \tag{2.1.75}$$

with regard for which, equation (2.1.71) may be written as follows:

$$\hat{\Phi} = \beta_0\hat{\delta} + \beta_1\hat{\varepsilon} + \beta_2\hat{\varepsilon}^2, \tag{2.1.76}$$

where

$$\beta_i = \beta_i(T(\tau - t), \lambda_1, \lambda_2, \lambda_3), \quad \hat{\varepsilon} = \hat{\varepsilon}(\tau).$$

For a large class of continuous processes,. the strain $\hat{\varepsilon}$ is an analytic function of time, therefore, the following expansion is valid:

$$\hat{\varepsilon}(\tau) = \hat{\varepsilon}(t) + \dot{\hat{\varepsilon}}(t)\cdot(\tau - t) + \frac{1}{2}\ddot{\hat{\varepsilon}}(t)\cdot(\tau - t)^2 + \cdots \tag{2.1.77}$$

Substituting (2.1.77) into (2.1.76), we obtain the expression:

$$\begin{aligned}\hat{\Phi} &= \beta_0\hat{\delta} + \beta_1\hat{\varepsilon}(t) + \beta_1(\tau - t)\,\dot{\hat{\varepsilon}}(t) + \beta_1\ddot{\hat{\varepsilon}}(t)\\ &+ \beta_2\hat{\varepsilon}^2(t) + \beta_2(\tau - t)^2\dot{\hat{\varepsilon}}^2(t) + 2\beta_2\,(\tau - t)\,\hat{\varepsilon}(t)\,\dot{\hat{\varepsilon}}(t) + \beta_2(\tau - t)^2\hat{\varepsilon}(t)\,\ddot{\hat{\varepsilon}}(t)\\ &+ \frac{1}{4}\beta_2(\tau - t)^4\,\ddot{\hat{\varepsilon}}^2(t) + \beta_2(\tau - t)^3\,\dot{\hat{\varepsilon}}(t)\,\ddot{\hat{\varepsilon}}(t) + \cdots,\end{aligned} \tag{2.1.78}$$

where the points denote the terms of the series, which contain the higher derivatives with respect to time $\dddot{\hat{\varepsilon}}, \ldots$. Substituting (2.1.78) into (2.1.70), we find the general form of the equation of state under the assumption that it is an analytic function:

$$\hat{\sigma}(\xi^i, t) = \omega_0\hat{\delta} + \omega_1\hat{\varepsilon} + \omega_2\,\hat{\varepsilon}^2 + \omega_3\,\dot{\hat{\varepsilon}} + \omega_4\hat{\varepsilon}\dot{\hat{\varepsilon}} + \omega_5\,\dot{\hat{\varepsilon}}^2 + \omega_6\,\ddot{\hat{\varepsilon}}, \tag{2.1.79}$$

where $\hat{\varepsilon} = \hat{\varepsilon}(\xi^i, t)$ and the variables ω_i denote the integrals of the form [the constants m, n, k, l are chosen in accordance with (2.1.78)]

$$\begin{aligned}\omega_i &= \int_{t_0}^{t}(1 + m(\tau - t) + n(\tau - t)^2 + k(\tau - t)^3\\ &+ l(\tau - t)^4)\beta_i(\lambda_1, \lambda_2, \lambda_3, T(\tau - t, \xi^i))\,d\tau.\end{aligned} \tag{2.1.80}$$

When writing formula (2.1.79), we have omitted the terms containing the derivatives of $\hat{\varepsilon}$ with respect to time t of the order higher than two. If there were in the closing relationship (2.1.79) the terms containing the derivatives of $\hat{\varepsilon}$ with respect to time of the order higher than two, then, by virtue of (1.2.29) and (1.2.30), there would also emerge in the motion equation (2.1.18) the terms involving the derivatives of the displacement having a higher order than $\frac{\partial^2 x^k}{\partial t^2}$. Since the motion equation for each individual ith molecule in the classical mechanics is a second-order differential equation $\frac{d^2 x_i^k}{dt^2} = F^k$, the equations of continuum mechanics should also have the second order with respect to time as functions of the displacement. This hypothesis appears to be well justified. The construction of the specific forms of the functionals ω_i can be carried out only on the basis of the experiments or by the methods of statistical physics and kinetic theory. We will consider in the following an example of a two-parametric equation of state for the ideal and viscous, heat-conducting gas.

2.1.5 Equations of an Ideal and Viscous, Heat-Conducting Gas

Consider at first the *ideal gas* representing a two-parametric medium with a spherical stress tensor

$$\hat{\sigma}^{ij} = -P\hat{g}^{ij}, \tag{2.1.81}$$

or in the Eulerian Cartesian coordinate system

$$\sigma^{ij} = -P\delta^{ij}. \tag{2.1.82}$$

The irreversible processes are absent in the ideal gas in the case of continuous motions; therefore, $dq' = 0$. Substituting this relation into (2.1.62) and the relation (2.1.81) into (2.1.51), let us rewrite with regard for the equations

$$\frac{1}{\rho}\sigma^{ij}\dot{\varepsilon}_{ij}\,dt = -\frac{1}{\rho}P\dot{\varepsilon}_i^i\,dt = -\frac{P}{\rho}\operatorname{div}\vec{v}\,dt = \frac{P}{\rho^2}\frac{d\rho}{dt}\,dt = -P\,d\Big(\frac{1}{\rho}\Big),$$

the first and the second laws of thermodynamics in the form

$$dE = -Pd\Big(\frac{1}{\rho}\Big) + dq_e, \qquad T\,dS = dq_e, \tag{2.1.83}$$

where the quantity dq_e is assumed to be known. Substituting the second equation into the first equation, we can rewrite (2.1.83) as follows:

$$dE = T\,dS - Pd\Big(\frac{1}{\rho}\Big), \tag{2.1.84}$$

from where it follows that the specific internal energy is a function of the density and the entropy $E = E(\rho, S)$; therefore, the ideal gas is called the two-parametric medium. Equation (2.1.84) coincides with (2.1.64) and is called the Gibbs formula in the literature. Since dE is a total differential, then

$$dE = \left(\frac{\partial E}{\partial S}\right)_\rho dS + \left(\frac{\partial E}{\partial \rho}\right)_S d\rho.$$

Substituting this expression into (2.1.84), we find with regard for the independence of the increments $d\rho$ and dS the remaining thermodynamic functions:

$$T = \left(\frac{\partial E}{\partial S}\right)_\rho, \quad P = \rho^2\left(\frac{\partial E}{\partial \rho}\right)_S. \tag{2.1.85}$$

It follows from here that the energy E is a thermodynamic potential in the variables ρ and S, the knowledge of which enables one to determine the remaining parameters P and T. If we choose ρ and T as independent variables, then we can determine with the aid of the Legendre transformation a new potential $F = E - TS$, which is called the *free energy* F and is a function of new variables $F = F(\rho, T)$. Assuming that

$$dE - T\,dS - S\,dT = \left(\frac{\partial F}{\partial T}\right)_\rho dT + \left(\frac{\partial F}{\partial \rho}\right)_T d\rho$$

we obtain with regard for (2.1.84):

$$-S\,dT - p\,d\left(\frac{1}{\rho}\right) = \left(\frac{\partial F}{\partial T}\right)_\rho dT + \left(\frac{\partial F}{\partial \rho}\right)_T d\rho,$$

from where we find:

$$S = -\left(\frac{\partial F}{\partial T}\right)_\rho, \quad P = \rho^2\left(\frac{\partial F}{\partial \rho}\right)_T. \tag{2.1.86}$$

Let us choose P and S as the independent variables and determine the thermodynamic potential $H = H(P, S)$, called the *enthalpy* by the formula $H = E + P/\rho$. Differentiating this relation:

$$\left(\frac{\partial H}{\partial S}\right)_P dS + \left(\frac{\partial H}{\partial P}\right)_S dP = dE + \frac{1}{\rho}dP + P\,d\left(\frac{1}{\rho}\right),$$

we can find with regard for (2.1.84) that

$$T = \left(\frac{\partial H}{\partial S}\right)_P, \quad \frac{1}{\rho} = \left(\frac{\partial H}{\partial P}\right)_S. \tag{2.1.87}$$

The *Gibbs thermodynamic potential* as a function of the variables P and T is defined in a similar way:

$$\mu = E - TS + P/\rho,$$

from where we obtain with regard for $\mu = \mu(P, T)$ and (2.1.84) that

$$S = -\left(\frac{\partial \mu}{\partial T}\right)_P, \quad \frac{1}{\rho} = \left(\frac{\partial \mu}{\partial P}\right)_T. \tag{2.1.88}$$

Note that, if the internal energy is given in the form of a function of some other variables, for example, $E = E(\rho, P)$, then the remaining parameters T and S are already not determined uniquely. Substituting

$$dE = \left(\frac{\partial E}{\partial P}\right)_\rho dP + \left(\frac{\partial E}{\partial \rho}\right)_P d\rho$$

into equation (2.1.84), we obtain:

$$dS = \frac{dE}{T} + \frac{P}{T}\frac{d}{dt}\left(\frac{1}{\rho}\right) = \left(\frac{1}{T}\frac{\partial E}{\partial P}\right) dP + \frac{1}{T}\left(\frac{\partial E}{\partial \rho} - \frac{P}{\rho^2}\right) d\rho, \tag{2.1.89}$$

where we have omitted for brevity the subscripts by the partial derivatives. According to the second law of thermodynamics (2.1.83), $dS = dq_e/T$ is a total differential, for which

$$dS = \frac{\partial S}{\partial P} dP + \frac{\partial S}{\partial \rho} d\rho.$$

Comparing this expression with (2.1.89), we find:

$$\frac{\partial S}{\partial P} = \frac{1}{T}\frac{\partial E}{\partial P}, \quad \frac{\partial S}{\partial \rho} = \frac{1}{T}\left(\frac{\partial E}{\partial \rho} - \frac{P}{\rho^2}\right). \tag{2.1.90}$$

From the condition of equality of the second derivatives

$$\frac{\partial}{\partial \rho}\left(\frac{\partial S}{\partial P}\right) = \frac{\partial}{\partial P}\left(\frac{\partial S}{\partial \rho}\right)$$

and from (2.1.90), we obtain the equation for the determination of $T = T(P, \rho)$:

$$\frac{\partial T}{\partial \rho}\frac{\partial E}{\partial P} + \frac{\partial T}{\partial P}\left(\frac{P}{\rho^2} - \frac{\partial E}{\partial \rho}\right) = \frac{T}{\rho^2}. \tag{2.1.91}$$

The partial differential equation (2.1.91) can be integrated by the method of characteristics (see Section 3.2) and contains some arbitrary function $\tilde{T}(\rho, P)$. If we choose some function $\tilde{T}(\rho, P)$ and consequently $T(\rho, P)$, then the entropy $S(\rho, P)$ is determined after integrating (2.1.89) with the accuracy up to a constant. The quantities ρ and S are called the eigenvariables of the internal energy E. Similar considerations are valid for other thermodynamic potentials.

A particular case of the ideal gas is the *perfect gas*, which is specified by the relations

$$E = C_V T, \qquad P = \rho R T, \tag{2.1.92}$$

where $R = \text{const}$ is the universal gas constant and $C_V = \text{const}$ is the specific heat under a constant volume. Substituting (2.1.92) into (2.1.89) we find:

$$\begin{aligned} dS &= \frac{dE}{T} + \frac{P}{T} d\Big(\frac{1}{\rho}\Big) = C_V \frac{dT}{T} + R\rho\, d\Big(\frac{1}{\rho}\Big) \\ &= d\Big(C_V \ln T + R \ln \frac{1}{\rho}\Big) = d\Big(C_V \ln T \Big(\frac{1}{\rho}\Big)^{(R/C_V)}\Big) \\ &= d\Big(C_V \ln T \Big(\frac{1}{\rho}\Big)^{(\gamma-1)}\Big), \end{aligned} \tag{2.1.93}$$

where we have assumed that $\frac{R}{C_V} = \frac{C_P - C_V}{C_V} = \frac{C_P}{C_V} - 1 = \gamma - 1$. The parameter $\gamma = \frac{C_P}{C_V}$ is called the adiabat exponent, and $C_P = (\frac{\partial q_e}{\partial T})_P$ is the specific heat at a constant pressure. For the perfect gas, the Meyer's formula $C_P - C_V = R$ is valid. Integrating (2.1.93), we obtain:

$$S = S_0 + C_V \ln \frac{T}{T_0} \Big(\frac{\rho_0}{\rho}\Big)^{\gamma-1}, \tag{2.1.94}$$

where S_0, T_0, ρ_0 are constants. Expressing from here the T and substituting into (2.1.92), we find the internal energy E as a function of the eigenvariables ρ and S

$$E = C_V\, T_0 \Big(\frac{\rho}{\rho_0}\Big)^{\gamma-1} e^{(S-S_0)/C_V}. \tag{2.1.95}$$

Consider the *adiabatic processes* $dq_e = 0$, for which equations (2.1.83) are to be rewritten in the form

$$dE = -P \frac{d}{dt}\Big(\frac{1}{\rho}\Big), \quad dS = 0, \tag{2.1.96}$$

from where it follows that the entropy remains constant at the adiabatic, reversible processes: $S = \text{const}$. Using (2.1.85) and (2.1.95), let us find the equation relating the pressure P to the density ρ at the adiabatic processes:

$$P = A\rho^\gamma, \tag{2.1.97}$$

which is called the *Poisson's adiabat*, where $A = A(S)$ is a constant.

Summarizing the above, let us write a complete system of equations of the ideal gas in the Eulerian coordinates $g_{ij} = \vec{e}_i \cdot \vec{e}_j$:

$$\begin{aligned} &\frac{d\rho}{dt} + \rho \operatorname{div} \vec{v} = 0, \qquad \frac{d}{dt} = \frac{\partial}{\partial t} + v^j \nabla_j, \\ &\frac{d\vec{v}}{dt} = -\frac{1}{\rho} \nabla P + \vec{F}, \qquad \nabla = \vec{e}_j \nabla^j, \quad \vec{v} = v^j \vec{e}_j \end{aligned} \tag{2.1.98}$$

$$\frac{dE}{dt} = -P\frac{d}{dt}\Big(\frac{1}{\rho}\Big) + \frac{dq_e}{dt}, \qquad T\frac{dS}{dt} = \frac{dq_e}{dt},$$
$$T = \Big(\frac{\partial E}{\partial S}\Big)_\rho, \qquad P = \rho^2\Big(\frac{\partial E}{\partial \rho}\Big)_S,$$

where $E = E(\rho, S)$, $\vec{F}$, $\frac{dq_e}{dt}$ are assumed to be given. At the adiabatic motions, the gas entropy in an individual volume remains constant. Note that the Eulerian coordinates are more often used for the description of the gas (or fluid) motions.

Consider a *viscous, heat-conducting gas* , for which

$$\sigma^{ij} = -Pg^{ij} + \tau^{ij}. \tag{2.1.99}$$

The tensor τ^{ij} corresponds to the viscous stresses, and as follows from the experiments, it is proportional to the rate-of-strain tensor $\dot{\varepsilon}_{ij} = \frac{1}{2}(\nabla_i v_j + \nabla_j v_i)$

$$\tau_{ij} = \lambda g_{ij}\operatorname{div}\vec{v} + 2\mu\dot{\varepsilon}_{ij}. \tag{2.1.100}$$

Employing the expansion

$$\dot{\varepsilon}_{ij} = \frac{1}{3}\dot{\varepsilon}_k^k\, g_{ij} + \dot{e}_{ij}, \quad \dot{\varepsilon}_k^k = \operatorname{div}\vec{v},$$

let us rewrite (2.1.99) and (2.1.100) in the form

$$\sigma_{ij} = (-P + \zeta\operatorname{div}\vec{v})g_{ij} + 2\mu\dot{e}_{ij}, \tag{2.1.101}$$

where $\zeta = \lambda + \frac{2}{3}\mu$. The viscosity coefficients λ and μ do not depend on the strain rates, but can depend on the temperature. They can be found from the experiment or the solution of the Boltzmann kinetic equation. Substituting (2.1.101) into (2.1.24), we obtain the *Navier-Stokes equation:*

$$\frac{d\vec{v}}{dt} = -\frac{1}{\rho}\nabla P + \Big(\frac{\zeta}{\rho} + \frac{\nu}{3}\Big)\nabla\operatorname{div}\vec{v} + \nu\Delta\,\vec{v}, \tag{2.1.102}$$

where $\nu = \mu/\rho$ is the kinematic viscosity coefficient and $\Delta = \nabla_i\nabla^i = g^{ij}\nabla_i\nabla_j$ is the Laplacian. Under the action of viscous stresses, the gas kinetic energy goes over to the thermal energy; therefore, the noncompensated heat dq' is different from zero. Subtracting the Gibbs equation (2.1.84) from the energy equation (2.1.51), we obtain:

$$\frac{1}{\rho}\sigma^{ij}\dot{\varepsilon}_{ij}dt + dq_e - T\,dS - \frac{P}{\rho^2}\,d\rho = 0. \tag{2.1.103}$$

Dividing equation (2.1.103) by dt, we find with regard for the second law of thermodynamics (2.1.62) and the continuity equation:

$$\frac{1}{\rho}\frac{d\rho}{dt} = -\operatorname{div}\vec{v} = -\dot{\varepsilon}_k^k$$

that

$$\frac{dq'}{dt} = \frac{1}{\rho}\sigma^{ij}\dot{\varepsilon}_{ij} + \frac{P}{\rho}\dot{\varepsilon}^k_k. \tag{2.1.104}$$

Employing (2.1.101), we obtain:

$$\sigma^{ij}\dot{\varepsilon}_{ij} = \sigma_{ij}\dot{\varepsilon}^{ij} = -P\dot{\varepsilon}^k_k + \zeta(\dot{\varepsilon}^k_k)^2 + 2\mu\dot{e}_{ij}\dot{e}^{ij}. \tag{2.1.105}$$

Substituting this expression into (2.1.104), we can find the rate of change of the noncompensated heat of a unit gas mass

$$\frac{dq'}{dt} = \frac{\zeta}{\rho}(\operatorname{div}\vec{v})^2 + 2\nu\dot{e}_{ij}\dot{e}^{ij}, \tag{2.1.106}$$

from where it follows that $\frac{dq'}{dt} \geq 0$ at $\zeta > 0$, $\nu > 0$, i.e., the closing relationships (2.1.101) agree with the second law of thermodynamics. An analysis of the expression (2.1.106) shows that the processes occurring with the finite strain rate $\dot{\varepsilon}_{ij} \neq 0$ are irreversible: $\frac{dq'}{dt} > 0$ and $\frac{dS}{dt} > 0$. For the infinitely slow processes when $\dot{\varepsilon}_{ij} \to 0$, the right-hand side of (2.1.106) will have the second order of smallness $\sim \dot{\varepsilon}_{ij}\dot{\varepsilon}^{ij}$, and one can assume with good accuracy these processes to be reversible (in the absence of heat conduction) $\frac{dq_e}{dt} \approx 0$, $\frac{dS}{dt} \approx 0$.

Let us determine the entropy variation $\frac{dS_e}{dt}$ due to the heat supply $\frac{dq_e}{dt}$. We neglect for a while the body heat supply $\dot{q}_V = 0$ and rewrite formula (2.1.48) in the Eulerian coordinates

$$\frac{dq_e}{dt} = -\frac{1}{\rho}\nabla_i q^i, \tag{2.1.107}$$

where $\vec{q} = q^i\vec{e}_i$ is the vector of the heat flux through a unit area. The experience shows that $\vec{q}$ is proportional to the temperature gradient

$$\vec{q} = -\kappa\nabla T, \tag{2.1.108}$$

where κ is the coefficient of heat conduction, which is a function of temperature. The minus sign is related to the fact that the heat always flows from hotter regions to colder regions. The coefficients of heat conduction κ and of the viscosity λ, μ depend on temperature, energy, and momentum at the thermal motion of molecules. The formula (2.1.108) is often called the *Fourier heat conduction law.* Substituting (2.1.108) into (2.1.107), we obtain:

$$\frac{dq_e}{dt} = \frac{\kappa}{\rho}\nabla_i\nabla^i T = \frac{\kappa}{\rho}\Delta T. \tag{2.1.109}$$

In the Cartesian coordinate system, equation (2.1.109) has the form

$$\frac{dq_e}{dt} = \frac{\kappa}{\rho}\left(\frac{\partial^2 T}{\partial x_1^2} + \frac{\partial^2 T}{\partial x_2^2} + \frac{\partial^2 T}{\partial x_3^2}\right). \tag{2.1.110}$$

Substituting (2.1.108) into (2.1.107) and then into the second law of thermodynamics, we can find the entropy variation $\frac{dS_T}{dt}$ due to the heat conduction

$$\begin{aligned}\rho\frac{dS_T}{dt} &= \frac{\rho}{T}\frac{dq_e}{dt} = -\frac{1}{T}\operatorname{div}\vec{q} = -\frac{1}{T^2}(\vec{q}\cdot\nabla T) - \operatorname{div}\Big(\frac{\vec{q}}{T}\Big)\\ &= \frac{\kappa}{T^2}(\nabla T)^2 - \operatorname{div}\Big(\frac{\vec{q}}{T}\Big).\end{aligned} \tag{2.1.111}$$

The action of a viscous dissipation of energy leads, according to (2.1.106), to the entropy increment

$$\rho\frac{dS_\mu}{dt} = \frac{1}{T}(\zeta(\operatorname{div}\vec{v})^2 + 2\mu\dot{e}_{ij}\dot{e}^{ij}).$$

The total entropy variation is given by the formula

$$\begin{aligned}\rho\frac{dS}{dt} &= \rho\Big(\frac{dS_T}{dt} + \frac{dS_\mu}{dt}\Big)\\ &= \frac{1}{T}(\zeta(\operatorname{div}\vec{v})^2 + 2\mu\dot{e}_{ij}\dot{e}^{ij}) + \frac{\kappa}{T^2}(\nabla T)^2 - \operatorname{div}\Big(\frac{\vec{q}}{T}\Big).\end{aligned} \tag{2.1.112}$$

Using (2.1.112), one can find the entropy variation in some fixed volume V

$$\begin{aligned}\frac{d}{dt}\int_V \rho S\,dV &= \int_V \frac{1}{T}(\zeta(\operatorname{div}\vec{v})^2 + 2\mu\dot{e}_{ij}\dot{e}^{ij}) + \frac{\kappa}{T^2}(\nabla T)^2)\,dV\\ &- \int_S \frac{1}{T}\vec{q}\cdot\vec{n}\,dS.\end{aligned} \tag{2.1.113}$$

It follows from here that the first integral over the volume yields the entropy production in the volume V at the expense of the irreversible and nonequilibrium processes related to the energy dissipation and the temperature equalization at the expense of the heat conduction. Since there should be $\frac{dS}{dt} > 0$, the heat conduction coefficient should also be positive: $\kappa > 0$.

Using formulas (2.1.51), (2.1.102), (2.1.105), and (2.1.109) let us write a complete system of equations for a viscous, heat conducting gas in the Eulerian coordinates:

$$\begin{aligned}&\frac{d\rho}{dt} + \rho\operatorname{div}\vec{v} = 0, \quad \frac{d}{dt} = \frac{\partial}{\partial t} + v^j\nabla_j,\\ &\frac{d\vec{v}}{dt} = -\frac{1}{\rho}\nabla P + \Big(\frac{\zeta}{\rho} + \frac{\nu}{3}\Big)\nabla\operatorname{div}\vec{v} + \nu\Delta\vec{v},\\ &\frac{dE}{dt} = -P\frac{d}{dt}\Big(\frac{1}{\rho}\Big) + \frac{\zeta}{\rho}(\operatorname{div}\vec{v})^2 + 2\nu\,\dot{e}_{ij}\dot{e}^{\,ij} + \frac{\kappa}{\rho}\Delta T,\\ &T = \Big(\frac{\partial E}{\partial S}\Big)_\rho, \quad P = \rho^2\Big(\frac{\partial E}{\partial \rho}\Big)_S, \quad E = E(\rho, S).\end{aligned} \tag{2.1.114}$$

The surface integral in (2.1.113) determines the entropy variation at the expense of the heat supply or removal and can therefore have any sign. The rate of the entropy production in the unit volume is called the dissipative function of the system σ

$$\rho \frac{dS}{dt} = \sigma, \quad \sigma = \zeta(\dot{\varepsilon}_k^k)^2 + 2\mu \dot{e}_{ij}\dot{e}^{ij} + \frac{\kappa}{T^2}(\nabla T)^2, \tag{2.1.115}$$

which is a homogeneous quadratic system function of $\nabla_i T$ and $\dot{\varepsilon}_{ij} = \frac{1}{2}(\nabla_i v_j + \nabla_j v_i)$. In the theory of irreversible processes, the gradients of temperature, velocity, etc., are called the thermodynamic forces X_α, and the irreversible phenomena caused by them in the form of a heat flux, momentum, etc., are called the fluxes j_α [11]. With the aid of the introduced notions, the dissipative function may be rewritten in the form

$$\sigma = j_i X_i, \tag{2.1.116}$$

where

$$j_k = q^k, \quad X_k = -\frac{1}{T^2}\nabla_k T, \quad j_{kl} = \tau^{kl}, \quad X_{kl} = \dot{\varepsilon}_{kl}, \tag{2.1.117}$$

and $\tau^{kl} = g^{ki}g^{lj}\tau_{ij}$, and the τ_{ij} are determined by formula (2.1.100). As follows from the heat conduction law (2.1.108) and the expression for the stress tensor (2.1.100), the fluxes j_α are proportional to the thermodynamic forces X_α. On the other hand, this result is a consequence of the fact that σ is a homogeneous quadratic function of X_α and can be generalized for a wide class of thermodynamic systems, which are near their equilibrium state. Consider a thermodynamic system whose state is characterized by the independent parameters μ_i. Let these parameters have in the equilibrium state the values μ_i^0. Let us introduce the notation $x_i = \mu_i - \mu_i^0$, so that the x_i will determine the deviation of the system from the equilibrium state $x_i = 0$. In the equilibrium state, the entropy S^0 reaches its maximum value; therefore, for the states different from the equilibrium one, but close to it, we will have that

$$S - S^0 = -\frac{1}{2}\beta_{ik}x_i x_k, \tag{2.1.118}$$

where the matrix β_{ik} is symmetric, i.e., $\beta_{ik} = \beta_{ki}$ (the substitution of the indices $i \leftrightarrow k$ should not change the scalar quantity S). The system deviation from the equilibrium state causes in the system the fluxes $\dot{x}_i = f_i(x_k)$, which tend to return the system to the equilibrium state $x_k = 0$, $\dot{x}_k = 0$. If the deviations are small, one can expand $f_i(x_k)$ into a Taylor series in the neighborhood of the equilibrium point $x_k = 0$ and retaining only the linear terms in the expansion, we obtain:

$$\dot{x}_i = -\lambda_{ik}x_k, \tag{2.1.119}$$

where $\lambda_{ik} = -|\partial f_i/\partial x_k|_{x_k=0}$. Let us introduce the thermodynamic forces X_i, which are conjugate to the velocities $\dot{x}_i$ in accordance with the formula

$$X_i = -\frac{\partial S}{\partial x_i}. \tag{2.1.120}$$

Substituting (2.1.118) into (2.1.120), we obtain:

$$X_i = \beta_{ik}x_k, \quad x_i = \beta_{ik}^{-1}X_k, \tag{2.1.121}$$

where the symbols β_{ik}^{-1} denote the matrix, which is inverse with respect to β_{ik}. Substituting the second equation from (2.1.121) into equation (2.1.119), we find:

$$\dot{x}_i = -\gamma_{ik}X_k, \tag{2.1.122}$$

where $\gamma_{ik} = \lambda_{ij}\beta_{jk}^{-1}$. There exists in the theory of nonequilibrium processes the Onsager's principle[11], according to which

$$\gamma_{ik} = \gamma_{ki}. \tag{2.1.123}$$

The proof of the Onsager's principle is based on the symmetry of the classical mechanics equations describing the evolution of the parameters $x_i = x_i(t)$ with respect to the inversion in time $t \to -t$ [11]. If the processes occur in a rotating medium with the angular velocity ω or in the magnetic field $\vec{H}$, then the symmetry with respect to the time inversion takes place under a simultaneous change of signs $\vec{\omega} \to -\vec{\omega}$, or $\vec{H} \to -\vec{H}$. In this case, the Onsager's principle will have the form

$$\gamma_{ik}(\vec{\omega}) = \gamma_{ki}(-\vec{\omega}), \quad \gamma_{ik}(\vec{H}) = \gamma_{ki}(-\vec{H}). \tag{2.1.124}$$

Differentiating (2.1.118) with respect to time, we can find the rate of the entropy production

$$\dot{S} = -\beta_{ik}\dot{x}_i x_k = -\dot{x}_i X_i. \tag{2.1.125}$$

A comparison of this formula with (2.1.116) shows that $j_i = -\dot{x}_i$, and equation (2.1.122), determining the kinetics of nonequilibrium processes, may be rewritten as

$$j_i = \gamma_{ik}X_k, \quad \gamma_{ik} = \gamma_{ki}. \tag{2.1.126}$$

The linear dependence (2.1.126) with symmetric coefficients determines a homogeneous quadratic function $\dot{S}$ of the thermodynamic forces. Substituting (2.1.122) into (2.1.125), we can find the form of this function:

$$\dot{S} = \gamma_{ik}X_i X_k. \tag{2.1.127}$$

It is seen that the dissipative function (2.1.116) is a particular case of (2.1.127). As the equilibrium is approached, the entropy increases: $\dot{S} >$

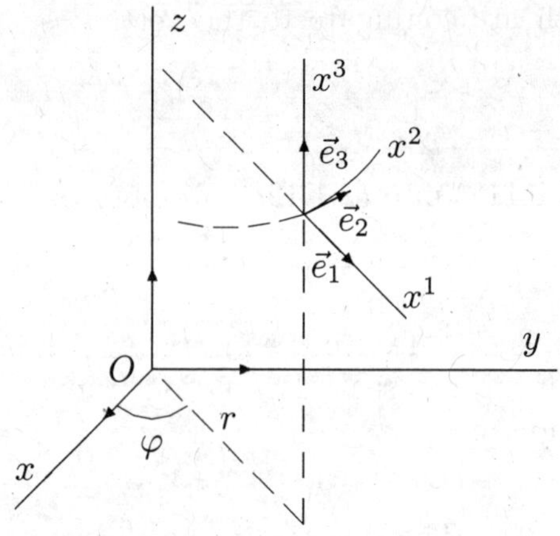

Figure 2.4: The cylindrical coordinate system.

0; therefore, the quadratic form should be positive definite. This is achieved in the case when all eigenvalues of the matrix γ_{ik} are positive. As an example, let us present the Fourier heat conduction law in an anisotropic medium

$$q_i = -\kappa_{ik} \nabla_k T. \tag{2.1.128}$$

It follows from the Onsager's principle (2.1.123) and the condition $\dot{S} > 0$ that

$$\kappa_{ij} = \kappa_{ji}, \tag{2.1.129}$$

and the eigenvalues of the matrix κ_{ij} should be positive.

Problem 2.1. Derive the equations of continuity, motion, and energy for an ideal gas in the Eulerian coordinates for three cases:
1) in the cylindrical coordinate system;
2) in the spherical coordinate system;
3) in the Cartesian coordinate system.

Solution: 1) Consider at first the *cylindrical coordinate system* specified with the aid of r, φ, and z, which we will denote for the unification of the tensor formulas by the symbols (see Fig. 2.4)

$$x^1 = r, \quad x^2 = \varphi, \quad x^3 = z.$$

The tangent vectors $\vec{e}_i$ to the coordinate lines x^i determine at each point an orthogonal basis of the cylindrical coordinate system (see Fig. 2.4). According to (1.1.77), the squared length element in the cylindrical coordinate system is equal to

$$ds^2 = (dx^1)^2 + (x^1)^2(dx^2)^2 + (dx^3)^2,$$

from where we obtain, with regard for the orthogonality of the coordinate system $g^{ii} = 1/g_{ii}$, the expressions for the components of the covariant and contravariant metric tensors

$$g_{11} = 1,\ g_{22} = (x^1)^2,\ g_{33} = 1,\ g^{11} = 1,\ g^{22} = \frac{1}{(x^1)^2},\ g^{33} = 1. \quad (2.1.130)$$

Let us introduce the physical components of the velocity u^i related to the coordinate components v^i by the formula $v^i = u^i/\sqrt{g_{ii}}$, with regard for which the continuity equation (2.1.8) can be written in the form

$$\sqrt{g_{11}g_{22}g_{33}}\,\frac{\partial \rho}{\partial t} + \frac{\partial}{\partial x^1}(\sqrt{g_{22}g_{33}}\,\rho u^1) + \frac{\partial}{\partial x^2}(\sqrt{g_{11}g_{33}}\,\rho u^2) + \frac{\partial}{\partial x^3}(\sqrt{g_{11}g_{22}}\,\rho u^3) = 0, \quad (2.1.131)$$

where we have used, for the determinant $g = \det \parallel g_{ij} \parallel$, the expression $g = g_{11}g_{22}g_{33}$. Substituting into the continuity equation the explicit form of the metric coefficients g_{ii} (the summation over i is absent) and going over to the notations $r = x^1$, $\varphi = x^2$, $z = x^3$, $u_r = u^1$, $u_\varphi = u^2$, $u_z = u^3$, let us rewrite the continuity equation in the form

$$r\frac{\partial \rho}{\partial t} + \frac{\partial}{\partial r}(r\rho u_r) + \frac{\partial}{\partial \varphi}(\rho u_\varphi) + r\frac{\partial}{\partial z}(\rho u_z) = 0. \quad (2.1.132)$$

Before proceeding to the motion equation, let us find the Christoffel symbols different from zero:

$$\Gamma^2_{12} = \Gamma^2_{21} = \frac{1}{2}g^{22}\frac{\partial g_{22}}{\partial x^1} = \frac{1}{x^1}, \quad \Gamma^1_{22} = -\frac{1}{2}g^{11}\frac{\partial g_{22}}{\partial x^1} = -x^1.$$

Let us determine the expressions for the substantive derivatives

$$\frac{dv^i}{dt} = \frac{\partial v^i}{\partial t} + v^j \nabla_j v^i = \frac{\partial v^i}{\partial t} + v^j\frac{\partial v^i}{\partial x^j} + v^j v^k \Gamma^i_{jk}. \quad (2.1.133)$$

Assuming $i = 1, 2, 3$ in this formula, we obtain:

$$\frac{dv^1}{dt} = \frac{\partial v^1}{\partial t} + v^j\frac{\partial v^1}{\partial x^j} + (v^2)^2\Gamma^1_{22},$$

$$\frac{dv^2}{dt} = \frac{\partial v^2}{\partial t} + v^j\frac{\partial v^2}{\partial x^j} + 2v^1v^2\Gamma^2_{12},$$

$$\frac{dv^3}{dt} = \frac{\partial v^3}{\partial t} + v^j\frac{\partial v^3}{\partial x^j}.$$

Taking into account the expressions for the Christoffel symbols and going over to the physical velocities u^i and the accelerations a^i by the formulas

$$v^i = \frac{u^i}{\sqrt{g_{ii}}}, \quad \frac{dv^i}{dt} = \frac{a^i}{\sqrt{g_{ii}}}$$

we obtain in the notation $r = x^1$, $\varphi = x^2$, $z = x^3$, $u_r = u^1$, $u_\varphi = u^2$, $u_z = u^3$, $a_r = a^1$, $a_\varphi = a^2$, $a_z = a^3$, the formulas for the accelerations:

$$\begin{aligned} a_r &= \frac{\partial u_r}{\partial t} + u_r \frac{\partial u_r}{\partial r} + \frac{u_\varphi}{r}\frac{\partial u_r}{\partial \varphi} + u_z \frac{\partial u_r}{\partial z} - \frac{u_\varphi^2}{r}, \\ a_\varphi &= \frac{\partial u_\varphi}{\partial t} + u_r \frac{\partial u_\varphi}{\partial r} + \frac{u_\varphi}{r}\frac{\partial u_\varphi}{\partial \varphi} + u_z \frac{\partial u_\varphi}{\partial z} + \frac{u_r u_\varphi}{r}, \\ a_z &= \frac{\partial u_z}{\partial t} + u_r \frac{\partial u_z}{\partial r} + \frac{u_\varphi}{r}\frac{\partial u_z}{\partial \varphi} + u_z \frac{\partial u_z}{\partial z}. \end{aligned} \tag{2.1.134}$$

Let us determine the physical components $(\nabla_j\sigma^{ij})$ of the vector $\nabla_j\sigma^{ij}\vec{e}_i$ by the formula $(\widetilde{\nabla_j\sigma^{ij}})\vec{E}_i = \nabla_j\sigma^{ij}\vec{e}_i$, where the unit vectors $\vec{E}_i = \vec{e}_i/|\vec{e}_i| = \vec{e}_i/\sqrt{g_{ii}}$. For the ideal gas $\sigma^{ij} = -Pg^{ij}$, therefore, with regard for $\vec{E}^i = \vec{E}_i$, $g_{ii} = 1/g^{ii}$, we have that

$$\begin{aligned} \nabla_j\sigma^{ij}\vec{e}_i &= -\nabla_j P g^{ij}\vec{e}_i = -\nabla_j P\vec{e}^j \\ &= -\nabla_j P\sqrt{g^{jj}}\frac{\vec{e}^j}{\sqrt{g^{jj}}} = -\nabla_j P\sqrt{g^{jj}}\vec{E}^j = -\frac{1}{\sqrt{g_{jj}}}\nabla_j P\vec{E}_j, \end{aligned}$$

from where it follows that

$$\widetilde{\nabla_j\sigma^{ij}} = -\frac{1}{\sqrt{g_{jj}}}\nabla_j P. \tag{2.1.135}$$

Going over to the notations r, φ, and z, let us rewrite this formula in a componentwise form

$$\widetilde{\nabla_j\sigma^{1j}} = -\frac{\partial P}{\partial r}, \quad \widetilde{\nabla_j\sigma^{2j}} = -\frac{1}{r}\frac{\partial P}{\partial \varphi}, \quad \widetilde{\nabla_j\sigma^{3j}} = -\frac{\partial P}{\partial z}. \tag{2.1.136}$$

Let us write the motion equations (2.1.24) in the physical coordinates. For this purpose, we multiply (2.1.24) by the basis vector $\vec{e}_i$ and write it in a vector form:

$$\rho\frac{dv^i}{dt}\vec{e}_i = \nabla_j\sigma^{ij}\vec{e}_i + \rho F^i\vec{e}_i.$$

Going over to the unit vectors, we can rewrite this equation as

$$\rho\frac{dv^i}{dt}\sqrt{g_{ii}}\vec{E}_i = (\widetilde{\nabla_j\sigma^{ij}})\vec{E}_i + \rho F^i\sqrt{g_{ii}}\vec{E}_i,$$

from where we obtain the motion equation in the physical coordinates:

$$\rho a^i = (\widetilde{\nabla_j\sigma^{ij}}) + \rho\tilde{F}^i, \tag{2.1.137}$$

where

$$\tilde{F}^i = F^i \sqrt{g_{ii}}, \quad a^i = \frac{dv^i}{dt}\sqrt{g_{ii}}.$$

Substituting (2.1.134) and (2.1.136) into (2.1.137), we obtain the motion equations in the cylindrical coordinates:

$$\begin{aligned}
&\frac{\partial u_r}{\partial t} + u_r\frac{\partial u_r}{\partial r} + \frac{u_\varphi}{r}\frac{\partial u_r}{\partial \varphi} + u_z\frac{\partial u_z}{\partial z} - \frac{u_\varphi^2}{r} = -\frac{1}{\rho}\frac{\partial P}{\partial r} + F_r,\\
&\frac{\partial u_\varphi}{\partial t} + u_r\frac{\partial u_\varphi}{\partial r} + \frac{u_\varphi}{r}\frac{\partial u_\varphi}{\partial \varphi} + u_z\frac{\partial u_\varphi}{\partial z} + \frac{u_r u_\varphi}{r} = -\frac{1}{\rho r}\frac{\partial P}{\partial \varphi} + F_\varphi,\\
&\frac{\partial u_z}{\partial t} + u_r\frac{\partial u_z}{\partial r} + \frac{u_\varphi}{r}\frac{\partial u_z}{\partial \varphi} + u_z\frac{\partial u_z}{\partial z} = -\frac{1}{\rho}\frac{\partial P}{\partial z} + F_z.
\end{aligned} \tag{2.1.138}$$

If there is no heat exchange in the ideal gas, i.e., $\frac{dq^e}{dt} = 0$, then the energy equation (2.1.83) leads to the condition for the constancy of the entropy S

$$\frac{dS}{dt} = \frac{\partial S}{\partial t} + v^j\nabla_j S = \frac{\partial S}{\partial t} + v^j\frac{\partial S}{\partial x^j} = \frac{\partial S}{\partial t} + \frac{u^j}{\sqrt{g_{jj}}}\frac{\partial S}{\partial x^j} = 0. \tag{2.1.139}$$

With regard for (2.1.130), this equation in the cylindrical coordinate system will have the form

$$\frac{\partial S}{\partial t} + u_r\frac{\partial S}{\partial r} + \frac{u_\varphi}{r}\frac{\partial S}{\partial \varphi} + \frac{\partial S}{\partial z} = 0. \tag{2.1.140}$$

In the particular case of an axisymmetric flow of an ideal gas, all of the functions ρ, u_r, u_φ, u_z, P do not depend on φ, and equations (2.1.132), and (2.1.138) have the form

$$\begin{aligned}
&\frac{\partial \rho}{\partial t} + \frac{\partial}{\partial r}(r\rho u_r) + \frac{\partial}{\partial z}(\rho u_z) = 0,\\
&\frac{\partial u_r}{\partial t} + u_r\frac{\partial u_r}{\partial r} + u_z\frac{\partial u_z}{\partial z} - \frac{u_\varphi^2}{r} = -\frac{1}{\rho}\frac{\partial P}{\partial r} + F_r,\\
&\frac{\partial u_\varphi}{\partial t} + u_r\frac{\partial u_\varphi}{\partial r} + u_z\frac{\partial u_\varphi}{\partial z} + \frac{u_r u_\varphi}{r} = F_\varphi,\\
&\frac{\partial u_z}{\partial t} + u_r\frac{\partial u_z}{\partial r} + u_z\frac{\partial u_z}{\partial z} = -\frac{1}{\rho}\frac{\partial P}{\partial z} + F_z.
\end{aligned}$$

If in addition the gas flow is irrotational, i.e., $u_\varphi = 0$, then this system of equations simplifies to the system

$$\begin{aligned}
&\frac{\partial \rho}{\partial t} + \frac{\partial}{\partial r}(r\rho u_r) + \frac{\partial}{\partial z}(\rho u_z) = 0,\\
&\frac{\partial u_r}{\partial t} + u_r\frac{\partial u_r}{\partial r} + u_z\frac{\partial u_z}{\partial z} = -\frac{1}{\rho}\frac{\partial P}{\partial r} + F_r,\\
&\frac{\partial u_z}{\partial t} + u_r\frac{\partial u_z}{\partial r} + u_z\frac{\partial u_z}{\partial z} = -\frac{1}{\rho}\frac{\partial P}{\partial z} + F_z.
\end{aligned} \tag{2.1.141}$$

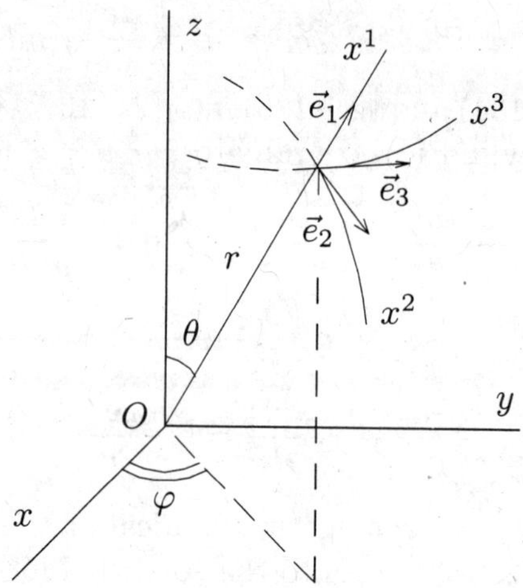

Figure 2.5: The spherical coordinate system.

2) Consider the *spherical coordinate system* (see Fig. 2.5). Introducing the tensor notations

$$x^1 = r, \quad x^2 = \theta, \quad x^3 = \varphi, \tag{2.1.142}$$

let us write the squared distance between two infinitesimally close points

$$ds^2 = (dx^1)^2 + (x^1)^2(dx^2)^2 + (x^1 \sin x^2)^2(dx^3)^2,$$

from where the formulas follow for the metric tensor components of the orthogonal spherical coordinate system

$$g_{11} = 1, \quad g_{22} = (x^1)^2, \quad g_{33} = (x^1 \sin x^2)^2, \quad g^{ii} = 1/g_{ii}, \tag{2.1.143}$$

where the summation over i is absent. Substituting expressions (2.1.143) into (2.1.131), we obtain the continuity equation in the spherical coordinates r, θ, φ:

$$\frac{\partial \rho}{\partial t} + \frac{1}{r^2}\frac{\partial}{\partial r}(r^2 \rho u_r) + \frac{1}{r \sin\theta}\frac{\partial}{\partial \theta}(\rho u_\theta \sin\theta) + \frac{1}{r\sin\theta}\frac{\partial}{\partial\varphi}(\rho u_\varphi) = 0, \tag{2.1.144}$$

where u_r, u_θ, and u_φ are the physical components of the velocity $u_i = u^i = v^i\sqrt{g_{ii}}$.

Let us find the expressions for the accelerations. We at first compute the Christoffel symbols, which are different from zero, with the aid of the *Mathematica* program `prog2-1.nb` as follows.

The Expressions for Metric Tensor Components

$$g = \begin{pmatrix} 1 & 0 & 0 \\ 0 & r^2 & 0 \\ 0 & 0 & r^2\sin^2(\theta) \end{pmatrix}$$

The Metric Tensor with Contravariant Components g^{ij}

$$g \text{ contravariant} = \begin{pmatrix} 1 & 0 & 0 \\ 0 & \frac{1}{r^2} & 0 \\ 0 & 0 & \frac{\csc^2(\theta)}{r^2} \end{pmatrix}$$

Calculation of the Christoffel Symbols $\Gamma^k{}_{ij}$, i, j, k = 1,2,3

The nonzero Christoffel symbols

$$\Gamma\_22\hat{}1 = -r$$

$$\Gamma\_33\hat{}1 = -r\sin^2(\theta)$$

$$\Gamma\_21\hat{}2 = \frac{1}{r}$$

$$\Gamma\_12\hat{}2 = \frac{1}{r}$$

$$\Gamma\_33\hat{}2 = -\cos(\theta)\sin(\theta)$$

$$\Gamma\_31\hat{}3 = \frac{1}{r}$$

$$\Gamma\_32\hat{}3 = \cot(\theta)$$

$$\Gamma\_13\hat{}3 = \frac{1}{r}$$

$$\Gamma\_23\hat{}3 = \cot(\theta)$$

Thus, we have obtained the following expressions for the nonzero Christoffel symbols:

$$\begin{aligned}
\Gamma^2_{21} &= \Gamma^2_{12} = \frac{1}{2} g^{22} \frac{\partial g_{22}}{\partial x^1} = \frac{1}{x^1}, \\
\Gamma^1_{22} &= -\frac{1}{2} g^{11} \frac{\partial g_{22}}{\partial x^1} = -x^1, \\
\Gamma^3_{13} &= \Gamma^3_{31} = \frac{1}{2} g^{33} \frac{\partial g_{33}}{\partial x^1} = \frac{1}{x^1},
\end{aligned}$$

$$\Gamma^1_{33} = -\frac{1}{2}g^{11}\frac{\partial g_{33}}{\partial x^1} = -x^1(\sin x^2)^2,$$
$$\Gamma^3_{32} = \Gamma^3_{23} = \frac{1}{2}g^{33}\frac{\partial g_{33}}{\partial x^2} = \operatorname{ctg} x^2,$$
$$\Gamma^2_{33} = -\frac{1}{2}g^{22}\frac{\partial g_{33}}{\partial x^2} = -\sin x^2 \cos x^2.$$

The substitution of these expressions into (2.1.133) and a passage to the physical components

$$a_r = \frac{dv^1}{dt}, \quad a_\theta = r\frac{dv^2}{dt}, \quad a_\varphi = \frac{dv^3}{dt} r\sin\theta$$

yields the formulas for accelerations:

$$\begin{aligned} a_r &= \frac{\partial u_r}{\partial t} + u_r\frac{\partial u_r}{\partial r} + \frac{u_\theta}{r}\frac{\partial u_r}{\partial \theta} + \frac{u_\varphi}{r\sin\theta}\frac{\partial u_r}{\partial \varphi} - \frac{1}{r}(u_\theta^2 + u_\varphi^2), \\ a_\theta &= \frac{\partial u_\theta}{\partial t} + u_r\frac{\partial u_\theta}{\partial r} + \frac{u_\theta}{r}\frac{\partial u_\theta}{\partial \theta} + \frac{u_\varphi}{r\sin\theta}\frac{\partial u_\theta}{\partial \varphi} + \frac{u_r u_\theta}{r} - \frac{\operatorname{ctg}\theta}{r}u_\varphi^2, \quad (2.1.145) \\ a_\varphi &= \frac{\partial u_\varphi}{\partial t} + u_r\frac{\partial u_\varphi}{\partial r} + \frac{u_\theta}{r}\frac{\partial u_\varphi}{\partial \theta} + \frac{u_\varphi}{r\sin\theta}\frac{\partial u_\varphi}{\partial \varphi} + \frac{u_r u_\varphi}{r} + \frac{\operatorname{ctg}\theta}{r}u_\theta u_\varphi. \end{aligned}$$

Substituting the metric coefficients g_{ii} (2.1.143) into formula (2.1.135), we can find the physical components of the vector $(\tilde{\nabla_j\sigma^{ij}})$:

$$(\tilde{\nabla_j\sigma^{1j}}) = -\frac{\partial P}{\partial r}, \quad (\tilde{\nabla_j\sigma^{2j}}) = -\frac{1}{r}\frac{\partial P}{\partial \theta}, \quad (\tilde{\nabla_j\sigma^{3j}}) = -\frac{1}{r\sin\theta}\frac{\partial P}{\partial \varphi}. \tag{2.1.146}$$

With regard for (2.1.146), the motion equations in the spherical coordinates will have the form

$$a_r = -\frac{1}{\rho}\frac{\partial P}{\partial r} + F_r, \quad a_\theta = -\frac{1}{\rho r}\frac{\partial P}{\partial \theta} + F_\theta, \quad a_\varphi = -\frac{1}{\rho r\sin\theta}\frac{\partial P}{\partial \varphi} + F_\varphi, \tag{2.1.147}$$

where the a_r, a_θ, and a_φ are determined by formulas (2.1.145). If there is no heat exchange (the adiabatic motion), then the motion equation will coincide with (2.1.139), which with regard for (2.1.143) will have the form

$$\frac{\partial S}{\partial t} + u_r\frac{\partial S}{\partial r} + \frac{u_\theta}{r}\frac{\partial S}{\partial \theta} + \frac{u_\varphi}{r\sin\theta}\frac{\partial S}{\partial \varphi} = 0. \tag{2.1.148}$$

Consider the particular case of a spherically symmetric flow of an ideal gas, then

$$u_\theta = u_\varphi = 0, \quad u_r = u_r(t, r), \quad \rho = \rho(t, r), \quad P = P(t, r); \quad S = S(t, r),$$

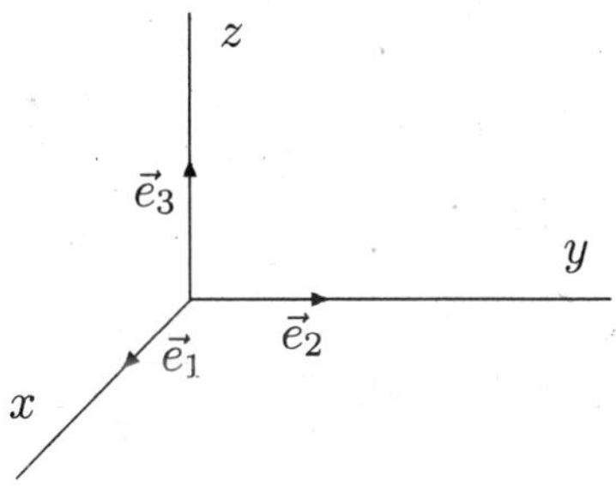

Figure 2.6: The Cartesian coordinate system.

and the system of equations (2.1.144), (2.1.145), (2.1.147), and (2.1.148) may be written in the form

$$\begin{aligned}
&\frac{\partial \rho}{\partial t} + \frac{1}{r^2}\frac{\partial}{\partial r}(r^2 \rho u_r) = 0, \\
&\frac{\partial u_r}{\partial t} + u_r \frac{\partial u_r}{\partial r} = -\frac{1}{\rho}\frac{\partial P}{\partial r} + F_r, \\
&\frac{\partial S}{\partial t} + u_r \frac{\partial S}{\partial r} = 0.
\end{aligned} \tag{2.1.149}$$

3) In the case of the *Cartesian coordinates* (see Fig. 2.6), the squared element of the length between two infinitesimally close points is equal to

$$ds^2 = (dx^1)^2 + (dx^2)^2 + (dx^3)^2,$$

where we have introduced the tensor notations

$$x^1 = x, \quad x^2 = y, \quad x^3 = z.$$

It follows from here that the metric tensor is a unit matrix, $g_{11} = g_{22} = g_{33} = g^{11} = g^{22} = g^{33} = 1$, and the remaining g_{ij} are equal to zero. For this reason, the coordinate and physical components coincide:

$$v^i = u^i = u_i, \quad a^i = a_i = \frac{dv^i}{dt}, \quad (\tilde{\nabla_j \sigma^{ij}}) = \nabla_j \sigma^{ij},$$

and the equations of continuity and motion will have the form

$$\frac{\partial \rho}{\partial t} + \frac{\partial}{\partial x}(\rho u_x) + \frac{\partial}{\partial y}(\rho u_y) + \frac{\partial}{\partial z}(\rho u_z) = 0,$$

$$\begin{aligned}
\frac{\partial u_x}{\partial t}+u_x\frac{\partial u_x}{\partial x}+u_y\frac{\partial u_x}{\partial y}+u_z\frac{\partial u_x}{\partial z}&=-\frac{1}{\rho}\frac{\partial P}{\partial x}+F_x,\\
\frac{\partial u_y}{\partial t}+u_x\frac{\partial u_y}{\partial x}+u_y\frac{\partial u_y}{\partial y}+u_z\frac{\partial u_y}{\partial z}&=-\frac{1}{\rho}\frac{\partial P}{\partial y}+F_y,\\
\frac{\partial u_z}{\partial t}+u_x\frac{\partial u_z}{\partial x}+u_y\frac{\partial u_z}{\partial y}+u_z\frac{\partial u_z}{\partial z}&=-\frac{1}{\rho}\frac{\partial P}{\partial z}+F_z.
\end{aligned} \tag{2.1.150}$$

For the adiabatic flow, it is necessary to augment the system (2.1.150) by the condition of the entropy constancy

$$\frac{\partial S}{\partial t}+u_x\frac{\partial S}{\partial x}+u_y\frac{\partial S}{\partial y}+u_z\frac{\partial S}{\partial z}=0. \tag{2.1.151}$$

In the case of a planar one-dimensional nonstationary flow, where

$$u_y=u_z=0,\;\; u_x=u_x(t,x),\;\; \rho=\rho(t,x),\;\; S=S(t,x),\;\; P=P(t,x),$$

the system (2.1.150), (2.1.151) simplifies to

$$\begin{aligned}
&\frac{\partial \rho}{\partial t}+\frac{\partial}{\partial x}(\rho u_x)=0,\\
&\frac{\partial u_x}{\partial t}+u_x\frac{\partial u_x}{\partial x}=-\frac{1}{\rho}\frac{\partial P}{\partial x}+F_x,\\
&\frac{\partial S}{\partial t}+u_x\frac{\partial S}{\partial x}=0.
\end{aligned} \tag{2.1.152}$$

In the planar stationary case $u_z=0$, $u_x=u_x(x,y)$, $u_y=u_y(x,y)$, $\rho=\rho(x,y)$, $P=P(x,y)$, and $S=S(x,y)$, and in the absence of the body forces $F_x=F_y=F_z=0$, we can find from (2.1.150) and (2.1.151) the system of equations for the ideal gas

$$\begin{aligned}
&\frac{\partial}{\partial x}(\rho u_x)+\frac{\partial}{\partial y}(\rho u_y)=0,\\
&u_x\frac{\partial u_x}{\partial x}+u_y\frac{\partial u_x}{\partial y}=-\frac{1}{\rho}\frac{\partial P}{\partial x},\\
&u_x\frac{\partial u_y}{\partial x}+u_y\frac{\partial u_y}{\partial y}=-\frac{1}{\rho}\frac{\partial P}{\partial y},\\
&u_x\frac{\partial S}{\partial x}+u_y\frac{\partial S}{\partial y}=0.
\end{aligned} \tag{2.1.153}$$

Problem 2.2. Assuming the body forces and the body heat sources to be equal to zero ($F^i=0,\; q_V=0$), write the equations of continuity (2.1.7), motion (2.1.24), and energy (2.1.50) in divergence form (the Eulerian coordinates). (Under the divergence form of equations, the following form of equations is meant: $\frac{\partial\varphi^{i\ldots s}}{\partial t}+\nabla_j\psi^{i\ldots sj}=0$, where $\varphi^{i\ldots s}$ and $\psi^{i\ldots sj}$ are some arbitrary tensor functions. Equations in divergence form are sometimes called the differential conservation laws.)

Solution: Let us multiply the continuity equation (2.1.7), which is already written in divergence form

$$\frac{\partial\rho}{\partial t}+\nabla_j\rho v^j=0, \tag{2.1.154}$$

by the velocity v^i and add to the motion equation (2.1.24):

$$\rho\Big(\frac{\partial v^i}{\partial t}+v^j\nabla_j v^i\Big)=\nabla_j\sigma^{ij}. \tag{2.1.155}$$

As a result, we obtain the motion equations in the divergence form:

$$\frac{\partial\rho v^i}{\partial t}+\nabla_j(\rho v^i v^j-\sigma^{ij})=0. \tag{2.1.156}$$

Assuming that $\dot\varepsilon_{ij}=\frac{1}{2}(\nabla_i v_j+\nabla_j v_i)$ in (2.1.50), we can rewrite the energy equation with regard for $\sigma^{ij}=\sigma^{ji}$ in the form

$$\rho\Big(\frac{\partial E}{\partial t}+v^j\nabla_j E\Big)=\sigma^{ij}\nabla_j v_i-\nabla_j\dot q^j. \tag{2.1.157}$$

Multiplying the continuity equation (2.1.154) by the specific internal energy E and adding to (2.1.157), we find:

$$\frac{\partial}{\partial t}(\rho E)+\nabla_j(\rho v^j E)+\nabla_j\dot q^j=\sigma^{ij}\nabla_j v_i. \tag{2.1.158}$$

Let us multiply the motion equation by the velocity v^i and sum over i. As a result, we obtain a single equation:

$$\rho\frac{\partial}{\partial t}\Big(\frac{v^2}{2}\Big)+\rho v^j\nabla_j\Big(\frac{v^2}{2}\Big)=v_i\nabla_j\sigma^{ij}, \tag{2.1.159}$$

where $v^2=v^i v_i=g_{ij}v^i v^j$. In the derivation of equation (2.1.159), we have used the following identities:

$$\begin{aligned}
\frac{\partial}{\partial t}\Big(\frac{v^2}{2}\Big) &\equiv \frac{1}{2}\frac{\partial}{\partial t}(g_{ij}v^i v^j)=\frac{g_{ij}}{2}\Big(v^i\frac{\partial v^j}{\partial t}+v^j\frac{\partial v^i}{\partial t}\Big)\\
&= \frac{1}{2}\Big(v_j\frac{\partial v^j}{\partial t}+v_i\frac{\partial v^i}{\partial t}\Big)=v_i\frac{\partial v^i}{\partial t};\\
\nabla_j\Big(\frac{v^2}{2}\Big) &\equiv \frac{1}{2}\nabla_j(g_{kl}v^k v^l)=\frac{g_{kl}}{2}(v^l\nabla_j v^k+v^k\nabla_j v^l)\\
&= \frac{1}{2}(v_k\nabla_j v^k+v_l\nabla_j v^l)=v_k\nabla_j v^k.
\end{aligned}$$

Multiplying the continuity equation by $\frac{v^2}{2}$ and adding to (2.1.159), we obtain the equation:

$$\frac{\partial}{\partial t}\Big(\rho\frac{v^2}{2}\Big) + \nabla_j\Big(\rho v^j\frac{v^2}{2}\Big) = v_i\nabla_j\sigma^{ij}. \qquad (2.1.160)$$

Adding equation (2.1.158) and (2.1.160), we find the energy equation in divergence form

$$\frac{\partial}{\partial t}\Big(\rho\Big(E+\frac{v^2}{2}\Big)\Big) + \nabla_j\Big(\rho v^j\,(E+\frac{v^2}{2}\Big) + \dot{q}^j - \sigma^{ij}v_i\Big) = 0. \qquad (2.1.161)$$

As will be shown in the following, the divergence form of the equations enables one to obtain the integral conservation laws and the relations at the discontinuities.

Problem 2.3. Find the difference of the specific heats $C_P - C_V$ for a given equation of state of an ideal gas $P = P(V,T)$ in the variables V and T, where $C_P = (\frac{\partial q_e}{\partial T})_P$ is the specific heat at a constant pressure, $C_V = (\frac{\partial q_e}{\partial T})_V$ is the specific heat under a constant volume, $V = 1/\rho$ is the specific volume.

Solution: For the ideal gas, the second law of thermodynamics has the form (2.1.83) (the second formula)

$$dq_e = T\,dS.$$

Substituting this equation into the definition of the specific heats, we obtain:

$$C_P = T\Big(\frac{\partial S}{\partial T}\Big)_P, \quad C_V = T\Big(\frac{\partial S}{\partial T}\Big)_V. \qquad (2.1.162)$$

Let us present the entropy S in the first equation (2.1.162) as a complex function $S = S(T, V(T,P))$. Then

$$\Big(\frac{\partial S}{\partial T}\Big)_P = \Big(\frac{\partial S}{\partial T}\Big)_V + \Big(\frac{\partial S}{\partial V}\Big)_T\Big(\frac{\partial V}{\partial T}\Big)_P.$$

Substituting this formula into the first equation in (2.1.162), we find:

$$C_P - C_V = T\Big(\frac{\partial S}{\partial V}\Big)_T\Big(\frac{\partial V}{\partial T}\Big)_P. \qquad (2.1.163)$$

Consider some isobaric process $P = \text{const}$. Then the differentiation of the equation $P = P(V,T) = \text{const}$ yields

$$0 = \Big(\frac{\partial P}{\partial T}\Big)_V + \Big(\frac{\partial P}{\partial V}\Big)_T\Big(\frac{\partial V}{\partial T}\Big)_P,$$

from where it follows that

$$\left(\frac{\partial V}{\partial T}\right)_P = -\left(\frac{\partial P}{\partial T}\right)_V \Big/ \left(\frac{\partial P}{\partial V}\right)_T. \tag{2.1.164}$$

To determine $(\frac{\partial S}{\partial V})_T$, let us make use of the condition for the total differential of the free energy F

$$dF = -S\,dT - P\,dV. \tag{2.1.165}$$

By virtue of

$$dF = \left(\frac{\partial F}{\partial T}\right)_V dT + \left(\frac{\partial F}{\partial V}\right)_T dV,$$

it follows from (2.1.165) that

$$S = -\left(\frac{\partial F}{\partial T}\right)_V, \quad P = -\left(\frac{\partial F}{\partial V}\right)_T. \tag{2.1.166}$$

From the equality of the mixed derivatives $\frac{\partial^2 F}{\partial T \partial V} = \frac{\partial^2 F}{\partial V \partial T}$ and equations (2.1.166), we obtain:

$$\left(\frac{\partial S}{\partial V}\right)_T = \left(\frac{\partial P}{\partial T}\right)_V. \tag{2.1.167}$$

Substituting (2.1.164) and (2.1.167) into (2.1.163), we obtain the desired relationship:

$$C_P - C_V = -T\left(\frac{\partial P}{\partial T}\right)_V^2 \Big/ \left(\frac{\partial P}{\partial V}\right)_T. \tag{2.1.168}$$

Since $(\frac{\partial P}{\partial V})_T < 0$, then $C_P > C_V$; i.e., the specific heat under a constant pressure is larger than the specific heat under a constant volume. For the perfect gas $PV = RT$, the Meyer's relation $C_P - C_V = R$ follows from (2.1.168).

2.2 The Hamilton–Ostrogradsky's Variational Principle in Continuum Mechanics

2.2.1 Euler–Lagrange Equations in Lagrangian Coordinates

It is well known[12–14] that the equations of the motion of a material point in the field of conservative forces can be obtained in the analytical mechanics from the Hamilton–Ostrogradsky's variational principle, which states that a certain functional termed the action reaches its extremum (local minimum) along the true trajectory. The action functional represents an integral in time of a difference of the kinetic and potential energy of a material point (particle). A similar variational principle can be formulated in continuum mechanics[7,14,15], and one can obtain the motion equations for a continuum. Unlike the standard approach based

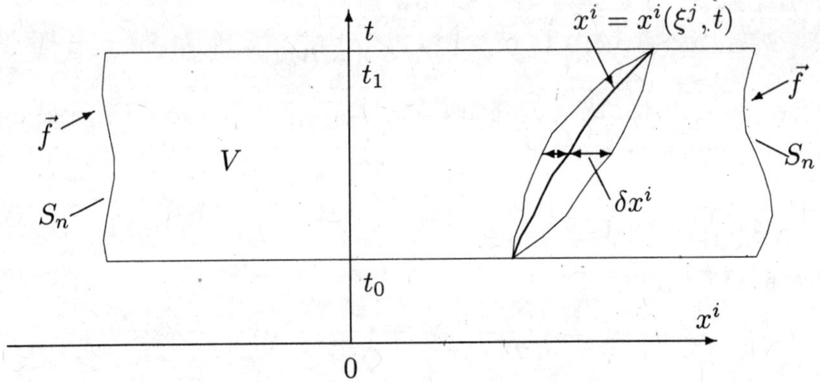

Figure 2.7: The true trajectory $x^i = x^i(\xi^j, t)$ and its variations δx^i in the action functional.

on the law of momentum conservation, the variational principle also enables one to obtain the boundary conditions and equations of state. In addition, at the derivation of equations for the media with internal microstructure, the variational principle is the simplest, and sometimes the only method for obtaining the motion equations, boundary conditions and equations of state (Section 7.7.1).

In the present section, the variational principle is applied for a formless continuum, mainly with the purpose of development of a general mathematical approach, which can be extended for the media with an internal microstructure. We now derive the motion equations, boundary conditions, and equations of state with the aid of the Hamilton–Ostrogradsky's principle. Let certain continuum perform an adiabatic motion in the field of potential forces. In this case, the Hamilton–Ostrogradsky's variational principle is valid, which asserts that the true continuum motion is determined from the condition of the minimum of action functional

$$\tilde{S} = \int_{t_0}^{t_1} \int_V L \, dV \, dt + \int_{t_0}^{t_1} A_s dt \tag{2.2.1}$$

on the set $x^i = x^i(\xi^j, t)$ determining the particles trajectories. As was noted above in Section 1.2, one can use the Eulerian and Lagrangian description. At the Lagrangian description, the accompanying coordinate system or the coordinate system of a reference frame is used. The quantity L is called the *density of the Lagrange function* or, briefly, the *Lagrangian* and is determined in classical continuum mechanics as a difference of the kinetic and potential and internal energies

$$L(x^i, \dot{x}^i, x^i_{,j}, S) = \frac{1}{2}\rho v_i v^i - U(x^i_{,j}, S) - \rho\,\Phi(x^i),$$

where $v^i = \frac{\partial x^i}{\partial t}$ is the velocity, $x^i_{,j} = \frac{\partial x^i}{\partial \xi^j}$ is the distortion, S is the entropy, U is the internal energy density, and Φ is the potential of bulk forces. The quantity A_s determines the work of surface forces at the displacement of the surface S_n by a vector $\Delta\vec{x} = \vec{x} - \vec{x}_0$

$$A_s = \int_{S_n} \vec{f} \cdot \Delta\vec{x}\, dS_n.$$

The particle trajectory $x^i = x^i(\xi^j, t)$ is varied in the functional (2.1.1) at a fixed Lagrangian coordinate $\xi^i = \text{const}$. The positions of all particles are assumed to be given at the initial moment of time t_0 and at the final moment of time t_1. Two cases are possible here. In the first case, the surface S_n bounding a volume is fixed: $\delta x^i\Big|_{S_n} = 0$, and in accordance with the Hamilton–Ostrogradsky principle (2.2.1), the variation of the particle trajectory leads to the equation (see Fig. 2.7)

$$\delta\tilde{S} = \int_{t_0}^{t_1}\int_V \delta L\, dV\, dt = 0. \tag{2.2.2}$$

In the second case, the particle trajectory is varied both inside the volume and on the boundary. The condition for the minimum of functional (2.2.1) under the trajectory variation δx^i leads to the equation

$$\delta\tilde{S} = \int_{t_0}^{t_1}\int_V \delta L\, dV\, dt + \int_{t_0}^{t_1} \delta A_s\, dt = 0. \tag{2.2.3}$$

Consider as an example the Lagrangian description in the coordinate system of the reference frame. Let the Lagrangian coordinates be given together with the Cartesian basis $\overset{o}{\vec{e}}_i$ in a reference frame $\overset{o}{\vec{e}}_i \cdot \overset{o}{\vec{e}}_j = \delta_{ij}$ and the Eulerian Cartesian coordinate system $\vec{e}_i \cdot \vec{e}_j = \delta_{ij}$. Then $\xi^i = x^i(\xi^k, t_0)$, and the displacement of particles of a continuum is equal to $u^i = x^i(\xi^k, t) - \xi^i$. It is well known that the work variation in the actual configuration is given by the formula $\delta A_s = \int_{S_n} \vec{f}_n \cdot \delta\vec{x}\, dS_n$. Using the relationships (1.3.23) $\vec{f}_n dS_n = \vec{P}_n dS^0_n$, we obtain the work variation in a reference frame

$$\delta A_s = \int_{S^0_n} \vec{P}_n \cdot \delta\vec{x}\; dS^0_n = \int_{S^0_n} P^i_n\, \delta x^i\, dS^0_n. \tag{2.2.4}$$

where the integration over the area S_0 is already performed in a reference frame. Substituting (2.2.4) in (2.2.3), we will have with regard for $dV =$

$d\xi^1 d\xi^2 d\xi^3$:

$$\int_{t_0}^{t_1}\int_{V_0} \delta L\, d\xi^1 d\xi^2 d\xi^3 dt + \int_{t_0}^{t_1}\int_{S_n^0} P_n^i \delta x^i dS_n^0\, dt = 0. \tag{2.2.5}$$

Let us vary the first item in (2.2.5):

$$\int_{t_0}^{t_1}\int_{V_0} \delta L\, dV\, dt = \int_{t_0}^{t_1}\int_{V_0}\left(\frac{\partial L}{\partial \dot{x}^i}\delta \dot{x}^i + \frac{\partial L}{\partial x^i_{,j}}\cdot \delta\left(\frac{\partial x^i}{\partial \xi^j}\right) + \frac{\partial L}{\partial x^i}\delta x^i\right) dV\, dt. \tag{2.2.6}$$

The derivatives and variations are commutative:

$$\delta \dot{x}^i = \delta\left(\frac{\partial x^i}{\partial t}\right) = \frac{\partial(\delta x^i)}{\partial t}, \quad \delta x^i_{,j} = \delta\left(\frac{\partial x^i}{\partial \xi^j}\right) = \frac{\partial(\delta x^i)}{\partial \xi^j}; \tag{2.2.7}$$

therefore,

$$\begin{aligned}
\int_{t_0}^{t_1}\int_{V_0} \frac{\partial L}{\partial \dot{x}_i}\delta \dot{x}^i dV\, dt &= \int_{t_0}^{t_1}\int_{V_0} \frac{\partial L}{\partial \dot{x}_i}\frac{\partial(\delta x^i)}{\partial t} dV\, dt \\
&= \int_{t_0}^{t_1}\int_{V_0} \frac{\partial}{\partial t}\left(\frac{\partial L}{\partial \dot{x}^i}\delta x^i\right) dV\, dt - \int_{t_0}^{t_1}\int_{V_0} \frac{\partial}{\partial t}\left(\frac{\partial L}{\partial \dot{x}^i}\right)\delta x^i dV\, dt \\
&= \int_{V_0}\left(\frac{\partial L}{\partial \dot{x}^i}\delta x^i\right)\Big|_{t_0}^{t_1} dV - \int_{t_0}^{t_1}\int_{V_0} \frac{\partial}{\partial t}\left(\frac{\partial L}{\partial \dot{x}^i}\right)\delta x^i dV\, dt.
\end{aligned}$$

Since the particle trajectory is fixed at the moments of time t_0, t_1, we have $\delta x\Big|_{t_0} = \delta x^i\Big|_{t_1} = 0$; therefore, the first integral vanishes, and we have

$$\int_{t_0}^{t_1}\int_{V_0} \frac{\partial L}{\partial \dot{x}^i}\delta \dot{x}^i dV\, dt = -\int_{t_0}^{t_1}\int_{V_0} \frac{\partial}{\partial t}\left(\frac{\partial L}{\partial \dot{x}^i}\right)\delta x^i dV\, dt. \tag{2.2.8}$$

We now transform the second integral in (2.2.6):

$$\begin{aligned}
\int_{t_0}^{t_1}\int_{V_0} \frac{\partial L}{\partial x^i_{,j}}\delta\left(\frac{\partial x^i}{\partial \xi^j}\right) dV\, dt &= \int_{t_0}^{t_1}\int_{V_0} \frac{\partial L}{\partial x^i_{,j}}\frac{\partial(\delta x^i)}{\partial \xi^j} dV\, dt \\
&= \int_{t_0}^{t_1}\int_{V_0} \frac{\partial}{\partial \xi^j}\left(\frac{\partial L}{\partial x^i_{,j}}\delta x^i\right) dV\, dt - \int_{t_0}^{t_1}\int_{V_0} \frac{\partial}{\partial \xi^j}\left(\frac{\partial L}{\partial x^i_{,j}}\right)\delta x^i dV\, dt \\
&= \int_{t_0}^{t_1}\int_{S_n^0} \frac{\partial L}{\partial x^i_{,j}} n_j^0 \delta x^i dS_n^0 dt - \int_{t_0}^{t_1}\int_{V_0} \frac{\partial}{\partial \xi^j}\left(\frac{\partial L}{\partial x^i_{,j}}\right)\delta x^i dV\, dt.
\end{aligned} \tag{2.2.9}$$

Substituting (2.2.6), (2.2.8), (2.2.9) into (2.2.5), we find:

$$\begin{aligned}
&\int_{t_0}^{t_1}\int_{V_0}\left(\frac{\partial L}{\partial x^i} - \frac{\partial}{\partial t}\left(\frac{\partial L}{\partial \dot{x}^i}\right) - \frac{\partial}{\partial \xi^j}\left(\frac{\partial L}{\partial x^i_{,j}}\right)\right)\delta x^i dV\, dt \\
&+ \int_{t_0}^{t_1}\int_{S_n^0}\left(P_n^i + \frac{\partial L}{\partial x^i_{,j}} n_j^0\right)\delta x^i\, dS_n^0\, dt = 0.
\end{aligned} \tag{2.2.10}$$

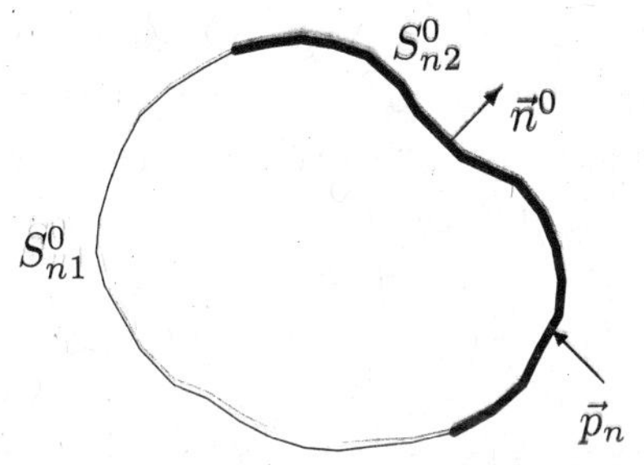

Figure 2.8: The boundary conditions for a mixed boundary-value problem.

Since the variations δx^i as well as V_0 and S^0_n are arbitrary, the *Euler-Lagrange equations*

$$\frac{\partial}{\partial t}\left(\frac{\partial L}{\partial \dot{x}^i}\right) + \frac{\partial}{\partial \xi^j}\left(\frac{\partial L}{\partial x^i_{,j}}\right) - \frac{\partial L}{\partial x^i} = 0, \tag{2.2.11}$$

boundary conditions on S_0, and equations of state

$$P^i_n = \overset{o}{\sigma}{}^{ij} n^0_j, \qquad \overset{o}{\sigma}{}^{ij} = -\frac{\partial L}{\partial x^i_{,j}}. \tag{2.2.12}$$

follow from (2.2.10). The quantity $\overset{o}{\sigma}{}^{ij}$ is called the *Piola-Kirchhoff stress tensor.*

The Euler–Lagrange equations (2.2.11) in fact represent the motion equations for a continuum. In the analytic mechanics, the corresponding Lagrange equations have the form[12–14]

$$\frac{\partial}{\partial t}\left(\frac{\partial L}{\partial \dot{x}^i}\right) - \frac{\partial L}{\partial x^i} = 0.$$

It follows from a comparison of this formula with (2.2.11) and (2.2.12) that the Euler–Lagrange equations for a continuum contain an additional item $\frac{\partial \overset{o}{\sigma}{}^{ij}}{\partial \xi^j}$ related to the action of internal stresses $\overset{o}{\sigma}{}^{ij}$. Equation (2.2.11) is the partial differential equation for the integration of which it is necessary to specify the boundary conditions on the surface S^0_n bounding the given material volume. One can specify on the surface S^0_n either the boundary positions (displacements) $\vec{x}_n = x^i_n \overset{o}{\vec{e}}_i$ – the first boundary-value problem – or the surface forces applied to the boundary $\vec{P}_n = p^i_n \overset{o}{\vec{e}}_i$ – the second boundary-value problem. A mixed problem is possible when the displacement $\vec{x}_n$ is specified on a part of the boundary

S^0_{n1} and the forces are specified on the remaining part S^0_{n2} (see Fig. 2.8). The given displacements $u^i = x^i - \xi^i$ are the boundary conditions for equations (2.2.11). If a vector of forces $\vec{P}_n$ is given on the boundary, then the boundary conditions for the Euler–Lagrange equations (2.2.11) are determined from the first equation (2.2.12).

Note that these boundary conditions refer to a reference frame in Lagrangian coordinates. The second equation in (2.2.12) is the equation of state and will be discussed in detail in the following.

As an example, let us derive the equations, which can be obtained from (2.2.11) for the Lagrange function L of a formless medium of the form

$$L = \rho_0\Big(\Big(\frac{\partial x^i}{\partial t}\Big)^2 - E(x^i_{,j}, S) - \Phi(x^i)\Big), \tag{2.2.13}$$

where $E(x^i_{,j}, S)$ is the specific internal energy. By computing the derivatives

$$\frac{\partial L}{\partial \dot{x}^i} = \rho_0 \frac{\partial x^i}{\partial t}, \quad \frac{\partial L}{\partial (x^i_{,j})} = -\rho_0 \frac{\partial E}{\partial x^i_{,j}}, \quad F^i = -\frac{\partial \Phi}{\partial x^i}$$

and substituting into (2.2.11) and (2.2.12), we obtain the motion equations:

$$\rho_0 \ddot{x}^i = \frac{\partial \overset{o}{\sigma}{}^{ij}}{\partial \xi^j} + \rho_0 F^i, \tag{2.2.14}$$

boundary conditions, and expression for the Piola–Kirchhoff tensor (equation of state) in terms of the derivatives of the *internal energy*

$$P^i_n = \overset{o}{\sigma}{}^{ij} n^0_j, \quad \overset{o}{\sigma}{}^{ij} = \rho_0 \frac{\partial E}{\partial x^i_{,j}}. \tag{2.2.15}$$

It can be seen from here that equations (2.2.14) coincide with motion equations (2.1.21) obtained from the second Newton's law written for a formless continuum.

Let us discuss in more detail the second equation in (2.2.15), which is the equation of state of a formless continuum (see Section 2.1.4). The internal energy E is usually specified as a function of the strain tensor $\hat{\varepsilon}_{ij}$, which in the case under consideration $g_{ij} = \delta_{ij}$, $g^0_{ij} = \delta_{ij}$ has the form (1.2.23)

$$\hat{\varepsilon}_{ij} = \frac{1}{2}\Big(\frac{\partial x^k}{\partial \xi^i}\frac{\partial x^k}{\partial \xi^j} - \delta_{ij}\Big). \tag{2.2.16}$$

We now rewrite the right-hand side of the second equation in (2.2.15) with the use of (2.2.16) in the form

$$\rho_0 \frac{\partial E}{\partial x^l_{,s}} = \frac{\rho_0}{2}\frac{\partial E}{\partial \hat{\varepsilon}_{ij}}\Big(\frac{\partial x^k}{\partial \xi^i}\delta_{kl}\,\delta_{js} + \frac{\partial x^k}{\partial \xi^j}\delta_{kl}\,\delta_{is}\Big) = \rho_0 \frac{\partial E}{\partial \hat{\varepsilon}_{is}}\frac{\partial x^l}{\partial \xi^i}.$$

Substituting this expression into the second equation in (2.2.15), we obtain the equation of state in a reference frame

$$\overset{o}{\sigma}{}^{ij} = \rho_0 \frac{\partial x^i}{\partial \xi^k} \frac{\partial E}{\partial \hat{\varepsilon}_{kj}}. \tag{2.2.17}$$

Multiplying (1.3.26) by $\frac{\partial \xi^l}{\partial x^k}$, we can find the expression for the Cauchy tensor in the actual configuration:

$$\hat{\sigma}^{li} = \frac{1}{\sqrt{\hat{g}}} \frac{\partial \xi^l}{\partial x^k} \overset{o}{\sigma}{}^{ki}. \tag{2.2.18}$$

Employing the continuity equation

$$\rho = \rho_0 / \sqrt{\hat{g}} \tag{2.2.19}$$

and the relationship (2.2.17), we write the equation of state in the actual configuration:

$$\hat{\sigma}^{ik} = \rho \frac{\partial E}{\partial \hat{\varepsilon}_{ik}}. \tag{2.2.20}$$

The equation of state (2.2.20) can be obtained independently from the energy equation. Let us define some process in the space of states $\hat{\varepsilon}_{ij}, S$ with the aid of infinitesimal increments (variations) of independent parameters $\delta\hat{\varepsilon}_{ij} = \dot{\hat{\varepsilon}}_{ij}\delta t$, $\delta q_e = T\,\delta S$, which lead in accordance with (2.1.49) to the variation of the specific internal energy:

$$\delta E = \frac{1}{\rho}\hat{\sigma}^{ij}\,\delta\hat{\varepsilon}_{ij} + T\,\delta S. \tag{2.2.21}$$

On the other hand, the quantity $E = E(\hat{\varepsilon}_{ij}, S)$ is a full differential of its variables

$$\delta E = \frac{\partial E}{\partial \hat{\varepsilon}_{ij}}\,\delta\hat{\varepsilon}_{ij} + \frac{\partial E}{\partial S}\,\delta S,$$

from where we obtain with regard for (2.2.21) and independence of $\delta\varepsilon_{ij}$ and δS:

$$\hat{\sigma}^{ij} = \rho \frac{\partial E}{\partial \hat{\varepsilon}_{ij}}, \qquad T = \frac{\partial E}{\partial S}. \tag{2.2.22}$$

Thus, the first equation of state in (2.2.22) coincides with equation (2.2.20) obtained from the variational principle.

2.2.2 Hamilton's Equations in Lagrangian Coordinates

As can be seen from (2.2.14), the motion equations are differential equations of the second order in time. By introducing new independent variables, coordinates, and momenta, instead of x^i, one can go over from (2.2.14) to a system of first-order equations in which the number of equations will be doubled in comparison with (2.2.14). This so-called *Hamiltonian* form of motion equations is often more convenient, since the momentum and the coordinate enter this system symmetrically. Within the framework of the Hamiltonian mechanics, the mathematical technique of Poisson brackets[12–14] has been developed, which enables one to find the integral conservation laws.

We will derive in what follows the Hamilton's equations of a continuum and the corresponding Poisson's brackets.

Let the continuum have the Lagrange function density

$$L = L(t, u^i, \dot{u}^i, u^i_{,j}, S),$$

and let us consider the adiabatic motions $S = \text{const}$, where $u^i = x^i - \xi^i$ are the components of the displacements vector $\dot{u}^i = \frac{\partial x^i}{\partial t}$, $u^i_{,j} = \frac{\partial u^i}{\partial \xi^j}$. In comparison with Section 2.2.1, we consider in what follows a more general Lagrangian, which may depend explicitly on the time t. By analogy with the analytical mechanics[12–14], we determine the momentum density

$$\pi_i = \frac{\partial L}{\partial \dot{u}^i}. \tag{2.2.23}$$

With the aid of the Legendre transformation, we can introduce the Hamilton's function density (the Hamiltonian)

$$H = \pi_i \dot{u}^i - L, \tag{2.2.24}$$

in which the independent variables are t, π_i, u^i, $u^i_{,j}$, and S:

$$H = H(t,\ \pi_i,\ u^i,\ u^i_{,j},\ S). \tag{2.2.25}$$

Assuming the volume in which the continuum is comprised to be invariable, we find the Hamilton's equations from the variational equation (2.2.2):

$$\int_{t_0}^{t_1} \delta(\pi_i \dot{u}^i - H)\, dV\, dt = 0. \tag{2.2.26}$$

It is necessary to augment equation (2.2.26) by the conditions for the absence of the variations at the moments of time t_0, t_1 and on the boundary S_n^0:

$$\delta u^i\Big|_{t_0} = \delta u^i\Big|_{t_1} = 0, \quad \delta \pi_i\Big|_{t_0} = \delta \pi_i\Big|_{t_1} = 0, \quad \delta u^i\Big|_{S_n^0} = \delta \pi_i\Big|_{S_n^0} = 0. \tag{2.2.27}$$

By varying (2.2.26) and omitting the notations of the integration limits, we obtain:

$$\iint \Big(\frac{\partial u^i}{\partial t} \delta \pi_i + \pi_i \delta \Big(\frac{\partial u^i}{\partial t} \Big) - \frac{\partial H}{\partial u^i} \delta u^i - \frac{\partial H}{\partial u^i_{,j}} \delta u^i_{,j} - \frac{\partial H}{\partial \pi_i} \delta \pi_i \Big) dV\, dt = 0. \tag{2.2.28}$$

The integration by parts similar to (2.2.7)–(2.2.9) yields

$$\iint \pi_i \delta\Big(\frac{\partial u^i}{\partial t}\Big) dV\, dt = -\iint \frac{\partial \pi_i}{\partial t} \delta u^i dV\, dt, \tag{2.2.29}$$

$$\iint \frac{\partial H}{\partial u^i_{,j}} \delta\Big(\frac{\partial u^i}{\partial \xi^j}\Big) dV\, dt = -\iint \frac{\partial}{\partial \xi^j}\Big(\frac{\partial H}{\partial u^i_{,j}}\Big) \delta u^i dV\, dt. \tag{2.2.30}$$

Substituting (2.2.29) and (2.2.30) into (2.2.28), we obtain:

$$\iint\Big(\Big(-\frac{\partial \pi_i}{\partial t} - \frac{\partial H}{\partial u^i} + \frac{\partial}{\partial \xi^j}\Big(\frac{\partial H}{\partial u^i_{,j}}\Big)\Big)\delta u^i + \Big(\frac{\partial u^i}{\partial t} - \frac{\partial H}{\partial \pi_i}\Big)\delta \pi_i\Big) dV\, dt = 0.$$

Since the variations δu^i and $\delta \pi_i$ are independent, it is necessary for the equality of this integral to zero that the expressions within the brackets affecting the variations δu^i and $\delta \pi_i$ are equal to zero separately, which implies the *Hamilton's equations*

$$\frac{\partial \pi_i}{\partial t} = -\frac{\partial H}{\partial u^i} + \frac{\partial}{\partial \xi^j}\Big(\frac{\partial H}{\partial u^i_{,j}}\Big), \quad \frac{\partial u^i}{\partial t} = \frac{\partial H}{\partial \pi_i}. \tag{2.2.31}$$

Let us introduce the following notations for the functional derivatives:

$$\frac{\delta H}{\delta u^i} = \frac{\partial H}{\partial u^i} - \frac{\partial}{\partial \xi^j}\Big(\frac{\partial H}{\partial u^i_{,j}}\Big), \quad \frac{\delta H}{\delta \pi_i} = \frac{\partial H}{\partial \pi_i}. \tag{2.2.32}$$

Now, rewrite the Hamilton's equations in the canonical form

$$\dot{\pi}_i = -\frac{\delta H}{\delta u^i}, \quad \dot{u}^i = \frac{\delta H}{\delta \pi_i}. \tag{2.2.33}$$

Hamilton's equations (2.2.33) have the first order in time; therefore, it is necessary for their integration to specify the coordinates u^i and the momentum density π_i at the initial moment of time. The boundary conditions for the Hamilton's equations coincide with the corresponding boundary conditions for the Euler–Lagrange equations (2.2.12).

Let us obtain with the aid of Hamilton's equations the energy conservation law of a continuum comprised within certain fixed volume V_0. For this purpose, let us find the total derivative with respect to time of the complete Hamiltonian $\tilde{H}$ of a continuum in a fixed volume V_0:

$$\tilde{H} = \int_{V_0} H\, dV.$$

For the Lagrange function density (2.2.13), the quantity $\tilde{H}$ is equal to the total continuum energy in the volume V_0. The total energy variation is determined by the expression

$$\frac{d\tilde{H}}{dt} = \int_{V_0}\Big(\frac{\partial H}{\partial u^i}\dot{u}^i + \frac{\partial H}{\partial u^i_{,j}}\frac{\partial}{\partial t}\Big(\frac{\partial u^i}{\partial \xi^j}\Big) + \frac{\partial H}{\partial \pi_i}\frac{\partial \pi_i}{\partial t} + \frac{\partial H}{\partial t}\Big) dV. \tag{2.2.34}$$

Integrating by parts the second item in (2.2.34), we obtain similarly to (2.2.30):

$$\int_{V_0} \frac{\partial H}{\partial u^i_{,j}} \frac{\partial \dot{u}^i}{\partial \xi^j} dV = -\int_{V_0} \dot{u}^i \frac{\partial}{\partial \xi^j}\Big(\frac{\partial H}{\partial u^i_{,j}}\Big) dV.$$

Substituting this expression into (2.2.34), we obtain:

$$\frac{d\tilde{H}}{dt} = \int_{V_0} \Big(\frac{\delta H}{\delta u^i}\dot{u}^i + \frac{\delta H}{\delta \pi_i}\dot{\pi}_i + \frac{\partial H}{\partial t}\Big) dV. \tag{2.2.35}$$

Substituting (2.2.33) into (2.2.35), we find:

$$\frac{d\tilde{H}}{dt} = \int_{V_0} \frac{\partial H}{\partial t} dV,$$

from where an important conclusion follows that the total energy of a continuum is conserved if the Lagrange function density L or the *Hamilton's density* H *of a closed system does not depend explicitly on time.* Since the law of energy conservation is confirmed by all experiments without any exception, the time cannot enter explicitly the set of the independent variables of the Lagrange function. This points to the fact that the time is homogeneous, and the motion equations are invariant with respect to the translations in time.

Let us generalize the above-developed method and find the conditions for the function $a = a(u^i, u^i_{,j}, \pi_i, t, S)$, at which it is the motion integral

$$A = \int_{V_0} a\, dV = \text{const}.$$

Computing the total time derivative $\dot{A}$ similarly to (2.2.35), we obtain the equation:

$$\frac{dA}{dt} = \int_{V_0} \Big(\frac{\partial a}{\partial t} + \frac{\delta a}{\delta u^i}\dot{u}^i + \frac{\delta a}{\delta \pi_i}\dot{\pi}_i\Big) dV = \int_{V_0} \Big(\frac{\partial a}{\partial t} + \frac{\delta a}{\delta u^i}\frac{\delta H}{\delta \pi_i} - \frac{\delta a}{\delta \pi_i}\frac{\delta H}{\delta u^i}\Big) dV. \tag{2.2.36}$$

By introducing the Poisson brackets

$$[AH] = \int_{V_0} \Big(\frac{\delta a}{\delta u^i}\frac{\delta H}{\delta \pi_i} - \frac{\delta a}{\delta \pi_i}\frac{\delta H}{\delta u^i}\Big) dV, \tag{2.2.37}$$

let us rewrite equation (2.2.36) as

$$\frac{dA}{dt} = \frac{\partial A}{\partial t} + [AH]. \tag{2.2.38}$$

If A does not depend explicitly on the time t, then it follows from (2.2.38) that

$$\frac{dA}{dt} = [AH],$$

from where we determine under the condition

$$[AH] = 0$$

that A is an integral $A = \text{const}$.

2.2.3 Euler–Lagrange Equations in Eulerian Coordinates and Murnaghan's Formula

Let us obtain the Euler–Lagrange equations from the variational principle (2.2.2) in Eulerian coordinates. The main purpose of the present section is the construction of the technique for the action variation in Eulerian coordinates and the derivation of the motion equations and equation of state of a formless medium in the case of nonlinear deformations. This equation of state is called the *Murnaghan's formula* in the literature and has a very nontrivial form. It can also be obtained by other techniques, which are discussed below. In the case of using the variational principle, however, the Murnaghan's equation is obtained in a natural way as a result of the action variation.

We will assume the Eulerian coordinate system to be a Cartesian system, i.e., $g_{ij} = \delta_{ij}$, and the Euler–Lagrange equations will follow from the variational principle

$$\delta\tilde{S} = \delta\int_{t_0}^{t_1}\int_V L(S, x^i, \dot{x}^i, x^i_{,j})\, dV\, dt = 0, \tag{2.2.39}$$

$$\delta x^i\Big|_{t_1} = \delta x^i\Big|_{t_2} = \delta x^i\Big|_{S_n} = 0,$$

where it is supposed that the volume is fixed; therefore, the variations are equal to zero on the boundary S_n. The Lagrange function density in Eulerian coordinates has the form

$$L = \rho(\frac{1}{2}v^2 - E(x^i_{,j}, S) - \Phi(x^i)), \quad v^i = \frac{\partial x^i}{\partial t}. \tag{2.2.40}$$

Let us choose the displacements $u^i(x^k, t)$ in (2.2.39) and (2.2.40) as independent functions to be determined. Assuming that the Eulerian coordinate system coincides with the coordinate system of a reference frame, we will have

$$u^i(x^k, t) = x^i(\xi^k, t) - \xi^i, \tag{2.2.41}$$

and we can rewrite the variational principle (2.2.40) in the new variables as follows:

$$\delta\tilde{S} = \delta\int_{t_0}^{t_1}\int_V L(S, x^i, u^i, \dot{u}^i, u^i_{,j})\, dV\, dt = 0, \tag{2.2.42}$$

$$\delta u^i\Big|_{t_1} = \delta u^i\Big|_{t_2} = \delta u^i\Big|_{S_n} = 0.$$

Let us express all functions in (2.2.40) in terms of the new variables. For this purpose, we define the direct b^i_j and inverse a^i_j distortion tensors

$$b^i_j = \frac{\partial x^i}{\partial \xi^j}, \ a^i_j = \frac{\partial \xi^i}{\partial x^j} = \delta^i_j - u^i_{,j} = \delta^i_j - \frac{\partial u^i}{\partial x^j}, \ b^i_j a^j_k = \delta^i_k. \qquad (2.2.43)$$

Differentiating equation (2.2.41) with respect to time, we obtain:

$$v^i = \frac{\partial u^i}{\partial t} + v^k \frac{\partial u^i}{\partial x^k},$$

from where we have

$$v^k (\delta^i_k - u^i_{,k}) = \frac{\partial u^i}{\partial t}.$$

Multiplying the both sides of this equation by b^j_i, we find with regard for (2.2.43):

$$v^j = b^j_i \frac{\partial u^i}{\partial t}. \qquad (2.2.44)$$

According to (1.4.3), the density ρ is determined by the equation

$$\rho = \rho_0 \frac{1}{\sqrt{g}} = \frac{\rho_0}{\left|\frac{\partial x}{\partial \xi}\right|} = \rho_0 \left|\frac{\partial \xi}{\partial x}\right| = \rho_0 |a^i_j|, \qquad (2.2.45)$$

where the vertical bars denote the determinant of a corresponding matrix. By varying (2.2.42) similarly to (2.2.6)–(2.2.11) we obtain:

$$\begin{aligned} \delta \tilde{S} &= \int_{t_0}^{t_1} \int_{V_0} \left(\frac{\partial L}{\partial u^i} \delta u^i + \frac{\partial L}{\partial \dot{u}^i} \delta \dot{u}^i + \frac{\partial L}{\partial u^i_{,j}} \delta u^i_{,j} \right) dV\, dt \\ &= \int_{t_0}^{t_1} \int_{V_0} \frac{\delta L}{\delta u^i} \delta u^i \, dV\, dt = 0, \end{aligned}$$

from where the Euler–Lagrange equations follow:

$$\frac{\delta L}{\delta u^i} = \frac{\partial L}{\partial u^i} - \frac{\partial}{\partial t}\left(\frac{\partial L}{\partial \dot{u}^i}\right) - \frac{\partial}{\partial x^k}\left(\frac{\partial L}{\partial u^i_{,k}}\right) = 0. \qquad (2.2.46)$$

The obtained equations are equivalent to the motion equations in Eulerian coordinates (1.4.24). To prove this fact, let us apply the technique used in[15]. Note that by virtue of the nondegeneracy of the matrix a^i_j ($|a^i_j| \neq 0$), equations (2.2.46) are equivalent to the system of equations

$$a^k_i \frac{\delta L}{\delta x^k} = 0, \qquad (2.2.47)$$

which can be written down as follows:

$$
\begin{aligned}
a_i^k \frac{\delta L}{\delta x^k} &= \frac{\partial L}{\partial u^i} - \frac{\partial}{\partial x^l}\left(\frac{\partial L}{\partial u^k_{,l}}\right)\delta_i^k - \frac{\partial}{\partial t}\left(\frac{\partial L}{\partial \dot{u}^k}\right)\delta_i^k \\
&- \frac{\partial u^k}{\partial x^i}\frac{\partial L}{\partial u^k} + \frac{\partial u^k}{\partial x^i}\frac{\partial}{\partial x^l}\left(\frac{\partial L}{\partial u^k_{,l}}\right) + \frac{\partial u^k}{\partial x^i}\frac{\partial}{\partial t}\left(\frac{\partial L}{\partial \dot{u}^i}\right) = 0. \quad (2.2.48)
\end{aligned}
$$

Using the expression for the total derivative of L with respect to the coordinate x^i

$$
\frac{\partial L}{\partial x^i} = \left.\frac{\partial L}{\partial x^i}\right|_{u^k} + \frac{\partial L}{\partial u^k}\frac{\partial u^k}{\partial x^i} + \frac{\partial L}{\partial u^k_{,l}}\frac{\partial u^k_{,l}}{\partial x^i} + \frac{\partial L}{\partial \dot{u}^k}\frac{\partial \dot{u}^k}{\partial x^i},
$$

let us rewrite (2.2.48) as follows:

$$
\left.\frac{\partial L}{\partial x^i}\right|_{u^k} + \frac{\partial L}{\partial u^i} - \frac{\partial}{\partial x^l}\left(a_i^k \frac{\partial L}{\partial u^k_{,l}} + L\delta_i^l\right) - \frac{\partial}{\partial t}\left(a_i^k \frac{\partial L}{\partial \dot{u}^k}\right) = 0. \quad (2.2.49)
$$

For the Lagrangian (2.2.40), we have:

$$
\left.\frac{\partial L}{\partial x^i}\right|_{u^k} = -\rho\frac{\partial \Phi}{\partial x^i}, \qquad \frac{\partial L}{\partial u^i} = 0, \quad (2.2.50)
$$

where we have taken into account the fact that ρ is a function of $u^i_{,j}$ (2.2.45). Using (2.2.44), let us find the derivative with respect to time:

$$
\begin{aligned}
\frac{\partial}{\partial t}\left(a_i^k \frac{\partial L}{\partial \dot{u}^k}\right) &= \frac{\partial}{\partial t}\left(a_i^k \frac{\partial}{\partial \dot{u}^k}\left(\frac{1}{2}\rho v_l v^l\right)\right) = \frac{\partial}{\partial t}\left(a_i^k \rho v_l \frac{\partial v^l}{\partial \dot{u}^k}\right) \\
&= \frac{\partial}{\partial t}(\rho v_l a_i^k b_k^l) = \frac{\partial}{\partial t}(\rho v_l \delta_i^l) = \frac{\partial}{\partial t}(\rho v_i). \quad (2.2.51)
\end{aligned}
$$

Before proceeding to the calculation of the derivative with respect to $u^k_{,l} = \frac{\partial u^k}{\partial x^l}$, let us find two auxiliary formulas. Differentiating (2.2.45) with respect to $u^k_{,j}$, we obtain:

$$
\frac{\partial \rho}{\partial u^k_{,j}} = \frac{\partial \rho}{\partial a_j^k}\frac{\partial a_j^k}{\partial u^k_{,j}} = -\frac{\partial \rho}{\partial a_j^k} = -\rho_0 \frac{\partial |a_i^s|}{\partial a_j^k}. \quad (2.2.52)
$$

The differential of the determinant $d|a_i^s|$ is equal to the differential of the element da_j^k multiplied by the corresponding cofactor. On the other hand, the inverse matrix $(a^{-1})_k^i$ is equal to the cofactors of the matrix $\| a_k^i \|$ divided by the determinant $|a_k^i|$:

$$
\frac{\partial |a_i^s|}{\partial a_j^k} = A_k^j = |a_i^s|(a^{-1})_k^j = |a_i^s| b_k^j, \quad (2.2.53)
$$

where $(a^{-1})^j_k = b^j_k$ is the inverse matrix and A^j_k is the matrix of cofactors. Substituting (2.2.53) into (2.2.52), we obtain with regard for (2.2.45):

$$\frac{\partial \rho}{\partial u^k_{,j}} = -\rho\, b^j_k. \tag{2.2.54}$$

We now determine the derivatives $\frac{\partial b^i_j}{\partial u^l_{,k}}, \frac{\partial v^2}{\partial u^k_{,j}}$. Let us differentiate equations in (2.2.43):

$$\frac{\partial b^i_j}{\partial u^l_{,s}}\, a^j_k + b^i_j\, \frac{\partial a^j_k}{\partial u^l_{,s}} = 0, \qquad \frac{\partial a^j_k}{\partial u^l_{,s}} = -\delta^j_l\, \delta^k_s.$$

We obtain from here the equation:

$$\frac{\partial b^i_j}{\partial u^l_{,s}}\, a^j_k = b^i_l\, \delta^s_k.$$

Multiplying both sides of this equation by the matrix b^k_n, we obtain:

$$\frac{\partial b^i_n}{\partial u^l_{,s}} = b^i_l\, b^s_n. \tag{2.2.55}$$

It is easy to find with the aid of (2.2.44) and (2.2.55) the second derivative:

$$\begin{aligned} a^k_i \frac{\partial}{\partial u^k_{,j}} \frac{v^2}{2} &= a^k_i\, v_l \frac{\partial v^l}{\partial u^k_{,j}} = a^k_i\, v_l\, \dot{u}^s \frac{\partial b^l_s}{\partial u^k_{,j}} \\ &= v_l\, a^k_i\, b^l_k\, b^j_s\, \dot{u}^s = v_l\, \delta^l_i\, v^j = v_i v^j. \end{aligned} \tag{2.2.56}$$

Using (2.2.40), (2.2.54), and (2.2.56), we obtain

$$\begin{aligned} a^k_i \frac{\partial L}{\partial u^k_{,l}} + L\,\delta^l_i &= a^k_i \frac{\partial \rho}{\partial u^k_{,l}} \Big(\frac{1}{2} v^2 - E - \Phi\Big) + \rho a^k_i \frac{\partial}{\partial u^k_{,l}} \frac{v^2}{2} \\ &- \rho a^k_i \frac{\partial E(u^k_{,l})}{\partial u^k_{,l}} + \rho\Big(\frac{1}{2} v^2 - E - \Phi\Big)\,\delta^l_i \\ &= -\rho a^k_i b^l_k L + \rho v_i v^l - \sigma^l_i + \rho L \delta^l_i = \rho v_i v^l - \sigma^l_i, \end{aligned} \tag{2.2.57}$$

where we have introduced the following notation for the stress tensor:

$$\sigma^l_i = a^k_i\, \rho \frac{\partial E}{\partial u^k_{,l}} = \rho(\delta^k_i - u^k_{,i}) \frac{\partial E}{\partial u^k_{,l}}. \tag{2.2.58}$$

Substituting (2.2.50), (2.2.51), and (2.2.57) into (2.2.49), we will find the motion equations

$$\frac{\partial}{\partial t}(\rho v_i) + \frac{\partial}{\partial x^l}(\rho v^s_i v^l - \sigma^l_i) + \rho \frac{\partial \Phi}{\partial x^i} = 0. \tag{2.2.59}$$

Multiplying the continuity equation (2.1.7)

$$\frac{\partial \rho}{\partial t} + \frac{\partial}{\partial x^l} \rho v^l = 0$$

by v_i^s and subtracting from (2.2.59), we obtain:

$$\rho\Big(\frac{\partial v_i}{\partial t} + v^l \frac{\partial v_i}{\partial x^l}\Big) = \frac{\partial \sigma_i^l}{\partial x^l} + \rho \frac{\partial \Phi}{\partial x^i}. \qquad (2.2.60)$$

Since the coordinates are Cartesian, $v_i = v^i$, $\sigma_i^l = \sigma^{il}$, $G^i = \rho\frac{\partial \Phi}{\partial x^i}$, and from (2.2.60) the equation follows:

$$\rho\Big(\frac{\partial v^i}{\partial t} + v^l \frac{\partial v^i}{\partial x^l}\Big) = \frac{\partial \sigma^{il}}{\partial x^l} + G^i, \qquad (2.2.61)$$

which coincides with equation (2.1.24).

For an isotropic medium, the specific internal energy is a function of the invariants of the strain tensor $E = E(I(\varepsilon_{ij}))$; therefore, it makes sense to rewrite formula (2.2.58) in such a form that it includes the derivatives of energy with respect to the components of the strain tensor ε_{ij}. In the Eulerian Cartesian coordinates, the strain tensor has the form (1.2.26)

$$\begin{aligned}\varepsilon_{ij} &= \frac{1}{2}\Big(\delta_{ij} - \frac{\partial \xi^k}{\partial x^i}\frac{\partial \xi^k}{\partial x^j}\Big) = \frac{1}{2}(\delta_{ij} - (\delta_i^k - u^k_{,i})(\delta_j^k - u^k_{,j})) \\ &= \frac{1}{2}(u^i_{,j} + u^j_{,i}) - \frac{1}{2}u^k_{,i}u^k_{,j},\end{aligned}$$

from where it follows that

$$\frac{\partial E}{\partial u^k_{,l}} = \frac{\partial E}{\partial \varepsilon_{ij}}\frac{\partial \varepsilon_{ij}}{\partial u^k_{,l}} = \frac{\partial E}{\partial \varepsilon_{kl}} - u^k_{,j}\frac{\partial E}{\partial \varepsilon_{jl}}. \qquad (2.2.62)$$

Substituting (2.2.62) into (2.2.58), we find the Murnaghan's formula

$$\sigma^{ij} = \rho\Big(\frac{\partial E}{\partial \varepsilon_{ij}} - 2\varepsilon_{ik}\frac{\partial E}{\partial \varepsilon_{kj}}\Big). \qquad (2.2.63)$$

In contrast with (2.2.58), the Murnaghan's formula is valid only for an isotropic medium, since the distortion tensor b_i^j enters the expression for E in Eulerian coordinates along with ε_{ij} for an anisotropic medium.

Since the motion equations were previously obtained from the conservation laws, it can seem at first sight that it is impossible to obtain with the aid of a variational principle anything new; however, this is not so. So far we have considered the "simplest" continuum, which does not possess the internal microstructure. While proceeding to the media with microstructure, internal dynamical variables appear, the differential equations for which can easily be obtained from the variational principle. This way of obtaining the equations is sometimes the only possible way. The variational principle is used in the following in a natural way for finding the equations in the media with microstructure of the type of a bubbly fluid. One can find with the aid of the variational principle the boundary conditions (of the Cauchy type) and expressions for the stress tensor in terms of the derivatives of the internal energy with respect to the tensor of strain or distortion. In addition, the variational principle enables one to elucidate a relation between the conservation laws and the isotropicity properties and the homogeneity of space and time, which is the subject of the next section.

Problem 2.3. Assume that the specific internal energy E depends on the strain deformation ε_{ij} and the entropy S. Obtain the Murnaghan's formula from the energy conservation law.

Solution: Let us write the energy conservation law in Eulerian coordinates (2.1.51) and the second law of thermodynamics (2.1.60) in the form of infinitesimal increments

$$\delta E = \frac{1}{\rho}\sigma^{ij}\,\dot{\varepsilon}_{ij}\,\delta t + T\,\delta S, \tag{2.2.64}$$

$$\delta E = \frac{\partial E}{\partial \varepsilon_{ij}}\delta\varepsilon_{ij} + \frac{\partial E}{\partial S}\,\delta S, \quad \delta x^l = v^l\,\delta t, \tag{2.2.65}$$

where δt and δS have the independent variations (increments). The strain variation $\delta\varepsilon_{ij} \neq \dot{\varepsilon}_{ij}\delta t$. This is related to the fact that the first equation in (1.3.54) has been obtained in the Lagrangian coordinate system; therefore, the relationship

$$\delta\hat{\varepsilon}_{ij} = \dot{\hat{\varepsilon}}_{ij}\,\delta t$$

is valid also in the Lagrangian coordinate system. In the Eulerian coordinates, however,

$$\dot{\varepsilon}_{ij}\delta t = \frac{1}{2}\Big(\frac{\partial v^i}{\partial x^j} + \frac{\partial v^j}{\partial x^i}\Big)\delta t = \frac{1}{2}\Big(\frac{\partial(\delta x^i)}{\partial x^j} + \frac{\partial(\delta x^j)}{\partial x^i}\Big), \tag{2.2.66}$$

where we have used the last equation in (2.2.65). Multiplying the left- and right-hand sides of equation (2.2.66) by $\frac{1}{\rho}\sigma^{ij}$, we find with regard for the symmetry $\sigma^{ij} = \sigma^{ji}$:

$$\frac{1}{\rho}\sigma^{ij}\dot{\varepsilon}_{ij}\,\delta t = \frac{1}{\rho}\sigma^{ij}\frac{\partial}{\partial x^j}(\delta x^i). \tag{2.2.67}$$

Let us express the variation of the strain tensor δ_{ij} in terms of the displacement variation δx^l. According to (1.2.26), we have in Eulerian coordinates that

$$\varepsilon_{ij} = \frac{1}{2}\Big(\delta_{ij} - \frac{\partial \xi^k}{\partial x^i}\frac{\partial \xi^k}{\partial x^j}\Big),$$

from where it follows that

$$\delta\varepsilon_{ij} = -\frac{1}{2}\Big(\Big(\frac{\partial \xi^k}{\partial x^i}\Big)\delta\Big(\frac{\partial \xi^k}{\partial x^j}\Big) + \Big(\frac{\partial \xi^k}{\partial x^j}\Big)\delta\Big(\frac{\partial \xi^k}{\partial x^i}\Big)\Big). \qquad (2.2.68)$$

At the particle motion, $\xi^k = \text{const}$; therefore,

$$\frac{\partial \xi^k}{\partial t} + v^l\frac{\partial \xi^k}{\partial x^l} = 0.$$

Differentiating this expression with respect to x^i, we obtain:

$$\frac{\partial}{\partial t}\Big(\frac{\partial \xi^k}{\partial x^i}\Big) + v^l\frac{\partial}{\partial x^l}\Big(\frac{\partial \xi^k}{\partial x^i}\Big) + \frac{\partial v^l}{\partial x^i}\frac{\partial \xi^k}{\partial x^l} = \frac{d}{dt}\Big(\frac{\partial \xi^k}{\partial x^i}\Big) + \frac{\partial \xi^k}{\partial x^l}\frac{\partial v^l}{\partial x^i} = 0,$$

from where we find that

$$\delta\Big(\frac{\partial \xi^k}{\partial x^i}\Big) = -\frac{\partial \xi^k}{\partial x^l}\frac{\partial}{\partial x^i}(\delta x^l). \qquad (2.2.69)$$

Substituting (2.2.69) into (2.2.68), we obtain:

$$\delta\varepsilon_{ij} = \frac{1}{2}\Big((\delta_{jl} - 2\varepsilon_{jl})\frac{\partial(\delta x^l)}{\partial x^i} + (\delta_{il} - 2\varepsilon_{il})\frac{\partial(\delta x^l)}{\partial x^j}\Big). \qquad (2.2.70)$$

Let us substitute (2.2.67) into (2.2.64) and (2.2.70) into (2.2.65). Equalling then (2.2.64) and (2.2.65), we find that

$$\frac{\partial E}{\partial \varepsilon_{ij}}(\delta_{il} - 2\varepsilon_{il})\frac{\partial(\delta x^l)}{\partial x^j} + \frac{\partial E}{\partial S}\delta S = \frac{\sigma^{lj}}{\rho}\frac{\partial(\delta x^l)}{\partial x^j} + T\,\delta S,$$

from where it follows by virtue of the independency of δx^l and δS that

$$\sigma^{lj} - \rho(\delta_{il} - 2\varepsilon_{il})\frac{\partial E}{\partial \varepsilon_{ij}}, \quad T = \frac{\partial E}{\partial S}. \qquad (2.2.71)$$

It is seen that the first equation in (2.2.71) coincides with Murnaghan's formula (2.2.63).

Problem 2.4. Let the specific internal energy be a function of the density ρ and the entropy S. Find the stress tensor σ^{ij} by using the Murnaghan's formula.

Solution: Let us write the Eulerian strain tensor in the form

$$\varepsilon_{ij} = \frac{1}{2}(\delta_{ij} - \tilde{g}_{ij}), \tag{2.2.72}$$

where

$$\tilde{g}_{ij} = \frac{\partial \xi^k}{\partial x^i}\frac{\partial \xi^k}{\partial x^j}. \tag{2.2.73}$$

Taking into account (2.2.72) and (2.2.73), we can rewrite formula (2.2.71) as follows:

$$\sigma^{ij} = -2\rho\tilde{g}_{ik}\frac{\partial E}{\partial \tilde{g}_{kj}}. \tag{2.2.74}$$

It follows from the mass conservation law written for an infinitesimal volume dV that

$$\rho\, dV = \rho_0\, dV_0, \quad dV = \left|\frac{\partial x^i}{\partial \xi^j}\right| dV_0, \tag{2.2.75}$$

where $|\frac{\partial x^i}{\partial \xi^i}|$ is the Jacobian of the transformation $x^i = x^i(\xi^j)$. Substituting (2.2.73) into (2.2.75), we obtain:

$$\rho = \rho_0 / \left|\frac{\partial x^j}{\partial \xi^i}\right| = \rho_0 \left|\frac{\partial \xi^i}{\partial x^j}\right| = \rho_0\sqrt{\tilde{g}}, \tag{2.2.76}$$

where the last equality follows from a linear algebra theorem, which states that the determinant of the product of matrices is equal to the product of the determinants of these matrices:

$$\tilde{g} = \left|\frac{\partial \xi^k}{\partial x^i}\frac{\partial \xi^k}{\partial x^j}\right| = \left|\frac{\partial \xi^k}{\partial x^j}\right|^2.$$

Thus, the derivative of the energy E is equal to

$$\frac{\partial E}{\partial \tilde{g}_{kj}} = \frac{1}{2}\frac{\partial E}{\partial \rho}\frac{\rho_0}{\sqrt{\tilde{g}}}\frac{\partial \tilde{g}}{\partial \tilde{g}_{kj}} = \frac{1}{2}\frac{\partial E}{\partial \rho}\rho_0\sqrt{\tilde{g}}\,\tilde{g}^{-1}_{kj}, \tag{2.2.77}$$

where we have used formula (2.2.53), which has the following form in the given notations:

$$\frac{\partial \tilde{g}}{\partial \tilde{g}_{kj}} = \tilde{g}\,\tilde{g}^{-1}_{kj}, \qquad \tilde{g}_{ik}\,\tilde{g}^{-1}_{kj} = \delta_{ij}.$$

Substituting (2.2.77) in (2.2.74), we finally obtain:

$$\sigma^{ij} = -\rho^2\frac{\partial E}{\partial \rho}\delta^{ij}. \tag{2.2.78}$$

Problem 2.5. Find the Euler equations in an accompanying coordinate system from the Hamilton–Ostrogradsky's variational principle.

Solution: The accompanying coordinate system is determined by the metric tensor $\hat{g}_{ij} = \hat{\vec{e}}_i \cdot \hat{\vec{e}}_j$ and the conditions $\xi^k =$ const. Let us determine the Eulerian Cartesian coordinate system $g_{ij} = \vec{e}_i \cdot \vec{e}_j = \delta_{ij}$, in which the motion law will have the form

$$x^i = x^i(\xi^j, t), \quad x^i\Big|_{t=0} = \xi^i. \tag{2.2.79}$$

The variational principle (2.2.3) in the accompanying coordinate system can be written in the form

$$\begin{aligned} \delta\tilde{S} &= \delta\tilde{S}' + \int_{t_0}^{t_1} \delta A_s \, dt \\ &= \delta \int_{t_0}^{t_1} \int_V \rho\sqrt{\hat{g}} L(x^i, \dot{x}^i, \hat{g}_{ij}, S) \, d\tau \, dt + \int_{t_0}^{t_1} \delta A_s \, dt = 0, \end{aligned} \tag{2.2.80}$$

where $\delta x^i|_{t_1} = \delta x^i|_{t_2} = 0$, $d\tau = d\xi^1 d\xi^2 d\xi^3$, $dV = \sqrt{\hat{g}}\, d\tau$. In the accompanying Lagrangian coordinates, the following relationships are valid:

$$\rho\sqrt{\hat{g}} = \rho_0, \; L(x^i, \dot{x}^i, \hat{g}_{ij}, S) = \frac{1}{2}\left(\frac{\partial x^k}{\partial t}\right)^2 - E(\hat{g}_{ij}(x^k), S) - \Phi(x^k). \tag{2.2.81}$$

The work of surface forces due to the variation on the boundary is equal to

$$\delta A_s = \int_{S_n} \vec{f} \cdot \delta\vec{x} \, dS_n = \int_{S_n} f^j \delta x^k \, \hat{\vec{e}}_j \cdot \vec{e}_k \, dS_n = \int_{S_n} f^j \frac{\partial x^k}{\partial \xi^j} \delta x^k \, dS_n. \tag{2.2.82}$$

With regard for (2.2.81), the variation of the first item in (2.2.80) is equal to

$$\delta\tilde{S}' = \int_{t_0}^{t_1} \int_V \rho_0 \left(\frac{\partial L}{\partial x^k} \delta x^k + \frac{\partial L}{\partial \dot{x}^k} \delta \dot{x}^k + \frac{\partial L}{\partial \hat{g}_{ij}} \delta \hat{g}_{ij} \right) d\tau \, dt. \tag{2.2.83}$$

Using the second equation in (2.2.81), we obtain:

$$\frac{\partial L}{\partial x^k} \delta x^k = -\frac{\partial \Phi}{\partial x^k} \delta x^k, \tag{2.2.84}$$

$$\frac{\partial L}{\partial \dot{x}^k} \delta \dot{x}^k = \frac{\partial}{\partial t}\left(\frac{\partial L}{\partial \dot{x}^k} \delta x^k \right) - \frac{\partial}{\partial t}\left(\frac{\partial L}{\partial \dot{x}^k} \right) \delta x^k = \frac{\partial}{\partial t}\left(\frac{\partial x^k}{\partial t} \delta x^k \right) - \frac{\partial^2 x^k}{\partial t^2} \delta x^k. \tag{2.2.85}$$

Since

$$\hat{g}_{ij} = \hat{\vec{e}}_i \cdot \hat{\vec{e}}_j = \frac{\partial x^k}{\partial \xi^i} \frac{\partial x^k}{\partial \xi^j},$$

we find with regard for the equation

$$\delta\frac{\partial x^k}{\partial \xi^i} = \frac{\partial(\delta x^k)}{\partial \xi^i}$$

that

$$\delta\hat{g}_{ij} = \frac{\partial(\delta x^k)}{\partial \xi^i}\frac{\partial x^k}{\partial \xi^j} + \frac{\partial x^k}{\partial \xi^i}\frac{\partial(\delta x^k)}{\partial \xi^j}.$$

By virtue of the symmetry of the tensor $\frac{\partial E}{\partial \hat{g}_{ij}}$, we obtain:

$$\frac{\partial L}{\partial \hat{g}_{ij}}\,\delta\hat{g}_{ij} = -\frac{\partial E}{\partial \hat{g}_{ij}}\,\delta\hat{g}_{ij} = -2\frac{\partial E}{\partial \hat{g}_{ij}}\frac{\partial x^k}{\partial \xi^j}\frac{\partial(\delta x^k)}{\partial \xi^i}$$
$$= -\frac{\partial}{\partial \xi^i}\Big(2\frac{\partial E}{\partial \hat{g}_{ij}}\frac{\partial x^k}{\partial \xi^j}\,\delta x^k\Big) + \frac{\partial}{\partial \xi^i}\Big(2\frac{\partial E}{\partial \hat{g}_{ij}}\frac{\partial x^k}{\partial \xi^j}\Big)\delta x^k. \qquad (2.2.86)$$

Since $\hat{\varepsilon}_{ij} = \frac{1}{2}(\hat{g}_{ij} - \delta_{ij})$, then

$$\frac{\partial E}{\partial \hat{\varepsilon}_{ij}} = 2\frac{\partial E}{\partial \hat{g}_{ij}}. \qquad (2.2.87)$$

Substituting (2.2.84)–(2.2.87) into (2.2.83), we find with regard for the equalities $\delta x^i|_{t_1} = \delta x^i|_{t_2} = 0$ that

$$\delta\tilde{S}' = \int_{t_0}^{t_1}\int_V \Big(-\rho_0\frac{\partial \Phi}{\partial x^k} - \rho_0\frac{\partial^2 x^k}{\partial t^2} + \frac{\partial}{\partial \xi^i}\Big(\rho_0\frac{\partial E}{\partial \hat{\varepsilon}_{ij}}\frac{\partial x^k}{\partial \xi^j}\Big)\Big)\delta x^k\,d\tau\,dt$$
$$- \int_{t_0}^{t_1}\int_{S_n} \rho\frac{\partial E}{\partial \hat{\varepsilon}_{ij}}\frac{\partial x^k}{\partial \xi^j} n_i\delta x^k\,dS_n\,dt. \qquad (2.2.88)$$

The last item in (2.2.88) has been obtained with the aid of the Ostrogradsky–Gauss theorem in the following way. Let us introduce for convenience the quantity

$$H^i = \frac{\partial E}{\partial \hat{\varepsilon}_{ij}}\frac{\partial x^k}{\partial \xi^j}\,\delta x^k.$$

Since the divergence $\frac{\partial H^i}{\partial \xi^i}$ does not depend on the coordinate system, we at first go over to the surface integral in a coordinate system of the reference frame, and by using the formula (1.3.22) $n_i\,dS_n = \sqrt{\hat{g}}\,n_i^0\,dS_n$, we then go over to the accompanying coordinate system:

$$\int_V \rho_0\frac{\partial H^i}{\partial \xi^i}\,d\tau\,dt = \int_{S_n^0} \rho_0 H^i n_i^0 dS_n^0 = \int_{S_n} \frac{\rho_0}{\sqrt{\hat{g}}} H^i n_i dS_n = \int_{S_n} \rho H^i n_i dS_n.$$

Substituting (2.2.82) and (2.2.88) into (2.2.80) and equalling to zero the surface integrals and the volume integrals, we obtain:

$$\frac{\partial^2 x^k}{\partial t^2} = \frac{\partial}{\partial \xi^i}\Big(\frac{\partial E}{\partial \hat{\varepsilon}_{ij}}\frac{\partial x^k}{\partial \xi^j}\Big) - \frac{\partial \Phi}{\partial x^k}, \qquad (2.2.89)$$

$$\hat{\sigma}^{ij} = \rho \frac{\partial E}{\partial \hat{\varepsilon}_{ij}}, \quad f^j = \hat{\sigma}^{ji} n_i. \tag{2.2.90}$$

Substituting (2.2.90) into (2.2.89), we obtain with regard for the continuity equation the motion equations

$$\rho_0 \frac{\partial^2 x^k}{\partial t^2} = \frac{\partial}{\partial \xi^i} \left(\sqrt{\hat{g}}\, \hat{\sigma}^{ij} \frac{\partial x^k}{\partial \xi^j} \right) + \rho_0 F^k, \tag{2.2.91}$$

which by virtue of the symmetry $\hat{\sigma}^{ij} = \hat{\sigma}^{ji}$ coincide with (1.3.20). Equation (2.2.90) enables one to find the stresses in a medium if the specific internal energy is given as a function of $\hat{\varepsilon}_{ij}$, S, and it also coincides with the corresponding equation (2.2.20).

2.3 Conservation Laws for Energy and Momentum in Continuum Mechanics

2.3.1 Conservation Laws in Cartesian Coordinates

The equations for the motion and energy of a continuum, which were obtained in Section 2.1, are a consequence of the integral conservation laws for momentum and energy. The existence of the conservation laws is related to the isotropicity and homogeneity of space and time and invariance of the action under the transformation of spatiotemporal coordinates.

Consider a continuum filling the overall space and representing a closed system, so that there is no external potential field. In this case, it follows from the condition of space and time homogeneity that the Lagrange function will not depend on the coordinates and time: $L = L(\dot{u}^i, u^i_{,j}, S)$, where $u^i = x^i - \xi^i$ are the components of the displacement vector in a Cartesian coordinate system (it is assumed similarly to Section 2.2 that the Eulerian Cartesian coordinate system coincides with the Cartesian coordinate system in the reference frame). The entropy S is assumed to be constant; therefore, it will not be included in what follows in the set of independent variables. [Note that, if a continuum is in an external potential field with a vector potential $A^i(t, x^k)$, the Lagrange function will also depend on the coordinate x^k and the time t; therefore, the conservation laws for a continuum will not be satisfied. The violation of the conservation laws for a continuum is related here to the fact that the continuum interacts with the external field and is no closed system. As was shown in[16,17], it is necessary to consider in this case a closed system field, continuum, for which the law of conservation of the sum of the energy and momentum of the field and continuum takes place.]

With regard for this remark, let us obtain the conservation laws for a formless continuum. The expression for the action in Lagrangian coordinates in a Cartesian basis of the reference frame has the form

$$\tilde{S} = \iint L(\dot{u}^i, u^i_{,j})\, d\xi^1\, d\xi^2\, d\xi^3\, dt, \tag{2.3.1}$$

where the integration is performed over the total space. It is necessary to require for the convergence of the integral that $L \to 0$ at $\xi^i \to \infty$. Let the coordinates and time be subject to small (infinitesimal) transformations

$$\xi'^i = \xi^i + \delta\xi^i(t, \xi^j), \quad t' = t + \delta t(t, \xi^j), \tag{2.3.2}$$

at which the functional $\tilde{S}$ is transformed into a certain functional $\tilde{S}'$. Consider only those functionals that do not change (remain invariable) under the transformations (2.3.2):

$$\iint L\Big(\frac{\partial u^i}{\partial t}, \frac{\partial u^i}{\partial \xi^j}\Big) d\xi^1\, d\xi^2\, d\xi^3\, dt = \iint L\Big(\frac{\partial u'^i}{\partial t'}, \frac{\partial u'^i}{\partial \xi'^j}\Big) d\xi'^1\, d\xi'^2\, d\xi'^3\, dt'. \tag{2.3.3}$$

Since the "new" volume $dV'\,dt'$ is equal to the product of the Jacobian and $dV\,dt$, we have:

$$\begin{aligned} d\xi'^1\, d\xi'^2\, d\xi'^3\, dt' &= \frac{\partial(\xi'^1, \xi'^2, \xi'^3, t')}{\partial(\xi^1, \xi^2, \xi^3, t)}\, d\xi^1\, d\xi^2\, d\xi^3\, dt \\ &\approx \Big(1 + \frac{\partial}{\partial t}\delta t + \frac{\partial}{\partial \xi^i}\delta\xi^i\Big) d\xi^1\, d\xi^2\, d\xi^3\, dt \\ &= \Big(1 + \frac{\partial}{\partial t}\delta t + \frac{\partial}{\partial \xi^i}\delta\xi^i\Big) dV\, dt. \end{aligned} \tag{2.3.4}$$

The displacement vector changes under the transformations (2.3.2) as follows:

$$u'^i(\xi'^j) = u^i(\xi^j) + \delta u^i(\xi^j). \tag{2.3.5}$$

Let us present the variations δu^i as

$$\delta u^i(\xi^j) = \bar{\delta} u^i(\xi^j) + \frac{\partial u^i}{\partial \xi^j}\delta\xi^j + \frac{\partial u^i}{\partial t}\delta t, \tag{2.3.6}$$

where $\bar{\delta} u^i(\xi^j) = u'^i(\xi^j) - u^i(\xi^j)$ is called the variation of the form of the function u^i or, briefly, the Lie variation. By introducing the notations

$$L'\Big(\frac{\partial u'^i}{\partial t'}, \frac{\partial u'^i}{\partial \xi'^j}\Big) = L'(\xi'^j), \quad L\Big(\frac{\partial u^i}{\partial t}, \frac{\partial u^i}{\partial \xi^j}\Big) = L(\xi^j),$$

let us find the variation of the Lagrange density function

$$L'(\xi'^j) = L(\xi^j) + \delta L(\xi^j), \tag{2.3.7}$$

where the variation of the Lagrangian is

$$\delta L(\xi^j) = \bar{\delta} L(\xi^j) + \frac{\partial L}{\partial \xi^j}\,\delta\xi^j + \frac{\partial L}{\partial t}\,\delta t \tag{2.3.8}$$

and the Lie variation is equal to

$$\bar{\delta} L = \frac{\partial L}{\partial u^i_{,j}}\,\bar{\delta} u^i_{,j} + \frac{\partial L}{\partial \dot{u}^i}\,\bar{\delta}\dot{u}^i. \tag{2.3.9}$$

Since the Lie differentiation and variation operators are commutative:

$$\bar{\delta} u^i_{,j} \neq \frac{\partial}{\partial \xi^j}\bar{\delta} u^i, \quad \bar{\delta}\dot{u}^i = \frac{\partial}{\partial t}\bar{\delta} u^i, \tag{2.3.10}$$

we have that

$$\frac{\partial L}{\partial u^i_{,j}}\bar{\delta} u^i_{,j} = \frac{\partial}{\partial \xi^j}\Big(\frac{\partial L}{\partial u^i_{,j}}\bar{\delta} u^i\Big) - \frac{\partial}{\partial \xi^j}\Big(\frac{\partial L}{\partial u^i_{,j}}\Big)\bar{\delta} u^i, \tag{2.3.11}$$

$$\frac{\partial L}{\partial \dot{u}^i}\bar{\delta}\dot{u}^i = \frac{\partial}{\partial t}\Big(\frac{\partial L}{\partial \dot{u}^i}\bar{\delta} u^i\Big) - \frac{\partial}{\partial t}\Big(\frac{\partial L}{\partial \dot{u}^i}\Big)\bar{\delta} u^i. \tag{2.3.12}$$

Substituting (2.3.11) and (2.3.12) into (2.3.9), we rewrite (2.3.9) in the form

$$\bar{\delta} L = \Big(-\frac{\partial}{\partial \xi^j}\Big(\frac{\partial L}{\partial u^i_{,j}}\Big) - \frac{\partial}{\partial t}\Big(\frac{\partial L}{\partial \dot{u}^i}\Big)\Big)\bar{\delta} u^i + \frac{\partial}{\partial \xi^j}\Big(\frac{\partial L}{\partial u^i_{,j}}\bar{\delta} u^i\Big) + \frac{\partial}{\partial t}\Big(\frac{\partial L}{\partial \dot{u}^i}\bar{\delta} u^i\Big). \tag{2.3.13}$$

Using (2.3.4), (2.3.7), and (2.3.8), we can rewrite equation (2.3.3) as follows:

$$\begin{aligned}
&\iint \Big\{(L+\delta L)\Big(1 + \frac{\partial}{\partial t}\delta t + \frac{\partial}{\partial \xi^i}\delta\xi^i\Big) - L\Big\}\,dV\,dt \\
&= \iint \Big\{\bar{\delta} L + \frac{\partial L}{\partial t}\delta t + \frac{\partial L}{\partial \xi^j}\delta\xi^j + L\frac{\partial}{\partial t}\delta t + L\frac{\partial}{\partial \xi^j}\delta\xi^j\Big\}\,dV\,dt \\
&= \iint \Big\{\bar{\delta} L + \frac{\partial}{\partial t}\,(L\delta t) + \frac{\partial}{\partial \xi^j}\,(L\delta\xi^j)\Big\}\,dV\,dt = 0.
\end{aligned}$$

Substituting here, instead of $\bar{\delta} L$, the right-hand side of equation (2.3.13), we obtain:

$$\begin{aligned}
&\iint \Big\{\Big(-\frac{\partial}{\partial \xi^j}\Big(\frac{\partial L}{\partial u^i_{,j}}\Big) - \frac{\partial}{\partial t}\Big(\frac{\partial L}{\partial \dot{u}^i}\Big)\Big)\bar{\delta} u^i \\
&+ \frac{\partial}{\partial t}\Big(L\,\delta t + \frac{\partial L}{\partial \dot{u}^i}\,\bar{\delta} u^i\Big) + \frac{\partial}{\partial \xi^j}\Big(L\,\delta\xi^j + \frac{\partial L}{\partial u^i_{,j}}\,\bar{\delta} u^i\Big)\Big\}\,dV\,dt = 0.
\end{aligned} \tag{2.3.14}$$

The motion equations of a continuum in the reference frame basis have been obtained in the foregoing section [see (2.2.11)]. Using the relations

$$u^i = x^i - \xi^i, \quad \dot{x}^i = \dot{u}^i, \quad x^i_{,j} = 1 + u^i_{,j}, \quad L = L(u^i, u^i_{,j}),$$

let us rewrite (2.2.11) as

$$\frac{\partial}{\partial t}\left(\frac{\partial L}{\partial \dot{u}^i}\right) + \frac{\partial}{\partial \xi^j}\left(\frac{\partial L}{\partial u^i_{,j}}\right) = 0. \tag{2.3.15}$$

If the real motions are considered, for which the Euler–Lagrange equations (2.3.15) are valid, then equation (2.3.14) takes the form

$$\iint \left\{\frac{\partial}{\partial t}\left(L\,\delta t + \frac{\partial L}{\partial \dot{u}^i}\,\bar{\delta} u^i\right) + \frac{\partial}{\partial \xi^j}\left(L\,\delta\xi^j + \frac{\partial L}{\partial u^i_{,j}}\,\bar{\delta} u^i\right)\right\} dV\,dt = 0,$$

from which it follows that

$$\frac{\partial}{\partial t}\left(L\,\delta t + \frac{\partial L}{\partial \dot{u}^i}\,\bar{\delta} u^i\right) + \frac{\partial}{\partial \xi^j}\left(L\,\delta\xi^j + \frac{\partial L}{\partial u^i_{,j}}\,\bar{\delta} u^i\right) = 0, \tag{2.3.16}$$

where in accordance with (2.3.6)

$$\bar{\delta} u^i = \delta u^i - \frac{\partial u^i}{\partial t}\,\delta t - \frac{\partial u^i}{\partial \xi^j}\,\delta\xi^j. \tag{2.3.17}$$

Formulas (2.3.16) and (2.3.17) can be extended for a more general class of transformations. Let us introduce for convenience the following notations:

$$x^\alpha = (x^0, x^1, x^2, x^3) = (t, \xi^1, \xi^2, \xi^3), \tag{2.3.18}$$

in which the Greek letters denote the components of the four-dimensional vector a^α $(\alpha = 0, 1, 2, 3)$ and the Latin letters denote the components of the three-dimensional vector b^i $(i = 1, 2, 3)$. In this notation, equations (2.3.16) and (2.3.17) can be rewritten in the form

$$\frac{\partial}{\partial x^\alpha}\left(L\,\delta x^\alpha + \frac{\partial L}{\partial(\partial u^i/\partial x^\alpha)}\,\bar{\delta} u^i\right) = 0, \tag{2.3.19}$$

$$\bar{\delta} u^i = \delta u^i - \frac{\partial u^i}{\partial x^\beta}\,\delta x^\beta. \tag{2.3.20}$$

Consider instead of (2.3.2) the transformations of a more general form depending on n small parameters ε^α

$$x'^\beta = X^\beta\left(x^\gamma, u^j, \frac{\partial u^j}{\partial x^\delta}, \varepsilon^\alpha\right), \quad u'^i = Y^i\left(x^\gamma, u^j, \frac{\partial u^j}{\partial x^\delta}, \varepsilon^\alpha\right). \tag{2.3.21}$$

Expand (2.3.21) into a series in the neighborhood of point $\varepsilon^\alpha = 0$:

$$x'^\beta = x^\beta + X^\beta_\alpha \varepsilon^\alpha, \quad u'^i = u^i + Y^i_\alpha \varepsilon^\alpha, \tag{2.3.22}$$

where

$$X^\beta_\alpha = \frac{\partial X^\beta}{\partial \varepsilon^\alpha}\bigg|_{\varepsilon^\alpha=0}, \quad Y^i_\alpha = \frac{\partial Y^i}{\partial \varepsilon^\alpha}\bigg|_{\varepsilon^\alpha=0}. \tag{2.3.23}$$

We obtain from here with regard for (2.3.20) the variations:

$$\delta x^\beta = X^\beta_\alpha \varepsilon^\alpha, \quad \delta u^i = Y^i_\alpha \varepsilon^\alpha, \quad \bar{\delta} u^i = \bar{Y}^i_\alpha \varepsilon^\alpha, \tag{2.3.24}$$

where

$$\bar{Y}^i_\alpha = Y^i_\alpha - \frac{\partial u^i}{\partial x^\beta} X^\beta_\alpha. \tag{2.3.25}$$

Substituting (2.3.24) into (2.3.19) and taking into account the independence of ε^α, we obtain n relationships ($\alpha = 1, \dots, n$):

$$\frac{\partial}{\partial x^\beta}\left(L X^\beta_\alpha + \frac{\partial L}{\partial u^i_{,\beta}} \bar{Y}^i_\alpha\right) = 0. \tag{2.3.26}$$

Thus, we have proved the *Noether's theorem*, which states that, if the action (2.3.1) is invariant under the infinitesimal transformations (2.3.22) depending on n small parameters, then n differential conservation laws (2.3.26) are valid in Cartesian coordinates, in which u^i are the solutions of the Euler-Lagrange equations (2.3.15), and X^β_α and $\bar{Y}^i_\alpha$ are determined from equations (2.3.23) and (2.3.25), respectively.

For each relationship (2.3.26), one can obtain an integral conservation law in an Euclidean space. The equations

$$\frac{\partial Q_\alpha}{\partial t} + \frac{\partial J^j_\alpha}{\partial \xi^j} = 0, \; Q_\alpha = L X^0_\alpha + \frac{\partial L}{\partial \dot{u}^i} \bar{Y}^i_\alpha, \; J^j_\alpha = L X^j_\alpha + \frac{\partial L}{\partial u^i_{,j}} \bar{Y}^i_\alpha \tag{2.3.27}$$

follow from (2.3.18) and (2.3.26), where the Q_α are the Noether's charges and J^j_α are the Noether's currents. Multiplying both sides of the first equation in (2.3.27) by dV and integrating with the use of the Ostrogradsky–Gauss theorem, we obtain the equation:

$$\frac{\partial}{\partial t} \int Q_\alpha dV = - \int_{S_n} J^j_\alpha n_j \, dS_n, \tag{2.3.28}$$

where the integral over an infinitely distant surface stands on the right-hand side. Since $L \to 0$ at infinity, the integral on the right-hand side of (2.3.28) is equal to zero. This implies that the integrals

$$\int Q_\alpha dV = \text{const}; \tag{2.3.29}$$

thus, they do not depend on time and are the integral conservation laws. Strictly speaking, only the integral conservation laws have a direct meaning and determine the conservation of the Noether's total charge Q_α in a finite (and infinite) volume, if the Noether's total current in (2.3.28) across the surface is equal to zero: $\int_{S_n} J^j_\alpha n_j \, dS_n = 0$. Note that the availability of the differential conservation laws does not lead in a general case to the existence of the integral conservation laws. As will be shown in the following, the existence of the solutions of Killing's equations is required. In the case of an Euclidean space such solutions exist; therefore, one can find from the differential conservation laws (2.3.26) and (2.3.27) the integral conservation laws (2.3.29).

Consider a particular case of the transformations (2.3.21), a translation in space and time

$$x'^\beta = x^\beta + \varepsilon^\beta, \quad u'(x'^\beta) = u^i(x^\beta). \tag{2.3.30}$$

Substituting (2.3.30) into (2.3.22) and (2.3.24), let us write the variatons $\delta x^\beta = \varepsilon^\beta$, $\delta u^i = 0$, from which we obtain the formulas:

$$X^\beta_\alpha = \delta^\beta_\alpha, \quad Y^i_\alpha = 0, \quad \bar{Y}^i_\alpha = -\frac{\partial u^i}{\partial x^\alpha}. \tag{2.3.31}$$

Taking into account these relations, we can write equation (2.3.26) in the form

$$\frac{\partial T^\beta_\alpha}{\partial x^\beta} = 0, \quad T^\beta_\alpha = \frac{\partial L}{\partial u^i_{,\beta}} \frac{\partial u^i}{\partial x^\alpha} - L\,\delta^\beta_\alpha. \tag{2.3.32}$$

The quantity T^β_α is called the *canonical tensor of energy and momentum.* Dividing the temporal and spatial components, we can rewrite (2.3.32) as follows:

$$\frac{\partial T^0_0}{\partial t} + \frac{\partial T^j_0}{\partial \xi^j} = 0, \quad \frac{\partial T^0_l}{\partial t} + \frac{\partial T^k_l}{\partial \xi^k} = 0, \tag{2.3.33}$$

where we have with regard for (2.1.12), (2.1.23), and (2.1.24) that

$$\begin{aligned} T^0_0 &= \frac{\partial L}{\partial \dot{u}^i}\dot{u}^i - L = H, \quad T^j_0 = \frac{\partial L}{\partial u^i_{,j}}\frac{\partial u^i}{\partial t} = -\overset{\circ}{\sigma}{}^{ij}\dot{u}^i, \\ T^0_l &= \frac{\partial L}{\partial \dot{u}^i}\frac{\partial u^i}{\partial \xi^l} = \pi_i \frac{\partial u^i}{\partial \xi^l} = \rho_0 \dot{u}^i \frac{\partial u^i}{\partial \xi^l}, \\ T^k_l &= \frac{\partial L}{\partial u^i_{,k}}\frac{\partial u^i}{\partial \xi^l} - L\delta^k_l = -\left(\overset{\circ}{\sigma}{}^{ik}\frac{\partial u^i}{\partial \xi^l} + L\delta^k_l\right), \end{aligned} \tag{2.3.34}$$

where H is the Hamilton's function density. Since the coordinates are Cartesian ($\overset{\circ}{g}_{ij} = \delta_{ij}$), we have in (2.3.34) that $T^k_l = T^{lk} = T_{lk}$. It can be seen from a comparison of formulas (2.3.33) and (2.3.34) that

H is the energy density, T_0^j determines the energy flux through a unit area, T_l^0 is the wave momentum density, and T_l^k is the tensor of the momentum flux through the unit area. It follows from (2.3.34) that the wave momentum density T_l^0 differs from the momentum density π^l equal to $\rho_0 \dot{u}^l$ for the Lagrangian of the form (2.2.13). To elucidate the meaning of the wave momentum, let us consider an infinitesimal particle of the mass dm moving at a speed $\vec{v}$ [18]. Expand $\vec{v}$ into the basis of the accompanying coordinate system $\hat{\vec{e}}_i$ and into the Cartesian basis of the reference frame $\overset{o}{\vec{e}}_i$:

$$\vec{v} = v^l \hat{\vec{e}}_l = \dot{u}^i \overset{o}{\vec{e}}_i \, .$$

Multiplying this equation by $\hat{\vec{e}}_s$, we determine similarly to (2.1.16) that

$$v^s = \hat{g}^{sl} \frac{\partial x^i}{\partial \xi^l} \dot{u}^i .$$

Going over to the covariant components $v_k = \hat{g}_{ks} v^s$, we can find with regard for formula $x^i = \xi^i + u^i$ that

$$v_k = \dot{u}_k + \frac{\partial u^i}{\partial \xi^k} \dot{u}^i . \tag{2.3.35}$$

Thus, the wave momentum components in the basis $\overset{o}{\vec{e}}_i$ are equal to $T_k^0 = \rho_0 \dot{u}^i \frac{\partial u^i}{\partial \xi^k}$ and represent the difference of the components of the momentum $\rho_0 v_k$ in the accompanying basis and the components of the momentum $\rho_0 \dot{u}_k$ in a fixed basis of the reference frame.

Comparing equations (2.3.33) and (2.3.27), we obtain by virtue of (2.3.29) the laws for the conservation of energy:

$$E = \int H \, dV = \text{const} \tag{2.3.36}$$

and of wave momentum

$$P_k = \int \frac{\partial L}{\partial \dot{u}^i} \frac{\partial u^i}{\partial \xi^k} \, dV = \text{const}, \tag{2.3.37}$$

the existence of which is related to the action invariance with respect to translations in time and space (2.3.30). Let us determine the momentum moment by the formula[16,19]

$$M_{ij} = \int (\xi_i T_j^0 - \xi_j T_i^0) \, dV, \tag{2.3.38}$$

from where we have with regard for (2.3.33):

$$\begin{aligned} \frac{dM_{ij}}{dt} &= \int \left(\xi_i \frac{\partial T_j^0}{\partial t} - \xi_j \frac{\partial T_i^0}{\partial t} \right) dV = - \int \left(\xi_i \frac{\partial T_j^k}{\partial \xi^k} - \xi_j \frac{\partial T_i^k}{\partial \xi^k} \right) dV \\ &= - \int \frac{\partial}{\partial \xi^k} (\xi_i T_j^k - \xi_j T_i^k) \, dV + \int (T_j^i - T_i^j) \, dV. \end{aligned}$$

Using the Ostrogradsky–Gauss' theorem and the relation $T_j^i = T_{ij}$, we now rewrite this expression in the form

$$\frac{dM_{ij}}{dt} = -\int (\xi_i T_j^k - \xi_j T_i^k)\, n_j\, dS_n + \int (T_{ij} - T_{ji})\, dV.$$

By assuming that the integration surface S_n tends to infinity, where the tensor $T_i^j = 0$, we obtain:

$$\frac{dM_{ij}}{dt} = \int (T_{ij} - T_{ji})\, dV, \tag{2.3.39}$$

where the integration is performed over the total space. It follows from here that the conservation law for the momentum moment $M_{ij} = \text{const}$ is valid under the condition of the symmetry of the tensor $T_{ij} = T_{ji}$. In the case under consideration (2.3.34), the tensor T_{ij} is not symmetric, which formally points to the presence of an internal moment. This is not so, however, since equation (2.3.38) does not determine the momentum moment in a reference frame[18]. According to[18,20], the shoulder for the momentum moment in formula (2.3.38) should serve $\xi^i + u^i$ rather than ξ^i; therefore, one should make in (2.3.38) a substitution $\xi^i \to \xi^i + u^i$. In this connection, we recall that the condition for the symmetry of the stress tensor $\hat{\sigma}^{ij} = \hat{\sigma}^{ji}$ was obtained from the conservation law for the momentum moment in an accompanying coordinate system, where the shoulder is equal to ξ^i (the shoulder vector is equal to $\xi^i \hat{\vec{e}}_i$) and the moment is determined by the formula (2.3.38). By a choice of the coordinate system in the reference frame, one can achieve the coincidence of the Eulerian and accompanying coordinate systems; therefore, in the Eulerian coordinate system, the equation $\sigma^{ij} = \sigma^{ji}$ also follows from the condition of the moment conservation.

Note that the Lagrange function in (2.2.1) and (2.2.5) and, correspondingly, the tensor of energy and momentum (2.3.32) are not determined uniquely with the accuracy up to an arbitrary function. Employing this arbitrariness, one can always ensure the symmetry of the tensor of energy and momentum[16,17].

Let us show that the motion equations (2.2.14) will not change if the total derivative of some vector function φ^j is added to the Lagrangian:

$$L' = L + \frac{\partial \varphi^\alpha}{\partial x^\alpha}. \tag{2.3.40}$$

Since the Lagrangian contains the derivatives of the displacements u^i of the order no higher than the first order, then φ^j must depend only on u^i, therefore, $\varphi^j = \varphi^j(u^i)$. Substituting (2.3.40) into (2.2.5), we rewrite the Hamilton–Ostrogradsky variational principle (2.2.1), (2.2.5) in the form

$$\int_{V_0}\int_{t_0}^{t_1} \delta L\, d\Omega + \int_{S_n^0}\int_{t_0}^{t_1} \overset{o}{\sigma}{}^{ij}\; n_j^0 \delta u^i\, dS_n^0\, dt + \int_{V_0}\int_{t_0}^{t_1} \delta\Big(\frac{\partial \varphi^\alpha}{\partial x^\alpha}\Big) d\Omega = 0, \tag{2.3.41}$$

where $d\Omega = dV\, dt$. Let us transform the third item:

$$\iint \delta \frac{\partial \varphi^\alpha}{\partial x^\alpha}\, d\Omega = \iint \frac{\partial}{\partial x^\alpha} \delta\varphi^\alpha d\Omega = \iint \frac{\partial}{\partial t}\delta\varphi^0 d\Omega + \iint \frac{\partial}{\partial \xi^j}\delta\varphi^j\, d\Omega$$
$$= \int \frac{\partial \varphi^0}{\partial u^k}\delta u^k \Big|_{t_1}^{t_2} dV + \iint \frac{\partial \varphi^j}{\partial u^i}\delta u^i n_j^0\, dt\, dS_n^0 = \int \frac{\partial \varphi^j}{\partial u^i} n_j^0\, \delta u^i\, dt\, dS_n^0,$$

where we have taken into account the fact that $\delta u^k(t_1) = \delta u^k(t_2) = 0$. Substituting the above expression into (2.3.41), we obtain the formula:

$$\int_{V_0}\int_{t_0}^{t_1} \frac{\delta L}{\delta u^i}\delta u^i\, d\Omega + \int_{S_n^0}\int_{t_1}^{t_2} \Big(\frac{\partial L}{\partial u^i_{,j}} + \frac{\partial \varphi^j}{\partial u^i} + \overset{o}{\sigma}{}^{ij}\Big)\, \delta u^i n_j^0\, dS_n^0\, dt = 0,$$

from where we find that the motion equations do not change [compare with (2.2.11)]:

$$\frac{\delta L}{\delta u^i} = -\frac{\partial}{\partial t}\Big(\frac{\partial L}{\partial \dot u^i}\Big) - \frac{\partial}{\partial \xi^j}\Big(\frac{\partial L}{\partial u^i_{,j}}\Big) = 0, \tag{2.3.42}$$

and the stresses are determined by the formula

$$\overset{o}{\sigma}{}^{ij} = -\frac{\partial L}{\partial u^i_{,j}} - \frac{\partial \varphi^j}{\partial u^i}. \tag{2.3.43}$$

It follows from the comparison of (2.2.12) and (2.3.43) that there exists a functional arbitrariness in the determination of the stress tensor. Since $\varphi^\beta = \varphi^\beta(u^i)$, we can rewrite (2.3.40) as follows:

$$L' = L + \frac{\partial \varphi^\beta}{\partial u^i}\frac{\partial u^i}{\partial x^\beta}. \tag{2.3.44}$$

Substituting expression (2.3.44) into the second equation (2.3.32) instead of L, we obtain an expression for the stress of energy and momentum $\tilde T_\alpha^\beta$, which corresponds to L':

$$\begin{aligned} \tilde T_\alpha^\beta &= T_\alpha^\beta + \frac{\partial \varphi^\beta}{\partial u^i}\frac{\partial u^i}{\partial x^\alpha} - \frac{\partial \varphi^\gamma}{\partial u^i}\frac{\partial u^i}{\partial x^\gamma}\delta_\alpha^\beta \\ &= T_\alpha^\beta + \frac{\partial \varphi^\beta}{\partial x^\alpha} - \frac{\partial \varphi^\gamma}{\partial x^\gamma}\delta_\alpha^\beta \\ &= T_\alpha^\beta - \frac{\partial}{\partial x^\gamma}(\delta_\alpha^\gamma\varphi^\beta - \delta_\alpha^\beta\varphi^\gamma) = T_\alpha^\beta + \frac{\partial}{\partial x^\gamma}\psi_\alpha^{\beta\gamma}, \end{aligned} \tag{2.3.45}$$

where we have introduced an antisymmetric tensor

$$\psi_\alpha^{\beta\gamma} = \varphi^\beta \delta_\alpha^\gamma - \varphi^\gamma \delta_\alpha^\beta, \quad \psi_\alpha^{\beta\gamma} = -\psi_\alpha^{\gamma\beta}. \tag{2.3.46}$$

By a direct substitution of (2.3.45) into the first equation (2.3.32), it is easy to be sure of the validity of the conservation laws

$$\frac{\partial \tilde{T}_\alpha^\beta}{\partial x^\beta} = 0.$$

By the choice of φ^α and correspondingly of $\psi_\alpha^{\beta\gamma}$, one can attain the symmetry of the tensor $\tilde{T}_\alpha^\beta = \tilde{T}_\beta^\alpha$; therefore, the tensor of energy and momentum for a formless continuum is assumed to be *symmetric* in the following.

2.3.2 Conservation Laws in an Arbitrary Coordinate System

As was shown above, the conservation laws (2.3.29) follow in a Cartesian coordinate system of the Euclidean space from the differential continuity equations (2.3.27). The majority of the authors usually believe that this is valid also in the general case for an arbitrary metric tensor $g_{ij}(\xi)$. For a four-dimensional Riemann space, this question has been discussed in detail, for example, in Logunov's work[17]. The results obtained by Logunov can be transferred to the three-dimensional Euclidean space of a classical mechanics.

Let a curvilinear coordinate system be defined at each point of space, which is specified by the tensor $g_{ij}(x)$. In this case, equations (2.3.27) and (2.3.32) can be written in the form

$$\frac{\partial H}{\partial t} + \nabla_j T_0^j = 0, \quad \frac{\partial Q_k}{\partial t} + \nabla_j J_k^j = 0, \tag{2.3.47}$$

where

$$Q_k = T_k^0, \quad J_k^j = T_k^j, \quad j, k = 1, 2, 3$$

and ∇_j is the covariant derivative (1.1.25), (1.1.27). The tensor T^{jk} is assumed to be symmetric below; therefore, the conditions $T^{jk} = T^{kj}$ are satisfied. By using the Weyl formula (1.1.73), we can rewrite the first equation in (2.3.47) as follows:

$$\frac{\partial H}{\partial t} + \frac{1}{\sqrt{g}} \frac{\partial}{\partial \xi^j} (\sqrt{g}\, T_0^j) = 0.$$

Multiplying this equation by an infinitesimal volume

$$dV = \sqrt{g}\, d\tau = \sqrt{g}\, d\xi^1\, d\xi^2\, d\xi^3, \tag{2.3.48}$$

we obtain with regard for the Gauss–Ostrogradsky's theorem:

$$\frac{\partial}{\partial t}\int H\,dV = -\int \frac{\partial}{\partial \xi^j}(\sqrt{g}\,T_0^j)\,d\xi^1 d\xi^2 d\xi^3 = -\int \sqrt{g}\,T_0^j\,n_j\,dS_n.$$

By assuming that the integration surface in the last integral tends to infinity, we find with regard for relation $T_0^j \to 0$ at $\xi^i \to \infty$ the energy conservation law

$$\int H\,dV = \text{const}, \tag{2.3.49}$$

where the integration is performed over the total space. Thus, the space properties $g_{ij}(x)$ do not affect the energy conservation law. This is related to the fact that the energy conservation law is related to the action invariance with respect to the time shifts (2.3.30) $x'^0 = x^0 + \varepsilon^0$, and the time in the classical mechanics does not depend on the spatial coordinates.

Multiplying the second equation in (2.3.47) by g^{ik} with regard for $\frac{\partial g^{ik}}{\partial t} = \nabla_j g^{ik} = 0$, we rewrite this equation for the contravariant components:

$$\frac{\partial Q^i}{\partial t} + \nabla_j J^{ij} = 0, \quad Q^i = g^{ik} Q_k, \quad J^{ij} = g^{ik} J_k^j, \tag{2.3.50}$$

$$\nabla_j J^{ij} = \frac{\partial J^{ij}}{\partial \xi^j} + \Gamma^j_{jl} J^{il} + \Gamma^i_{jl} J^{lj}, \quad J^{ij} = J^{ji}.$$

Multiplying the first equation in (2.3.50) by $dV = \sqrt{g}\,d\tau$, we obtain with regard for the Weyl formula (1.1.73) and the condition $J^{ij} \to 0$ at $x^j \to \infty$:

$$\frac{\partial}{\partial t}\int Q^i dV = -\int \frac{\partial}{\partial \xi^j}(\sqrt{g}\,J^{ij})\,d\tau - \int \sqrt{g}\,\Gamma^i_{jl} J^{lj}\,d\tau$$

$$= -\int \sqrt{g}\,J^{ij} n_j\,dS_n - \int \sqrt{g}\,\Gamma^i_{jl} J^{lj}\,d\tau = -\int \sqrt{g}\,\Gamma^i_{jl} J^{lj}\,d\tau.$$

It follows from this formula that the conservation laws will take place provided that

$$\int \sqrt{g}\,\Gamma^i_{jl} J^{lj}\,d\xi^1\,d\xi^2\,d\xi^3 = 0.$$

Since $\Gamma^i_{jl} = \Gamma^i_{jl}(g_{kl})$, $g = g(g_{kl})$, and J^{lj} is an arbitrary function of x^i, it is indeed required to find the conditions for g_{kl} under which the identities

$$\frac{\partial}{\partial t}\int Q^i\,dV = 0 \tag{2.3.51}$$

are satisfied.

Let us find the conditions for g_{kl} at which the conditions (2.3.51) are satisfied. It turns out that the metric tensor g_{ij} should satisfy for this purpose the forminvariance condition

$$\bar{\delta} g_{ij} = 0 \tag{2.3.52}$$

under the infinitesimal coordinate transformations

$$\xi'^i = \xi^i + \eta^i(\xi^j), \tag{2.3.53}$$

where $\eta^i(\xi^j)$ is an infinitesimal vector. Let us rewrite for convenience equation (2.3.52) for the contravariant components of the metric tensor

$$g'^{ij}(\xi^k) - g^{ij}(\xi^k) = 0, \tag{2.3.54}$$

where

$$g'^{ij}(\xi^k) = g'^{ij}(\xi'^k) - \frac{\partial g^{ij}}{\partial \xi^k}\eta^k. \tag{2.3.55}$$

The quantity $g'^{ij}(\xi'^k)$ is determined in accordance with the rules for the transformation of contravariant components under the transformations (2.3.53):

$$\begin{aligned} g'^{ij}(\xi'^s) &= \frac{\partial \xi'^i}{\partial \xi^k}\frac{\partial \xi'^j}{\partial \xi^l} g^{kl}(\xi^s) = \left(\delta^i_k + \frac{\partial \eta^i}{\partial \xi^k}\right)\left(\delta^j_l + \frac{\partial \eta^j}{\partial \xi^l}\right) g^{kl} \\ &\approx g^{ij}(\xi^s) + g^{il}\frac{\partial \eta^j}{\partial \xi^l} + g^{kj}\frac{\partial \eta^i}{\partial \xi^k}. \end{aligned} \tag{2.3.56}$$

Substituting (2.3.56) into (2.3.55), we find

$$g'^{ij}(\xi^k) = g^{ij}(\xi^k) - \frac{\partial g^{ij}}{\partial \xi^k}\eta^k + g^{il}\frac{\partial \eta^j}{\partial \xi^l} + g^{kj}\frac{\partial \eta^i}{\partial \xi^k} = g^{ij}(\xi^k) + \nabla^i \eta^j + \nabla^j \eta^i. \tag{2.3.57}$$

Substituting (2.3.57) into (2.3.54), we obtain:

$$\bar{\delta} g^{ij} = \nabla^i \eta^j + \nabla^j \eta^i = 0. \tag{2.3.58}$$

It follows from the equality $g_{ij} g^{jk} = \delta^k_i$ that $\bar{\delta} g_{ij} g^{jk} + g_{ij}\bar{\delta} g^{jk} = 0$. Multiplying this equation by g_{kl}, we find:

$$\bar{\delta} g_{il} = -g_{ij}\, g_{kl}\, \bar{\delta} g^{jk}. \tag{2.3.59}$$

Multiplying equation (2.3.59) by $g_{ki}\, g_{jl}$, we obtain:

$$\nabla_k \eta_l + \nabla_l \eta_k = 0. \tag{2.3.60}$$

It follows from here that the metric tensor g_{ij} will be forminvariant under the infinitesimal transformations (2.3.59) if η_i will satisfy equations

(2.3.60). The solutions of equations (2.3.60) are called the *Killing's vectors* η_i. In the Cartesian coordinates x^i, the equations (2.3.60) simplify to

$$\frac{\partial \eta_i}{\partial x^j} + \frac{\partial \eta_j}{\partial x^i} = 0$$

and have the solutions

$$\eta_i = \varepsilon_i + \omega_{ij} x^j, \quad \omega_{ij} = -\omega_{ji}, \tag{2.3.61}$$

where ε_i are three infinitesimal constant translation vectors and ω_{ij} is an infinitesimal constant rotation tensor containing three independent components. Using (2.3.61), one can find the Killing's vectors η_i' in arbitrary coordinates $\xi^i = \xi^i(x^j)$, which satisfy equations (2.3.60)

$$\eta_i'(\xi) = \frac{\partial x^j}{\partial \xi^i} \eta_j(x(\xi)). \tag{2.3.62}$$

Thus, the system of equations (2.3.60) is integrable in the Euclidean space, and the solution depends on six parameters ε_i, ω_{ij} determining the availability of six integral conservation laws for momentum and momentum moment. [In the case of an arbitrary space, the system of equations (2.3.60) is integrable only in the space of constant curvature[17] $R = g^{ij} R^k_{ikj} =$ const, where R^k_{ikj} is the Riemann–Christoffel curvature tensor (1.2.52).]

To obtain the integral conservation laws in arbitrary coordinates, let us multiply similarly to [17,19] equation (2.3.50) by the Killing's vector $\eta_i(\xi^j)$ and sum over the index i:

$$\frac{\partial \eta_i Q^i}{\partial t} + \eta_i \nabla_j J^{ij} = 0. \tag{2.3.63}$$

Let us transform the second item in (2.3.63):

$$\begin{aligned} \eta_i \nabla_j J^{ij} &= \nabla_j(\eta_i J^{ij}) - J^{ij} \nabla_j \eta_i \\ &= \nabla_j(\eta_i J^{ij}) - \frac{1}{2} J^{ij}(\nabla_j \eta_i - \nabla_i \eta_j) - \frac{1}{2} J^{ij}(\nabla_j \eta_i + \nabla_i \eta_j). \end{aligned}$$

The last two items are equal to zero by virtue of equality $J^{ij} = J^{ji}$ and equation (2.3.60); therefore,

$$\eta_i \nabla_j J^{ij} = \nabla_j(\eta_i J^{ij}). \tag{2.3.64}$$

Substituting (2.3.64) into (2.3.63), we obtain the equation:

$$\frac{\partial \eta_i Q^i}{\partial t} + \nabla_j(\eta_i J^{ij}) = 0. \tag{2.3.65}$$

Multiplying (2.3.65) by $dV = \sqrt{g}\,d\tau$ and integrating over the total volume with regard for $J^{ij} \to 0$ at $\xi^i \to \infty$, we find:

$$\frac{\partial}{\partial t}\int \eta_i Q^i\,dV = -\int \sqrt{g}\,\nabla_j(\eta_i J^{ij})\,d\tau$$

$$= -\int \frac{\partial}{\partial \xi^j}(\sqrt{g}\,\eta_i J^{ij})\,d\xi^1 d\xi^2 d\xi^3 = -\int \sqrt{g}\,\eta_i J^{ij}\,n_j\,dS_n = 0,$$

from where the integral conservation law in arbitrary curvilinear coordinates follows:

$$\int \sqrt{g}\,\eta_i\,Q^i\,d\tau = \text{const}. \qquad (2.3.66)$$

Let us find the conservation laws for the momentum and momentum moment. Let the relation between the Cartesian and curvilinear coordinates be determined by the formula $x^i = y^i(\xi^j)$. Then in accordance with (2.3.61) and (2.3.62) the Killing's vectors in the curvilinear coordinate system ξ^i will have the form

$$\eta_i = y^j_{,i}\varepsilon_i + y^j_{,i}y^k\,\omega_{ik}, \qquad (2.3.67)$$

where $y^j_{,i} = \frac{\partial y^j}{\partial x^i}$. Consider at first the translation transformation (2.3.53) for which

$$\eta_i = y^j_{,i}\,\varepsilon_i, \qquad (2.3.68)$$

where $\varepsilon_i =$const is an infinitesimal constant vector. Substituting (2.3.68) into (2.3.66), we find with regard for (2.3.47) and (2.3.50):

$$\int y^j_{,i}\,g^{ik}\,T^0_k\,\varepsilon_j\,dV = \text{const},$$

from where we obtain by virtue of $\varepsilon_j = $ const the conservation law for the momentum P^j in an arbitrary coordinate system:

$$P^j = \int \sqrt{g}\,g^{ik}T^0_k y^j_{,i}\,d\xi^1\,d\xi^2\,d\xi^3 = \text{const}. \qquad (2.3.69)$$

In the Cartesian coordinate system $g = 1$, $g^{ik} = \delta^{ik}$, $y^j_{,i} = \delta^j_i$, and we find from (2.3.69) that $P_j = \int T^0_k dV = $ const, which coincides with (2.3.37) with regard for (2.3.34).

Let us obtain the conservation law for the momentum moment. For the transformations of rotation (2.3.53) by a constant infinitesimal vector $\delta\varphi_k = \omega_{ij}$, we have from (2.3.67):

$$\eta_i = y^j_{,i}y^k\omega_{jk}. \qquad (2.3.70)$$

Substituting (2.3.70) into (2.3.66), we find:

$$\int y^j_{,i}y^k Q^i\omega_{ik}dV = \frac{1}{2}\int (y^j_{,i}y^k - y^k_{,i}y^j)\,Q^i\omega_{jk}\,dV = \text{const},$$

from where we obtain by virtue of $\omega_{jk} = \text{const}$ the conservation law for the moment of momentum M^{jk} in an arbitrary curvilinear coordinate system:

$$M^{jk} = \int \sqrt{g}\, g^{is} T^0_s \,(y^j_{,i} y^k - y^k_{,i} y^j)\, d\xi^1\, d\xi^2\, d\xi^3 = \text{const}. \qquad (2.3.71)$$

In the particular case of Cartesian coordinates, $g = 1$, $g^{is} = \delta^{is}$, $y^k = x^k = \xi^k$, and $y^j_{,i} = \delta^j_i$. Substituting these relationships into (2.3.71), we obtain the conservation law for the momentum moment in Cartesian coordinates:

$$M^{jk} = \int (T^0_j \xi^k - T^0_k \xi^j)\, d\xi^1 d\xi^2 d\xi^3 = \text{const}$$

[cf. (2.3.38)].

Thus, we have proved that the conservation laws for momentum and energy follow from the homogeneity of space and time and the conservation law for the momentum moment follows from the space isotropicity.

The relativity principle of Galilei plays an important role in classical mechanics. According to this principle, all phenomena in different inertial reference frames proceed in the same way. The Galilei's principle postulates the existence of many inertial reference frames possessing the property that the isolated material point is either at rest with respect to these frames or moves at a constant speed. The relation between the Cartesian coordinates in two reference frames, one of which x'^i moves with respect to the other one, x^i, at a constant speed v^i_0, is given by the Galilean transformations

$$x'^i = x^i - v^i_0\, t, \qquad t' = t. \qquad (2.3.72)$$

It follows from the Galilei's relativity principle that the action $\tilde{S}$ should be invariant under the Galilean transformations. As a result, the conservation law for the momentum of the center of mass of a moving continuum emerges.

Let us derive the conservation law for the momentum of the center of mass. Differentiating (2.3.72) with respect to time:

$$v'^i = v^i - v^i_0,$$

and multiplying the left- and right-hand sides of this equation by $\rho_0 dV$, we integrate it over the total space in a reference frame:

$$\tilde{P}'^i = \int_{V_0} \rho_0 v'^i\, dV = \int_{V_0} \rho_0\, v^i\, dV - v^i_0 \int_{V_0} \rho_0\, dV.$$

The system of the center of mass is determined by the condition $\tilde{P}'^i = 0$, from where the velocity v_0^i will be determined as well as the momentum $\tilde{P}_0^i = v_0^i \int_{V_0} \rho_0 \, dV$ of the system of the mass center:

$$v_0^i = \int_{V_0} \rho_0 v^i \, dV \Big/ \int_{V_0} \rho_0 \, dV, \quad \tilde{P}_0^i = \int_{V_0} \rho_0 v^i \, dV.$$

Differentiating the second equation with respect to time, we obtain with regard for the motion equation:

$$\begin{aligned} \frac{\partial}{\partial t} \int_{V_0} \rho_0 v^i dV &= \int_{V_0} \frac{\partial}{\partial t} \left(\frac{\partial L}{\partial \dot{u}^i} \right) dV = - \int_{V_0} \frac{\partial}{\partial \xi^j} \left(\frac{\partial L}{\partial u^i_{,j}} \right) dV \\ &= - \int_{S_n^0} \frac{\partial L}{\partial u^i_{,j}} n_j^0 \, dS_n^0 = 0, \end{aligned}$$

from where the conservation law for the momentum of the center of mass follows:

$$\tilde{P}_0^i = \int_{V_0} \rho_0 \, v^i \, dV = \text{const.} \tag{2.3.73}$$

The theory of Lie groups[21,22] is concerned with a study of the invariance properties of some system with respect to certain continuous transformations. This general theory gives an algorithm for the identification of the invariance properties of the differential equations and determination of the corresponding integrals on their basis. In particular, we have studied in the present section the Galilean group, consisting of the translation and rotation transformations. The other important group describing the tension transformations will be considered in the following at the proof of the Pi theorem.

Problem 2.6. Construct in Cartesian coordinates a tensor of energy and momentum, which is symmetric in spatial indices, i.e., $\tilde{T}_i^j = \tilde{T}_j^i$, by using the Killing's vectors η^i as the variations of δx^i in (2.3.19) (in the Cartesian coordinates $\tilde{T}_i^j = \tilde{T}^{ij} = \tilde{T}_{ij}$).

Solution: Since time is not involved in the Killing's transformations (2.3.61), let us consider the stationary case in which the equations (2.3.19) and (2.3.20) may be written down in the form

$$\frac{\partial}{\partial x^j} \left(L\eta^j + \frac{\partial L}{\partial u^i_{,j}} \bar{\delta} u^i \right) = 0, \tag{2.3.74}$$

$$\bar{\delta} u^i = \delta u^i - \frac{\partial u^i}{\partial x^j} \eta^j .$$

Under the infinitesimal coordinate transformations $x'^i = x^i + \eta^i$, the components of the displacement vector change as follows:

$$u'^i(x') = \frac{\partial x'^i}{\partial x^k} u^k(x) \approx u^i(x) + \frac{\partial \eta^i}{\partial x^k} u^k(x).$$

On the other hand,

$$u'^i(x') = u'^i(x) + \frac{\partial u^i}{\partial x^k} \eta^k,$$

from where it follows that

$$\bar{\delta} u^i = u'^i(x) - u^i(x) = \frac{\partial \eta^i}{\partial x^k} u^k - \frac{\partial u^i}{\partial x^k} \eta^k. \qquad (2.3.75)$$

Consider the rotation transformations. In the Cartesian coordinates, we have, according to (2.3.61):

$$\eta^i = \omega^i_j x^j, \qquad (2.3.76)$$

where $\omega^j_i = -\omega^i_j$ is an antisymmetric rotation tensor whose components determine three constant rotation vectors $\omega^j_i = \text{const}$. Substituting (2.3.76) into (2.3.75), we find the formula for the Lie variation:

$$\bar{\delta} u^i = \left(u^j \delta^i_s - x^j \frac{\partial u^i}{\partial x^s} \right) \omega^s_j. \qquad (2.3.77)$$

Substituting (2.3.76) and (2.3.77) into the first equation (2.3.74), we obtain:

$$\frac{\partial}{\partial x^j} \left(\frac{\partial L}{\partial u^i_{,j}} u^k \delta^i_s - T^j_s x^k \right) \omega^s_k = 0,$$

from where we find with regard for (2.3.32) $\frac{\partial T^j_i}{\partial x^j} = 0$ that

$$\left(\frac{\partial}{\partial x^j} \left(\frac{\partial L}{\partial u^i_{,j}} u^k \right) - T^k_i \right) \omega^k_i = 0. \qquad (2.3.78)$$

Since the tensor ω^k_i is antisymmetric, the tensor standing before it will be symmetric; therefore,

$$\frac{\partial}{\partial x^j} \left(\frac{\partial L}{\partial u^i_{,j}} u^k \right) - T^k_i = \frac{\partial}{\partial x^j} \left(\frac{\partial L}{\partial u^k_{,j}} u^i \right) - T^i_k,$$

from where

$$\begin{aligned} T^k_i - T^i_k &= -\frac{\partial}{\partial x^j} \left(\frac{\partial L}{\partial u^k_{,j}} u^i - \frac{\partial L}{\partial u^i_{,j}} u^k \right) \\ &= -\frac{\partial}{\partial x^j} (S^{ij}_k - S^{kj}_i) = -\frac{\partial}{\partial x^j} B^{kj}_i, \end{aligned} \qquad (2.3.79)$$

where the quantities

$$S_k^{ij} = \frac{\partial L}{\partial u^k_{,j}} u^i \tag{2.3.80}$$

are called the *spin tensor* in the literature[17] and

$$B_i^{kj} = S_k^{ij} - S_i^{kj} \tag{2.3.81}$$

is a tensor of rank three, which is antisymmetric with respect to the indices i and k:

$$B_i^{kj} = -B_k^{ij}. \tag{2.3.82}$$

With the aid of B_i^{kj}, one can construct a tensor $\tilde{S}_i^{kj}$, which is antisymmetric with respect to the indices j and k:

$$\tilde{S}_i^{kj} = \frac{1}{2}(B_j^{ik} + B_j^{ki} - B_k^{ij}). \tag{2.3.83}$$

Following Belinfante[7], we now determine the symmetric tensor of energy and momentum

$$\tilde{T}_i^k = T_i^k + \frac{\partial}{\partial x^j} \tilde{S}_i^{kj}. \tag{2.3.84}$$

It is easy to verify that this tensor satisfies the conservation laws (2.3.32):

$$\frac{\partial}{\partial x^k} \tilde{T}_i^k = \frac{\partial}{\partial x^k} T_i^k + \frac{\partial^2}{\partial x^k \partial x^j} \tilde{S}_i^{kj} = 0,$$

and is symmetric with respect to i and k:

$$\begin{aligned} \tilde{T}_i^k - \tilde{T}_k^i &= T_i^k - T_k^i + \frac{\partial}{\partial x^j}(\tilde{S}_i^{kj} - \tilde{S}_k^{ij}) \\ &= -\frac{\partial}{\partial x^j} B_i^{kj} + \frac{\partial}{\partial x^j} B_i^{kj} = 0. \end{aligned}$$

References

1. **Eringen, A.C.,** *Mechanics of Continua*, John Wiley and Sons, Inc., New York, 1967.
2. **Mindlin, R.D.,** Micro-structure in linear plasticity, *Archive of Rational Mechanics and Analysis*, 1:51, 1964.
3. **Palmov, V.A.,** Basic equations of the nonsymmetric elasticity, *Prikladnaya Matematika i Mekhanika* (in Russian), 28(3):401, 1964.

4. **Cosserat, E. and Cosserat, F.,** *Théorie des corps déformables*, Hermann, Paris, 1909.
5. **Uhlenbeck, G.E. and Ford, G.W.,** *Lectures in Statistical Mechanics*, American Mathematical Society, Providence, Rhode Island, 1963.
6. **Hirshfelder, J.O., Curtiss, C.F., and Bird, R.B.,** *The Molecular Theory of Gases and Liquids*, John Wiley and Sons, Inc., New York, 1954.
7. **Sedov, L.I.,** *Continuum Mechanics*, Vol. 1 (in Russian), Fifth Edition, Nauka, Moscow, 1994.
8. **Nicolis, G. and Prigogine, I.,** *Exploring Complexity: An Introduction*, W.H. Freeman and Company, New York, 1989.
9. **Nicolis, J.S.,** *Dynamics of Hierarchical Systems. An Evolutionary Approach*, Springer-Verlag, Berlin, 1986.
10. **Ilyushin, A.A.,** *Continuum Mechanics* (in Russian), Moscow State University, Moscow, 1990.
11. **Landau, L.D. and Lifschitz, E.M.,** *Statistical Physics, Pt. I* (in Russian), Nauka, Moscow, 1976.
12. **Goldstein, H.,** *Classical Mechanics*, Addison-Wesley, Cambridge, Massachusetts, 1950.
13. **Lanszos, C.,** *The Variational Principles of Mechanics*, Univ. Toronto Press, 1949.
14. **Leech, J.W.,** *Classical Mechanics*, John Wiley and Sons, Inc., New York, 1958.
15. **Berdichevskii, V.L.,** *Variational Principles of Continuum Mechanics* (in Russian), Nauka, Moscow, 1983.
16. **Landau, L.D. and Lifschitz, E.M.,** *Field Theory* (in Russian), Vol. 2, Nauka, Moscow, 1988.
17. **Logunov, A.A.,** *Lectures on the Theories of Relativity and Gravitation (Up-to-Date Analysis of the Problem)* (in Russian), Vol. 2, Nauka, Moscow, 1988.
18. **Eshelby, J.D.,** Continual Theory of Defects, *Solid State Phys.*, Vol. 3, p. 79, 1956.
19. **Warsi, Z.U.A.,** *Fluid Dynamics. Theoretical and Computational Approaches*, CRC Press, Boca Raton, 1993.
20. **Landau, L.D. and Lifschitz, E.M.,** *Elasticity Theory* (in Russian), Nauka, Moscow, 1987.
21. **Ovsyannikov, L.V.,** *The Group Analysis of Differential Equations* (in Russian), Nauka, Moscow, 1978.
22. **Richtmyer, R.D.,** *Principles of Advanced Mathematical Physics*, Vol. 2, Springer-Verlag, New York, 1978.

3

The Features of the Solutions of Continuum Mechanics Problems

In this chapter, we consider the similarity and dimensional methods, including the construction of self-similar solutions. We also present the theory of weak discontinuities (the characteristics) and strong discontinuities (shock waves and tangential discontinuities).

3.1 Similarity and Dimension Theory in Continuum Mechanics

The theory of similarity and dimension plays an important role in continuum mechanics. It enables one to validate the principles of the physical modeling of actual processes under laboratory conditions and to construct the self-similar solutions of continuum mechanics equations[1,2]. The Pi theorem is the cornerstone of this theory. Before we proceed to its formulation and proof, we introduce the basic notions of the similarity theory.

The equations of continuity, momentum, energy, and state introduced in Section 2.1 establish some relationships between the dimensional quantities: the velocity, density, pressure, temperature, etc. One can use different systems of units to specify these quantities. One can draw here an analogy with the way in which we have used the different coordinate systems for the description of a continuum. The most widespread systems of units are the CGS system, in which the units of length L (centimeter), mass M (gramm), and time T (second) are used as the primary measurement units; the MKS system with the primary measurement units for the length L (meter), force K (the kilogramm-force), time T (second); the SI system with the primary units, such as meter, kilogramm-mass, second, ampere, the Kelvin degree, and lumen. The primary quantities and their primary measurement units are introduced experimentally with the aid of special standards. The primary measurement units are the independent units and cannot be expressed in terms of each other. The remaining quantities have the dependent dimensions in the sense that their measurement units are expressed in terms of the primary independent measurement units. For example, the velocity has the dimension $[v] = L/T$, the kinetic energy $[E] = ML^2/T^2$, the stress $[\sigma_{ij}] = ML^{-1}T^{-2}$, etc. The dimension q of any quantity a in a certain basis of the independent measurement units $q_1, q_2, \ldots, q_n$ may generally be presented by the formula

$$q = q_1^{\alpha_1} q_2^{\alpha_2} \ldots q_n^{\alpha_n}. \tag{3.1.1}$$

We immediately note that the number of independent measurement units can increase or decrease depending on the problem formulation. So in the kinematics problems, it is sufficient to specify two independent measurement units with the dimensions of time T and length L. In the thermodynamics problems, it is on the contrary reasonable to extend the basis by introducing the temperature in Kelvin as an additional independent measurement unit. In this case, the kinetic energy of a monoatomic molecule is determined by the formula $\mathcal{E} = \frac{3}{2}kT$, and one should include an additional parameter, the Boltzmann constant $k = 1.38 \cdot 10^{-23} J/K^o$, in the set of the determining parameters of a problem. Thus, the extension of the basis at the expense of new independent measurement units leads to the appearance of new problem parameters.

Let us prove the Pi theorem following[1,2]. Let us assume that there is a certain physical process that is characterized completely with the aid of n independent finite dimensional and nondimensional quantities $a_1, a_2, \ldots, a_n$. Then, any quantity a describing a given process is a function of these n quantities

$$a = f(a_1, a_2, \ldots, a_k, a_{k+1}, \ldots, a_n). \tag{3.1.2}$$

Let the first k quantities $a_1, a_2, \ldots, a_k$ $(k \le n)$ have the independent dimensions $[a_1] = A_1,\ [a_2] = A_2, \ldots, [a_k] = A_k$. (For the case $k = 3$ these may be any three quantities having the independent dimensions of the form $L^\alpha M^\beta T^\gamma$.) Then, the dimensions of the remaining quantities may be expressed in terms of the first k quantities:

$$\begin{aligned} [a] &= A_1^{\alpha_1} A_2^{\alpha_2} \ldots A_k^{\alpha_k}, \quad [a_{k+1}] = A_1^{\beta_1} A_2^{\beta_2} \ldots A_k^{\beta_k}, \ldots, \\ [a_n] &= A_1^{\gamma_1} A_2^{\gamma_2} \ldots A_k^{\gamma_k}. \end{aligned} \tag{3.1.3}$$

Let us change the measurement units of the quantities $a_1, a_2, \ldots, a_k$ by the factors $\lambda_1, \lambda_2, \ldots, \lambda_k$, so that the numerical values of the quantities $a, a_1, , \ldots, a_n$ in a new system of units are equal to

$$\tilde{a} = \lambda_1^{\alpha_1} \lambda_2^{\alpha_2} \ldots \lambda_k^{\alpha_k} a, \quad \tilde{a}_1 = \lambda_1 a_1, \ldots, \tilde{a}_k = \lambda_k a_k,$$
$$\tilde{a}_{k+1} = \lambda_1^{\beta_1} \lambda_2^{\beta_2} \ldots \lambda_k^{\beta_k} a_{k+1}, \ldots, \tilde{a}_n = \lambda_1^{\gamma_1} \lambda_2^{\gamma_2} \ldots \lambda_k^{\gamma_k} a_n.$$

In the new system of measurement units, the formula (3.1.2) takes the form

$$\begin{aligned}\tilde{a} &= \lambda_1^{\alpha_1}\lambda_2^{\alpha_2}\ldots\lambda_k^{\alpha_k}a = f(\lambda_1 a_1, \lambda_2 a_2 \ldots \lambda_k a_k,\\ &\lambda_1^{\beta_1}\lambda_2^{\beta_2}\ldots\lambda_k^{\beta_k}a_{k+1},\ldots,\lambda_1^{\gamma_1}\lambda_2^{\gamma_2}\ldots\lambda_k^{\gamma_k}a_n).\end{aligned} \tag{3.1.4}$$

Assuming $\lambda_i = 1/a_i$, where $1 \le i \le k$, and substituting into equation (3.1.4), we can rewrite it as

$$\Pi = f(\Pi_1, \Pi_2, \ldots, \Pi_{n-k}), \tag{3.1.5}$$

where

$$\begin{aligned}\Pi &= \frac{a}{a_1^{\alpha_1}a_2^{\alpha_2}\cdots a_k^{\alpha_k}}, \quad \Pi_1 = \frac{a_{k+1}}{a_1^{\beta_1}a_2^{\beta_2}\cdots a_k^{\beta_k}}, \ldots,\\ \Pi_{n-k} &= \frac{a_n}{a_1^{\gamma_1}a_2^{\gamma_2}\cdots a_k^{\gamma_k}},\end{aligned} \tag{3.1.6}$$

and the first k parameters $\lambda_i a_i = 1$. As follows from (3.1.3), the quantities $\Pi, \Pi_1, \ldots, \Pi_{n-k}$ are nondimensional and they do naturally not depend on the choice of the nondimensional measurement units A_i, $1 \le i \le k$. The relationship (3.1.5) makes up the contents of the Pi theorem, which asserts that the relation (3.1.2) between $n+1$ dimensional quantities, of which k quantities have the independent dimensions, may be presented in the form of a relation between $n+1-k$ nondimensional quantities determined by formulas (3.1.6). The practical usefulness of the Pi theorem consists of the reduction of the independent problem parameters by k parameters at a passage from a dimensional form of equations to their nondimensional form. In particular, at $n = k$, it follows that the functional relationship is determined with the accuracy up to the constant:

$$a = Ba_1^{\alpha_1}a_2^{\alpha_2}\cdots a_k^{\alpha_k},$$

where the constant B is determined from a complete solution of a problem or from the experiment. It appears at first sight that one can reduce the difference $n+1-k$ by increasing the number of independent measurement units. It is not so, however, in the general case. As was noted above, the introduction of the additional independent measurement units is accompanied as a rule by the introduction of the same number of additional parameters, so that the difference $n+1-k$ will be the same. The cases are nevertheless possible when the new introduced parameters do not play a significant role in the problem under consideration and can be eliminated from the list of the independent quantities a_i, which leads to the diminution of the difference $n+1-k$. The choice of the group of independent dimensional quantities $a_1, \ldots, a_n$ determining a given physical process is a nontrivial operation and requires the understanding of a

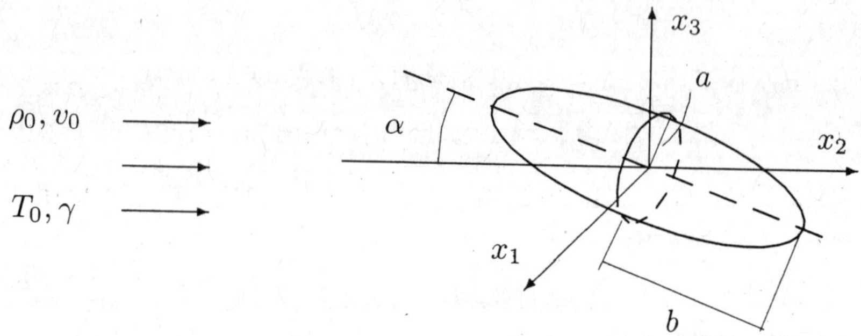

Figure 3.1: The gas flow around an ellipsoid of revolution.

physical essence of the problem. This choice is simplified if it is possible to perform a strict mathematical formulation of the problem. A physical process is usually described in continuum mechanics by a partial differential equation or a system of partial differential equations, which are valid in some spatiotemporal domain (t, x_k). In addition, it is necessary to specify the initial and boundary conditions for a given system of equations.

If one succeeds in such a formulation of the problem, then all independent variables and physical constants entering the system of equations, initial and boundary conditions, and geometric parameters of the region will enter the set of independent parameters. A complete set of the independent parameters should characterize completely a given physical process.

The nondimensional formulas (3.1.5), which make up the contents of the Pi theorem, indeed determine a *diversity of physically similar processes*, in which the nondimensional combinations $\Pi_1, \Pi_2, \ldots, \Pi_{n-k}$ are the same. It is clear that the dimensional parameters may differ significantly in these processes; however, the equality of $\Pi_1, \ldots, \Pi_{n-k}$ enables one to obtain the parameters of some process from another process by a simple alteration of the corresponding scales. Thus, the Pi theorem serves as a background for modeling of the actual processes in experiments conducted on the corresponding models. In particular, the experiments are carried out in the wind tunnels, in which the aerodynamic drag of the models flowed past is measured. Then the results of these experiments are presented in a nondimensional form (3.1.5). After that one finds the drag of an actual body in the flow by specifying the corresponding dimensional parameters.

Let us show how the similarity criteria work using the example of a problem of the flow past an ellipsoid of revolution with the main semi-axes a and b, which is inclined at the angle α to the free stream having at "infinity" the density ρ_0, temperature T_0, velocity v_0, and adiabatic exponent γ (see Fig. 3.1). We have obtained above the equations (2.1.114), which govern the flow of a viscous, heat-conducting gas. Neglecting the viscous terms proportional to $\operatorname{div}\vec{v}$ in these equations and omitting the items $\frac{\partial\rho}{\partial t}$, $\frac{\partial\vec{v}}{\partial t}$, and $\frac{\partial E}{\partial t}$ (the flow is stationary), let us write the governing equations in the Cartesian coordinates x_i (see Fig. 3.1):

$$\begin{aligned}
\frac{\partial}{\partial x_i}\rho v_i &= 0, \quad v_i\frac{\partial\vec{v}}{\partial x_i} = -\frac{1}{\rho}\frac{\partial P}{\partial x_i} + \nu\sum_{i=1}^{3}\frac{\partial^2\vec{v}}{\partial x_i^2} - g\vec{e}_3, \\
v_i\frac{\partial E}{\partial x_i} &= -Pv_i\frac{\partial}{\partial x_i}\Big(\frac{1}{\rho}\Big) + 2\nu\Big(\dot{\varepsilon}_{ij}\dot{\varepsilon}_{ij} - \frac{1}{3}(\dot{\varepsilon}_{ii})^2\Big) + c_v\omega\sum_{i=1}^{3}\frac{\partial^2 T}{\partial x_i^2}, \quad (3.1.7) \\
E &= c_vT, \quad P = (\gamma-1)\rho E, \quad \dot{\varepsilon}_{ij} = \frac{1}{2}\Big(\frac{\partial v_i}{\partial x_j} + \frac{\partial v_j}{\partial x_i}\Big),
\end{aligned}$$

where $\nu = \mu/\rho$ is the kinematic viscosity, $\omega = \kappa/c_v\rho$ is the temperature conduction, and g is the gravity acceleration. It follows from the form of equations (3.1.7) and the boundary conditions (Fig. 3.1) that the ρ, $\vec{v}, T, P, E$ will depend on the following parameters: x_i, α, a, b, ρ_0, v_0, T_0, γ, c_v, ν, ω, g. The force acting on the ellipsoid is an integral of the stress tensor (1.3.10) over the ellipsoid surface:

$$f_i = \iint \sigma_{ij}n_j\,dS,$$

where n_j is a normal to the ellipsoid surface. The coordinates x_i will disappear from the parameters set, and

$$f_i = f_i(\alpha, a, b, \rho_0, v_0, T_0, \gamma, c_v, \nu, \omega, g). \quad (3.1.8)$$

We choose a, ρ_0, v_0, and T_0 as the parameters with independent dimensions. Then, we can obtain the following expression for the force by using the Pi theorem and constructing the nondimensional combinations of the form (3.1.6):

$$f_i = \rho_0 v_0^2 a^2 A_i\Big(\alpha, \gamma, \frac{b}{a}, \frac{c_vT_0}{v_0^2}, \frac{\nu}{v_0a}, \frac{\omega}{v_0a}, \frac{gb}{v_0^2}\Big). \quad (3.1.9)$$

As will be shown below, the derivative of the pressure with respect to density at a constant entropy $c = (\frac{\partial P}{\partial\rho})_S$ determines the velocity of the propagation of small disturbances and is usually called the sound speed. Differentiating the Poisson adiabat (2.1.97), we obtain:

$$c^2 = \gamma\frac{P}{\rho} = \gamma(\gamma-1)c_vT. \quad (3.1.10)$$

With regard for (3.1.10), formula (3.1.9) may be written as

$$f_i = \rho_0 v_0^2 a^2 A_i(\alpha,\ \gamma,\ \frac{b}{a},\ M,\ \mathrm{Re},\ \mathrm{Pe},\ \mathrm{Fr}), \tag{3.1.11}$$

where $M = v_0/c_0$ is the Mach number, $\mathrm{Re} = \frac{av_0}{\nu}$ is the Reynolds number, $\mathrm{Pe} = \frac{av_0}{\omega}$ is the Péclet number, and $\mathrm{Fr} = v_0/\sqrt{ga}$ is the Froude number. The flow parameters, for example, the density, will be represented similarly by the formula

$$\rho = \rho_0 H(\alpha,\ \gamma,\ \frac{b}{a},\ \frac{x_i}{a},\ M,\ \mathrm{Re},\ \mathrm{Pe},\ \mathrm{Fr}). \tag{3.1.12}$$

Note that the similarity criteria can be obtained by a different technique: by reducing the system of equations (3.1.7) to a nondimensional form. For this purpose, let us divide the both sides of the continuity equation in (3.1.7) by $\rho_0 v_0/a$, both sides of the momentum equation by v_0^2/a and both sides of the energy equation by $c_v T_0 v_0/a$. As a result, we obtain the equations:

$$\begin{aligned}
\frac{\partial}{\partial \tilde{x}_i}\tilde{\rho}\tilde{v}_i &= 0,\\
\tilde{v}_i\frac{\partial \tilde{\vec{v}}}{\partial \tilde{x}_i} &= -\frac{1}{\gamma M^2}\left(\frac{\tilde{T}}{\tilde{\rho}}\frac{\partial \tilde{\rho}}{\partial \tilde{x}_i} + \frac{\partial \tilde{T}}{\partial \tilde{x}_i}\right) + \frac{1}{\mathrm{Re}}\sum_{i=1}^{3}\frac{\partial^2 \tilde{\vec{v}}}{\partial \tilde{x}_i^2} - \frac{1}{\mathrm{Fr}^2}\vec{e}_3,\\
\tilde{v}_i\frac{\partial \tilde{E}}{\partial \tilde{x}_i} &= -(\gamma-1)\tilde{P}\tilde{v}_i\frac{\partial}{\partial \tilde{x}_i}\left(\frac{1}{\tilde{\rho}}\right) + 2\gamma(\gamma-1)\frac{M^2}{\mathrm{Re}}\left(\tilde{\varepsilon}_{ij}\tilde{\varepsilon}_{ij} - \frac{1}{3}(\tilde{\varepsilon}_{ii})^2\right)\\
&+ \frac{1}{\mathrm{Pe}}\sum_{i=1}^{3}\frac{\partial^2 \tilde{T}}{\partial \tilde{x}_i^2},
\end{aligned} \tag{3.1.13}$$

where

$$\begin{aligned}
\tilde{x}_i &= \frac{x_i}{a},\quad \tilde{v}_i = \frac{v_i}{v_0},\quad \tilde{\vec{v}} = \frac{\vec{v}}{v_0},\quad \tilde{\rho} = \frac{\rho}{\rho_0},\quad \tilde{T} = \frac{T}{T_0},\\
\tilde{E} &= \frac{E}{c_v T_0},\quad \tilde{P} = \tilde{\rho}\tilde{T},\quad \tilde{\varepsilon}_{ij} = \frac{1}{2}\left(\frac{\partial \tilde{v}_i}{\partial \tilde{x}_i} + \frac{\partial \tilde{v}_j}{\partial \tilde{x}_i}\right).
\end{aligned}$$

The formula (3.1.12) follows from the form of equations (3.1.13) and the boundary conditions $\tilde{\vec{v}} = 0$ on the ellipsoid surface.

As was noted above, two flows will be similar if their similarity criteria, i.e., the arguments of the function (3.1.12), coincide:

$$\begin{aligned}
\alpha_1 &= \alpha_2,\ \gamma_1 = \gamma_2,\ \frac{b_1}{a_1} = \frac{b_2}{a_2},\ \frac{v_1}{c_1} = \frac{v_2}{c_2},\ \frac{a_1 v_1}{\omega_1} = \frac{a_2 v_2}{\omega_2},\\
\frac{a_1 v_1}{\nu_1} &= \frac{a_2 v_2}{\nu_2},\ \frac{v_1}{\sqrt{a_1}} = \frac{v_2}{\sqrt{a_2}}.
\end{aligned} \tag{3.1.14}$$

It follows from the analysis of equations (3.1.14) that it is practically impossible to achieve a complete similarity of the two flows. The size of the experimental model a_2 is usually considerably smaller than the size a_1 of the actual body. In order to achieve the equality of the Reynolds number, the flow velocity v_2 in a model experiment should be much larger than the velocity v_1 of the actual flow ($v_2 \gg v_1$). At the same time, from the equality of the Froude numbers, an opposite inequality $v_2 \ll v_1$ follows. How can one fit these contradictory requirements to each other? The nature itself comes here to the aid. While considering the gas flows around actual bodies, one can neglect the gravity effects and eliminate the Froude number from the set of the similarity criteria. It is also very difficult in practice to ensure a simultaneous equality of the Mach numbers M and the Reynolds numbers Re, since the inequalities $v_2 \gg v_1$ and $c_2 \gg c_1$ should be simultaneously satisfied here. The latter inequality can be ensured only by a strong heating of gas in the experiment, which is very difficult to make in practice. One indeed studies in the experiment the dependence of the drag force only on the Mach number or on the Reynolds number, and the dependences on other parameter are taken from the theoretical computations or from independent experiments. In addition, it should be kept in mind that the new physical phenomena and processes may arise in the model experiments, which are absent in the actual flow.

It follows from the above that the researcher can face serious problems in the process of physical modeling, and considerable efforts are required to overcome these problems. At present, a large number of various wind tunnels exist nevertheless in the world, the experimental studies on which constitute a basis for the development of aerospace technology. A similar situation takes place in other applied branches of mechanics.

Among other important applications of the Pi theorem and of the *similarity methods*, we point to a possibility of constructing the *classes of self-similar solutions* of the partial differential equations with their aid[1,2]. Let us illustrate this fact using a classical example[2] about the heat propagation from a point source in a medium with a constant temperature conduction ω.

Let some quantity of heat q be released at the time $t = 0$ at the coordinate origin. Assuming $v_i = 0$, $\dot{\varepsilon}_{ij} = 0$, $E = c_v T$, we obtain the heat equation:

$$\frac{\partial T}{\partial t} = \omega \, \Delta T. \tag{3.1.15}$$

Choosing a spherical coordinate system, we can write the Laplacian in the form

$$\Delta T = \frac{\partial^2 T}{\partial r^2} + \frac{2}{r}\frac{\partial T}{\partial r}. \tag{3.1.16}$$

Since the total heat quantity is conserved, the desired solution $T(r, t)$

in the region $0 \leq r < \infty$ should satisfy at each moment of time the condition

$$q = 4\pi \int_0^\infty T\, r^2\, dr. \tag{3.1.17}$$

It follows from the physical formulation of the problem that the temperature T will depend on the following parameters:

$$T = f(\omega,\ q,\ t,\ r). \tag{3.1.18}$$

Let us choose the first three parameters as the quantities with independent dimensions $[\omega] = m^2/s$, $[q] = K^o/m^3$, $[t] = s$, where s is the second, m is the meter, and K^o is the temperature in Kelvin. Using the Pi theorem (3.1.5), (3.1.6), let us rewrite (3.1.18) in a nondimensional form

$$\frac{T(\omega t)^{\frac{3}{2}}}{q} = f\Big(\frac{r^2}{\omega t}\Big),$$

from where it follows that the solution of the problem (3.1.15)–(3.1.17) can be written as

$$T = \frac{q}{(\omega t)^{\frac{3}{2}}} f(\xi), \tag{3.1.19}$$

where $\xi = r^2/(\omega t)$ is a self-similar variable. It is seen from (3.1.19) that the solutions at each moment of time are similar to each other and can be obtained one from another with the aid of a compression and expansion along the axes r and t, respectively. The solutions of equations (in the variables r and t), which depend on the combination r/At^α, are generally called self-similar[1,2]. The exponent α and the quantity A are determined from the similarity considerations, and in a given problem $A = \omega^{1/2}$, and $\alpha = 1/2$.

The existence of self-similar solutions is ensured by the invariance of the equations and corresponding boundary and initial conditions under a group of the similarity transformations. In particular, equations (3.1.15) and (3.1.16) and condition (3.1.17) are invariant under the group of transformations $r' = ar$, $t' = a^2 t$, and $T' = a^{-3}T$. At present, the methods have been developed on the basis of the Lie group theory, which enables one to construct the analytical solutions of the partial differential equations. These solutions reflect the internal properties of symmetry of the governing equations. The presentation of these methods goes beyond the scope of this book, and we refer the interested reader to the relevant literature[2,3].

Before proceeding to the determination of the form of the function $f(\xi)$, we would like to make one more remark. The self-similar solutions of the second kind, in which the similarity exponent α and the dimensional constant A in the self-similar variable r/At^α cannot be determined from the dimensional considerations, exist along with the above-discussed self-similar solutions. The problem of a converging shock wave[4] is such an example. In this case, the total energy of gas in the considered flow region is not constant, which does not enable one to find the value of α. The parameter α in this problem is found from the condition of a passage of the solution through a singular point and is an irrational number. A feature of this solution is that, at the initial moment of time, it is not self-similar (it depends on the character of the motion of a piston creating the shock wave) and becomes self-similar as the shock wave approaches the coordinate origin (the notion of shock wave will be introduced below in this section). The dimensional constant A is determined from the condition of matching of non-self-similar and self-similar solutions.

Let us return to our problem and find the function $f(\xi)$ in (3.1.19). Substituting the solution (3.1.19) into equations (3.1.15) and (3.1.16), we obtain an ordinary differential equation:

$$\xi(4f' + f)' + \frac{3}{2}(4f' + f) = 0,$$

where $f' = \frac{df}{d\xi}$. From the condition $f \to 0$, $f' \to 0$ at $r \to \infty$, $\xi \to \infty$, we obtain from the above equation the equation $4f' + f = 0$, whose solution has the form

$$f = Ce^{-\xi/4}. \tag{3.1.20}$$

Substituting (3.1.20) into (3.1.19) and then into (3.1.17), we can easily find the constant $C = 1/(2\sqrt{\pi})^3$. As a result, we obtain the solution of the given problem in the form

$$T = \frac{q}{(4\pi\omega t)^{3/2}} e^{-\frac{r^2}{4\omega t}}. \tag{3.1.21}$$

3.2 The Characteristics of Partial Differential Equations

As was shown above, the motions of continua are governed by partial differential equations, the variables of which are the coordinates x_i and the time t. For the problem formulation, it is necessary to specify the domain of variation D of the variables x_i and t and the initial and boundary conditions. As the initial conditions for equations $\hat{L}y = 0$, the dependences $y(x_i)$ are taken at $t = 0$. If the initial conditions are specified at $t = 0$ in the overall space $-\infty < x_i < +\infty$, then one says that the Cauchy problem is defined in the hyperplane $t = 0$. In a boundary-value problem, it is necessary to specify the boundary conditions on the surfaces bounding the region D at $t > 0$. As will be shown below, the formulation of a boundary-value problem depends on the type of equation or system of equations. The type of system of equations is determined by its characteristics. There are sufficiently many books in which the theory of characteristics is presented[5–8]. Let us give a short overview of this theory, keeping in mind its applications to the continuum mechanics problems.

Following[7], let at first consider a system of n linear equations

$$A\frac{\partial u}{\partial t} + B\frac{\partial u}{\partial x} = g, \tag{3.2.1}$$

where

$$u = \begin{pmatrix} u_1 \\ u_2 \\ \vdots \\ u_n \end{pmatrix}, \quad A = \begin{Vmatrix} a_{11} & \dots & a_{1n} \\ \vdots & & \vdots \\ a_{n1} & \dots & a_{nn} \end{Vmatrix}, \quad B = \begin{Vmatrix} b_{11} & \dots & b_{1n} \\ \vdots & & \vdots \\ b_{n1} & \dots & b_{nn} \end{Vmatrix},$$

$$g = \begin{pmatrix} g_1 \\ g_2 \\ \vdots \\ g_n \end{pmatrix},$$

$$\det \| A \| \neq 0, \quad a_{ij} = a_{ij}(t,x), \quad b_{ij} = b_{ij}(t,x), \quad g_i = g_i(t,x).$$

Let us multiply both sides of equation (3.2.1) by the inverse matrix A^{-1}; then we obtain a system of equations, which is more convenient for a subsequent study:

$$E\frac{\partial u}{\partial t} + C\frac{\partial u}{\partial x} = f, \tag{3.2.2}$$

$$C = A^{-1}B, \quad f = A^{-1}g, \quad E = \begin{pmatrix} 1 & \dots & 0 \\ \vdots & \ddots & \vdots \\ 0 & \dots & 1 \end{pmatrix}$$

Let the system (3.2.2) have the solution in region D. Define in D some line l (Fig. 3.2), on which we specify a vector $u_i(t,x)$ and find $u_i(t,x)$ in some neighborhood of l (the Cauchy problem). The Cauchy problem will be solvable if the derivatives $\frac{\partial u_i}{\partial t}$, $\frac{\partial u_i}{\partial x}$ are known on l. Let us find the $\frac{\partial u_i}{\partial t}$, $\frac{\partial u_i}{\partial x}$ on l. For this purpose, let us specify the displacements dt and dx along l, to which the change $du_i = \frac{\partial u_i}{\partial t}dt + \frac{\partial u_i}{\partial x}dx$ corresponds, or in the matrix form:

$$du = E\,dt\frac{\partial u}{\partial t} + E\,dx\,\frac{\partial u}{\partial x}. \tag{3.2.3}$$

Uniting equations (3.2.2) and (3.2.3), we obtain the system of equations for determining the $\frac{\partial u}{\partial t}$, $\frac{\partial u}{\partial x}$ on l:

$$E\frac{\partial u}{\partial t} + C\frac{\partial u}{\partial x} = f,$$

$$E\,dt\frac{\partial u}{\partial t} + E\,dx\frac{\partial u}{\partial x} = du. \tag{3.2.4}$$

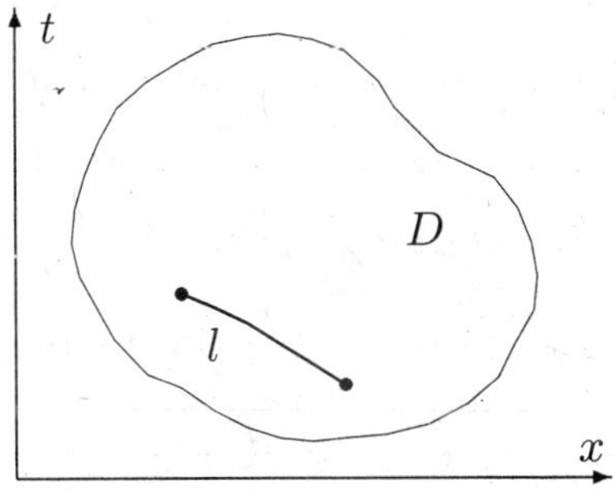

Figure 3.2: To the formulation of the Cauchy problem for system (3.2.2).

Let us define the *characteristics* of system (3.2.2) as the lines on which *the Cauchy problem cannot be posed.* The derivatives $\frac{\partial u}{\partial t}$, $\frac{\partial u}{\partial x}$ cannot be found on such lines; therefore, we can find from the system (3.2.4) the equation for the characteristics:

$$\det \begin{Vmatrix} E & C \\ E\,dt & E\,dx \end{Vmatrix} = 0. \tag{3.2.5}$$

Thus, the derivatives $\frac{\partial u}{\partial t}$, $\frac{\partial u}{\partial x}$ undergo a discontinuity on the characteristics. Since the solution of (3.2.2) in D exists, the rank of the augmented matrix in (3.2.4) should be equal to the rank of a matrix in the left-hand side of (3.2.4):

$$\operatorname{rank} \begin{Vmatrix} E & C & f \\ E\,dt & E\,dx & du \end{Vmatrix} = \operatorname{rank} \begin{Vmatrix} E & C \\ E\,dt & E\,dx \end{Vmatrix}. \tag{3.2.6}$$

Equation (3.2.6) determines the relations of the characteristics (3.2.5). Computing the determinant in (3.2.5), we obtain[7]:

$$\det \begin{Vmatrix} E & C \\ E\,dt & E\,dx \end{Vmatrix} = dt^n \det \begin{Vmatrix} E & C \\ E & \frac{dx}{dt}E \end{Vmatrix}$$
$$= dt^n \det \begin{Vmatrix} 0 & C - \frac{dx}{dt}E \\ E & \frac{dx}{dt}E \end{Vmatrix} = (-1)^n dt^n \det \| C - \frac{dx}{dt}E \| = 0,$$

from where the equations of the characteristics follow:

$$\frac{dx}{dt} = \lambda_i, \tag{3.2.7}$$

where $\lambda_i = \lambda_i(t, x)$ are the eigenvalues of the matrix C:

$$\det \| C - \lambda E \| = 0. \tag{3.2.8}$$

If *all of the zeroes* λ_i in (3.2.8) are *real and different*, then the system (3.2.2) is called *hyperbolic.*

An eigenvector S^i corresponds to each eigenvalue λ_i. It is found from the system of equations

$$(C_{kl} - \lambda_i \delta_{kl}) S_l^i = 0. \tag{3.2.9}$$

The quantities S_l^i form a matrix S, in which there are n rows corresponding to the numbering of l in the range $1 \leq l \leq n$ and n columns formed by numbering with respect to i. With regard for this remark, one can rewrite (3.2.9) in the matrix form $CS = S\lambda$, from where it follows that the matrix S performs a linear transformation, which reverts the matrix C to a diagonal form:

$$\Lambda = S^{-1}CS, \quad \Lambda = \begin{pmatrix} \lambda_1 & 0 & 0 & 0 \\ 0 & \lambda_2 & 0 & 0 \\ 0 & 0 & \ddots & 0 \\ 0 & 0 & 0 & \lambda_n \end{pmatrix} \tag{3.2.10}$$

As a result of this transformation, the vector u goes over to v by the formula $u = Sv$. Substituting this formula into (3.2.2), we find:

$$\frac{\partial}{\partial t}(Sv) + C\frac{\partial}{\partial x}(Sv) = f,$$

from where we obtain:

$$S\frac{\partial v}{\partial t} + CS\frac{\partial v}{\partial x} + \left(\frac{\partial}{\partial t}S + C\frac{\partial}{\partial x}S\right)v = f.$$

Multiplying this equation by the matrix S^{-1}, we obtain with regard for (3.2.10) a canonical form of the hyperbolic system of equations (3.2.2):

$$\frac{\partial v}{\partial t} + \Lambda\frac{\partial v}{\partial x} + Gv = g, \tag{3.2.11}$$

where

$$G = S^{-1}\left(\frac{\partial}{\partial t}S + C\frac{\partial}{\partial x}S\right), \quad g = S^{-1}f.$$

One can say that, if the system of equations (3.2.2) can be reduced to the form (3.2.11), then it is called hyperbolic (note that the requirement of the absence of multiple roots λ_i is not necessary in this definition).

Let us formulate an initial- and boundary-value problem for the hyperbolic system (3.2.11) in the strip $0 \leq x \leq l$, $t > 0$. We at first elucidate the basic features using a simple example of the system of equations

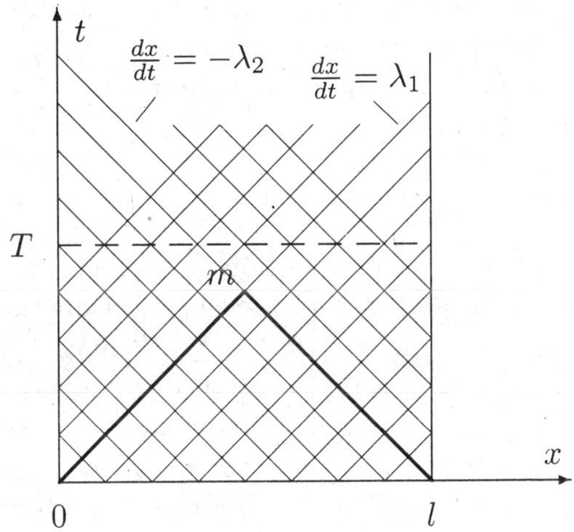

Figure 3.3: To the formulation of a boundary-value problem for system (3.2.12).

$$
\begin{aligned}
&\frac{\partial v_1}{\partial t} + \lambda_1 \frac{\partial v_1}{\partial x} + g_{11}v_1 + g_{12}v_2 = f_1, \\
&\frac{\partial v_2}{\partial t} - \lambda_2 \frac{\partial v_2}{\partial x} + g_{21}v_1 + g_{22}v_2 = f_2,
\end{aligned}
\tag{3.2.12}
$$

where $\lambda_1 > 0,\ \lambda_2 > 0$. The system of equations (3.2.12) has the characteristics

$$\frac{dx}{dt} = \lambda_1, \qquad \frac{dx}{dt} = -\lambda_2,$$

which are shown in Fig. 3.3 by straight lines. To determine the solution in the strip $0 \le x \le l,\ t > 0$, one must specify the initial conditions at $t = 0$:

$$v_1 = \varphi_1(x), \qquad v_2 = \varphi_2(x)$$

and the boundary conditions at $x = 0$ and at $x = l$. The sum of the two partial derivatives in the first equation of system (3.2.12)

$$\frac{\partial v_1}{\partial t} + \lambda_1 \frac{\partial v_1}{\partial x} = \frac{\partial v_1}{\partial t} + \frac{\partial v_1}{\partial x}\frac{dx}{dt} = \frac{d_1 v_1}{dt}$$

represents a derivative along the characteristic $\frac{dx}{dt} = \lambda_1$. Therefore, one must pose on the left boundary one condition, which will be "carried" into the region by the characteristics $\frac{dx}{dt} = \lambda_1$. One can show in the same way that in the second equation of (3.2.12) the expression

$$\frac{\partial v_2}{\partial t} - \lambda_2 \frac{\partial v_2}{\partial x} = \frac{d_2 v_2}{dt}$$

is the derivative along the characteristic $\frac{dx}{dt} = -\lambda_2$, and one must specify on the right boundary one condition. One can generally formulate such a *rule* that the number of the boundary conditions should be equal to the number of the characteristics emanating from the boundary. Let us specify the boundary conditions in our case in the form

$$v_1\Big|_{x=0} = \alpha_{12}v_2 + \psi_1(t), \quad v_2\Big|_{x=l} = \alpha_{21}v_1 + \psi_2(t).$$

After that, one can construct the solution in the rectangle $0 \le x \le l,\ 0 \le t \le T$, where T is determined as the time when the characteristic with the maximum slope intersects the strip $0 \le x \le l$ ($T \max|\frac{dx}{dt}| < l$). It can be seen that the solution in the triangle Olm is determined completely from the initial conditions $\varphi_1(x)$, $\varphi_2(x)$. It is conventional to call this triangle the *characteristic* triangle. The value of v_2 on the right boundary is determined by the initial conditions along the characteristics $\frac{dx}{dt} = -\lambda_2$. After that, one can find $v_1|_{x=0}$. The value $v_2|_{x=0}$ is determined in the same way. By using the given values $v_1|_{x=0}$ and $v_2|_{x=0}$, one restores the solution in the remaining part of the region $0 \le x \le l,\ 0 \le t \le T$. The solution obtained at $t = T$ serves as an initial condition for constructing the solution in the next rectangle $0 \le x \le l,\ T \le t \le 2T$, where the overall procedure is repeated. To ensure the necessary solution smoothness, it is necessary to satisfy the matching conditions. If one requires the continuity of the function sought for, then one must satisfy the conditions

$$\varphi_1(0) = \alpha_{12}\,\varphi_2(0) + \psi_1(0), \quad \varphi_2(l) = \alpha_{21}\,\varphi_1(l) + \psi_2(0).$$

If the continuity of the first derivatives is required, then the following relations must be satisfied, for example, at point $t = 0,\ x = 0$:

$$\lambda_1 \frac{d\varphi_1(0)}{dx} + \frac{d\psi_1(0)}{dt} + \frac{d\alpha_{12}}{dt}\varphi_2(0) + \alpha_{12}\frac{dv_2(0)}{dt} + g_{11}\varphi_1(0)$$
$$+ g_{12}\,\varphi_2(0) = f_1(0),$$
$$\frac{dv_2(0)}{dt} - \lambda_2\frac{d\varphi_2(0)}{dx} + g_{12}\,\varphi_1(0) + g_{22}\,\varphi_2(0) = f_2(0).$$

These equalities should be augmented by the dissipativity conditions, which ensure the uniqueness of the solution of the given problem. The uniqueness theorem has been proved in[7] on the basis of the energy integral for a linear hyperbolic system of equations. We present this theorem without proof as follows.

The system of linear equations

$$\frac{\partial v_i}{\partial t} + \lambda_i\frac{\partial v_i}{\partial x} + g_{ij}v_i = 0, \quad (i = 1, \dots, n_0),$$
$$\frac{\partial v_i}{\partial t} - \lambda_i\frac{\partial v_i}{\partial x} + g_{ij}v_i = 0, \quad (i = n_0 + 1, \dots, n),$$

where $\lambda_i(x,t) > 0$ in the region $0 \le x \le l$, $0 \le t \le T$, with the boundary conditions on the left boundary

$$v_i = \alpha_{ij} v_j, \qquad (i = 1, \ldots, n_0),$$

and with the boundary conditions on the right boundary

$$v_i = \beta_{ij} v_j, \qquad (i = n_0 + 1, \ldots, n),$$

$[\alpha_{ij}(t),\ \beta_{ij}(t)$ are smooth functions$]$, with the initial condition

$$v_i\Big|_{t=0} = \varphi_i(x),$$

which are matched with the boundary conditions, *will have a unique solution* if the boundary conditions satisfy the dissipativity conditions expressed in the form of the inequalities

$$-\sum_{i=1}^{n_0} \lambda_i v_i^2 + \sum_{i=n_0+1}^{n} \lambda_i v_i^2 \le -\lambda_0 \quad \text{on the right boundary},$$

$$-\sum_{i=n_0+1}^{n} \lambda_i v_i^2 + \sum_{i=1}^{n_0} \lambda_i v_i^2 \le -\lambda_0 \quad \text{on the left boundary},$$

where $\lambda_0 > 0$ is a constant. If α_{ij}, β_{ij} are constant, one can assume $\lambda_0 = 0$.

The above definition of the characteristics (3.2.5)–(3.2.8) and the principle of the formulation of the boundary conditions (the number of boundary conditions should be equal to the number of characteristics emanating from the boundary) are also valid for quasilinear equations in which the matrix C and the vector f in (3.2.2) [or the matrices A, B and the vector g in (3.2.1)] depend on the solution itself:

$$\frac{\partial u}{\partial t} + C(t,x,u)\frac{\partial u}{\partial x} = f(t,x,u). \tag{3.2.13}$$

Note, however, that the above-presented uniqueness theorem can already be violated in this case. The characteristics of the system of equations (3.2.13) will here depend on the solution itself: $\frac{dx}{dt} = \lambda(x,t,u)$. Such situations are possible here when several characteristics $(\frac{dx}{dt})_i = \lambda(t,x,u_i)$ corresponding to several values $u_i = u_i(t,x)$ will pass through the same (t,x) point in certain subdomain. Looking ahead, let us show in Fig. 3.4 a picture of the C_+ characteristics in the Riemann wave created in the gas by a moving piston (shown in Chapter 6). It can be seen that at $t > t_A$ the solution uniqueness is violated, and two characteristics determining the two-valued property of the solution pass through

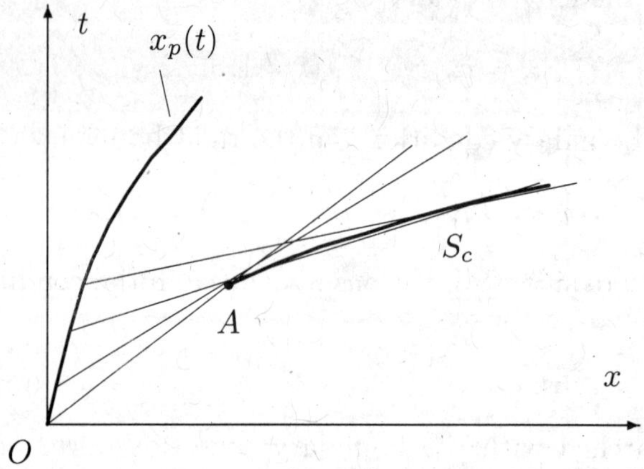

Figure 3.4: The picture of the C_+ characteristics in the Riemann wave. The line $x = x_p(t)$ is the piston trajectory.

each (t, x) point above the line S_c. The line S_c is called the *caustic* and is an envelope of a family of the characteristics. There arises indeed a *discontinuity of the first kind* at point A, which is called a *shock wave* in mechanics. As will be shown below, the characteristics (3.2.5)–(3.2.8) of the continuum mechanics are the lines along which the disturbances propagate. Just this fact conditions their big role in continuum mechanics problems.

In the case of a larger number of the variables, the characteristics are defined in a similar way and are the hypersurfaces in the corresponding space. For example, for the system of equations[3,7]

$$A\frac{\partial u}{\partial t} + B\frac{\partial u}{\partial x} + C\frac{\partial u}{\partial y} = f(t, x, y, u), \tag{3.2.14}$$

the vector of a normal (τ, ξ, η) to the characteristic surface is determined by the equation

$$\det \| \tau A + \xi B + \eta C \| = 0. \tag{3.2.15}$$

In concluding this section, we would like to note that, if the *characteristics are complex*, then such a system of equations is called *elliptic*. As was shown by Hadamard, it is impossible to formulate a Cauchy problem for the elliptic system of equations. As a matter of fact, the small disturbances given at $t = 0$ grow exponentially. As a result, the problem solution does not already depend continuously on the initial data. As the mathematicians say, the problem becomes *illposed.* For an elliptic system, the Dirichlet problem of the function restoration in a region on

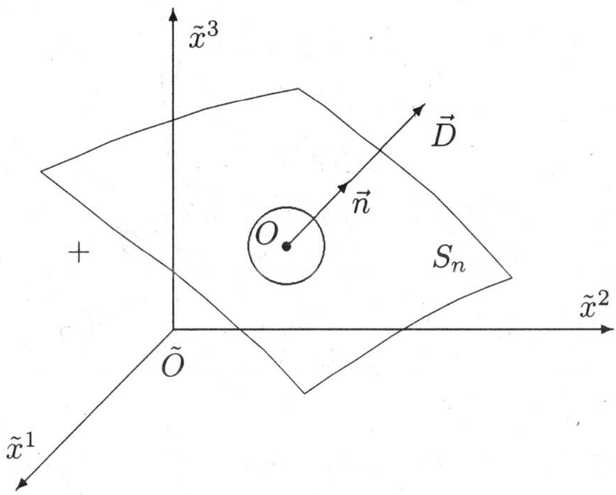

Figure 3.5: The discontinuity surface.

the basis of the given values on the boundary is a well-posed problem. For example, in the case of the elliptic Cauchy-Riemann system

$$\frac{\partial u}{\partial t} + \frac{\partial v}{\partial x} = 0, \quad \frac{\partial v}{\partial t} - \frac{\partial u}{\partial x} = 0, \tag{3.2.16}$$

a well-posed problem is the Dirichlet problem of determining the function $u(t, x)$ [and $v(t, x)$ with the accuracy up to a constant] in a region $D(x, t)$ bounded by a closed curve γ: $x = x(\xi)$, $t = t(\xi)$ on the basis of a function $u|_\gamma = \Phi(\xi)$ given on the boundary γ.

3.3 Discontinuity Surfaces in Continuum Mechanics

We have studied in the foregoing section the characteristics, which represent the lines or the surfaces on which some derivatives cannot be determined, i.e., they undergo a discontinuity. A part of them are also called the *weak discontinuities*, because the functions themselves remain continuous on these discontinuities. In addition to them, the strong discontinuities on which the functions themselves undergo a discontinuity: the velocity, density, stress, etc., can take place in continuum mechanics. Since the differential equations governing the continuum motions assume the existence of continuous functions and their derivatives, they are inapplicable on the discontinuities. Therefore, it is necessary to find the algebraic relationships on these discontinuities.

Let the functions $\varphi(\tilde{x}^i)$ sought for undergo a discontinuity on some surface S_n (see Fig. 3.5), which is specified in the Eulerian coordinate system $\tilde{x}^i$ by the equation

$$\tilde{f}(t, \tilde{x}^i) = 0, \tag{3.3.1}$$

with the normal

$$\vec{n} = \nabla \tilde{f} / |\nabla \tilde{f}|. \tag{3.3.2}$$

The displacement velocity of this surface in the space is determined by the vector $\vec{D} = D\vec{n}$, which in each point is perpendicular to the surface S_n. Differentiating both sides of (3.3.1) with respect to time

$$\frac{\partial \tilde{f}}{\partial t} + \frac{\partial \tilde{f}}{\partial \tilde{x}^i} \frac{d\tilde{x}^i}{dt} = 0,$$

we obtain with regard for the definition (3.3.2) and the relationship

$$\frac{\partial \tilde{f}}{\partial \tilde{x}^i} \frac{d\tilde{x}^i}{dt} = n_i \frac{d\tilde{x}^i}{dt} \left| \frac{\partial \tilde{f}}{\partial \tilde{x}^i} \right| = (\vec{n} \cdot \tilde{\vec{v}}) \left| \frac{\partial \tilde{f}}{\partial \tilde{x}^i} \right| = D \left| \frac{\partial \tilde{f}}{\partial \tilde{x}^i} \right|$$

the formula:

$$D = -\frac{\partial \tilde{f}}{\partial t} \Big/ \left| \frac{\partial \tilde{f}}{\partial \tilde{x}^i} \right|, \tag{3.3.3}$$

where

$$\left| \frac{\partial \tilde{f}}{\partial \tilde{x}^i} \right| = \sqrt{\tilde{g}^{kl} \frac{\partial \tilde{f}}{\partial \tilde{x}^k} \frac{\partial \tilde{f}}{\partial \tilde{x}^l}}.$$

To derive the relations on a discontinuity at point O (Fig. 3.5), let us go over to the inertial reference frame K', in which the point O is at rest. The velocity $\vec{u}$ of the continuum motion in the system K' is related to the velocity $\vec{v}$ in a rest-system K by the Galilean transformation

$$\vec{u} = \vec{v} - \vec{D}. \tag{3.3.4}$$

Introduce the Eulerian coordinate system x^i in the reference frame K' with the center at point O (see Fig. 3.6). Since continuum mechanics equations retain their form in any inertial coordinate system (they are invariant under the Galilean transformation), the continuity equation (2.1.154), momentum equation (2.1.156), and energy equation (2.1.161) will have in the Eulerian coordinates of the system K' the form

$$\begin{aligned} &\frac{\partial \rho}{\partial t} + \nabla_j \rho u^j = 0, \\ &\frac{\partial}{\partial t}(\rho u^i) + \nabla_j (\rho u^i u^j - \sigma^{ij}) = 0, \\ &\frac{\partial}{\partial t}\left(\rho\left(E + \frac{u^2}{2}\right)\right) + \nabla_j \left(\rho u^j \left(E + \frac{u^2}{2}\right) - \sigma^{ij} u_j + q^j\right) = 0. \end{aligned} \tag{3.3.5}$$

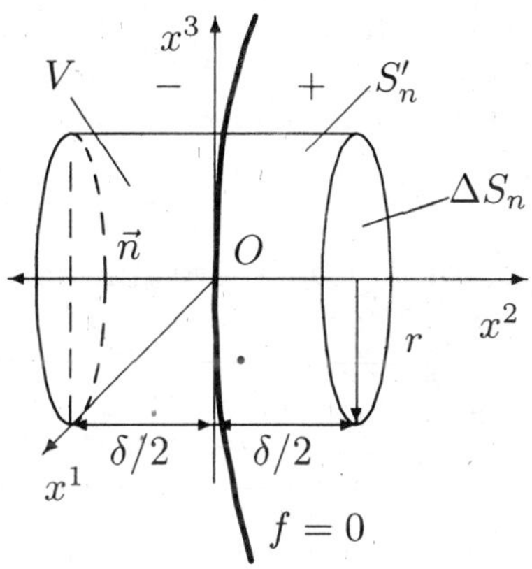

Figure 3.6: The coordinate system and the integration region in a reference frame in which the velocity of point O is equal to zero.

The body forces and the body heat sources are not taken into account in equations (3.3.5). As was noted above, these equations cannot be used on a discontinuity; therefore, let us go over from the differential equations to the integral equations.

For this purpose, let us consider a divergence equation of the general form

$$\frac{\partial \varphi}{\partial t} + \nabla_j \psi^j = 0. \tag{3.3.6}$$

Multiply both sides of equation (3.3.6) by dV and integrate over a finite volume V. Using the Gauss–Ostrogradsky theorem, (1.1.61) we obtain:

$$\frac{\partial}{\partial t} \int_V \varphi \, dV + \int_S \psi_n \, dS_n = 0, \tag{3.3.7}$$

where $\psi_n = \psi^j n_j$. In this equation, the φ and ψ^j cannot only be continuous, but also discontinuous functions of the coordinates x^i. In the particular case of the continuous functions, equation (3.3.6) follows from it. In a general case, however, it describes a wider class of the solutions. We recall that equations (3.3.5) were obtained in Section 2.1 from the integral equations.

Let us integrate equation (3.3.7) on a discontinuity in our case. As a volume V, we choose a cylinder whose generatrix has the length δ and is parallel with the vector of normal $\vec{n}$ at point O (see Fig. 3.6). A part of the cylinder $\delta/2$ in length along the normal $\vec{n}$ lies on the one side of the discontinuity surface $f(t, x^i) = 0$, and the remaining part lies on the other side. We will mark our functions in the first region by the subscript "minus," i.e., φ_-, ψ^j_-, and we will mark the functions in the second region by the plus sign, i.e., φ_+, ψ^j_+. The point O will be always at rest; however, the other points of the surface $f(t,x^i)=0$ have different velocities $D=D(t,x^i)$. Therefore, the surface $f=0$ will shift with time. If we choose a sufficiently small radius r of the cylinder base and consider the surface $f=0$ only in a small neighborhood of the point O, however, then we can neglect this shift. After these remarks, let us calculate the integral in (3.3.7) over the volume and the cylinder surface:

$$\frac{\partial}{\partial t}\int_V \varphi\, dV = \frac{\partial}{\partial t}\Big(\frac{1}{V}\int \varphi\, dV\Big)\cdot V = \frac{\partial}{\partial t}\langle\varphi\rangle\pi r^2\,\delta, \tag{3.3.8}$$

$$\int_{S_n}\psi_n\, dS_n = \int_{S_n}\psi^j n_j\, dS_n = (\psi^j_- - \psi^j_+)\, n_j\pi r^2 + \int_{S'_n}\psi_n\, dS_n, \tag{3.3.9}$$

where we have taken into account the fact that the normals to the opposite cylinder bases have the opposite signs. Let us write the integral over the lateral cylinder surface in the form

$$\int_{S'_n}\psi_n\, dS_n = \frac{1}{S'_n}\int_{S'_n}\psi_n\, dS_n\cdot S'_n = \langle\psi_n\rangle S'_n = \langle\psi_n\rangle 2\pi r\,\delta. \tag{3.3.10}$$

Substituting (3.3.8)–(3.3.10) into (3.3.7), we obtain the equation:

$$\frac{\partial}{\partial t}\langle\varphi\rangle\pi r^2\delta + (\psi_{n-} - \psi_{n+})\pi r^2 + \langle\psi_n\rangle 2\pi r\delta = 0. \tag{3.3.11}$$

It is assumed that the functions φ and ψ^j are bounded and continuous, with respect to t and x^i, together with their derivatives on both sides of the discontinuity surface $f(t,x^i)=0$. In this case, the functions $\langle\varphi\rangle$, $\frac{\partial\langle\varphi\rangle}{\partial t}$, and $\langle\psi_n\rangle$ are also bounded and continuous functions of the time t. Therefore, as the length of the cylinder generatrix tends to zero, i.e., $\delta\to 0$, we obtain from (3.3.11) the relation on the discontinuity:

$$[\psi_n] = 0, \tag{3.3.12}$$

where the square brackets denote the quantity

$$[\psi_n] = \psi_{n+} - \psi_{n-}, \qquad \psi_n = \psi^j n_j. \tag{3.3.13}$$

Let us apply the obtained formulas to the divergence system of equations (3.3.5). In particular, for the continuity equation $\varphi=\rho$, $\psi^j=\rho u^j$, and from (3.3.12), we obtain the relation:

$$[\rho u_n] = 0. \tag{3.3.14}$$

Let us multiply both sides of equation (3.3.4) by $\vec{n}$ as a scalar product. With regard for the relations

$$\vec{D} = D\cdot\vec{n}, \quad \vec{n}\cdot\vec{n} = 1, \quad v_n = \vec{v}\cdot\vec{n}, \quad u_n = \vec{u}\cdot\vec{n} = u^j n_j,$$

we can find the quantity u_n:

$$u_n = v_n - D. \tag{3.3.15}$$

For the momentum equation (3.3.5), one must consider three components:

$$\varphi^j = \rho u^i, \quad \psi_n^i = \rho u^i u_n - f_n^i, \quad f_n^i = \sigma^{ij} n_j, \tag{3.3.16}$$

from where we obtain the following three relations ($i = 1, 2, 3$):

$$[\rho u_n u^i - f_n^i] = 0. \tag{3.3.17}$$

Multiplying both sides of (3.3.17) by the basis vectors $\vec{e}_i$ and adding to each other, we obtain with regard for the formulas $\vec{u} = u^i \vec{e}_i$, $\vec{f}_n = f_n^i \vec{e}_i$ one vector relationship:

$$[\rho u_n \vec{u} - \vec{f}_n] = 0. \tag{3.3.18}$$

While considering the energy equation in (3.3.5), let us set

$$\varphi = \rho\Big(E + \frac{u^2}{2}\Big), \quad \psi_n = \rho u_n\Big(E + \frac{u^2}{2}\Big) - \vec{f}_n \cdot \vec{u} + q_n, \tag{3.3.19}$$
$$\vec{f}_n \cdot \vec{u} = \sigma^{ij} u_i n_j, \quad q_n = q^j n_j.$$

Substituting (3.3.19) into (3.3.12), we find the last relationship:

$$[\rho u_n \Big(E + \frac{u^2}{2}\Big) - \vec{f}_n \cdot \vec{u} + q_n] = 0. \tag{3.3.20}$$

The algebraic equations (3.3.14), (3.3.18), and (3.3.20) determine completely the relations on the discontinuity surface $f(t, x^i) = 0$ and are consequences of the conservation laws of mass, momentum, and energy.

Let us identify two classes of the discontinuities: the *tangential discontinuities*, which are determined by the condition $u_n = 0$, and the *shock waves* at $u_n \neq 0$. In the case of tangential discontinuities, it follows from (3.3.15) that $v_n = D$; i.e., the particles of a continuum do not intersect the tangential discontinuity surface. Assuming $u_n = 0$ in the obtained relations, wc can find the relations on the tangential discontinuities:

$$[\vec{f}_n] = 0, \quad [\vec{f}_n \cdot \vec{u} - q_n] = 0. \tag{3.3.21}$$

It follows from (3.3.14) and (3.3.18) that the density jumps $[\rho]$ and the velocity component tangential to the discontinuity surface $[u_\tau]$ may be arbitrary. At the shock waves, the quantities $[u_n] \neq 0$ and the jump of the normal velocity component $[u_n] \neq 0$. In the literature, it is conventional to supply the parameters ahead of the shock wave with the subscript 1, and the parameters behind the shock wave are supplied with the subscript 2 instead of the subscripts $+, -$. With regard for this remark, let us write the continuity equation (3.3.14) in the form

$$\rho_1 u_{n1} = \rho_2 u_{n2}. \tag{3.3.22}$$

Here, two substantially different cases are possible depending on the sign of $[u_n]$. The first case $u_{n1} > u_{n2}$, $\rho_1 < \rho_2$ corresponds to the compression shock wave, and the second case $u_{n1} < u_{n2}$, $\rho_1 > \rho_2$ corresponds to the rarefaction shock wave.

The choice of the needed solution is determined with the aid of the second law of thermodynamics (2.1.62), which in the Eulerian coordinates x^i in the reference frame K' has the form

$$\frac{\partial S}{\partial t} + u^j \nabla_j S = \frac{1}{T}(\dot{q}_e + \dot{q}'), \quad \dot{q}' \geq 0. \tag{3.3.23}$$

Multiplying both sides of (3.3.23) by ρ and both sides of the continuity equation by S and adding the obtained equations, we can rewrite equation (3.3.23) in the divergence form

$$\frac{\partial}{\partial t}(\rho S) + \nabla_j(\rho u^j S) = \frac{\rho}{T}(\dot{q}_e + \dot{q}'). \tag{3.3.24}$$

Integrating this equation over the volume V (see Fig. 3.6), we obtain similarly to (3.3.11):

$$[\rho u_n S] = \frac{1}{\pi r^2} \int_V \frac{\rho}{T}(\dot{q}_e + \dot{q}')\, dV. \tag{3.3.25}$$

In contrast with (3.3.12), the integral over the volume has emerged on the right-hand side.

Let us show that this integral is different from zero and that it always has a positive sign. Before proceeding to this proof, we note that the notion of the discontinuity surface is a mathematical idealization. The flow parameters at a shock wave rapidly change within narrow transitional layers, the width of which h reduces with increasing the shock wave strength down to the order of smallness of several free paths λ. To estimate the integral in (3.3.24), we shall assume the shock wave to be weak, so that $h \gg \lambda$. In this case, continuum mechanics hypotheses will be valid in the transitional layer. A specific heat flux is determined by formulas (2.1.107) and (2.1.108):

$$\dot{q}_e = -\frac{1}{\rho}\nabla_i q^i, \quad q_i = -\kappa \nabla_i T, \tag{3.3.26}$$

and with their regard, we obtain:

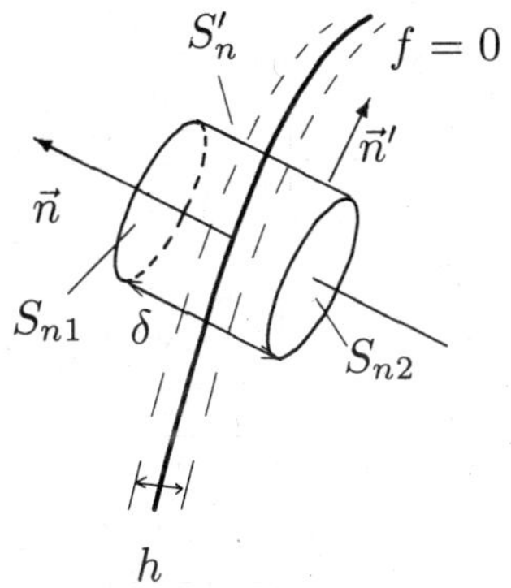

Figure 3.7: To the estimation of the entropy growth at a shock wave.

$$\begin{aligned} \int_V \frac{\rho}{T} \dot{q}_e \, dV &= -\int_V \nabla_i \Big(\frac{q^i}{T}\Big) \, dV + \int_V q^i \nabla_i \Big(\frac{1}{T}\Big) \, dV \\ &= -\int_S \frac{q_n}{T} \, dS_n + \int_V \kappa \frac{(\nabla_i T)^2}{T^2} \, dV. \end{aligned} \tag{3.3.27}$$

Since the gradients outside a narrow transitional layer are small, the integrals taken over the cylinder bases S_{n1} and S_{n2} will be small. In addition, we note that the gradients of all quantities in the transitional layer are large along the normal $\vec{n}$ to the surface $f(t, x^i) = 0$, and they are small along all other directions lying in a plane tangent to the surface $f = 0$. A normal-to-the-lateral cylinder surface $\vec{n}'$ is perpendicular to the normal $\vec{n}$ (see Fig. 3.7); therefore, the integral along the lateral surface $\int_{S'_n} \frac{1}{T} q_{n'} \, dS_n$ will also be small. One can consequently neglect in (3.3.26) the integral over the cylinder surface S_n and assume that

$$\int_V \frac{\rho}{T} \dot{q}_e \, dV \approx \int_V \kappa \frac{(\nabla_i T)^2}{T^2} \, dV > 0. \tag{3.3.28}$$

According to the second law of thermodynamics, $\dot{q}' > 0$ in the transitional layer; therefore,

$$\int_V \frac{\rho}{T} \dot{q}' \, dV > 0. \tag{3.3.29}$$

Substituting (3.3.27) and (3.3.28) into (3.3.24) and taking into account the relations $j = \rho_1 u_{n1} = \rho_2 u_{n2} > 0$ [see formula (3.3.14)], we obtain a criterion for the solution choice:

$$[S] > 0. \tag{3.3.30}$$

Thus, the entropy S_2 behind the shock wave should always be larger than the entropy S_1 ahead of the shock wave: $S_2 > S_1$. Let us estimate

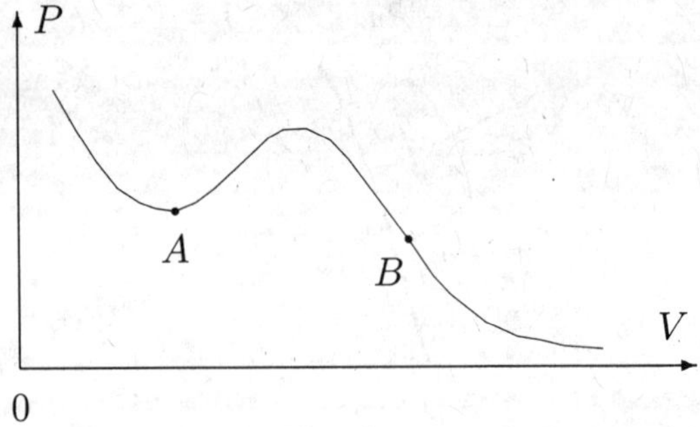

Figure 3.8: The adiabat of the van der Waals' gas.

the entropy increment behind the shock wave. The magnitude of a non-compensated heat is determined by the work of viscous forces. Using formula (2.1.106) for $\dot{q}'$ and formula (3.3.27), we can find the entropy change for a viscous, heat-conducting gas:

$$[S] = \frac{1}{j\pi r^2} \int_V \left(\kappa \frac{|\nabla_i T|^2}{T^2} + \frac{1}{T}((\operatorname{div} \vec{u})^2 + 2\mu \dot{e}_{ij} \dot{e}^{ij} \right) dV.$$

Assuming in this formula that $\nabla_i T \sim [T]/h$, $\dot{e}_{ij} \sim \operatorname{div} \vec{u} \sim \frac{[u]}{h}, j \sim \rho u$, we obtain:

$$[S] \sim \frac{\delta}{\rho u h^2} \left(\kappa \frac{[T]^2}{T^2} + \mu \frac{[u]^2}{T} \right). \tag{3.3.31}$$

Since $h \leq \delta$, $[T] \neq 0$, and $[u] \neq 0$, the entropy jump in (3.3.30) will be a finite quantity. One can find from the conditions at a shock wave (3.3.14), (3.3.18), (3.3.20) the entropy jump $[S]$, which does not depend on the transitional layer structure and is determined by shock wave strength specified, for example, by the velocity jump $[u_n]$. As will be shown below in Problem 3.1, it follows from (3.3.29) that, for the "normal" media for which the adiabat is convex downwards, the compression shock waves are stable. For the media having the intervals that are convex upwards on the adiabat (the interval AB in Fig. 3.8), the existence of the rarefaction shock waves is possible.

Problem 3.1. Determine with the aid of the Pi theorem the drag force of a sphere as it enters at a constant speed v, a half-space filled with a viscous, incompressible fluid (see Fig. 3.9). The sphere radius is equal to a. Consider the limiting cases $v \to 0$ and $v \to \infty$.

Solution: The drag force F depends on the following parameters:

$$F = F(\rho,\ v,\ \mu,\ t,\ a), \tag{3.3.32}$$

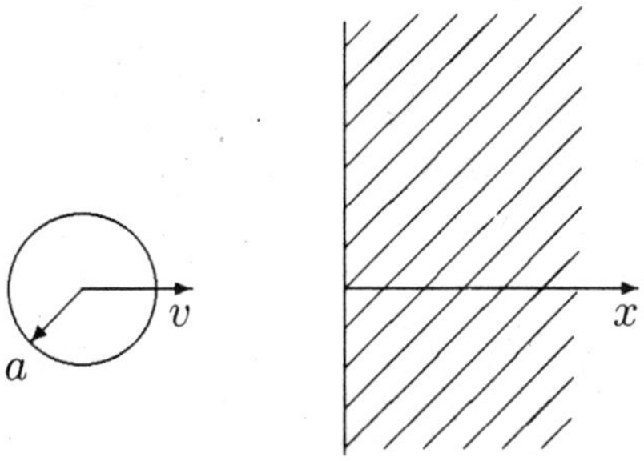

Figure 3.9: Penetration of a rigid sphere into a semi-infinite obstacle.

where ρ and μ are the fluid density and viscosity and t is the time. Choosing ρ, v, and t as the independent dimensional quantities, let us rewrite equation (3.3.32) in the nondimensional variables:

$$\Pi_F = \Pi_F(\Pi_1,\ \Pi_2), \tag{3.3.33}$$

where

$$\Pi_F = \frac{F}{\rho^k v^l t^n}, \quad \Pi_1 = \frac{\mu}{\rho^\alpha v^\beta t^\gamma}, \quad \Pi_2 = \frac{a}{\rho^s v^m t^h}.$$

The quantity Π_F is nondimensional; therefore,

$$\begin{aligned} [\Pi_F] &= 1 = \frac{n}{\left(\frac{kg}{m^3}\right)^k \left(\frac{m}{s}\right)^l s^n} = \frac{kg \cdot m}{s^2 \left(\frac{kg}{m^3}\right)^k \left(\frac{m}{s}\right)^l s^n} \\ &= 1/\left(kg^{(k-1)} \cdot m^{(l-3k-1)} \cdot s^{(2+n-l)}\right), \end{aligned}$$

from where it follows that

$$k - 1 = 0, \quad l - 3k - 1 = 0, \quad 2 + n - l = 0.$$

The solution of this system is

$$k = 1, \qquad l = 4, \qquad n = 2;$$

therefore,

$$\Pi_F = \frac{F}{\rho v^4 t^2}.$$

We have in a similar way for the Π_1 that

$$[\Pi_1] = 1 = \frac{\frac{n}{m^2} s}{\left(\frac{kg}{m^3}\right)^\alpha \left(\frac{m}{s}\right)^\beta s^\gamma} = \frac{1}{kg^{(\alpha-1)} m^{(1-3\alpha+\beta)} s^{(1-\beta+\gamma)}},$$

from where it follows that

$$\alpha - 1 = 0, \quad 1 - 3\alpha + \beta = 0, \quad 1 - \beta + \gamma = 0.$$

The solution of this system of equation is

$$\alpha = 1, \qquad \beta = 2, \qquad \gamma = 1;$$

therefore,

$$\Pi_1 = \mu/(\rho v^2 t).$$

We obtain in a similar way that $\Pi_2 = a/(vt)$. Substituting the expressions for Π_F, Π_1, and Π_2 into equation (3.3.33), we obtain the relation:

$$\frac{F}{\rho v^4 t^2} = \Pi_F(\frac{\rho v^2 t}{\mu}, \frac{vt}{a}). \tag{3.3.34}$$

Denoting the unknown function Π_F by φ, we can rewrite equation (3.3.34) in the form

$$F = \rho v^4 t^2 \varphi\Big(\frac{\rho v^2 t}{\mu}, \frac{vt}{a}\Big). \tag{3.3.35}$$

The nondimensional variable $\Pi_1 = \frac{\rho v^2 t}{\mu}$ is an analog of the Reynolds number, and $\Pi_2 = \frac{vt}{a}$ is an analog of the Strouhal number.

In the limit of small velocities $v \to 0$, the force F is determined by the viscosity μ and does not depend on the inertial properties ρ; therefore,

$$F = F(v, \ \mu, \ t, \ a).$$

Choosing the μ, v, and t as the independent quantities, we obtain:

$$\Pi_F = \Pi_F(\Pi_1),$$
$$\Pi_F = F/(\mu v^2 t), \quad \Pi_1 = vt/a,$$

from where it follows that

$$F = \mu v^2 t\, \Phi(vt/a). \tag{3.3.36}$$

In the limit of large velocities $v \to \infty$, the force F does not depend on μ; therefore, we find from equation (3.3.35) that

$$F = \rho v^4 t^2 \psi\Big(\frac{vt}{a}\Big). \tag{3.3.37}$$

The obtained expressions (3.3.35), (3.3.36), and (3.3.37) do not depend on the choice of the independent dimensional parameters. Taking ρ, v, and a as the independent quantities, we obtain:

$$\Pi_F = \frac{F}{\rho v^2 a^2}; \quad \Pi_1 = \frac{\rho v a}{\mu}; \quad \Pi_2 = \frac{vt}{a},$$

from where it follows that

$$F = \rho v^2 a^2 \varphi_2\Big(\frac{\rho v a}{\mu},\ \frac{vt}{a}\Big). \tag{3.3.38}$$

Let us rewrite equation (3.3.34) in the form

$$F = \rho v^2 a^2 \Big(\frac{vt}{a}\Big)^2 \varphi\Big(\frac{\rho v a}{\mu}\Big(\frac{vt}{a}\Big),\ \frac{vt}{a}\Big).$$

It follows from here that the formulas (3.3.35) and (3.3.38) coincide at

$$\varphi_2\Big(\frac{\rho v a}{\mu},\ \frac{vt}{a}\Big) = \Big(\frac{vt}{a}\Big)^2 \varphi\Big(\Big(\frac{\rho v a}{\mu}\Big)\Big(\frac{vt}{a}\Big),\ \frac{vt}{a}\Big),$$

where $\mathrm{Re} = \frac{\rho v a}{\mu}$ is the Reynolds number and $St = \frac{vt}{a}$ is the Strouhal number.

Problem 3.2. Find the characteristics for the following second-order equations:

1) $\dfrac{\partial^2 u}{\partial t^2} - c^2 \dfrac{\partial^2 u}{\partial x^2} = 0$ the wave equation;

2) $\dfrac{\partial^2 u}{\partial t^2} + \dfrac{\partial^2 u}{\partial x^2} = 0$ the Laplace equation;

3) $\dfrac{\partial u}{\partial t} - \dfrac{\partial^2 u}{\partial x^2} = 0$ the heat equation.

Solution: Consider at first the general second-order equation

$$A\frac{\partial^2 u}{\partial t^2} + 2B\frac{\partial^2 u}{\partial t \partial x} + C\frac{\partial^2 u}{\partial x^2} = f, \tag{3.3.39}$$

where A, B, and C are the functions of t, x, and u, and

$$f = f\Big(t,\ x,\ u,\ \frac{\partial u}{\partial x},\ \frac{\partial u}{\partial t}\Big).$$

Following[7], we go over in (3.3.39) to the new coordinates

$$\alpha = \alpha(x,t), \quad \beta = \beta(x,t), \quad \frac{\partial(\alpha,\beta)}{\partial(t,x)} \neq 0,$$

in which equation (3.3.39) will have the form

$$L\frac{\partial^2 u}{\partial \alpha^2} + 2K\frac{\partial^2 u}{\partial \alpha \partial \beta} + M\frac{\partial^2 u}{\partial \beta^2} + N\frac{\partial u}{\partial \alpha} + P\frac{\partial u}{\partial \beta} = f, \tag{3.3.40}$$

where

$$
\begin{aligned}
f &= f\Big(t,\ x,\ u,\ \frac{\partial u}{\partial \alpha}\frac{\partial \alpha}{\partial x} + \frac{\partial u}{\partial \beta}\frac{\partial \beta}{\partial x},\ \frac{\partial u}{\partial \alpha}\frac{\partial \alpha}{\partial t} + \frac{\partial u}{\partial \beta}\frac{\partial \beta}{\partial t}\Big),\\
L &= A\Big(\frac{\partial \alpha}{\partial t}\Big)^2 + 2B\Big(\frac{\partial \alpha}{\partial t}\Big)\Big(\frac{\partial \alpha}{\partial x}\Big) + C\Big(\frac{\partial \alpha}{\partial x}\Big)^2,\\
K &= A\Big(\frac{\partial \alpha}{\partial t}\Big)\Big(\frac{\partial \beta}{\partial t}\Big) + B\bigg(\Big(\frac{\partial \alpha}{\partial t}\Big)\Big(\frac{\partial \beta}{\partial x}\Big) + \Big(\frac{\partial \alpha}{\partial x}\Big)\Big(\frac{\partial \beta}{\partial t}\Big)\bigg) + C\frac{\partial \alpha}{\partial x}\frac{\partial \beta}{\partial x},\\
M &= A\Big(\frac{\partial \beta}{\partial t}\Big)^2 + 2B\Big(\frac{\partial \beta}{\partial t}\Big)\Big(\frac{\partial \beta}{\partial x}\Big) + C\Big(\frac{\partial \beta}{\partial x}\Big)^2,\\
N &= A\frac{\partial^2 \alpha}{\partial t^2} + 2B\frac{\partial^2 \alpha}{\partial t \partial x} + C\frac{\partial^2 \alpha}{\partial x^2},\\
P &= A\frac{\partial^2 \beta}{\partial t^2} + 2B\frac{\partial^2 \beta}{\partial t \partial x} + C\frac{\partial^2 \beta}{\partial x^2}.
\end{aligned}
$$

While moving along the coordinate line $\alpha = \text{const}$, the β changes, and conversely. Let us formulate a Cauchy problem on the coordinate line $\alpha = \text{const}$ for the second-order equation (3.3.40). For this purpose, we specify on the line $\alpha = \text{const}$ the $u(\beta)$ and $\frac{\partial u(\beta)}{\partial \alpha}$. Differentiating these functions with respect to β, we can find that, in addition to them, the values $\frac{\partial u}{\partial \beta}$, $\frac{\partial^2 u}{\partial \beta^2}$, $\frac{\partial^2 u}{\partial \alpha \partial \beta}$ are specified on the given coordinate line. We determine the characteristics from the condition that the given Cauchy problem for equation (3.3.39) is unsolvable [although the solution $u(\alpha, \beta)$ itself exists]. It follows from the form of (3.3.40) that the coefficient L affecting the second derivative $\frac{\partial^2 u}{\partial \alpha^2}$ should vanish for this purpose. Thus, the *lines* $\alpha(t, x) = \text{const}$ on which

$$A\Big(\frac{\partial \alpha}{\partial t}\Big)^2 + 2B\Big(\frac{\partial \alpha}{\partial t}\Big)\Big(\frac{\partial \alpha}{\partial x}\Big) + C\Big(\frac{\partial \alpha}{\partial x}\Big)^2 = 0 \tag{3.3.41}$$

are called the characteristics of equation (3.3.39). From (3.3.40), (3.3.41), and the condition for the solution existence, it follows a relation along the characteristic $\alpha(t, x) = \text{const}$:

$$2K\frac{\partial^2 u}{\partial \alpha \partial \beta} + M\frac{\partial^2 u}{\partial \beta^2} + N\frac{\partial u}{\partial \alpha} + P\frac{\partial u}{\partial \beta} = f. \tag{3.3.42}$$

Using the obtained formulas, it is easy to obtain the characteristics for the equations 1), 2), 3) given in the problem formulation. For the wave equation, the characteristics are determined from (3.3.41) at $A = 1$, $B = 0$, and $C = -c^2$:

$$\Big(\frac{\partial \alpha}{\partial t}\Big)^2 - c^2\Big(\frac{\partial \alpha}{\partial x}\Big)^2 = \Big(\frac{\partial \alpha}{\partial t} - c\Big(\frac{\partial \alpha}{\partial x}\Big)\Big)\Big(\frac{\partial \alpha}{\partial t} + c\Big(\frac{\partial \alpha}{\partial x}\Big)\Big) = 0,$$

from where we obtain the equations $\alpha(ct - x) = \text{const}$, and $\alpha(ct + x) = \text{const}$. These equations are valid if

$$ct - x = \text{const}, \qquad ct + x = \text{const}, \tag{3.3.43}$$

which determine two families of lines, along which the disturbances propagate at the speed c to the right and to the left.

For the Laplace equation, we set $A = 1,\ B = 0$, and $C = 1$, and the characteristics obey the equation

$$\left(\frac{\partial\alpha}{\partial t}\right)^2 + \left(\frac{\partial\alpha}{\partial x}\right)^2 = \left(\frac{\partial\alpha}{\partial t} + i\frac{\partial\alpha}{\partial x}\right)\left(\frac{\partial\alpha}{\partial t} - i\frac{\partial\alpha}{\partial x}\right) = 0,$$

from which it follows that the characteristics are complex:

$$it - x = \text{const}, \qquad it + x = \text{const},$$

i.e., the Laplace equation has the elliptic type.

For the heat equation, we have $A = B = 0,\ C = 1$, and the equation for the characteristics

$$\left(\frac{\partial\alpha}{\partial x}\right)^2 = 0$$

has the solution $\alpha = \alpha(t)$. It follows from here that the characteristics are represented by the straight lines $\alpha(t) = \text{const}$ parallel with the x axis. It follows from here that the Cauchy problem for the heat equation on the line $t = 0$ is illposed. Only one function is specified as an initial condition for the heat equation, although the equation itself has the second order. Such a problem is no Cauchy problem.

Problem 3.3. Write the relations on a shock wave in an ideal gas, and find the entropy jump in a weak shock wave.

Solution: According to (2.1.81), we have in the ideal gas that

$$\sigma^{ij} = -Pg^{ij}, \quad \dot{q}' = 0, \quad q^j = 0;$$

therefore,

$$\begin{aligned} f_n^i &= \sigma^{ij} n_j = -Pn^i, \quad \vec{f}_n \cdot \vec{u} = \sigma^{ij} u_i n_j = -Pu^j n_j = -Pu_n, \\ u_n &= \vec{u} \cdot \vec{n} = u^j n_j, \qquad q_n = q^j n_j = 0. \end{aligned}$$

Substituting these formulas into (3.3.14), (3.3.18), and (3.3.20), we obtain:

$$[\rho u_n] = 0, \ [\rho u_n \vec{u} + P\vec{n}] = 0, \ \left[\rho u_n\left(E + \frac{u^2}{2}\right) + Pu_n\right] = 0. \tag{3.3.44}$$

Multiplying the second equation by $\vec{n}$ in accordance with a scalar product and by the vector $\vec{\tau}$ perpendicular to $\vec{n}$, we find with regard for equation $\vec{n} \cdot \vec{\tau} = 0$:

$$[\rho u_n] = 0, \ [\rho u_n^2 + P] = 0, \ [u_\tau] = 0, \ \left[\left(H + \frac{u_n^2}{2}\right)\right] = 0, \qquad (3.3.45)$$

where $[\varphi] = \varphi_2 - \varphi_1$, u_τ is the velocity component parallel with a tangent plane to the discontinuity surface at point O, and $H = E + P/\rho$ is the enthalpy (2.1.87).

Let us find the entropy jump in a weak shock wave. For this purpose, we combine the relations (3.3.45) with each other and rewrite them in a slightly different form

$$\begin{aligned} u_{n1} = jV_1, \quad u_{n2} = jV_2, \quad P_2 - P_1 &= j^2(V_1 - V_2), \\ H_2 - H_1 &= \frac{1}{2}(V_1 + V_2)(P_2 - P_1), \end{aligned} \qquad (3.3.46)$$

where $j = \rho u_n$ is the mass flux, and $V = 1/\rho$ is the specific volume. The last equation in (3.3.46) is called the *Hugoniot adiabat* in the literature. Since $H = H(P, S)$, the difference $H_2 - H_1$ will depend on $S_2 - S_1$ and $(P_2 - P_1)$. Expanding $H_2 - H_1$ into a Taylor series in $(S_2 - S_1)$ up to the first-order terms, and in $(P_2 - P_1)$ up to the third-order terms, we obtain the formula:

$$\begin{aligned} H_2 - H_1 &= \left(\frac{\partial H}{\partial S}\right)_P (S_2 - S_1) + \left(\frac{\partial H}{\partial P}\right)_S (P_2 - P_1) \\ &+ \frac{1}{2}\left(\frac{\partial^2 H}{\partial P^2}\right)_S (P_2 - P_1)^2 + \frac{1}{6}\left(\frac{\partial^3 H}{\partial P^3}\right)_S (P_2 - P_1)^3. \end{aligned} \qquad (3.3.47)$$

All derivatives are taken in (3.3.47) at point S_1, P_1. Using (2.1.87), let us rewrite (3.3.47) in the form

$$\begin{aligned} H_2 - H_1 &= T_1(S_2 - S_1) + V_1(P_2 - P_1) + \frac{1}{2}\left(\frac{\partial V}{\partial P_1}\right)_S (P_2 - P_1)^2 \\ &+ \frac{1}{6}\left(\frac{\partial^2 V}{\partial P_1^2}\right)_S (P_2 - P_1)^3. \end{aligned} \qquad (3.3.48)$$

Since the product $(V_1 + V_2)(P_2 - P_1)$ in the Hugoniot adiabat [the last equation in (3.3.46)] already has the first order of smallness with respect to $(P_2 - P_1)$, the specific volume V_2 in this expression should be expanded up to the second-order terms with respect to $(P_2 - P_1)$:

$$V_2 - V_1 = \left(\frac{\partial V}{\partial P_1}\right)_S (P_2 - P_1) + \frac{1}{2}\left(\frac{\partial^2 V}{\partial P_1^2}\right)_S (P_2 - P_1)^2. \qquad (3.3.49)$$

Substituting (3.3.48) and (3.3.49) into the equation for the Hugoniot adiabat, we obtain:

$$S_2 - S_1 = \frac{1}{12T_1}\Big(\frac{\partial^2 V}{\partial P_1^2}\Big)_S (P_2 - P_1)^3. \qquad (3.3.50)$$

For the normal gases $(\frac{\partial^2 V}{\partial P^2})_S > 0$; therefore, the condition $S_2 > S_1$ will be satisfied if $P_2 > P_1$, which corresponds to a compression shock wave. At the phase transitions, the regions can arise, where $(\frac{\partial^2 V}{\partial P^2})_S < 0$; therefore, the entropy will increase $(S_2 > S_1)$ in the rarefaction shock waves, where $P_2 < P_1$.

References

1. **Sedov, L.I.,** *Similarity and Dimensional Methods in Mechanics* (in Russian), Ninth Edition, Nauka, Moscow, 1987 [English translation made by A.G. Volkovets: **Sedov, L.I.,** *Similarity and Dimensional Methods in Mechanics*, CRC Press, Boca Raton, 1993].
2. **Birkhoff, G.,** *Hydrodynamics. A Study in Logic, Fact and Similitude*, Princeton University Press, Princeton, 1960.
3. **Ovsyannikov, L.V.,** *Lectures on the Fundamentals of Gas Dynamics* (in Russian), Nauka, Moscow, 1981.
4. **Landau, L.D. and Lifschitz, E.M.,** *Hydrodynamics* (in Russian), Nauka, Moscow, 1986.
5. **Smirnov, V.I.,** *A Course of Higher Mathematics, Vol. IV*, Birkhäuser Verlag, Basel and Stuttgart, 1961.
6. **Chorin, A.J. and Marsden, J.E.,** *Mathematical Introduction to Fluid Mechanics.* Second Edition, Springer-Verlag, New York, 1990.
7. **Godunov, S.K.,** *Equations of the Mathematical Physics* (in Russian), Nauka, Moscow, 1979.
8. **Roždestvenskii, B.L. and Janenko, N.N.,** *Systems of Quasilinear Equations and Their Applications to Gas Dynamics.* Second Edition (in Russian), Nauka, Moscow, 1978. English transl.: *Systems of Quasilinear Equations and Their Applications to Gas Dynamics*, Translations of Mathematical Monographs, Vol. 55, American Mathematical Society, Providence, Rhode Island, 1983.

4

Ideal Fluid

This chapter deals with incompressible ideal fluid flows. While choosing the material, the authors aimed at presenting those reults that have now become the classical results and are widely used in the current research work of the aerohydrodynamicists[1–9]. In particular, the Bernoulli and Lagrange integrals are derived in Section 4.1. They enable one to find the pressure distribution in the fluid from a given velocity field.

In Section 4.2, we consider planar irrotational fluid motions. The method of the theory of functions of a complex variable is presented in detail. This has enabled one to solve a variety of problems on the fluid flow around sufficiently arbitrary planar profiles. The computation of a lift force of a wing, which is known in the literature as the Joukowskii force, should be considered as the most important result of this method.

In sections 4.3 and 4.4, we study the three-dimensional fluid flows. We demonstrate here with the aid of *Mathematica* programs the efficiency of the method for computation of three-dimensional flows around the bodies, which is based on a substitution of the body surface by the distributed sources and sinks or dipoles.

In Section 4.5, we consider the general properties of vortex flows and derive the Biot–Savart formula, which enables one to find the fluid flow velocity from a given distribution of vortices.

4.1 Integrals of Motion Equations of Ideal Fluid and Gas

We will call the fluid or gas *ideal* if there are no irreversible processes related to viscosity and heat conduction. The shear stresses are equal to zero in the ideal fluid and gas, and there are only the normal stresses. In the real fluids or gases, the shear stresses are different from zero by virtue of the action of viscous forces. Often the cases take place, where the shear stresses are small, however, as compared with the normal stresses. Under such conditions, it is convenient to represent the fluids or gases as ideal fluids, for which we have in the Eulerian Cartesian coordinates that

$\sigma^{ij} = -P\delta^{ij}$, where P is the pressure. The basic equations governing the ideal fluid or gas motion have been discussed in sufficient detail in Section 2.1.5. We will consider in what follows some exact solutions, which enable one to understand deeper the flow of ideal fluids or gases.

4.1.1 Motion Equations in the Gromeka–Lamb Form

Let us write the momentum equation of an ideal gas (2.1.96) in the Gromeka–Lamb form[1–4]. For this purpose, we make use of a well-known identity

$$(\vec{v} \cdot \nabla)\vec{v} = \nabla\left(\frac{v^2}{2}\right) - \vec{v} \times \operatorname{rot} \vec{v}$$

and obtain:

$$\frac{\partial \vec{v}}{\partial t} + \nabla\left(\frac{v^2}{2}\right) - \vec{v} \times \operatorname{rot} \vec{v} = \vec{F} - \frac{1}{\rho}\nabla P. \tag{4.1.1}$$

The equations in the Gromeka–Lamb form explicitly contain the *curl vector* $\vec{\Omega} = \operatorname{rot} \vec{v}$, which is equal to the doubled angular velocity of the liquid particle rotation $\vec{\Omega} = 2\vec{\omega}$ (1.2.75). Using equation (4.1.1), we can obtain the Bernoulli and Lagrange integrals.

4.1.2 The Bernoulli Integral

We will assume the gas motion to be steady ($\frac{\partial \vec{v}}{\partial t} = 0$), barotropic [$P = P(\rho)$], and that the body forces have the potential $\vec{F} = -\nabla \Pi$. In this case equation (4.1.1) may be presented in the form

$$\nabla\left(\frac{v^2}{2} + \mathcal{P} + \Pi\right) = \vec{v} \times \operatorname{rot} \vec{v},$$

where

$$\mathcal{P} = \int \frac{dP}{\rho}.$$

Let us multiply the left- and right-hand sides of this equation by $\vec{v}$ as in a scalar product:

$$\nabla\left(\frac{v^2}{2} + \mathcal{P} + \Pi\right) \cdot \vec{v} = (\vec{v} \times \operatorname{rot} \vec{v}) \cdot \vec{v}.$$

Since $(\vec{v} \times \operatorname{rot} \vec{v}) \cdot \vec{v} = 0$, we have:

$$\frac{d}{dl}\left(\frac{v^2}{2} + \mathcal{P} + \Pi\right) = 0,$$

where $\vec{v} \cdot \nabla = \frac{d}{dl}$ is a derivative along the streamline. Integrating the last equation along a streamline, we obtain:

$$\frac{v^2}{2} + \mathcal{P} + \Pi = C(l). \tag{4.1.2}$$

Here $C(l)$ is a constant along a streamline, but it changes at a passage to another streamline. It should be noted that, in the case of *irrotational* flows, in which $\operatorname{rot}\vec{v} = 0$, the $C(l)$ will be constant in the overall flow region. Equation (4.1.2) is called the *Bernoulli integral*[1−6].

Consider *homogeneous incompressible fluid in the gravity force field.* In this case, $\rho = \rho_0$ is a given constant, $\mathcal{P} = \frac{P}{\rho_0}$, and $\Pi = g_0 z$, where g_0 is the acceleration of gravity. The Bernoulli integral (4.1.2) takes the form

$$\frac{v^2}{2} + g_0 z + \frac{P}{\rho_0} = C$$

or

$$\frac{v^2}{2g_0} + z + \frac{P}{\rho_0 g_0} = \frac{C}{g_0} = C'. \tag{4.1.3}$$

The individual items in (4.1.3) have the dimension of length and are called, respectively: $\frac{v^2}{2g_0}$ the velocity height, z the geometric height, and $\frac{P}{\rho_0 g_0}$ the piezometric height.

In the case of adiabatic flow of the *perfect gas*, the equation of state has the form (2.1.97) $P = A \cdot \rho^\gamma$, where A is a constant, and γ is the exponent of the Poisson's adiabat. Then, $\mathcal{P} = \frac{\gamma}{\gamma-1} \cdot \frac{P}{\rho}$ and equation (4.1.2) takes the form

$$\frac{v^2}{2} + z + \frac{c^2}{\gamma - 1} + \Pi = C(l), \tag{4.1.4}$$

where $c^2 = \frac{dP}{d\rho} = \frac{\gamma P}{\rho}$ is the squared sound velocity.

In order to determine the constant $C(l)$ in (4.1.4), it is sufficient to know the gas characteristics at some point of the streamline.

4.1.3 The Lagrange Integral

In the case where the flow of an ideal and barotropic fluid or gas is irrotational and the body forces are conservative, equation (4.1.1) may be presented in the form

$$\frac{\partial \vec{v}}{\partial t} + \nabla\left(\frac{v^2}{2}\right) + \nabla\left(\int \frac{dP}{d\rho}\right) + \nabla\Pi = 0.$$

Taking into account the fact that $\vec{v} = \nabla\varphi$, $\frac{\partial \vec{v}}{\partial t} = \nabla\left(\frac{\partial\varphi}{\partial t}\right)$, we obtain:

$$\nabla\left(\frac{\partial\varphi}{\partial t} + \frac{v^2}{2} + \mathcal{P} + \Pi\right) = 0.$$

It follows from this equation that the expression in the brackets does not depend on the coordinates but can depend on time:

$$\frac{\partial\varphi}{\partial t} + \frac{v^2}{2} + \Pi = f(t), \tag{4.1.5}$$

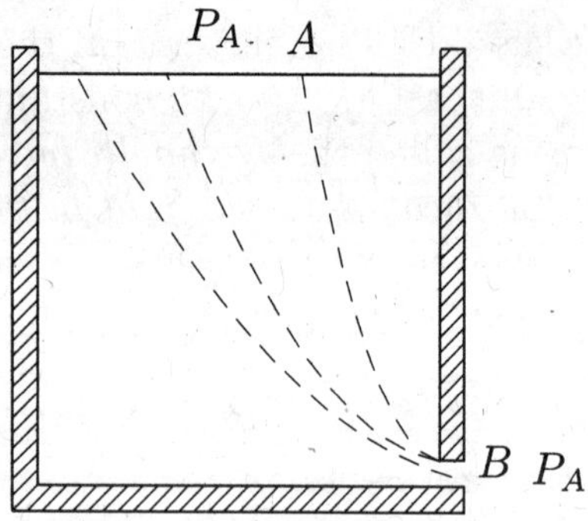

Figure 4.1: A qualitative picture of the fluid outflow from a vessel.

where $f(t)$ is an arbitrary function of time. Equation (4.1.5) is called *the Lagrange integral.* The Bernoulli and Lagrange integrals are the exact solutions of the Euler equations under corresponding conditions. If the domain of applicability of these conditions coincides, we can obtain the *Euler–Bernoulli integral* in the form

$$\frac{v^2}{2} + \Pi + \int \frac{dP}{\rho} = C.$$

The constant C is here the same for the overall flow in contrast with the Bernoulli integral (4.1.2), in which the constant $C_{(l)}$ is different on different streamlines.

The Bernoulli and Lagrange integrals play an important role in fluid and gas mechanics. They enable one to find the pressure distribution on the basis of the known velocity modulus. The velocity $\vec{v}$ can be found from the continuity and momentum equations. In a particular case of a potential flow $\vec{v} = \nabla\varphi$, and upon substitution into the continuity equation at $P = P_0$ we obtain the equation $\Delta\varphi = 0$. This circumstance is illustrated in the following on a number of simple problems.

Problem 4.1. Find the outflow velocity of an incompressible fluid with the density ρ_0 from a vessel through a small orifice (see Fig. 4.1) under the gravity force. The open fluid area in the vessel is equal to S_0, and the orifice area is equal to S_1, where $S_0 \gg S_1$ and H is the fluid height in the vessel.

Solution: When the fluid escapes from the vessel, the fluid height reduces; however, under the condition $S_1/S_0 \ll 1$, this reduction will be slow. Therefore, the given fluid motion can be considered approximately as a sequential change of stationary states. It is conventional to call such an approach to the nonstationary flow the *quasistationary regime.* Let us write the Bernoulli integral along the streamline AB, where the point A is located on the surface S_0 and B is located on the surface S_1; that is,

$$\frac{v_A^2}{2} + \frac{P_A}{\rho_0} + g_0 H = \frac{v_B^2}{2} + \frac{P_B}{\rho_0},$$

where ρ_0 is the fluid density. Since the pressure at points A and B is equal to the atmospheric pressure $P_A = P_B = P_a$, the Bernoulli integral simplifies greatly:

$$\frac{v_A^2}{2} + g_0 H = \frac{v_B^2}{2}.$$

We find from the equation for the constancy of flow rates $v_A \cdot S_0 = v_B \cdot S_1$ and the last relationship that

$$v_B = \sqrt{\frac{2g_0 H}{1 - S_1^2/S_0^2}} \approx \sqrt{2g_0 H},$$

since $S_1/S_0 \ll 1$ by virtue of the problem condition. The obtained formula is known as the Torricelli formula, and it does not differ from a formula describing the velocity of a material point falling from height H.

Problem 4.2. Find the outflow velocity of an adiabatic gas obeying the Poisson equation. The gas escapes from a vessel with given parameters P_0, ρ_0, and T_0 inside the vessel (see Fig. 4.2).

Solution: Let the gas escape into the atmosphere through a small orifice. We assume that the gas parameters inside the vessel change little, and consequently, we can use the quasistationary approach. Let us take the streamline AB, where A is a point inside the vessel and B is a point in the section of an escaping jet. Write the Poisson's adiabat and the Bernoulli integral for the points A and B as

$$\frac{P_A}{\rho_A^\gamma} = \frac{P_B}{\rho_B^\gamma}, \quad \frac{v_B^2}{2} + \frac{\gamma}{\gamma - 1}\frac{P_A}{\rho_A} = \frac{v_B^2}{2} + \frac{\gamma}{\gamma - 1}\frac{P_B}{\rho_B}.$$

Since the vessel is large and the orifice is small, $v_A \ll v_B$, and one can assume that $v_A = 0$, $P_A = P_0$, and $\rho_A = \rho_0$. This enables us to find from the Bernoulli integral:

$$\frac{v_B^2}{2} = \frac{\gamma}{\gamma - 1}\left(\frac{P_0}{\rho_0} - \frac{P_B}{\rho_B}\right).$$

Since the flow is adiabatic, we have:

$$\rho_B = \rho_0 \left(\frac{P_B}{P_0}\right)^{\frac{1}{\gamma}},$$

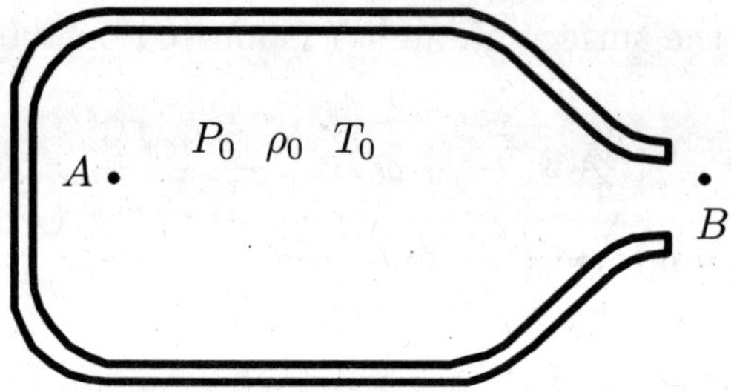

Figure 4.2: The adiabatic gas outflow from a vessel through a small orifice.

and it follows from the foregoing formula that

$$v_B = \sqrt{\frac{2\gamma}{\gamma - 1} \cdot \frac{P_0}{\rho_0} \left[1 - \left(\frac{P_B}{P_0}\right)^{\frac{\gamma-1}{\gamma}}\right]}. \tag{4.1.6}$$

Consider the physical meaning of this formula. Introduce the flow rate per unit area $Q = \rho_B v_B$. We find from formula (4.1.6):

$$Q = \sqrt{\frac{2\gamma}{\gamma - 1} \cdot P_0 \rho_0 \xi^{\frac{2}{\gamma}} (1 - \xi^{\frac{\gamma-1}{\gamma}})},$$

where $\xi = P_B/P_0$, and P_B is the pressure of a medium into which the gas escapes. At $\xi = 1$, i.e., $P_B = P_0$, we have an equilibrium; consequently, the gas does not escape from the vessel. At the diminution of P_B, i.e., at the diminution of ξ, the flow rate increases and reaches its maximum at certain $\xi = \xi_*$. At a further diminution of ξ, the value of Q reduces and vanishes at $\xi = 0$. It should be noted, however, that the experiment confirms the validity of the dependence $Q(\xi)$ only for $\xi > \xi_*$. At $\xi < \xi_*$ Q indeed remains constant and is equal to its maximum value, and the outflow velocity is equal to the sound velocity. A further diminution of P_B already does not affect the outflow from the orifice, since the disturbances from the ambient medium do not penetrate the vessel interior. At $\xi < \xi_*$, when the jet becomes supersonic, the assumption on the one-dimensional character of the flow proves to be invalid, and it is necessary to take into account the multidimensional character of the flow.

4.2 Planar Irrotational Steady Motions of an Ideal Incompressible Fluid

We will call the flow *planar* if all particles move in parallel with some plane chosen in advance. The particles' velocities at the corresponding points of the planes parallel with this chosen plane have the same moduli and directions. This leads to the fact that it is sufficient to consider the flow of an ideal incompressible fluid in such a single plane, which we will denote by the (x, y) plane. At such a choice of the coordinate system, all gasdynamic quantities will depend only on the coordinates (x, y) and they do not depend on z, i.e., $\vec{v} = \vec{v}(v_x, v_y, 0)$ and $\frac{\partial v_x}{\partial z} = \frac{\partial v_y}{\partial z} = 0$. We will assume the flow to be *steady*; consequently, the gasdynamic parameters do not depend on time, i.e., $\frac{\partial \vec{v}}{\partial t} = 0$. The availability of only two independent variables x and y enables one to apply, for the investigation of the planar flows, the mathematical apparatus of the theory of functions of complex variable and to solve a large number of problems on the flow around the bodies. Kirchhoff[7] was one of the first researchers who actively applied the methods of complex variable functions to problems of planar flow around the body. These methods have enjoyed the most significant development in the works of Joukowskii and Chaplygin. These methods are presented in detail in Russian literature[1,8,9] as well as other literature[4,10,11]. Note that, in order to avoid confusion with the potential φ, we will denote the polar angle in this section by the letter θ.

4.2.1 The Governing Equations of Planar Flows

Since the fluid is incompressible and the flow is steady, the density is constant and is known, $\rho = \rho_0 = \text{const}$, and the functions v_x, v_y, P, which are sought for, are the functions of the coordinates x and y only. With regard for the above assumptions, the system of Euler equations can be written in projections onto the x- and y-axes as

$$\begin{aligned} &\frac{\partial v_x}{\partial x} + \frac{\partial v_y}{\partial y} = 0, \\ &\frac{\partial v_y}{\partial x} - \frac{\partial v_x}{\partial y} = 0, \\ &\frac{v_x^2 + v_y^2}{2} + \frac{P}{\rho_0} + \Pi = C, \end{aligned} \tag{4.2.1}$$

where Π is the potential of body forces. Here the first equation is the continuity equation, and instead of the vector equation for the momentum, we have written the Euler–Bernoulli integral and the condition of the curl absence $\operatorname{rot} \vec{v} = 0$ for a planar motion. As was shown in Section 4.1, such a form of equations is possible, since the passage from one system to another one was carried out with the aid of identical transformations.

The energy equation for an incompressible fluid in the absence of heat supply yields

$$\frac{dE}{dt} = 0;$$

that is, the energy is conserved in a particle for an incompressible fluid. The first two equations (4.2.1) contain only the functions v_x and v_y, and the last equation can be used for finding the pressure on the basis of the known values of v_x and v_y. Thus, the system (4.2.1) completely describes the flow.

The condition for the curl absence enables us to introduce the *velocity potential* φ by the formulas

$$v_x = \frac{\partial \varphi}{\partial x}, \quad v_y = \frac{\partial \varphi}{\partial y}.$$

Then the second equation in (4.2.1) is satisfied automatically, since $\operatorname{rot}(\nabla \varphi) \equiv 0$. Substituting these formulas into the first equation of system (4.2.1), we find:

$$\frac{\partial^2 \varphi}{\partial x^2} + \frac{\partial^2 \varphi}{\partial y^2} = 0. \tag{4.2.2}$$

The solution of equation (4.2.2) should satisfy certain boundary conditions. In the case of an unbounded flow around a body at rest, the solution should be such that the flow velocity at infinity be equal to a given quantity $\vec{v}_\infty$. On the body surface S, the slip condition is satisfied, that is,

$$\left.\frac{\partial \varphi}{\partial x}\right|_\infty = v_{x\infty}; \quad \left.\frac{\partial \varphi}{\partial y}\right|_\infty = v_{y\infty}; \quad \left.\frac{\partial \varphi}{\partial n}\right|_S = 0, \tag{4.2.3}$$

where $\vec{n}$ is the vector of a normal to the surface S. Thus, the problem on an unbounded fluid flow around a body reduces to the solution of the Laplace equation (4.2.2) under the boundary conditions (4.2.3) and is called the *external Neumann problem.* The pressure distribution is determined from the Euler–Bernoulli integral.

Basing on the continuity equation of system (4.2.1), let us introduce the *stream function* $\psi = \psi(x, y)$ by the formulas

$$v_x = \frac{\partial \psi}{\partial y}, \quad v_y = -\frac{\partial \psi}{\partial x},$$

where $d\psi = \frac{\partial \psi}{\partial x}dx + \frac{\partial \psi}{\partial y}dy$. Write the equation of a streamline for a planar flow as

$$\frac{dx}{v_x} = \frac{dy}{v_y} \text{ or } v_x\, dy - v_y\, dx = 0.$$

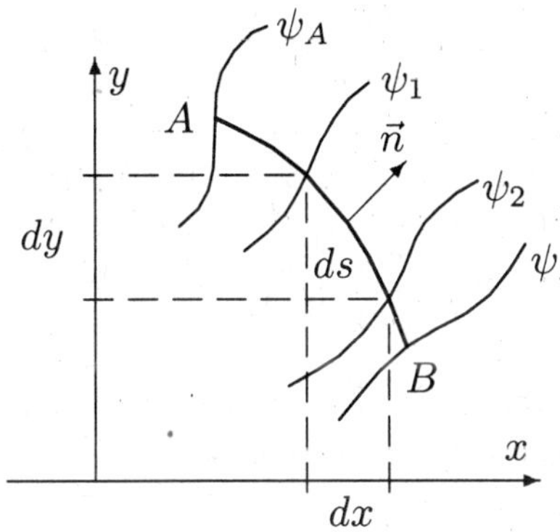

Figure 4.3: The streamlines intersecting an arbitrary curve AB.

Comparing this relation with the definition $d\psi = v_x dy - v_y dx$, we can see that $d\psi = 0$ along the streamline; that is, $\psi = \text{const}$ along the streamline. It is conventional to call the function $\psi(x, y)$ the *stream function* and the equation $\psi(x, y) = \text{const}$ the equation of a streamline. The different values of a constant correspond to different streamlines. One can compute in terms of the stream function the flow rate across a curve intersecting different streamlines (see Fig. 4.3). Let us calculate the flow rate per unit time across the curve AB (a piece of a cylindrical surface with the height $\Delta z = 1$ with the generatrix AB) by the formula

$$Q = \int_B^A v_n\, ds = \int_B^A (v_x n_x + v_y n_y)\, ds,$$

where $\vec{n}$ is a normal to the curve AB, $n_x = \cos(\hat{nx})$, $n_y = \cos(\hat{ny})$. If dx and dy are the projections of the curve element ds, then it is obvious that

$$\frac{dy}{ds} = n_x, \quad \frac{dx}{ds} = -n_y,$$

and we obtain:

$$Q = \int_B^A (v_x\, dy - v_y\, dx) = \int_B^A d\psi = \psi_A - \psi_B.$$

Thus, the flow rate across the curve AB is equal to the difference of the values of the stream function at the ends of this curve.

It is easy to be sure of the fact that the equipotential lines ($\varphi = \text{const}$) and the streamlines ($\psi = \text{const}$) are perpendicular to each other in any flow subregion. For this purpose, it is sufficient to show that $\nabla\varphi \cdot \nabla\psi = 0$; that is

$$\frac{\partial\varphi}{\partial x} \cdot \frac{\partial\psi}{\partial x} + \frac{\partial\varphi}{\partial y} \cdot \frac{\partial\psi}{\partial y} = \frac{\partial\varphi}{\partial x}\left(-\frac{\partial\varphi}{\partial y}\right) + \frac{\partial\varphi}{\partial y} \cdot \frac{\partial\varphi}{\partial x} \equiv 0.$$

We have used here the fact that

$$v_x = \frac{\partial \varphi}{\partial x} = \frac{\partial \psi}{\partial y} \quad \text{and} \quad v_y = \frac{\partial \varphi}{\partial y} = -\frac{\partial \psi}{\partial x}.$$

If the flow is irrotational, then we have for the planar flows that

$$\frac{\partial v_y}{\partial x} - \frac{\partial v_x}{\partial y} = 0.$$

Substituting into this relation the expressions for the velocity components $v_x = \frac{\partial \psi}{\partial y}$ and $v_y = -\frac{\partial \psi}{\partial x}$, we obtain:

$$\frac{\partial^2 \psi}{\partial x^2} + \frac{\partial^2 \psi}{\partial y^2} = 0. \tag{4.2.4}$$

When solving the problem on a flow around the bodies in terms of the stream function, it is necessary to have the boundary conditions written also for this function. Consider the fluid flow around an impermeable body surface S. It is known that $\vec{v}_n = 0$ for an ideal fluid on this surface. Let us write this condition in terms of the stream function ψ, i.e.,

$$\vec{v}_n = v_x n_x + v_y n_y = \frac{\partial \psi}{\partial y} \cdot \frac{dy}{ds} + \frac{\partial \psi}{\partial x} \cdot \frac{dx}{ds} = \frac{\partial \psi}{\partial s}.$$

Consequently, we determine on the body contour that $\frac{\partial \psi}{\partial s}|_s = 0$, that is, $\psi|_s = \text{const}$, and the stream function conserves a constant value on the contour s. This means that the body boundary should be a streamline.

Thus, the problem on an unbounded fluid flow past a body is reduced in terms of the stream function to the solution of the Laplace equation (4.2.4) under the following boundary conditions:

$$\psi(x, y)\Big|_s = 0, \quad \frac{\partial \psi}{\partial x}\Big|_\infty = -v_{y\infty}, \quad \frac{\partial \psi}{\partial y}\Big|_\infty = v_{x\infty}.$$

It is conventional to call such a problem formulation the *Dirichlet problem.* The pressure distribution is determined from the Euler–Bernoulli integral.

Comparing the expressions for v_x and v_y computed in terms of φ and ψ, one can write that

$$\frac{\partial \varphi}{\partial x} = \frac{\partial \psi}{\partial y}, \quad \frac{\partial \varphi}{\partial y} = -\frac{\partial \psi}{\partial x}. \tag{4.2.5}$$

These are the Cauchy–Riemann conditions, which are well known in the theory of the functions of complex variable; they ensure that the function

$$w = \varphi(x, y) + i\psi(x, y)$$

will be a function of one complex variable $z = x + iy$, which is conventional to call the *complex potential.*

Problem 4.3. Prove that, if the functions φ and ψ satsfy the relations (4.2.5), then the function $w(x, y) = w(z)$.

Solution: Let $w(x, y) = \varphi(x, y) + i\psi(x, y)$ and introduce $z = x + iy$, $\bar{z} = x - iy$. By definition,

$$w(x, y) = w(\frac{z+\bar{z}}{2}, \frac{z-\bar{z}}{2i}) = w(z, \bar{z}).$$

We have that

$$\frac{\partial w}{\partial \bar{z}} = \frac{\partial \varphi}{\partial x} \cdot \frac{dx}{d\bar{z}} + \frac{\partial \varphi}{\partial y} \cdot \frac{dy}{d\bar{z}} + i\frac{\partial \psi}{\partial x} \cdot \frac{dx}{d\bar{z}} + i\frac{\partial \psi}{\partial y} \cdot \frac{dy}{d\bar{z}}.$$

Since

$$x = \frac{1}{2}(z + \bar{z}), \quad y = \frac{1}{2i}(z - \bar{z}),$$

we have that

$$\frac{dx}{d\bar{z}} = \frac{1}{2}, \quad \frac{dy}{d\bar{z}} = \frac{i}{2}.$$

Substituting these values into the expression for $\frac{\partial w}{\partial \bar{z}}$, we find:

$$\frac{\partial w}{\partial \bar{z}} = \frac{1}{2} \cdot \frac{\partial \varphi}{\partial x} + \frac{1}{2} i \frac{\partial \varphi}{\partial y} + \frac{\partial \psi}{\partial x} \frac{i}{2} - \frac{1}{2} \frac{\partial \psi}{\partial y}.$$

Using the Cauchy–Riemann conditions, we obtain:

$$\frac{\partial w}{\partial \bar{z}} \equiv 0,$$

which proves that $w = w(z)$.

If the complex potential is known, it is easy to find the functions characterizing the flow by the formulas

$$\varphi = \mathrm{Re}\,(w(z)), \quad \psi = \mathrm{Im}(w(z)).$$

While presenting the theory of planar ideal incompressible fluid motions, we shall denote by V the *complex velocity* with the components (u, v), i.e.,

$$V = u + iv, \quad (u = v_x, \ v = v_y),$$

and the modulus is $|V| = \sqrt{u^2 + v^2}$. We will denote the complex conjugate velocity by $\bar{V}$,

$$\bar{V} = u - iv.$$

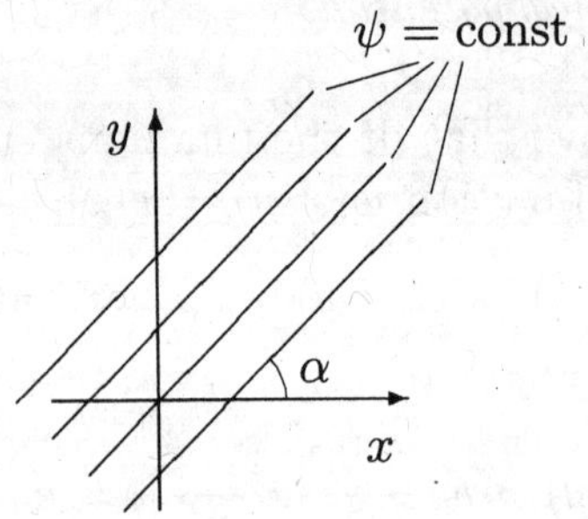

Figure 4.4: The streamlines in a plane-parallel flow.

On the other hand, we have that

$$\frac{dw}{dz} = \frac{dw}{dx} = \frac{dw}{d(iy)} = \frac{\partial\varphi}{\partial x} + i\frac{\partial\psi}{\partial x}.$$

By the definition $\frac{\partial\varphi}{\partial x} = v_x = u$ and $\frac{\partial\psi}{\partial x} = -v_y = -v$, consequently, $\frac{dw}{dz} = u - iv = \bar{V}(z)$; that is

$$u = \mathrm{Re}\left(\frac{dw}{dz}\right), \quad v = -\mathrm{Im}\left(\frac{dw}{dz}\right).$$

If θ is the angle between the vector V and the x-axis, then

$$\begin{aligned} V &= u + iv = |V|(\cos\theta + i\sin\theta) = |V|e^{i\theta}, \\ \bar{V} &= u - iv = |V|e^{-i\theta}. \end{aligned}$$

Computing a contour integral of $\bar{V}$ over a closed contour c in the flow plane, we find:

$$\oint_c \bar{V}\,dz = \oint_c \frac{dw}{dz}dz = \oint_c (d\varphi + i\,d\psi).$$

Determining from here the imaginary and real parts, we obtain:

$$\Gamma = \mathrm{Re}\oint_c \bar{V}\,dz, \quad Q = \mathrm{Im}\oint_c \bar{V}\,dz;$$

that is, the real part determines the velocity circulation, and the imaginary part determines the volume flow rate of fluid per second. Thus, the introduction of a complex potential enables one to apply a well-developed apparatus of the theory of the functions of complex variable for finding the solutions of many boundary-value problems on the steady planar potential flows of an ideal incompressible fluid. As already stated

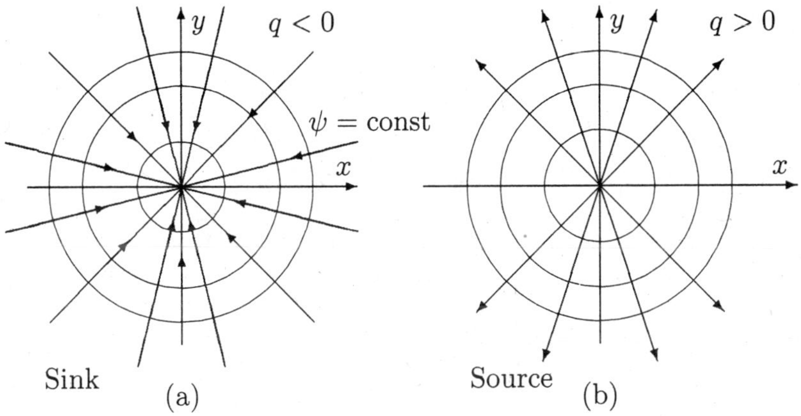

Figure 4.5: The picture of streamlines for the (a) sink and the (b) source.

above, the knowledge of $w(z)$ immediately enables one to determine the basic gasdynamic functions characterizing the flow.

Let $w(z) = az$, where $a > 0$ is real. Then

$$\varphi = ax, \quad \psi = ay.$$

The velocity components are $u = \frac{\partial\varphi}{\partial x} = a$ and $v = \frac{\partial\varphi}{\partial y} = 0$. The streamlines $\psi = \text{const}$ are the straight lines $y = \text{const}$, and the lines of equal potential $\varphi = \text{const}$ are the lines $x = \text{const}$. Consequently, we have a flow along the x-axis at a constant velocity a. Let $w(z) = ae^{-i\alpha} \cdot z$, where α and a are real and positive. We have that $\varphi + i\psi = a(\cos\alpha - i\sin\alpha)(x + iy)$, or

$$\varphi = a(x\cos\alpha + y\sin\alpha), \quad \psi = a(y\cos\alpha - x\sin\alpha).$$

The velocity components are $u = a\cos\alpha$ and $v = a\sin\alpha$. The streamlines $\psi = \text{const}$ are the straight lines $y = (\tan\alpha)x + c$, which make the angle α with the x-axis. Thus, we have a translational flow at a constant velocity, which makes the angle α with the x-axis (see Fig. 4.4).

Consider the complex potential of the form

$$w = \frac{q}{2\pi}\ln z,$$

where $z = re^{i\theta}$, $r = |z|$, and $\theta = \arg z$. We find that $\varphi + i\psi = (\ln r + i\theta) \cdot \frac{q}{2\pi}$ or

$$\varphi = \frac{q}{2\pi}\ln r, \quad \psi = \frac{q\theta}{2\pi}.$$

The streamlines will be the beams emanating from the coordinate origin. The equipotential lines are the circles $r = \text{const}$ (see Fig. 4.5 (a) and

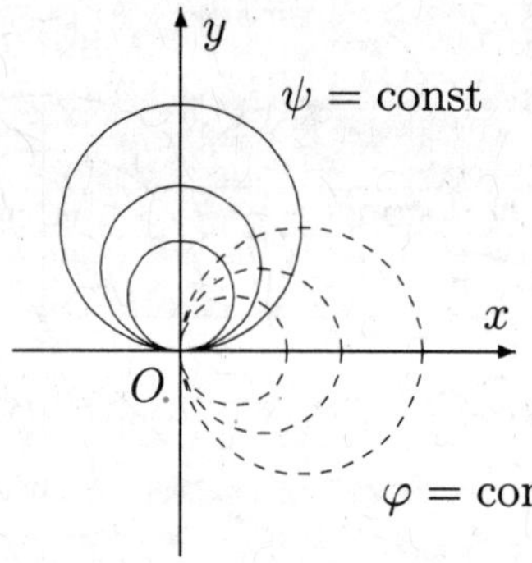

Figure 4.6: The streamlines from a dipole.

(b)). For the complex potential $w(z) = \frac{q}{2\pi}\ln z$ the pojections of the velocities on the axes of polar coordinates will be $v_r = \frac{q}{2\pi} \cdot \frac{1}{r}$, $v_\theta = 0$. It can be seen from here that the velocity has a constant magnitude on each circle with a fixed radius and with a center at the coordinate origin; the velocity vector is directed along the radius and reduces as r^{-1} when r increases. At $q > 0$ the velocity is directed away from the coordinate origin [see Fig. 4.5 (b)], and at $q < 0$ the velocity vector is directed to the coordinate origin and $v_r < 0$ [see Fig. 4.5 (a)].

Let us compute the flow rate of a fluid across a contour, enclosing the coordinate origin by using the formula $Q = \text{Im} \oint \bar{V}\, dz = \oint d\psi = q$. Thus, q is the source strength. At $q > 0$, we have a source, and at $q < 0$, we have a sink. If the source is located at point $z = a$ rather than in the coordinate origin, then the complex potential will have the form $w(z) = \frac{q}{2\pi}\ln(z - a)$.

Let a source of strength q be located at point A of the (x, y) plane, let a sink of strength q be located at point B, and let the complex coordinates of points A and B be $z_A = \frac{l}{2}e^{i\alpha}$, $z_B = -\frac{l}{2}e^{i\alpha}$. The complex potential of the flow from these two sources has the form

$$w_A(z) = \frac{q}{2\pi}\ln(z - \frac{l}{2}e^{i\alpha}), \quad w_B(z) = \frac{-q}{2\pi}\ln(z + \frac{l}{2}e^{i\alpha}),$$

and the complex potential of the total flow is

$$w(z) = w_A(z) + w_B(z), \quad w(z) = \frac{q}{2\pi}\ln\left(\frac{z - \frac{l}{2}e^{i\alpha}}{z + \frac{l}{2}e^{i\alpha}}\right). \tag{4.2.6}$$

We will consider such a point z that $|z| \gg l$. Expanding (4.2.6) into a series with respect to $\frac{l}{z}$, we obtain the formula:

$$w(z) \approx \frac{q}{2\pi}\Big(-\frac{l}{2z}e^{i\alpha} - \frac{l}{2z}e^{i\alpha} + \ldots\Big) = -\frac{q}{2\pi}\frac{l}{z}e^{i\alpha} + \ldots.$$

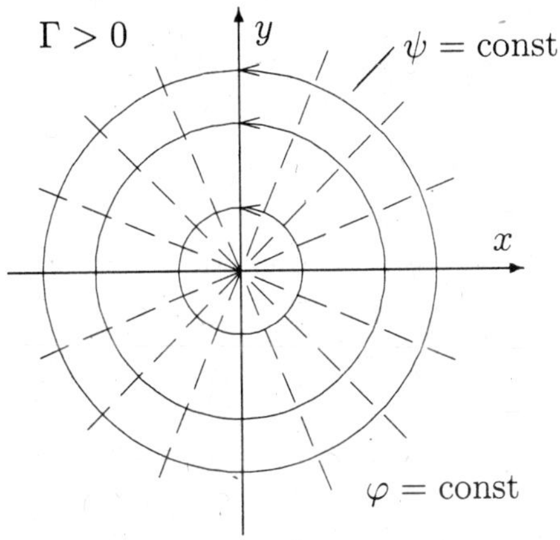

Figure 4.7: The streamlines from a curl.

Let $l \to 0$ and the intensity $q \to \infty$, so that the product $l \cdot q = M$ remains constant. Then, the complex potential will have, for such a limiting flow, the form

$$w(z) = -\frac{M}{2\pi} \cdot \frac{e^{i\alpha}}{z}. \qquad (4.2.7)$$

Let us study the flow pattern determined by the complex potential (4.2.7). For the sake of simplicity, we assume $\alpha = 0$; i.e., the dipole is located at the coordinate origin and its axis coincides with the x-axis. In this case, the functions φ and ψ are presented by formulas

$$\varphi = -\frac{M}{2\pi} \cdot \frac{x}{x^2 + y^2}, \quad \psi = \frac{M}{2\pi} \cdot \frac{y}{x^2 + y^2}.$$

The streamlines $\psi = \text{const}$ are the lines on which

$$\frac{y}{x^2 + y^2} = \frac{1}{2c}, \quad x^2 + (y - c)^2 = c^2, \quad c = \frac{\psi \cdot \pi}{M}$$

and they represent the circles passing through the coordinate origin and the centers are on the y-axis (see Fig. 4.6), where the equipotential lines $\varphi = \text{const}$ are shown by the dashed lines. It is conventional to call the flow described by such a complex potential the *dipole*.

Consider a complex potential of the form

$$w(z) = \frac{\Gamma}{2\pi i} \ln z. \qquad (4.2.8)$$

In the polar coordinates, we have:

$$\varphi = \frac{\Gamma}{2\pi} \theta, \quad \psi = -\frac{\Gamma}{2\pi} \ln r.$$

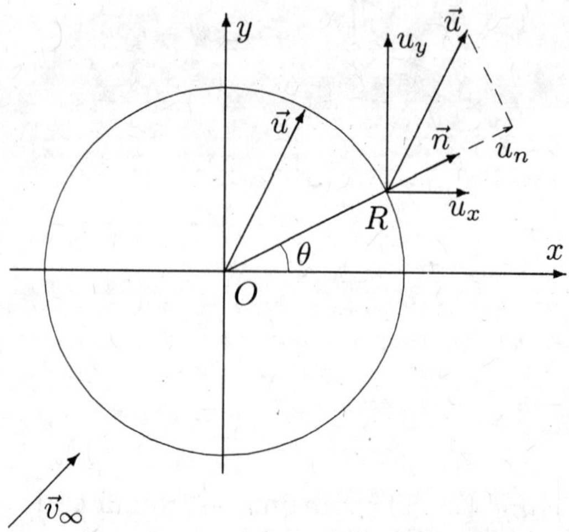

Figure 4.8: The flow past a cylinder moving at a velocity $\vec{u}$.

The streamlines $\psi = \text{const}$ are the circles with a center at the coordinate origin, and the lines $\varphi = \text{const}$ are the beams $\theta = \text{const}$ (the dashed lines in Fig. 4.7). The coordinate origin $r = 0$ is a singular point, since

$$v_r = \frac{\partial \varphi}{\partial r} = 0, \quad v_\theta = \frac{1}{r}\frac{\partial \varphi}{\partial \theta} = \frac{\Gamma}{2\pi} \cdot \frac{1}{r}.$$

The velocity $v_\theta > 0$ at $\Gamma > 0$; that is, the motion along a circle in a counter-clockwise direction corresponds to a positive circulation value. Let us take a contour l, comprising the coordinate origin, and compute the velocity circulation over this contour by the formula

$$\text{Re} \oint_l \bar{V}\, dz = \oint_l d\varphi = \Gamma.$$

Thus, Γ is the velocity circulation over a closed contour enclosing the coordinate origin.

4.2.2 The Potential Flow past the Cylinder

Let a circular cylinder with the radius R move at the velocity $\vec{u}$ having the value $\vec{v}_\infty$ at the infinity (see Fig. 4.8). The $\vec{u}$ and $\vec{v}_\infty$ are perpendicular to the cylinder axis; that is

$$\vec{u} = \vec{u}(u_x, u_y, 0), \quad \vec{v}_\infty = \vec{v}_\infty(v_x, v_y, 0).$$

We assume the motion to be planar and irrotational; therefore, the function $w(z)$ exists such that $\frac{dw}{dz} = \bar{V}(z)$. It is clear from the physical considerations that the function $\bar{V} = v_x - iv_y$ should be determined at all points of the (x, y) plane besides the circle with the radius R. It should be single-valued, bounded, and take a given value at the infinity. It is known from the theory of the functions of complex variable that such a function can be presented in the form of the Laurent series in negative powers of z:

$$\bar{V}(z) = c_0 + \frac{c}{z} + \frac{c_1}{z^2} + \frac{c_2}{z^2} + \cdots, \tag{4.2.9}$$

$$\text{at } z \to \infty \quad \bar{V}(z)|_{\infty} = v_x - iv_y = c_0 = \text{const.}$$

It follows from (4.2.9) that

$$\frac{dw}{dz} = \bar{V}(z) = v_x - iv_y + \frac{c}{z} + \frac{c_1}{z^2} + \cdots.$$

Integrating with respect to z, we obtain:

$$w(z) = (v_x - iv_y)z + c\ln z + \sum_{n=1}^{\infty} \frac{c_n}{z^n}.$$

From the boundary condition on the circle, we have that $v_n|_S = u_n$ or $v_n = \frac{\partial\varphi}{\partial n}$, since the flow is irrotational. In the polar coordinates (r, θ) on the cylinder curface $r = R$, we have:

$$\left.\frac{\partial\varphi}{\partial r}\right|_{r=R} = u_x \cos\theta + u_y \sin\theta.$$

This relation serves for finding $c, c_1, \ldots, c_n$. Going over to the polar coordinates in $w(z)$, subdividing into φ and ψ and differentiating φ with respect to r, we substitute the obtained expression into the boundary condition and determine c and c_n. This yields

$$\begin{aligned}
c &= A + iB, \quad c_n = A_n + iB_n, \quad z = re^{i\theta}, \\
w(z) &= \varphi + i\psi = (v_x - iv_y)\, r\,(\cos\theta + i\sin\theta) + (A + iB)\,(\ln r + i\theta) \\
&+ (A_1 + iB_1)\frac{1}{r}\,(\cos\theta - i\sin\theta) \\
&+ \sum_{n=2}^{\infty}(A_n + iB_n)\frac{1}{r^n}(\cos n\theta - i\sin n\theta).
\end{aligned}$$

We can determine from here the real part in the form

$$\begin{aligned}
\varphi &= \left(v_x r + \frac{A_1}{r}\right)\cos\theta + \left(v_y r + \frac{B_1}{r}\right)\sin\theta + A\ln r - B\theta \\
&+ \sum_{n=2}^{\infty}\left(\frac{A_n}{r^n}\cos n\theta + \frac{B_n}{r^n}\sin n\theta\right),
\end{aligned}$$

which enables us to find $\frac{\partial \varphi}{\partial r}$. Equalling the coefficients in the obtained equality at equal powers of the sine and cosine functions, we have:

$$A = 0,\ v_x - \frac{A_1}{R^2} = u_x,\ v_y - \frac{B_1}{R^2} = u_y,\ A_k = B_k = 0\ (k = 2, 3, \ldots);$$

that is

$$A = 0,\ A_1 = (v_x - u_x)R^2,\ B_1 = (v_y - u_y)R^2,\ A_k = B_k = 0,\ k = 2, 3, \ldots.$$

The coefficient B still remains indeterminate. We denote it by $B = -\frac{\Gamma}{2\pi}$ and finally obtain:

$$w(z) = (v_x - iv_y)z + \frac{\Gamma}{2\pi i}\ln z + \frac{(v_x - u_x)R^2 + i(v_y - u_y)R^2}{z}, \tag{4.2.10}$$

where $v_x + iv_y = V_\infty$ and $u_x + iu_y = u$, or

$$w(z) = \bar{V}_\infty z + (V_\infty - u)\frac{R^2}{z} + \frac{\Gamma}{2\pi i}\ln z.$$

This is the general form of a complex potential for the flow past the cylinder, where there are the following complex potentials:
1) $\bar{V}_\infty z$ coresponds to the translational flow;
2) $(V_\infty - u)\frac{R^2}{z}$ corresponds to the dipole;
3) $\frac{\Gamma}{2\pi i} \cdot \ln z$ corresponds to a point vortex.

Let us write the complex potential in the following particular cases:
1) The cylinder is at rest, i.e., $u \equiv 0$. Then

$$w(z) = \bar{V}_\infty z + V_\infty\frac{R^2}{z} + \frac{\Gamma}{2\pi i}\ln z. \tag{4.2.11}$$

2) The fluid is at rest, i.e., $V_\infty \equiv 0$. Then

$$w(z) = -u\frac{R^2}{z} + \frac{\Gamma}{2\pi i}\ln z.$$

3) The fluid at infinity is at rest and the cylinder is also at rest. Then

$$w(z) = \frac{\Gamma}{2\pi i}\ln z;$$

this is the circulatory flow past the cylinder.

Let us study in detail the flow pattern for a stationary flow of a cylinder, which is at rest, that is, Case 1 when $u = 0$ and the flow at infinity is directed along the x-axis. The complex potential at $v_x = v_\infty, v_y = 0$ takes the form

$$w(z) = v_\infty\left(z + \frac{R^2}{z}\right) + \frac{\Gamma}{2\pi i}\ln z.$$

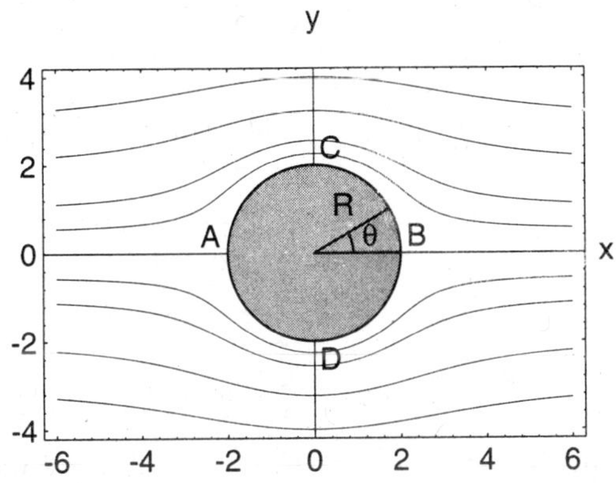

Figure 4.9: The flow pattern around the cylinder at $\Gamma = 0$.

The following two cases are possible here.
$\Gamma = 0$: in this case, we have the noncirculatory flow past the cylinder,

$$w(z) = v_\infty\Big(z + \frac{R^2}{z}\Big)$$

or

$$\varphi + i\psi = v_\infty\Big(x + iy + R^2 \cdot \frac{x - iy}{x^2 + y^2}\Big).$$

From here, we find:

$$\varphi = v_\infty x\Big(1 + \frac{R^2}{x^2 + y^2}\Big), \quad \psi = v_\infty y\Big(1 - \frac{R^2}{x^2 + y^2}\Big). \tag{4.2.12}$$

The streamlines $\psi =$ const are the lines

$$y\Big(1 - \frac{R^2}{x^2 + y^2}\Big) = \text{const},$$

which are the third-order curves symmetric with respect to the y-axis. The lines $\psi = c_1$ and $\psi = -c_1$ are symmetric with respect to the x-axis. At $\psi = 0$, the equation for the streamline is split into two equations: $y = 0$ is the x-axis and $x^2 + y^2 = R^2$ is the circle. The flow pattern past the cylinder is presented in Fig. 4.9. We have obtained Fig. 4.9 with the aid of the *Mathematica* Notebook `prog4-1.nb` (see also Appendix B).

Consider the field of the velocity vector. Let us go over in (4.2.12) to the polar coordinates (r, θ) by the formulas $x = r\cos\theta$, $y = r\sin\theta$. Then

$$\varphi = v_\infty r\cos\theta\Big(1 + \frac{R^2}{r^2}\Big),$$

$$\begin{aligned} v_r &= \frac{\partial \varphi}{\partial r} = v_\infty \cos\theta\left(1 - \frac{R^2}{r^2}\right), \\ v_\theta &= \frac{1}{r}\frac{\partial \varphi}{\partial \theta} = -v_\infty \sin\theta\left(1 + \frac{R^2}{r^2}\right). \end{aligned} \tag{4.2.13}$$

The formulas (4.2.13) yield the velocity components at any point of the flow. Assuming $r = R$ in (4.2.13), we obtain the magnitude of the velocity on the surface:

$$v_r = 0, \quad v_\theta = -2v_\infty \sin\theta.$$

At the cylinder points $\theta_A = \pi$, $\theta_B = 0$, the velocity is equal to zero; that is, these points are critical. At points $\theta_C = \frac{\pi}{2}$, $\theta_D = -\frac{\pi}{2}$, the velocity has the largest magnitude equal to $2v_\infty$. We find from the Bernoulli integral $\frac{v^2}{2} + \frac{P}{\rho} = C$ that $P = \rho(C - 2v_\infty^2 \sin^2\theta)$. It can be seen from this relation that, at points $\pm\theta$, $\pm(\pi - \theta)$, the pressure is the same; therefore, the main vector of the forces acting on the cylinder will be equal to zero. This result, which consists of the fact that the body, past which there is an ideal fluid flow does not experience any drag, bears the name of *d'Alembert's paradox.*

In the case of the circulatory flow past the cylinder, $w(z)$ has the form

$$w(z) = v_\infty\left(z + \frac{R^2}{z}\right) + \frac{\Gamma}{2\pi i}\ln z \tag{4.2.14}$$

and the complex velocity is

$$\bar{V}(z) = \frac{dw}{dz} = v_\infty\left(1 - \frac{R^2}{z^2}\right) + \frac{\Gamma}{2\pi i}\cdot\frac{1}{z}.$$

Let us find the critical points of the flow, where $v_x = 0$, $v_y = 0$; that is,

$$v_\infty z^2 + \frac{\Gamma}{2\pi i}z - v_\infty R^2 = 0$$

or

$$z_{1,2} = \frac{1}{2v_\infty}\left(-\frac{\Gamma}{2\pi i} \pm \sqrt{-\frac{\Gamma^2}{4\pi^2} + 4v_\infty^2 R^2}\right). \tag{4.2.15}$$

Analyzing (4.2.15), we have the following:

1) $-\frac{\Gamma^2}{4\pi^2} + 4v_\infty^2 R^2 > 0$. The critical points are located on the cylinder $|z_{1,2}| = R$ symmetrically with respect to the y-axis; that is, $\mathrm{Im}\, z_1 = \mathrm{Im}\, z_2$ and $\mathrm{Re}\, z_1 = -\mathrm{Re}\, z_2$. The flow pattern is shown for this case in Fig. 4.10 (a). We have obtained Fig. 4.10 (a) with the aid of the *Mathematica* Notebook `prog4-2.nb`. The parameter g$= \frac{\Gamma}{2\pi v_\infty}$. Therefore, the above case is obtained at g$< 2R$. We have chosen for Fig. 4.10 (a) the value $R = 1$.

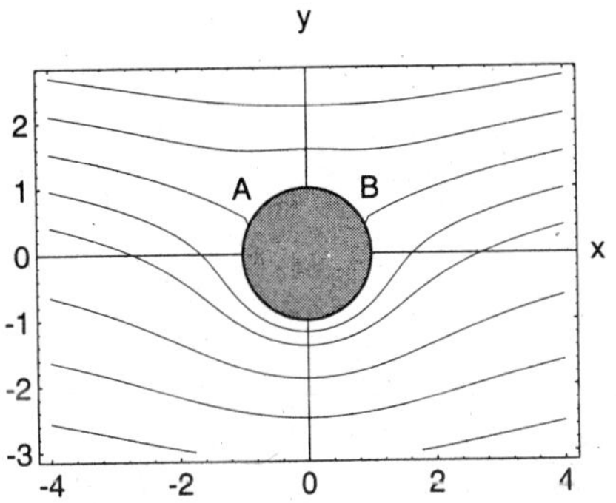

(a)

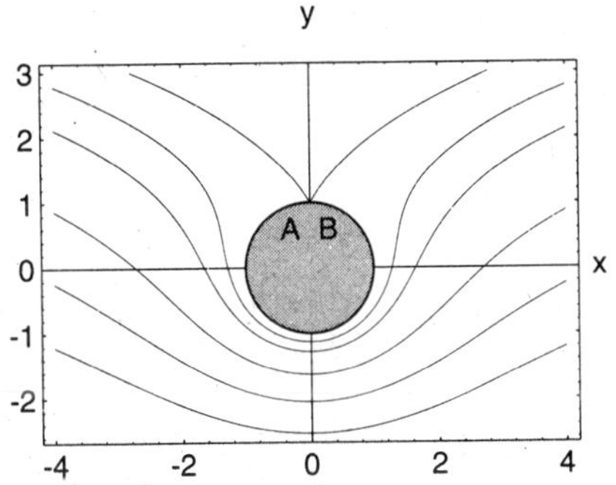

(b)

Figure 4.10: The flow patterns around the cylinder at (a) $\Gamma \neq 0$ and $4v_\infty^2 R^2 > \frac{\Gamma^2}{4\pi^2}$ and at (b) $\Gamma \neq 0$ and $4v_\infty^2 R^2 = \frac{\Gamma^2}{4\pi^2}$.

2) $-\frac{\Gamma^2}{4\pi^2} + 4v_\infty^2 R^2 = 0$. The critical points merge into one point located on the imaginary axis: $|z_{1,2}| = R$, $z_1 = z_2 = \frac{\Gamma}{4\pi v_\infty} \cdot i$ [see Fig. 4.10 (b)]. In order to obtain Fig. 4.10 (b) with the aid of *Mathematica*, it is sufficient to replace the value `g=1` in the above program for Fig. 4.10 (a) with the value `g = 2`.

3) $-\frac{\Gamma^2}{4\pi^2} + 4v_\infty^2 R^2 < 0$. Both of the zeroes are imaginary, $|z_1| < R$, and $|z_2| > R$. There is one critical point in the flow region outside the cylinder on the imaginary axis (see Fig. 4.11). We have obtained Fig. 4.11 by using our *Mathematica* Notebook `prog4-3.nb`.

In the case of the circulatory flow past the cylinder, the streamlines are symmetric with respect to the y-axis. The pressures at the cylinder points symmetrical with respect to the y-axis have equal magnitudes. There is already here no flow symmetry with respect to the x-axis. Therefore, a force acting on the cylinder in the direction of the y-axis arises, and the force in the direction of the x-axis is equal to zero as in the flow without circulation; that is, the D'alembert's paradox takes place here also. Thus, in the presence of circulation, different flow patterns can take place and, therefore, it is necessary for the solution uniqueness to specify the circulation magnitude or some additional conditions for its determination. This fact is substantial while solving many practical problems, which we will show later. Thus, it follows from the considered examples that, if the complex potential is given, the physical flow pattern is determined easily.

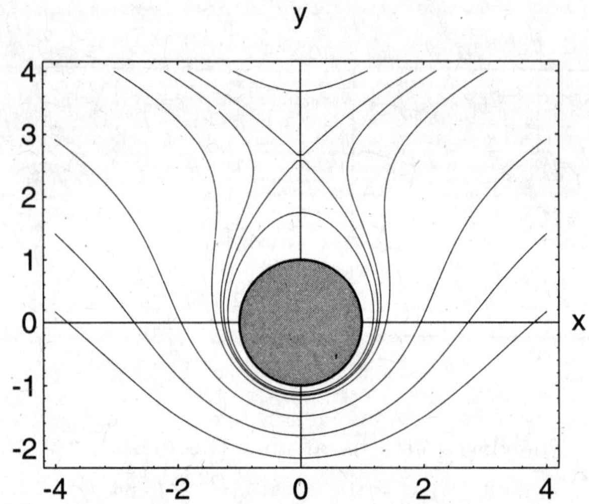

Figure 4.11: The flow pattern around the cylinder at $\Gamma \neq 0$ and $4v_\infty^2 R^2 < \frac{\Gamma^2}{4\pi^2}$.

4.2.3 The Method of Conformal Mappings

Consider the solution of the problem of a flow past a contour of an arbitrary shape l in the plane of complex variable $z = x + iy$. We will denote the region of the z-plane outside the contour l by D. We also introduce an auxiliary plane $\zeta = \xi + i\eta$ and a circle with the radius R with a center at the coordinate origin in the ζ-plane. We will denote a region in the ζ-plane outside the circle l' with the radius R by D' (see Fig. 4.12).

In accordance with the Riemann's theorem about the conformal mapping, an analytical function $z = f(\zeta)$ exists, which maps the region $D' \to D$ in such a way that the points of the contour $l' \to l$. The point $z \to \infty$ is mapped onto the point $\zeta \to \infty$, and the direction of contour tracing is not changed at this point. We assume that

$$\left.\frac{df}{d\zeta}\right|_{\zeta\to\infty} = \left.\frac{dz}{d\zeta}\right|_{\zeta\to\infty} = k > 0,$$

where k is a real number. We will consider the problem of the potential fluid flow past the contour l in the plane D. The flow velocity at infinity is given:

$$V_\infty = v_{\infty x} + iv_{\infty y},$$

and we will denote the corresponding complex potential by $w(z)$ (see Fig. 4.12). In accordance with the Riemann's theorem, an inverse transformation $\zeta = F(z)$ should also exist. Thus, $w(z) = w[f(\zeta)] = w(\zeta)$, where $w(z)$ is determined in D outside l, and $w(\zeta)$ is determined in D'

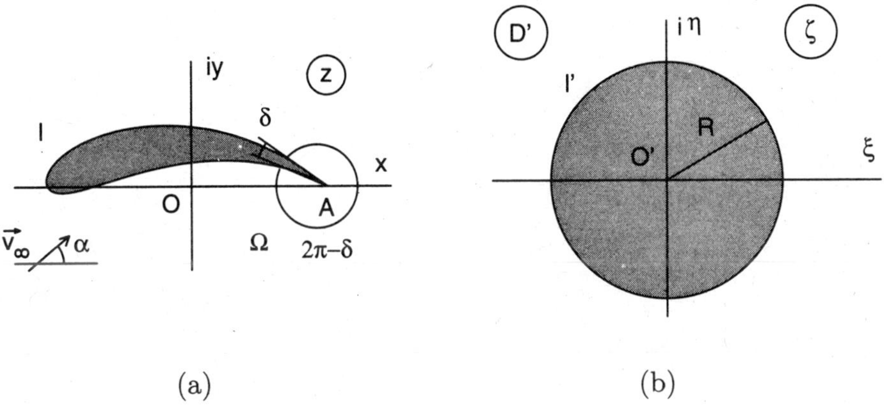

Figure 4.12: The (a) physical and (b) auxiliary planes in the problem of the flow around the contour.

outside l'. We will consider the function $w(\zeta)$ as a complex potential in the ζ-plane. One can associate with each flow in the z-plane a flow in the ζ-plane; that is

$$w(z) = \varphi(x, y) + i\psi(x, y), \quad w(\zeta) = \Phi(\xi, \eta) + i\Psi(\xi, \eta),$$

where the equalities

$$\varphi(x, y) = \Phi(\xi, \eta), \quad \psi(x, y) = \Psi(\xi, \eta)$$

take place at the corresponding points of the z-plane and of the ζ-plane. The function $w(z)$ is a complex potential of the flow past a contour l at rest in the z-plane. Therefore, the function $\psi(x, y)$ is constant on l. The circle l' in the ζ-plane corresponds to the contour l, but since $\psi(x, y) = \Psi(\xi, \eta)$, the function $\Psi(\xi, \eta)$ will also be constant on l'; that is, the circle is a streamline of the flow whose complex potential is $w(\zeta)$. Let us find the conditions at infinity for this flow. The complex velocity is

$$\bar{V}(z) = \frac{dw}{dz} = \frac{dw}{dz} \cdot \frac{dz}{d\zeta} = \bar{V}(z) \cdot \frac{dz}{d\zeta}.$$

At point $z \to \infty$, the value $\bar{V}_\infty$ is known, consequently

$$\left(\frac{dw}{d\zeta}\right)_\infty = \bar{V}\Big|_{\zeta\to\infty} = k\bar{V}\Big|_{z\to\infty} = k\bar{V}_\infty.$$

Thus, the $w(\zeta)$ determines in ζ-plane a flow outside the circle with the velocity at infinity $k\bar{V}_\infty$, which corresponds to formula (4.2.11):

$$w(\zeta) = k\bar{V}_\infty\zeta + \frac{kV_\infty R^2}{\zeta} + \frac{\Gamma}{2\pi i}\ln\zeta.$$

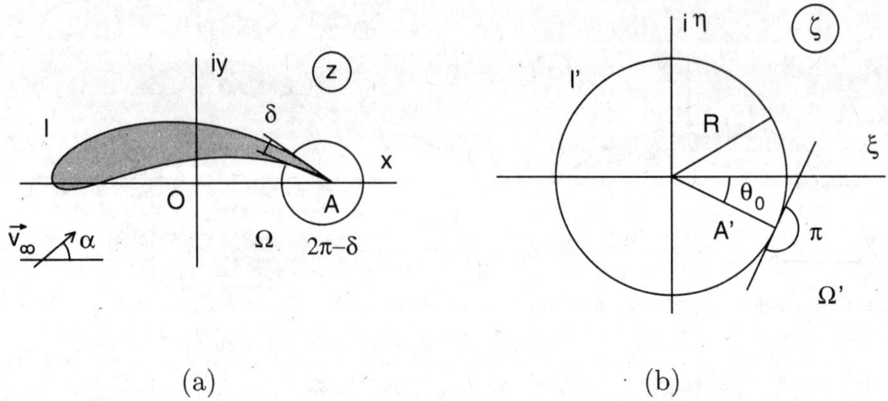

Figure 4.13: A profile with (a) one corner point and (b) its mapped form in the auxiliary plane ζ.

Making the substitution $\zeta = F(z)$ in the obtained equation, we can write that

$$w(z) = k\bar{V}_\infty F(z) + \frac{kV_\infty R^2}{F(z)} + \frac{\Gamma}{2\pi i}\ln F(z). \tag{4.2.16}$$

Formula (4.2.16) gives the solution of the problem of the potential flow past an arbitrary contour, if the conformal mapping of the region outside l onto the exterior of the circle is known; that is, when the function $\zeta = F(z)$ is known. The value k is found by the formula

$$k = \left.\frac{dz}{d\zeta}\right|_\infty = [F'(z)_\infty]^{-1}.$$

It should be noted that the circulation Γ in the solution (4.2.16) remains indeterminate; for its determination, the Chaplygin–Joukowskii–Kutta postulate is used.

Assume that there is in the z-plane a profile with one corner point and the angle $\delta < \pi$ at this point as shown in Fig. 4.13. Let us introduce an auxiliary plane ζ, and the function $z = f(\zeta)$ maps the region Ω' of the ζ-plane outside the circle with the radius R, with the contour l' onto the exterior of the profile l of the region Ω of the z-plane. Let us compute the velocity magnitude at the corner point A, which goes over at the transformation to the point A' of the circle l'; that is

$$\bar{V}_A = \left.\frac{dw(z)}{dz}\right|_A = \left.\frac{dw(\zeta)}{d\zeta}\right|_{A'} \cdot \left.\frac{d\zeta}{dz}\right|_A = \left.\frac{dw(\zeta)}{d\zeta}\right|_{A'} \cdot \frac{1}{\left.\frac{dz}{d\zeta}\right|_{A'}}.$$

The function $z = f(\zeta)$ transforms the angle π into the angle $2\pi - \delta$. This points to the fact that the transformation conformity is violated, and in the vicinity of point A, the function $z(\zeta)$ should have the expansion of the form

$$z - z_A = M(\zeta - \zeta_{A'})^{\frac{2\pi-\delta}{\pi}} + \cdots.$$

We find from here that

$$\left.\frac{dz}{d\zeta}\right|_{A'} = \frac{2\pi - \delta}{\pi} \cdot M(\zeta - \zeta_{A'})^{\frac{\pi-\delta}{\pi}}\bigg|_{\zeta=\zeta_{A'}} = 0,$$

i.e., at $\zeta < \pi$ at point A' $\frac{dz}{d\zeta}|_{A'} = 0$; consequently, if $\frac{dw(\zeta)}{d\zeta}|_{A'} \neq 0$, then $\bar{V}_A \to \infty$, which is physically inadmissible.

The requirement that the velocity $\bar{V}_A$ at the sharp trailing edge be finite makes the contents of the *Chaplygin–Joukowskii–Kutta postulate*[1,4,8,9,10,11]. The satisfaction of this condition is possible if $\frac{dw}{d\zeta}|_{A'} = 0$; i.e., A' is a critical point of the flow past a cylinder, and as is already known it depends on the magnitude of the circulation Γ. From here, it follows the second formulation of the Chaplygin–Joukowskii–Kutta postulate: the circulation in the flow past the profile with a sharp trailing edge A is such that the point of a circle onto which A is mapped at the transformation should be a critical point in the flow past the cylinder.

This postulate enables one to determine the value of Γ. So we have:

$$w(\zeta) = k\bar{V}_\infty \zeta + k\frac{V_\infty R^2}{\zeta} + \frac{\Gamma}{2\pi i} \ln \zeta,$$

where the complex velocity is expressed by the formula $\frac{dw}{d\zeta}|_{\zeta\to\infty} = k\bar{V}_\infty$. Let the flow impinging on the profile have the inclination angle α to the x-axis; that is, $\bar{V}_\infty = |v_\infty|e^{-i\alpha}$, $V_\infty = |v_\infty|e^{i\alpha}$. Let us compute the quantity

$$\left.\frac{dw}{d\zeta}\right|_{A'} = k\bar{V}_\infty - \frac{kV_\infty R^2}{\zeta_{A'}^2} + \frac{\Gamma}{2\pi i} \cdot \frac{1}{\zeta_{A'}} = 0.$$

We find from here that

$$\Gamma = 2\pi i k\Big(\frac{V_\infty R^2}{\zeta_{A'}} - \bar{V}_\infty \cdot \zeta_{A'}\Big).$$

Taking into account that $\zeta_{A'} = Re^{i\theta_0}$ and $V_\infty = |v_\infty|e^{i\alpha}$, where θ_0 is the angle determining the position of point A' on the circle, we obtain:

$$\Gamma = 2\pi i k R|v_\infty| \cdot (e^{i(\alpha-\theta_0)} - e^{-i(\alpha-\theta_0)})$$

or

$$\Gamma = 4\pi k R|v_\infty| \sin(\theta_0 - \alpha). \tag{4.2.17}$$

It is conventional to call the angle $(\alpha - \theta_0)$ the angle of attack. It is seen that, if $\theta_0 - \alpha = 0$ then $\Gamma = 0$ automatically. This enables one to ensure the solution uniqueness and formulate a question on the calculation of the forces acting on the profile from the flow. If the contour is smooth or has the angle $\delta > \pi$ or several corner points, then the question of circulation cannot be solved without imposing additional conditions.

Let us calculate the magnitude of the main vector and of the main moment of the pressure forces acting on the body at a planar attached steady ideal incompressible fluid flow past it. The main vector of the forces acting on the body is equal to

$$\vec{F} = -\oint_l P \cdot \vec{n}\, dl,$$

where l is the body contour and $\vec{n}$ is the vector of the normal to it. Let us rewrite this formula in the projections onto the coordinate axes:

$$F_x = -\oint_l P n_x\, dl = -\oint_l P\, dy, \quad F_y = -\oint_l P n_y\, dl = +\oint_l P\, dx. \tag{4.2.18}$$

Introduce the quantity $\bar{R} = F_x - iF_y$; that is

$$\bar{R} = -\oint_l P\, dy - i\oint_l P\, dx = -i\oint_l P\, d\bar{z}.$$

Along the contour l, the Bernoulli integral is valid, which may be written in the absence of the body forces as

$$P = \rho_0 C - \rho_0 \frac{v^2}{2}, \tag{4.2.19}$$

where C is a constant in the Bernoulli integral. Substituting (4.2.19) into $\bar{R}$, we find:

$$\bar{R} = -i\oint_l \rho_0 C\, d\bar{z} + i\frac{\rho_0}{2}\oint_l v^2\, d\bar{z}.$$

Since the contour l is closed, then $\oint_l \rho_0 C\, d\bar{z} = 0$; consequently

$$\bar{R} = i\frac{\rho_0}{2}\oint_l v^2\, d\bar{z}. \tag{4.2.20}$$

Consider the element of the body contour dl and denote by θ the angle between the tangent to the contour l and the x-axis. Then

$$dz = e^{i\theta} dl, \quad d\bar{z} = e^{-i\theta} dl, \quad d\bar{z} = e^{-2i\theta} dz,$$

and we can rewrite formula (4.2.20) as

$$\bar{R} = i\frac{\rho_0}{2}\oint_l v^2 e^{-2i\theta}\, dz.$$

At the contour l points, the velocity is directed along the tangent; that is, $ve^{-i\theta} = v\cos\theta - iv\sin\theta = v_x - iv_y = \bar{V}$, which enables us to write that

$$\bar{R} = i\frac{\rho_0}{2}\oint_l \bar{V}^2\, dz.$$

Taking into account the fact that $\bar{V} = \frac{dw}{dz}$, we have:

$$\bar{R} = F_x - iF_y = i\frac{\rho_0}{2}\oint_l \left(\frac{dw}{dz}\right)^2 dz. \tag{4.2.21}$$

Formula (4.2.21) is called the *first formula of Chaplygin–Blasius.* Let us compute the magnitude of the main moment of the pressure forces. The contour element dl is affected by the force whose projection is

$$dF_x = -P\,dy, \quad dF_y = P\,dx.$$

The moment of this force with respect to the coordinate origin is

$$dL = dF_y\cdot x - dF_x\cdot y = P(x\,dx + y\,dy),$$

from where we obtain the moment of the forces acting on the body in the form:

$$L = \oint_l P(x\,dx + y\,dy). \tag{4.2.22}$$

Substituting the Bernoulli integral in (4.2.22), we find:

$$L = C\rho_0 \oint_l (x\,dx + y\,dy) - \frac{\rho_0}{2}\oint_l v^2(x\,dx + y\,dy) = -\frac{\rho_0}{2}\oint_l v^2(x\,dx + y\,dy).$$

Consider the expression $z\,d\bar{z}$:

$$z\,d\bar{z} = (x+iy)(dx - i\,dy) = x\,dx + y\,dy + i(y\,dx - x\,dy).$$

Hence

$$x\,dx + y\,dy = \mathrm{Re}\,(z\cdot d\bar{z}),$$

and consequently

$$L = -\frac{\rho_0}{2}\oint_l v^2 \mathrm{Re}\,(z\,d\bar{z}) = \mathrm{Re}\left(-\frac{\rho_0}{2}\oint_l v^2 z\,d\bar{z}\right).$$

Taking into account the fact that $ve^{-i\theta} = \bar{V}$ and $d\bar{z} = e^{-2i\theta}dz$, we can rewrite the *second formula of Chaplygin–Blasius* in the form

$$L = \mathrm{Re}\left(-\frac{\rho_0}{2}\oint_l \left(\frac{dw}{dz}\right)^2 z\,dz\right). \tag{4.2.23}$$

It is conventional to call formulas (4.2.21) and (4.2.23) the *Chaplygin–Blasius formulas.* They enable one to compute the main vector of the pressure forces and the main moment if the complex potential of the flow $w(z)$ is known.

Using the obtained formulas, we now prove the *Joukowskii theorem.*

The main vector of the pressure forces acting on the profile is equal numerically to the product of the density and the absolute values of the velocity and circulation and has the direction obtained by the rotation of the velocity vector v_∞ by the angle $\frac{\pi}{2}$ in the direction opposite to circulation.

Proof. Consider an irrotational steady planar flow of an ideal incompressible fluid past some profile l. Let the complex potential $w(z)$ correspond to this flow. Compute $\bar{R}$, the complex force by the Chaplygin–Blasius formula

$$\bar{R} = F_x - iF_y = \frac{i\rho_0}{2}\oint_l \left(\frac{dw}{dz}\right)^2 dz.$$

From the theory of functions of complex variable, we have for the region outside the profile l the expansion

$$\bar{V}(z) = \frac{dw}{dz} = A_0 + \frac{A_1}{z} + \frac{A_2}{z^2} + \cdots.$$

Let us find the coefficients A_0 and A_1 from the condition that $\bar{V} = \bar{V}_\infty$ for $z \to \infty$, i.e.,

$$A_0 = \left(\frac{dw}{dz}\right)_{z\to\infty} = \bar{V}_\infty.$$

Since the function $\bar{V} = \frac{dw}{dz}$ is bounded outside l and does not have singularities in the overall exterior of the z-plane with respect to l, then in accordance with the residue theorem, we have:

$$\oint_l \frac{dw}{dz}\, dz = 2\pi A_1.$$

Since $\oint_l \frac{dw}{dz}\, dz = \Gamma + iQ$ and the profile is impermeable, we have:

$$2\pi i A_1 = \Gamma \quad \text{or} \quad A_1 = \frac{\Gamma}{2\pi i}.$$

Then

$$\bar{V}(z) = \frac{dw}{dz} = \bar{V}_\infty + \frac{\Gamma}{2\pi i}\cdot\frac{1}{z} + \frac{A_2}{z^2} + \cdots.$$

We have that

$$\left(\frac{dw}{dz}\right)^2 = \bar{V}_\infty^2 + 2\bar{V}_\infty\cdot\frac{\Gamma}{2\pi i}\cdot\frac{1}{z} + \left(2A_2\bar{V}_\infty - \frac{\Gamma^2}{4\pi^2}\right)\frac{1}{z^2} + \cdots.$$

The integration of both sides of this equation and the use of the residue theorem yields

$$\oint \left(\frac{dw}{dz}\right)^2 dz = 2\bar{V}_\infty \Gamma.$$

Then we obtain from (4.2.21) that $\bar{R} = i\rho_0 \bar{V}_\infty \Gamma$, where $\bar{R} = F_x - iF_y$ and $\bar{V}_\infty = v_{x\infty} - iv_{y\infty}$. Taking this formula into account, we have:

$$\bar{R} = F_x - iF_y = i\rho_0\Gamma(v_{x\infty} - iv_{y\infty}) \quad \text{or} \quad F_x = +\rho_0\Gamma v_{y\infty}, \ F_y = -\rho_0\Gamma v_{x\infty}.$$

Since the main vector of the pressure forces is equal to $F = F_x + iF_y$, we find that $F = -i\rho_0\Gamma V_\infty$, where $V_\infty = v_{x\infty} + iv_{y\infty}$. Calculating the modulus of both sides of the equations, we find:

$$|F| = \rho_0 |V_\infty|\,|\Gamma|,$$

which proves the *Joukowskii theorem*.

It is important that the main vector of forces is perpendicular to the velocity direction at infinity. Let us direct the x-axis along the velocity direction $\vec{v}_\infty$; then the force F_y, which is perpendicular to the velocity direction at infinity, is called the lift force, and the force F_x in the flow direction is called the drag. It follows from the Joukowskii theorem that only the lift force F_y arises at a potential flow past a profile, and it exists only in the presence of circulation. Using the formula (4.2.17), we find:

$$|F| = 4\pi k R\rho_0 |v_\infty|^2 \cdot |\sin(\theta_0 - \alpha)|.$$

If we make use of the drag coefficients

$$C_x = \frac{F_x}{\frac{1}{2}\rho_0 v_\infty^2 S}, \quad C_y = \frac{F_y}{\frac{1}{2}\rho_0 v_\infty^2 S},$$

(S is the area of a reference section of the body), then the D'Alembert's paradox will be valid for the ideal fluid: $C_x = 0$.

It should be noted that the formula for C_y at the flow around the wings, where F_y is computed on the basis of the Joukowskii theorem, coincides sufficiently well with experimental data.

Using the second Chaplygin–Blasius formula (4.2.23), we have

$$L = \mathrm{Re}\left(-\frac{\rho_0}{2}\oint z\left(\frac{dw}{dz}\right)^2 dz\right). \qquad (4.2.24)$$

Using the expansion into the Laurent series of $\frac{dw}{dz}$ in a region outside the body, we find:

$$z\left(\frac{dw}{dz}\right)^2 = z\bar{V}_\infty^2 + 2\bar{V}_\infty \frac{\Gamma}{2\pi i} + \left(2A_2\bar{V}_\infty - \frac{\Gamma^2}{4\pi^2}\right)\frac{1}{z} + \cdots.$$

Substituting this expression into (4.2.24) and using the residue theorem, we get:

$$L = \mathrm{Re}\left[-\frac{\rho_0}{2}2\pi i\left(2A_2\bar{V}_\infty - \frac{\Gamma}{4\pi^2}\right)\right]$$

or

$$L = -\mathrm{Re}\,(2\rho_0\pi i\bar{V}_\infty \cdot A_2).$$

Thus, if the complex velocity expansion into the Laurent series is known, i.e., A_2 is known, then the magnitude of the moment can easily be computed. It is often convenient to use the expansion of the mapping function $z = f(\zeta)$ in the neighborhood of a point at infinity

$$z = k\zeta + k_0 + \frac{k_1}{\zeta} + \frac{k_2}{\zeta^2} + \cdots.$$

Let us go over to the variable ζ in the integral (4.2.24) by using the formula

$$L = \mathrm{Re}\left[-\frac{\rho_0}{2}\oint z(\zeta)\left(\frac{dw}{d\zeta}\right)^2 \cdot \left(\frac{d\zeta}{dz}\right)d\zeta\right]. \tag{4.2.25}$$

Let us calculate

$$\begin{aligned}
\left(\frac{dw}{d\zeta}\right)^2 &= k^2\bar{V}_\infty^2 + 2k\bar{V}_\infty\frac{\Gamma}{2\pi i}\cdot\frac{1}{\zeta} \\
&- \left(\frac{\Gamma^2}{4\pi^2} + 2k^2\,V_\infty\bar{V}_\infty R^2\right)\frac{1}{\zeta^2} + \cdots, \\
z\frac{d\zeta}{dz} &= z\cdot\frac{1}{\frac{dz}{d\zeta}} = \zeta + \frac{k_0}{k} + \frac{2k_1}{k}\cdot\frac{1}{\zeta} + \cdots; \\
z(\zeta)\left(\frac{dw}{d\zeta}\right)^2\cdot\frac{d\zeta}{dz} &= c_1\zeta + c_0 + \left[2k_0\bar{V}_\infty\frac{\Gamma}{2\pi i} + 2k\cdot k_1\bar{V}_\infty^2\right. \\
&- \left.\left(\frac{\Gamma^2}{4\pi^2} + 2k^2V_\infty\bar{V}_\infty R^2\right)\right]\frac{1}{\zeta} + \frac{c_2}{\zeta^2} + \cdots.
\end{aligned}$$

Substituting the obtained formulas into (4.2.25) and applying the residue theorem, we find:

$$L = \mathrm{Re}\left\{-\frac{\rho_0}{2}2\pi i\left[2k_0\bar{V}_\infty\frac{\Gamma}{2\pi i} + 2k_1k\bar{V}_\infty^2 - \left(\frac{\Gamma}{4\pi^2} + 2k^2V_\infty\bar{V}_\infty R^2\right)\right]\right\},$$

from where we finally have:

$$L = -\mathrm{Re}\,(k_0\rho_0\bar{V}_\infty\Gamma + 2\pi ikk_1\rho_0\bar{V}_\infty^2). \tag{4.2.26}$$

As an example of the application of the above-presented method, we consider the problem of the flow around a plate. Take in the z-plane a segment $[-a,\ a]$ on the x-axis. A plane-parallel flow comes on this

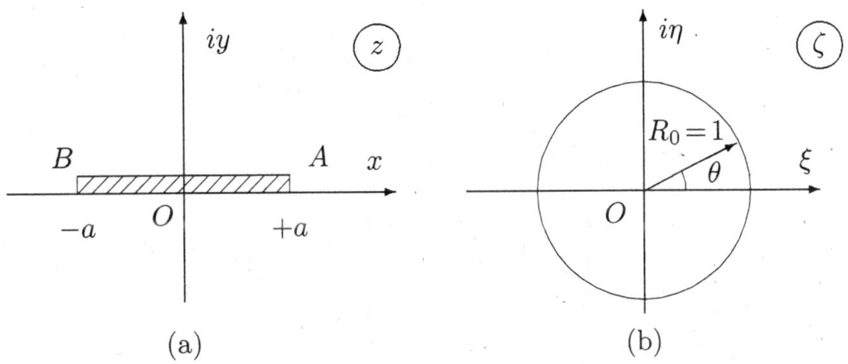

Figure 4.14: The (a) physical and (b) auxiliary planes in the problem of the flow around the plate.

segment and has the velocity $V_\infty = |V_\infty|e^{i\alpha}$ at infinity. It is required to determine the fluid velocity near the plate.

We will search for the solution by using the method of conformal mapping. We take the ζ-plane as a canonical plane in which the solution of a problem on the flow past a circular cylinder of unit radius is known (see Fig. 4.14). Thus, the problem reduces to the determination of a mapping function that maps conformally the exterior of a circle in the ζ-plane onto the exterior of the interval $[-a, a]$ of the z-plane. The function

$$z = \frac{a}{2}\left(\zeta + \frac{1}{\zeta}\right) \tag{4.2.27}$$

proposed by Joukowskii will be such a function. We indeed have on the circle $R_0 = 1$ that $\zeta = e^{i\theta}$. Substituting this value of ζ into (4.2.26), we obtain:

$$z = x + iy = \frac{a}{2}(e^{i\theta} + e^{-i\theta}) = a\cos\theta, \quad x = a\cos\theta, \quad y = 0;$$

that is, the circle is mapped onto the interval in such a way that the upper semicircle is mapped onto the upper side of the segment and the lower semicircle is mapped onto the lower side of the segment. The contrary assertion is also valid, i.e., the mapping of the exterior of a segment onto the exterior of a circle. Resolving equations (4.2.27) with respect to ζ, we find:

$$\zeta = \frac{z + \sqrt{z^2 - a^2}}{a}. \tag{4.2.28}$$

Knowing (4.2.28) we can write a complex potential for the flow around a plate. We have in our case that

$$z \approx k\zeta + k_0 + \frac{k_1}{\zeta} + \frac{k_2}{\zeta^2} + \dots$$

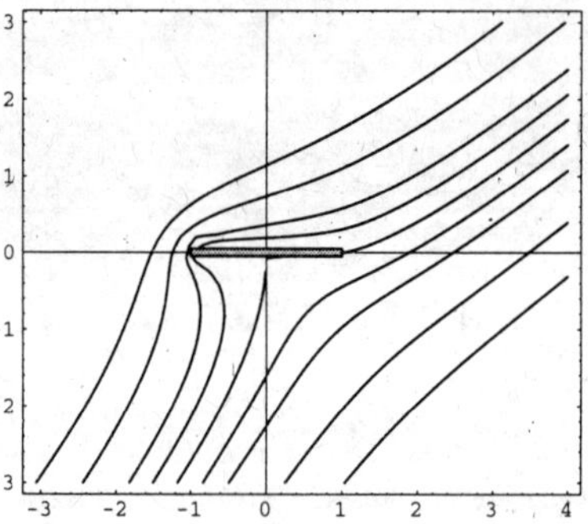

Figure 4.15: The streamlines in the incompressible fluid flow around the plate at $\alpha = \pi/4$.

and

$$k = \frac{dz}{d\zeta}\bigg|_{\infty} = \frac{a}{2}, \quad k_0 = 0, \quad k_1 = \frac{a}{2}, \quad R_0 = 1.$$

Therefore,

$$\begin{aligned} w(z) &= \frac{1}{2}\bar{V}_{\infty}(z + \sqrt{z^2 - a^2}) + \frac{1}{2}V_{\infty}(z - \sqrt{z^2 - a^2}) \\ &+ \frac{\Gamma}{2\pi i}\ln\Big(\frac{z + \sqrt{z^2 - a^2}}{a}\Big), \end{aligned} \tag{4.2.29}$$

where $\bar{V}_{\infty} = |V_{\infty}|e^{-i\alpha}$ and $V_{\infty} = |V_{\infty}|e^{+i\alpha}$. We can determine the circulation magnitude from the condition of the velocity finiteness at the trailing sharp edge in accordance with the Chaplygin–Joukowskii–Kutta postulate by the formula (4.2.17). Here α is the angle between the freestream direction and the x-axis, and θ_0 is the angle determining the location in the ζ-plane of point A', onto which the trailing sharp edge A is mapped. In our case, $\theta_0 = 0$, $k = \frac{a}{2}$, $R_0 = 1$, and the expression for the circulation will then be determined by the formula

$$\Gamma = -2\pi a|V_{\infty}|\sin\alpha. \tag{4.2.30}$$

Thus, having the complex potential (4.2.29) and (4.2.30), we can find the complex velocity V and its components V_x and V_y in the region, including also the plate points. We have made the *Mathematica* program `prog4-4.nb`, which solves the problem of the incompressible fluid flow around a plate. We show in Fig. 4.15 a picture of the streamlines obtained with the aid of this program.

Let us determine the main vector of the pressure forces acting on the plate with the aid of the Joukowskii theorem:

$$\bar{R} = F_x - iF_y = i\rho_0\bar{V}_{\infty}\Gamma = -2\pi i\rho_0 a|V_{\infty}|^2 e^{-i\alpha}\sin\alpha.$$

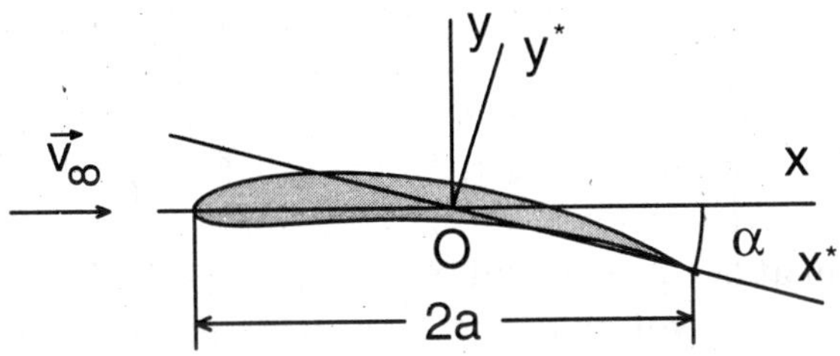

Figure 4.16: The flow around a slender profile.

We find from here that

$$F_x = -2\pi\rho_0|V_\infty|a\sin^2\alpha, \quad F_y = 2\pi\rho_0|V_\infty|^2 a\sin\alpha\cos\alpha.$$

The magnitude of the Joukowskii force for the plate is

$$F = |R| = 2\pi a\rho_0|V_\infty|^2|\sin\alpha|, \tag{4.2.31}$$

and the lift force coefficient is

$$C_P = \frac{F}{\frac{1}{2}\rho_0|V_\infty|^2 S} = 2\pi\sin\alpha,$$

where $S = 2a$. At small angles of attack, this formula simplifies to

$$C_P \approx 2\pi\alpha,$$

since $\sin\alpha \approx \alpha$. Taking into account the expression for the moment L (4.2.26), one can write that

$$L = -\frac{a}{2}\cos\alpha \cdot F.$$

The experimental data show that the obtained results can be used as the solution of a problem of the attached flow around slender profiles at small angles of attack. The profile is assumed to be slender if the ratio of the profile thickness to its chord length is much smaller than unity, and the need in an attached flow requires a smallness of the angle between the tangent direction at any point of the profile and the chord as well as a small angle of attack.

4.2.4 The Problem of the Flow around a Slender Profile

Consider the problem of the ideal incompressible fluid flow around a slender profile at a given freestream velocity. Denote by h the profile thickness and by $2a$ its chord length. The assumption on the slender profile means that the ratio $\frac{h}{2a}$ is small, $\frac{h}{2a} \ll 1$, and that the angle between the direction of a tangent at any profile point and the chord is small. Let us choose the coordinate system x, y in such a way that the flow velocity V at infinity is directed in parallel with the x-axis and the angle between $\vec{v}_\infty$ and the profile chord is small ($\alpha \approx 0$). We place the coordinate origin at the profile mid-chord (see Fig. 4.16). Let $y_t = F_1(x)$ and $y_b = F_2(x)$ be the equations for the upper (top) and bottom surfaces of the profile. The above-introduced limitations are expressed in the form of the following inequalities:

$$\left|\frac{F_1(x)}{2a}\right| \ll 1, \quad \left|\frac{F_2(x)}{2a}\right| \ll 1, \quad \left|\frac{dF_1(x)}{dx}\right| \ll 1, \quad \left|\frac{dF_2(x)}{dx}\right| \ll 1. \tag{4.2.32}$$

It should be noted that the limitations (4.2.31) for the subsonic profiles may not be satisfied in the vicinity of a rounded leading edge. In this region, one must use with extreme care the solutions that we construct below.

Introduce a coordinate system x^*Oy^* fixed in a profile by directing the x^*-axis along the profile chord $(-a,\ a)$. The angle between the velocity $\vec{v}_\infty$ direction of the Ox-axis and the chord (the Ox^*-axis) is the angle of attack α. Let $y_t^* = F_t(x^*)$, $y_b^* = F_b(x^*)$ be the profile equations in this coordinate system. The relation between the coordinates (x, y) and (x^*, y^*) is given by the formulas

$$x^* = x\cos\alpha - y\sin\alpha, \quad y^* = x\sin\alpha + y\cos\alpha.$$

If we take into account the fact that $\alpha \sim 0$, then we have:

$$x^* = x, \qquad y^* = x\alpha + y, \tag{4.2.33}$$

since $\sin\alpha \approx \alpha$ and $\cos\alpha \approx 1$. The profile equations y_t^* and y_b^* in the coordinate system (x, y) with regard for (4.2.33) take the form

$$x\alpha + y_t = F_t(x), \qquad x\alpha + y_b = F_b(x)$$

or

$$y_t = F_t(x) - x\alpha, \quad y_b = F_b(x) - x\alpha. \tag{4.2.34}$$

The problem of the flow around the profile will be solved if we find a function $w(z)$ satisfying the conditions at infinity, the conditions for the flow around the profile, and the Chaplygin–Joukowskii postulate. Let us present the complex potential $w(z)$ in the form

$$w(z) = V_\infty \cdot z + w'(z),$$

where $V_\infty \cdot z$ is a complex potential of a plane-parallel flow having the velocity $\vec{v}_\infty$ directed along the x-axis, and $w'(z)$ is a complex potential of disturbances. It is obviously necessary to require that

$$\left.\frac{dw}{dz}\right|_\infty = V_\infty, \qquad \left.\frac{dw'}{dz}\right|_\infty = 0.$$

Taking into account the complex potential definition and subdividing the real and imaginary parts, we find:

$$\varphi(x,y) = V_\infty \cdot x + \varphi'(x,y), \quad \psi(x,y) = V_\infty \cdot y + \psi'(x,y). \qquad (4.2.35)$$

Here φ' and ψ' are the velocity potential and the stream function of the disturbed flow. Let us obtain the condition for the function ψ'. Since the profile contour S is a streamline, one can set without loss of generality that $\psi|_S = 0$. Substituting (4.2.33) and (4.2.34) into this equation, we find:

$$\psi'(x,y_t) = -V_\infty(F_t(x) - \alpha x); \quad \psi'(x,y_b) = -V_\infty(F_b(x) - \alpha x). \quad (4.2.36)$$

Taking into account the fact that a slender profile introduces small disturbances into the flow, we expand the functions $\psi'(x,y_t)$ and $\psi'(x,y_b)$ into the Taylor series in powers of y_t and y_b in the neighborhood of $y_t = y_b = 0$; that is

$$\begin{aligned}\psi'(x,y_t) &= \psi'(x,+0) + \left.\frac{\partial\psi'}{\partial y}\right|_{+0} y_t + \ldots,\\ \psi'(x,y_b) &= \psi'(x,-0) + \left.\frac{\partial\psi'}{\partial y}\right|_{-0} y_b + \ldots\end{aligned}$$

Substituting the obtained expressions into (4.2.35) and retaining only the terms of the first-order smallness, we find:

$$\psi'(x,+0) = -V_\infty(F_t(x) - \alpha x); \quad \psi'(x,-0) = -V_\infty(F_b(x) - \alpha x). \qquad (4.2.37)$$

Thus, the problem reduces to the determination of $w'(z)$ outside the cut $(-a,a)$ by the given values (4.2.36) for ψ' on the cut. The conditions at infinity as well as the Joukowskii–Chaplygin postulate should be satisfied. Let us make use of the Joukowskii transformation of the form

$$z(\zeta) = \frac{a}{2}\left(\zeta + \frac{1}{\zeta}\right),$$

which maps the exterior of a unit circle in the ζ-plane onto the exterior of the cut $(-a,a)$ in the z-plane. Let us set $w'(z) = w'(z(\zeta)) = w'(\zeta)$ and search for the function $w'(z)$ determined in the exterior of the unit circle in the ζ-plane and satisfying the condition

$$\left.\frac{dw'}{d\zeta}\right|_\infty = 0 \qquad (4.2.38)$$

and corresponding to the no-slip condition on the unit circle. Let us find this condition. Set $\zeta = \rho e^{i\theta}$, and introduce the functions $\Phi'(\rho, \theta)$ and $\Psi'(\rho, \theta)$ such that

$$w'(\zeta) = \Phi'(\rho, \theta) + i\Psi'(\rho, \theta).$$

Taking into account the formula $\psi'(x, y) = \Psi'(\rho, \theta)$, we have on the circle $\rho = 1$ that

$$\Psi'(1, \theta) = \begin{cases} -V_\infty[F_t(a\cos\theta) - \alpha a\cos\theta], & 0 \le \theta \le \pi, \\ -V_\infty[F_b(a\cos\theta) - \alpha a\cos\theta], & \pi \le \theta \le 2\pi. \end{cases}$$

We will search for the function $w'(\zeta)$ given in the exterior of the circle $\rho = 1$ and satisfying the condition (4.2.38) in the form of the series

$$w'(\zeta) = \frac{\Gamma}{2\pi i}\ln\zeta + \sum_{n=0}^{\infty}\frac{c_n}{\zeta^n}, \tag{4.2.39}$$

where $c_n = a_n + ib_n$ are the unknown constants to be determined. Determining the real and imaginary parts in (4.2.39), we find the expressions:

$$\begin{aligned} \Phi'(\rho, \theta) &= \frac{\Gamma}{2\pi}\theta + \sum_{n=0}^{\infty}\frac{1}{\rho^n}(a_n\cos n\theta + b_n\sin n\theta), \\ \Psi'(\rho, \theta) &= -\frac{\Gamma}{2\pi}\ln\rho + \sum_{n=0}^{\infty}\frac{1}{\rho^n}(-a_n\sin n\theta + b_n\cos n\theta). \end{aligned}$$

We have on the circle $\rho = 1$ that

$$\Psi'(1, 0) = \sum_{n=0}^{\infty}(b_n\cos n\theta - a_n\sin n\theta). \tag{4.2.40}$$

Substituting (4.2.40) into the boundary condition for $\Psi'(1, \theta)$, we find:

$$\sum_{n=0}^{\infty}(b_n\cos n\theta - a_n\sin n\theta) = aV_\infty(\alpha\cos\theta - f(\theta)), \tag{4.2.41}$$

where

$$f(\theta) = \begin{cases} \dfrac{F_t(a\cos\theta)}{a}, & 0 \le \theta \le \pi, \\ \dfrac{F_b(a\cos\theta)}{a}, & \pi \le \theta \le 2\pi. \end{cases}$$

We now expand the known function $f(\theta)$ into the Fourier series:

$$f(\theta) = \sum_{n=0}^{\infty}(\alpha_n\cos n\theta + \beta_n\sin n\theta)$$

and substitute this series into (4.2.41). Then we obtain:

$$\sum_{n=0}^{\infty}(b_n \cos n\theta - a_n \sin n\theta) = aV_\infty[\alpha \cos\theta - \sum_{n=0}^{\infty}(\alpha_n \cos n\theta + \beta_n \sin n\theta)].$$

Comparing the coefficients at equal $\sin n\theta$ and $\cos n\theta$, we find:

$$\begin{aligned} a_n &= V_\infty a\beta_n, \quad n \geq 1, \\ b_0 &= -aV_\infty\alpha_0, \quad b_1 = aV_\infty(\alpha - \alpha_1), \quad b_n = -aV_\infty\alpha_n, \ n \geq 2. \end{aligned}$$

To determine Γ, let us make use of the Chaplygin–Joukowskii postulate, i.e., the velocity finiteness at point $z = a$ and $\frac{\partial\varphi'}{\partial x}|_{x=0} = 0$. By virtue of the fact that

$$\frac{d\varphi'}{dx} = \frac{d\Phi'}{d\theta} \cdot \frac{1}{\frac{dx}{d\theta}} = -\frac{d\Phi'}{d\theta} \cdot \frac{1}{a\sin\theta},$$

the condition

$$\left.\frac{d\Phi'}{d\theta}\right|_{\theta=0} = 0$$

should be satisfied at the trailing edge $\theta = 0$. We now use the formula for $\Phi'(\rho, \theta)$ and write the values of $\frac{d\Phi'}{d\theta}$ on the circle $\rho = 1$, i.e.,

$$\left.\frac{d\Phi'}{d\theta}\right|_{\rho=1} = \frac{\Gamma}{2\pi} + \sum_{n=0}^{\infty}(-na_n \sin n\theta + nb_n \cos n\theta).$$

Taking into account the condition at point $\theta = 0$, we have:

$$\Gamma = -2\pi \sum_{n=0}^{\infty} nb_n.$$

Thus, we have determined in the complex potential $w'(\zeta)$ all coefficients of the series, which enables us to perform a complete construction of the problem solution by the presented algorithm.

4.3 Axisymmetric and Three-Dimensional Potential Ideal Incompressible Fluid Flows

4.3.1 Axially Symmetric Flows

These are such flows, which will be the same in all planes passing through some fixed line l. The liquid particles trajectories lie in the halfplanes passing through the l. The axially symmetric flows are often encountered in practice: for example, the flows in cylindrical tubes and channels, and the flow around the bodies of revolution at zero incidence. It is convenient to describe the axially symmetric flows both in the cylindrical coordinates (r, φ, z) (see Fig. 2.4) and in the spherical coordinates (r, θ, λ) (see Fig. 2.5). (In contrast with Fig. 2.5, the axial angle is denoted here by the letter λ. This is related to the fact that the letter φ is used in the present chapter to denote the velocity potential.) In the case of an axially symmetric flow, all hydrodynamic quantities depend in the cylindrical coordinates only on r and z, and in the spherical coordinates, they depend on r and θ and do not depend on λ.

Consider a fluid flow from a source (sink) of strength q located at the coordinate origin. Such a flow is a particular case of an axially symmetric flow in which all hydrodynamic functions depend only on r in the spherical coordinates. Since the flow is potential, then $\vec{v} = \nabla\varphi$, and in the spherical coordinate system, we obtain, for the physical components of the velocity vector, the expressions:

$$v_r = \frac{\partial \varphi}{\partial r}, \quad v_\theta = \frac{1}{r}\frac{\partial \varphi}{\partial \theta}, \quad v_\lambda = \frac{1}{r \sin\theta} \cdot \frac{\partial \varphi}{\partial \lambda}.$$

The continuity equation for an incompressible fluid in the spherical coordinates follows from formula (2.1.144) if we assume $\frac{\partial \rho}{\partial t} = 0$:

$$\frac{1}{r^2 \sin\theta}\left\{\frac{\partial}{\partial r}\left(r^2 \sin\theta \frac{\partial \varphi}{\partial r}\right) + \frac{\partial}{\partial \theta}\left(\sin\theta \frac{\partial \varphi}{\partial \theta}\right) + \frac{\partial}{\partial \lambda}\left(\frac{\partial \varphi}{\partial \lambda}\right)\right\} = 0.$$

Let us make use of the condition that $\varphi = \varphi(r)$. Then we find from the continuity equation:

$$\frac{\partial}{\partial r}\left(r^2 \frac{\partial \varphi}{\partial r}\right) = 0.$$

The integration yields

$$\varphi = -\frac{c}{r} + c_1,$$

where c and c_1 are constants. Since the velocity potential is determined with the accuracy up to an arbitrary constant, we can assume without loss of generality that $c_1 = 0$.

The source strength q is the fluid quantity, which flows through a surface of a sphere of radius r per unit time, i.e.,

$$q = 4\pi r^2 v_r = 4\pi c \quad \text{or} \quad c = \frac{q}{4\pi}.$$

This enables one to write down the velocity potential in the case of a flow from a source placed at the coordinate origin in the form

$$\varphi = -\frac{q}{4\pi r},$$

where $r = \sqrt{x^2 + y^2 + z^2}$. If the source is placed not in the coordinate origin but in a fixed point with the coordinates $x = a$, $y = b$, and $z = c$, then

$$\varphi = -\frac{q}{4\pi\sqrt{(x-a)^2 + (y-b)^2 + (z-c)^2}}. \tag{4.3.1}$$

Consider a flow from a source and sink located at a distance l from one another and having the same strengths q but having the opposite signs. Since the equations governing the flows are linear, the sum of the two solutions is also the solution $\varphi = \varphi_1 + \varphi_2$, where

$$\varphi_1 = -\frac{q}{4\pi}\frac{1}{\sqrt{x^2 + y^2 + (z - \frac{l}{2})^2}}$$

is a potential corresponding to the source and

$$\varphi_2 = \frac{q}{4\pi}\frac{1}{\sqrt{x^2 + y^2 + (z + \frac{l}{2})^2}}$$

is a potential corresponding to the sink. That is, we have the solution

$$\varphi = -\frac{q}{4\pi}\left\{\frac{1}{\sqrt{x^2 + y^2 + (z - \frac{l}{2})^2}} - \frac{1}{\sqrt{x^2 + y^2 + (z + \frac{l}{2})^2}}\right\}. \tag{4.3.2}$$

Consider the limiting case where $q \to \infty$, $l \to 0$, and $q \cdot l = M = \text{const.}$ In this case, it is conventional to call the flow the *flow from a spatial dipole*. Expand the expressions in (4.3.2) in the brackets into the Taylor series in powers of l and go over to the limit as $l \to 0$. As a result, we obtain:

$$\varphi = -\frac{M \cdot z}{4\pi r^3}, \tag{4.3.3}$$

where $M = q \cdot l$ is called the dipole moment. The obtained formula (4.3.3) may also be written in the form

$$\varphi = \frac{M}{4\pi} \cdot \frac{\partial}{\partial z}\left(\frac{1}{r}\right). \tag{4.3.4}$$

If the dipole axis l does not coincide with the coordinate axis, then the potential of a flow from the dipole will in a general case have the form

$$\varphi = \frac{M}{4\pi} \cdot \frac{\partial}{\partial l}\left(\frac{1}{r}\right),$$

where the

$$\frac{\partial}{\partial l} = \frac{\partial}{\partial x} \cdot \cos(l,\hat{} x) + \frac{\partial}{\partial y} \cdot \cos(l,\hat{} y) + \frac{\partial}{\partial z} \cdot \cos(l,\hat{} z)$$

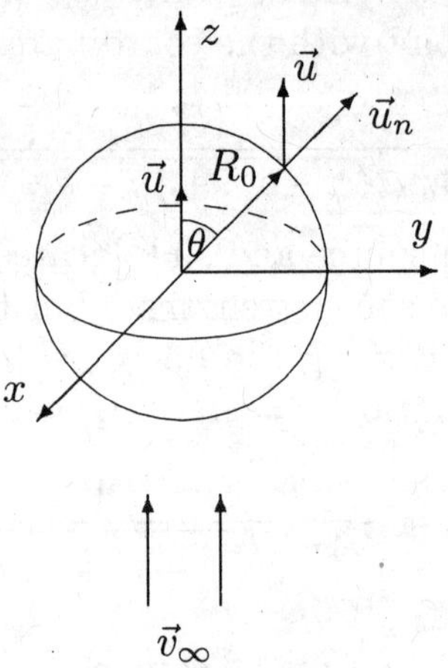

Figure 4.17: The flow around a sphere moving at a velocity $\vec{u} \parallel \vec{v}_\infty$.

is the derivative taken along the direction of the dipole axis.

Consider a problem of the flow around a sphere of radius R_0 moving at a velocity $\vec{u}$ along the z axis (see Fig. 4.17). The velocity vector of a freestream flow $\vec{v}_\infty$ is directed along the Oz-axis. It is required to find the flow. In this case, the velocity potential φ will satisfy the Laplace equation

$$\Delta\,\varphi = 0,$$

the boundary conditions on the sphere surface

$$\left.\frac{\partial \varphi}{\partial n}\right|_S = u_n,$$

and the condition at infinity

$$\left.\frac{\partial \varphi}{\partial x}\right|_\infty = 0, \quad \left.\frac{\partial \varphi}{\partial y}\right|_\infty = 0, \quad \left.\frac{\partial \varphi}{\partial z}\right|_\infty = v_\infty.$$

Writing the Laplace equation in the spherical coordinates and taking into account the fact that the flow is axisymmetric, we have:

$$\frac{\partial}{\partial r}\left(r^2 \sin\theta \frac{\partial \varphi}{\partial r}\right) + \frac{\partial}{\partial \theta}\left(\sin\theta \frac{\partial \varphi}{\partial \theta}\right) = 0. \tag{4.3.5}$$

Rewrite the boundary conditions in the form

$$\left.\frac{\partial \varphi}{\partial n}\right|_{r=R_0} = u\cos\theta, \quad \left. v_r \right|_{r\to\infty} = \left.\frac{\partial \varphi}{\partial r}\right|_{r\to\infty} = v_\infty \cos\theta,$$

$$\left. v_\theta \right|_{r\to\infty} = \left.\frac{1}{r}\frac{\partial \varphi}{\partial \theta}\right|_{r\to\infty} = -v_\infty \sin\theta, \quad \left. v_\lambda \right|_{r\to\infty} = 0. \tag{4.3.6}$$

It is convenient to search for the solution of the formulated problem in the form of the sum of two solutions

$$\varphi = \varphi_1 + \varphi_2;$$

where φ_1 is a potential of a plane-parallel flow having the velocity $\vec{v}_\infty$ in the direction of the Oz-axis, and φ_2 is the potential of a flow near the sphere moving at a velocity $-(\vec{v}_\infty - \vec{u})$. The representation of the solution in the form of a sum is possible since equation (4.3.5) is linear. The boundary conditions (4.3.6) may easily be rewritten for the functions φ_1 and φ_2 in the form

$$\left.\frac{\partial\varphi_1}{\partial r}\right|_{r\to\infty} = v_\infty cos\theta, \quad \frac{1}{r}\cdot\left.\frac{\partial\varphi_1}{\partial\theta}\right|_{r\to\infty} = -v_\infty \sin\theta,$$
$$\left.\varphi_2\right|_{r\to\infty} = 0, \quad \left.\frac{\partial\varphi_2}{\partial r}\right|_{r=R_0} = -(v_\infty - u)\cos\theta. \qquad (4.3.7)$$

The potential φ_1 is determined by the formula

$$\varphi_1 = v_\infty z = v_\infty r\cos\theta$$

since $z = r\cos\theta$. For the construction of the potential φ_2, we make use of the flow from a dipole with an axis parallel with the Oz-axis and located at the coordinate origin O; i.e.,

$$\varphi_2 = -A\frac{\partial}{\partial z}\left(\frac{1}{r}\right) = A\cdot\frac{\cos\theta}{r^2},$$

where A is a constant, which we find by using the boundary condition (4.3.7) $A = \frac{(v_\infty-u)R_0^3}{2}$. Thus, the general solution will have the form

$$\varphi(r,\theta) = \left(v_\infty r + \frac{v_\infty - u}{2}\cdot\frac{R_0^3}{r^2}\right)\cos\theta \qquad (4.3.8)$$

or

$$\varphi = v_\infty z + \frac{1}{2}\left(\frac{R}{r}\right)^3 (v_\infty - u)\, z.$$

The first item is the potential of a plane-parallel flow at a velocity v_∞, and the second item is the potential of a dipole with the moment $M = 2\pi R^3(u - v_\infty)$.

Consider the particular cases. Let the sphere be at rest. Then $u = 0$, and

$$\varphi = v_\infty\left(r + \frac{1}{2}\frac{R_0^3}{r^2}\right)\cos\theta. \qquad (4.3.9)$$

If the fluid at infinity is at rest, then $v_\infty = 0$ and $\varphi = -\frac{R_0^3}{2r^2} \cdot u\cos\theta$. Let us study the distribution of the velocities and the pressure on the surface of a fixed sphere. We have from (4.3.9):

$$v_r = v_\infty\left(1 - \frac{R_0^3}{r^3}\right)\cos\theta, \quad v_\theta = -v_\infty\left(1 + \frac{R_0^3}{2r^3}\right)\sin\theta,$$

and on the surface $r = R_0$, we have:

$$v_r\Big|_{r=R_0} = 0, \quad v_\theta = -\frac{3}{2}v_\infty \sin\theta.$$

The maximum value of the velocity magnitude on the sphere surface is equal to $\frac{3}{2}v_\infty$ and is reached at the points $\theta = \pm\frac{\pi}{2}$. It should be noted that for the case of the flow past a cylinder the velocity maximum on the surface is equal to $2v_\infty$.

We have from the Bernoulli integral on the sphere surface that

$$\frac{P - P_\infty}{\rho_0} = \frac{v_\infty^2}{2}\left(1 - \frac{9}{4}\sin^2\theta\right),$$

where the magnitude of a constant was found from the condition at infinity. It follows from the symmetry of the pressure distribution that the main vector of all pressure forces is equal to zero; that is, the D'Alembert's paradox takes place.

Following[1,9], we now consider the *axially symmetric flows in the cylindrical coordinates*, where the z-axis is taken as the symmetry axis. All hydrodynamic quantities do not depend in this case on φ. Therefore, the continuity equation (2.1.139) in the case $\rho_0 = \text{const}$ can be written in the form

$$\frac{\partial}{\partial r}(rv_r) + \frac{\partial}{\partial z}(rv_z) = 0.$$

If we introduce the function ψ by the formulas

$$v_r = \frac{1}{r}\frac{\partial\psi}{\partial z}, \quad v_z = -\frac{1}{r}\frac{\partial\psi}{\partial r}, \tag{4.3.10}$$

then the continuity equation is satisfied automatically. From the equation for the streamlines $\frac{dr}{v_r} = \frac{dz}{v_z}$, the expression $rv_r\,dz = rv_z\,dr$ follows. By virtue of (4.3.10), it is a total differential

$$d\psi = \frac{\partial\psi}{\partial r}\,dr + \frac{\partial\psi}{\partial z}\,dz,$$

and by the definition of the streamline, the function ψ will be constant on a streamline. Thus, we will call the function ψ, introduced by formulas (4.3.10), the stream function.

If the flow is potential, then a velocity potential exists, which for the flow with the axial symmetry in the cylindrical coordinates is related to the velocity components via the formulas

$$v_r = \frac{\partial \varphi}{\partial r}, \quad v_z = \frac{\partial \varphi}{\partial z}.$$

Taking into account these equations and formulas (4.3.10), we find:

$$\frac{\partial \varphi}{\partial r} = \frac{1}{r}\frac{\partial \psi}{\partial z}, \quad \frac{\partial \varphi}{\partial z} = -\frac{1}{r}\frac{\partial \psi}{\partial r}.$$

These relations differ from the Cauchy–Riemann conditions, which took place in the planar flow, by the presence of the factor $1/r$. If one of the functions φ or ψ is known, then the computation of them in terms of one another reduces to the quadratures

$$\psi = \psi(r_0, z_0) + \int_l r\left(\frac{\partial \varphi}{\partial r}dz - \frac{\partial \varphi}{\partial z}dr\right)$$

or (4.3.11)

$$\varphi = \varphi(r_0, z_0) + \int_l \frac{1}{r}\left(\frac{\partial \psi}{\partial z}dr - \frac{\partial \psi}{\partial r}dz\right).$$

Let us construct the stream functions for some specific simplest flows.

1) *The translational flow* $\varphi = v_\infty z$. We find by formula (4.3.11) that

$$\psi = -v_\infty \frac{r^2}{2} + c.$$

If the flow axis $r = 0$ is the streamline $\psi = 0$, then $c = 0$ and $\psi = -v_\infty \frac{r^2}{2}$.

2) *The flow from a source* $\varphi = -\frac{q}{4\pi} \cdot \frac{1}{\sqrt{r^2+z^2}}$. It is clear that

$$\frac{\partial \varphi}{\partial r} = \frac{q}{4\pi} \cdot \frac{r}{(\sqrt{r^2+z^2})^3}, \quad \frac{\partial \varphi}{\partial z} = \frac{q}{4\pi} \cdot \frac{z}{(\sqrt{r^2+z^2})^3}.$$

Using the relation between φ and ψ, we find:

$$\psi = \frac{q}{4\pi} \cdot \frac{z}{\sqrt{r^2+z^2}} + f(z).$$

Computing the quantity $\frac{\partial \psi}{\partial z}$, we obtain:

$$\frac{\partial \psi}{\partial z} = \frac{q}{4\pi} \cdot \frac{r^2}{(\sqrt{r^2+z^2})^3} + \frac{df}{dz}.$$

It follows from the equality $\frac{\partial\psi}{\partial z} = r \cdot \frac{\partial\varphi}{\partial r}$, however, that $\frac{df}{dz} = 0$; that is, $f = \text{const}$. Consequently the stream function describing a flow from the source has the form

$$\psi = \frac{q}{4\pi} \cdot \frac{z}{\sqrt{r^2 + z^2}} + c.$$

3) *The flow from a dipole.* The velocity potential in this case is represented by a well-known formula

$$\varphi = -\frac{Mz}{4\pi r^3} = \frac{M}{4\pi} \cdot \frac{\partial}{\partial z}\Big(\frac{1}{\sqrt{r^2 + z^2}}\Big).$$

By using the relation

$$\frac{\partial\psi}{\partial z} = r\frac{\partial\varphi}{\partial r} = \frac{\partial}{\partial z}\Big(\frac{M}{4\pi}r\frac{\partial}{\partial r}\Big(\frac{1}{\sqrt{r^2 + z^2}}\Big)\Big)$$

and integrating it with respect to z, we find:

$$\psi = \frac{M}{4\pi}r\frac{\partial}{\partial r}\Big(\frac{1}{\sqrt{r^2 + z^2}}\Big) + f(r).$$

Computing the quantity $\frac{\partial\psi}{\partial r}$ from this formula and comparing it with the expression for the $\frac{\partial\psi}{\partial r}$ found from the condition $\frac{\partial\psi}{\partial r} = -r\frac{\partial\varphi}{\partial z}$, we obtain $\frac{df}{dr} = 0$; that is, $f = \text{const}$. Consequently, the stream function describing the flow from a dipole has the form

$$\psi = -\frac{M}{4\pi} \cdot \frac{r^2}{(\sqrt{r^2 + z^2})^3} + c.$$

In the general case of the potential axially symmetric flows of an ideal incompressible fluid, the formulation of problems on the flows around the bodies in terms of φ reduces to the solution of the equation

$$\frac{\partial^2\varphi}{\partial r^2} + \frac{\partial^2\varphi}{\partial z^2} + \frac{1}{r}\frac{\partial\varphi}{\partial r} = 0$$

and the satisfaction of the boundary conditions

$$\frac{\partial\varphi}{\partial n}\Big|_S = 0, \quad \frac{\partial\varphi}{\partial r}\Big|_{r\to\infty} = 0 \quad \text{and} \quad \frac{\partial\varphi}{\partial z}\Big|_{z\to\infty} = V_\infty,$$

where S is the body surface. If we consider the same problem in terms of ψ, then it is necessary to search for the solution of the equation

$$\frac{\partial^2\psi}{\partial r^2} + \frac{\partial^2\psi}{\partial z^2} - \frac{1}{r}\frac{\partial\psi}{\partial r} = 0$$

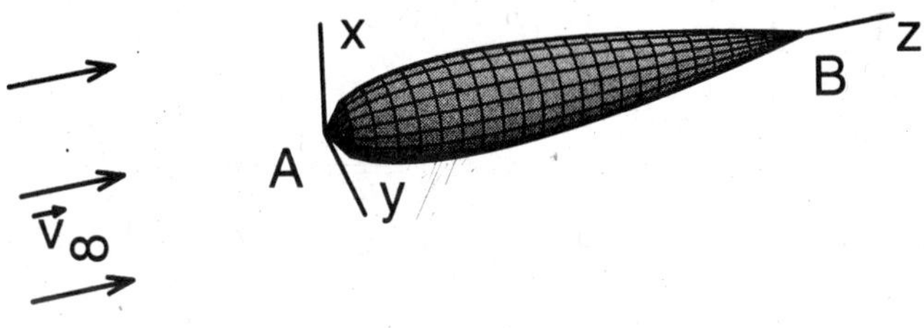

Figure 4.18: The flow around an axisymmetric body.

under the boundary conditions

$$\psi|_S = 0, \quad \frac{1}{r}\frac{\partial \psi}{\partial z}\bigg|_{z\to\infty} = 0, \quad -\frac{1}{r}\frac{\partial \psi}{\partial r}\bigg|_{r\to\infty} = V_\infty.$$

The obtained equation differs from the Laplace equation by the item $-\frac{1}{r}\frac{\partial \psi}{\partial r}$; therefore, the well-developed methods of the theory of functions of complex variable for the given class of problems will already be inapplicable.

4.3.2 The Method of Sources and Sinks

Consider a longitudinal fluid flow around the body of revolution AB (see Fig. 4.18) at a velocity $\vec{v}_\infty$ directed along the Oz-axis. We will solve the problem by the method of the sources and sinks, whose idea consists in a substitution of the body AB under consideration by a system of sources and sinks on the axis of revolution. One of the stream surfaces for a flow formed by this system of singularities must coincide with the surface of the body of revolution; that is, a distribution of the sources' strength is chosen on the basis of a given body of revolution. This method was applied for the first time by Rankine[1].

Let us distribute the sources and sinks continuously on the Oz-axis with the density $\mu(\zeta)$. The total strength of the sources (sinks) located in the interval $[\zeta, \zeta + d\zeta]$ is equal to $\mu(\zeta)\,d\zeta$. At a small $d\zeta$, we can write the stream function from this singularity in the form

$$d\psi_1 = -\frac{\mu(\zeta)\,d\zeta}{4\pi}\Big(1 - \frac{z-\zeta}{\sqrt{r^2 + (z-\zeta)^2}}\Big). \tag{4.3.12}$$

Integrating (4.3.12) from A to B, we obtain:

$$\psi_1 = -\frac{1}{4\pi}\int_a^B \mu(\zeta)\Big(1 - \frac{z-\zeta}{\sqrt{r^2 + (z-\zeta)^2}}\Big)\,d\zeta. \tag{4.3.13}$$

Let us present the overall flow around the body of revolution in the form of a sum of two flows: the translational flow $\psi_2 = -r^2\frac{V_\infty}{2}$ and the flow determined by ψ_1; that is

$$\psi = -r^2\frac{V_\infty}{2} - \frac{1}{4\pi}\int_A^B \mu(\zeta)\Big(1 - \frac{z-\zeta}{\sqrt{r^2+(z-\zeta)^2}}\Big)\,d\zeta. \tag{4.3.14}$$

Since the body is impenetrable, we have:

$$\int_A^B \mu(\zeta)\,d\zeta = 0;$$

that is, the total strength of the sources (sinks) located inside the body should be equal to zero. Under satisfaction of this condition, equation (4.3.14) has the form

$$\psi = -r^2\frac{V_\infty}{2} + \frac{1}{4\pi}\int_A^B \frac{(z-\zeta)\mu(\zeta)}{\sqrt{r^2+(z-\zeta)^2}}\,d\zeta.$$

We shall assume that $r = \tilde{r}(z)$ is the equation of the body contour and, consequently, $\psi = 0$ on the contour. Taking the boundary condition into account, we find:

$$\frac{1}{4\pi}\int_A^B \frac{(z-\zeta)\mu(\zeta)\,d\zeta}{\sqrt{\tilde{r}^2(z)+(z-\zeta)^2}} = \frac{1}{2}V_\infty\tilde{r}^2(z). \tag{4.3.15}$$

Thus, the problem on determining $\mu(z)$ reduces to the solution of the Fredholm integral equation of the first kind. Equation (4.3.15) is usually solved by the method of the reduction to a system of linear algebraic equations. The interval AB is partitioned into subintervals, and in each of them, a point ζ_i is chosen and the integral is replaced with the Riemann sum

$$I = \sum_{i=1}^{n} \frac{(z-\zeta_i)\mu(\zeta_i)\,\Delta_i}{\sqrt{\tilde{r}^2(z)+(z-\zeta_i)^2}},$$

where the $\mu(\zeta_i)$ at points ζ_i $(i = 1, 2, \ldots, n)$ are the unknown quantities. Replace the integral in (4.3.15) with this sum, and require that the obtained equation be satisfied at points z_k belonging to the interval δ_k, where $k = 1, 2, \ldots, n$. In this way, we obtain a system of linear algebraic equations for $\mu(\zeta_i)$ in the form:

$$\frac{1}{4\pi}\sum_{i=1}^{n} \frac{(z_k-\zeta_i)\,\mu(\zeta_i)\,\Delta_i}{\sqrt{\tilde{r}^2(z_k)+(z_k-\zeta_i)^2}} = \frac{1}{2}V_\infty\,\tilde{r}^2\,(z_k).$$

4.3.3 The Program prog4-5.nb

The above-presented method of sources and sinks proved to be very convenient for its implementation with *Mathematica 3.0* because this software system enables the user to easily combine the symbolic and numerical computations in the same program. In this way, we have developed the *Mathematica* Notebook `prog4-5.nb`, which solves the problem of the incompressible axisymmetric fluid flow around a body of revolution by the method of sources and sinks. We present in what follows the numerical values of the source strengths $\mu(\zeta_i)$ and the approximate analytic expression for the stream function $\psi(z, x)$, which were obtained by this Notebook.

Computation of the Entries of Matrix A of the System Aμ = b

$$\{0.20944, 0.167643, 0.0508645, -0.00681594, -0.0458145, -0.0645419, -0.0717302, -0.0704602, -0.0634079, -0.0517386, -0.0359175, -0.0163396\}$$

Computation of Stream Function ψ(x, y)

$$\psi(z,x) = -\frac{x^2}{2} + \frac{1}{4\pi}\left(-\frac{0.00136163(-0.958333+z)}{\sqrt{x^2+(-0.958333+z)^2}} - \frac{0.00299313(-0.875+z)}{\sqrt{x^2+(-0.875+z)^2}} - \frac{0.00431155(-0.791667+z)}{\sqrt{x^2+(-0.791667+z)^2}} - \frac{0.00528399(-0.708333+z)}{\sqrt{x^2+(-0.708333+z)^2}} - \frac{0.00587169(-0.625+z)}{\sqrt{x^2+(-0.625+z)^2}} - \frac{0.00597752(-0.541667+z)}{\sqrt{x^2+(-0.541667+z)^2}} - \frac{0.00537849(-0.458333+z)}{\sqrt{x^2+(-0.458333+z)^2}} - \frac{0.00381788(-0.375+z)}{\sqrt{x^2+(-0.375+z)^2}} - \frac{0.000567995(-0.291667+z)}{\sqrt{x^2+(-0.291667+z)^2}} + \frac{0.00423871(-0.208333+z)}{\sqrt{x^2+(-0.208333+z)^2}} + \frac{0.0139702(-0.125+z)}{\sqrt{x^2+(-0.125+z)^2}} + \frac{0.0174533(-0.0416667+z)}{\sqrt{x^2+(-0.0416667+z)^2}}\right)$$

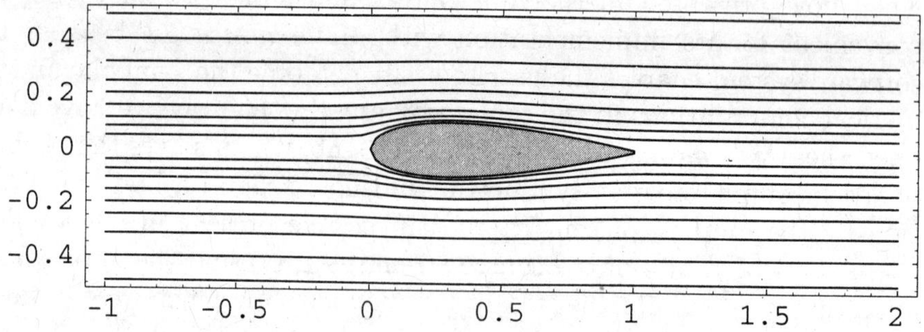

Figure 4.19: The picture of streamlines obtained by the method of sources and sinks at $n = 12$.

In this program, we have used as the function $r = \tilde{r}(z)$ the function describing the surface of the airfoil NACA0020 (see Fletcher[12]). The accuracy of a solution obtained by the method of sources and sinks depends substantially on the number n of the subintervals into which the interval AB is partitioned. In addition, one should take a larger value of the number n in the case of a complex geometry of the body of revolution, when, for example, the body surface has the areas with sign-changing curvature. In the case of the NACA0020 profile (see Fig. 4.18), we have tried several values of n: $n = 4$, $n = 6$, $n = 8$, $n = 10$, and $n = 12$ (the number n is a positive integer; i.e., we could also take $n = 13$, etc.). The behavior of the streamlines obtained at $n = 10$ and $n = 12$ is nearly the same. This points to the fact that, for the body of revolution formed by the profile NACA0020 it is sufficient to take the value $n = 12$ to achieve a sufficiently high accuracy. We show in Fig. 4.19 the picture of the streamlines around the body of revolution whose section in the plane passing through the axis of revolution coincides with the NACA0020 profile. This picture was obtained by the *Mathematica* program `prog4-5.nb` at $n = 12$. The interested reader can perform a number of runs for increasing values of n ($n = 4, 5, 6, \ldots$) to observe the solution convergence and the improvement of the local behavior of the streamlines with increasing n.

It should be noted that the method of singularities was used actively for the solution of problems on the flow around the bodies, especially in the case of three-dimensional flows. The main difficulty of the solution in this case reduces to the difficulty of the solution of a linear system of algebraic equations whose condition number depends substantially on

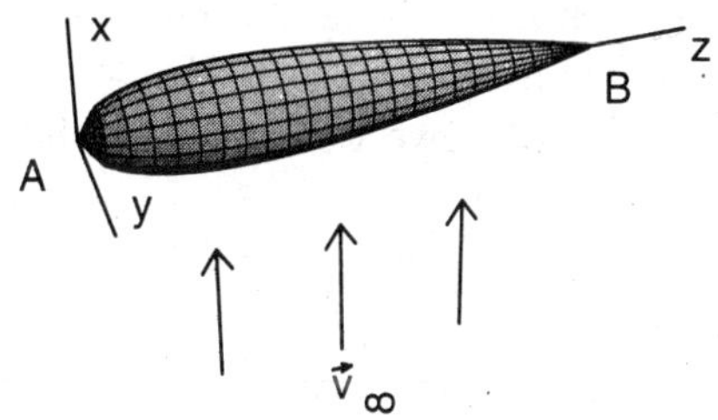

Figure 4.20: The flow around the body of revolution.

the choice of the singularities replacing the body. In a general case, other methods for the solution of equation (4.3.15) also exist.

4.3.4 The Transverse Flow around the Body of Revolution: The Program prog4-6.nb

Consider a problem of the fluid flow around the body of revolution AB in the case in which the fluid velocity $\vec{v}_\infty$ is perpendicular to the axis of revolution. Let us assume that the vector $\vec{v}_\infty$ is parallel with the x-axis (see Fig. 4.20). In this case, we already have a three-dimensional flow. We will construct the solution by the method of singularities, assuming that the dipoles with the axes parallel with the x-axis are located continuously in the interval AB with the density $\mu(\zeta)$. The total moment of the dipoles located in the interval $(\zeta, \zeta + d\zeta)$ is equal to $\mu(\zeta)\,d\zeta$, and if $d\zeta$ is small, then the velocity potential for the flow from such a dipole is equal to

$$d\varphi_1 = -\frac{\mu(\zeta)\,d\zeta}{4\pi} \cdot \frac{x}{(\sqrt{x^2+y^2+(z-\zeta)^2})^3}.$$

All dipoles located in the interval AB form a flow with the velocity potential

$$\varphi_1 = -\frac{1}{4\pi}\int_A^B \frac{\mu(\zeta)x\,d\zeta}{(\sqrt{x^2+y^2+(z-\zeta)^2})^3}$$

or in the cylindrical coordinates

$$\varphi_1 = -\frac{r\cos\theta}{4\pi}\int_A^B \frac{\mu(\zeta)\,d\zeta}{(\sqrt{r^2+(z-\zeta)^2})^3}.$$

Let us present the flow near the body in the form of the sum

$$\varphi = V_\infty r\cos\theta - \frac{r\cos\theta}{4\pi}\int_A^B \frac{\mu(\zeta)\,d\zeta}{(\sqrt{r^2+(z-\zeta)^2})^3}, \tag{4.3.16}$$

where $x = r\cos\theta$. For the determination of $\mu(\zeta)$, we use the boundary condition on the body surface, since the condition at infinity is satisfied automatically at such a choice of the velocity potential. Let us write the equations for the streamlines in cylindrical coordinates (to avoid confusion with the potential φ, we denote here the polar angle by the letter θ):

$$\frac{dr}{v_r} = \frac{dz}{v_z} = \frac{r\,d\theta}{v_\theta}$$

and find the expressions for v_r, v_z, and v_θ by using (4.3.16) in the form

$$\begin{aligned} v_r &= \frac{\partial\varphi}{\partial r} = v_\infty\cos\theta - \frac{\cos\theta}{4\pi}\frac{\partial}{\partial r}\left[r\int_A^B \frac{\mu(\zeta)\,d\zeta}{(\sqrt{r^2+(z-\zeta)^2})^3}\right], \\ v_z &= \frac{\partial\varphi}{\partial z} = -\frac{r\cos\theta}{4\pi}\frac{\partial}{\partial z}\left[r\int_A^B \frac{\mu(\zeta)\,d\zeta}{(\sqrt{r^2+(z-\zeta)^2})^3}\right], \\ v_\theta &= \frac{1}{r}\frac{\partial\varphi}{\partial\theta} = -v_\infty\sin\theta + \frac{\sin\theta}{4\pi}\int_A^B \frac{\mu(\zeta)\,d\zeta}{(\sqrt{r^2+(z-\zeta)^2})^3}. \end{aligned}$$

Substituting these expressions into the equation for the streamline $\frac{dr}{dz} = \frac{v_r}{v_z}$ we obtain an ordinary differential equation of the form

$$\frac{dr}{dz} = f(r,z). \tag{4.3.17}$$

Since the equation of the body is given, $r = \Phi(z)$, then $\frac{dr}{dz} = \Phi'(z)$ will be a known function. Taking this condition, as well as (4.3.17), into account, we find:

$$\Phi'(z) = \frac{v_\infty - \frac{1}{4\pi}\frac{\partial}{\partial r}\left[r\int_A^B \frac{\mu(\zeta)\,d\zeta}{(\sqrt{r^2+(z-\zeta)^2})^3}\right]}{-\frac{1}{4\pi}r\frac{\partial}{\partial z}\int_A^B \frac{\mu(\zeta)\,d\zeta}{(\sqrt{r^2+(z-\zeta)^2})^3}}. \tag{4.3.18}$$

Thus, we have reduced the problem of the flow past a body to the solution of an integral equation. This equation can in practice be easily reduced to a system of linear algebraic equations, as this has been done in the foregoing case.

Let us now describe the computer implementation of the above method of singularities. Since the computational results in the problems of three-dimensional flow around a body are usually presented in the Cartesian coordinates x, y, z, it is desirable to go over to these coordinates from the cylindrical coordinates r, θ, z. For this purpose, we must find the Cartesian velocity components u_x, u_y, u_z of the velocity vector $\vec{v}$ from the components v_r, v_θ, v_z in cylindrical coordinates. Let us denote the basis vectors of the cylindrical coordinate system by $\vec{e}_1, \vec{e}_2, \vec{e}_3$. Then $\vec{v} = v^j \vec{e}_j$, where $v^1 = v_r$, $v^2 = v_\theta$, and $v^3 = v_z$. While solving Problem 1.2 (see Section 1.1), we have found that

$$\vec{e}_1 = \cos\theta \vec{E}_1 + \sin\theta \vec{E}_2, \ \vec{e}_2 = -\sin\theta \vec{E}_1 + \cos\theta \vec{E}_2, \ \vec{e}_3 = \vec{E}_3,$$

where $\vec{E}_1, \vec{E}_2$, and $\vec{E}_3$ are the basis vectors of the Cartesian coordinate system. Therefore, we can write:

$$\vec{v} = u_x \vec{E}_1 + u_y \vec{E}_2 + u_z \vec{E}_3 = v^j \vec{e}_j.$$

We can find from here that

$$\begin{aligned} u_x &= u_x(\vec{E}_1 \cdot \vec{E}_1) = v^j(\vec{e}_j \cdot \vec{E}_1) = v^1[\cos\theta(\vec{E}_1 \cdot \vec{E}_1) + \sin\theta(\vec{E}_2 \cdot \vec{E}_1)] \\ &+ v^2[-\sin\theta(\vec{E}_1 \cdot \vec{E}_1) + \cos\theta(\vec{E}_2 \cdot \vec{E}_1)] \\ &= v^1 \cos\theta - v^2 \sin\theta = \cos\theta\, v_r - \sin\theta\, v_\theta. \end{aligned}$$

We can find in a similar way that $u_y = \sin\theta\, v_r + \cos\theta\, v_\theta$ and $u_z = v_z$. Let us now write the equations for the streamlines in the Cartesian coordinates:

$$\frac{dx}{u_x} = \frac{dy}{u_y} = \frac{dz}{u_z}.$$

We can obtain the following ordinary differential equation (ODE) of streamline from these equations:

$$\frac{dx}{dz} = \frac{u_x}{u_z}.$$

Since the streamlines at infinity coincide with the free stream, it is clear that $|dx/dz| \to \infty$ as $|x| \to \infty$. Therefore, the above ODE cannot be used directly for numerical integration. In this connection, we use a parametric representation for each streamline: $x = x(t), \quad y = y(t), \quad z = z(t)$, where t is a parameter, which changes along the streamline. Then we can write the following ODEs for the streamline:

$$\begin{aligned} \frac{dx}{dt} &= u_x(x(t), y(t), z(t)); \\ \frac{dy}{dt} &= u_y(x(t), y(t), z(t)); \\ \frac{dz}{dt} &= u_z(x(t), y(t), z(t)). \end{aligned} \tag{4.3.19}$$

The integral equation (4.3.18) for μ can easily be rewritten in the form

$$\frac{3\Phi(z)\Phi'(z)}{4\pi} \int_A^B \frac{(z-\zeta)\mu(\zeta)\, d\zeta}{(\sqrt{\Phi(z)^2 + (z-\zeta)^2})^3}$$

$$+ \frac{1}{4\pi}\left[\int_A^B \frac{\mu(\zeta)\,d\zeta}{(\sqrt{\Phi(z)^2+(z-\zeta)^2})^3}\right.$$

$$\left. - 3\Phi^2(z)\int_A^B \frac{\mu(\zeta)\,d\zeta}{(\sqrt{\Phi(z)^2+(z-\zeta)^2})^5}\right] = v_\infty. \qquad (4.3.20)$$

The approximation of the integrals standing on the left-hand side of this equation leads to a linear algebraic system for determining the values of μ. To write a discrete analog of equation (4.3.20), let us assume that the body AB (see Fig. 4.20) is located in the interval $0 \le z \le a$, where $a > 0$. Let us subdivide the segment $[0, a]$ into n equal subintervals, each of which has the length $h = a/n$ and n is a positive integer specified by the program user. We now choose the point $\zeta_i = (i - 0.5)h$, $i = 1, \ldots, n$ in each subinterval. In addition, we also take the point

$$z_k = kh, \quad k = 1, \ldots, n$$

in each subinterval. If we approximate the integrals in (4.3.20) by the formula of rectangles, we obtain the following linear algebraic system for determining the values $\mu(\zeta_i)$, $i = 1, \ldots, n$:

$$\frac{3\Phi(z_k)\Phi'(z_k)}{4\pi}\sum_{i=1}^{n}\frac{(z_k-\zeta_i)\mu(\zeta_i)h}{(\sqrt{\Phi(z_k)^2+(z_k-\zeta_i)^2})^3}$$

$$+ \frac{1}{4\pi}\left[\sum_{i=1}^{n}\frac{\mu(\zeta_i)\,h}{(\sqrt{\Phi(z_k)^2+(z_k-\zeta_i)^2})^3}\right.$$

$$\left. - 3\Phi^2(z_k)\sum_{i=1}^{n}\frac{\mu(\zeta_i)\,h}{(\sqrt{\Phi(z_k)^2+(z_k-\zeta_i)^2})^5}\right] = v_\infty,$$

$$i = 1, \ldots, n; \qquad k = 1, \ldots, n.$$

We have used the built-in *Mathematica* function `LinearSolve[A,b]` for the numerical solution of this system of equations. This enables us to write the approximations of the velocity components v_r, v_θ, and v_z by the method of singularities. For example,

$$v_\theta = -v_\infty \sin\theta + \frac{\sin\theta}{4\pi}\sum_{i=1}^{n}\frac{\mu(\zeta_i)\,h}{(\sqrt{r^2+(z-\zeta_i)^2})^3}.$$

While passing from the cylindrical coordinates r, θ, z to the Cartesian coordinates x, y, z, we have used the definition of the cylindrical coordinates: $x = r\cos\theta$, $y = r\sin\theta$, and $z = z$. We find from here that

$$\cos\theta = x/r, \quad \sin\theta = y/r, \quad r = \sqrt{x^2+y^2}.$$

Since the right-hand sides of the ODEs (4.3.19) are very complex, these equations cannot be integrated in analytic form. Therefore, one must apply some numerical method to solve these equations. We have used the classical fourth-order Runge–Kutta method[13] for the numerical integration of the system (4.3.19). The integration step $\Delta t = 0.02$ along the t-axis proved to be sufficient to obtain a good accuracy of the numerical results. For each streamline we have specified the Cauchy data, i.e., the initial points (x_0, y_0, z_0) of a streamline:

$$x(0) = x_0, \quad y(0) = y_0, \quad z(0) = z_0.$$

These initial points were chosen for each streamline at a sufficiently large distance from the body, where the flow differs little from the free stream. We have implemented the above-presented variant of the method of sources and sinks for three-dimensional problems in the *Mathematica* Notebook `prog4-6.nb`. In what follows, we present the following output of this program:
1) the numerical values of the source strengths $\mu(\zeta_i)$;
2) the approximate analytic expression for the velocity component v_z obtained by the method of sources and sinks.

Computation of the Entries of Matrix A of the System $\mathbf{A}\mu = \mathbf{b}$

$$\mu = \{0.00495366, -0.0385903, -0.0442002, -0.0619501, -0.0566708, -0.0632327, -0.0507179, -0.0531756, -0.0366697, -0.0405405, -0.0215066, -0.0356558, -0.0162334, -0.174031, 0.0156886\}$$

$$\begin{aligned} uz(x,y,z) = \frac{1}{4\pi}\Bigg(x\Bigg(&\frac{0.000970804(-0.975+z)}{(x^2+y^2+(-0.975+z)^2)^{5/2}} \\ &- \frac{0.00387169(-0.925+z)}{(x^2+y^2+(-0.925+z)^2)^{5/2}} \\ &- \frac{0.00164998(-0.875+z)}{(x^2+y^2+(-0.875+z)^2)^{5/2}} \\ &- \frac{0.00207601(-0.825+z)}{(x^2+y^2+(-0.825+z)^2)^{5/2}} \\ &- \frac{0.00241959(-0.775+z)}{(x^2+y^2+(-0.775+z)^2)^{5/2}} \\ &- \frac{0.00326743(-0.725+z)}{(x^2+y^2+(-0.725+z)^2)^{5/2}} \end{aligned}$$

$$- \frac{0.00397453(-0.675+z)}{(x^2+y^2+(-0.675+z)^2)^{5/2}}$$
$$- \frac{0.00495214(-0.625+z)}{(x^2+y^2+(-0.625+z)^2)^{5/2}}$$
$$- \frac{0.0057672(-0.575+z)}{(x^2+y^2+(-0.575+z)^2)^{5/2}}$$
$$- \frac{0.00672378(-0.525+z)}{(x^2+y^2+(-0.525+z)^2)^{5/2}}$$
$$- \frac{0.00746622(-0.475+z)}{(x^2+y^2+(-0.475+z)^2)^{5/2}}$$
$$- \frac{0.00823441(-0.425+z)}{(x^2+y^2+(-0.425+z)^2)^{5/2}}$$
$$- \frac{0.00869796(-0.375+z)}{(x^2+y^2+(-0.375+z)^2)^{5/2}}$$
$$- \frac{0.00905145(-0.325+z)}{(x^2+y^2+(-0.325+z)^2)^{5/2}}$$
$$- \frac{0.00895613(-0.275+z)}{(x^2+y^2+(-0.275+z)^2)^{5/2}}$$
$$- \frac{0.00857577(-0.225+z)}{(x^2+y^2+(-0.225+z)^2)^{5/2}}$$
$$- \frac{0.00753456(-0.175+z)}{(x^2+y^2+(-0.175+z)^2)^{5/2}}$$
$$- \frac{0.00600065(-0.125+z)}{(x^2+y^2+(-0.125+z)^2)^{5/2}}$$
$$- \frac{0.0035396(-0.075+z)}{(x^2+y^2+(-0.075+z)^2)^{5/2}}$$
$$- \frac{0.000519049(-0.025+z)}{(x^2+y^2+(-0.025+z)^2)^{5/2}}\Bigg)\Bigg)$$

The above-presented algorithm proved to be very efficient in terms of a needed computer time. We show in Fig. 4.21 the picture of streamlines in the flow around the body of revolution whose section by the plane $y=0$ represents the NACA0020 profile[12]. This picture was obtained at

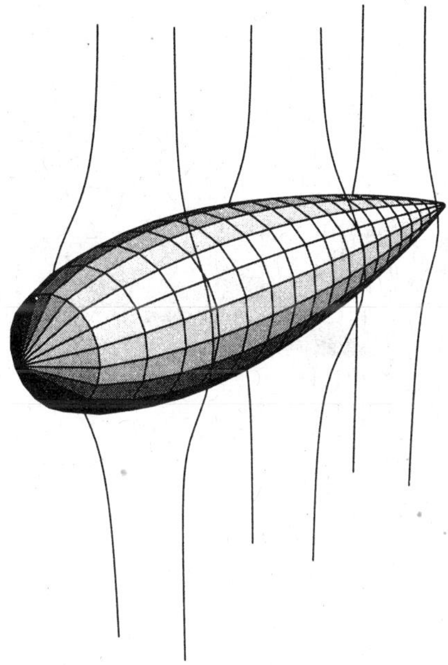

Figure 4.21: Streamlines in a three-dimensional incompressible fluid flow around the body of revolution.

$n = 20$. We show in Fig. 4.21 three pairs of the streamlines. In each pair of the streamlines, the initial points (x_0, y_0, z_0) were taken symmetrically with respect to the plane $y = 0$. The interested reader can easily obtain other streamlines by specifying his own values of (x_0, y_0, z_0) (see the lists `z01, y01, x01` in the main program `ThreeDimFlow[...]` of our *Mathematica* Notebook `prog4-6.nb`).

In Fig. 4.22, we show the distribution of the fluid velocity vectors in the three-dimensional flow around the same body of revolution as in Fig. 4.21. To obtain this figure, we have used the built-in *Mathematica* function `PlotVectorField3D[_]` of the software system *Mathematica 3.0* (see a description of this function in Appendix A).

As compared with the finite difference or finite element methods for the numerical solution of three-dimensional problems, the above-presented method of sources and sinks is much more efficient because it does not need any spatial computing mesh in the three-dimensional space around the body of revolution.

The approach we have used above is similar to the *boundary element method.* In contrast with the latter method, however, the above presented method has the advantage, that it enables one to obtain the analytic formulas, with the aid of which it is possible to study the solution behavior at infinity.

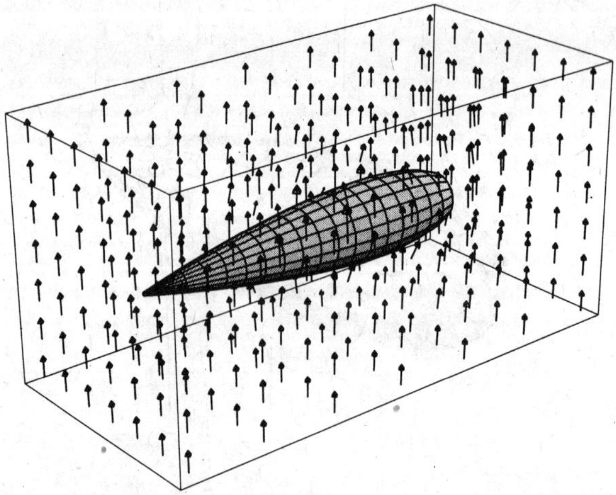

Figure 4.22: The distribution of the fluid velocity vectors in the flow around the body of revolution.

It should be noted that the solution of a problem of the flow around the body of revolution for the case in which the velocity vector $\vec{v}_\infty$ is located in an arbitrary plane can be constructed in the form of a superposition of two solutions, where one solution is the solution of a problem of the longitudinal flow around the body and the second one is the solution of a problem of the transverse flow around the body.

4.4 Nonstationary Motion of a Solid in the Fluid

4.4.1 Formulation of a Problem on Nonstationary Body Motion in Ideal Fluid

Let a solid bounded by a convex smooth surface S move in the fluid. The fluid is ideal, incompressible, and the body forces have a potential. The perturbed fluid motion will then be potential. The moving body will perturb the surrounding fluid, therein creating a velocity field $\vec{v}(x, y, z, t)$, which reduces with a distance from the body and vanishes at infinity. Choose a moving coordinate system (x, y, z) fixed in a body. Then the body velocity in a fixed coordinate system (x_0, y_0, y_0) will consist of the two velocities: the velocity of the translational motion $\vec{u}_0$ and the rotational motion $\vec{u}_\omega = (\vec{\omega} \times \cdot \vec{r})$; that is, $\vec{u} = \vec{u}_0 + (\vec{\omega} \times \vec{r})$. The instantaneous angular velocity of the body is measured with respect to the center O in which the moving coordinate system is placed (see Fig. 4.23). Since the

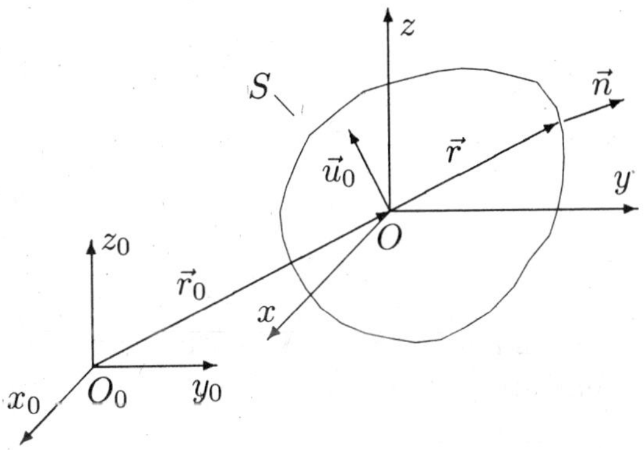

Figure 4.23: The motion of a solid body in an incompressible fluid.

fluid is incompressible and the flow is potential, the velocity potential in the fixed coordinate system will satisfy the Laplace equation

$$\Delta\,\varphi = 0$$

and the slip condition on the body surface

$$v_n = \left(\frac{\partial\tilde{\varphi}}{\partial n}\right)_S = (\vec{u}_0 + \vec{u}_\omega)\cdot\vec{n},$$

where $\vec{n}$ is a normal to the surface S of the body (Fig. 4.23). If the motion law of the body is known, then the coordinates (x_0, y_0, z_0) for each given time t can be expressed in terms of the coordinates (x, y, z) and consequently the value $\tilde{\varphi}(t, x_0, y_0, z_0)$ will be expressed in terms of (x, y, z), that is $\tilde{\varphi}(t, x_0, y_0, z_0) = \varphi(t, x, y, z)$. Such a passage from one coordinate system to another is made with the aid of a translation of the coordinate origin and rotation of the coordinates. As is known, the Laplace operator retains its form in this case so that

$$\Delta\varphi\,(t, x, y, z) = 0. \tag{4.4.1}$$

The condition at infinity also retains its form since the relations

$$(x_0^2 + y_0^2 + z_0^2) \to \infty \quad \text{and} \quad (x^2 + y^2 + z^2) \to \infty$$

are equivalent; i.e., during the time Δt, the body will pass only a finite interval. The condition on a body surface will have in this case the following representation:

$$\left.\frac{\partial\varphi}{\partial n}\right|_S = (u_{ox} + u_{\omega x})\alpha + (u_{oy} + u_{\omega y})\beta + (u_{oz} + u_{\omega z})\gamma,$$

where $\alpha = \cos(\hat{nx})$, $\beta = \cos(\hat{ny})$, $\gamma = \cos(\hat{nz})$,

$$u_{\omega x} = \omega_y z - \omega_z y, \quad u_{\omega y} = \omega_z x - \omega_x z, \quad u_{\omega z} = \omega_x y - \omega_y x.$$

Consequently,

$$\begin{aligned} \left.\frac{\partial \varphi}{\partial n}\right|_S &= u_{0x}\alpha + u_{0y}\beta + u_{0z}\gamma + \omega_x(y\gamma - z\beta) \\ &+ \omega_y(z\alpha - x\gamma) + \omega_z(x\beta - y\alpha). \end{aligned} \tag{4.4.2}$$

It follows from this formula that the potential φ should linearly depend on the velocities that are variable in time and will have the following form[7]:

$$\varphi = u_{0x}\varphi_1 + u_{0y}\varphi_2 + u_{0z}\varphi_3 + \omega_x\varphi_4 + \omega_y\varphi_5 + \omega_z\varphi_6, \tag{4.4.3}$$

where the functions φ_i ($i = 1, \dots, 6$) are the functions of the coordinates (x, y, z). Such a form for the representation of the velocity potential was proposed for the first time by Kirchhoff. Thus, if the body form and the law of its motion are given, then the determination of φ reduces to the solution of the external Neumann problem for the Laplace equation.

By virtue of the linearity of problem (4.4.1), all functions $\varphi_i(x, y, z)$ must satisfy the Laplace equation

$$\Delta\varphi_i = 0 \qquad (i = 1, 2, \dots, 6),$$

the conditions at infinity

$$\left.\frac{\partial \varphi_i}{\partial x}\right|_\infty = \left.\frac{\partial \varphi_i}{\partial y}\right|_\infty = \left.\frac{\partial \varphi_i}{\partial z}\right|_\infty = 0,$$

and the conditions on the body surface S

$$\left.\frac{\partial \varphi_1}{\partial n}\right|_S = \alpha, \quad \left.\frac{\partial \varphi_2}{\partial n}\right|_S = \beta, \quad \left.\frac{\partial \varphi_3}{\partial n}\right|_S = \gamma, \quad \left.\frac{\partial \varphi_4}{\partial n}\right|_S = y\gamma - z\cdot\beta,$$
$$\left.\frac{\partial \varphi_5}{\partial n}\right|_S = z\cdot\alpha - x\cdot\gamma, \quad \left.\frac{\partial \varphi_6}{\partial n}\right|_S = x\cdot\beta - y\cdot\alpha.$$

The determination of each function φ_i reduces to the solution of the corresponding Neumann problem, and the dependence on time will only be realized in terms of the functions $\vec{u}_0$ and $\vec{\omega}$.

4.4.2 The Hydrodynamic Reactions at the Body Motion

For the main vector of the pressure forces and the main moment with respect to the coordinate origin, one can write the formulas

$$\vec{F} = -\iint_S P\cdot\vec{n}\,dS, \qquad \vec{L} = -\iint_S P(\vec{r}\times\vec{n})\,dS, \tag{4.4.4}$$

where $\vec{r}$ is the radius vector of a surface point with respect to the coordinate origin (see Fig. 4.23). Let us write the Lagrange integral in the coordinate system (x_0, y_0, z_0):

$$\frac{\partial \tilde{\varphi}}{\partial t} + \frac{v^2}{2} + \frac{P}{\rho_0} = f(t),$$

which has the following form at $v_\infty = 0$:

$$\left.\frac{\partial \tilde{\varphi}}{\partial t}\right|_\infty + \frac{P_\infty}{\rho_0} = f(t).$$

Introduce the notation

$$\left.\frac{\partial \tilde{\varphi}}{\partial t}\right|_\infty = f(t) - \frac{P_\infty}{\rho_0} = f_1(t),$$

which enables us to write the Lagrange integral in the form

$$\frac{\partial \tilde{\varphi}'}{\partial t} + \frac{v^2}{2} + \frac{P}{\rho_0} = \frac{P_\infty}{\rho_0}, \tag{4.4.5}$$

where $\tilde{\varphi}' = \tilde{\varphi} - \int f_1(t)\, dt$. Omitting the primes in (4.4.5) and resolving with respect to P, we find:

$$P = P_\infty - \rho_0 \frac{\partial \tilde{\varphi}}{\partial t} - \rho_0 \frac{v^2}{2}. \tag{4.4.6}$$

Substituting (4.4.6) into (4.4.4), we obtain:

$$\vec{F} = \rho_0 \iint_S \vec{n}\left(\frac{\partial \tilde{\varphi}}{\partial t} + \frac{v^2}{2}\right) dS, \quad \vec{L} = \rho_0 \iint_{S'} (\vec{r} \times \vec{n}) \cdot \left(\frac{\partial \tilde{\varphi}}{\partial t} + \frac{v^2}{2}\right) dS. \tag{4.4.7}$$

Similar results can be obtained if we make use of the conservation laws for the momentum and for the momentum moment written in integral form. Let us take an arbitrary surface Σ, which is fixed in space and encloses the body S. By definition, the momentum $\vec{K}$ available within a volume τ between the surfaces S and Σ is equal to

$$\vec{K} = \rho_0 \iiint_V \vec{v}\, dV = \rho_0 \iiint_V \nabla \tilde{\varphi}\, dV.$$

Using the Gauss–Ostrogradsky formula, we can find:

$$\vec{K} = \rho_0 \iint_\Sigma \tilde{\varphi} \cdot \vec{n}\, dS - \rho_0 \iint_S \tilde{\varphi} \cdot \vec{n}\, dS.$$

Applying the theorem on the momentum change to the fluid mass within the volume V, we have:

$$\frac{d\vec{K}}{dt} = \vec{F}' - \vec{F}, \tag{4.4.8}$$

where $\vec{F}'$ is the main vector of the forces acting on the surface Σ from the fluid located outside the volume V. With regard for (4.4.7), we have the following expression for $\vec{F}'$:

$$\vec{F}' = -\iint_{\Sigma} P \cdot \vec{n}\, dS' = \rho_0 \iint_{\Sigma} \vec{n}\Big(\frac{\partial \tilde{\varphi}}{\partial t} + \frac{v^2}{2}\Big)\, dS. \tag{4.4.9}$$

The total variation of the momentum during the time dt within a volume is equal to

$$d\vec{K} = d\left[\rho_0 \iint_{\Sigma} \tilde{\varphi} \cdot \vec{n}\, dS - \rho_0 \iint_{S} \tilde{\varphi} \cdot \vec{n}\, dS\right] + \rho_0 \iint_{\Sigma} \vec{v} \cdot v_n\, dS \cdot dt,$$

where the last item corresponds to the momentum variation at the expense of a fluid that has flown into the volume V or has left it during the time dt; i.e.,

$$\frac{d\vec{K}}{dt} = \frac{d}{dt} \iint_{\Sigma} \rho_0 \tilde{\varphi} \cdot \vec{n}\, dS - \frac{d}{dt} \iint_{S} \rho_0 \tilde{\varphi} \cdot \vec{n}\, dS + \iint_{\Sigma} \rho_0 \vec{v} \cdot v_n\, dS. \tag{4.4.10}$$

Substituting (4.4.10) and (4.4.9) into (4.4.8), we obtain:

$$\begin{aligned} \vec{F} &= \rho_0 \iint_{\Sigma} \vec{n}\Big(\frac{\partial \tilde{\varphi}}{\partial t} + \frac{v^2}{2}\Big)\, dS - \frac{d}{dt} \iint_{\Sigma} \rho \tilde{\varphi} \cdot \vec{n}\, dS \\ &+ \frac{d}{dt} \iint_{S} \rho_0\, \tilde{\varphi} \cdot \vec{n}\, dS - \iint_{\Sigma} \rho_0\, \vec{v} \cdot v_n\, dS. \end{aligned}$$

Since the surface Σ is at rest, then

$$\frac{d}{dt} \iint_{S} \rho_0\, \tilde{\varphi} \cdot \vec{n}\, dS = \iint_{\Sigma} \rho_0\, \vec{n} \frac{d\tilde{\varphi}}{dt}\, dS,$$

therefore,

$$\vec{F} = \frac{d}{dt} \iint_{S} \rho_0\, \tilde{\varphi} \cdot \vec{n}\, dS + \rho_0 \iint_{\Sigma} \Big(\vec{n} \cdot \frac{v^2}{2} - \vec{v} \cdot v_n\Big)\, dS. \tag{4.4.11}$$

It is known from the theory of elliptic equations that, at $R^2 = x^2 + y^2 + z^2 \to \infty$ the potential $\tilde{\varphi} \sim \frac{1}{R^2}$, and $V \sim \frac{1}{R^3}$ [see, for example, formula (4.3.8) at $v_\infty = 0$]. Therefore, we obtain:

$$\lim_{R\to\infty} \iint_{\Sigma} \Big(\vec{n} \cdot \frac{v^2}{2} - \vec{v} \cdot v_n\Big)\, dS = 0.$$

Thus, formula (4.4.11) simplifies to

$$\vec{F} = \frac{d}{dt} \iint_{S} \rho_0 \tilde{\varphi} \cdot \vec{n}\, dS. \tag{4.4.12}$$

Let us compute in a similar way the main moments of the pressure forces that act on the surfaces S and Σ within the volume V. Then the law of the variation of moments is equal to

$$\frac{d\vec{l}}{dt} = \vec{L}' - \vec{L}.$$

By the definition,

$$\vec{l} = \rho_0 \iiint_V (\vec{r} \times \vec{v})\, dV = \rho_0 \iiint (\vec{r} \times \nabla\tilde{\varphi})\, dV.$$

With the aid of the Ostrogradsky–Gauss theorem, we obtain:

$$\vec{l} = \rho_0 \iint_\Sigma \tilde{\varphi}(\vec{r} \times \vec{n})\, dS - \rho_0 \iint_S \tilde{\varphi}(\vec{r} \times \vec{n})\, dS.$$

The expressions for $\vec{L}'$ and $\frac{d\vec{l}}{dt}$ will be similar to the above-obtained formulas (4.4.9) and (4.4.10); i.e.,

$$\begin{aligned} \vec{L}' &= \rho_0 \iint_\Sigma (\vec{r} \times \vec{n})\Big(\frac{\partial\tilde{\varphi}}{\partial t} + \frac{v^2}{2}\Big)\, dS, \\ \frac{d\vec{l}}{dt} &= \frac{d}{dt}\iint_\Sigma \rho_0\, \tilde{\varphi}(\vec{r} \times \vec{n})\, dS - \frac{d}{dt}\iint_S \rho_0\, \tilde{\varphi}(\vec{r} \times \vec{n})\, dS \\ &+ \iint_\Sigma \rho_0\, (\vec{r} \times \vec{v}) \cdot v_n\, dS. \end{aligned}$$

Taking into account the fact that the surface Σ is fixed in space and turning R to infinity, we obtain:

$$\vec{L} = \frac{d}{dt}\iint_S \rho_0\, \tilde{\varphi}(\vec{r} \times \vec{n})\, dS. \qquad (4.4.13)$$

4.4.3 Equations of Solid Motion in a Fluid under the Action of Given Forces

Denote by $\vec{G}$ the main momentum vector, and denote by $\vec{H}$ the main momentum moment of a solid. The external forces different from the pressure forces are reduced to the main vector $\vec{R}$ and the moment $\vec{Q}$. Applying the law of the moment and momentum variation, we can write:

$$\frac{d\vec{G}}{dt} = \vec{F} + \vec{R}, \quad \frac{d\vec{H}}{dt} = \vec{L} + \vec{Q}.$$

The quantities $\vec{R}$ and $\vec{Q}$ are assumed to be given here. Substituting into these equations the expressions for $\vec{F}$ and $\vec{L}$ determined by formulas (4.4.12) and (4.4.13), we obtain:

$$\frac{d}{dt}\left(\vec{G} - \rho_0 \iint_S \tilde{\varphi} \cdot \vec{n}\, dS\right) = \vec{R}, \quad \frac{d}{dt}\left(\vec{H} - \rho_0 \iint_S \tilde{\varphi}(\vec{r} \times \vec{n})\, dS\right) = \vec{Q}. \tag{4.4.14}$$

It is conventional to call the integrals of the form

$$\vec{B} = -\rho_0 \iint_S \tilde{\varphi} \cdot \vec{n}\, dS, \quad \vec{I} = -\rho_0 \iint_S \tilde{\varphi}\,(\vec{r} \times \vec{n})\, dS$$

the *virtual momentum* and the *virtual momentum moment*, respectively.

Let us present the formulas (4.4.14) in a moving coordinate system comoving with a body. Using the representation for the velocity potential in the form (4.4.3)

$$\varphi = \sum_{k=1}^{6} u_k \cdot \varphi_k,$$

where

$$u_1 = u_{0x};\ u_2 = u_{0y}; u_3 = u_{0z};\ u_4 = \omega_x;\ u_5 = \omega_y;\ u_6 = \omega_z,$$

we can present the formulas for $\vec{B}(B_1, B_2, B_3)$ and $\vec{I}(B_4, B_5, B_6)$ in the form

$$\begin{aligned}
B_1 &= -\rho_0 \iint_S \varphi \cdot \alpha\, dS = -\rho_0 \iint_S \varphi \cdot \frac{\partial \varphi_1}{\partial n}\, dS,\\
B_2 &= -\rho_0 \iint_S \varphi \cdot \beta\, dS = -\rho_0 \iint_S \varphi \cdot \frac{\partial \varphi_2}{\partial n}\, dS,\\
B_3 &= -\rho_0 \iint_S \varphi \cdot \gamma\, dS = -\rho_0 \iint_S \varphi \cdot \frac{\partial \varphi_3}{\partial n}\, dS,\\
B_4 &= -\rho_0 \iint_S \varphi \cdot (y\gamma - z\beta)\, dS = -\rho_0 \iint_S \varphi \cdot \frac{\partial \varphi_4}{\partial n}\, dS,\\
B_5 &= -\rho_0 \iint_S \varphi \cdot (z\alpha - x\gamma)\, dS = -\rho_0 \iint_S \varphi \cdot \frac{\partial \varphi_5}{\partial n}\, dS,\\
B_6 &= -\rho_0 \iint_S \varphi \cdot (x\beta - y\alpha)\, dS = -\rho_0 \iint_S \varphi \cdot \frac{\partial \varphi_6}{\partial n}\, dS.
\end{aligned}$$

The obtained formulas can be presented in a compact form

$$B_i = -\rho_0 \iint_S \varphi \cdot \frac{\partial \varphi_i}{\partial n}\, dS \quad (i = 1, 2, \ldots, 6)$$

or

$$B_i = \sum_{k=1}^{6} \lambda_{ik} u_k, \tag{4.4.15}$$

where

$$\lambda_{ik} = -\rho_0 \iint_S \varphi_k \cdot \frac{\partial \varphi_i}{\partial n}\, dS \quad (i = 1, 2, \ldots, 6;\ k = 1, 2, \ldots, 6).$$

It follows from these formulas that all B_i are expressed in terms of u_k, that is, in terms of the components of the solid body velocity $\vec{u}_0$ and the angular velocity ω. The coefficients λ_{ik} having the dimension of mass are essentially determined by the body geometry, and it is conventional to call them the *virtual masses*. There are 36 such coefficients λ_{ik}, and if the values of φ_k are known, then their computation reduces to the numerical quadratures. One can show that the tensor λ_{ik} is symmetric: $\lambda_{ik} = \lambda_{ki}$. Therefore, the number of different coefficients λ_{ik} is no more than 21.

As an example, let us consider the above-solved problem of the flow past a sphere of radius R_0 moving in a fluid at a velocity V_∞ under the action of the force $\vec{R}$ applied to the sphere center. The velocity potential for the flow past a sphere moving at a unit velocity along the Oz-axis has the form

$$\varphi_3 = -\frac{R_0^3 \cos\theta}{2r^2},$$

where the θ, r, and λ are the spherical coordinates with the origin at the sphere center. We find that

$$\left.\frac{\partial \varphi_3}{\partial n}\right|_S = \left.\frac{\partial \varphi_3}{\partial r}\right|_{r=R_0} = \cos\theta, \quad \left.\varphi_3\right|_{r=R_0} = -\frac{R_0}{2}\cos\theta$$

and, consequently,

$$\begin{aligned}\lambda_{33} &= -\rho_0 \iint_S \varphi_3 \frac{\partial \varphi_3}{\partial n}\, dS = \rho_0 \frac{R_0}{2} \iint_S \cos^2\theta\, dS \\ &= \frac{\rho_0 R_0^3}{2} \int_0^{\pi}\int_0^{2\pi} \cos\theta \sin\theta\, d\theta\, d\lambda = \frac{2}{3}\rho_0 \pi R_0^3.\end{aligned}$$

A similar computation yields $\lambda_{11} = \lambda_{22} = \lambda_{33} = \frac{2}{3}\rho_0\pi R_0^3$,

$$\lambda_{44} = \lambda_{55} = \lambda_{66} = 0 \quad \text{and} \quad \lambda_{ik} = 0 \quad \text{at} \quad i \neq k.$$

Thus, both of the formulas (4.4.15) take the form

$$\begin{aligned}B_i &= \frac{2}{3}\rho_0 \pi R_0^3 u_i, \qquad & i = 1, 2, 3, \\ B_i &= 0, & i = 4, 5, 6,\end{aligned}$$

or in the vector form:

$$\vec{B} = \frac{2}{3}\rho_0 \pi R_0^3\, \vec{u}, \quad \vec{I} = 0,$$

and in accordance with formulas (4.4.12) and (4.4.13), we have:

$$\vec{F} = -\frac{2}{3}\rho_0 \pi R_0^3 \frac{d\vec{u}}{dt}, \quad \vec{L} = 0.$$

It follows from the obtained formulas that the forces are reduced to a single resultant force applied to the sphere center, and the main moment will be equal to zero. If the sphere mass is equal to m and the force $\vec{R}$ applied to its center acts on the sphere, then the motion equations (4.4.14) for the sphere may be written in the form

$$m\frac{d\vec{u}}{dt} + \frac{2}{3}\rho_0 \pi R^3 \frac{d\vec{u}}{dt} = \vec{R} \text{ or } \left(m + \frac{2}{3}\rho_0 \pi R_0^3\right)\frac{d\vec{u}}{dt} = \vec{R}.$$

Thus, the sphere motion occurs in such a way as if it had occurred in a vacuum, and the sphere mass is increased by the amount $\frac{2}{3}\rho_0 \pi R_0^3$, equal to a half mass of a fluid displaced by the sphere.

4.5 Vortical Motions of Ideal Fluid

4.5.1 The Theorems of Thomson, Lagrange, and Helmholtz

We will consider in this section the vortical flows of an ideal fluid, in which the curl vector $\Omega = \operatorname{rot} \vec{v} \neq 0$. The velocity circulation along some line L consisting of the fluid particles

$$\Gamma = \int_L \vec{v} \cdot d\vec{r}$$

is an important characteristic of the vortex flows. If the line L is closed and $\vec{v}(\vec{r})$ is a smooth differentiable function, then one can introduce, with the aid of the Stokes theorem (1.1.67), a vortex flux across the surface S spanned on the closed contour L:

$$\Gamma = \oint_L \vec{v} \cdot d\vec{r} = \iint_S \vec{\Omega} \cdot \vec{n}\, dS.$$

This quantity is termed the *strength of a vortex tube* passing through a given closed contour L.

Let us prove the following *Thomson theorem*:

In the ideal barotropic fluid moving in an external potential field, the velocity circulation over any closed contour does not depend on time.

Proof. Consider the variation of the circulation $\Gamma(t) = \int_{\breve{AB}} \vec{v} \cdot d\vec{r}$ calculated along a "liquid" contour AB, consisting of the same particles all the time. We show in Fig. 4.24 such a liquid contour at the initial

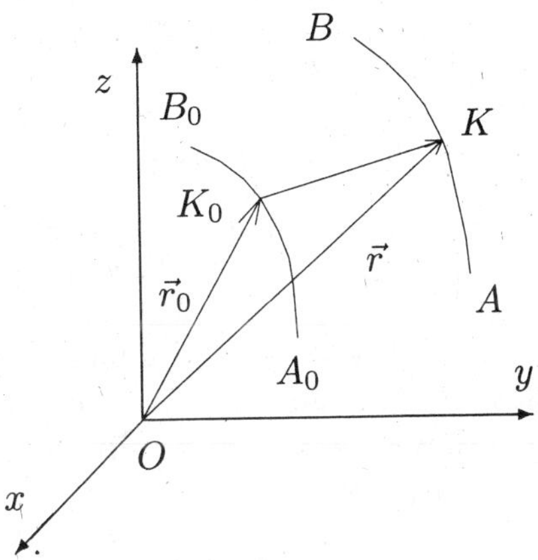

Figure 4.24: A "liquid" integration contour at two moments of time $t = 0$ and t.

moment of time $t = 0$ and at time t. Since the length of contour AB changes with time, the integration limits also depend on time. Let us go over in the integral $\Gamma(t)$ from the Eulerian coordinates to the Lagrangian coordinates. We will assume that the contour A_0B_0 is given at time $t = 0$, and as the Lagrangian coordinates of the particles on the contour A_0B_0, we choose the arc length s, which we will measure from point A_0. Let the particle K_0 be characterized at the initial moment of time by the vector $\vec{r}_0 = \vec{r}_0(s)$. At the time t, the particle K_0 will move to point K with a radius vector $\vec{r} = \vec{r}(s,t)$, the velocity $\vec{v} = \frac{\partial \vec{r}}{\partial t}$, and the acceleration $\vec{a} = \frac{\partial^2 \vec{r}}{\partial t^2}$. Since $d\vec{r} = \frac{\partial \vec{r}}{\partial s}\, ds$, the circulation magnitude will be given by the integral

$$\Gamma(t) = \int_{A_0 \breve{} B_0} \vec{v} \cdot \frac{\partial \vec{r}}{\partial s}\, ds \tag{4.5.1}$$

in which the integration limits do not already depend on time. Differentiating both sides of (4.5.1) with respect to time, we find with regard for the relation

$$\frac{\partial^2 \vec{r}}{\partial t\, \partial s} = \frac{\partial^2 \vec{r}}{\partial s\, \partial t} = \frac{\partial \vec{v}}{\partial s}$$

that

$$\frac{d\Gamma}{dt} = \int_{A_0 \breve{} B_0} \vec{a} \cdot \frac{\partial \vec{r}}{\partial s}\, ds + \int_{A_0 \breve{} B_0} \vec{v} \cdot \frac{\partial \vec{v}}{\partial s}\, ds.$$

Since

$$\vec{v}\frac{\partial \vec{v}}{\partial s}\, ds = d\Big(\frac{v^2}{2}\Big), \qquad \vec{a} \cdot \frac{\partial \vec{r}}{\partial s}\, ds = \vec{a} \cdot d\vec{r},$$

the final expression for the derivative of circulation will have the form

$$\frac{d\Gamma}{dt} = \int_{\breve{AB}} \vec{a} \cdot d\vec{r} + \frac{v_B^2}{2} - \frac{v_A^2}{2}.$$

If the curve AB is closed, then $A = B$ and

$$\frac{d\Gamma}{dt} = \oint \frac{d\vec{v}}{dt} \cdot d\vec{r}. \tag{4.5.2}$$

Thus, the time derivative of the velocity circulation over a closed contour is equal to the circulation of acceleration over the same contour. For the ideal fluid and gas, we have the following equation for the momentum change in the form (2.1.98):

$$\frac{d\vec{v}}{dt} = -\frac{1}{\rho}\nabla P + \vec{F}. \tag{4.5.3}$$

Since the fluid is barotropic $[\rho = \rho(P)]$, a function $\mathcal{P}(P)$ exists, such that $\frac{1}{\rho}\nabla P = \nabla\mathcal{P}$. The body forces are conservative by the definition; therefore, $\vec{F} = -\nabla U$. Taking into account the Thomson's theorem conditions, we can rewrite formula (4.5.3) as

$$\frac{d\vec{v}}{dt} = -\nabla(\mathcal{P} + U).$$

Substituting the obtained equation into (4.5.2), we find:

$$\frac{d\Gamma}{dt} = -\oint d(\mathcal{P} + U) \equiv 0,$$

from where it follows that $\Gamma(t) = \text{const}$.

Thus, the velocity circulation over any closed contour moving with a fluid remains constant for this contour at any time in the motion. The proof of the Thomson's theorem is then completed.

Now, let us prove the following *Lagrange theorem:*

If the conditions of the Thomson's theorem are satisfied and there are no vortices in a fixed fluid mass at some moment of time $t = t_0$, then there will also be no vortices at the subsequent moments of time.

Proof. Assume that at some moment of time $t = t_0$ there are no vortices in the fluid mass under consideration lying within a volume V; that is, $\vec{\Omega} \equiv 0$. Consequently, the fluid flow will be potential and $\vec{v} = \nabla\varphi$, where φ is the velocity potential. The velocity circulation Γ_0 over an arbitrary closed contour l_0 will be equal to zero by the definition:

$$\Gamma_0 = \oint_{l_0} \vec{v} \cdot d\vec{r} = \oint_{l_0} \nabla\varphi \cdot d\vec{r} = 0. \tag{4.5.4}$$

Consider a specified fluid mass at some other moment of time t confined by an arbitrary contour l. The fluid particles, which were located on the contour l_0 at the moment of time t_0 for which the formula (4.5.4) is valid, also correspond to any contour l at the time t. By virtue of the Thomson's theorem, the circulation Γ over the contour l will also be equal to zero. Using the Stokes formula, we obtain for any moment of time:

$$\iint \vec{\Omega} \cdot \vec{n}\, dS = 0, \tag{4.5.5}$$

where S is a surface bounded by the contour l and located completely within a volume occupied by the fluid. Since the above integral is equal to zero for any surface S, it follows from here that $\vec{\Omega} = 0$.

The Lagrange theorem is of fundamental importance for studying the irrotational flows of ideal fluid since it implies that, if the ideal fluid flow is irrotational at an initial moment of time, it will also be irrotational at subsequent times.

One can define a *vortex filament* (1.2.77) as such a line that the tangent at each point of this line is directed either along the curl vector or opposite to the curl vector:

$$\frac{dx}{\Omega_x} = \frac{dy}{\Omega_y} = \frac{dz}{\Omega_z}.$$

Now, let us prove the following

Helmholtz theorem:
If the liquid particles satisfy at their motion the conditions of the Thomson theorem and form a vortex filament at some moment of time, then these particles form a vortex filament at all subsequent and previous moments of time, and the vortex tube strength Γ *will be constant along its length and will not change with time.*

Proof. Let us choose a line l, which is not a vortex filament and draw the vortex filaments through each point of l. As a result, we obtain a *vortex surface* S, a normal to which satisfies the condition:

$$\Omega_n = \vec{\Omega} \cdot \vec{n} = 0. \tag{4.5.6}$$

In this particular case, where the vortex surface has a tube shape, it is called the *vortex tube*. Let the liquid particles form a vortex surface l_0 at the moment of time t_0. Let us choose on this surface an arbitrary closed contour l_0 bounding a piece of the surface σ_0. From the Stokes formula, we have:

$$\Gamma = \oint_{l_0} \vec{v} \cdot d\vec{r} = \iint_{\sigma_0} \vec{\Omega} \cdot \vec{n}\, dS = 0.$$

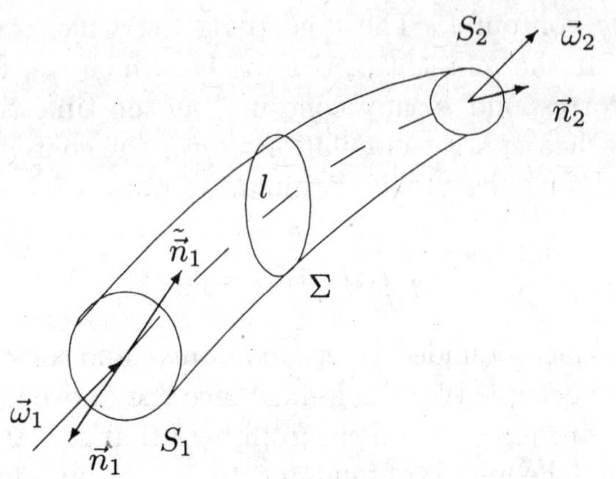

Figure 4.25: The vortex tube.

At the moment of time t, the fluid particles, which were located at time t_0 on l_0, will pass to the contour l bounding the area σ of the surface S. Since the motion of liquid particles obeys the conditions of Thomson's theorem, we have:

$$\Gamma = \oint_l \vec{v} \cdot d\vec{r} = 0$$

or, by virtue of the Stokes formula,

$$\iint_\sigma \vec{\Omega} \cdot \vec{n}\, dS = 0. \tag{4.5.7}$$

Since the surface σ in (4.5.7) is arbitrary, the condition (4.5.6) is satisfied at any point on the surface S; consequently, the surface S is a vortex surface.

Let a vortex filament l_0 be given at time $t = 0$. The line l_0 can be presented as an intersection of two vortex surfaces σ_1^0 and σ_2^0. As was shown above, at time t, the vortex surfaces σ_1^0 and σ_2^0 will again go over to the vortex surfaces σ_1 and σ_2. Since the corresponding points of σ_i^0 and σ_i consist of the same particles, the points of a vortex filament l formed by the intersection of σ_1 and σ_2 will correspond to the points on the vortex filament l_0. The curl vector Ω at any point of l is tangent to σ_1 and σ_2; i.e., it is directed along a tangent to the line of intersection l. Thus, we have proved that the vortex filament remains a vortex filament in the process of its motion.

Let us now prove the second part of the theorem, which states that the vortex tube strength will be constant along its length and will not change with time. Consider a vortex tube with a curvilinear axis (see

Fig. 4.25). By definition, the vortex tube strength is the quantity

$$\Gamma = \oint_l \vec{v}\, d\vec{r} = \iint_\sigma \Omega_n d\sigma,$$

where l is a contour bounding the tube and σ is a surface intersecting the vortex tube. Let l_1 and l_2 be two arbitrary contours bounding the vortex tube. Consider a volume τ bounded by the surface $S = S_1 + S_2 + \Sigma$ (see Fig. 4.25), where S_1 and S_2 are the tube sections bounded by contours l_1 and l_2, respectively, and Σ is a lateral tube surface between l_1 and l_2. Write the vortex flux across the surface S by the formula

$$\iint_S \Omega_n\, dS = \iiint_\tau \operatorname{div} \Omega\, d\tau = 0, \tag{4.5.8}$$

since $\operatorname{div} \Omega = \operatorname{div}(\operatorname{rot} \vec{v}) = 0$. It follows from (4.5.8) that

$$\iint_S \Omega_n\, dS = \iint_{S_1+S_2+\Sigma} \Omega_n\, dS = 0.$$

Since $\Omega_n = 0$ on Σ, then

$$\iint_{S_1} \Omega_n\, dS + \iint_{S_2} \Omega_n\, dS = 0, \tag{4.5.9}$$

where $\vec{n}$ is an outer normal to the surface bounding the volume τ (see Fig. 4.25). Taking into account the fact that $\tilde{\vec{n}}_1 = -\vec{n}_1$ and using the Stokes formula, we obtain:

$$\iint_{S_2} \Omega_n\, dS = \oint_{l_2} \vec{v} \cdot d\vec{r} = \Gamma_2, \quad \iint_{S_1} \Omega_n\, dS = \oint_{l_1} \vec{v} \cdot d\vec{r} = \Gamma_1' = -\Gamma_1, \tag{4.5.10}$$

where Γ_1 and Γ_2 are the velocity circulations calculated by a passage along the contours l_1 and l_2 in the same direction. It follows from formulas (4.5.9) and (4.5.10) that

$$\oint_{l_1} \vec{v} \cdot d\vec{r} = \oint_{l_2} \vec{v}\, d\vec{r}, \quad \Gamma_1 = \Gamma_2.$$

Since the conditions of Thomson's theorem are satisfied and the contours l_1 and l_2 have been chosen arbitrarily, the circulation along any liquid contour does not depend on time and, consequently, the vortex tube strength does not change with time. The Helmholtz theorem has been proved.

If the vortex tube ends in the fluid, then $S_2 \to 0$, and according to (4.5.7) the angular velocity $\frac{1}{2}\vec{\Omega}$ in this section tends to infinity, which is physically impossible. By virtue of the Helmholtz theorem, the vortex tubes can begin on rigid or free surfaces, extend to infinity, or be closed onto themselves.

The vortex helices leaving the surfaces flowed past and extending to infinity behind these surfaces as well as the vortex formations of the type of tornados and whirlpools or the vortex rings can serve as the examples of such vortex formations.

Let us study the case in which the ideal fluid flow does not satisfy the barotropicity and nonconservativity of the body forces. It is interesting to elucidate a question on the possibility of the onset and decay of vortices in such motions. We have obtained, at the proof of the Thomson's theorem, the equality

$$\frac{d\Gamma}{dt} = \oint_l \frac{d\vec{v}}{dt} \cdot d\vec{r},$$

which can be rewritten with the aid of the momentum equation in the Euler form

$$\frac{d\vec{v}}{dt} = \vec{F} - \frac{1}{\rho}\nabla P$$

as

$$\frac{d\Gamma}{dt} = \oint_l \vec{F}\, d\vec{r} - \oint_l \frac{1}{\rho}\nabla P\, d\vec{r}. \tag{4.5.11}$$

Consider the case of a barotropic fluid; i.e., $\rho = \rho(P)$, but the body forces are nonconservative. Equation (4.5.11) can be simplified with regard for this condition to the form

$$\frac{d\Gamma}{dt} = \oint_l \vec{F}\, d\vec{r}.$$

The work $\oint_l \vec{F} \cdot d\vec{r}$ for a nonconservative force $\vec{F}$ is not equal to zero at a passage along a contour l; therefore, $\frac{d\Gamma}{dt} \neq 0$ and the Thomson's theorem is invalid; that is the vortices may arise and vanish. Let us consider the case in which the body forces are conservative, i.e., $\vec{F} = -\nabla U$, but the fluid is baroclinic. In this case, ρ depends not only on pressure, but also on temperature, humidity (of air), or salinity (water). Equation (4.5.11) takes the form

$$\frac{d\Gamma}{dt} = -\oint_l \frac{1}{\rho}\nabla P\, d\vec{r} = -\oint_l \frac{1}{\rho} dP = -\oint_l \omega\, dP,$$

where $\omega = 1/\rho$. Let us construct two families of surfaces: $P =$ const (the isobaric surfaces) and $\omega =$ const (the isosteric surfaces). The four surfaces $\omega = \omega_0$, $\omega = \omega_1$; $P = P_0$, and $P = P_1$ form a tube, which is called the *isobaric isosteric tube*. Let $\omega_1 = \omega_0 + 1$ and $P_1 = P_0 + 1$, and the contour $ABCD$ enclosing this tube is shown in Fig. 4.26. Then

$$\frac{d\Gamma}{dt} = -\int_A^B \omega\, dP - \int_B^C \omega\, dP - \int_C^D \omega\, dP - \int_D^A \omega\, dP = \omega_0 - (\omega_0 + 1) = -1.$$

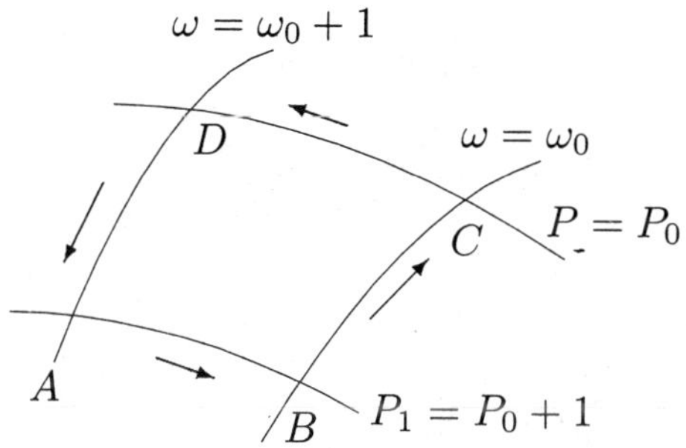

Figure 4.26: A contour enclosing the isobaric/isosteric tube.

At a different disposition of the surfaces, one can obtain the equality $\frac{d\Gamma}{dt} = +1$. If the contour encloses K^+ unit positive tubes and K^- negative unit tubes, then

$$\frac{d\Gamma}{dt} = K^+ - K^-$$

and, consequently, the vortices can arise and vanish in a baroclinic fluid since $K^+ - K^- \neq 0$. This makes the contents of the *Bjerknes theorem.*

4.5.2 Motion Equations in Friedmann's Form

Write the momentum equation for ideal fluid in the Gromeka–Lamb form (4.1.1):

$$\frac{\partial \vec{v}}{\partial t} + \nabla\left(\frac{\vec{v}^2}{2}\right) - \vec{v} \times \operatorname{rot} \vec{v} = \vec{F} - \frac{1}{\rho}\nabla P.$$

Let us apply the rot operation to both sides of this equation. Then we get:

$$\frac{\partial \vec{\Omega}}{\partial t} + \operatorname{rot}\left(\nabla\left(\frac{\vec{v}^2}{2}\right)\right) - \operatorname{rot}(\vec{v} \times \operatorname{rot} \vec{v}) = \operatorname{rot} \vec{F} - \operatorname{rot}\left(\frac{1}{\rho}\nabla P\right).$$

Since

$$\begin{aligned}
\operatorname{rot}(\vec{v} \times \operatorname{rot} \vec{v}) &= (\vec{\Omega} \cdot \nabla)\vec{v} - (\vec{v} \cdot \nabla)\vec{\Omega} + \vec{\Omega} \operatorname{div} \vec{v}, \\
\operatorname{rot}\left(\nabla \cdot \frac{\vec{v}^2}{2}\right) &= 0, \\
\operatorname{rot}\left(\frac{1}{\rho}\nabla P\right) &= -\frac{1}{\rho^2}\nabla\rho \times \nabla P, \quad \operatorname{rot} \vec{v} = \vec{\Omega},
\end{aligned}$$

we have:

$$\frac{\partial \vec{\Omega}}{\partial t} + (\vec{v} \cdot \nabla)\vec{\Omega} - (\vec{\Omega} \cdot \nabla)\vec{v} - \vec{\Omega} \operatorname{div} \vec{v} = \operatorname{rot} \vec{F} + \frac{1}{\rho^2}\nabla\rho \times \nabla P,$$

or

$$\frac{d\vec{\Omega}}{dt} = (\vec{\Omega} \cdot \nabla)\,\vec{v} + \vec{\Omega}\,\mathrm{div}\,\vec{v} + \mathrm{rot}\,\vec{F} + \frac{1}{\rho^2}\nabla\rho \times \nabla P. \tag{4.5.12}$$

It is conventional to call equation (4.5.12) the *Friedmann's equation.* If we assume that the field of the body forces is conservative that is $\vec{F} = -\nabla U$, and the fluid is barotropic, then equation (4.5.12) simplifies to the form

$$\frac{d\vec{\Omega}}{dt} - (\vec{\Omega} \cdot \nabla)\vec{v} - \vec{\Omega} \cdot \mathrm{div}\,\vec{v} = 0. \tag{4.5.13}$$

If we assume that the fluid is incompressible, then we obtain the equation:

$$\frac{d\vec{\Omega}}{dt} = (\vec{\Omega} \cdot \nabla)\vec{v}. \tag{4.5.14}$$

These equations were first obtained by Helmholtz. It is convenient to use them while solving the meteorology problems. The Helmholtz theorems can also be proved by using equations (4.5.13) or (4.5.14).

4.5.3 The Biot–Savart Formulas and the Straight Vortex Filament

If the velocity field $\vec{v}\,(x, y, z)$ is given, then it is sufficiently simple to find its divergence $\theta = \mathrm{div}\,\vec{v}$ and the curl vector $\vec{\Omega} = \mathrm{rot}\,\vec{v}$. Consider an inverse problem. Let the functions $\theta(x, y, z)$ and $\vec{\Omega}(x, y, z)$ be given, and it is required to find the velocity field $\vec{v}(x, y, z)$, which satisfies the equations

$$\mathrm{div}\,\vec{v} = \theta\,(x, y, z), \quad \mathrm{rot}\,\vec{v} = \vec{\Omega}\,(x, y, z). \tag{4.5.15}$$

We also assume that the fluid occupies the overall space and is at rest at infinity; that is, $\vec{v}|_\infty = 0$.

It is clear that the system (4.5.15) does not always have a solution because the number of equations is four, and the number of the functions to be determined is only three ($\vec{v}(v_x, v_y, v_z)$). Since $\mathrm{div}(\mathrm{rot}\,\vec{v}) = 0$, one of the necessary solvability conditions is the satisfaction of the equality

$$\mathrm{div}\,\vec{\Omega} = 0.$$

We will search for the solution of equations (4.5.15) in the form of the sum

$$\vec{v} = \vec{v}_1 + \vec{v}_2,$$

where the functions $\vec{v}_1$ and $\vec{v}_2$ satisfy the following equation and the boundary conditions, respectively:

$$\mathrm{div}\,\vec{v}_1 = \theta, \quad \mathrm{rot}\,\vec{v}_1 = 0, \quad \vec{v}_1\Big|_\infty = 0, \tag{4.5.16}$$

$$\operatorname{div} \vec{v}_2 = 0, \quad \operatorname{rot} \vec{v}_2 = \vec{\Omega}, \quad \vec{v}_2 \Big|_{\infty} = 0. \tag{4.5.17}$$

Since the original problem is linear, the sum of these two solutions will be the solution of problem (4.5.15).

Let us construct the solution of problem (4.5.16). Introduce the function φ by the formula

$$\vec{v}_1 = \nabla \varphi.$$

In this case, the second equation of system (4.5.16) is satisfied automatically and the substitution into the first equation yields

$$\Delta \varphi = \theta\,(x, y, z). \tag{4.5.18}$$

Thus, the solution of problem (4.5.16) reduces to the solution of the Poisson equation in an unbounded (x, y, z) space. Using the theory of potential, we can write the solution of (4.5.18) as

$$\varphi = -\frac{1}{4\pi} \iiint_{-\infty}^{\infty} \frac{\theta(\xi, \eta, \zeta)}{r}\, d\xi\, d\eta\, d\zeta, \tag{4.5.19}$$

where $r^2 = (x-\xi)^2 + (y-\eta)^2 + (z-\zeta)^2$. As is known from mathematical physics courses, the constructed solution (4.5.19) is a unique solution of the Poisson equation and tends to zero at infinity if the function $\theta(x, y, z)$ is pieacewise continuous and reduces at infinity as $\frac{1}{(x^2+y^2+z^2)^{\frac{2+\alpha}{2}}}$, where α is a positive constant. Thus, the solution of problem (4.5.16) determines a vector

$$\vec{v}_1 = \nabla \varphi = -\frac{1}{4\pi} \nabla \iiint \frac{\theta(x, y, z)}{r}\, d\xi\, d\eta\, d\zeta. \tag{4.5.20}$$

We will search for the solution of problem (4.5.17) in the form

$$\vec{v}_2 = \operatorname{rot} \vec{A},$$

where thc following equality is valid for the vector $\vec{A}$:

$$\operatorname{div} \operatorname{rot} \vec{A} = 0.$$

At such a choice of the solution, the first equation of system (4.5.17) is satisfied identically and the second equation in this case has the form

$$\operatorname{rot} \operatorname{rot} \vec{A} = \vec{\Omega}. \tag{4.5.21}$$

Using the equality

$$\operatorname{rot} \operatorname{rot} \vec{A} = \nabla \cdot (\operatorname{div} \vec{A}) - \Delta \vec{A},$$

let us write equation (4.5.21) as follows:

$$\Delta\vec{A} - \nabla\cdot(\operatorname{div}\vec{A}) = -\vec{\Omega}. \tag{4.5.22}$$

One can assume without loss of generality that $\operatorname{div}\vec{A} = 0$. Let us prove the validity of this assertion. If we indeed assume that $\operatorname{div}\vec{A} = f \neq 0$, then, assuming $\vec{A}_1 = \vec{A} + \nabla\varphi$, we get:

$$\operatorname{div}\vec{A}_1 = f + \operatorname{div}(\nabla\varphi) = f + \Delta\varphi.$$

Choosing φ as the solution of the Poisson equation $\Delta\,\varphi = -f$ [see problem (4.5.16)], we obtain $\operatorname{div}\vec{A}_1 = 0$, $\vec{v}_2 = \operatorname{rot}\vec{A} = \operatorname{rot}\vec{A}_1$. Thus, the use of the vectors $\vec{A}$ and $\vec{A}_1$ for the computation of the velocity $\vec{v}_2$ will lead to the equal results and $\operatorname{div}\vec{A}_1 = 0$. These considerations enable us to assume that $\operatorname{div}\vec{A} = 0$, and consequently, equation (4.5.22) takes the form

$$\Delta\vec{A} = -\vec{\Omega},$$

which can be written in projections onto the coordinate axes as

$$\Delta A_x = -\Omega_x; \quad \Delta A_y = -\Omega_y; \quad \Delta A_z = -\Omega_z.$$

Each of these equations is the Poisson equation, and the solution of these equations has the form

$$\begin{aligned} \vec{A} &= \frac{1}{4\pi}\iiint_{-\infty}^{\infty}\frac{\vec{\Omega}}{r}\,d\xi\,d\eta\,d\zeta, \\ \vec{v}_2 &= \frac{1}{4\pi}\operatorname{rot}\iiint_{-\infty}^{\infty}\frac{\vec{\Omega}}{r}\,d\xi\,d\eta\,d\zeta. \end{aligned} \tag{4.5.23}$$

The obtained formulas enable us to construct the general solution of problem (4.5.15) in the form

$$\vec{v} = -\frac{1}{4\pi}\nabla\cdot\iiint\frac{\theta}{r}\,d\xi\,d\eta\,d\zeta + \frac{1}{4\pi}\operatorname{rot}\iiint_{-\infty}^{\infty}\frac{\vec{\Omega}}{r}\,d\xi\,d\eta\,d\zeta. \tag{4.5.24}$$

A direct calculation shows that $\operatorname{div}\vec{A} = 0$. Consider the expression $\Delta\cdot(\operatorname{div}\vec{A})$. Taking (4.5.23) into account, we obtain:

$$\Delta\cdot(\operatorname{div}\vec{A}) = \operatorname{div}(\Delta\vec{A}) = -\operatorname{div}\vec{\Omega} = -\operatorname{div}\operatorname{rot}\vec{v} = 0.$$

Thus, $\operatorname{div}\vec{A}$ is a harmonic function with the property $\lim_{r\to\infty}(\operatorname{div}\vec{A}) = 0$ and, consequently, it is equal to zero in the overall space, what was to be proved.

We now establish the uniqueness of the solution of problem (4.5.15). Let us assume that there exist two solutions $\vec{v}$ and $\vec{v}_1$ of the given problem. Then, the difference $\vec{u} = \vec{v} - \vec{v}_1$ satisfies the equations

$$\operatorname{div} \vec{u} = 0, \quad \operatorname{rot} \vec{u} = 0, \quad u\Big|_{\infty} = 0.$$

It is obvious that $\vec{u}$ is a potential field $\vec{u} = \nabla\varphi$, but $\operatorname{div} \vec{u} = \operatorname{div} \cdot (\nabla\varphi) = 0$. Consequently, φ and $\vec{u}$ are the harmonic functions. Since $\vec{u}|_{\infty} = 0$, however, then $\vec{u} \equiv 0$ in the overall space and $\vec{v}_1 = \vec{v}$. Thus, we have proved the uniqueness of the solution of problem (4.5.15).

Let a closed vortex tube with a finite volume V exist in a fluid filling the overall space. The velocity field induced by such a vortex tube is determined by formula (4.5.24). In our case, $\vec{\Omega}(x, y, z) = 0$ outside the region V. Since we assume that there are no sources in the fluid, then $\theta(x, y, z) = 0$ everywhere. Therefore,

$$\vec{v} = \frac{1}{4\pi} \operatorname{rot} \iiint_V \frac{\vec{\Omega}}{r}\, dV. \tag{4.5.25}$$

Denote the tube section by σ, and denote the mean tube line by l. Let $\vec{t}$ be a unit vector of the tangent to the mean line. Assuming the velocity vortex $\vec{\Omega}$ to be constant in each section, we can write for the length element dl of a vortex tube that

$$\vec{\Omega} \cdot dV = \vec{\Omega}\, \sigma\, dl = \Omega\, \sigma \vec{t}\, dl.$$

Then we can transform the integral in (4.5.25) to the form

$$\vec{v} = \frac{1}{4\pi} \operatorname{rot} \int_l dl \iint_\sigma \frac{\vec{\Omega}}{r}\, d\sigma \approx \frac{1}{4\pi} \operatorname{rot} \int_l \frac{\Omega\, \sigma \vec{t}}{r}\, dl.$$

Turning σ to zero and $\Omega \to \infty$ in such a way that the product $\Omega \cdot \sigma = \Gamma$ remains constant, we obtain:

$$\vec{v} = \frac{\Gamma}{4\pi} \cdot \operatorname{rot} \int_l \frac{\vec{t}}{r}\, dl \tag{4.5.26}$$

or in the projections onto the axes of a Cartesian coordinate system

$$\begin{aligned}
v_x &= \frac{\Gamma}{4\pi} \left(\frac{\partial}{\partial y} \int_l \frac{t_z}{r}\, dl - \frac{\partial}{\partial z} \int_l \frac{t_y}{r}\, dl \right), \\
v_y &= \frac{\Gamma}{4\pi} \left(\frac{\partial}{\partial z} \int_l \frac{t_x}{r}\, dl - \frac{\partial}{\partial x} \int_l \frac{t_z}{r}\, dl \right), \\
v_z &= \frac{\Gamma}{4\pi} \left(\frac{\partial}{\partial x} \int_l \frac{t_y}{r}\, dl - \frac{\partial}{\partial y} \int_l \frac{t_x}{r}\, dl \right).
\end{aligned}$$

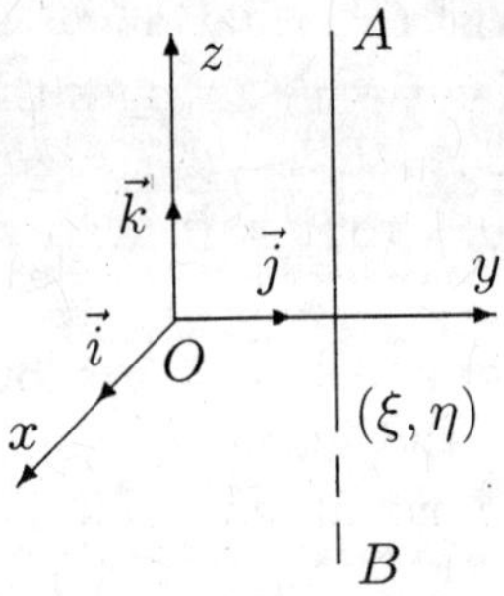

Figure 4.27: A rectilinear vortex filament.

The vector $\vec{t}$ does not depend on the coordinates x, y, and z. Performing the differentiation under the integral sign and taking into account the fact that $\nabla(\frac{1}{r}) = -\frac{\vec{r}}{r^3}$, where $\vec{r} = (x-\xi)\vec{i} + (y-\eta)\vec{j} + (z-\zeta)\vec{k}$, we obtain:

$$\begin{aligned} v_x &= -\frac{\Gamma}{4\pi}\int_l \left(\frac{y-\eta}{r}t_z - \frac{z-\zeta}{r}t_y\right)\frac{dl}{r^2}, \\ v_y &= -\frac{\Gamma}{4\pi}\int_l \left(\frac{z-\zeta}{r}t_x - \frac{x-\xi}{r}t_z\right)\frac{dl}{r^2}, \\ v_z &= -\frac{\Gamma}{4\pi}\int_l \left(\frac{x-\xi}{r}t_y - \frac{y-\eta}{r}t_x\right)\frac{dl}{r^2}. \end{aligned} \tag{4.5.27}$$

We have in (4.5.27) under the integral sign a vector product of the two vectors $\vec{t}$ and $\vec{n} = \frac{\vec{r}}{r}$; that is,

$$\vec{v} = \frac{\Gamma}{4\pi}\int_l (\vec{t} \times \vec{n})\frac{dl}{r^2} = \frac{\Gamma}{4\pi}\int_l (\vec{t} \times \vec{r})\frac{dl}{r^3}. \tag{4.5.28}$$

It follows from the obtained formula that the vortex filament element dl engenders at point $M(\vec{r})$ the velocity $\Delta\vec{v}$, which is computed by the formula

$$\Delta\vec{v} = \frac{\Gamma}{4\pi}(\vec{t} \times \vec{r})\frac{dl}{r^3}. \tag{4.5.29}$$

The numerical value of $|\Delta\vec{v}|$ is equal to

$$|\Delta\vec{v}| = \frac{\Gamma}{4\pi}\sin\alpha \cdot \frac{dl}{r^2},$$

where α is the angle between the vectors $\vec{t}$ and $\vec{r}$. The formulas (4.5.28) or (4.5.29) are similar to the Biot–Savart formulas in electrodynamics.

Consider a particular case in which the vortex filament is rectilinear and infinite (see Fig. 4.27). Let a rectilinear vortex filament AB (Fig.

4.27) pass through point (ξ, η) in parallel with the z-axis. Then $\vec{t}_x = \vec{t}_y = 0$, $\vec{t}_z = \vec{k}$, $dl = d\zeta$, and the formulas (4.5.28) simplify to

$$\vec{v} = \frac{\Gamma}{4\pi} \int_{-\infty}^{\infty} \frac{\vec{k} \times \vec{r}}{r^3}\, d\zeta.$$

Mapping this equality onto the coordinate axes and calculating the integrals, we obtain:

$$v_x = -\frac{\Gamma}{2\pi} \frac{y - \eta}{\rho^2}, \quad v_y = -\frac{\Gamma}{2\pi} \frac{x - \xi}{\rho^2}, \quad v_z = 0, \tag{4.5.30}$$

where

$$\rho^2 = (x - \xi)^2 + (y - \eta)^2 \quad \text{and} \quad \int_{-\infty}^{\infty} \frac{d\zeta}{r^3} = \frac{2}{\rho^2}.$$

The formulas (4.5.30) describe a planar fluid flow, where at each point perpendicular to the vortex the particles move along a circle at the center of which the vortex is located. The velocity magnitude is equal to $v = \frac{\Gamma}{2\pi} \cdot \frac{1}{\rho}$. The counter-clockwise motion along a circle, ρ in radius, corresponds to the positive values of Γ, and the clockwise motion along this circle corresponds to the negative values of Γ. As a consequence of the symmetry of the fluid flow around a point vortex it is obvious that the vortex will be at rest. Using the functions of a complex variable $z = x + iy$, let us write the formulas (4.5.30) in the form

$$v_x - iv_y = \frac{\Gamma}{2\pi i} \cdot \frac{1}{z - z_0},$$

where $z_0 = \xi + i\eta$, $\bar{z} = x - iy$, and $\bar{z}_0 = \xi - i\eta$. We recall that the complex potential of the vortex $w = \frac{\Gamma}{2\pi i} \ln(z - z_0)$, and we have by definition that $v_x - iv_y = \frac{dw}{dz}$, where Γ is the vortex intensity, or nothing more or less than the velocity circulation along any closed contour enclosing the point z_0.

For a qualitative explanation of a variety of the phenomena occurring in nature one can introduce the concept of the discontinuity surface, which is the surface on which some hydrodynamic parameters undergo a discontinuity. The velocity is usually chosen as such a quantity. Such is, for example, the discontinuity surface in a cyclone, along which the cold and warm air get in touch, and where a jump in the wind velocity takes place. Let us show that the surface of a discontinuity in the tangential velocity component may be considered as a limiting case of a vortex layer, that is, a space between two close surfaces that is filled by vortices, and in this case, a continuous although rapid velocity variation takes place. We assume for simplicity that the discontinuity surface is the plane S

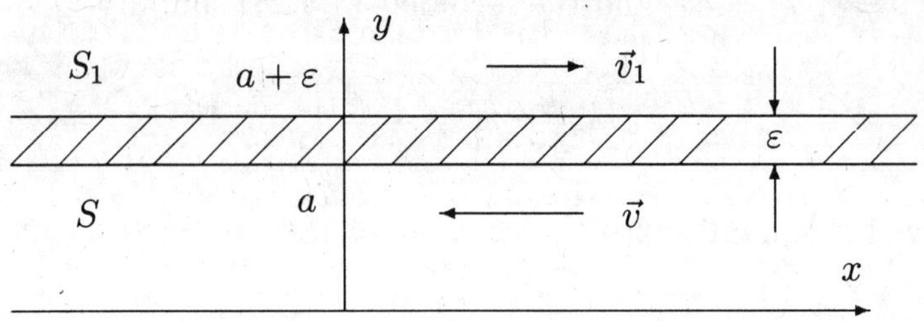

Figure 4.28: A vortex layer of thickness ε.

parallel with the Oxy plane so that its equation is $y = a$ (see Fig. 4.28). Introduce the plane S_1 with equation $y = a + \varepsilon$, which lies at a distance ε from the plane S. Let the fluid move at a velocity $\vec{v}$ on the one side of S and at a velocity $\vec{v}_1$ on the other side S_1. Both velocities are constant and parallel with the x-axis. Assume that the components of the velocities $\vec{v}$ and $\vec{v}_1$ along the Ox-axis are u and u_1. Thus, we have the discontinuity only in the tangential velocity component. Assume that, in a layer between S and S_1, the velocity components are determined by the formulas

$$v_x = (u - u_1)\left(1 - \frac{y-a}{\varepsilon}\right) + u_1, \quad v_y = v_z = 0.$$

Then we will have in the plane S, for which $y - a = 0$, that $v_x = u$, and in the plane S_1, where $y - a = \varepsilon$, that $v_x = u_1$.Consequently, at such a specification of the velocity, its magnitude will change continuously from u to u_1 while passing to the plane S_1 from the plane S. The curl vector in the layer SS_1 has a direction perpendicular to the Oxy plane, and its component along the Oz-axis is equal to

$$\Omega_z = \frac{\partial v_y}{\partial x} - \frac{\partial v_x}{\partial y} = \frac{u - u_1}{\varepsilon}.$$

The remaining curl components $\Omega_x = \Omega_y = 0$. At a small ε, the curl may be very large. Inside the layer, the Ω_z has a constant value different from zero, and outside the layer, $\Omega = 0$. Therefore, the layer SS_1 may be termed a *vortex layer*. Let us identify in the layer a vortex tube of a rectangular section, $\Delta x = 1$ in width and ε in height, and calculate its intensity by the formula

$$\Gamma = \iint_S \Omega_z \, dx \, dy = +\int_0^1 dx \cdot \int_0^\varepsilon \frac{u - u_1}{\varepsilon} dy,$$

which yields

$$\Gamma = +(u - u_1).$$

It follows from the last formula that the vortex tube strength $\Gamma = \Omega_z \cdot \varepsilon$ does not depend on the layer thickness ε. In the limit, as $\varepsilon \to 0$, $\Omega_z \to \infty$, and the intensity $\Gamma = u - u_1$ remains constant, we will have a flow with a surface of the discontinuity in the tangential velocity component. Such a flow with a tangential discontinuity may be interpreted as a flow engendered by a vortex layer in which the vortices of a sufficiently large intensity are located. The introduction of the vortex layer concept gives a possibility to explain the origin of the vortices in a fluid. By virtue of the Lagrange theorem, if there are no vortices in the ideal fluid at the initial moment of time, then there will be no vortices at all times of motion. In reality, we have that, under conditions close to the conditions of the Lagrange theorem (the constancy of density, small viscosity of fluid, and the availability of a potential of the acting forces), the vortices in the fluid arise. If we assume that a vortex layer on the surface of a body flowed past emerges, then it is not difficult to imagine that, in the case of instability of this layer, the vortices can separate from it, as this often takes place in reality at the motion of a body in the fluid.

References

1. **Kochin, N.E., Kibel, I.A., and Rose, N.V.,** *Theoretical Hydromechanics* (in Russian), Vol. I, 6th Edition; Vol. II, 4th Edition, Fizmatgiz, Moscow, 1963.
2. **Milne-Thompson, L.M.,** *Theoretical Hydrodynamics*, 5th Edition, MacMillan, New York, 1967.
3. **Gromeka, I.S.,** *Some Cases of Incompressible Fluid Flow* (in Russian), Kazan, 1881 (Reprinted in: **Gromeka, I.S.,** *Collected Works* (in Russian), USSR Academy of Sciences, Moscow, 1952, p. 76.
4. **Lamb, H.,** *Hydrodynamics*, 6th Edition, Cambridge University Press, London, 1932; Dover Publications, New York, 1945.
5. **Germain, P.,** *Cours de Mécanique des Milieux Continus. Tome I. Théorie Générale*, Masson et Cie, Éditeurs, Paris, 1973.
6. **Sedov, L.I.,** *Continuum Mechanics*, Vols. I and II (in Russian), Fifth Edition, Nauka, Moscow, 1994.
7. **Kirchhoff, G.R.,** *Mechanics* (in Russian; translated from German), USSR Academy of Sciences, Moscow, 1962.

8. **Sedov, L.I.,** *Planar Problems of Hydrodynamics and Aerodynamics* (in Russian), Second Edition, Nauka, Moscow, 1966.
9. **Vallander, S.V.,** *Lectures in Hydroaeromechanics* (in Russian), Leningrad State University, Leningrad, 1978.
10. **Warsi, Z.U.A.,** *Fluid Dynamics. Theoretical and Computational Approaches*, CRC Press, Boca Raton, 1993.
11. **Batchelor, G.K.,** *An Introduction to Fluid Dynamics*, Cambridge University Press, London, 1967.
12. **Fletcher, C.A.J.,** *Computational Techniques for Fluid Dynamics*, Vols. I, II, 3rd Edition, Springer-Verlag, Berlin, 1996.
13. **Strampp, W., Ganzha, V., and Vorozhtsov, E.,** *Höhere Mathematik mit Mathematica. Band 3: Differentialgleichungen und Numerik*, Verlag Vieweg, Braunschweig/Wiesbaden, 1997.

5

Viscous Fluid

This chapter is devoted to the viscous fluid flows, which are described by the Navier–Stokes equations. We derive the Navier–Stokes equations in the Cartesian, cylindrical, and spherical coordinate systems and consider their exact solutions at small Reynolds numbers. We present the Prandtl's theory of boundary layer, which is valid at large Reynolds numbers. This theory enables one to calculate the drag force acting on a plate in the viscous fluid flow. We also outline the theory for the transition from laminar viscous fluid flow to turbulent flow and discuss a number of the semiempirical theories of turbulence.

As was noted above, in the ideal fluid, the surface forces applied to the surface elements of any fluid volume represent the normal pressures directed inside the volume. Any actual fluid possesses, however, the viscous property, which gives rise to *shear stresses.* It is this property that is one of the reasons causing the drag of the fluid flow in pipes and channels or the drag of bodies moving in the fluid. In this case, there are, between the layers in the viscous fluid flow, the forces tangent to the direction of motion of these layers.

The classical viscous fluid is an isotropic medium whose shear drag is different from zero and linearly depends on the shear strain rate [see (2.1.101)]. The equations governing the viscous fluid flows were derived for the first time by Navier (1822), who applied a simplified molecular model for the gases. This has led to the introduction of a positive viscosity $\mu > 0$, which in the opinion of Navier desribes the molecular diffusion of the momentum. It has been generally recognized at present, however, that simple laws for the molecular forces describe inadequately the actual fluids. Therefore, a continual approach proposed by Stokes (1845) is considered to be more preferable, and we will follow this approach.

5.1 General Equations of Viscous Incompressible Fluid

5.1.1 The Navier–Stokes Equations

We will assume that the coefficients of shear viscosity μ and heat conduction κ are constant. Assuming $E = c_V T$, $\rho = \rho_0 = \text{const}$ in formulas (2.1.114), we obtain the system of Navier–Stokes equations governing the flow of an incompressible viscous, heat conducting fluid in the form

$$\begin{aligned} &\operatorname{div} \vec{v} = 0, \\ &\frac{\partial \vec{v}}{\partial t} + (\vec{v}\,\nabla)\vec{v} = -\frac{1}{\rho_0}\nabla P + \nu \Delta\, \vec{v}, \\ &c_V\, \rho_0 \frac{dT}{dt} = \kappa\, \Delta T + \Phi, \end{aligned} \tag{5.1.1}$$

where $\Phi = 2\mu\, \dot{\varepsilon}_{ij}\dot{\varepsilon}_{ij}$, $\dot{\varepsilon}_{ij} = \frac{1}{2}(\frac{\partial v_i}{\partial x^j} + \frac{\partial v_j}{\partial x^i})$ is the rate-of-strain tensor. The function Φ entering the system of equations (5.1.1) is nonnegative and vanishes only in the case in which the fluid is at rest or moves as an absolutely rigid body. For the ideal fluid, $\Phi = 0$ since $\mu = 0$. It is conventional to call the function Φ the *dissipative function.*

It should be noted that the first two equations in (5.1.1) do not contain the temperature; therefore, the solution of this system can be subdivided into two stages, in which at the first stage the unknown functions $\vec{v}$ and P are determined, and then the temperature is found from the last equation.

Using the formulas of covariant differentiation (1.1.23) and (1.1.25), we can write the Navier–Stokes equations (5.1.1) in any orthogonal coordinate system. The continuity equations and the expressions for convective derivatives $(\vec{v}\,\nabla)\vec{v}$ were obtained above while solving Problem 2.1 [see formulas (2.1.131), (2.1.132), (2.1.138), (2.1.144), (2.1.145)]. Therefore, it the expression for the Laplacian $\Delta\vec{v}$ remains to be found. We at first determine the physical components of the velocity u_i, which should enter the equations. Similarly to (1.1.43), let us write in an orthogonal coordinate system $g_{ij} = 0$, if $i \neq j$, $g_i = g_{ii}$, and $g^i = 1/g_i$:

$$\begin{aligned} \vec{v} &= v^j \vec{e}_j = \sqrt{g_j} v^j \frac{\vec{e}_j}{\sqrt{g_j}} = \sqrt{g_j} v^j \vec{E}_j = u^j \vec{E}_j = u_j \vec{E}_j, \\ \vec{v} &= v_j \vec{e}^{\,j} = \sqrt{g^j}\, v_j \frac{\vec{e}^{\,j}}{\sqrt{g^j}} = \frac{1}{\sqrt{g_j}} v_j \vec{E}^j = u_j \vec{E}^j = u_j \vec{E}_j, \end{aligned}$$

where $\vec{E}_i \vec{E}_j = \delta_{ij}$, $\vec{E}_i = \vec{E}^i$, and $u_i = u^i$. We obtain from these formulas the relation between the physical velocity components u_i and the coordinate velocity components v_i, v^i:

$$v^j = \frac{u_j}{\sqrt{g_j}}, \qquad v_j = \sqrt{g_j}u_j, \tag{5.1.2}$$

where $g_j \equiv g_{jj}$ and there is no summation in j. By introducing the physical components of the vectors

$$\begin{aligned} \Delta\vec{v} &= (\Delta\vec{v})_{\tilde{j}}\vec{E}_j, \quad \nabla\operatorname{div}\vec{v} = (\nabla\operatorname{div}\vec{v})_{\tilde{j}}\vec{E}_j, \\ \operatorname{rot}\vec{v} &= (\operatorname{rot}\vec{v})_{\tilde{j}}E_j, \quad \operatorname{rot}\operatorname{rot}\vec{v} = (\operatorname{rot}\operatorname{rot}\vec{v})_{\tilde{j}}\vec{E}_j, \end{aligned}$$

we can rewrite the identity

$$\Delta\vec{v} = \nabla\operatorname{div}\vec{v} - \operatorname{rot}\operatorname{rot}\vec{v}$$

in the physical components as

$$(\Delta\vec{v})_{\tilde{k}} = (\nabla\operatorname{div}\vec{v})_{\tilde{k}} - (\operatorname{rot}\operatorname{rot}\vec{v})_{\tilde{k}}. \tag{5.1.3}$$

Let us find the form of the individual items entering this expression.

The definition of a gradient of an arbitrary scalar function Φ

$$\nabla\Phi = \nabla_j\Phi\vec{e}^{\,j} = \sqrt{g^j}\nabla_j\Phi\vec{E}^j = \frac{1}{\sqrt{g_j}}\nabla_j\Phi\vec{E}_j = (\nabla_j\Phi)\tilde{}\,\vec{E}_j$$

implies the formulas for the physical components:

$$(\nabla_j\Phi)\tilde{} = \frac{1}{\sqrt{g_j}}\nabla_j\Phi. \tag{5.1.4}$$

While deriving formula (5.1.4), we have used the equality $g^j = 1/g_j$, which is valid in an orthogonal coordinate system.

Using the definition of the rot operation, we can write with (1.1.36) and (1.1.42) in view that

$$\operatorname{rot}\vec{v} = \frac{1}{\sqrt{g}}\left(\nabla_j v_k - \nabla_k v_j\right)\vec{e}_i.$$

With the use of the covariant derivative definition (1.1.25), this formula can be rewritten with regard for the symmetry of the Christoffel symbols $\Gamma^l_{jk} = \Gamma^l_{kj}$, as follows:

$$\operatorname{rot}\vec{v} = \frac{1}{\sqrt{g}}\left(\frac{\partial v_k}{\partial x^j} - \frac{\partial v_j}{\partial x^k}\right)\vec{e}_i.$$

Expressing the v_j in terms of the u_j with the aid of (5.1.2), we obtain with regard for relations $g = \det \| g_{ij} \| = g_i g_j g_k$, $\vec{e}_i = \sqrt{g_i}\vec{E}_i$:

$$(\operatorname{rot}\vec{v})_{\tilde{i}} = \frac{1}{\sqrt{g_j g_k}}\left(\frac{\partial}{\partial x^j}(\sqrt{g_k}\,u_k) - \frac{\partial}{\partial x^k}(\sqrt{g_j}\,u_j)\right), \tag{5.1.5}$$

where there is no summation over the repeating indices. Substituting in (1.1.74) the formulas $v^j = u_j/\sqrt{g_j}$, $g = g_1 g_2 g_3$, we obtain the following expression for divergence:

$$\operatorname{div}\vec{v} = \frac{1}{\sqrt{g_i g_j g_k}}\Big(\frac{\partial}{\partial x^i}(\sqrt{g_j g_k}\, u_i) + \frac{\partial}{\partial x^j}(\sqrt{g_i g_k}\, u_j) + \frac{\partial}{\partial x^k}(\sqrt{g_i g_j}\, u_k)\Big). \tag{5.1.6}$$

Substituting (5.1.6) into (5.1.4) and applying twice the operation rot (5.1.5), we can write expression (5.1.3) as

$$\begin{aligned}(\Delta\vec{v})_k &= \frac{1}{\sqrt{g_k}}\frac{\partial}{\partial x^k}\left[\frac{1}{\sqrt{g_i g_j g_k}}\left(\frac{\partial}{\partial x^i}(\sqrt{g_j g_k} u_i) + \frac{\partial}{\partial x^j}(\sqrt{g_i g_k} u_j)\right.\right.\\ &+ \left.\left.\frac{\partial}{\partial x^k}(\sqrt{g_i g_j} u_k)\right)\right] - \frac{1}{\sqrt{g_i g_j}}\left[\frac{\partial}{\partial x^i}\left(\sqrt{\frac{g_j}{g_i g_k}}\frac{\partial}{\partial x^k}(\sqrt{g_i} u_i)\right)\right.\\ &- \frac{\partial}{\partial x^i}\left(\sqrt{\frac{g_j}{g_i g_k}}\frac{\partial}{\partial x^i}(\sqrt{g_k} u_k)\right) - \frac{\partial}{\partial x^j}\left(\sqrt{\frac{g_i}{g_j g_k}}\frac{\partial}{\partial x^j}(\sqrt{g_k} u_k)\right)\\ &+ \left.\frac{\partial}{\partial x^j}\left(\sqrt{\frac{g_i}{g_j g_k}}\frac{\partial}{\partial x^k}(\sqrt{g_j} u_j)\right)\right].\end{aligned} \tag{5.1.7}$$

Using a cyclic permutation of indices, one must set in expression (5.1.7) at $k = 1$ $i = 2, j = 3$, $k = 2$ $i = 3, j = 1$, and $k = 3$ $i = 1, j = 2$.

Before we write the Navier–Stokes equations in the Cartesian, cylindrical, and spherical coordinate systems, we obtain the expressions for the physical components of stresses. Using formulas (1.1.25) and (5.1.2), we obtain:

$$\begin{aligned}\dot{\varepsilon}_{ij} &= \frac{1}{2}(\nabla_i v_j + \nabla_j v_i)\\ &= \frac{1}{2}\Big(\frac{\partial}{\partial x^i}(\sqrt{g_j} u_j) + \frac{\partial}{\partial x^j}(\sqrt{g_i} u_i)\Big) - \sqrt{g_\alpha}\, u_\alpha\, \Gamma^\alpha_{ij},\end{aligned} \tag{5.1.8}$$

where there is a summation over the Greek index α and there is no summation over the Latin indices. In the orthogonal coordinate system, $g_{ij} = 0$ at $i \neq j$; therefore, it follows from the definition of the Christoffel symbols that the following symbols will be different from zero:

$$\begin{aligned}\Gamma^i_{ij} &= \frac{1}{2}g^i\frac{\partial g_i}{\partial x^j} = \frac{1}{\sqrt{g_i}}\frac{\partial\sqrt{g_i}}{\partial x^j}, \quad \Gamma^i_{ii} = \frac{1}{\sqrt{g_i}}\frac{\partial\sqrt{g_i}}{\partial x^i},\\ \Gamma^i_{jj} &= -\frac{1}{2}g^i\frac{\partial g_j}{\partial x^i} = -\frac{1}{2g_i}\frac{\partial g_j}{\partial x^i}.\end{aligned} \tag{5.1.9}$$

Substituting (5.1.9) in (5.1.8), we obtain at $i = j$:

$$\dot{\varepsilon}_{ii} = \frac{\partial}{\partial x^i}(\sqrt{g_i} u_i) - u_i\frac{\partial\sqrt{g_i}}{\partial x^i} + \sqrt{\frac{g_i}{g_j}}u_j\frac{\partial\sqrt{g_i}}{\partial x^j} + \sqrt{\frac{g_i}{g_k}}u_k\frac{\partial\sqrt{g_i}}{\partial x^k}; \tag{5.1.10}$$

and at $i \neq j$:

$$\dot{\varepsilon}_{ij} = \frac{1}{2}\Big(\frac{\partial}{\partial x^i}(\sqrt{g_j}u_j) + \frac{\partial}{\partial x^j}(\sqrt{g_i}u_i) - 2u_i\frac{\partial\sqrt{g_i}}{\partial x^j} - 2u_j\frac{\partial\sqrt{g_j}}{\partial x^i}\Big). \quad (5.1.11)$$

According to (2.1.101), the stress tensor in a viscous incompressible fluid ($\operatorname{div}\vec{v} = 0$) has the form

$$\sigma_{ij} = -Pg_{ij} + 2\mu\dot{\varepsilon}_{ij}. \quad (5.1.12)$$

Knowing the σ_{ij}, one can find the physical components of the stress tensor $\tilde{\sigma}_{ij}$ by formulas (1.1.45):

$$\tilde{\sigma}_{ij} = \frac{1}{\sqrt{g_i g_j}}\sigma_{ij}. \quad (5.1.13)$$

Substituting (5.1.13) in (5.1.12), we obtain in an orthogonal coordinate system ($g_{ij} = 0$ at $i \neq j$):

$$\tilde{\sigma}_{ij} = -P\delta_{ij} + \frac{2\mu}{\sqrt{g_i g_j}}\,\dot{\varepsilon}_{ij}. \quad (5.1.14)$$

Substituting (5.1.10) and (5.1.11) in (5.1.14), we find the desired expression for the physical components $\tilde{\sigma}_{ij}$:

$$\begin{aligned} \tilde{\sigma}_{ii} &= -P + 2\mu\Big(\frac{1}{\sqrt{g_i}}\frac{\partial u_i}{\partial x^i} + \frac{u_i}{\sqrt{g_i g_j}}\frac{\partial\sqrt{g_i}}{\partial x^j} + \frac{u_k}{\sqrt{g_i g_k}}\frac{\partial\sqrt{g_i}}{\partial x^k}\Big), \\ \tilde{\sigma}_{ij} &= \mu\Big(\frac{1}{\sqrt{g_j}}\frac{\partial u_i}{\partial x^j} + \frac{1}{\sqrt{g_i}}\frac{\partial u_j}{\partial x^i} - \frac{u_i}{\sqrt{g_i g_j}}\frac{\partial\sqrt{g_i}}{\partial x^j} - \frac{u_j}{\sqrt{g_i g_j}}\frac{\partial\sqrt{g_j}}{\partial x^i}\Big), \\ i &\neq j. \end{aligned} \quad (5.1.15)$$

Note that the coefficients $\sqrt{g_i} \equiv H_i$ are sometimes called the Lamé coefficients in the literature[1].

In the *Cartesian coordinate system*, we set $g_1 = g_2 = g_3 = 1$, $x^i = x$, $x^j = y$, $x^k = z$, $u_i = v_x$, $u_j = v_y$, and $u_k = v_z$. Substituting these values in (5.1.6), (5.1.7), (5.1.15) and (5.1.1), respectively, we obtain the Navier–Stokes equations in the *Cartesian coordinates*:

$$\frac{\partial v_x}{\partial x} + \frac{\partial v_y}{\partial y} + \frac{\partial v_z}{\partial z} = 0,$$

$$\frac{\partial v_x}{\partial t} + v_x\frac{\partial v_x}{\partial x} + v_y\frac{\partial v_x}{\partial y} + v_z\frac{\partial v_x}{\partial z} = -\frac{1}{\rho_0}\frac{\partial P}{\partial x} + \nu\Big(\frac{\partial^2 v_x}{\partial x^2} + \frac{\partial^2 v_x}{\partial y^2} + \frac{\partial^2 v_x}{\partial z^2}\Big);$$

$$\frac{\partial v_y}{\partial t} + v_x\frac{\partial v_y}{\partial x} + v_y\frac{\partial v_y}{\partial y} + v_z\frac{\partial v_y}{\partial z} = -\frac{1}{\rho_0}\frac{\partial P}{\partial y} + \nu\Big(\frac{\partial^2 v_y}{\partial x^2} + \frac{\partial^2 v_y}{\partial y^2} + \frac{\partial^2 v_y}{\partial z^2}\Big);$$

$$\frac{\partial v_z}{\partial t} + v_x\frac{\partial v_z}{\partial x} + v_y\frac{\partial v_z}{\partial y} + v_z\frac{\partial v_z}{\partial z} = -\frac{1}{\rho_0}\frac{\partial P}{\partial z} + \nu\Big(\frac{\partial^2 v_z}{\partial x^2} + \frac{\partial^2 v_z}{\partial y^2} + \frac{\partial^2 v_z}{\partial z^2}\Big).$$

$$\sigma_{xx} = -P + 2\mu\frac{\partial v_x}{\partial x}, \quad \sigma_{xy} = \sigma_{yx} = \mu\Big(\frac{\partial v_x}{\partial y} + \frac{\partial v_y}{\partial x}\Big);$$
$$\sigma_{yy} = -P + 2\mu\frac{\partial v_y}{\partial y}, \quad \sigma_{xz} = \sigma_{zx} = \mu\Big(\frac{\partial v_x}{\partial z} + \frac{\partial v_z}{\partial x}\Big);$$
$$\sigma_{zz} = -P + 2\mu\frac{\partial v_z}{\partial z}, \quad \sigma_{yz} = \sigma_{zy} = \mu\Big(\frac{\partial v_y}{\partial z} + \frac{\partial v_z}{\partial y}\Big). \tag{5.1.16}$$

The energy equation is not considered here for the reason noted above.

In the *cylindrical coordinate system* $x^i = r$, $x^j = \varphi$, $x^k = z$ the metric tensor components are [see (2.1.130)]

$$g_i = 1, \quad g_j = r^2, \quad g_k = 1. \tag{5.1.17}$$

Introduce the notations

$$u_i = v_r, \quad u_j = v_\varphi, \quad u_k = v_z, \quad \tilde{\sigma}_{ii} = \sigma_{rr},$$
$$\tilde{\sigma}_{ij} = \sigma_{r\varphi}, \; \tilde{\sigma}_{ik} = \sigma_{rz}, \; \tilde{\sigma}_{jj} = \sigma_{\varphi\varphi}, \; \tilde{\sigma}_{jk} = \sigma_{\varphi z}, \; \tilde{\sigma}_{kk} = \sigma_{zz}.$$

With regard for (2.1.134), we obtain the system of the Navier–Stokes equations in the cylindrical coordinates:

$$\frac{\partial v_r}{\partial r} + \frac{1}{r}\frac{\partial v_\varphi}{\partial \varphi} + \frac{\partial v_z}{\partial z} + \frac{v_r}{r} = 0,$$
$$\frac{\partial v_r}{\partial t} + v_r\frac{\partial v_r}{\partial r} + \frac{v_\varphi}{r}\frac{\partial v_r}{\partial \varphi} + v_z\frac{\partial v_r}{\partial z} - \frac{v_\varphi^2}{r} = -\frac{1}{\rho_0}\frac{\partial P}{\partial r}$$
$$+ \nu\Big(\frac{\partial^2 v_r}{\partial r^2} + \frac{1}{r^2}\frac{\partial^2 v_r}{\partial \varphi^2} + \frac{\partial^2 v_r}{\partial z^2} + \frac{1}{r}\frac{\partial v_r}{\partial r} - \frac{2}{r^2}\frac{\partial v_\varphi}{\partial \varphi} - \frac{v_r}{r^2}\Big),$$
$$\frac{\partial v_\varphi}{\partial t} + v_r\frac{\partial v_\varphi}{\partial r} + \frac{v_\varphi}{r}\frac{\partial v_\varphi}{\partial \varphi} + v_z\frac{\partial v_\varphi}{\partial z} + \frac{v_r v_\varphi}{r} = -\frac{1}{r\rho_0}\frac{\partial P}{\partial \varphi}$$
$$+ \nu\Big(\frac{\partial^2 v_\varphi}{\partial r^2} + \frac{1}{r^2}\frac{\partial^2 v_\varphi}{\partial \varphi^2} + \frac{\partial^2 v_\varphi}{\partial z^2} + \frac{1}{r}\frac{\partial v_\varphi}{\partial r} + \frac{2}{r^2}\frac{\partial v_r}{\partial \varphi} - \frac{v_\varphi}{r^2}\Big), \tag{5.1.18}$$
$$\frac{\partial v_z}{\partial t} + v_r\frac{\partial v_z}{\partial r} + \frac{v_\varphi}{r}\frac{\partial v_z}{\partial \varphi} + v_z\frac{\partial v_z}{\partial z} = -\frac{1}{r\rho_0}\frac{\partial P}{\partial z}$$
$$+ \nu\Big(\frac{\partial^2 v_z}{\partial r^2} + \frac{1}{r^2}\frac{\partial^2 v_z}{\partial \varphi^2} + \frac{\partial^2 v_z}{\partial z^2} + \frac{1}{r}\frac{\partial v_z}{\partial r}\Big),$$
$$\sigma_{rr} = -P + 2\mu\frac{\partial v_r}{\partial r}, \quad \sigma_{r\varphi} = \mu\Big(\frac{1}{r}\frac{\partial v_r}{\partial \varphi} + \frac{\partial v_\varphi}{\partial r} - \frac{v_\varphi}{r}\Big),$$
$$\sigma_{\varphi\varphi} = -P + 2\mu\Big(\frac{1}{r}\frac{\partial v_\varphi}{\partial \varphi} + \frac{v_r}{r}\Big), \quad \sigma_{\varphi z} = \mu\Big(\frac{\partial v_\varphi}{\partial z} + \frac{1}{r}\frac{\partial v_z}{\partial \varphi}\Big),$$
$$\sigma_{zz} = -P + 2\mu\frac{\partial v_z}{\partial z}, \quad \sigma_{zr} = \mu\Big(\frac{\partial v_z}{\partial r} + \frac{\partial v_r}{\partial z}\Big).$$

In the *spherical coordinate system* $x^i = r$, $x^j = \theta$, and $x^k = \varphi$, the metric tensor components have the form (2.1.143):

$$g_i = 1, \quad g_j = r^2, \quad g_k = r^2\sin^2\theta. \tag{5.1.19}$$

Introduce the following notations for the physical components:

$$
\begin{aligned}
u_i &= v_r, \quad u_j = v_\theta, \quad u_k = v_\varphi, \\
\tilde{\sigma}_{ii} &= \sigma_{rr}, \quad \tilde{\sigma}_{ij} = \sigma_{r\theta}, \quad \tilde{\sigma}_{ik} = \sigma_{r\varphi}, \\
\tilde{\sigma}_{jj} &= \sigma_{\theta\theta}, \quad \tilde{\sigma}_{jk} = \sigma_{\theta\varphi}, \quad \tilde{\sigma}_{kk} = \sigma_{\varphi\varphi}.
\end{aligned}
$$

In order to obtain the desired equations, one must substitute (5.1.19) in (5.1.4), (5.1.6), (5.1.15), and the differentiation results in (5.1.1), respectively, and take into account (1.4.145). We have implemented these analytic computations with the aid of the *Mathematica* program `prog5-1.nb`. For the purpose of brevity, we restrict ourselves in our computer program to the computation of the continuity equation and the expressions for the Laplace operator entering the momentum equations (the "inviscid parts" of these equations were presented above in Section 2.1).

For the purpose of convenience, we denote the velocity components while computing the left-hand side of continuity equation as

$$v_1[r,\theta,\varphi] \equiv v_r, \quad v_2[r,\theta,\varphi] \equiv v_\theta, \quad v_3[r,\theta,\varphi] \equiv v_\varphi.$$

When writing the program `prog5-1.nb` for calculating the Laplace operator components in three momentum equations, it was convenient to denote the velocity components as

$$u[r,\theta,\varphi,1] \equiv v_r, \quad u[r,\theta,\varphi,2] \equiv v_\theta, \quad u[r,\theta,\varphi,3] \equiv v_\varphi.$$

This has enabled us to write a single function `laplace[k_]`, which finds the expressions for the components of $\Delta\vec{v}$ in the spherical coordinates at $k = 1, 2, 3$. In what follows, we present the output of the *Mathematica* Notebook `prog5-1.nb`, which contains the obtained expressions for $\operatorname{div}\vec{v}$ and the components of $\Delta\vec{v}$ in the spherical coordinates.

The Expressions for Metric Tensor Components

$$g = \begin{pmatrix} 1 & 0 & 0 \\ 0 & r^2 & 0 \\ 0 & 0 & r^2\sin^2(\theta) \end{pmatrix}$$

The Continuity Equation

$$
\begin{aligned}
\frac{1}{r}(2v_1[r,\Theta,\varphi] &+ \cot(\theta)v_2[r,\Theta,\varphi] \\
&+ \csc(\theta)v_3^{(0,0,1)}(r,\Theta,\varphi) + rv_1^{(1,0,0)}(r,\Theta,\varphi)) = 0
\end{aligned}
$$

The Laplace Operator in the Momentum Equations

$$\frac{1}{r^2}(-2u[r,\theta,\varphi,1]-2\cot[\theta]u[r,\theta,\varphi,2]$$
$$-2\csc[\theta]u^{(0,0,1,0)}[r,\theta,\varphi,3]+\csc[\theta]^2u^{(0,0,2,0)}[r,\theta,\varphi,1]$$
$$+\cot[\theta]u^{(0,1,0,0)}[r,\theta,\varphi,1]-2u^{(0,1,0,0)}[r,\theta,\varphi,2]$$
$$+u^{(0,2,0,0)}[r,\theta,\varphi,1]+2ru^{(1,0,0,0)}[r,\theta,\varphi,1]$$
$$+r^2u^{(2,0,0,0)}[r,\theta,\varphi,1])$$

$$\frac{1}{r^2}(-\csc[\theta]^2u[r,\theta,\varphi,2]-2\cot[\theta]\csc[\theta]u^{(0,0,1,0)}[r,\theta,\varphi,3]$$
$$+\csc[\theta]^2u^{(0,0,2,0)}[r,\theta,\varphi,2]+2u^{(0,1,0,0)}[r,\theta,\varphi,1]$$
$$+\cot[\theta]u^{(0,1,0,0)}[r,\theta,\varphi,2]+u^{(0,2,0,0)}[r,\theta,\varphi,2]$$
$$+2ru^{(1,0,0,0)}[r,\theta,\varphi,2]+r^2u^{(2,0,0,0)}[r,\theta,\varphi,2])$$

$$\frac{1}{r^2}(-\csc[\theta]^2u[r,\theta,\varphi,3]+2\csc[\theta]u^{(0,0,1,0)}[r,\theta,\varphi,1]$$
$$+2\cot[\theta]+\csc[\theta]u^{(0,0,1,0)}[r,\theta,\varphi,2]$$
$$+\csc[\theta]^2u^{(0,0,2,0)}[r,\theta,\varphi,3]+\cot[\theta]u^{(0,1,0,0)}[r,\theta,\varphi,3]$$
$$+u^{(0,2,0,0)}[r,\theta,\varphi,3]+2ru^{(1,0,0,0)}[r,\theta,\varphi,3]$$
$$+r^2u^{(2,0,0,0)}[r,\theta,\varphi,3])$$

With regard for the above expressions for components of $\Delta\vec{v}$, we obtain the system of Navier–Stokes equations in the spherical coordinates:

$$\frac{\partial v_r}{\partial r}+\frac{1}{r}\frac{\partial v_\theta}{\partial\theta}+\frac{1}{r\sin\theta}\frac{\partial v_\varphi}{\partial\varphi}+\frac{2v_r}{r}+\frac{v_\theta \mathrm{ctg}\theta}{r}=0,$$

$$\frac{\partial v_r}{\partial t}+v_r\frac{\partial v_r}{\partial r}+\frac{v_\theta}{r}\frac{\partial v_r}{\partial\theta}+\frac{v_\varphi}{r\sin\theta}\frac{\partial v_r}{\partial\varphi}-\frac{v_\theta^2+v_\varphi^2}{r}=-\frac{1}{\rho_0}\frac{\partial P}{\partial r}$$
$$+\nu\Big(\frac{\partial^2 v_r}{\partial r^2}+\frac{1}{r^2}\frac{\partial^2 v_r}{\partial\theta^2}+\frac{1}{r^2\sin^2\theta}\frac{\partial^2 v_r}{\partial\varphi^2}+\frac{2}{r}\frac{\partial v_r}{\partial r}+\frac{\mathrm{ctg}\,\theta}{r^2}\frac{\partial v_r}{\partial\theta}$$
$$-\frac{2}{r^2}\frac{\partial v_\theta}{\partial\theta}-\frac{2}{r^2\sin\theta}\frac{\partial v_\varphi}{\partial\varphi}-\frac{2v_r}{r^2}-\frac{2\mathrm{ctg}\,\theta}{r^2}v_\theta\Big),$$

$$\frac{\partial v_\theta}{\partial t}+v_r\frac{\partial v_\theta}{\partial r}+\frac{v_\theta}{r}\frac{\partial v_\theta}{\partial\theta}+\frac{v_\varphi}{r\sin\theta}\frac{\partial v_\theta}{\partial\varphi}+\frac{v_rv_\theta}{r}-\frac{v_\varphi^2\mathrm{ctg}\,\theta}{r}=-\frac{1}{\rho_0 r}\frac{\partial P}{\partial\theta}$$
$$+\nu\Big(\frac{\partial^2 v_\theta}{\partial r^2}+\frac{1}{r^2}\frac{\partial^2 v_\theta}{\partial\theta^2}+\frac{1}{r^2\sin^2\theta}\frac{\partial^2 v_\theta}{\partial\varphi^2}+\frac{2}{r}\frac{\partial v_\theta}{\partial r}+\frac{\mathrm{ctg}\,\theta}{r^2}\frac{\partial v_\theta}{\partial\theta}$$
$$-\frac{2\cos\theta}{r^2\sin^2\theta}\frac{\partial v_\varphi}{\partial\varphi}+\frac{2}{r^2}\frac{\partial v_r}{\partial\theta}-\frac{v_\theta}{r^2\sin^2\theta}\Big),\tag{5.1.20}$$

$$\frac{\partial v_\varphi}{\partial t}+v_r\frac{\partial v_\varphi}{\partial r}+\frac{v_\theta}{r}\frac{\partial v_\varphi}{\partial\theta}+\frac{v_\varphi}{r\sin\theta}\frac{\partial v_\varphi}{\partial\varphi}+\frac{v_rv_\varphi}{r}+\frac{v_\theta v_\varphi\mathrm{ctg}\,\theta}{r}$$

$$= -\frac{1}{\rho_0 r\sin\theta}\frac{\partial P}{\partial \varphi} + \nu\Big(\frac{\partial^2 v_\varphi}{\partial r^2} + \frac{1}{r^2}\frac{\partial^2 v_\varphi}{\partial \theta^2} + \frac{1}{r^2\sin^2\theta}\frac{\partial^2 v_\varphi}{\partial \varphi^2} + \frac{2}{r}\frac{\partial v_\varphi}{\partial r}$$

$$+ \frac{\operatorname{ctg}\theta}{r^2}\frac{\partial v_\varphi}{\partial \theta} + \frac{2}{r^2\sin^2\theta}\frac{\partial v_r}{\partial \varphi} + \frac{2\cos\theta}{r^2\sin^2\theta}\frac{\partial v_\theta}{\partial \varphi} - \frac{v_\varphi}{r^2\sin^2\theta}\Big),$$

$$\sigma_{rr} = -P + 2\mu\frac{\partial v_r}{\partial r}, \quad \sigma_{r\theta} = \mu\Big(\frac{1}{r}\frac{\partial v_r}{\partial \theta} + \frac{\partial v_\theta}{\partial r} - \frac{v_\theta}{r}\Big),$$

$$\sigma_{\theta\theta} = -P + 2\mu\Big(\frac{1}{r}\frac{\partial v_\theta}{\partial \theta} + \frac{v_r}{r}\Big),$$

$$\sigma_{\theta\varphi} = \mu\Big(\frac{1}{r\sin\theta}\frac{\partial v_\theta}{\partial \varphi} + \frac{1}{r}\frac{\partial v_\varphi}{\partial \theta} - \frac{v_\varphi\operatorname{ctg}\theta}{r}\Big),$$

$$\sigma_{\varphi\varphi} = -P + 2\mu\Big(\frac{1}{r\sin\theta}\frac{\partial v_\varphi}{\partial \varphi} + \frac{v_r}{r} + \frac{v_\theta\operatorname{ctg}\theta}{r}\Big),$$

$$\sigma_{\varphi r} = \mu\Big(\frac{\partial v_\varphi}{\partial r} + \frac{1}{r\sin\theta}\frac{\partial v_r}{\partial \varphi} - \frac{v_\varphi}{r}\Big).$$

5.1.2 Formulation of Problems for the System of the Navier–Stokes Equations

In order to determine the solutions of the system of Navier–Stokes equations, it is necessary to specify the initial and boundary conditions. The no-slip condition is specified on the surfaces of the bodies, past which the fluid moves. *This condition means that the fluid velocity is equal to the body velocity.* The initial conditions and the conditions at infinity depend on the problem character.

Consider a problem of the steady viscous fluid flow around a body at rest. As follows from (5.1.1), this problem reduces to the solution of the system of equations

$$\operatorname{div}\vec{v} = 0,$$
$$(\vec{v}\,\nabla)\vec{v} = -\frac{1}{\rho_0}\nabla P + \nu\,\Delta\vec{v}, \tag{5.1.21}$$

in an infinite region, which satisfies the boundary conditions

$$\vec{v}\Big|_S = 0, \quad \vec{v}\Big|_\infty = \vec{v}_\infty, \quad P\Big|_\infty = P_\infty,$$

where S is the body surface. If one searches for the solution of (5.1.1) in the class of irrotational flows, then $\vec{v} = \nabla\varphi$, and by virtue of the continuity equation in (5.1.1), we have $\Delta\varphi = 0$. It follows from here that $\Delta\vec{v} = \Delta(\nabla\varphi) = \nabla(\Delta\varphi) = 0$. In the presence of this relationship, the Navier–Stokes equations will coincide with the Euler equations; that is, the solutions of the Euler equations under the condition $\vec{v} = \nabla\varphi$ satisfy equations (5.1.2) and are the solutions of the Navier–Stokes equations. As shown above, however, the solution of equation $\Delta\varphi = 0$ for the problem of a flow around the body is determined with the accuracy up to circulation under the following boundary conditions:

$$v_n\Big|_S = \frac{\partial\varphi}{\partial n}\Big|_S = 0, \quad \nabla\varphi\Big|_\infty = \vec{v}_\infty.$$

The tangent velocity component v_τ on the body surface S will be different from zero, that is $v_\tau\Big|_S = \frac{\partial\varphi}{\partial\tau}\Big|_S \neq 0$. This means that the potential flow in the case of a viscous fluid does not satisfy the no-slip condition $\vec{v}|_S = 0$ at points of a solid body. Thus, the class of the potential flows cannot be used for solving the problems of the viscous incompressible fluid flow around the bodies. The viscous fluid flows should always be rotational in this case.

An inherent nonlinearity of the system of Navier–Stokes equations makes impossible the application of the superposition of the solutions, which enabled us to construct the solutions for a variety of practical ideal fluid flows, as shown in the foregoing sections. As is known, any streamline in the inviscid fluid can be replaced with a rigid wall and thereby the fluid flow will not be disturbed. This is related to the fact that the boundary condition for the normal velocity component on a streamline and on a rigid wall is the same: $v_n = 0$. In the case of a viscous fluid, one more boundary condition $v_\tau = 0$ appears on the wall, which is absent on a streamline. It follows from here that in the viscous fluid case a substitution of an arbitrary streamline for a rigid wall is invalid. This should be taken into account while constructing the solutions.

5.2 Viscous Fluid Flows at Small Reynolds Numbers

The majority of specialists in the field of fluid mechanics believe that the theoretical hydrodynamics based on the Navier–Stokes equations gives a reasonably accurate approximation of the actual fluid dynamics if the Mach number is small, so that one can neglect the compressibility effects. They are sure that, if the Navier–Stokes equations were integrable, then one would be able to determine completely all of the fluid motions. Therefore, it is important to obtain the exact solutions of these equations, to carry out a correct comparison with experimental data, and outline the domain of applicability of the conventional Navier–Stokes model.

5.2.1 Exact Solutions of the System of Equations for a Viscous Fluid

Problem formulation. We will assume that the flow is one-dimensional if the velocities are parallel with some chosen direction in space. Let us take the x axis as this direction. Then $\vec{v} = \vec{v}(v_x, 0, 0)$, and we have from the system of equations governing the viscous fluid flow (5.1.16) that

$$\frac{\partial v_x}{\partial x} = 0, \quad \frac{\partial P}{\partial y} = \frac{\partial P}{\partial z} = 0, \quad \rho_0 = \text{const},$$
$$\frac{\partial v_x}{\partial t} + v_x \frac{\partial v_x}{\partial x} = -\frac{1}{\rho_0}\frac{\partial P}{\partial x} + \nu\Big(\frac{\partial^2 v_x}{\partial x^2} + \frac{\partial^2 v_x}{\partial y^2} + \frac{\partial^2 v_x}{\partial z^2}\Big). \quad (5.2.1)$$

It follows from the first three equations of system (5.2.1) that

$$v_x = v_x(y, z, t), \qquad P = P(x, t).$$

Taking this into account, we can present the last equation of system (5.2.1) as

$$\frac{\partial v_x}{\partial t} - \nu\Big(\frac{\partial^2 v_x}{\partial y^2} + \frac{\partial^2 v_x}{\partial z^2}\Big) = -\frac{1}{\rho_0}\frac{\partial P}{\partial x}. \quad (5.2.2)$$

The left-hand side of (5.2.2) does not depend on x. Consequently, $\frac{\partial P}{\partial x}$ can depend only on time; that is,

$$\frac{\partial P}{\partial x} = f(t), \qquad P = f(t) \cdot x + f_1(t).$$

Thus, in a one-dimensional flow the pressure is a linear function of x and the functions $f(t)$ and $f_1(t)$ can be determined if the pressure P is given in two sections x_1 and x_2, namely,

$$P(x_1, t) = \mathcal{F}_1(t), \qquad P(x_2, t) = \mathcal{F}_2(t).$$

Then

$$\frac{\partial P}{\partial x} = \frac{\mathcal{F}_2(t) - \mathcal{F}_1(t)}{x_2 - x_1} = \frac{\Delta P}{\Delta x}(t),$$

and the velocity is found from equation (5.2.2):

$$\rho_0 \frac{\partial v_x}{\partial t} = \mu\Big(\frac{\partial^2 v_x}{\partial y^2} + \frac{\partial^2 v_x}{\partial z^2}\Big) - \frac{\Delta P}{\Delta x}. \quad (5.2.3)$$

Equation (5.2.3) coincides with a well-studied heat equation. To find its solution, we must specify the initial and boundary conditions of the form

$$v_x\Big|_{t=t_0} = \mathcal{F}(y, z), \quad v_x\Big|_{x=l_k} = u_k(t).$$

The problem is simplified if the flow is steady. In this case, the pressure jump is constant and equation (5.2.3) reduces to the solution of the Poisson equation

$$\mu\left(\frac{\partial^2 v_x}{\partial y^2}+\frac{\partial^2 v_x}{\partial z^2}\right)=\frac{\Delta P}{\Delta x} \tag{5.2.4}$$

under the boundary conditions $v_x|_{l_k}=u_k$.

Consider at first the case of steady flow between two parallel planes. Assume that the space between two parallel planes $y=\pm h$ is filled by a viscous fluid. It is necessary to find possible one-dimensional steady flows. It follows from the physical meaning of problem that the flow is planar. The function v_x does not depend on z; that is, $v=v(x,y)$. Equation (5.2.4) simplifies in this case to

$$\frac{d^2 v_x}{dy^2}=\frac{1}{\mu}\frac{\Delta P}{\Delta x}, \tag{5.2.5}$$

where $\frac{\Delta P}{\Delta x}=\frac{P_2-P_1}{x_2-x_1}$ is a given constant, $\mu=\rho_0\nu$. The solution of (5.2.5) should satisfy the boundary conditions $v_x|_{y=h}=v_1$, and $v_x|_{y=-h}=v_2$. Let us present the general solution of equation (5.2.5) in the form

$$v_x=\left(\frac{1}{\mu}\frac{\Delta P}{\Delta x}\right)\frac{y^2}{2}+C_1 y+C_2,$$

where C_1 and C_2 are the integration constants. Determining the C_1 and C_2 from the boundary conditions, we obtain:

$$v_x=\frac{1}{2\mu}\frac{\Delta P}{\Delta x}(y^2-h^2)+\frac{v_1-v_2}{2h}y+\frac{v_1+v_2}{2}.$$

If both channel walls are at rest, then $v_1=v_2\equiv 0$, and the obtained solution simplifies to

$$v_x=-\frac{1}{2\mu}\frac{\Delta P}{\Delta x}(h^2-y^2). \tag{5.2.6}$$

The expression within the brackets is nonnegative by virtue of $|y|\le h$ consequently, the fluid always moves in the direction of the pressure reduction. The maximum value of v_x is achieved on the axis $y=0$, and the dependence $v_x=v_x(y)$ has the form of a parabola (Fig. 5.1). Let us calculate the flow rate across a section between the plates when a fluid layer has the thickness 1 along the z-axis:

$$Q=\iint_S v_x dS=\int_0^1 dz\int_{-h}^{h} v_x\,dy=-\frac{2}{3\mu}\frac{\Delta P}{\Delta x}h^3.$$

Thus, the flow rate is directly proportional to the pressure drop and to the cubed distance between the plates, and it is inversely proportional to the viscosity coefficient.

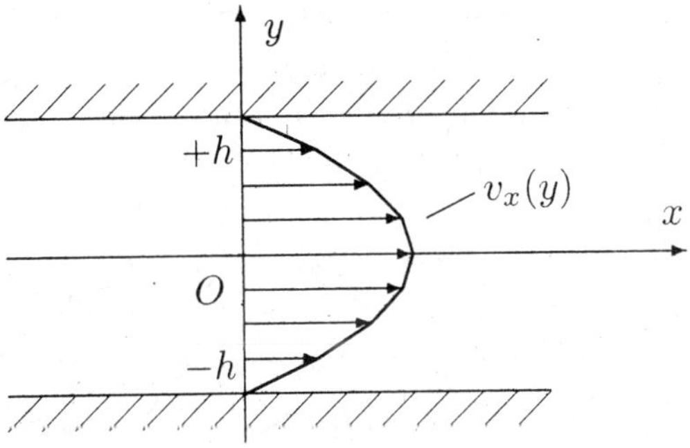

Figure 5.1: The velocity profile $v_z(y)$ in the steady flow between two parallel planes.

Problem 5.1. Find a steady viscous fluid flow between two parallel plates $y = \pm h$ when one of the plates moves at a velocity v_0 and the second plate is at rest.

Consider a viscous fluid flow in a circular tube of radius R. The tube is at rest, and the z-axis coincides with the tube axis. Assuming that only one velocity component $v_z = v_z(r)$ is different from zero, we obtain from system (5.1.18) the equation for v_z:

$$\frac{d^2 v_z}{dr^2} + \frac{1}{r}\frac{dv_z}{dr} = \frac{1}{\mu}\frac{\Delta P}{\Delta z},$$

where $\mu = \rho_0 \nu$. Let us present the solution of this equation as follows:

$$v_z = \frac{1}{\mu}\frac{\Delta P}{\Delta z} \cdot \frac{r^2}{4} + C_1 \ln r + C_2.$$

Taking into account the boundary no-slip condition on the tube contour $v_z|_R = 0$ and the finiteness of the velocity magnitude along the axis, we find *the Poiseuille formula*

$$v_z = -\frac{1}{4\mu} \cdot \frac{\Delta P}{\Delta z}(R^2 - r^2). \tag{5.2.7}$$

Let us calculate the flow rate across the tube crosssection by the formula

$$Q = \int_0^R \int_0^{2\pi} v_z\, r\, d\varphi\, dr = -\frac{2\pi}{4\mu}\frac{\Delta P}{\Delta z}\int_0^R r(R^2 - r^2)\, dr = -\frac{\pi}{8\mu} \cdot \frac{\Delta P}{\Delta z} R^4.$$

The obtained solution does not always agree well with experimental data and depends substantially on the Reynolds number $\mathrm{Re} = \frac{vl}{\nu}$. If $\mathrm{Re} \le 1000 - 1100$, then there is a good agreement with the experiment. At

Re > 1100, a drastical change in the flow pattern occurs. At small Re, each fluid particle moves closely to a straight line: the flow is stratified and quiet. It is conventional to call such a flow the *laminar flow.* At Re > 10^3, each fluid particle performs a chaotic motion and it is no longer a one-dimensional and steady flow. Such a flow is called *turbulent.*

The formulas (5.2.6) and (5.2.7) are valid only for laminar flows. The number Re, at which a flow passage from a laminar to turbulent regime takes place, is called the *critical Reynolds number.* The numerical value of the critical Reynolds number depends substantially on the quality of the polishing of the material of tubes, the inlet section, and many other parameters.

Problem 5.2. Extend the solution (5.2.7) obtained for the axisymmetric flows in a circular tube for the case of axisymmetric steady fluid flows inside an annular tube $R_1^2 \leq y^2 + z^2 \leq R_2^2$, where $v|_{r=R_1} = v_1$ and $v|_{r=R_2} = v_2$ are the velocities at which the tubes move in parallel with their axes.

5.2.2 Viscous Fluid Motion between Two Rotating Coaxial Cylinders

Let the viscous incompressible fluid fill the space between two circular coaxial cylinders c_1 and c_2 with the radii r_1 and r_2 (see Fig. 5.2). The cylinders rotate around a common axis at constant angular velocities ω_1 and ω_2, respectively. We assume that the fluid flow is stationary and the external forces are absent. By introducing the cylindrical coordinates (r, φ, z), we can obviously assume that the fluid motion occurs along a circle with a center on the z-axis; that is,

$$v_z = v_r = 0, \quad v_\varphi = v(r), \quad P = P(r).$$

The system of motion equations (5.1.18) simplifies respectively and has the form

$$\frac{1}{\rho_0} \cdot \frac{dP}{dr} = \frac{v^2}{r}, \quad \frac{d^2v}{dr^2} + \frac{1}{r} \cdot \frac{dv}{dr} - \frac{v}{r^2} = 0. \qquad (5.2.8)$$

Determine the solution of the second equation of system (5.2.8) in the class of the functions $v = r^k$. Substituting this solution, we find that $k(k-1) + k - 1 = 0$, which gives two roots $k_1 = 1$ and $k_2 = -1$. Consequently, we have two particular solutions $v_1 = r$ and $v_2 = \frac{1}{r}$. The general solution will have the form

$$v = Ar + \frac{B}{r},$$

where A and B are arbitrary constants, which are determined from the boundary conditions

$$v = \omega_1 r_1 \text{ at } r = r_1, \quad v = \omega_2 r_2 \text{ at } r = r_2.$$

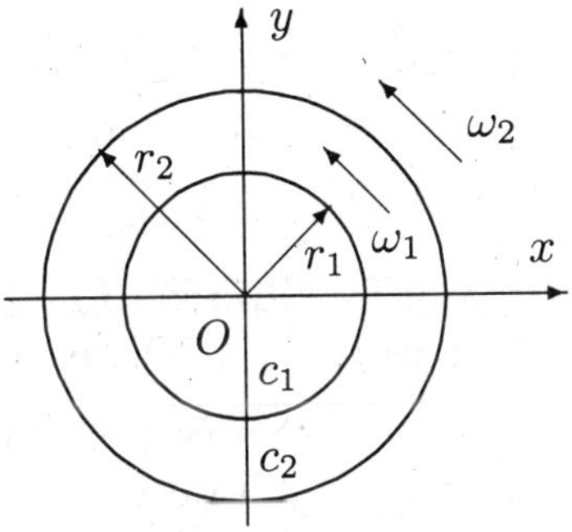

Figure 5.2: The viscous fluid flow between two rotating coaxial cylinders.

A simple calculation of A and B gives the final expression for v as

$$v = \frac{(\omega_2 r_2^2 - \omega_1 r_1^2)\, r^2 + (\omega_1 - \omega_2)\, r_1^2 r_2^2}{r\,(r_2^2 - r_1^2)},$$

and from the first equation of system (5.1.10), we find:

$$P = P_1 + \rho_0 \int_{r_1}^{r} \frac{v^2}{r} \cdot dr.$$

Let us calculate the friction force that acts on the elements of the internal and external cylinders by the formula (5.1.18)

$$\sigma_{r\varphi} = \mu\left(\frac{\partial v}{\partial r} - \frac{v}{r}\right) = -\frac{2\,\mu\,(\omega_1 - \omega_2)\, r_1^2 r_2^2}{r^2\,(r_2^2 - r_1^2)}.$$

The particular cases.
1) The both cylinders rotate at the same angular velocity $\omega_1 = \omega_2 = \omega$; that is,

$$v = \omega r.$$

The viscous fluid rotates as a solid at the same angular velocity.
2) One of the cylinders is at rest, for example, $\omega_1 = 0,\ \omega_2 = \omega$. Then

$$v = \frac{r_2^2}{r_2^2 - r_1^2}\,\omega r - \frac{r_1^2 r_2^2}{r_2^2 - r_1^2} \cdot \frac{\omega}{r}.$$

3) Let $r_2 \to \infty, \omega_2 = 0$. We then obtain the fluid motion outside the cylinder at a velocity

$$v = \frac{\omega_1 r_1^2}{r}.$$

As is known, the vortices are absent in such a fluid motion.

It should be noted that the experiments show a satisfactory agreement in the velocities calculated by the above-obtained formulas only in the case of laminar flows, that is, when the angular velocities of the cylinders rotation remain sufficiently small and do not exceed the critical values, after which a turbulent flow regime begins.

5.2.3 The Viscous Incompressible Fluid Flow around a Sphere at Small Reynolds Numbers

Consider a steady fluid flow around a sphere of radius $r = a$. The fluid velocity at infinity is equal to $\vec{v}_\infty$ and directed in parallel with the Ox-axis. The coordinate origin coincides with the sphere center. Thus, it is necessary to solve the system of equations (5.1.16) at small Reynolds numbers $\mathrm{Re} = \frac{v_\infty \cdot a}{\nu} \ll 1$ under the following boundary conditions at infinity:

$$v_x\Big|_\infty = v_\infty, \quad v_y\Big|_\infty = 0, \quad v_z\Big|_\infty = 0, \quad P\Big|_\infty = P_\infty$$

and on the sphere surface

$$v_x\Big|_{r=a} = 0, \quad v_y\Big|_{r=a} = 0, \quad v_z\Big|_{r=a} = 0,$$

where $r^2 = x^2+y^2+z^2$. Since the solution is sought for at small Reynolds numbers, it is necessary that the following two conditions be satisfied: the freestream velocity v_∞ is sufficiently small or the sphere radius a is small. Introducing the nondimensional variables by the formulas

$$\xi = \frac{x}{a}, \; \eta = \frac{y}{a}, \; \zeta = \frac{z}{a}, \; \vec{w} = \frac{a}{\nu}\vec{v}, \; \mathcal{P} = \left(\frac{a}{\nu}\right)^2 \cdot \frac{P}{\rho_0}, \tag{5.2.9}$$

we can rewrite the system (5.1.16) in the new variables as

$$\frac{\partial w_x}{\partial \xi} + \frac{\partial w_y}{\partial \eta} + \frac{\partial w_z}{\partial \zeta} = 0,$$
$$w_x\frac{\partial \vec{w}}{\partial \xi} + w_y\frac{\partial \vec{w}}{\partial \eta} + w_z\frac{\partial \vec{w}}{\partial \zeta} = -\nabla\mathcal{P} + \Delta\vec{w}. \tag{5.2.10}$$

The nondimensional velocity $\vec{w}$ and the pressure $\mathcal{P}$ satisfy the following conditions at infinity:

$$w_x\Big|_\infty = \mathrm{Re}_\infty, \quad w_y\Big|_\infty = w_z\Big|_\infty = 0, \quad \mathcal{P}\Big|_\infty = \left(\frac{a}{\nu}\right)^2\frac{P_\infty}{\rho_0}. \tag{5.2.11}$$

It follows from here that the nondimensional velocity will have the order $|\vec{w}| \sim \mathrm{Re}$ in the flow region. Since the Reynolds number is small, $\mathrm{Re} \ll 1$, the following inequalities will be valid:

$$|w_x| \ll 1, \quad |w_y| \ll 1, \quad |w_z| \ll 1, \quad \left|\frac{\partial w_i}{\partial \xi_i}\right| \ll 1.$$

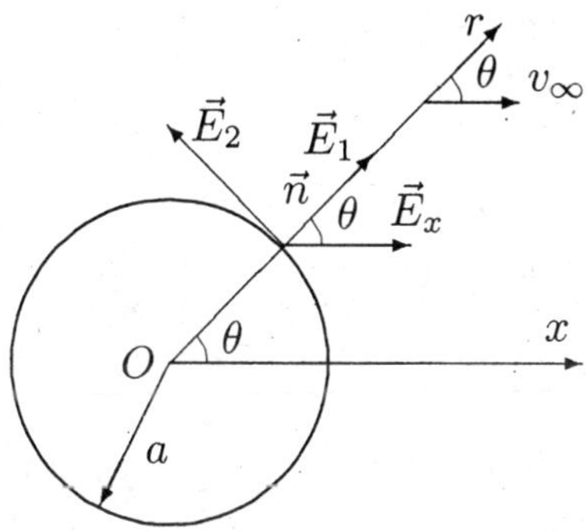

Figure 5.3: To the problem of incompressible fluid flow around a sphere.

Thus, the convective derivatives $w_i \frac{\partial w_j}{\partial \xi_i}$ in the system (5.2.10) are the quantities of the second-order smallness. Neglecting in (5.2.10) the terms of the second-order smallness in comparison with the terms of first-order smallness and returning to the dimensional variables by formulas (5.2.9), we obtain the *system of the Stokes equations*:

$$\operatorname{div} \vec{v} = 0, \quad \frac{1}{\rho_0} \nabla P = \nu \, \Delta \vec{v}. \tag{5.2.12}$$

Thus, the system of the Stokes equations (5.2.12) governs the incompressible fluid flow at small Reynolds numbers Re $\ll 1$ and is obtained from the complete system of the Navier–Stokes equations (5.1.16) by neglecting the convective derivatives.

The Stokes equations are linear, which facilitates greatly the construction of analytic solutions. It is in particular possible to solve with its aid the above-formulated problem of the flow past a sphere. For this purpose, we introduce a spherical coordinate system whose origin is placed at the sphere center (see Fig. 5.3). By virtue of the flow symmetry with respect to the Ox-axis, the desired functions will depend only on r and θ:

$$v_r = v_r(r, \theta), \quad v_\theta = v_\theta(r, \theta), \quad P = P(r, \theta), \quad v_\varphi = 0. \tag{5.2.13}$$

Neglecting the second-order smallness terms proportional to v_i^2, $v_i \frac{\partial v_j}{\partial x^i}$ in the system of the Navier–Stokes equations (5.1.20), we obtain with (5.2.13) in view the system of the Stokes equations (5.2.12) in the spherical coordinate system:

$$\frac{\partial v_r}{\partial r} + \frac{1}{r} \frac{\partial v_\theta}{\partial \theta} + \frac{2 v_r}{r} + \frac{v_\theta \operatorname{ctg} \theta}{r} = 0,$$

$$\begin{aligned}\frac{\partial^2 v_r}{\partial r^2} &+ \frac{1}{r^2}\frac{\partial^2 v_r}{\partial \theta^2} + \frac{2}{r}\frac{\partial v_r}{\partial r} + \frac{\operatorname{ctg}\theta}{r^2}\frac{\partial v_r}{\partial \theta} - \frac{2}{r^2}\frac{\partial v_\theta}{\partial \theta} - \frac{2v_r}{r^2} \\ &- \frac{2\operatorname{ctg}\theta}{r^2}v_\theta = \frac{1}{\mu}\frac{\partial P}{\partial r}, \\ \frac{\partial^2 v_\theta}{\partial r^2} &+ \frac{1}{r^2}\frac{\partial^2 v_\theta}{\partial \theta^2} + \frac{2}{r}\frac{\partial v_\theta}{\partial r} + \frac{\operatorname{ctg}\theta}{r^2}\frac{\partial v_\theta}{\partial \theta} \\ &+ \frac{2}{r^2}\frac{\partial v_r}{\partial \theta} - \frac{v_\theta}{r^2\sin^2\theta} = \frac{1}{\mu r}\frac{\partial P}{\partial \theta}.\end{aligned} \tag{5.2.14}$$

On the sphere surface, the solution of the system of equations (5.2.14) should satisfy the conditions

$$v_r(a,\theta) = 0, \qquad v_\theta(a,\theta) = 0, \tag{5.2.15}$$

and at infinity it should satisfy the conditions

$$v_r\Big|_{r\to\infty} = v_\infty\cos\theta, \quad v_\theta\Big|_{r\to\infty} = -v_\infty\sin\theta, \quad P\Big|_{r\to\infty} = P_\infty. \tag{5.2.16}$$

Taking into account the form of the boundary conditions (5.2.16), we will search for the solution similar to[1] in the form

$$v_r = f(r)\cos\theta, \quad v_\theta = -g(r)\sin\theta, \quad P(r,\theta) = P_\infty + \mu\, b(r)\cos\theta. \tag{5.2.17}$$

Substituting these functions into system (5.2.14), we obtain the following system of ordinary differential equations:

$$\begin{aligned} b' &= f'' + \frac{2}{r}f' - \frac{4(f-g)}{r^2}, \\ \frac{b}{r} &= g'' + \frac{2}{r}g' + \frac{2(f-g)}{r^2}, \\ f' &+ \frac{2(f-g)}{r} = 0, \end{aligned} \tag{5.2.18}$$

where the prime denotes the derivative with respect to r. The boundary conditions for system (5.2.18) follow from (5.2.15) and (5.2.16):

$$f(a) = 0, \quad g(a) = 0, \quad f(\infty) = v_\infty, \quad g(\infty) = v_\infty, \quad b(\infty) = 0. \tag{5.2.19}$$

Expressing the function g from the last equation of system (5.2.18) and eliminating the b' and g, we can rewrite (5.2.18) as

$$\begin{aligned} &r^3 f'''' + 8r^2 f''' + 8rf'' - 8f' = 0, \\ &b = \frac{1}{2}f'''r^2 + 3rf'' + 2f', \\ &g = \frac{1}{2}f'r + f. \end{aligned} \tag{5.2.20}$$

We search for the solution of the first equation in (5.2.20) in the form $f = r^n$. As a result of substitution, we obtain an equation for n: $n(n-2)(n+1)(n+3) = 0$, which has the solution $n = -1$, $n = -3$, $n = 0$, and $n = 2$. Thus, the general solution of the first equation in (5.2.20) has the form

$$f = \frac{C_1}{r^3} + \frac{C_2}{r} + C_3 + C_4 r^2, \tag{5.2.21}$$

where C_1, C_2, C_3, and C_4 are the integration constants. Substituting (5.2.21) in the second and third equations of system (5.2.20), we obtain:

$$g = -\frac{C_1}{2r^3} + \frac{C_2}{2r} + C_3 + 2C_4 r^2, \quad b = \frac{C_2}{r^2} + 10C_4 r. \tag{5.2.22}$$

From the boundary conditions (5.2.19), we find the constants C_i:

$$C_1 = \frac{1}{2} v_\infty a^3, \quad C_2 = -\frac{3}{2} v_\infty a, \quad C_3 = v_\infty, \quad C_4 = 0. \tag{5.2.23}$$

Substituting (5.2.21), (5.2.22), and (5.2.23) in (5.2.17), we obtain the solution sought for:

$$\begin{aligned} v_r &= v_\infty \cos\theta \Big(1 - \frac{3}{2}\frac{a}{r} + \frac{1}{2}\Big(\frac{a}{r}\Big)^3 \Big), \\ v_\theta &= -v_\infty \sin\theta \Big(1 - \frac{3}{4}\frac{a}{r} - \frac{1}{4}\Big(\frac{a}{r}\Big)^3 \Big), \\ P &= P_\infty - \frac{3}{2}\mu v_\infty \frac{a}{r} \cos\theta. \end{aligned} \tag{5.2.24}$$

Knowing the velocity distribution in space (5.2.24), one can find the force acting on the sphere. Substituting (5.2.24) in the last six formulas (5.1.20), we determine that the following physical components of the stress tensor will be different from zero on the sphere surface $r = a$:

$$\sigma_{rr} = -P = \frac{3}{2}\mu \frac{v_\infty}{a} \cos\theta, \quad \sigma_{r\theta} = \mu \frac{\partial v_\theta}{\partial r} = -\frac{3}{2}\mu \frac{v_\infty}{a} \sin\theta. \tag{5.2.25}$$

Write the Cauchy formula in the orthonormal spherical basis $\vec{E}_i = \vec{e}_i / \sqrt{g_i}$:

$$\tilde{f}_i = \tilde{\sigma}_{ij} n_j, \tag{5.2.26}$$

where $\vec{f} = \tilde{f}_i \vec{E}_i$, $\vec{E}_i \cdot \vec{E}_j = \delta_{ij}$, $\tilde{\sigma}_{11} = \sigma_{rr}$, and $\tilde{\sigma}_{12} = \sigma_{r\theta}$. The unit vector $\vec{E}_1$ is directed along a normal to the sphere, and the vector $\vec{E}_2$ is directed along the tangent to the sphere (see Fig. 5.3). It is clear that the resultant force $\vec{F}$ is directed along the Ox-axis. Therefore, to compute $\vec{F}$, it is sufficient to find on each area with a normal $\vec{n}$ the projection of force $\vec{f}$ onto the unit vector $\vec{E}_x$ directed along the Ox-axis (see Fig. 5.3):

$$f_x = \vec{f} \cdot \vec{E}_x = \tilde{f}_i \vec{E}_i \cdot \vec{E}_x = \tilde{\sigma}_{ij} n_j \vec{E}_i \cdot \vec{E}_x.$$

It follows from Fig. 5.3 that

$$\vec{E}_1 = \vec{n},\ \vec{E}_1 \cdot \vec{E}_x = \cos\theta, \quad \vec{E}_2 \cdot \vec{E}_x = -\sin\theta, \quad n_1 = 1, \quad n_2 = 0;$$

therefore, we have with regard for these relations that

$$f_x = \tilde{\sigma}_{11}\cos\theta - \tilde{\sigma}_{12}\sin\theta = \sigma_{rr}\cos\theta - \sigma_{r\theta}\sin\theta.$$

The resultant force is equal to the integral $F = \int f_x\, dS$, and assuming $dS = 2\pi a^2 \sin\theta\, d\theta$, we write:

$$F = \int_0^\pi (\sigma_{rr}\cos\theta - \sigma_{r\theta}\sin\theta) 2\pi a^2 \sin\theta\, d\theta.$$

Substituting here the stresses from (5.2.25) and integrating over θ, we obtain the *Stokes formula*

$$F = 6\pi\mu v_\infty a. \tag{5.2.27}$$

We make two remarks about the obtained solution (5.2.24), (5.2.27).

First, the Stokes formula (5.2.27) can easily be obtained with the accuracy up to the numerical coefficient with the aid of the Pi theorem (3.1.5), (3.1.6). It follows from the system of equations (5.2.14) and the boundary conditions (5.2.15), (5.2.16) that the force F can depend only on μ, v_∞, and a. According to the Pi theorem, this dependence has the nondimensional form

$$\Pi_F = C_0, \quad \Pi_F = \frac{F}{\mu v_\infty a},$$

where C_0 is a constant, from where it follows that $F = C_0 \mu v_\infty a$. According to formula (5.2.27), $C_0 = 6\pi$. Note that the Stokes formula (5.2.27) describes well the results of numerous experiments at $\mathrm{Re} \ll 1$.

Secondly, it turns out that the solution (5.2.24) is inapplicable at sufficiently large distances from the sphere $r \gg a$, where $|\vec{v}| \sim v_\infty$. Computing the convective derivative with the aid of (5.2.24), we obtain $|(\vec{v}\,\nabla)\vec{v}| \sim v_\infty^2 a/r^2$. The retained terms in the Stokes motion equation (5.2.12) $\nu\Delta\vec{v}$ have the order $\nu\, v_\infty a/r^3$. Thus, the condition $|\nu\,\Delta\vec{v}| \gg |(\vec{v}\,\nabla)\vec{v}|$ leads to the inequality $\nu v_\infty a/r^3 \gg v_\infty^2 a/r^2$, which is satisfied only for $r \ll \nu/v_\infty$. At larger distances, the Stokes approximation (5.2.12) is invalid. At $r > \nu/v_\infty$, it is necessary to retain the convective terms in the Navier–Stokes equations (5.1.21). Since at these distances in the general case $\vec{v} \approx \vec{v}_\infty$, one can assume in equations (5.1.21) $(\vec{v}\,\nabla)\vec{v} \approx (\vec{v}_\infty\nabla)\vec{v}$, and as a result, we obtain the *Oseen equatons*:

$$\operatorname{div}\vec{v} = 0, \quad (\vec{v}_\infty\nabla)\vec{v} = -\frac{1}{\rho_0}\nabla P + \nu\,\Delta\vec{v},$$

which are also linear. One can compute with the aid of these equations the velocity field and find a refined formula for the drag force of a sphere[1]

$$F = 6\pi\mu v_\infty a\left(1 + \frac{3}{8}\mathrm{Re}\right), \quad \mathrm{Re} = \frac{av_\infty}{\nu} \ll 1. \tag{5.2.28}$$

Note that, while solving the problem of the viscous incompressible fluid flow moving perpendicularly to the cylinder, one must immediately use the Oseen equations. [The Stokes equations (5.2.12) in this case have no solution satisfying the boundary conditions on the cylinder and at infinity[2].] The force acting on the unit cylinder length was calculated for the first time by Lamb and is equal to[1]

$$F = \frac{8\pi\mu\, v_\infty}{1 - 2\ln\left(\frac{\gamma \mathrm{Re}}{4}\right)}, \qquad \gamma \approx 1.78. \tag{5.2.29}$$

5.3 Viscous Fluid Flows at Large Reynolds Numbers

We will consider the viscous fluid flows at $\mathrm{Re} \gg 1$. It is to be expected that such fluid flows will be close to the ideal fluid flows because $\mathrm{Re} = \infty$ in the ideal fluid. As was shown in the foregoing section, however, although the potential solutions for ideal fluid are also the solutions of the Navier–Stokes equations they do not ensure the satisfaction of the boundary conditions on the body surface around which the fluid flow is considered. Therefore, the proximity of these results may take place only at a large distance from the body. Prandtl[1–6] has supposed that the viscous fluid flows at $\mathrm{Re} \gg 1$ will be close to the ideal fluid flows everywhere except for a layer near the boundary of a rigid surface. The viscosity effects are significant inside this layer. The experimental results confirm the Prandtl's hypothesis about the existence of a thin transition layer, which is conventionally termed the *boundary layer*. Thus, the flow region around the body can be subdivided into two subregions (Fig. 5.4):

1) the boundary layer region, which is located near the body;

2) the region outside the boundary layer, where the flow is close to the ideal fluid flow.

In region (1) (see Fig. 5.4), where $\delta/L \ll 1$, the viscous forces have a significant effect; δ is the boundary layer thickness and L is a reference length of the body. In region (2) outside the boundary layer, one can assume that the flow coincides with a potential flow of ideal fluid. This enables one to simplify substantially in region (1) the system of the Navier–Stokes equations, and as a consequence of this, the solution of the problem of the fluid flow around the body.

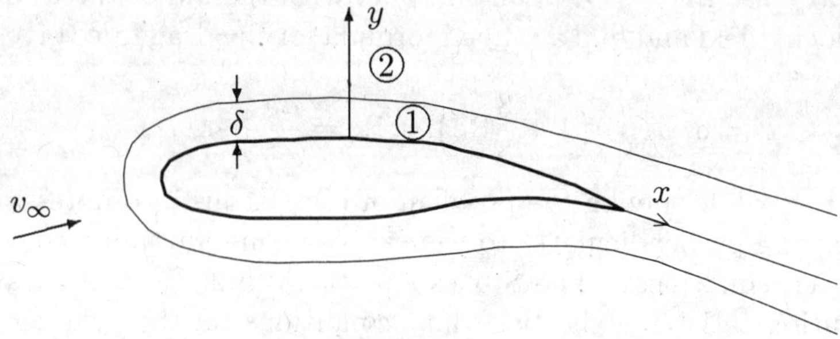

Figure 5.4: Fluid flow region around a body.

5.3.1 Prandtl's Theory of Boundary Layers

We will assume that Re $\gg 1$, which enables us to carry out a simplification of the Navier–Stokes equations if we assume that $\delta/L \ll 1$. We will assume that the fluid flow is laminar. Consider a problem of the viscous incompressible fluid flow around some planar contour. Consider the flow inside the boundary layer, that is, at $0 \le y \le \delta(x)$, where $\delta(x)$ is the boundary layer thickness. Since the boundary layer thickness δ is much less than the curvature radius of the body, one can neglect the body contour curvature. Introduce the Cartesian coordinates x and y, where x is the arc length measured from the point O of flow divergence (see Fig. 5.4) and y is a distance along a normal to the contour. Neglecting the body forces, let us write the system of the Navier–Stokes equations (5.1.16) in the plane-parallel case $v_x = v_x(x,y)$, $v_y = v_y(x,y)$, $P = P(x,y)$, $v_z = 0$:

$$\begin{aligned} &\frac{\partial v_x}{\partial x} + \frac{\partial v_y}{\partial y} = 0, \\ &\frac{\partial v_x}{\partial t} + v_x \frac{\partial v_x}{\partial x} + v_y \frac{\partial v_x}{\partial y} = -\frac{1}{\rho_0}\frac{\partial P}{\partial x} + \nu\,\Delta x, \\ &\frac{\partial v_y}{\partial t} + v_x \frac{\partial v_y}{\partial x} + v_y \frac{\partial v_y}{\partial y} = -\frac{1}{\rho_0}\frac{\partial P}{\partial y} + \nu\,\Delta x. \end{aligned} \tag{5.3.1}$$

Let us estimate the terms entering system (5.3.1), assuming that $\delta/L \ll 1$, where L is a reference length of the body. The velocity component v_x can have different values at the outer edge of the boundary layer, but all of these values will have the same order of magnitude, which we denote by $v_x \approx O(v)$, where v is a reference flow velocity at infinity. While x changes from 0 to L, the velocity varies by the quantity of the order v; therefore,

$$\frac{\partial v_x}{\partial x} \approx O(\frac{v}{L}), \quad \frac{\partial^2 v_x}{\partial x^2} \approx O(\frac{v}{L^2}). \tag{5.3.2}$$

As y varies from 0 to δ, the velocity magnitude changes from zero to the value of the order v; therefore,

$$\frac{\partial v_x}{\partial y} \approx O(\frac{v}{\delta}), \quad \frac{\partial^2 v_x}{\partial y^2} \approx O(\frac{v}{\delta^2}). \tag{5.3.3}$$

Using the above-introduced hypothesis $\delta/L \ll 1$ and comparing the orders of the terms $\frac{\partial^2 v_x}{\partial x^2}$ and $\frac{\partial^2 v_x}{\partial y^2}$ we find that $\frac{\partial^2 v_x}{\partial y^2} \gg \frac{\partial^2 v_x}{\partial x^2}$. Estimating the remaining terms entering the second equation of system (5.3.1), we find that

$$v_x \cdot \frac{\partial v_x}{\partial x} \approx O(\frac{v^2}{L}), \quad \frac{\partial v_x}{\partial t} \approx O(\frac{v^2}{L}).$$

We can estimate the order of magnitude of the quantity v_y by using the continuity equation in (5.3.1):

$$\frac{\partial v_y}{\partial y} \approx -\frac{\partial v_x}{\partial x} \approx O(\frac{v}{L}), \quad v_y = \int_0^y \frac{\partial v_y}{\partial y} dy \approx O(\frac{v\delta}{L});$$

consequently,

$$v_y \frac{\partial v_x}{\partial y} \approx O(\frac{v\delta}{L} \cdot \frac{v}{\delta}).$$

Thus, the terms of the left-hand side of the second equation of system (5.3.1) will have the order of magnitude $O(\frac{v^2}{L})$.

Prandtl assumed that the inertia forces and the viscous friction forces in the boundary layer have the same order of magnitude; that is,

$$\nu \frac{\partial^2 v_x}{\partial y^2} \approx O(\frac{v^2}{L}).$$

Taking (5.3.3) into account, we find that

$$\nu \frac{\partial^2 v_x}{\partial y^2} \approx O(\nu \frac{v}{\delta^2}) \approx O(\frac{v^2}{L})$$

or

$$\left(\frac{\delta}{L}\right)^2 \approx O\left(\frac{\nu}{vL}\right), \quad \left(\frac{\delta}{L}\right) \approx O\left(\frac{1}{\sqrt{\mathrm{Re}}}\right),$$

where $\mathrm{Re} = \frac{vL}{\nu}$. It follows from the obtained estimate that the relative thickness of the boundary layer is inversely proportional to $\sqrt{\mathrm{Re}}$; that is, the larger the number Re, the thinner the boundary layer.

We will estimate the terms involving the pressure by using the Bernoulli integral, which will be valid at the outer edge of the boundary layer; that is,

$$\frac{v^2}{2} + \frac{P}{\rho_0} = \text{const}.$$

We find from here that

$$\frac{\partial}{\partial x}\left(\frac{v^2}{2}\right) = -\frac{1}{\rho_0}\cdot\frac{\partial P}{\partial x} \approx O\left(\frac{v^2}{L}\right).$$

If we compare the orders of the terms entering the second equation of system (5.3.1), we have:

$$\frac{\partial v_x}{\partial t} \approx v_x\frac{\partial v_x}{\partial x} \approx v_y\frac{\partial v_x}{\partial y} \approx \frac{1}{\rho_0}\frac{\partial P}{\partial x} \approx \nu\frac{\partial^2 v_x}{\partial y^2} \gg \nu\frac{\partial^2 v_x}{\partial x^2}.$$

Consequently, one can neglect the term $\nu\frac{\partial^2 v_x}{\partial x^2}$ in this equation, and it takes the form

$$\frac{\partial v_x}{\partial t} + v_x\frac{\partial v_x}{\partial x} + v_y\frac{\partial v_x}{\partial y} = -\frac{1}{\rho_0}\frac{\partial P}{\partial x} + \nu\frac{\partial^2 v_x}{\partial y^2}.$$

The continuity equation remains unchanged since the order of its terms is the same.

Consider the third equation of system (5.3.1). Estimating the orders of terms, we have:

$$v_x\frac{\partial v_y}{\partial x} \approx O\left(\frac{\delta}{L}\cdot\frac{v^2}{L}\right), \quad v_y\left(\frac{\partial v_y}{\partial y}\right) \approx O\left(\frac{\delta}{L}\cdot\frac{v^2}{L}\right),$$
$$\frac{\partial^2 v_y}{\partial x^2} \approx O\left(\frac{v}{L^2}\cdot\frac{\delta}{L}\right), \quad \frac{\partial^2 v_y}{\partial y^2} \approx O\left(\frac{v}{L\,\delta}\right).$$

It is clear that $\frac{\partial^2 v_y}{\partial y^2} \gg \frac{\partial^2 v_y}{\partial x^2}$. Estimating the term $\frac{1}{\rho_0}\frac{\partial P}{\partial y}$, we have:

$$\frac{1}{\rho_0}\frac{\partial P}{\partial y} \approx O\left(\frac{\delta}{L}\cdot\frac{v^2}{L}\right).$$

Comparing the terms $\frac{1}{\rho_0}\frac{\partial P}{\partial y}$ and $\frac{1}{\rho_0}\frac{\partial P}{\partial x}$ in the boundary layer region, we find:

$$O\left(\frac{1}{\rho_0}\frac{\partial P}{\partial y}\right) \approx \frac{\delta}{L}O\left(\frac{1}{\rho_0}\frac{\partial P}{\partial x}\right);$$

that is, the pressure changes along the y-axis much slower than along the x-axis. The remaining terms of the third equation of system (5.3.1) have a much lesser order of magnitude than the corresponding terms of the second equation of the system; therefore, this equation can be replaced with equation $\frac{\partial P}{\partial y} = 0$ whose integral is $P = P(x,t)$. It follows from here that the pressure across the boundary layer does not change and is a known function found from the solution of a problem of the ideal fluid flow around the body.

Thus, taking the above into account, we can simplify the system of the Navier-Stokes equations (5.3.1) and write it as

$$\begin{aligned} &\frac{\partial v_x}{\partial t}+v_x\frac{\partial v_x}{\partial x}+v_y\frac{\partial v_x}{\partial y}=-\frac{1}{\rho_0}\frac{\partial P}{\partial x}+\nu\frac{\partial^2 v_x}{\partial y^2},\\ &\frac{\partial v_x}{\partial x}+\frac{\partial v_y}{\partial y}=0,\\ &P=P(x,t). \end{aligned} \tag{5.3.4}$$

Here v_x and v_y are the unknown functions of x, y, and t in the boundary layer region; P is a given function of x and t. If the flow is steady, then the Bernoulli integral $\frac{u_0^2}{2}+\frac{P}{\rho_0}=\text{const}$ will be valid in the region outside the boundary layer, and the differentiation of this integral yields

$$u_0\frac{\partial u_0}{\partial x}+\frac{1}{\rho_0}\frac{\partial P}{\partial x}=0,$$

where u_0 is the flow velocity at the outer edge of the boundary layer. With regard for this relation, we can simplify the system of equations (5.3.4) and write it in the form

$$\begin{aligned} &\frac{\partial v_x}{\partial t}+v_y\frac{\partial v_x}{\partial y}-\nu\frac{\partial^2 v_x}{\partial y^2}=u_0\frac{\partial u_0}{\partial x},\\ &\frac{\partial v_x}{\partial x}+\frac{\partial v_y}{\partial y}=0. \end{aligned} \tag{5.3.5}$$

Since at $y=0$ the quantity $\frac{1}{\rho_0}\frac{\partial P}{\partial x}=-(u_0\frac{\partial u_0}{\partial x})_{y=0}$, one can take as the function u_0 the solution of ideal fluid motions at $y=0$. The function $u_0=u_x$ depends only on x. The functions v_x and v_y are determined as a solution of the system of equations (5.3.5). To solve these equations, it is necessary to specify the initial and boundary conditions. The first equation in (5.3.5) has the parabolic type (see Problem 3.2 in Chapter 3); therefore, it is necessary to specify for its solution the initial condition at $x=0$ and the boundary condition at $y=0$ and $y=\delta(x)$ as follows:
1) the initial condition at $x=0$: $v_x=u_0(0)$;
2) the no-slip conditions are satisfied on the body wall: $v_x|_{y=0}=0$ and $v_y|_{y=0}=0$;
3) at the outer edge of the boundary layer,

$$v_x|y\to\delta(x)=u_0(x). \tag{5.3.6}$$

It should be noted that the condition (5.3.6) is indeed no boundary condition because the function $\delta(x)$ is unknown therein. To determine this function, it is necessary to have an additional condition. It is mathematically correct to formulate the problem of the flow around a body

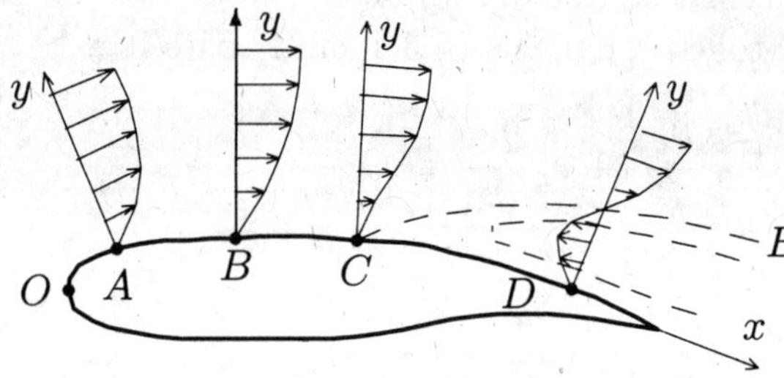

Figure 5.5: Flow separation from the body surface.

as a conjugate problem with an unknown boundary, which separates the boundary layer region from the region filled with ideal fluid. The solution of such a problem, however, is very difficult. Prandtl has proposed to write the condition (5.3.6) at the outer edge of the boundary layer at infinity. The possibility of such an approximation is related to the fact that the velocity v_x changes rapidly inside the boundary layer, and outside, it rapidly tends to the limiting values as the distance from the body increases. Thus, the initial and boundary conditions for the Prandtl boundary layer equations are taken as

$$v_x\Big|_{x=0,y>0} = u_0(0), \quad v_x\Big|_{y=0} = 0, \quad v_y\Big|_{y=0} = 0,$$
$$v_x\Big|_{y\to\infty,0\le x\le L} = u_0(x). \tag{5.3.7}$$

Having the velocity distribution in the boundary layer, one can find the outer edge of the boundary layer $\delta(x)$ by using the equality $v_x(x,\delta) = (1-\varepsilon)u_0(x)$, where ε is a given small quantity.

Finally, in the case of an unsteady motion, one must also specify an initial condition at $t = 0$: $v_x = v_x^0(x,y)$, where $v_x^0(x,y)$ is a given function of its arguments.

The obtained equations (5.3.5) describe a laminar flow; therefore, they are often termed the equations of a laminar boundary layer in the literature. The use of the Cartesian coordinates at the derivation of (5.3.5) is of no fundamental importance. A similar derivation can also be performed for an arbitrary curvilinear coordinate system[1,7].

Note that the initial condition $v_x|_{x=0} = u_0(0)$ does not significantly affect the solution in the boundary layer. It affects the solution in a thin layer $\Delta x \sim \delta$, and at $\Delta x \gg \delta$, its influence rapidly decays.

The boundary layer theory enables one to explain the phenomenon of the flow separation from a smooth surface of the body (see Fig. 5.5). Let the derivative $\frac{dP}{dx} < 0$ on the body surface (Fig. 5.5) from point O to point B. At point B, the pressure achieves its minimum $\frac{dP}{dx}|_B =$

0, and after that, we have $\frac{dP}{dx} > 0$ at $x > x_B$. The presence of a positive pressure gradient at $x > x_B$ leads to the flow deceleration. If the fluid were ideal, the available amount of kinetic energy would suffice by virtue of the Bernoulli integral to surmount the decelerating effect of the pressure gradient. Just this fact takes place outside the boundary layer. Inside the boundary layer, however, the flow pattern will be different. By virtue of the viscous effects, the flow velocity in it is small; therefore, the pressure gradient downstream the point B causes at first the stagnation and then the reverse fluid flow (point D in Fig. 5.5). The interaction of a freestream with the fluid moving in the opposite direction leads to a significant displacement of the streamline CE from the body surface, the thickening of the boundary layer, and its separation from the body surface at point C. As can be seen from Fig. 5.5, the derivative $\frac{\partial v_x}{\partial y}|_{y=0} > 0$, and at the separation point C, we have $\frac{\partial v_x}{\partial y}|_{y=0} = 0$, and downstream of the separation point, we have $\frac{\partial v_x}{\partial y}|_{y=0} < 0$. It follows from the above explanation that the separation point C is always located more downstream than the point of the pressure minimum B. The boundary layer separation leads to a considerable increase of its thickness; after that, the approximation of the boundary layer is already not so good. In this case, it is necessary to perform a computation of a viscous fluid flow with the aid of a complete system of the Navier–Stokes equations.

Another limitation is related to the fact that the flow can become turbulent even upstream from the separation point C, and the approximation of a laminar boundary layer will not correspond to the physics of the phenomenon.

Note that the boundary layer arises not only in the fluid, but also in the gas flow. The temperature and the density also rapidly change in the gas flow in the direction across the boundary layer. Therefore, the problem of a heat exchange between the body and gas is closely related to the problems of the boundary layer theory (see, for example, [6]).

5.3.2 Boundary Layer of a Flat Plate

Consider the problem on a stationary fluid flow around a thin semi-infinite plate at a freestream velocity v_∞ directed along the Ox-axis (see Fig. 5.6). We assume the plate to be thin, so that we can neglect the disturbances, which it introduces into the ideal fluid flow. In this case, $P = P(x) = P_\infty = \text{const}$, and the velocity at the outer edge of the boundary layer is equal to v_∞. Consequently, the problem reduces to the solution of a system of equations (5.3.5) in which one must set $\frac{du_0}{dx} = 0$:

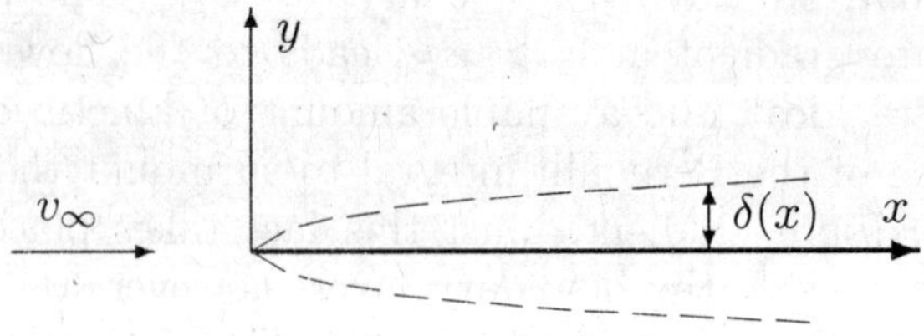

Figure 5.6: Stationary fluid flow around a thin semi-infinite plate.

$$v_x\frac{\partial v_x}{\partial x}+v_y\frac{\partial v_x}{\partial y}=\nu\frac{\partial^2 v_x}{\partial y^2},\qquad \frac{\partial v_x}{\partial x}+\frac{\partial v_y}{\partial y}=0 \tag{5.3.8}$$

under the following boundary conditions:

$$v_x\Big|_{y=0,x>0}=0,\quad v_y\Big|_{y=0,x>0}=0,\quad v_x\Big|_{y=\delta(x),x\geq 0}=v_\infty. \tag{5.3.9}$$

Following Prandtl, we replace the condition at the outer edge of the boundary layer with a condition of the form

$$v_x\Big|_{x=0,y>0}=v_\infty,\quad v_x\Big|_{y=\infty,x>0}=v_\infty. \tag{5.3.10}$$

We find from the first equation of (5.3.8) that

$$v_y=\left(\nu\frac{\partial^2 v_x}{\partial y^2}-v_x\frac{\partial v_x}{\partial x}\right)\Big/\frac{\partial v_x}{\partial y}. \tag{5.3.11}$$

The substitution of this expression into the second equation of system (5.3.8) yields

$$\frac{\partial v_x}{\partial x}+\frac{\partial}{\partial y}\left(\frac{\nu\frac{\partial^2 v_x}{\partial y^2}-v_x\frac{\partial v_x}{\partial x}}{\frac{\partial v_x}{\partial y}}\right)=0. \tag{5.3.12}$$

Under the satisfaction of the boundary condition $v_x=0$ at $y=0, x>0$, we obtain from (5.3.11) that

$$\frac{\partial^2 v_x}{\partial y^2}\Big|_{y=0,x>0}=0.$$

Thus, the solution of the problem reduces to the solution of equation (5.3.12) under the following boundary conditions:

$$\begin{aligned} v_x\Big|_{y=0,x>0}&=0, & \frac{\partial^2 v_x}{\partial y^2}\Big|_{y=0,x>0}&=0,\\ v_x\Big|_{x=0,y>0}&=v_\infty, & v_x\Big|_{x>0,y\to\infty}&=v_\infty. \end{aligned} \tag{5.3.13}$$

Since the problem under consideration does not involve any reference length, the solution will be self-similar. The functions to be determined, for example, the velocity v_x, will depend on the variables x and y and the viscosity ν. Using the Pi theorem, we will search for the solution of equation (5.3.12) in the form $v_x = \mathcal{F}(\xi)$, where $\xi = \frac{1}{\sqrt{2\nu}} \cdot \frac{y}{\sqrt{x}}$ is a self-similar variable. We can calculate that

$$\frac{\partial v_x}{\partial x} = \mathcal{F}' \left[\frac{1}{\sqrt{2\nu}} \left(-\frac{1}{2} \right) y x^{-\frac{3}{2}} \right],$$

$$\frac{\partial v_x}{\partial y} = \mathcal{F}' \left[\frac{1}{\sqrt{2\nu}} \cdot \frac{1}{\sqrt{x}} \right], \quad \frac{\partial^2 v_x}{\partial y^2} = \mathcal{F}'' \cdot \frac{1}{2\nu x}.$$

The substitution of these expressions in (5.3.12) yields

$$\mathcal{F}' \cdot \mathcal{F}''' - (\mathcal{F}'')^2 + \mathcal{F} \cdot (\mathcal{F}')^2 = 0, \tag{5.3.14}$$

where the prime denotes the differentiation with respect to ξ. Substituting the solution in the boundary conditions (5.3.13), we have:

$$\mathcal{F}(0) = 0, \quad \mathcal{F}''(0) = 0 \text{ and } \mathcal{F}(\infty) = v_\infty. \tag{5.3.15}$$

The obtained equation (5.3.14) under the boundary conditions (5.3.15) cannot be integrated in closed form.

Let us construct the solution as follows. Instead of problems (5.3.14) and (5.3.15), we at first consider the Cauchy problem for the function $\mathcal{F}_1(\xi)$, satisfying equation (5.3.14) and the following initial conditions:

$$\mathcal{F}_1(0) = 0, \quad \mathcal{F}_1'(0) = 1, \quad \mathcal{F}_1''(0) = 0.$$

Such a problem can be solved with the aid of any numerical method. Assume that we have found such a function and, in particular, we know its value at $\xi \to \infty$; that is, $\mathcal{F}_1(\infty) = C_0$. Having the function $\mathcal{F}_1(\xi)$, we construct the function $\mathcal{F}(\xi)$ by the formula

$$\mathcal{F}(\xi) = k^2 \mathcal{F}_1(k\xi),$$

where k is a constant. It is easy to be sure by a direct substitution that the function $\mathcal{F}(\xi)$ satisfies equation (5.3.14) if the function $\mathcal{F}_1(\xi)$ satisfies this equation. Let us choose k in such a way that $\mathcal{F}(\xi)$ is the solution of problem (5.3.14) with conditions (5.3.15). We find from the third condition in (5.3.15) that $k = \sqrt{\frac{v_\infty}{C_0}}$. Consequently,

$$v_x = \frac{v_\infty}{C_0} \mathcal{F}_1 \left(\sqrt{\frac{v_\infty}{C_0}} \cdot \frac{1}{\sqrt{2\nu}} \cdot \frac{y}{\sqrt{x}} \right), \tag{5.3.16}$$

where C_0 and $\mathcal{F}_1$ are the known quantities as the solution of the auxiliary Cauchy problem.

Let us calculate the drag R_x of the plate of a finite length l and width b by formula

$$R_x = 2\int_0^b dz \int_0^l \sigma_{yx}\Big|_{y=0} dx.$$

The appearance of the coefficient 2 is related to the fact that both sides of the plate give the same contribution to the drag force. Since $\sigma_{yx}|_{y=0}$ does not depend on z, then

$$R_x = 2b\int_0^l \sigma_{yx}\Big|_{y=0} dx. \tag{5.3.17}$$

Since

$$\sigma_{yx}\Big|_{y=0} = \mu\left(\frac{\partial v_x}{\partial y} + \frac{\partial v_y}{\partial x}\right)\Big|_{y=0} = \mu\frac{\partial v_x}{\partial y}\Big|_{y=0},$$

we find with regard for (5.3.16) that

$$\sigma_{xy}\Big|_{y=0} = \mu\frac{\partial}{\partial y}\left(\frac{v_\infty}{C_0}\mathcal{F}_1\left(\frac{v_\infty}{C_0}\cdot\frac{y}{\sqrt{2\nu x}}\right)\right)\Big|_{y=0} = \mu\left(\frac{v_\infty}{C_0}\right)^{3/2}\frac{1}{\sqrt{2\nu x}},$$

where we have taken into account that $\mathcal{F}_1'(0) = 1$. Substituting the obtained expression into formula (5.3.17) and computing the integral, we obtain:

$$R_x = \frac{2\sqrt{2}}{C_0^{3/2}}\cdot\rho_0\sqrt{\nu}\,v_\infty^{3/2}b\cdot\sqrt{l},$$

where we have assumed that $\mu = \rho_0\nu$. The drag coefficient

$$C_x = \frac{R_x}{\frac{1}{2}\rho_0 v_\infty^2 S_0},$$

which at the total area of both plate sides $S_0 = 2b\cdot l$ is equal to

$$C_x = 2\sqrt{2}\left(\frac{1}{C_0}\right)^{\frac{3}{2}}\cdot\frac{1}{\sqrt{\mathrm{Re}}},$$

where $\mathrm{Re} = \frac{lv_\infty}{\nu}$ is the Reynolds number.

We have used the *Mathematica 3.0* system to solve numerically the auxiliary problem

$$\mathcal{F}_1'\cdot\mathcal{F}_1''' - (\mathcal{F}_1'')^2 + \mathcal{F}_1\cdot(\mathcal{F}_1')^2 = 0, \tag{5.3.18}$$

$$\mathcal{F}_1(0) = 0, \quad \mathcal{F}_1'(0) = 1, \quad \mathcal{F}_1''(0) = 0. \tag{5.3.19}$$

For the computer implementation, it is convenient to reduce equation (5.3.18) to a system of three first-order ordinary differential equations (ODEs)

$$\mathcal{F}_1' = Y, \quad Y' = Z, \quad Z' = \frac{Z^2}{Y} - \mathcal{F}_1 Y. \tag{5.3.20}$$

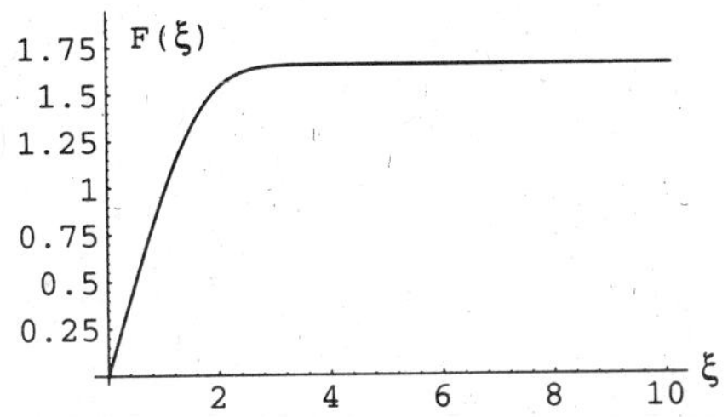

Figure 5.7: The graph of the function $\mathcal{F}_1(\xi)$.

With regard for (5.3.19), the initial conditions for system (5.3.20) have the form

$$\mathcal{F}_1(0) = 0, \quad Y(0) = 1, \quad Z(0) = 0. \tag{5.3.21}$$

We have used in our *Mathematica* Notebook **prog5-2.nb** the standard fourth-order Runge–Kutta method[8] to solve the initial-value problem (5.3.20), (5.3.21). The software system *Mathematica 3.0* contains a large library of the methods for the analytical solution of the ODEs. Therefore, we hoped to solve the nonlinear ODE (5.3.14) with its aid in analytic form. The methods available, however, in the corresponding library of *Mathematica 3.0* are still insufficient to solve this task in analytic form.

The obtaining of a numerical solution of problem (5.3.20), (5.3.21) with the aid of our Notebook **prog5-2.nb** takes only about a second. We show in Fig. 5.7 the graph of the function $\mathcal{F}_1(\xi)$ obtained as a result of the numerical integration of system (5.3.20). It can be seen from the numerical results presented in the above *Mathematica* Notebook that

$$2\sqrt{2}\Big(\frac{1}{C_0}\Big)^{3/2} \approx 1.32822;$$

therefore, the drag coefficient is

$$C_x \approx \frac{1.328}{\sqrt{\mathrm{Re}}}. \tag{5.3.22}$$

Thus, at large numbers Re, the drag coefficient of a plate is inversely proportional to $\sqrt{\mathrm{Re}}$. The obtained formula (5.3.22) is confirmed well by the experimental data for $\mathrm{Re} \le 3 \cdot 10^5$. At larger Reynolds numbers, a transition of the flow regime from laminar to turbulent regime takes place. The boundary layer thickness has the order

$$\delta(x) \sim \sqrt{\frac{\nu x}{v_\infty}},$$

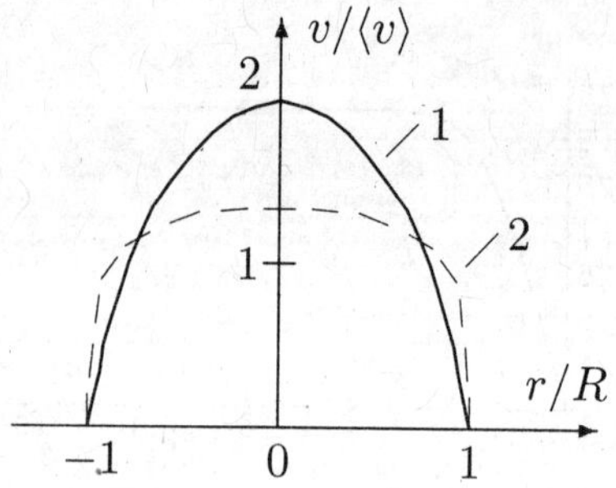

Figure 5.8: The Poiseuille profile 1 and the mean velocity profile of the turbulent flow 2[6].

from where it follows that $\delta(x)$ increases with x and the Prandtl assumptions on which the boundary layer theory is based are violated. The experiments show that, at some distance from the plate leading edge, the laminar boundary layer passes to a turbulent boundary layer.

5.4 Turbulent Fluid Flows

5.4.1 Basic Properties of Turbulent Flows

So far we studied the laminar fluid flows. The experience shows, however, that they are not always stable. At a certain critical Reynolds number Re^*, the laminar flow loses its stability, destroys, and a turbulent flow arises. The liquid particle velocity in turbulent flow chaotically changes in the neighborhood of certain mean flow, or it undergoes fluctuations.

The general properties of turbulent motions were discovered by Reynolds in 1883 when he studied the water motion in a circular cylindrical pipe. For the flow visualization, Reynolds used a painted liquid jet. At increasing laminar flow velocity in a rectilinear jet, at a certain distance from the tube inlet, the wavy disturbances appear. As the flow velocity increases, the amplitude of these disturbances grows. As a result, the jet destroys and mixes with a liquid and uniformly paints it in the overall section. Thus, a transition from laminar to turbulent flow takes place. Reynolds has discovered that this passage occurred in the case in which the Reynolds number $\mathrm{Re} = \langle u\rangle d/\nu$ achieved certain critical value $\mathrm{Re}^* \approx 2 \cdot 10^3$, where $\langle u\rangle$ is the mean flow velocity and $d = 2R$ is the tube diameter. The given value $\mathrm{Re}^* \approx 2 \cdot 10^3$ is the lower critical value. If the disturbances at the tube inlet are removed, for example, at the expense of a smooth flow entry into the tube, then one can increase the critical Reynolds number up to $\mathrm{Re}^* \approx 1.3 \cdot 10^4$. The passage from a laminar flow to the turbulent one leads to a significant alteration of the velocity

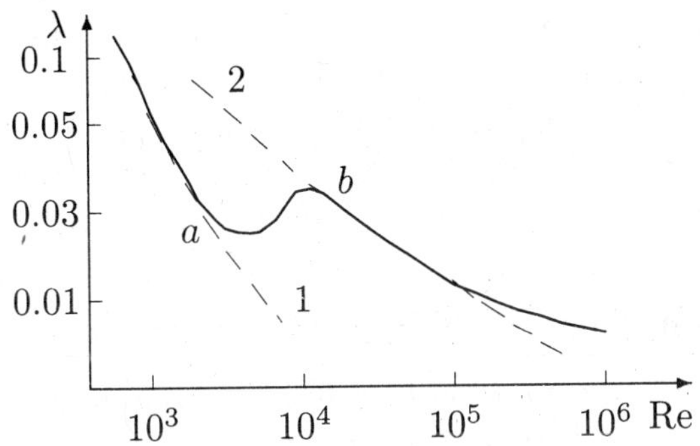

Figure 5.9: The dependence of the coefficient of loading loss λ on the Re number in the logarithmic coordinates[6].

profile $v(r)$ (see Fig. 5.8) and to an increase of the skin-friction drag. The skin-friction drag is characterized by the coefficient of the loading loss $\lambda = -\frac{\Delta P}{\Delta z}/(\frac{\rho\langle v\rangle^2}{2d})$. At a laminar flow, the dependence $v = v_z(r)$ is determined by the Poiseuille formula (5.2.7), with the aid of which, one can compute the mean fluid velocity in the tube:

$$\langle v\rangle = \frac{2}{R^2}\int_0^R v_z(r)\,r\,dr = -\frac{1}{8\mu}\frac{\Delta P}{\Delta z}R^2.$$

Substituting this value into the expression for λ, we obtain the dependence $\lambda = 64/\mathrm{Re}$, which is shown by a dashed line 1 in Fig. 5.9. In the turbulent flow regime, the dependence $\lambda(\mathrm{Re})$ within a wide range of the Re values is given by an empirical formula $\lambda = 0.316/\mathrm{Re}^{1/4}$ shown in Fig. 5.9 by a dashed curve 2. The actual dependence $\lambda(\mathrm{Re})$ observed in the experiment is shown by a solid line in Fig. 5.9. The transitional regime from a laminar flow to a turbulent one corresponds to the interval "*ab*" in Fig. 5.9. In the transitional regime at the values of Re close to Re^*, an alternation of laminar and turbulent regimes (the intermittency) can take place. This is related to the fact that the turbulence at first forms in some bounded regions, representing the liquid tubes filling a part of the tube crosssection. The experience shows that at $\mathrm{Re} > \mathrm{Re}^*$ the front end of the liquid tube moves faster than the rear end. This leads to the extension of tubes, their merging, and the flow turbulization. At $\mathrm{Re} < \mathrm{Re}^*$, the picture is opposite, which leads to the disappearance of the turbulent tubes and the flow laminarization. Returning to Fig. 5.9, we note that the quantity λ for a turbulent flow regime is larger than the value of λ for the laminar flow (line 1) at the same Reynolds number.

The above-noted *instability* of the laminar flow at $\mathrm{Re} > \mathrm{Re}^*$ and the *presence of fluctuations* in the turbulent flow are the *general properties* inherent in the viscous fluid flows. Attempts at a mathematical description of these phenomena, however, face insurmountable obstacles. We can immediately summarize that there is at present no consistent theory of turbulence. Therefore, we present in what follows only a brief survey of the stability theories and of semiempirical theories of turbulence, which are closely related to the specific flow diagrams.

5.4.2 Laminar Flow Stability and Transition to Turbulence

The mathematical theory of linear stability of laminar flows is the most developed theory[1,5,9,10]. It enables one to determine the stability boundary for laminar flows. We now briefly outline the main points of this theory at the example of the problem of the *stability of a planar flow between two parallel planes* (see Section 5.2.1). This problem was solved correctly for the first time by Lin[9] in 1945. Let us present the solution of the Navier-Stokes equations (5.1.21) in the form of a sum of a stationary solution $\vec{v}_0$, P_0 and the disturbances $\vec{v}\,'$, P':

$$\vec{v} = \vec{v}_0(x_i) + \vec{v}\,'(x_i, t), \quad P = P_0 + P'(x_k, t),$$

where $\vec{v}\,' \ll \vec{v}_0$, $P' \ll P_0$. Substituting this solution in (5.1.21) and neglecting the quadratic disturbance terms, we obtain a linear system of equations:

$$\operatorname{div} \vec{v}\,' = 0, \quad \frac{\partial \vec{v}\,'}{\partial t} + (\vec{v}_0 \nabla)\vec{v}\,' + (\vec{v}\,'\nabla)\vec{v}_0 = -\frac{1}{\rho}\nabla P' + \nu\Delta\vec{v}\,', \tag{5.4.1}$$

where the stationary solution $\vec{v}_0(x_i)$, $P_0(x_i)$ satisfies the stationary equations

$$\operatorname{div} \vec{v}_0 = 0, \quad (\vec{v}_0 \nabla)\vec{v}_0 = -\frac{1}{\rho}\nabla P_0 + \nu\,\Delta\vec{v}_0. \tag{5.4.2}$$

In our case $\vec{v}_0 = (U(y), 0, 0)$, where $U(y)$ was found above [see (5.2.6)]. Assuming in (5.4.1) that $v_x' = \frac{\partial \psi'}{\partial y}$ and $v_y' = -\frac{\partial \psi'}{\partial x}$, one can find the equation for the disturbances of the stream function ψ' (see also[1]):

$$\frac{\partial}{\partial t}(\Delta\psi') + U(y)\frac{\partial \Delta\psi'}{\partial x} - U''(y)\frac{\partial \psi'}{\partial x} - \nu\Delta^2\psi' = 0. \tag{5.4.3}$$

Representing the solution of (5.4.3) as

$$\psi' = f(y)e^{i(kx - \omega t)} \tag{5.4.4}$$

we obtain the *Orr-Sommerfeld equation* for f:

$$\left(U - \frac{\omega}{k}\right)(f'' - k^2 f) - U''f = -\frac{i\nu}{k}(f^{\text{IV}} - 2k^2 f'' + k^4 f). \tag{5.4.5}$$

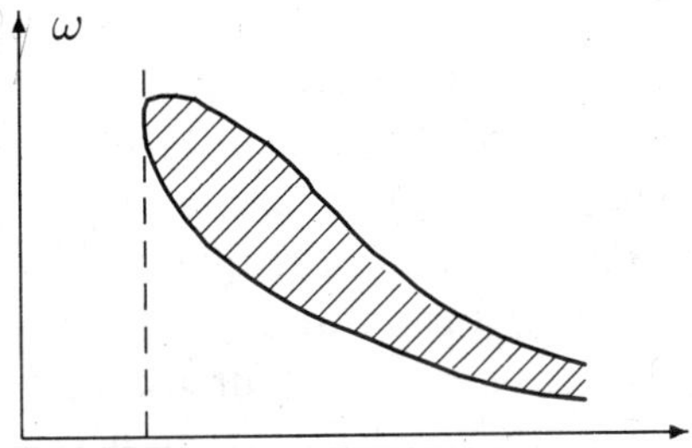

Figure 5.10: The neutral curve for the planar flow between the plates.

From the no-slip conditions $v'_x = v'_y = 0$ at $y = -h$ and $y = h$, we have the relations

$$f(-h) = f'(-h) = f(h) = f'(h) = 0. \qquad (5.4.6)$$

Four linearly independent solutions f_i of (5.4.5) must be related to each other by the homogeneous conditions (5.4.6). Equation $F(\omega, k, \mathrm{Re}) = 0$ follows from the condition for the existence of nonzero solutions of system (5.4.5) and (5.4.6). Resolving this equation with respect to k, we obtain $k = k(\omega, \mathrm{Re})$. Separating the real and imaginary parts $k = k_r + ik_i$, we obtain the equation of the neutral curve:

$$k_i(\omega, \mathrm{Re}) = 0, \qquad (5.4.7)$$

which determines the boundary of the stability region. If $k_i < 0$, then the factor $\exp(|k_i|x)$ leads to the growth of disturbances (5.4.4) as x increases. We present in Fig. 5.10 the neutral curve obtained by Lin. The dashed region is the instability region, where $\mathrm{Re} = U_{max}h/\nu$ and $U_{max} = U(0)$ is the maximum velocity. The wave vector $k_* \approx h$ corresponds to the minimal critical number $\mathrm{Re}^* = 5772$. At $\mathrm{Re} \to \infty$, both branches of the neutral curve are described by the equations

$$\omega \approx \frac{U(0)}{2h}\mathrm{Re}^{-3/11}, \quad \omega \approx \frac{U(0)}{2h}\mathrm{Re}^{-3/7}; \qquad (5.4.8)$$

the ω and k are related to each other on both curves (5.4.8) by relation $\omega \approx 4\frac{U}{h}(kh)^3$. It follows from Fig. 5.10 that, for any $\mathrm{Re} > \mathrm{Re}^*$, the disturbances exist that amplify in the flow. This is valid, in particular, for large numbers Re also. Thus, a small viscosity has a destabilizing effect. It follows from here that the term $\nu\Delta\vec{v}$ in the Navier–Stokes equations (5.1.21) cannot be neglected for a correct flow description even at small ν. This is related to the fact that the small parameter ν affects the higher (second) derivative and can lead to a qualitative change of the solution.

The above-considered linear theory describes only the initial stage of the small disturbance growth at the neutral curve intersection. After the disturbance $\vec{v}\,', P'$ becomes sufficiently large, it is necessary to take into account the nonlinear terms in the Navier–Stokes equations (5.1.21). At the expense of nonlinearity, the disturbance amplitude is stabilized and achieves certain value $\vec{v}_1\,', P_1'$. The subsequent growth of the Re number leads to the fact that this solution also ceases to be stable and a new qualitative solution modification (the bifurcation) occurs. As a result, in the flow the periodic solutions with new frequencies arise. After several bifurcations, the flow becomes complex and intricate; therefore, it is called turbulent. The character of the solution modification or, as it is said, the scenario of the turbulence development should be determined from the solution of a complete system of the Navier–Stokes equations. At present, such a complete solution is unfortunately impossible. The most promising direction is here a substitution of the Navier–Stokes equations for a system of ODEs in time $\dot{a}_i = F_i(a_k)$ for the coefficients in the solution expansion in the eigenfunctions $\vec{\varphi} = \sum_{i=1}^{N} a_i(t)\varphi_i(x_k)$. After that, one can use the classical apparatus of the theory of disturbances for the ODEs, which has its origin in the works of Lyapunov. Several scenarios of the turbulence development were predicted in this way[10–13]. In addition, it was shown that, in the phase space of a_i, there after several bifurcations a strange attractor arises (an attracting manifold of unstable trajectories is called the strange attractor). The mathematical aspects of these questions are very complex, and we refer the interested reader to the relevant literature[10–14].

5.4.3 Turbulent Fluid Flow

The first attempt at constructing the equations for turbulent flows belongs to Reynolds. He proposed to write the velocity (and the pressure) as a sum of the mean velocity $\bar{v}_i$ and the fluctuation velocity v_i': $v_i = \bar{v}_i + v_i'$, $P = \bar{P} + P'$. The mean values of the velocity and pressure were determined by him as the mean in time values:

$$\bar{v}_i = \frac{1}{T}\int_{t-T/2}^{t+T/2} v_i\,dt', \quad \bar{P} = \frac{1}{T}\int_{t-T/2}^{t+T/2} P\,dt', \tag{5.4.9}$$

where T is the averaging period. We immediately note that the averaging in time is not uniquely possible because one can perform the averaging over a volume or over an ensemble of realizations (as this is being done in the statistical physics). It is clear that the result should not depend on the averaging technique. The averaging period T is chosen to be such that it is substantially larger than the fluctuations time, but it should be less than a characteristic time for the variation of the mean flow parameters.

Differentiating the first equation in (5.4.9) with respect to time, we obtain:

$$\frac{\partial \bar{v}_i}{\partial t} = \frac{1}{T}(v_i(x^j, t+T/2) - v_i(x^j, t-T/2)) = \frac{1}{T}\int_{t-T/2}^{t+T/2} \frac{\partial v_i}{\partial t'}\, dt',$$

from where it follows that

$$\frac{\partial \bar{v}_i}{\partial t} = \overline{\frac{\partial v_i}{\partial t}}. \tag{5.4.10}$$

Since the t and x_i are the independent variables (the Eulerian description is considered), then

$$\frac{\partial \bar{v}_i}{\partial x^j} = \overline{\frac{\partial v_i}{\partial x^j}}, \qquad \frac{\partial \bar{P}}{\partial x^j} = \overline{\frac{\partial P}{\partial x^j}}. \tag{5.4.11}$$

Let us apply the averaging procedure (5.4.9) to the Navier–Stokes equations (5.1.21) in the Cartesian coordinates in the form

$$\frac{\partial v_j}{\partial x^j} = 0, \quad \rho\Big(\frac{\partial v_i}{\partial t} + v_j\frac{\partial v_i}{\partial x^j}\Big) = -\frac{\partial P}{\partial x^i} + \frac{\partial \tau_{ij}}{\partial x^j},$$
$$\tau_{ij} = 2\mu\dot{\varepsilon}_{ij}, \quad \dot{\varepsilon}_{ij} = \frac{1}{2}\Big(\frac{\partial v_i}{\partial x^j} + \frac{\partial v_j}{\partial x^i}\Big). \tag{5.4.12}$$

Using the first equation in (5.4.11), we obtain the continuity equation for the mean velocity

$$\frac{\partial \bar{v}_j}{\partial x^j} = 0. \tag{5.4.13}$$

Before proceeding to the averaging of the second equation in (5.4.12), we rewrite it with the aid of the continuity equation as follows:

$$\rho\Big(\frac{\partial v_i}{\partial t} + \frac{\partial}{\partial x^j}(\rho v_i v_j)\Big) = -\frac{\partial P}{\partial x^i} + \frac{\partial \tau_{ij}}{\partial x^j}.$$

Averaging this equation with regard for the relation $\overline{v_i v_j} = \bar{v}_i\bar{v}_j + \overline{v'_i v'_j}$ and the formulas (5.4.10), (5.4.11), we obtain the *Reynolds equations*:

$$\frac{\partial \rho\bar{v}_i}{\partial t} + \frac{\partial}{\partial x^j}(\rho\bar{v}_i\bar{v}_j) = -\frac{\partial \bar{P}}{\partial x^i} + \frac{\partial}{\partial x^j}(\bar{\tau}_{ij} - \rho\overline{v'_i v'_j}), \tag{5.4.14}$$

where

$$\bar{\tau}_{ij} = \mu\Big(\frac{\partial \bar{v}_i}{\partial x^j} + \frac{\partial \bar{v}_j}{\partial x^i}\Big). \tag{5.4.15}$$

The symmetric tensor

$$\pi_{ij} = -\rho\overline{v'_i v'_j} = -\frac{1}{T}\int_{t-T/2}^{t+T/2} \rho v'_i v'_j\, dt \tag{5.4.16}$$

is called the *tensor of the Reynolds stresses.* It is easy to see that the system of equations (5.4.13)–(5.4.16) is not closed since the tensor of the Reynolds stresses (5.4.16) is not determined. Therefore, it is necessary to invoke the additional hypotheses, which enable one to find the form of this tensor[6,10,15,16].

Let us present below some most widespread semiempirical hypotheses. Using an analogy between the molecular and fluctuating fluid motion in turbulent flow, Boussinesq has proposed for π_{ij} the formula

$$\pi_{ij} = 2A\overline{\varepsilon_{ij}}, \quad \overline{\varepsilon_{ij}} = \frac{1}{2}\left(\frac{\partial \bar{v}_i}{\partial x^j} + \frac{\partial \bar{v}_j}{\partial x^i}\right). \tag{5.4.17}$$

In this formula, A is the turbulent viscosity coefficient for which one can write by analogy with molecular viscosity:

$$A = \rho\overline{v'l'}. \tag{5.4.18}$$

Prandtl proposed to consider the quantity l', an analog of the mean free path for the molecules, and termed l' the mixing length. According to Prandtl, one can write for a planar flow along the x-axis, for example, between the plates or in a boundary layer in which $\bar{v}_x = \bar{v}_x(y)$, that $v'_x \sim l'\frac{\partial \bar{v}_x}{\partial y}$. Substituting this formula in expression (5.4.18), we obtain:

$$A = \rho l^2\left|\frac{\partial \bar{v}_x}{\partial y}\right|. \tag{5.4.19}$$

The quantity l is also called the mixing length and is determined as a rule from experiment. Note that also more complex formulas exist for the turbulent viscosity, which take into account the velocity profile curvature. Let the general dependence of the mixing length l on the first and second derivatives of the velocity have the form $l = l(\frac{\partial \bar{v}_x}{\partial y}, \frac{\partial^2 \bar{v}_x}{\partial y^2})$. Assuming that both parameters have the independent dimensions, we find from the Pi theorem (3.1.5), (3.1.6) that $l\frac{\partial^2 \bar{v}_x}{\partial y^2}/\frac{\partial \bar{v}_x}{\partial y} = \kappa$, where κ is a constant. Expressing from here l and substituting in formula (5.4.19) we obtain:

$$l = -\kappa\left(\frac{\partial \bar{v}_x}{\partial y}\right)/\left(\frac{\partial^2 \bar{v}_x}{\partial y^2}\right), \quad A = \kappa^2\rho\left(\frac{\partial \bar{v}_x}{\partial y}\right)^3/\left(\frac{\partial^2 \bar{v}_x}{\partial y^2}\right)^2. \tag{5.4.20}$$

The obtained expressions are called the von Kármán formulas. The turbulent viscosity A usually exceeds by a factor of several dozens the molecular viscosity μ; therefore, one can neglect the terms $\mu\bar{\varepsilon}_{ij}$ in the Reynolds equations. It should be noted that the above-considered hypotheses enable one to qualitatively describe the near-wall flows and freestreams in the far field from the bodies. Since there are no reference dimensions in the problem in these cases, the mixing length will be proportional to the corresponding coordinate: $l \sim Cx_i$.

Similarly to the Reynolds equations, one can obtain an equation for the heat propagation in a turbulent motion. Neglecting the viscous dissipation of energy, we can rewrite the third equation of system (5.1.1) as

$$C_V\rho\frac{\partial T}{\partial t}+\frac{\partial}{\partial x^j}(C_V\rho v_j T)=\kappa\frac{\partial^2 T}{\partial x^j\partial x^j}, \tag{5.4.21}$$

where κ is the heat conduction coefficient (which is different from the von Kármán constant). Applying the averaging procedure to (5.4.21), we obtain with regard for formulas $v_j=\bar{v}_j+v'_j$, $T=\bar{T}+T'$, and (5.4.10) and (5.4.11):

$$C_V\rho\Big(\frac{\partial\bar{T}}{\partial t}+\bar{v}_j\frac{\partial\bar{T}}{\partial x^j}\Big)=\frac{\partial}{\partial x^j}(Q_j-\bar{q}_j), \tag{5.4.22}$$

where

$$\bar{q}_j=-\kappa\frac{\partial\bar{T}}{\partial x^j},\quad Q_j=-C_V\rho\overline{v'_jT'}. \tag{5.4.23}$$

The quantity Q_j is called the turbulent heat flux and is proportional to the mean value of the product of the velocity pulsations v'_j and the temperature T'

$$\overline{v'_jT'}=\frac{1}{\Delta t}\int_{t-\Delta t/2}^{t+\Delta t/2}v'_jT'\,dt',$$

where we have denoted the interval of averaging by Δt to avoid confusion with temperature. Replacing C_VT in equations (5.4.22) and (5.4.23) with the concentration of the admixture of some substance, we obtain the equation for the turbulent diffusion of this substance:

$$\rho\Big(\frac{\partial\bar{c}}{\partial t}+\bar{v}_j\frac{\partial\bar{c}}{\partial x^j}\Big)=\frac{\partial}{\partial x^j}\big(-\bar{m}_j+M_j\big),$$
$$\bar{m}_j=-D\frac{\partial\bar{c}}{\partial x^j},\quad M_j=-\rho\overline{c'v'_j}, \tag{5.4.24}$$

where D is the molecular diffusion coefficient, $\bar{m}_j$ is the molecular diffusion vector, and M_j is the turbulent diffusion vector. By analogy with the Boussinesq formula (5.4.17), one can assume that

$$Q_j=-C_VA_Q\frac{\partial\bar{T}}{\partial x^j},\qquad M_j=-A_M\frac{\partial\bar{c}}{\partial x^j}. \tag{5.4.25}$$

Since the physical nature of the turbulent transport of the momentum, energy, and admixture is the same (at the expense of the velocity fluctuations), we can find by analogy with formula (5.4.19) for a planar flow with $\bar{v}_x=\bar{v}_x(y)$ that[6]

$$A_Q=\rho l_Q^2\left|\frac{\partial\bar{v}_x}{\partial y}\right|,\quad A_M=\rho l_M^2\left|\frac{\partial\bar{v}_x}{\partial y}\right|, \tag{5.4.26}$$

where the l_Q and l_M are the mixing lengths for the processes of turbulent transport of heat and admixture.

The coefficients of turbulent viscosity A, energy transport A_Q, and diffusion A_M are determined by the nature of the velocity fluctuations taking place in the turbulent flow. As already said above, there is no consistent theory here. There are a number of important qualitative results, however, explaining some properties of the velocity fluctuations v' (since in the results presented below the direction of the velocity vector is insignificant, we will omit the vector symbol over the velocity notation and we will mean the velocity module)[10,17,18].

It follows from the experiments that there are in turbulent flow the fluctuations with different spatial dimensions λ. Denote the magnitude order of the velocity fluctuation on the scale λ by v'_λ, and introduce the corresponding number $\mathrm{Re}_\lambda = v'_\lambda \lambda/\nu$. The large-scale fluctuations have the size $\lambda \sim L$ and the fluctuations velocity $v'_L \sim \Delta\bar{v}$, where L and $\Delta\bar{v}$ are the reference scales of the body and the mean velocity alteration, respectively. The least scale is determined from the condition $\mathrm{Re}_\lambda \sim 1$ and has the order $\lambda_0 \sim \nu/v'_0$. For the smaller scales $\lambda \ll \lambda_0$, the inequality $\mathrm{Re}_\lambda \ll 1$ is valid and the pulsations decay at the expense of viscosity occurs. Since $\mathrm{Re}_L = \Delta\bar{v}L/\nu \gg 1$ in a turbulent flow, then $\lambda_0 \ll L$, and there is in the interval from λ_0 to L a wide spectrum of pulsations scales $\lambda_0 \ll \lambda \ll L$, the *inertial interval.* According to Richardson[1], in the flow past a body the generation of turbulent flow energy at the expense of large-scale pulsations occurs. After a disintegration of these pulsations into smaller ones, this energy passes to the pulsations with smaller scales. Since $\mathrm{Re}_\lambda \gg 1$ therein, there is practically no energy dissipation. Thus, a constant specific energy flux $\dot{\mathcal{E}}$ from large-scale pulsations to small-scale pulsations takes place. This flux dissipates into the heat at the small-scale pulsations of the order λ_0. Using the constancy of a specific energy flux $\dot{\mathcal{E}}$, one can find the energy distribution $\mathcal{E}$ and the velocity v'_λ in the interval $\lambda_0 \ll \lambda \ll L$ for the *homogeneous isotropic turbulence.* In this case, the turbulence properties on the scales $\lambda \ll L$ do not depend on the mean velocity $\bar{v}$ and the pulsations velocity v'_λ will depend on $\dot{\mathcal{E}}$ and on λ; that is, $v_\lambda = f(\dot{\mathcal{E}}, \lambda)$. Since the dimension of the specific heat flux $[\dot{\mathcal{E}}] = [E]/(kg \cdot s) = J/(kg \cdot s) = m^2/s^3$, then we obtain with the aid of the Pi theorem the *Kolmogorov–Obukhov law*[10,17]

$$v'_\lambda = (\dot{\mathcal{E}}\lambda)^{1/3} \tag{5.4.27}$$

Formula (5.4.27) is valid within the *inertial interval* $\lambda_0 \ll \lambda \ll L$. Taking into account the fact that on the scale L the velocity pulsation $v'_\lambda \sim \Delta\bar{v}$, we can rewrite (5.4.27) in the form

$$\frac{v'_\lambda}{\Delta\bar{v}} \sim \left(\frac{\lambda}{L}\right)^{\frac{1}{3}}. \tag{5.4.28}$$

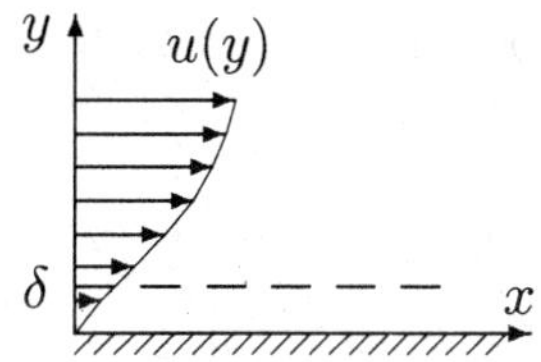

Figure 5.11: The velocity profile of turbulent flow near the wall.

Thus, the similarity law (5.4.28) is valid for the homogeneous isotropic turbulence. Applying (5.4.28) to the least scale $v_0' \sim \Delta\bar{v}(\lambda_0/L)^{1/3}$, we can find with regard for $\text{Re}_0 = \lambda_0 v_0'/\nu \sim 1$ the order of the least scale λ_0 and the velocity v_0':

$$\lambda_0 \sim L/\text{Re}_L^{3/4}, \quad v_0' \sim \Delta\bar{v}/\text{Re}_L^{1/4},$$

where $\text{Re}_L = L\,\Delta\bar{v}/\nu \gg 1$. Since the wavenumbers of pulsations $k = 2\pi/\lambda$ correspond to the pulsations scale λ, one can determine the specific kinetic energy of the pulsations with a given value of k and in the interval dk by the formula $\mathcal{E}(k)\,dk$. It follows from this formula that $\mathcal{E}(k)$ has the dimension $[\mathcal{E}(k)] = [E]/[k] = J \cdot m/kg = m^3/s^2$. In the inertial interval, $\mathcal{E}(k)$ is a function of $\dot{\mathcal{E}}$ and k only; that is, $\mathcal{E}(k) = f(\dot{\mathcal{E}}, k)$. Using the Pi theorem, we obtain with regard for relations $[\dot{\mathcal{E}}] = m^2/s^3$, $[k] = 1/m$:

$$\mathcal{E}(k) \sim \dot{\mathcal{E}}^{2/3} k^{-5/3}. \tag{5.4.29}$$

It follows from this formula that the main portion of energy is located in the large-scale pulsations with the large $\lambda \sim L$ and $k \sim 1/L$. These scales play the main role in the transport of energy and momentum; therefore, the turbulent viscosity $A \sim \rho\Delta\bar{v}L$.

Problem 5.3. Find the dependence of the mean velocity $\bar{v}_i(x^j)$ on coordinate x^j in the steady ($\frac{\partial \bar{v}_i}{\partial t} = 0$) plane-parallel turbulent flow along an unbounded flat surface (see Fig. 5.11).

Solution: Choose the x-axis along the flow and the y-axis along a normal to the wall. Only one velocity component will be different from zero: $\bar{v}_x = u(y)$, $\bar{v}_y = \bar{v}_z = 0$. The pressure gradient in the flow is equal to zero: $\frac{\partial P}{\partial x^i} = 0$. Under these assumptions and with regard for relations $\frac{\partial \bar{v}_i}{\partial t} = 0$, the Reynolds equations (5.4.14) reduce to a single ODE

$$\frac{d}{dy}(\tau_{xy} + \pi_{xy}) = 0,$$

whose integral is

$$\bar{\tau}_{xy} + \pi_{xy} = c_0. \tag{5.4.30}$$

Since the transverse velocity fluctuations on the wall are equal to zero, $v'_y|_{y=0} = 0$, we obtain from equation (5.4.16):

$$\pi_{xy}\Big|_{y=0} = \rho\overline{v'_x v'_y}\Big|_{y=0} = 0,$$

so that $c_0 = \tau_{xy}(0) = \tau_w$, the skin-friction stress on the wall. As a result, equation (5.4.30) can be rewritten as

$$\bar{\tau}_{xy} + \pi_{xy} = \tau_w. \tag{5.4.31}$$

The stresses conditioned by the molecular friction $\bar{\tau}_{xy}$ and the Reynolds stresses π_{xy} are found by formulas (5.4.15), (5.4.17), and (5.4.18) and in our case are equal to

$$\bar{\tau}_{xy} = \mu\frac{du}{dy}, \quad \pi_{xy} = A\frac{du}{dy}, \quad A = \rho l^2\left|\frac{du}{dy}\right|. \tag{5.4.32}$$

Since there is in the problem no reference length, it follows from the analysis of dimensions that $l = \kappa y$ (the Prandtl formula), where κ is a nondimensional constant. Substituting the formulas $l = \kappa y$, (5.4.32) in (5.4.31), we find the equation for u:

$$\mu\frac{du}{dy} + \rho\kappa^2 y^2\left(\frac{du}{dy}\right)^2 = \tau_w. \tag{5.4.33}$$

The solution of equation (5.4.33) should satisfy the condition $u|_{y=0} = 0$ on the wall. It is impossible to integrate equation (5.4.33) in a general case. Therefore, we find its solutions far from the wall and close to the wall, and then we match them. One can neglect the molecular friction in the region far from the wall since $\bar{\tau}_{xy} \ll \pi_{xy}$ and equation (5.4.33) will have the form

$$\rho\kappa^2 y^2\left(\frac{du}{dy}\right)^2 = \tau_w;$$

the solution of which is found by a simple integration:

$$u = \frac{1}{\kappa}\sqrt{\frac{\tau_w}{\rho}}\ln y + c. \tag{5.4.34}$$

On the contrary, near the wall the molecular friction plays the predominant role, $\bar{\tau}_{xy} \gg \pi_{xy}$, and we obtain from (5.4.33) the equation:

$$\mu\frac{du}{dy} = \tau_w$$

the solution of which is with regard for the boundary condition $u|_{y=0} = 0$

$$u = \frac{\tau_w}{\rho\nu}y, \tag{5.4.35}$$

where we have set $\mu = \rho\nu$. The near-wall region, where formula (5.4.35) is valid, is called the *viscous sublayer*. We can find the viscous sublayer thickness δ from the condition $\mathrm{Re} = \frac{v_*\delta}{\nu} \approx 1$, where $v_* = u(\delta) = \frac{\tau_w}{\rho\nu}\delta$. Substituting the expression for v_* into the condition $\mathrm{Re} \approx 1$, we obtain:

$$\delta \approx \frac{\nu}{v_*}, \quad v_* \approx \sqrt{\frac{\tau_w}{\rho}}. \tag{5.4.36}$$

Note that such a subdivision of the flow into a viscous sublayer and the turbulent flow (the turbulent kernel of the flow) is sufficiently rough. There is indeed a transitional region in which the molecular viscosity effect gradually decreases as the distance from the wall increases. In many problems, however, such a rough approximation is sufficient.

Assuming $u = v_*$, $y = \delta$ in formula (5.4.34), we find the constant $c \approx v_* - \frac{v_*}{\kappa}\ln\delta$. Substituting this expression in (5.4.34), we find the *Prandtl formula*:

$$u \approx v_* + \frac{v_*}{\kappa}\ln\frac{yv_*}{\nu}. \tag{5.4.37}$$

At large distances, one can neglect v_* and we obtain:

$$u = \frac{v_*}{\kappa}\ln\frac{yv_*}{\nu}. \tag{5.4.38}$$

The velocity distribution (5.4.37), (5.4.38) is called the *logarithmic velocity profile*.

References

1. **Kochin, N.E., Kibel, I.A., and Rose, N.V.,** *Theoretical Hydromechanics* (in Russian), Vol. II, 4th Edition, Fizmatgiz, Moscow, 1963.
2. **Birkhoff, G.,** *Hydrodynamics. A Study in Logic, Fact and Similitude*, Second Edition, Princeton University Press, Princeton, NJ, 1960.
3. **Prandtl, L.,** *The Mechanics of Viscous Fluids. Vol. III, Aerodynamic Theory*, Springer, Berlin, 1935.
4. **Prandtl, L. and Tietjens, O.G.,** *Applied Hydro- and Aeromechanics* (Translated from the German edition, Springer, Berlin, 1931), McGraw-Hill, New York, 1934.

5. **Schlichting, H.**, *Boundary Layer Theory*, Seventh Edition, McGraw-Hill, New York, 1979.
6. **Loitsyanskii, L.G.**, *Mechanics of Liquids and Gases*, Pergamon Press, Oxford, 1966.
7. **Mises, R.**, Bemerkungen zur Hydrodynamik, *Zeitschrift für angewandte Mathematik und Mechanik*, 7:425, 1927.
8. **Strampp, W., Ganzha, V., and Vorozhtsov, E.**, *Höhere Mathematik mit Mathematica. Band 3: Differentialgleichungen und Numerik*, Verlag Vieweg, Braunschweig/Wiesbaden, 1997.
9. **Lin, C.C.**, *The Theory of Hydrodynamic Stability*, University Press, Cambridge, 1955.
10. **Landau, L.D. and Lifschitz, E.M.**, *Hydrodynamics* (in Russian), Nauka, Moscow, 1986.
11. **Swinney, H.L. and Gollub, J.P.** (Editors), *Hydrodynamic Instabilities and the Transition to Turbulence*, Vol. 45, Springer-Verlag, Berlin, 1981.
12. **Ruelle, D. and Takens, F.**, On the nature of turbulence, *Communications on Mathematical Physics*, 20:167, 1971; 23:343, 1971.
13. **Feigenbaum, M.J.**, Universality in the behavior of nonlinear systems, *Uspekhi Fizicheskikh Nauk* (in Russian), 141:343, 1983 [translated from the Los Alamos Science (Summer 1980)], p. 4.
14. **Lorenz, E.N.**, Deterministic nonperiodic flow, *J. Atmos. Sci.*, 20:130, 1963.
15. **Libby, P.A. and Williams, F.** (Editors), *Turbulent Reacting Flows*, Topics in Applied Physics, vol. 44, Springer-Verlag, Berlin, 1980.
16. **Bradshaw, P.** (Editor), *Turbulence*, Second Edition, Topics in Applied Physics, Vol. 12, Springer-Verlag, Berlin, 1978.
17. **Monin, A.S. and Yaglom, A.M.**, *Statistical Fluid Mechanics*, Vols. I and II, MIT Press, Cambridge, MA, 1971.
18. **Hinze, J.O.**, *Turbulence, an Introduction to Its Mechanism and Theory*, McGraw-Hill, New York, 1959.

6

Gas Dynamics

This chapter deals with one-dimensional stationary and nonstationary as well as planar and three-dimensional stationary gas flows. The theories for the Laval nozzle and normal and oblique shock waves are presented. The Becker's solution for the shock wave structure is given. The solution of a simple wave type is obtained with the aid of the method of characteristics. The onset of a gradient catastrophe as well as discontinuous solutions in continuous flows is shown. These effects are caused by the nonlinear terms in the gas dynamics equations. The Chaplygin's method for the transformation of the solution of stationary gas dynamics equations to the plane of the hodograph variables is presented in detail. A new method is presented for the aerodynamic design of three-dimensional solutions with the aid of the solutions of a lesser dimension.

6.1 One-Dimensional Stationary Gas Flows

6.1.1 Governing Equations for Quasi-One-Dimensional Gas Flow

Consider a problem of the gas flow in a duct whose crosssection $F(x)$ changes slowly along the duct axis Ox (see Fig. 6.1). Assuming the gas to be barotropic and neglecting the body forces, let us write the system of the Euler equations in Cartesian coordinates (2.1.150) as follows:

$$\begin{aligned}\frac{\partial\rho}{\partial t} &+ v_x\frac{\partial\rho}{\partial x}+v_y\frac{\partial\rho}{\partial y}+v_z\frac{\partial\rho}{\partial z}+\rho\Big(\frac{\partial v_x}{\partial x}+\frac{\partial v_y}{\partial y}+\frac{\partial v_z}{\partial z}\Big)=0,\\ \frac{dv_x}{dt} &= \frac{\partial v_x}{\partial t}+v_x\frac{\partial v_x}{\partial x}+v_y\frac{\partial v_x}{\partial y}+v_z\frac{\partial v_x}{\partial z}=-\frac{1}{\rho}\frac{\partial P}{\partial x},\\ \frac{dv_y}{dt} &= -\frac{1}{\rho}\frac{\partial P}{\partial y},\quad \frac{dv_z}{dt}=-\frac{1}{\rho}\frac{\partial P}{\partial z},\quad \rho=\Phi(P).\end{aligned}\tag{6.1.1}$$

Since $F(x)$ changes slowly with x, it is natural to assume that $v_y \sim v_z \ll v_x$, $\frac{dv_y}{dt} \sim \frac{dv_z}{dt} \ll \frac{dv_x}{dt}$; therefore, we obtain $\frac{\partial P}{\partial y} \approx 0$, $\frac{\partial P}{\partial z} \approx 0$. It follows

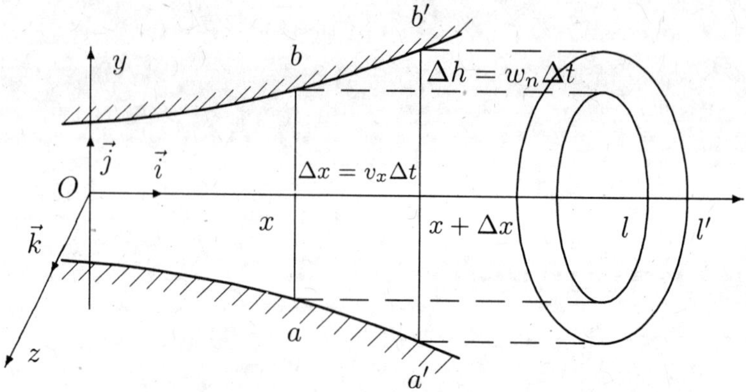

Figure 6.1: The duct with a variable crosssection $F(x)$.

from here that $P = P(t, x)$, $\rho = \rho(t, x)$. Assume that $v_x = v_x(t, x)$. Neglecting the terms $v_y \frac{\partial v_x}{\partial y}$ and $v_z \frac{\partial v_x}{\partial z}$ in the second equation (6.1.1) as compared with $v_x \frac{\partial v_x}{\partial x}$, we can rewrite (6.1.1) in a simplified form

$$\begin{aligned} \frac{\partial v_x}{\partial t} &+ v_x \frac{\partial v_x}{\partial x} = -\frac{1}{\rho}\frac{\partial P}{\partial x}, \\ \frac{\partial \rho}{\partial t} &+ v_x \frac{\partial \rho}{\partial x} + \rho\Big(\frac{\partial v_x}{\partial x} + \frac{\partial v_y}{\partial y} + \frac{\partial v_z}{\partial z}\Big) = 0, \\ \rho &= \Phi(P). \end{aligned} \tag{6.1.2}$$

Let us integrate the continuity equation over the cross-section area F of the tube[1]:

$$\iint_F \Big(\frac{\partial \rho}{\partial t} + v_x \frac{\partial \rho}{\partial x} + \rho \frac{\partial v_x}{\partial x}\Big)\, dS + \iint_F \rho\Big(\frac{\partial v_y}{\partial y} + \frac{\partial v_z}{\partial z}\Big)\, dS = 0,$$

and taking into account the fact that the relation $\Big(\frac{\partial \rho}{\partial t} + v_x \frac{\partial \rho}{\partial x} + \rho \frac{\partial v_x}{\partial x}\Big)$ does not depend on y and z, we find:

$$\Big(\frac{\partial \rho}{\partial t} + v_x \frac{\partial \rho}{\partial x} + \rho \frac{\partial v_x}{\partial x}\Big) F + \iint_F \rho\Big(\frac{\partial v_y}{\partial y} + \frac{\partial v_z}{\partial z}\Big)\, dS = 0.$$

Since $\rho = \rho(x, t)$, by introducing the vector $\vec{w} = v_y \cdot \vec{j} + v_z \cdot \vec{k}$ with the unit vectors $\vec{j}$ and $\vec{k}$, we obtain:

$$\iint_F \rho\Big(\frac{\partial v_y}{\partial y} + \frac{\partial v_z}{\partial z}\Big)\, dS = \rho \iint_F \operatorname{div} \vec{w}\, dS = \rho \oint_l w_n\, dl,$$

where l is the boundary of the area F. The particles displacement during a time Δt may be presented as a displacement in the transverse plane

$\vec{w}\,\Delta t$ and a displacement along the x-axis by a distance $\Delta x = v_x \Delta t$. Then the particles will go over during the time Δt from the contour l to the contour l' along a normal by a distance $\Delta h = w_n \Delta t$ (see Fig. 6.1). The area change ΔF will be equal to the area of a ring:

$$\Delta F \approx \oint \Delta h \, dl = \Delta t \oint w_n \, dl,$$

from where we find:

$$\oint w_n dl = \lim_{\Delta t \to 0} \frac{\Delta F}{\Delta t} = \frac{dF}{dt}.$$

Using the obtained formula, we can rewrite the continuity equation as

$$F\left(\frac{\partial \rho}{\partial t} + v_x \frac{\partial \rho}{\partial x} + \rho \frac{\partial v_x}{\partial x}\right) + \rho \frac{dF}{dt} = 0.$$

Assuming that the tube does not deform, that is $\frac{dF}{dt} = v_x \frac{\partial F}{\partial x}$, we can write the above equation as

$$F\left(\frac{\partial \rho}{\partial t} + v_x \frac{\partial \rho}{\partial x} + \rho \frac{\partial v_x}{\partial x}\right) + \rho v_x \frac{\partial F}{\partial x} = 0$$

or

$$F \cdot \frac{\partial \rho}{\partial t} + \frac{\partial}{\partial x}(\rho v_x F) = 0. \tag{6.1.3}$$

In the stationary flow case, we have from equations (6.1.2) and (6.1.3):

$$\frac{d}{dx}(F \rho v_x) = 0, \quad v_x \frac{dv_x}{dx} + \frac{1}{\rho}\frac{dP}{dx} = 0, \quad \rho = \Phi(P)$$

The integration with respect to x yields:

$$F \rho v_x = Q, \quad \frac{v_x^2}{2} + \int \frac{dP}{\Phi(P)} = B, \quad \rho = \Phi(P), \tag{6.1.4}$$

where Q and B are the integration constants. The second equation in these relations is the Bernoulli integral. The neglection of the transverse acceleration assumed in the above derivation is equivalent to the neglection of term $v_y^2 + v_z^2$ in the expression for v^2 as compared to v_x^2.

6.1.2 Gas Motion in a Variable Section Duct: Elementary Theory of the Laval Nozzle

Since the motion is steady, the formulas (6.1.4) give the general solution of the problem of the gas flow in a variable section duct, where the constant Q represents the gas flow rate in a fixed duct section. These

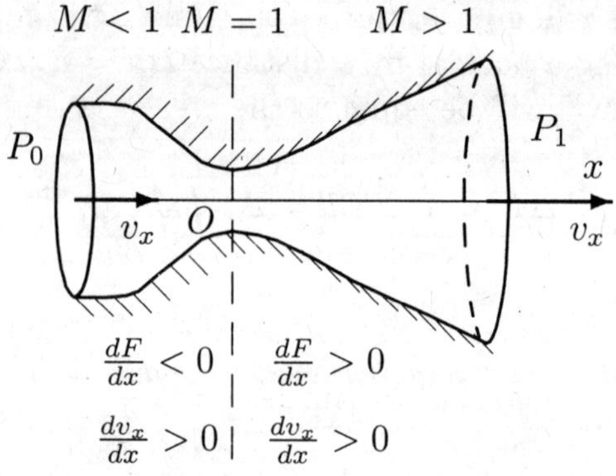

Figure 6.2: The Laval nozzle.

equations determine the flow completely if the values Q_0 and P_0 or ρ_0 are known in one duct section $x = x_0$.

For a subsequent analysis, it is convenient to present these formulas in another form. We at first take a logarithm, and then differentiate the first equality in (6.1.4). We then write the second equality in (6.1.4) in the differential form; that is,

$$\frac{dF}{F} + \frac{d\rho}{\rho} + \frac{dv_x}{v_x} = 0, \quad v_x\, dv_x + \frac{dP}{\rho} = 0. \tag{6.1.5}$$

It is well known that $\frac{dP}{d\rho} = c^2$; therefore, we can rewrite the second equation in (6.1.5) as

$$v_x dv_x + \frac{c^2 d\rho}{\rho} = 0 \text{ and } \frac{d\rho}{\rho} = -\frac{v_x}{c^2} \cdot dv_x.$$

Substituting this relation in the first equation in (6.1.5), we have:

$$\frac{dF}{F} + \frac{dv_x}{v_x} - \frac{v_x}{c^2} dv_x = 0$$

or

$$\frac{dv_x}{dx} = \frac{v_x}{(M^2 - 1)F} \frac{dF}{dx}, \tag{6.1.6}$$

where $M^2 = \frac{v_x^2}{c^2}$ and the boundary conditions are $x = x_0$, $v_x = v_x^0$, and $P = P_0$.The relation (6.1.6) enables us to draw a number of important conclusions. We will assume for definiteness that $v_x > 0$. The sign of $\frac{dv_x}{dx}$ in (6.1.6) depends on the flow type.

1) Let $M < 1$; that is, $v_x < c$. Then in the case in which the area F reduces, i.e., $\frac{dF}{dx} < 0$, then $\frac{dv_x}{dx} > 0$, that is the velocity increases. If

$\frac{dF}{dx} > 0$, i.e., the section area increases, then $\frac{dv_x}{dx} < 0$; that is, the velocity decreases.

2) Let $M > 1$, that is $v_x > c$. In this case, $\frac{dv_x}{dx} < 0$ if $\frac{dF}{dx} < 0$; that is, the velocity decreases as the duct area decreases. If $\frac{dF}{dx} > 0$ then $\frac{dv_x}{dx} > 0$; i.e., the duct area increase leads to the velocity increase.

3) At $M = 1$, it is necessary for the existence of a regular flow regime that $\frac{dF}{dx} = 0$; that is, $v_x = c$ in the minimal section. The satisfaction of this condition is related to the need for surmounting a singularity at point $M = 1$ of equation (6.1.6).

These considerations were put by Laval into the basis of the elementary nozzle theory, on the basis of which it is possible to revert a subsonic flow into a supersonic flow in a continuous way. The duct section should at first have a convergent part, in which the flow velocity increases up to the sound velocity in the minimal section, and then the duct should be divergent; in this duct part, the flow accelerates. In the minimal section $v_x = c$, that is $M = 1$ (see Fig. 6.2).

We now consider the one-dimensional stationary isentropic gas motion in the variable section duct. We will assume that the gas flow is adiabatic and that the gas is perfect. Under these conditions, we have with (2.1.97) in view that $P = A\rho^\gamma$, $\frac{dP}{d\rho} = c^2$, and $c^2 = \gamma \cdot \frac{P}{\rho} = \gamma R_0 T$. We find from the Bernoulli integral (6.1.4): $\frac{v^2}{2} + \frac{A\gamma}{\gamma-1}\rho^{\gamma-1} = B$ or

$$\frac{v^2}{2} + \frac{\gamma}{\gamma - 1} \cdot \frac{P}{\rho} = B.$$

We have assumed here $v = v_x$. It follows from the equation of state $P = (c_P - c_V)\rho T$ that $\frac{P}{\rho} \cdot \frac{1}{\gamma-1} = c_V T$. This enables us to write the Bernoulli integral as

$$\frac{v^2}{2} + c_P T = B.$$

Choosing the constant B from the condition that $T = T_0$ at $v = 0$, we have: $\frac{v^2}{2} + c_P T = c_P T_0$, where T_0 is the temperature in the adiabatically decelerated gas. Divide both sides of this equality by $c_P T$. As a result, we obtain:

$$\frac{T_0}{T} = 1 + \frac{v^2}{2c_P T}.$$

Observing that $\frac{v^2}{2c_P T} = \frac{\gamma-1}{2} \cdot \frac{v^2}{c^2}$, we can rewrite the above formula as

$$\frac{T_0}{T} = 1 + \frac{\gamma - 1}{2} M^2. \tag{6.1.7}$$

Recalling that the sound velocity is proportional to the square root of temperature, we obtain:

$$\frac{c_0}{c} = \left(1 + \frac{\gamma - 1}{2} M^2\right)^{1/2}. \tag{6.1.8}$$

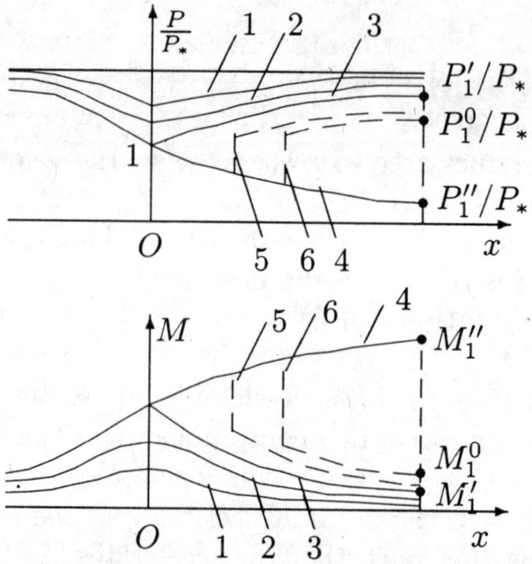

Figure 6.3: The distributions of the pressure $\frac{P}{P_*}$ and the Mach number M along the Laval nozzle (see Fig. 6.2).

Using the Poisson adiabat and the Clapeyron formula $P = \rho RT$, it is easy to get:

$$\frac{\rho_0}{\rho} = (1 + \frac{\gamma - 1}{2} M^2)^{\frac{1}{\gamma - 1}}, \quad \frac{P_0}{P} = (1 + \frac{\gamma - 1}{2} M^2)^{\frac{\gamma}{\gamma - 1}}. \tag{6.1.9}$$

Using the continuity equation for two different tube crosssections, we can write:

$$\frac{F}{F_1} = \frac{\rho_1 v_1}{\rho v} = \left(\frac{1 + \frac{\gamma-1}{2} M^2}{1 + \frac{\gamma-1}{2} M_1^2} \right)^{\frac{1}{\gamma-1}} \cdot \frac{M_1}{M} \cdot \frac{c_1}{c}$$

or after simple transformations

$$\frac{F}{F_1} = \frac{M_1}{M} \cdot \left(\frac{1 + \frac{\gamma-1}{2} M^2}{1 + \frac{\gamma-1}{2} M_1^2} \right)^{\frac{\gamma+1}{2(\gamma-1)}} \tag{6.1.10}$$

The relation (6.1.10) together with formulas (6.1.7) – (6.1.9) transformed to the form

$$\frac{P}{P_1} = \left(\frac{1 + \frac{\gamma-1}{2} M_1^2}{1 + \frac{\gamma-1}{2} M^2} \right)^{\frac{\gamma}{\gamma-1}}, \quad \frac{\rho}{\rho_1} = \left(\frac{1 + \frac{\gamma-1}{2} M_1^2}{1 + \frac{\gamma-1}{2} M^2} \right)^{\frac{1}{\gamma-1}}$$

$$\frac{T}{T_1} = \left(\frac{1 + \frac{\gamma-1}{2} M_1^2}{1 + \frac{\gamma-1}{2} M^2} \right), \quad \frac{v}{v_1} = \frac{M}{M_1} \left(\frac{1 + \frac{\gamma-1}{2} M_1^2}{1 + \frac{\gamma-1}{2} M^2} \right)^{\frac{1}{2}} \tag{6.1.11}$$

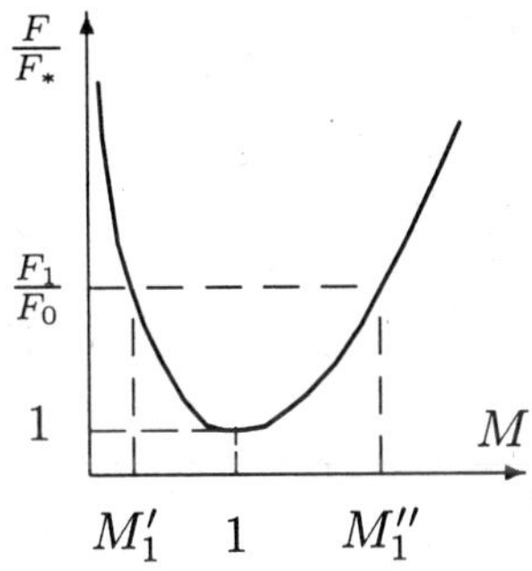

Figure 6.4: The curve $F/F_* = f(M)$ corresponding to equation (6.1.12).

yield a parametric solution of a problem of the quasi-one-dimensional gas motion in a variable section duct, where the Mach number plays the role of a parameter[2]. Specifying the function $F(x)$, we can determine $M(x)$ by (6.1.10) and then the desired functions $P(x)$, $\rho(x)$, $T(x)$ and $v(x)$ in accordance with (6.1.11). At present, the functions (6.1.11) have been tabulated and there are corresponding tables for them, which simplify greatly the specific computations.

Let us study the gas flow in the Laval nozzle by using the obtained formulas (6.1.10) and (6.1.11) (see Fig. 6.2). Assume that the gas escapes from a large reservoir with constant parameters P_0, ρ_0, T_0 through the Laval nozzle shown in Fig. 6.2. Let the pressure at the nozzle exit be equal to P_1. If the pressure P_1 at the nozzle exit is equal to the pressure P_0 at the inlet, then there will be no gas outflow and the pressure remains constant everywhere. If the pressure jump $P_0 - P_1$ is small, then the gas flow in the nozzle will be subsonic. The pressure distribution and the Mach numbers along the x coordinate are decribed by formulas (6.1.10) and (6.1.11) and are shown in Fig. 6.3 by curves 1 and 2. Note that the pressure P_1 in Fig. 6.3 is related to the critical pressure P_*, the pressure at a point where $M_1 = 1$. Assuming that $M_1 = 1, P = P_*$ in (6.1.11), we find:

$$\frac{P}{P_*} = \left(\frac{2}{\gamma+1}\left(1 + \frac{\gamma-1}{2}M^2\right)\right)^{-\frac{\gamma}{\gamma-1}},$$

where $M(x)$ is determined from formula (6.1.10), in which $F(x)$ is a known function. A further reduction of P_1 will lead to the gas flow acceleration. Finally, at a certain pressure jump $P_0 - P_1$, the gas velocity in the minimal section $x = 0$ will become equal to the sound velocity $M = 1$. In this case, the minimal section becomes critical and, assuming $F_1 = F_*, M_1 = 1$ in (6.1.10), we obtain:

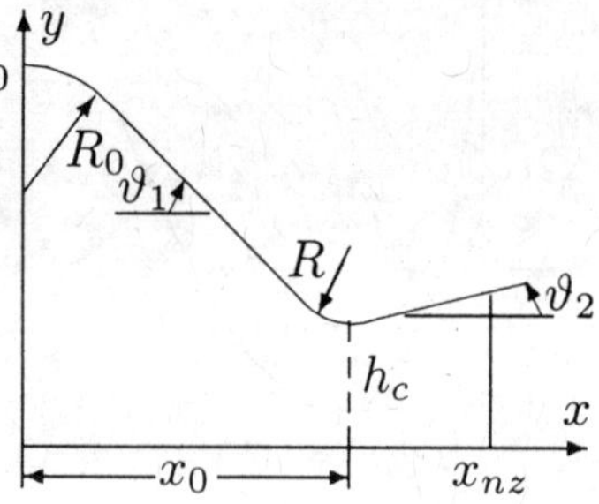

Figure 6.5: Axisymmetric nozzle ($R/h_c = 0.64$).

$$\frac{F}{F_*} = \frac{1}{M}\left(\frac{2}{\gamma+1}\right)^{\frac{\gamma+1}{2(\gamma-1)}}\left(1+\frac{\gamma-1}{2}M^2\right)^{\frac{\gamma+1}{2(\gamma-1)}} \tag{6.1.12}$$

A qualitative dependence of F/F_* on the Mach number, which corresponds to formula (6.1.12), is shown in Fig. 6.4. It follows from (6.1.12) at $M \to 0$ that $F/F_* \sim 1/M$ and at $M \to \infty$ that $F/F_* \sim M^{2/(\gamma-1)}$. It can be seen from Fig. 6.4 that, at a given nozzle cross-section at the exit F_1/F_0, the existence of two flow regimes is possible: the subsonic flow with the Mach number $M_1' < 1$ and the supersonic flow with the Mach number $M_1'' > 1$. The dependences $P(x)/P_*$ and $M(x)$ corresponding to these two flow regimes are shown in Fig. 6.3 by curves 3 and 4. It is seen that, in order to go over in the Laval nozzle from subsonic flow to supersonic flow, it is necessary to reduce the pressure at the exit from P_1' to P_1'' by a finite amount equal to $P_1' - P_1''$. In this connection, a question arises on the flow regime if the exit pressure P_1^0 lies in the interval $P_1'' < P_1^0 < P_1'$. It is impossible to give an answer to this question within the framework of the above-considered elementary theory. In this case, the flow will not be stationary and continuous, and it will contain the discontinuities of the shock wave type, the flow behind which will be subsonic (see curves 5 and 6 in Fig. 6.3). Such flow regimes are called the off-design regimes. Since the design supersonic flow is unique only one pressure value P_1'' corresponds to it. At $P_1 < P_1''$, the off-design regimes will take place also, at which the gas will continue to expand adiabatically after escaping from the nozzle exit down to the pressure P_1, but this flow will be intrinsically multidimensional.

We have implemented with *Mathematica 3.0* the solution process described by equations (6.1.10)–(6.1.11) (see our *Mathematica* Notebook `prog6-1.nb`). As the variable section duct, we have used, for the computational examples presented in Fig. 6.6, the axisymmetric Laval nozzle from[3] (see also[4]). We show in Fig. 6.5 the upper half of this nozzle, together with the explanation of the parameters $R_0, \theta_1, \theta_2, R, h_c, x_0$, gov-

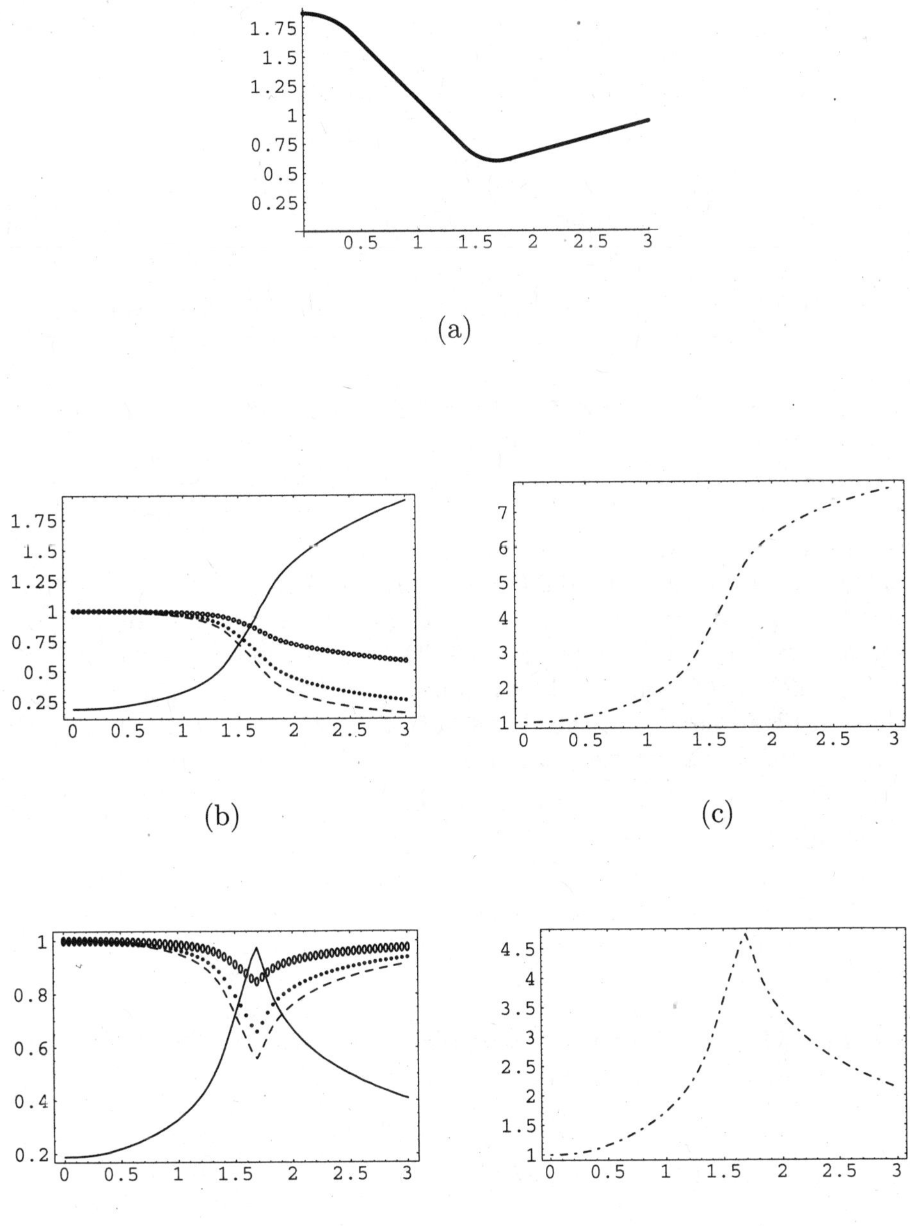

Figure 6.6: The quasi-one-dimensional nozzle flows: (a) the Laval nozzle geometry; (b) the profiles of the Mach number $M(x)$ (solid line), the pressure ratio $P(x)/P_1$ (dashed line), the density ratio $\rho(x)/\rho_1$ (dotted line), the temperature ratio $T(x)/T_1$ (white circles), supersonic flow regime; (c) the profile of the gas velocity ratio $v(x)/v_1$, supersonic flow regime; (d) the profiles of $M, P/P_1, \rho/\rho_1, T/T_1$, subsonic flow regime; and (e) the profile of v/v_1, subsonic flow regime.

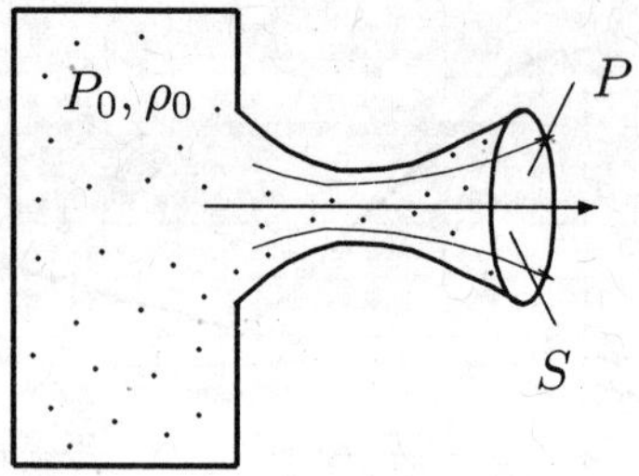

Figure 6.7: The gas outflow from a reservoir through the Laval nozzle.

erning the nozzle wall geometry. One can obtain different nozzle shapes by the variation of these parameters.

We show in Fig. 6.6 (a) the actual nozzle shape for which the quasi-one-dimensional gas flow was computed with the aid of our program `prog6-1.nb` [see Figs. 6.6 (b)–(e)]. The profiles of M, P/P_1, ρ/ρ_1, v/v_1, and T/T_1, shown in Figs. 6.6 (b),(c), correspond to the supersonic flow regime in the diverging nozzle part. The program `prog6-1.nb` has the input parameter `reg`, which enables the user to choose the subsonic or supersonic flow regime in the diverging part of the Laval nozzle. We show in Figs. 6.6 (d),(e) the profiles of M, P/P_1, ρ/ρ_1, v/v_1, and T/T_1 for the same Laval nozzle in the case of a subsonic flow regime.

Problem 6.1. A reservoir is filled with an ideal gas at rest with $\gamma = 1.4$ under the pressure P_0 and the gas density is ρ_0; $v_0 = 0$. Find the dependence of the gas flow rate Q on the pressure P at the nozzle exit for the gas escaping from the reservoir through a Laval nozzle (see Fig. 6.7). The area of the nozzle exit crosssection is equal to S.

Solution: The gas flow rate $Q = \rho v S$, where $\rho = \rho_0 (P/P_0)^{1/\gamma}$. The quantity v is determined from (6.1.8) and (6.1.9):

$$
\begin{aligned}
P_0 &= P\left(1+\frac{(\gamma-1)}{2}M^2\right)^{\frac{\gamma}{\gamma-1}}, \quad \frac{c_0}{c} = \left(1+\frac{\gamma-1}{2}M^2\right)^{1/2}, \\
M^2 &= \frac{v^2}{c^2} = \frac{v^2}{c_0^2}\frac{c_0^2}{c^2},
\end{aligned}
$$

and equals

$$
v^2 = \frac{2\gamma}{(\gamma-1)}\frac{P_0}{\rho_0}\left(1-\left(\frac{P}{P_0}\right)^{\frac{\gamma-1}{\gamma}}\right).
$$

Substituting v and ρ in the formula for Q, we obtain:

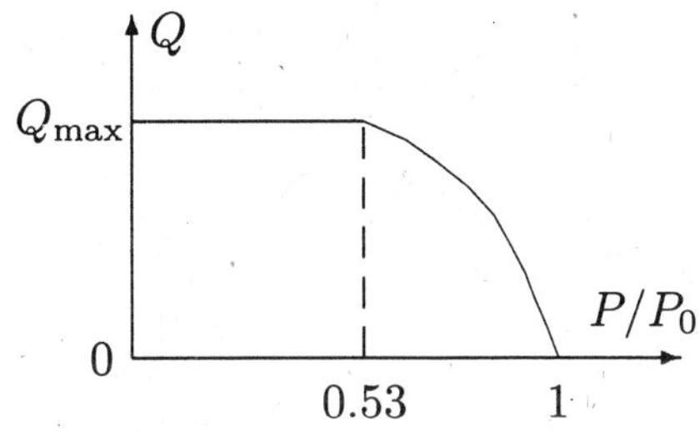

Figure 6.8: The dependence of the flow rate Q on the ratio of the pressure at the nozzle exit to the pressure inside the reservoir.

$$Q = \left(\frac{P}{P_0}\right)^{\frac{1}{\gamma}} S \sqrt{\frac{2\gamma}{(\gamma-1)} P_0\, \rho_0 \left(1 - \left(\frac{P}{P_0}\right)^{\frac{(\gamma-1)}{\gamma}}\right)},$$

where S is the area of the nozzle exit. The obtained formula is valid at $Q < Q_{max}$, after that $Q = Q_{max} = \text{const}$ (Fig. 6.8) at $\gamma = 1.4$. The appearance of this limitation $Q \leq Q_{max}$ is related to the flow choking. In this case, at $P/P_0 < 0.53$ ($\gamma = 1.4$), we have $M = 1$ in the minimal crosssection and $M > 1$ to the right of this crosssection, therefore, the disturbances from the nozzle exit propagating upstream at the velocity $\frac{dx}{dt} = v - c$ cannot enter the reservoir and increase the gas flow rate.

6.1.3 Planar Shock Wave in Ideal Gas

As was shown in Section 3.3, the shock wave in ideal gas represents a discontinuity surface, the mass flow rate j across which is different from zero. In a coordinate system comoving with the shock wave front, the jump relations have the form (3.3.45):

$$[\rho u_n] = 0, \ [\rho u_n^2 + P] = 0, \ [u_\tau] = 0, \ \left[H + \frac{u_n^2}{2}\right] = 0, \qquad (6.1.13)$$

where $H = E + P/\rho$ is the specific enthalpy, $\vec{u} = \vec{v} - \vec{D}$ is the gas velocity in the coordinate system comoving with the shock wave front, $u_n = \vec{u} \cdot \vec{n}$ is a normal velocity component, u_τ is the velocity component tangential to the discontinuity surface, $\vec{D} = D\vec{n}$ is the shock wave speed, and $\vec{n}$ is a normal to the surface of the shock wave front. The square brackets denote the jumps of the corresponding quantities, i.e., $[\varphi] = \varphi_2 - \varphi_1$, where the subscript 1 corresponds to the gas parameters before and subscript 2 corresponds behind the shock wave. The first relation in (6.1.13) is a consequence of the mass conservation law. The second and third relations are the consequences of the conservation laws for the momentum. The fourth relation is a consequence of the conservation law of the total energy.

Consider a planar shock wave in the case in which the gas velocity is directed along a normal to the discontinuity surface i.e., $u_\tau = 0$ and $u_n = u$. Defining the gas specific volume $V = 1/\rho$ let us rewrite the first two equations in (6.1.13) as follows:

$$u_1 = jV_1, \quad u_2 = jV_2, \quad P_2 - P_1 = j^2(V_1 - V_2), \tag{6.1.14}$$

where $j = \rho_1 u_1 = \rho_2 u_2$ is the flux of the gas mass inflowing across the discontinuity. Using (6.1.14), let us transform the last equation in (6.1.12):

$$H_2 - H_1 = \frac{u_1^2}{2} - \frac{u_2^2}{2} = \frac{1}{2}j^2(V_1^2 - V_2^2) = \frac{1}{2}(V_1 + V_2)(P_2 - P_1). \tag{6.1.15}$$

Substituting $H = E + PV$ in (6.1.15), we obtain the equation of the *Hugoniot adiabat* [5,6]

$$E_2 - E_1 = \frac{1}{2}(P_2 + P_1)(V_1 - V_2). \tag{6.1.16}$$

It is necessary to augment relation (6.1.16) by the ideal gas equation of state written in the variables P and V:

$$E = E(P, V). \tag{6.1.17}$$

Substituting (6.1.17) in (6.1.16), we obtain the equation of the Hugoniot adiabat in the form:

$$E(P_2, V_2) - E(P_1, V_1) = \frac{1}{2}(P_2 + P_1)(V_1 - V_2), \tag{6.1.18}$$

which determines the dependence of P_2 on V_2 at the given P_1 and V_1

$$P_2 = f(V_2, P_1, V_1). \tag{6.1.19}$$

At a given j, equation (6.1.19) [or (6.1.18)] together with the last equation in (6.1.14) enables one to find the gas state behind the shock wave. Assuming in (6.1.19) that $P_2 = P$ and $V_2 = V$, let us present the Hugoniot adiabat (6.1.19) passing through the point P_1, V_1 by the line H in Fig. 6.9. The last equation in (6.1.14) describes a straight line R (the Rayleigh line) passing through the point P_1, V_1. The intersection of the Rayleigh line R and the Hugoniot adiabat H determines the state behind the shock wave P_2, V_2 (point 2 in Fig. 6.9).

Let us determine the mutual disposition of the Hugoniot adiabat H and the Poisson adiabat P_s passing through the same point P_1, V_1 (the initial gas state before the compression). The equation for the Poisson

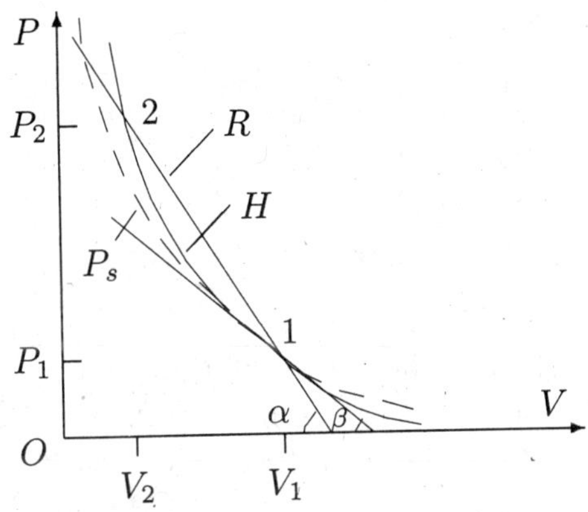

Figure 6.9: The Hugoniot adiabat H, the Poisson adiabat P_s, and the Rayleigh line R.

adiabat P_s follows from the ideal gas equation of state $V = V(P, S)$, in which one must set $S =$ const. Expanding the right-hand side of this equation in the neighborhood of P_1, V_1, let us rewrite the equation of the Poisson adiabat in the form

$$\begin{aligned} V - V_1 &= \left(\frac{\partial V}{\partial P}\right)_S (P - P_1) + \frac{1}{2}\left(\frac{\partial^2 V}{\partial P^2}\right)_S (P - P_1)^2 \\ &+ \frac{1}{6}\left(\frac{\partial^3 V}{\partial P^3}\right)_S (P - P_1)^3 + \cdots, \end{aligned} \tag{6.1.20}$$

where the derivatives are taken at $P = P_1$ and we have deleted the terms of a higher order of smallness.

At a gas compression in the shock wave, the final gas state P, V lies at the Hugoniot adiabat (6.1.18) and the gas entropy increases (see Section 3.3), i.e., $S > S_1$. Consequently, we obtain for a weak shock wave from the equation of state $V = V(P, S)$:

$$\begin{aligned} V - V_1 &= \left(\frac{\partial V}{\partial P}\right)_S (P - P_1) + \frac{1}{2}\left(\frac{\partial^2 V}{\partial P^2}\right)_S (P - P_1)^2 \\ &+ \frac{1}{6}\left(\frac{\partial^3 V}{\partial P^3}\right)_S (P - P_1)^3 + \left(\frac{\partial V}{\partial S}\right)_P (S - S_1) + \cdots. \end{aligned} \tag{6.1.21}$$

It can be seen that the first three items in (6.1.21) coincide with (6.1.20), and the last item related to the entropy growth is absent in (6.1.20). We have found previously in Problem 3.3 for the Hugoniot adiabat (6.1.18) the entropy change (3.3.50) in a weak shock wave. Substituting (3.3.50) in (6.1.21), we obtain the equation for the Hugoniot adiabat in the neighborhood of P_1, V_1:

$$\begin{aligned} V - V_1 &= \left(\frac{\partial V}{\partial P}\right)_S (P-P_1) + \frac{1}{2}\left(\frac{\partial^2 V}{\partial P^2}\right)_S (P-P_1)^2 \\ &+ \frac{1}{6}\left(\frac{\partial^3 V}{\partial P^3}\right)_S (P-P_1)^3 \\ &+ \frac{1}{12T_1}\left(\frac{\partial V}{\partial S}\right)_P \left(\frac{\partial^2 V}{\partial P^2}\right)_S (P-P_1)^3 + \cdots, \end{aligned} \qquad (6.1.22)$$

where we have neglected the terms of the higher order of smallness. The Hugoniot adiabat H (6.1.22) differs from the Poisson adiabat P_s (6.1.20) by the presence of the last term in (6.1.22). For the normal gases, $(\frac{\partial V}{\partial P})_S < 0$, $(\frac{\partial V}{\partial S})_P > 0$, and $(\frac{\partial^2 V}{\partial P^2})_S > 0$; therefore, the comparison of (6.1.20) and (6.1.22) yields that, at $P > P_1$, there will be $V_H(P) > V_{P_s}(P)$ and the Hugoniot adiabat H will lie higher than the Poisson adiabat P_s, and at $P < P_1$, it will lie lower, $V_H(P) < V_{P_s}(P)$ (see Fig. 6.9). At point P_1, V_1, both adiabats have the second-order tangent

$$\left(\frac{dV}{dP}\right)_{P_s} = \left(\frac{dV}{dP}\right)_H = \left(\frac{\partial V}{\partial P}\right)_S, \quad \left(\frac{d^2V}{dP^2}\right)_{P_s} = \left(\frac{d^2V}{dP^2}\right)_H = \left(\frac{\partial^2 V}{\partial P^2}\right)_S. \qquad (6.1.23)$$

Let us find the propagation speed of a weak shock wave D through the gas particles. In this case, we have $u_1 = -D$ ahead of the shock wave and we obtain from the first and third equations in (6.1.14):

$$D^2 = V_1^2\left(\frac{P_2 - P_1}{V_1 - V_2}\right). \qquad (6.1.24)$$

Using (6.1.23), we can find in a weak shock wave:

$$D^2 = -V_1^2\left(\frac{dP}{dV}\right)_H = -V_1^2\left(\frac{\partial P}{\partial V}\right)_S = \left(\frac{\partial P}{\partial \rho}\right)_S = c_1^2, \qquad (6.1.25)$$

where the derivatives are taken at point $V = V_1$. It follows from (6.1.25) that a weak shock wave propagates through the gas particles at a sound speed. In the case of a finite amplitude shock wave, the velocity of its propagation over the gas particles is given by formula (6.1.24) from which it follows that $D^2 = V_1^2 \operatorname{tg}\alpha$, where α is the angle between the Rayleigh line R and the OV axis (see Fig. 6.9). The squared sound velocity (6.1.25) is $c_1^2 = -V_1^2 \operatorname{tg}\beta$, where β is the angle between the tangent to the adiabats at point P_1, V_1 and the OV-axis [see Fig. 6.9)]. It follows from the concavity condition for the adiabats $(\frac{\partial^2 V}{\partial P^2})_S > 0$ that $\tan\alpha > \tan\beta$ (see Fig. 6.9). Therefore, for the finite amplitude shock wave $P_2 > P_1$ the inequality $D^2 > c_1^2$ is valid. Since $D = -u_1$, it follows from here that $u_1^2 > c_1^2$; that is, the flow ahead of the shock wave is supersonic.

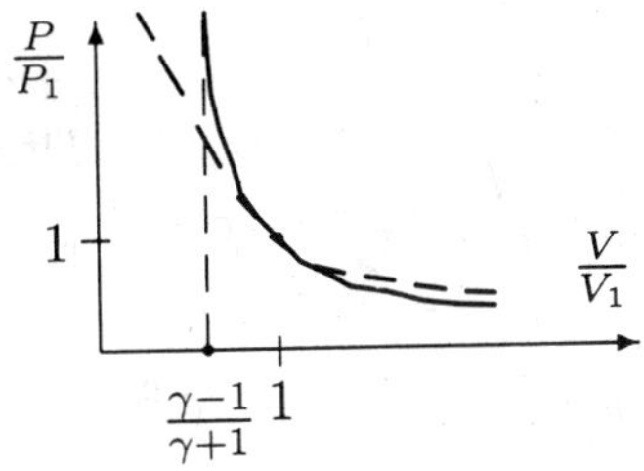

Figure 6.10: The adiabats of Hugoniot and Poisson for a perfect gas.

Let us find the gas velocity u_2 behind the shock wave front. We obtain from the second and third equations of system (6.1.14):

$$u_2^2 = -V_2^2\left(\frac{P_2 - P_1}{V_2 - V_1}\right). \tag{6.1.26}$$

The squared sound velocity at point P_2, V_2 is equal to

$$c_2^2 = -V_2^2\left(\frac{\partial P}{\partial V}\right)_{S_2}. \tag{6.1.27}$$

Comparing the inclinations of the tangent to the adiabat and to the Rayleigh line at point V_2, P_2, we obtain the inequality $u_2^2 < c_2^2$, from where it follows that the flow is subsonic behind the shock wave. Thus, we have proved the Zemplen's theorem[5–9], which asserts that the shock wave moves in the normal gas at a supersonic velocity in a gas ahead of the shock wave and at a subsonic velocity in a gas behind the shock wave: $u_1^2 > c_1^2$, $u_2^2 < c_2^2$. These inequalities are sometimes termed the stability conditions of a shock wave.

Consider a particular case of a *perfect gas* with the equation of state (2.1.92), which we rewrite in the form

$$E = \frac{PV}{\gamma - 1}, \quad \gamma = \frac{c_P}{c_V}, \quad c_P - c_V = R. \tag{6.1.28}$$

Substituting (6.1.28) in (6.1.18), we find the equation for the Hugoniot adiabat of a perfect gas

$$\frac{P}{P_1} = \frac{(\gamma+1)V_1 - (\gamma-1)V}{(\gamma+1)V - (\gamma-1)V_1}. \tag{6.1.29}$$

We show in Fig. 6.10 the Hugoniot adiabat (6.1.29) by a solid line, the Poisson adiabat $PV^\gamma = P_1V_1^\gamma$ for a perfect gas by a dashed line. It can be seen from Fig. 6.10 that the Hugoniot adiabat has a vertical asymptote

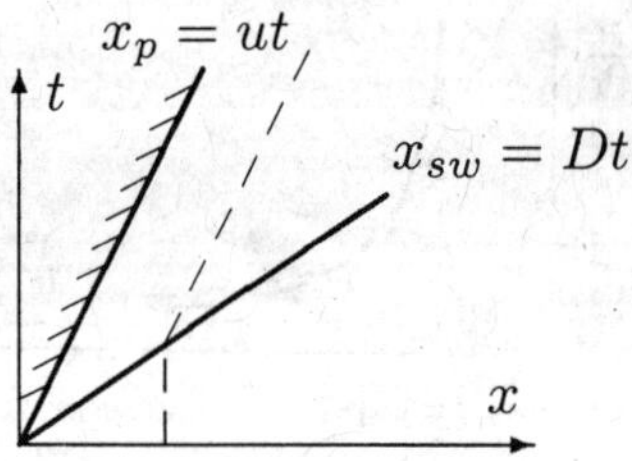

Figure 6.11: The trajectory of the piston $x_p(t)$ and of the shock wave $x_{sw}(t)$ in the t, x plane.

$\frac{V}{V_1} = \frac{\gamma-1}{\gamma+1}$, at which $P \to \infty$. This means that independently of the shock wave strength it is impossible to compress the gas in the shock wave by a factor more than $\frac{\gamma+1}{\gamma-1}$. In particular, $\gamma = 1.4$ for air and the maximum compression is equal to 6. At the same time, by compressing the gas adiabatically along the Poisson adiabat, one can compress it to an infinitesimal volume $V \to 0$ at $P \to \infty$. Such a considerable difference is related to irreversible processes at the shock wave front, as a result of which a considerable part of the gas kinetic energy passes in an irreversible way to heat. The thermal pressure arising there hinders the further compression of gas.

In the applications, the formulas are useful, which express the gas parameters behind the shock wave as functions of the shock wave Mach number $M_1 = u_1/c_1$. These formulas are easily found from (6.1.14), (6.1.28), (6.1.29), and $c_1^2 = \gamma P_1 V_1$:

$$\begin{aligned} \frac{V_1}{V_2} &= \frac{u_1}{u_2} = \frac{(\gamma+1)M_1^2}{(\gamma-1)M_1^2+2}, \\ \frac{P_2}{P_1} &= \frac{2\gamma}{\gamma+1}M_1^2 - \frac{\gamma-1}{\gamma+1}, \\ \frac{T_2}{T_1} &= \frac{(2\gamma M_1^2-(\gamma-1))((\gamma-1)M_1^2+2)}{(\gamma+1)^2 M_1^2}, \\ M_2^2 &= \frac{2+(\gamma-1)M_1^2}{2\gamma M_1^2-(\gamma-1)}; \qquad M_2^2 = \frac{u_2^2}{c_2^2}. \end{aligned} \tag{6.1.30}$$

Problem 6.2. A piston begins to move at a velocity u into a tube filled with an ideal gas at rest. The gas is under the pressure P_1 and has the specific volume V_1. Find the shock wave velocity D and the pressure behind the shock wave front P_2.

Solution: Let us represent the given process on the t, x diagram. We show in Fig. 6.11 by a solid line the shock wave trajectory and by a dashed line the trajectory of a gas particle. It can be seen that the gas velocity behind the shock wave front is equal to the piston velocity in the laboratory frame. In the coordinate system comoving with the shock wave front, we can use the jump relations (6.1.14) and the equation of the Hugoniot adiabat (6.1.16):

$$v_1 = jV_1, \quad v_2 = jV_2, \quad P_2 - P_1 = j^2(V_1 - V_2),$$
$$E_2 - E_1 = \frac{1}{2}(P_1 + P_2)(V_1 - V_2). \tag{6.1.31}$$

The specific internal energy of the ideal gas is

$$E = C_V T = \frac{PV}{\gamma - 1}.$$

Substituting this equation in the last equation of system (6.1.31), we obtain the shock adiabat of the polytropic gas:

$$\frac{V_2}{V_1} = \frac{(\gamma+1)P_1 + (\gamma-1)P_2}{(\gamma-1)P_1 + (\gamma+1)P_2}. \tag{6.1.32}$$

Using the first three equations of (6.1.31), we can find the jump of the gas velocity at the shock wave front:

$$v_1 - v_2 = j(V_1 - V_2) = \sqrt{\frac{P_2 - P_1}{V_1 - V_2}}\,(V_1 - V_2) = \sqrt{(P_2 - P_1)(V_1 - V_2)}.$$

Since the gas velocity jump is equal to u, we have $u^2 = (P_2 - P_1)(V_1 - V_2)$. Dividing both sides of this equation by P_1V_1 and using the relation

$$P_1V_1 = \frac{P_1}{\rho_1} = \frac{\gamma P_1}{\rho_1}\frac{1}{\gamma} = \frac{c_1^2}{\gamma},$$

we obtain:

$$\left(\frac{P_2}{P_1} - 1\right)\left(1 - \frac{V_2}{V_1}\right)\frac{c_1^2}{\gamma} = u^2. \tag{6.1.33}$$

Using the shock adiabat (6.1.32), we find:

$$1 - \frac{V_2}{V_1} = \frac{2(P_2 - P_1)}{(\gamma-1)P_1 + (\gamma+1)P_2} = \frac{2\left(\frac{P_2}{P_1} - 1\right)}{\gamma - 1 + (\gamma+1)\frac{P_2}{P_1}}. \tag{6.1.34}$$

Substituting (6.1.34) in (6.1.33), we obtain:

$$2\frac{c_1^2}{\gamma}\left(\frac{P_2}{P_1} - 1\right)^2 - (\gamma+1)\,u^2\left(\frac{P_2}{P_1} - 1\right) - 2\gamma u^2 = 0.$$

Introducing the new variable $x = \frac{P_2}{P_1} - 1$, we find:

$$x^2 - \frac{\gamma(\gamma+1)M^2}{2}x - \gamma^2 M^2 = 0, \quad M = \frac{u}{c_1},$$

from where

$$x_{1,2} = \frac{1}{4}(\gamma+1)\gamma M^2 \pm \sqrt{\frac{1}{16}(\gamma+1)^2\gamma^2 M^4 + \gamma^2 M^2}.$$

Since $P_2/P_1 > 1$, it is necessary to choose the "+" sign; therefore,

$$\frac{P_2}{P_1} = 1 + \frac{1}{4}(\gamma+1)\gamma M^2 + \sqrt{\frac{1}{16}(\gamma+1)^2\gamma^2 M^4 + \gamma^2 M^2}. \qquad (6.1.35)$$

In order to find D, we make use of the fact that the gas velocity before the shock wave front in the coordinate system of the shock wave front is equal to D. It follows from here that

$$D^2 = j^2 V_1^2 = \frac{(P_2 - P_1)V_1^2}{(V_1 - V_2)} = \frac{\left(\frac{P_2}{P_1} - 1\right)P_1 V_1}{\left(1 - \frac{V_2}{V_1}\right)} = \frac{c_1^2}{\gamma}\frac{\left(\frac{P_2}{P_1} - 1\right)}{\left(1 - \frac{V_2}{V_1}\right)}.$$

Taking into account equations (6.1.32) and (6.1.35), we obtain:

$$\begin{aligned} D^2 &= \frac{c_1^2}{2\gamma}\left((\gamma-1) + (\gamma+1)\frac{P_2}{P_1}\right) = \frac{c_1^2}{2\gamma}\Big(\gamma - 1 + \gamma + 1 \\ &+ \frac{1}{4}(\gamma+1)^2\gamma M^2 + \gamma(\gamma+1)M\sqrt{1 + \frac{1}{16}(\gamma+1)^2 M^2}\Big), \end{aligned}$$

from where

$$D^2 = c_1^2\left(1 + \frac{(\gamma+1)^2}{8}M^2 + \frac{(\gamma+1)}{2}M\sqrt{1 + \frac{(\gamma+1)^2}{16}M^2}\right). \qquad (6.1.36)$$

In the limit of a strong shock wave, i.e., at $M \gg 1$, we have from expressions (6.1.35), (6.1.36), (6.1.32):

$$\frac{\rho_2}{\rho_1} = \frac{\gamma+1}{\gamma-1}, \quad P_2 = \frac{(\gamma+1)}{2}\rho_1 u^2, \quad D = \frac{\gamma+1}{2}u. \qquad (6.1.37)$$

In the limit of a weak shock wave $M \ll 1$, it follows from (6.1.36) that

$$D = c_1 + \frac{(\gamma+1)}{4}u; \quad M = \frac{u}{c_1}.$$

6.1.4 Shock Wave Structure in Gas

So far, we have considered the shock wave as a discontinuity surface having a zero thickness. As was noted above, this is only a mathematical idealization. The shock wave indeed has a finite thickness, which is determined by the processes of viscosity and heat conduction in gas. There is a thin transitional layer in which the gas parameters vary along a normal $\vec{n}$ to the layer and vary weakly in the direction perpendicular to the normal $\vec{n}$ (see Fig. 6.12). Consider the shock wave structure in a transitional layer of thickness Δx. We will describe the gas flow by the equations of a viscous, heat-conducting gas. Choose the Eulerian Cartesian coordinate system $g_{ij} = \delta_{ij}$ and direct the x^1-axis along the direction of the shock wave propagation $\vec{n}$. In this case, only one velocity component v^1 will be different from zero, which can be assumed to be a function of the time t and the x^1 coordinate (we neglect the variation of all the flow parameters in the plane x^2x^3). For the brevity of notation, we will omit in this subsection the index "1" by the velocity and by the coordinate and assume that $v^1 \equiv v(x,t)$, and $x^1 \equiv x$. From the definition of the rate-of-strain tensor (1.2.62) $\dot{\varepsilon}_{ij}$ and the deviator of the strain rate $\dot{e}_{ij} = \dot{\varepsilon}_{ij} - \frac{1}{3}\dot{\varepsilon}_{kk}$, we obtain:

$$\dot{\varepsilon}_{11} = \frac{\partial v}{\partial x}, \quad \dot{e}_{11} = \frac{2}{3}\frac{\partial v}{\partial x}, \quad \operatorname{div}\vec{v} = \dot{\varepsilon}_{11} = \frac{\partial v}{\partial x}, \tag{6.1.38}$$

the remaining components $\dot{\varepsilon}_{ij} = 0$. Substituting (6.1.38) in (2.1.101), we find with regard for $g_{ij} = \delta_{ij}$ a component of the stress tensor, which is different from zero:

$$\sigma_{11} = -P + \eta\frac{\partial v}{\partial x}, \quad \eta = \zeta + \frac{4}{3}\mu. \tag{6.1.39}$$

Taking into account the temperature T variation only along a normal to the transitional layer, we obtain:

$$\dot{q}_1 = -\kappa\frac{\partial T}{\partial x}. \tag{6.1.40}$$

Substituting (6.1.39) and (6.1.40) in the general divergence form system of equations (2.1.154), (2.1.156), and (2.1.161), we obtain the system of equations for a viscous, heat-conducting gas in the one-dimensional nonstationary case, which is written in the divergence form:

$$\begin{gathered}\frac{\partial \rho}{\partial t} + \frac{\partial}{\partial x}(\rho v) = 0,\\ \frac{\partial}{\partial t}(\rho v) + \frac{\partial}{\partial x}\left(\rho v^2 + P - \eta\frac{\partial v}{\partial x}\right) = 0,\\ \frac{\partial}{\partial t}\left(\rho\left(E + \frac{v^2}{2}\right)\right) + \frac{\partial}{\partial x}\left(\rho v\left(H + \frac{v^2}{2}\right)\right) - \eta v\frac{\partial v}{\partial x} - \kappa\frac{\partial T}{\partial x}\bigg) = 0,\end{gathered} \tag{6.1.41}$$

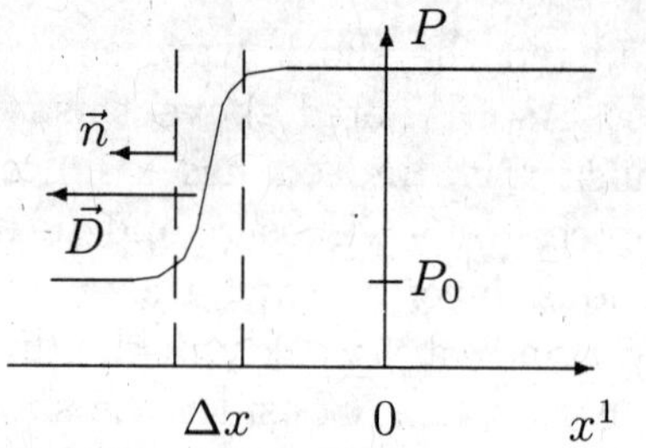

Figure 6.12: The profile of the pressure $P(x^1)$ in the shock wave propagating to the left: $\vec{D} = D\vec{n}$.

where $H = E + v^2/2$ is the enthalpy. It is necessary to augment the system (6.1.41) by the equation of state, which we write in the form $P = P(\rho, E)$. Using the Galilean transformation (2.3.72)

$$x' = x - Dt, \quad u = v - D, \quad P' = P,\ T' = T,\ \rho' = \rho,\ E' = E \tag{6.1.42}$$

we can go over to the coordinate system K' comoving with the shock wave front. Assume that in the coordinate system K' all parameters depend only on the coordinate x' and do not depend on the time t. Since the system (6.1.41) is invariant under the Galilean transformations, it will have the same form in the system K'. Omitting in (6.1.41) all partial derivatives with respect to time and replacing $x \to x'$, $v \to u$, $\rho \to \rho'$, and $E \to E'$, we can write the equations of a viscous, heat-conducting gas in the coordinate system K' comoving with the shock wave front as

$$\begin{aligned}
&\frac{d}{dx'}(\rho' u) = 0, \quad \frac{d}{dx'}\Big(\rho' u^2 + P' - \eta\frac{du}{dx'}\Big) = 0,\\
&\frac{d}{dx'}\Big(\rho' u\Big(H' + \frac{u^2}{2}\Big) - \eta u\frac{du}{dx'} - \kappa\frac{dT'}{dx'}\Big) = 0,\\
&P' = P'(\rho', E').
\end{aligned} \tag{6.1.43}$$

These equations have the following three first integrals:

$$\begin{aligned}
&\rho' u = C_1, \quad \rho' u^2 + P' - \eta\frac{du}{dx'} = C_2,\\
&C_1\Big(H' + \frac{u^2}{2}\Big) - \eta u\frac{du}{dx'} - \kappa\frac{dT'}{dx'} = C_3.
\end{aligned} \tag{6.1.44}$$

In the coordinate system K', we have $u > 0$, therefore, the integration constants $C_1 > 0$, $C_2 > 0$, and $C_3 > 0$. The first integral in (6.1.44) expresses the mass conservation law. The second integral expresses the momentum conservation law The third integral expresses the energy conservation law. After simple transformations, we can present the system (6.1.44) in the form

$$
\begin{aligned}
\eta\frac{dV'}{dx'} &= \frac{1}{C_1}\Big(C_1^2\Big(V'-\frac{C_2}{C_1}\Big)+P'(V',T')\Big)=\mathcal{M}_1'(V',T'),\\
\kappa\frac{dT'}{dx'} &= C_1\left(E'-\frac{1}{2}C_1^2\Big(V'-\frac{C_2}{C_1}\Big)^2-\frac{C_3}{C_1}+\frac{1}{2}\Big(\frac{C_2}{C_1}\Big)^2\right) \qquad (6.1.45)\\
&= \mathcal{M}_2'(V',T'),
\end{aligned}
$$

where $V'=1/\rho'$. Introducing the nondimensional variables $V=V'\frac{C_1}{C_2}$, $P=P'/C_2$, $E=E'(\frac{C_1}{C_2})^2$, $T=R(\frac{C_1}{C_2})^2T'$ and omitting the prime by x', we rewrite the system (6.1.45) in the nondimensional form:

$$
\begin{aligned}
\tilde{\eta}\frac{dV}{dx} &= V+P(V,T)-1=\mathcal{M}_1(V,T)\\
\tilde{\kappa}\frac{dT}{dx} &= E(V,T)-\frac{1}{2}(V-1)^2-\beta=\mathcal{M}_2(V,T), \qquad (6.1.46)
\end{aligned}
$$

where we have introduced the notations

$$\tilde{\eta}=\eta/C_1, \quad \tilde{\kappa}=\kappa/(C_1R), \quad \beta=C_3C_1/C_2^2-1.$$

Let us formulate for system (6.1.46) the following boundary-value problem.

Find the solution $V(x)$ and $T(x)$ satisfying system (6.1.46) at $x\in(-\infty,+\infty)$, which tends at $x\to\pm\infty$ to the constant values; that is,

$$
\begin{aligned}
&\text{at } x\to-\infty \quad V(x)\to V_1, \quad T(x)\to T_1,\\
&\text{at } x\to+\infty \quad V(x)\to V_2, \quad T(x)\to T_2. \qquad (6.1.47)
\end{aligned}
$$

A necessary condition for the existence of such a solution is the requirement that the points (V_1,T_1) and (V_2,T_2) are the stationary points of system (6.1.46); that is,

$$\mathcal{M}_1(V_1,T_1)=\mathcal{M}_2(V_1,T_1)=\mathcal{M}_1(V_2,T_2)=\mathcal{M}_2(V_2,T_2)=0.$$

This means that the points (V_1,T_1) and (V_2,T_2) should be the points of intersection of the curves $\mathcal{M}_1(V,T)=0$ and $\mathcal{M}_2(V,T)=0$. If such points exist, then the conditions $\tilde{\eta}\frac{dV}{dx}=\tilde{\kappa}\frac{dT}{dx}=0$ are satisfied at these points according to (6.1.46). Taking into account the satisfaction of these conditions at $x\to\pm\infty$, we find from system (6.1.44):

$$
\begin{aligned}
&\rho_1'u_1=\rho_2'u_2=C_1,\\
&\rho_1'u_1^2+P_1'=\rho_2'u_2^2+P_2'=C_2,\\
&\rho_1'u_1\Big(E_1'+\frac{P_1'}{\rho_1'}+\frac{u_1^2}{2}\Big)=\rho_2'u_2\Big(E_2'+\frac{P_2'}{\rho_2'}+\frac{u_2^2}{2}\Big)=C_3.
\end{aligned}
$$

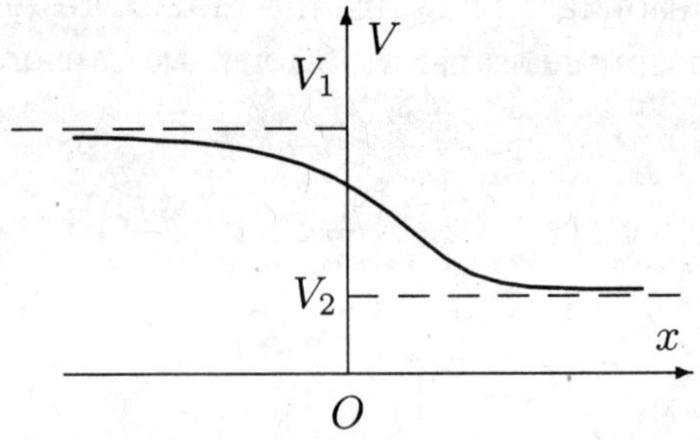

Figure 6.13: Shock wave structure in a viscous, non-heat-conducting gas.

It follows from these relations that the functions u_1, ρ_1', P_1', T_1'; u_2, ρ_2', P_2', and T_2' should satisfy the Hugoniot conditions. Resolving them, we find that the points (V_1, T_1) and (V_2, T_2) exist in the (V, T) plane and lie on the Hugoniot shock adiabat.

A detailed qualitative analysis of the solution of system (6.1.46) may be found in[9]. The system of equations (6.1.46) with boundary conditions (6.1.47) has an analytic solution for some particular cases. Consider a well-known solution found by Becker. Let the equation of state of a viscous, heat-conducting gas coincide with the ideal gas equation of state. Then we have in the nondimensional variables:

$$T = PV, \quad E = \frac{1}{\gamma - 1}T,$$

and the functions $\mathcal{M}_1$ and $\mathcal{M}_2$ my be presented by formulas

$$\mathcal{M}_1 = \frac{T}{V} + V - 1, \quad \mathcal{M}_2 = \frac{1}{\gamma - 1}T - \frac{1}{2}(V-1)^2 - \beta.$$

In the case in which there is no heat conduction, i.e., $\tilde{\kappa} = 0$, and the viscosity $\tilde{\eta} = \text{cons} \neq 0$, we find:

$$T = \frac{\gamma - 1}{2}[V(x) - 1]^2 + \beta(\gamma - 1),$$

and $V(x)$ satisfies the equation

$$\tilde{\eta}\frac{dV}{dx} = \frac{\gamma + 1}{2}\frac{(V - V_1)(V - V_2)}{V} \tag{6.1.48}$$

at $V_2 \leq V \leq V_1$. The integration of this equation yields:

$$x + \text{const} = \frac{2\tilde{\eta}}{(\gamma + 1)}\ln\left[(V_1 - V)^{\frac{V_1}{V_1 - V_2}} \cdot (V - V_2)^{-\frac{V_2}{V_1 - V_2}}\right]. \tag{6.1.49}$$

Without loss of generality, we can assume the integration constant to be equal to zero. Its value characterizes a topologically similar translation of the curve (6.1.49) with respect to the coordinate origin. A qualitative picture of the behavior of curve (6.1.49) is presented in Fig. 6.13. The curve $V = V(x)$ presented in this figure was obtained with the aid of our *Mathematica* Notebook `prog6-2.nb`. Following Prandtl, let us define the shock wave thickness by formula

$$\Delta x = \frac{V_1 - V_2}{\max|\frac{dV}{dx}|}. \tag{6.1.50}$$

It follows from (6.1.48) that the maximum of the derivative $|\frac{dV}{dx}|$ takes place at a point where

$$\frac{2\tilde{\eta}}{\gamma+1}\frac{d^2V}{dx^2} = \frac{V^2 - V_1V_2}{V^2} = 0,$$

from where we obtain $V = \sqrt{V_1V_2}$. Substituting this value in (6.1.48), we find:

$$\left|\frac{dV}{dx}\right|_{\max} = \frac{(\gamma+1)}{2\tilde{\eta}}(\sqrt{V_1} - \sqrt{V_2})^2.$$

Substituting this expression in (6.1.50), we obtain:

$$\Delta x = \frac{2\tilde{\eta}}{\gamma+1}\left(\frac{\sqrt{V_1}+\sqrt{V_2}}{\sqrt{V_1}-\sqrt{V_2}}\right). \tag{6.1.51}$$

Using this formula, we can estimate the thickness Δx in the case of a weak shock wave $\Delta V = V_1 - V_2 \ll V_2$. Assuming $V_1 = V_2 + \Delta V$, we have:

$$\sqrt{V_1} = \sqrt{V_2 + \Delta V} \approx \sqrt{V_2} + \frac{1}{2}\Delta V/\sqrt{V_2},$$

and with regard for the mass conservation law $u_1/V_1 = u_2/V_2$, we obtain: $\sqrt{V_1} = \sqrt{V_2}(1 + \frac{u_1-u_2}{2u_2})$. Substituting this expression in (6.1.51), we find with regard for the formula $\tilde{\eta} = \eta/(u_2\rho_2)$:

$$\Delta x = \frac{8\eta}{(\gamma+1)\rho_2(u_1-u_2)}. \tag{6.1.52}$$

We can estimate the magnitude of molecular viscosity η by formulas[3]

$$\eta = \zeta + \frac{4}{3}\mu \approx 2\mu, \quad \mu \approx 0.5\rho\bar{v}\lambda_0, \quad \bar{v} = \sqrt{\frac{8}{\pi\gamma}}\,c, \quad c = \sqrt{\frac{\gamma P}{\rho}},$$

where λ_0 is the mean free path of the molecules in gas, and c is the sound velocity. Taking these formulas into account, we can write the final expression for Δx at $\gamma = 1.4$ in the form

$$\Delta x \approx 1.6\frac{c_2}{u_1-u_2}\lambda_0. \tag{6.1.53}$$

It can be seen from here that $\lambda_0 < \Delta x \ll L$; therefore, from the viewpoint of continuum mechanics, the substitution of the transitional layer of width Δx by a discontinuity surface is justified. The constructed solution (6.1.49) is valid for a weak shock wave $u_1 - u_2 \ll c_1, \Delta x \gg \lambda_0$, when the approximation of continuum is applicable. For a strong shock wave, the thickness $\Delta x \sim \lambda_0$ and one must use the methods of the kinetic theory of gases for the flow description in a transitional layer.

6.2 Nonstationary One-Dimensional Flows of Ideal Gas

6.2.1 Planar Isentropic Waves

Consider a continuous nonstationary flow of ideal gas along the x-axis in which the entropy remains constant in the overall flow region $-\infty < x < \infty$: $S = \text{const}$. As follows from (2.1.152), the entropy is conserved in each particle at a continous flow of ideal gas: $\frac{dS}{dt} = 0$; therefore, to insure the isentropicity, it is sufficient to require that $S = \text{const}$ at the initial moment of time $t = 0$. The equations of continuity, motion, and entropy constancy for such flows were derived above in Chapter 2 [see (2.1.152)]. Neglecting the body force $F_x = 0$ and using the equation of the Poisson adiabat instead of the third equation in (2.1.152), we rewrite (2.1.152) in the form

$$\begin{aligned} &\frac{\partial \rho}{\partial t} + \frac{\partial}{\partial x}(\rho v) = 0, \\ &\frac{\partial v}{\partial t} + v\frac{\partial v}{\partial x} + \frac{1}{\rho}\frac{\partial P}{\partial x} = 0, \\ &P = A\rho^\gamma, \end{aligned} \tag{6.2.1}$$

where we have denoted the velocity along the x-axis by letter $v \equiv u_x$.

Consider at first the *propagation of a wave of small amplitude.* We will search for the solution of system (6.2.1) in this case by the method of small disturbances:

$$v = v_0 + v' + \cdots; \quad P = P_0 + P' + \cdots; \quad \rho = \rho_0 + \rho' + \cdots,$$

where v_0, P_0, ρ_0 are constants and $v'(t,x), P'(t,x), \rho'(t,x), \ldots$ are the disturbances, which are small, together with their derivatives with respect to t and x. Without loss of problem generality, we assume $v_0 = 0$, which corresponds to the propagation of disturbances in a medium at rest. Substituting the given values in system of equations (6.2.1) and retaining the terms of the first order of smallness, we find with regard for relations $\rho' \ll \rho_0, \ P' \ll P_0$ that

$$\rho_0\frac{\partial v'}{\partial t} = -c_0^2\frac{\partial \rho'}{\partial x}; \quad \rho_0\frac{\partial v'}{\partial x} = -\frac{\partial \rho'}{\partial t}; \quad P' = c_0^2\rho', \tag{6.2.2}$$

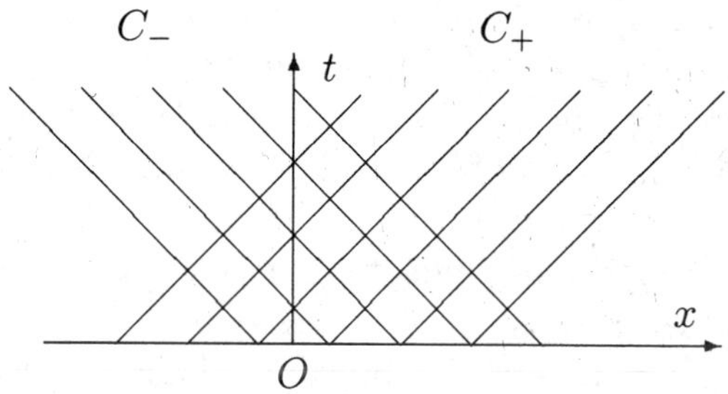

Figure 6.14: The characteristics $C_+ : x - c_0 t = \xi_2$, and $C_- : x + c_0 t = \xi_1$.

where $c_0^2 = (\frac{dP}{d\rho})_0$ is the sound velocity in a medium at rest. The system of equations (6.2.2) is already linear, and it can easily be reduced by a crossdifferentiation to a single equation

$$\frac{\partial^2 v'}{\partial t^2} - c_0^2 \frac{\partial^2 v'}{\partial x^2} = 0, \tag{6.2.3}$$

or

$$\frac{\partial^2 \rho'}{\partial t^2} - c_0^2 \frac{\partial^2 \rho'}{\partial x^2} = 0.$$

The general solution of any of these equations may be presented as a sum of two arbitrary functions:

$$v' = f_1(x + c_0 t) + f_2(x - c_0 t), \tag{6.2.4}$$

which are determined from the initial and boundary conditions. Let us introduce the new coordinates by formulas

$$\xi_1 = x + c_0 t, \quad \xi_2 = x - c_0 t. \tag{6.2.5}$$

The solution of (6.2.3) with regard for (6.2.4) has the form

$$v' = f_1(\xi_1) + f_2(\xi_2). \tag{6.2.6}$$

Each of the functions $f_1(\xi_1)$ and $f_2(\xi_2)$ entering (6.2.6) represents a corresponding distribution of the velocity disturbance in the moving coordinates $O_1\xi_1$ or $O_1\xi_2$. Assuming $\xi_1 = \text{const}$ or $\xi_2 = \text{const}$ in the obtained solutions, we obtain the equations:

$$x + c_0 t = \text{const}, \quad x - c_0 t = \text{const}, \tag{6.2.7}$$

describing two planes moving in the opposite directions at a velocity c_0, which are perpendicular to the Ox-axis. Each of the planes carries constant values of the disturbances of the velocity, pressure, and density, which are determined by the initial or boundary conditions. Thus, the general solution (6.2.4) is composed of two solutions corresponding to two planar waves propagating in the opposite directions. From the geometric viewpoint, the obtained solution (6.2.4) may be interpreted as the presence of two families of straight lines (6.2.7) in the (x,t) plane with the angular coefficients $\pm c_0$ (see Fig. 6.14). One of the families of lines $x - c_0 t = \xi_2$ corresponds to the characteristics C_+, and the other family of lines $x + c_0 t = \xi_1$ corresponds to the characteristics C_-. Along the C_- characteristic, the value of the disturbance $f_1(\xi_1)$ is constant, and along each C_+ characteristic the value of $f_2(\xi_2)$ is constant. Consider, for example, the wave propagating to the right, for which by definition $f_1(\xi_1) = 0$. Then we obtain from (6.2.5) and (6.2.6) that $v' = f_2(\xi_2)$. It follows from (6.2.2)–(6.2.4) that the disturbances of the density ρ' and the pressure P' will depend on the same variable: $\rho' = \rho'(\xi_2)$ and $P' = P'(\xi_2)$. Eliminating from here the variable ξ_2, we can write: $v' = v'(\rho')$ and $P' = P'(\rho')$. It is conventional to call such waves the *simple waves.* They will be considered in more detail below in the case of the finite amplitude waves. The velocity of the propagation of a simple wave with a small amplitude is constant and equal to the sound velocity in the medium at rest $c_0 = \sqrt{(\frac{dP}{d\rho})_0}$. This condition is not satisfied for a finite amplitude wave.

Consider the propagation of *finite amplitude waves.* It is necessary to find in this case an exact solution of the system of equations (6.2.1). For this purpose we use the method of characteristics presented in Chapter 3, Section 3.2. Using the relations

$$\frac{\partial \rho}{\partial t} = \left(\frac{\partial \rho}{\partial P}\right)_S \frac{\partial P}{\partial t} = \frac{1}{c^2}\frac{\partial P}{\partial t}, \quad \frac{\partial \rho}{\partial x} = \frac{1}{c^2}\frac{\partial P}{\partial x},$$

we can eliminate the derivatives $\frac{\partial \rho}{\partial t}$ and $\frac{\partial \rho}{\partial x}$ in the system (6.2.1):

$$\begin{aligned} &\frac{\partial P}{\partial t} + v\frac{\partial P}{\partial x} + \rho c^2 \frac{\partial v}{\partial x} = 0, \\ &\frac{\partial v}{\partial t} + v\frac{\partial v}{\partial x} + \frac{1}{\rho}\frac{\partial P}{\partial x} = 0, \end{aligned} \tag{6.2.8}$$

where c^2 is the squared sound velocity, which is found by formula

$$c^2 = \left(\frac{\partial P}{\partial \rho}\right)_S = \gamma A \rho^{\gamma-1} = \gamma \frac{P}{\rho}. \tag{6.2.9}$$

Present the system of equations (6.2.8) in matrix form:

$$\frac{\partial u}{\partial t} + C\frac{\partial u}{\partial x} = 0, \quad u = \begin{pmatrix} P \\ v \end{pmatrix}, \quad C = \begin{Vmatrix} v & \rho c^2 \\ \frac{1}{\rho} & v \end{Vmatrix} \tag{6.2.10}$$

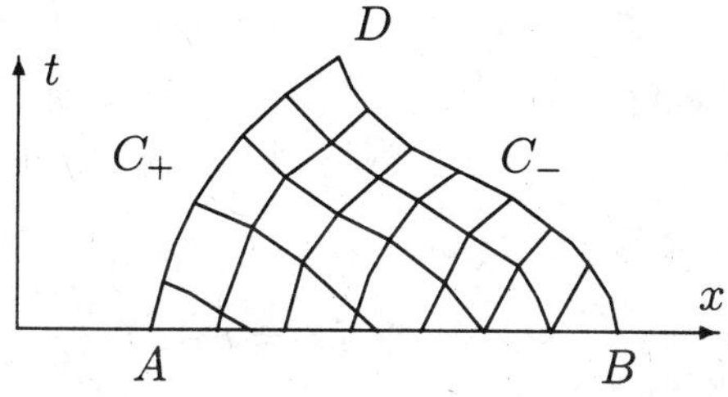

Figure 6.15: The characteristic triangle.

According to (3.2.5), the characteristics are determined from the condition of the equality to zero of the determinant

$$\begin{vmatrix} 1 & 0 & v & \rho c^2 \\ 0 & 1 & 1/\rho & v \\ dt & 0 & dx & 0 \\ 0 & dt & 0 & dx \end{vmatrix} = 0. \tag{6.2.11}$$

Calculating the determinant (6.2.11), we find two families $C_{\pm}$ of the characteristics:

$$C_+ : \quad \frac{dx}{dt} = v + c; \qquad C_- : \quad \frac{dx}{dt} = v - c. \tag{6.2.12}$$

It can be seen that the characteristics represent the lines along which the finite amplitude disturbances propagate. The velocity of disturbances propagation $v \pm c$ is not constant and depends on the amplitude of disturbances. The relations on the characteristics are found from the condition (3.2.6)

$$\text{rank} \begin{vmatrix} 1 & 0 & v & \rho c^2 & 0 \\ 0 & 1 & 1/\rho & v & 0 \\ dt & 0 & dx & 0 & dp \\ 0 & dt & 0 & dx & dv \end{vmatrix} = \text{rank} \begin{vmatrix} 1 & 0 & v & \rho c^2 \\ 0 & 1 & 1/\rho & v \\ dt & 0 & dx & 0 \\ 0 & dt & 0 & dx \end{vmatrix}$$

Since the determinant of the matrix standing on the right-hand side is equal to zero, the determinant of a matrix formed of any four columns of the matrix standing on the left-hand side of the given equation should be equal to zero. For example,

$$\begin{vmatrix} 1 & 0 & v & 0 \\ 0 & 1 & 1/\rho & 0 \\ dt & 0 & dx & dP \\ 0 & dt & 0 & dv \end{vmatrix} = dx\,dv + \frac{1}{\rho}\,dt\,dP - v\,dt\,dv = 0.$$

Substituting here the equations for characteristics (6.2.12), we obtain that the relation $dv + \frac{dP}{\rho c} = 0$ takes place along the C_+ characteristic $\frac{dx}{dt} = v + c$ and the relation $dv - \frac{dP}{\rho c} = 0$ takes place along the C_- characteristic $\frac{dx}{dt} = v - c$. Integrating these relations, we can find the *Riemann invariants*:

$$\begin{aligned} C_+ : \quad \frac{dx}{dt} &= v + c, \qquad \mathcal{I}_+ = v + \int \frac{dP}{\rho c} = \text{const}, \\ C_- : \quad \frac{dx}{dt} &= v - c, \qquad \mathcal{I}_- = v - \int \frac{dP}{\rho c} = \text{const}. \end{aligned} \tag{6.2.13}$$

Using the equation for the Poisson adiabat $P = A\rho^\gamma$, we now compute the integrals in (6.2.13) and obtain the Riemann invariants in a perfect gas:

$$\begin{aligned} C_+ : \quad \frac{dx}{dt} &= v + c, \qquad \mathcal{I}_+ = v + \frac{2c}{\gamma - 1} = \text{const}, \\ C_- : \quad \frac{dx}{dt} &= v - c, \qquad \mathcal{I}_- = v - \frac{2c}{\gamma - 1} = \text{const}. \end{aligned} \tag{6.2.14}$$

Note that the trajectory equation $\frac{dx}{dt} = v$ is the C_0 characteristic along which the entropy S is constant ($\frac{\partial S}{\partial t} + v\frac{\partial S}{\partial x} = 0$). This characteristic proved to be outside the scope of our consideration here since $S = \text{const}$ in the overall region, and the characteristics are the lines of weak discontinuity.

The knowledge of the Riemann invariants (6.2.14) enables us to construct the solution in a characteristic triangle ADB (see Fig. 6.15) by using the given values of v and P in the interval AB. The side AD of the characteristic triangle coincides with a C_+ characteristic, and the side DB coincides with a C_- characteristic. The interval AB determines the domain of influence for point D. The value of $\mathcal{I}_+$ at point D coincides with the corresponding $\mathcal{I}_+$ at point A, and the value of $\mathcal{I}_-$ at point D coincides with the value of $\mathcal{I}_-$ at point B. By using the known $\mathcal{I}_+$ and $\mathcal{I}_-$, it is easy to determine the v and c at point D.

Consider a particular class of flows in which *one of the Riemann invariants is identically constant* in the overall flow region all the time. It is easy to see that this flow is a *simple wave*. To prove this fact, let us consider a wave propagating to the right $\frac{dx}{dt} = v + c$ and require that the invariant $\mathcal{I}_- = v - \frac{2c}{\gamma-1}$ be constant in the overall flow region. Then we obtain from the equality $\mathcal{I}_- = \text{const}$ that $c = \frac{\gamma-1}{2}v + \text{const}$, and with regard for (6.2.9) and $P = A\rho^\gamma$, we can express all flow parameters as the functions of the velocity $c = c(v)$, $P = P(v)$, and $\rho = \rho(v)$, which coincides with the simple wave definition. Since the invariant $\mathcal{I}_+$ is constant along each C_+ characteristic, we obtain from (6.2.14):

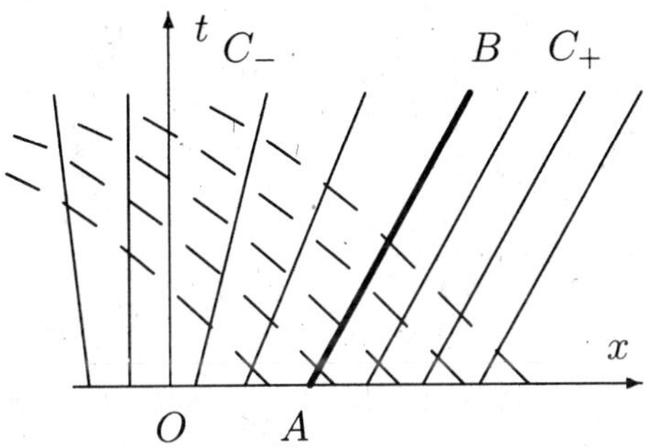

Figure 6.16: The solid lines are the C_+ characteristics; the dashed lines are the C_- characteristics.

$v + \frac{2c(v)}{\gamma-1} = \text{const}$. It follows from here that $v = \text{const}$ and $c = \text{const}$ along the C_+ characteristic, and we find from the first equation (6.2.14):

$$x = (v + c)t + f(v), \quad v - \frac{2c}{\gamma - 1} = \text{const}. \tag{6.2.15}$$

Since $v + c = \text{const}$ and $v = \text{const}$, the C_+ characteristics in a simple wave are the straight lines (see Fig. 6.16). An important property of simple waves is that they *adjoin continuously the region of constant flow* $v = \text{const}$ and $P = \text{const}$. Let us prove this assertion. Let the wave propagating to the right adjoin a region of constant flow along the line AB coinciding with some C_+ characteristic (see Fig. 6.16). The region of constant flow is located to the right of AB, both Riemann invariants $\mathcal{I}_+$ and $\mathcal{I}_-$ are constant in it, and the characteristics C_+ and C_- are the straight lines. Then the C_- characteristics will intersect the line AB and transfer a constant value of $\mathcal{I}_-$ into a region to the left of the line AB. Thus, the flow to the left of AB will be a simple wave.

It is easy to show in the same way that the invariant $\mathcal{I}_+$ should be constant in a simple wave propagating to the left at a velocity $\frac{dx}{dt} = v - c$. Integrating the third equation in (6.2.14), we obtain an equation for a simple wave propagating to the left:

$$x = (v - c)\,t + f(v), \quad v + \frac{2c}{\gamma - 1} = \text{const}. \tag{6.2.16}$$

The first equation in (6.2.16) determines the C_- characteristics, which are also the straight lines.

In a particlular case $f(v) = 0$, a simple *centered* wave (6.2.15) propagating to the right takes place:

$$v + c = \frac{x}{t}, \qquad v - \frac{2c}{\gamma - 1} = \text{const}, \tag{6.2.17}$$

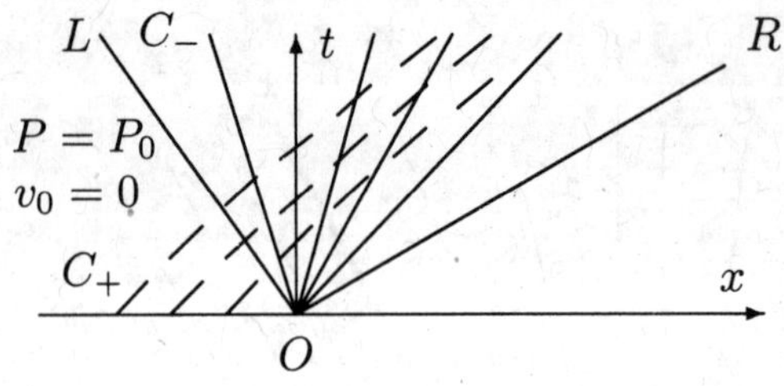

Figure 6.17: The solid lines are the C_- characteristics; the dashed lines are the C_+ characteristics in the problem of the gas outflow into vacuum.

and to the left (6.2.16):

$$v - c = \frac{x}{t}, \qquad v + \frac{2c}{\gamma - 1} = \text{const.} \tag{6.2.18}$$

It can be seen from (6.2.17) and (6.2.18) that the v and c depend on a self-similar variable $\xi = x/t$. Thus, a simple centered wave is described by a self-similar solution of equations (6.2.1). Note that an opposite assertion is also valid. In the one-dimensional case, a self-similar planar wave depending on a self-similar variable $\xi = x/t$ is a simple wave. The waves (6.2.17) and (6.2.18) are centered at point $t = 0$ and $x = 0$. If the wave is centered at point t^*, x^*, then it will also be described by formulas (6.2.17) and (6.2.18), if one replaces therein $t \to t - t^*$, $x \to x - x^*$ (one makes a shift transformation in time and in coordinate).

A method of a passage from the variables t, x to the variables $\mathcal{I}_+, \mathcal{I}_-$ is used in a general case while studying the interaction of planar waves. As a result, one obtains a linear Darboux equation with variable coefficients. One can construct for a perfect gas a solution, which is expressed in terms of the elementary functions. A region occupied by such a solution may adjoin either a simple wave or a wall. The reader can find the details of the formulation of this problem in the well-known monographs[8–10].

Problem 6.3. A perfect gas with the parameters $P_0, \rho_0, v_0 = 0$ fills a tube at time $t = 0$ at $x < 0$. The vacuum $P = 0$ is to the right at $x > 0$. Find $v(t, x)$, $\rho(t, x)$, and $P(t, x)$ at the gas outflow into vacuum.

Solution: After a gas outflow into vacuum begins, a rarefaction wave will propagate to the left (see Fig. 6.17). Since it adjoins a stagnation region $v_0 = 0$, $P = P_0$ along the characteristic L, this will be a simple wave. It is generally described by equations (6.2.16). There is no reference length in the formulation of the given problem; therefore, $f(v) = 0$ and a centered rarefaction wave (6.2.18) takes place. The second equation in

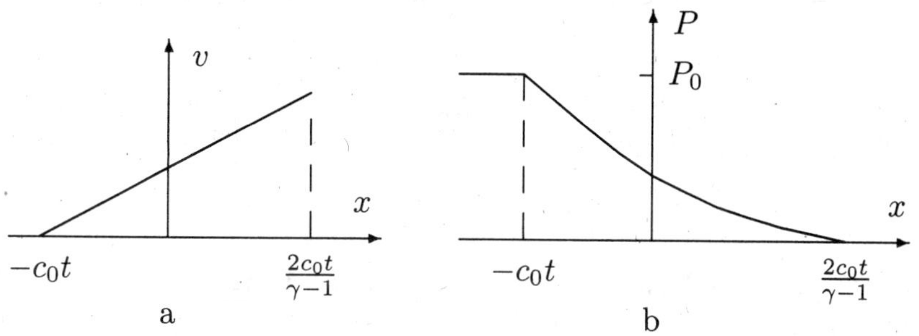

Figure 6.18: The dependencies $v(x)$ and $P(x)$ at the gas outflow into vacuum.

(6.2.18) is valid in the overall flow region, hence, at $t = 0$, from where we find:

$$v + \frac{2c}{\gamma - 1} = \frac{2c_0}{\gamma - 1}, \quad v - c = \frac{x}{t}. \tag{6.2.19}$$

Solving these equations with respect to v and c, we obtain a self-similar solution depending on the self-similar variable x/t:

$$v = \frac{2}{\gamma+1}\Big(c_0 + \frac{x}{t}\Big), \quad c = \frac{2}{\gamma+1}\Big(c_0 - \frac{\gamma-1}{2}\frac{x}{t}\Big). \tag{6.2.20}$$

Since the gas expansion is adiabatic, we find from the Poisson adiabat:

$$\frac{T}{T_0} = \Big(\frac{c}{c_0}\Big)^2, \quad \frac{\rho}{\rho_0} = \Big(\frac{c}{c_0}\Big)^{\frac{2}{(\gamma-1)}}, \quad \frac{P}{P_0} = \Big(\frac{c}{c_0}\Big)^{\frac{2\gamma}{(\gamma-1)}} \tag{6.2.21}$$

Combining equations (6.2.20) and (6.2.21), we obtain the distributions of density $\rho(t,x)$ and pressure $P(t,x)$:

$$\begin{aligned} \rho &= \rho_0\Big(\frac{2}{\gamma+1}\Big)^{\frac{2}{(\gamma-1)}}\Big(1 - \frac{(\gamma-1)}{2}\frac{x}{c_0t}\Big)^{\frac{2}{(\gamma-1)}}, \\ P &= P_0\Big(\frac{2}{\gamma+1}\Big)^{\frac{2\gamma}{(\gamma-1)}}\Big(1 - \frac{(\gamma-1)}{2}\frac{x}{c_0t}\Big)^{\frac{2\gamma}{(\gamma-1)}}. \end{aligned} \tag{6.2.22}$$

The left characteristic L is found from the condition $v = 0$, the substitution of which in the first equation of (6.2.20) yields $x = -c_0t$. The right characteristic R adjoins the vacuum $\rho = 0$; therefore, we find from (6.2.22) its equation $x = \frac{2c_0}{(\gamma-1)}t$. It follows from here that the maximum velocity of the gas outflow into vacuum is equal to $v_* = \frac{2c_0}{\gamma-1}$. It is interesting to note that this velocity is by a factor of $\sqrt{\frac{2}{\gamma-1}}$ larger than the maximum velocity of a stationary gas outflow into vacuum

$v_m = \sqrt{\frac{2}{\gamma-1}}c_0$. The latter formula for v_m may be obtained from the Bernoulli integral $\frac{v^2}{2} + \frac{c^2}{\gamma-1} = \frac{c_0^2}{\gamma-1}$ if we set $v = v_m$, and $c = 0$ therein. We show in Fig. 6.18 the qualitative dependencies $v(x)$ and $P(x)$ at some fixed moment of time t, which are described by formulas (6.2.20) and (6.2.22), respectively.

6.2.2 Gradient Catastrophe and Shock Wave Formation

A continuous gas flow in the simple compression waves cannot exist as long as is wished. As a consequence of the nonlinearity of equations (6.2.1), in a compression wave an intersection of the characteristics of the same family takes place. An unbounded growth of the gradients of all quantities occurs at the point of intersection of the characteristics; therefore, this phenomenon is called the *gradient catastrophe.*

To prove this, let us consider a simple wave propagating to the right, in which

$$\frac{dx}{dt} = k = v + c, \quad \mathcal{I}_- = v - \int \frac{dP}{\rho c} \equiv \text{const}, \tag{6.2.23}$$

where the coefficient k determines the inclination of a rectilinear C_+ characteristic in the t, x plane. It follows from these equations that

$$\frac{dk}{d\rho} = \frac{dv}{d\rho} + \frac{dc}{d\rho} = \frac{1}{\rho c}\frac{dP}{d\rho} + \frac{dc}{d\rho} = \frac{c}{\rho} + \frac{dc}{d\rho} = \frac{1}{\rho}\frac{d(\rho c)}{d\rho}. \tag{6.2.24}$$

Using the expression

$$\rho c = \rho\sqrt{\frac{\partial P}{\partial \rho}} = \sqrt{-\frac{\partial P}{\partial V}} = 1/\sqrt{-\frac{\partial V}{\partial P}},$$

let us rewrite (6.2.24) in the form

$$\frac{d(\rho c)}{d\rho} = c^2\frac{d(\rho c)}{dP} = \frac{\rho^3 c^5}{2}\left(\frac{\partial^2 V}{\partial P^2}\right)_S. \tag{6.2.25}$$

Substituting (6.2.25) in (6.2.24), we obtain for a normal gas the inequality

$$\frac{dk}{d\rho} = \frac{\rho^2 c^5}{2}\left(\frac{\partial^2 V}{\partial P^2}\right)_S = G > 0. \tag{6.2.26}$$

Differentiating the second equation in (6.2.23) with respect to x:

$$\frac{\partial v}{\partial x} = \frac{\partial P}{c\rho\partial x} = \frac{c}{\rho}\frac{\partial \rho}{\partial x}$$

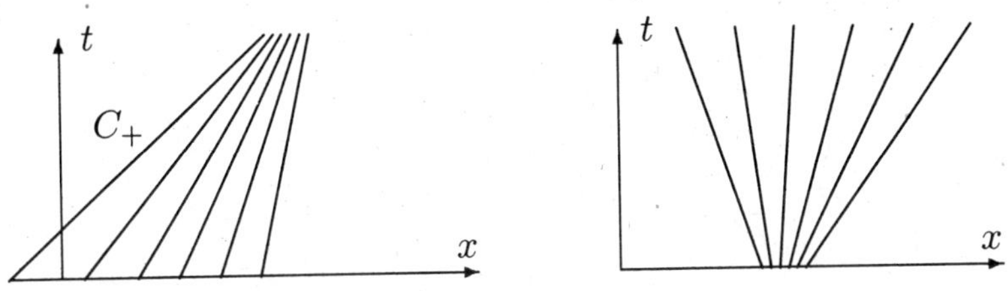

Figure 6.19: The characteristics (a) in a simple compression wave and (b) in a simple rarefaction wave.

and using the continuity equation [the first equation in (6.2.1)], we write: $\frac{\partial \rho}{\partial x} = -\frac{1}{c}\frac{d\rho}{dt}$. Combining this equation with (6.2.26), we find:

$$\frac{\partial k}{\partial x} = -\frac{G}{c}\frac{d\rho}{dt}, \tag{6.2.27}$$

where $G > 0$. One can show in a similar way that formula (6.2.27) holds also for a simple wave propagating to the left. Since $\frac{d\rho}{dt} > 0$ in a *compression wave*, and $\frac{d\rho}{dt} < 0$ in a *rarefaction wave*, the characteristics in the compression wave and in the rarefaction wave will behave differently (see Fig. 6.19). The characteristics approach one another in a simple compression wave as t increases; therefore, the gradient $|k_x|$ increases and, consequently, the $|v_x|$, $|\rho_x|$, $|P_x|$ increase. At the point of the characteristics intersection, these derivatives turn to infinity, which means the *gradient catastrophe*. At this time, a discontinuity arises in the solution, which is called the *shock wave*. The shock waves were predicted for the first time theoretically in this way by Riemann. Riemann, however, used as the jump relations at a shock wave the conservation laws for the gas mass, momentum, and entropy. It turned out that the gas entropy is not conserved but increases at a passage across the shock wave front as a consequence of the viscosity and heat conduction action. Hugoniot proposed to use the energy conservation law instead of the entropy conservation. The relations obtained in this way were analyzed in Section 6.1.3. Note that the characteristics in the rarefaction wave diverge with increasing time t [see Fig. 6.19 (b)]; therefore, the shock waves do not arise in a normal gas in the rarefaction wave.

Problem 6.4. A piston moves at a constant acceleration a into a tube filled with a perfect gas at rest ($v_0 = 0$, $\rho = \rho_0$, $P = P_0$) with the adiabatic exponent γ (see Fig. 6.20). Find the time of the shock wave formation.

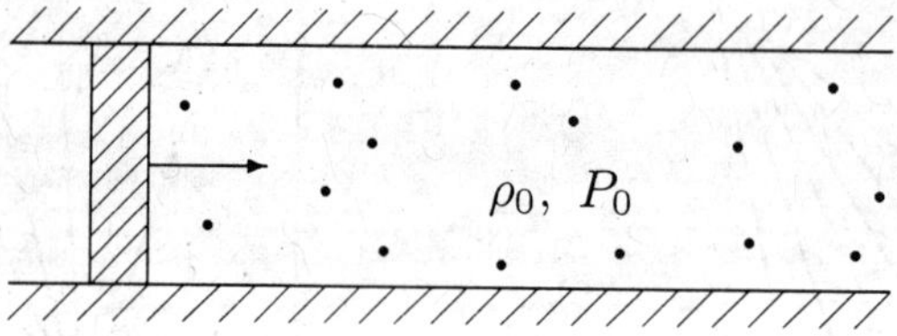

Figure 6.20: The piston motion in a tube filled with a gas at rest.

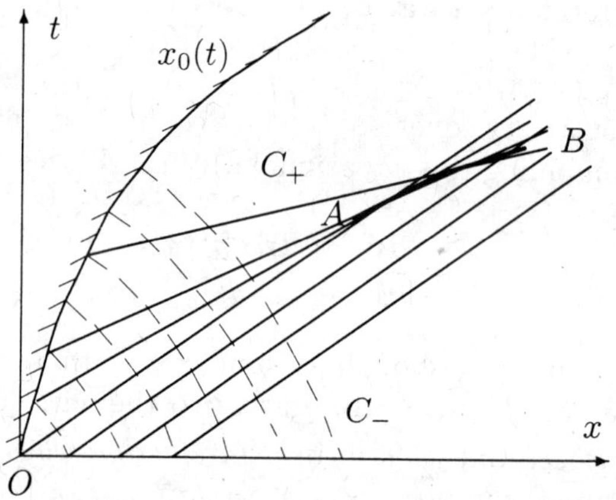

Figure 6.21: The picture of characteristics in the piston problem.

Solution: The piston sends ahead of itself a compression wave, the flow behind which adjoins the stagnation region. Therefore, this wave is simple (see Fig. 6.21). Solid lines in Fig. 6.21 show the C_+ characteristics, the dashed lines show the C_- characteristics. A solid line with dashes corresponds to the piston trajectory, which is described by the equation $x_0 = at^2/2$. All C_+ characteristics in the simple wave are rectilinear, and all gas parameters retain their constant values along them. The flow region adjoins the rest of the region along the characteristic OA, on which by virtue of the continuity condition the equalities $v = 0$ and $c = c_0$ are valid, where $c_0 = \sqrt{\gamma P_0/\rho_0}$ is the sound velocity in the nondisturbed gas. The solution in a simple wave is determined by formulas (6.2.15). We will rewrite these formulas in the form

$$x = (v + c)t + f(v), \quad v - \frac{2c}{\gamma - 1} = -\frac{2c_0}{\gamma - 1}. \tag{6.2.28}$$

Eliminating from here the sound velocity c, we obtain:

$$x = \Big(\frac{(\gamma - 1)}{2}v + c_o\Big)t + f(v). \tag{6.2.29}$$

The function $f(v)$ is determined from the boundary condition: the gas velocity on the piston surface is equal to the piston velocity $v(t, x_0(t)) = \dot{x}_0(t)$. Substituting this condition in (6.2.29), we obtain:

$$f(\dot{x}_0) = x_0(t(\dot{x}_0)) - \left(\frac{\gamma+1}{2}\dot{x}_0 + c_0\right)t(\dot{x}_0). \tag{6.2.30}$$

Substituting here the relations

$$x_0(t(\dot{x}_0)) = \frac{\dot{x}_0^2}{2a}, \qquad t(\dot{x}_0) = \frac{\dot{x}_0}{a},$$

we arrive at the expression

$$f(\dot{x}_0) = -\left(c_0\frac{\dot{x}_0}{a} + \frac{\gamma}{2a}\dot{x}_0^2\right),$$

from where we obtain the formula:

$$f(v) = -\left(c_0\frac{v}{a} + \frac{\gamma}{2a}v^2\right).$$

Substituting this formula in (6.2.29), we obtain a quadratic equation for determining the gas velocity:

$$v^2 - \frac{2}{\gamma}\left(-c_0 + \frac{\gamma+1}{2}at\right)v + \frac{2a}{\gamma}(x - c_0t) = 0$$

the solution of which has the form

$$v_{1,2} = \frac{1}{\gamma}\left(-c_0 + \frac{\gamma+1}{2}at\right) \pm \frac{1}{\gamma}\sqrt{\left(-c_0 + \frac{\gamma+1}{2}at\right)^2 + 2a\gamma(c_0t - x)}.$$

As was noted above, the velocity $v = 0$ and $x = c_0t$ on the line OA; therefore, one must take the minus sign before the square root:

$$v_{1,2} = \frac{1}{\gamma}\left(-c_0 + \frac{\gamma+1}{2}at\right) - \frac{1}{\gamma}\sqrt{\left(-c_0 + \frac{\gamma+1}{2}at\right)^2 + 2a\gamma(c_0t - x)}. \tag{6.2.31}$$

Let us turn to the analysis of the obtained solution. We show in Fig. 6.22 a qualitative dependence $v(x)$ described by equation (6.2.31). At time t_1, the velocity v is a smooth and single-valued function of x. At time $t_2 = t_A$, the derivative $\frac{\partial v}{\partial x}|_{xA} \to -\infty$, and a shock wave arises. At $t_3 > t_A$, the solution (6.2.31) is ambiguous, and therefore it does not describe the actual gas flow, which already contains a shock wave marked by a thick line in Fig. 6.21. Differentiating (6.2.31) with respect to x, we find the expression:

$$\frac{\partial v}{\partial x} = -a\gamma/\sqrt{\left(-c_0 + \frac{\gamma+1}{2}at\right)^2 + 2a\gamma(c_0t - x)}. \tag{6.2.32}$$

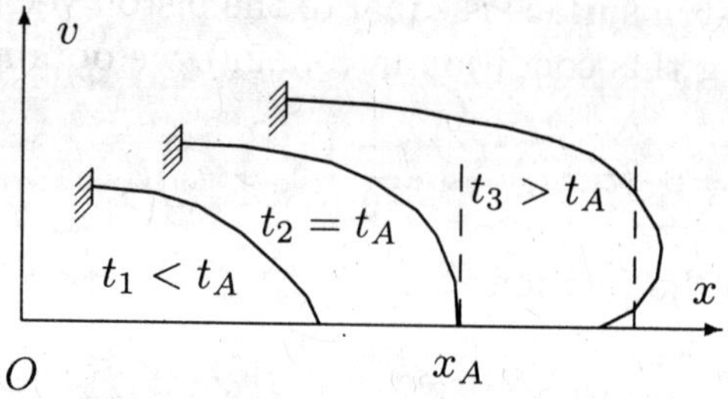

Figure 6.22: The dependence $v(x)$ described by (6.2.31) at three moments of time $t_1 < t_2 < t_3$.

In the simple wave $c_0 t \geq x$, where the equality is achieved on the characteristic OA. With regard for the above, the denominator in (6.2.32) vanishes at a simultaneous vanishing of both items:

$$c_0 - \frac{\gamma+1}{2}at = 0, \qquad x - c_0 t = 0.$$

It follows from here that the shock wave forms at the first C_+ characteristic OA emanating from the piston at time $t_A = \frac{2c_0}{(\gamma+1)a}$. Note that, if the piston acceleration $a \to \infty$, then the shock wave arises immediately at a piston motion: $t_A \to 0$.

6.3 Planar Irrotational Ideal Gas Motion (Linear Approximation)

6.3.1 Governing Equations and Their Linearization

Consider a planar motion of an ideal barotropic gas. We will assume that the gas motion is stationary and potential, in which the gas-dynamic functions will depend only on two variables x and y. Using the results presented in Chapter 2, we write the original system of the Euler equations in the form (2.1.153):

$$\begin{gathered}\rho\left(\frac{\partial u}{\partial x} + \frac{\partial v}{\partial y}\right) + u\frac{\partial \rho}{\partial x} + v\frac{\partial \rho}{\partial y} = 0,\\ \rho\left(u\frac{\partial u}{\partial x} + v\frac{\partial u}{\partial y}\right) = -c^2\frac{\partial \rho}{\partial x},\\ \rho\left(u\frac{\partial v}{\partial x} + v\frac{\partial v}{\partial y}\right) = -c^2\frac{\partial \rho}{\partial y}, \qquad P = A\rho^\gamma.\end{gathered} \tag{6.3.1}$$

Here A is a constant quantity, $\gamma = \frac{C_P}{C_V}$, $c^2 = \frac{dP}{d\rho}$, and from the barotropicity condition, we have

$$c^2 \frac{\partial \rho}{\partial x} = \frac{\partial P}{\partial x}, \quad c^2 \frac{\partial \rho}{\partial y} = \frac{\partial P}{\partial y}$$

and the velocity vector $\vec{v}\,(u, v)$, $u \equiv u_x$, and $v \equiv u_y$. Substituting the values of $\frac{\partial \rho}{\partial x}$ and $\frac{\partial \rho}{\partial y}$ from the momentum equations of (6.3.1) in the continuity equation, we obtain:

$$(c^2 - u^2)\frac{\partial u}{\partial x} - uv\Big(\frac{\partial u}{\partial y} + \frac{\partial v}{\partial x}\Big) + (c^2 - v^2)\frac{\partial v}{\partial y} = 0. \qquad (6.3.2)$$

In the case of an irrotational flow, we can write:

$$\frac{\partial u}{\partial y} - \frac{\partial v}{\partial x} = 0, \quad \frac{v^2}{2} + \frac{c^2}{\gamma - 1} = \frac{1}{2} v_\infty^2 + \frac{c_\infty^2}{\gamma - 1}. \qquad (6.3.3)$$

As follows from the above consideration, these equations can be used instead of the momentum equations (6.3.1). Introducing the velocity potential by formulas $u = \frac{\partial \varphi}{\partial x}$ and $v = \frac{\partial \varphi}{\partial y}$, we can transform (6.3.2) to the form

$$\Big(c^2 - \Big(\frac{\partial \varphi}{\partial x}\Big)^2\Big) \cdot \frac{\partial^2 \varphi}{\partial x^2} - 2\frac{\partial \varphi}{\partial x}\frac{\partial \varphi}{\partial y} \cdot \frac{\partial^2 \varphi}{\partial x \partial y} + \Big(c^2 - \Big(\frac{\partial \varphi}{\partial y}\Big)^2\Big)\frac{\partial^2 \varphi}{\partial y^2} = 0,$$

where the squared sound velocity is determined from the Bernoulli integral:

$$c^2 = \frac{\gamma - 1}{2} v_\infty^2 + c_\infty^2 - \frac{\gamma - 1}{2}\Big(\Big(\frac{\partial \varphi}{\partial x}\Big)^2 + \Big(\frac{\partial \varphi}{\partial y}\Big)^2\Big).$$

The obtained system of equations is nonlinear, and its solution generally involves considerable mathematical difficulties. We will study the flow around slender bodies for which the thickness is much less than their length. In this case, the solution of system (6.3.2) and (6.3.3) may be presented in the form

$$u = u_\infty + \tilde{u}, \; v = \tilde{v}, \; P = P_\infty + \tilde{P}, \; \rho = \rho_\infty + \tilde{\rho}, \; c = c_\infty + \tilde{c}.$$

Here $u_\infty, P_\infty, \rho_\infty$, and c_∞ are the constant quantities and $u_\infty \gg \tilde{u}$, $\tilde{u} \sim \tilde{v}$, $\tilde{P} \ll P_\infty$, $\tilde{\rho} \ll \rho_\infty$, $c_\infty \gg \tilde{c}$, including also their derivatives with respect to spatial variables. Substituting these values in equations (6.3.2) and (6.3.3) and neglecting the terms of the second order of smallness, we find the linearized equations:

$$\begin{gathered}
(c_\infty^2 - u_\infty^2)\frac{\partial \tilde{u}}{\partial x} + c_\infty^2 \frac{\partial \tilde{v}}{\partial y} = 0, \\
\frac{\partial \tilde{u}}{\partial y} - \frac{\partial \tilde{v}}{\partial x} = 0, \quad u_\infty \tilde{u} + \frac{c_\infty^2}{\rho_\infty}\tilde{\rho} = 0, \; c_\infty^2 = \gamma \frac{P_\infty}{\rho_\infty}, \qquad (6.3.4) \\
\tilde{P} = \gamma \frac{P_\infty}{\rho_\infty} \cdot \tilde{\rho}.
\end{gathered}$$

Dividing in the first equation of (6.3.4) the left- and right-hand sides by $c_\infty^2 \neq 0$, we obtain the system of equations:

$$(1 - M_\infty^2)\frac{\partial \tilde{u}}{\partial x} + \frac{\partial \tilde{v}}{\partial y} = 0, \quad \frac{\partial \tilde{u}}{\partial y} - \frac{\partial \tilde{v}}{\partial x} = 0, \tag{6.3.5}$$

where $M_\infty = \frac{u_\infty}{c_\infty}$. Knowing the $\tilde{u}$ and $\tilde{v}$, we can find the quantity $\tilde{\rho}$ from the equation $\tilde{\rho} = -\frac{M_\infty \tilde{u} \rho_\infty}{c_\infty}$. Thus, a complex nonlinear system (6.3.2) and (6.3.3) has been reduced to a linear system, for the solution of which one can apply the well-known methods of the mathematical physics.

6.3.2 The Problem of the Flow around a Slender Profile

Consider a problem of the subsonic gas flow around a slender body at a given velocity u_∞. We will assume that the profile contour is given by the equation $y = h_{1,2}(x)$, $a \leq x \leq b$, where the subscript 1 refers to the upper contour part and subscript 2 refers to the lower contour part. We consider the profile to be slender, and consequently, the angle Θ of a tangent to the x axis is determined by the equality $|\tan \Theta| \approx |\Theta| \approx |h'_{1,2}(x)|$. Let us present the velocity potential φ and the stream function ψ of the disturbed motion in the form of a sum

$$\varphi = \varphi_\infty + \tilde{\varphi}, \quad \psi = \psi_\infty + \tilde{\psi},$$

where $\tilde{\varphi}$ and $\tilde{\psi}$ are small perturbations of φ_∞ and ψ_∞; the latter quantities characterize the uniform flow. By definition of the velocity potential, we have:

$$u_\infty + \tilde{u} = \frac{\partial \varphi_\infty}{\partial x} + \frac{\partial \tilde{\varphi}}{\partial x}, \quad \tilde{v} = \frac{\partial \varphi_\infty}{\partial y} + \frac{\partial \tilde{\varphi}}{\partial y}.$$

The continuity equation [the first equation in (6.3.1)] will be identically satisfied if we introduce the stream function ψ by formulas

$$\rho u = \rho_\infty \frac{\partial \psi}{\partial y}, \quad \rho v = -\rho_\infty \frac{\partial \psi}{\partial x},$$

from which it follows that

$$(\rho_\infty + \tilde{\rho})(u_\infty + \tilde{u}) = \rho_\infty \left(\frac{\partial \psi_\infty}{\partial y} + \frac{\partial \tilde{\psi}}{\partial y} \right),$$

$$(\rho_\infty + \tilde{\rho})\tilde{v} = -\rho_\infty \left(\frac{\partial \psi_\infty}{\partial x} + \frac{\partial \tilde{\psi}}{\partial x} \right).$$

Comparing the quantities of the same orders in the obtained formulas, we obtain:

$$\varphi_\infty = u_\infty x, \quad \psi_\infty = u_\infty y,$$

$$\tilde{u} = \frac{\partial \tilde{\varphi}}{\partial x}, \quad \tilde{v} = \frac{\partial \tilde{\varphi}}{\partial y}, \quad \tilde{\rho} u_\infty + \rho_\infty \tilde{u} = \rho_\infty \frac{\partial \tilde{\psi}}{\partial y}, \quad \tilde{v} = -\frac{\partial \tilde{\psi}}{\partial x}. \tag{6.3.6}$$

Substituting (6.3.6) in formulas (6.3.4) and (6.3.5), we find the basic linearized equations for the determination of the perturbed velocity potential φ:

$$(1 - M_\infty^2)\frac{\partial^2 \tilde{\varphi}}{\partial x^2} + \frac{\partial^2 \tilde{\varphi}}{\partial y^2} = 0 \tag{6.3.7}$$

and of the stream function of the perturbation $\tilde{\psi}$:

$$(1 - M_\infty^2)\frac{\partial^2 \tilde{\psi}}{\partial x^2} + \frac{\partial^2 \tilde{\psi}}{\partial y^2} = 0. \tag{6.3.8}$$

The magnitude of the disturbance pressure can be determined from the linearized Bernoulli integral by formula

$$\tilde{P} = -\rho_\infty u_\infty \tilde{u}, \tag{6.3.9}$$

which enables us to obtain for the pressure coefficient C_p the equation:

$$C_p = \frac{2\tilde{P}}{\rho_\infty u_\infty^2} = -\frac{2\tilde{u}}{u_\infty}.$$

It follows from the obtained equations (6.3.6) and (6.3.7) that, at $M_\infty < 1$, they will have the elliptic type and at $M_\infty > 1$, they will have the hyperbolic type.

Let us write the boundary condition on the profile contour in two different forms: either in terms of the velocity potential or in terms of the stream function. The slip condition for the profile contour will have the form

$$v_n = u_{\infty n} + \tilde{v}_n = 0 \quad \text{at} \quad y = h_{1,2}(x), \quad x \in [a, b]$$

or with regard for the expansion powers

$$\tilde{v} = \frac{\partial \tilde{\varphi}}{\partial y} = u_\infty h'(x) \quad \text{at} \quad y = h_{1,2}(x), \quad x \in [a, b].$$

If we use formulas (6.3.6), then the condition $\psi = \psi_\infty + \tilde{\psi} = u_\infty y + \tilde{\psi} = 0$ on the contour may be transformed to the form

$$\tilde{\psi} = -u_\infty h_{1,2}(x), \quad y = h_{1,2}(x), \quad x \in [a, b]. \tag{6.3.10}$$

Since we assume the profile to be slender, we will require the satisfaction of the condition (6.3.10), not on the contour itself but on its projection onto the Ox-axis, that is on a segment of the Ox axis between $x = a$ and $x = b$, and the upper and lower sides of the segment correspond to the condition $y = \pm 0$. Thus, the condition (6.3.10) for the stream function of the disturbances has the form

$$\tilde{\psi} = -u_\infty h_{1,2}(x), \quad y = \pm 0, \quad a \leq x \leq b. \tag{6.3.11}$$

Consider at first the subsonic gas flow ($M_\infty < 1$) around a slender profile. Then the condition at an infinite distance from the profile is

$$\tilde{\psi} \to 0 \quad \text{at } y \to \pm\infty. \tag{6.3.12}$$

If we go over from the x, y coordinates to the new coordinates ξ, η by formulas

$$\xi = x, \qquad \eta = \omega y,$$

where $\omega^2 = 1 - M_\infty^2$, then the mathematical formulation of the problem of the subsonic gas flow around a slender profile reduces to the following. Find the solution of equation

$$\frac{\partial^2 \tilde{\psi}}{\partial \xi^2} + \frac{\partial^2 \tilde{\psi}}{\partial \eta^2} = 0 \tag{6.3.13}$$

in the (ξ, η) plane under the conditions

$$\begin{aligned} &\tilde{\psi} = -u_\infty h_{1,2}(\xi), \quad \eta = \pm 0, \quad a \le \xi \le b, \\ &\tilde{\psi} \to 0, \quad \eta \to \pm\infty. \end{aligned}$$

Consider a problem formulation for the incompressible fluid flow at a velocity u_∞ around the same profile; that is, we will have:

$$\begin{aligned} &\frac{\partial^2 \tilde{\psi}_0}{\partial x^2} + \frac{\partial^2 \tilde{\psi}_0}{\partial y^2} = 0, \\ &\tilde{\psi}_0 = -u_0 h_{1,2}(x) \qquad \text{at } y = \pm 0, \ a \le x \le b, \\ &\tilde{\psi}_0 \to 0 \ \text{ at infinity,} \end{aligned} \tag{6.3.14}$$

where $\tilde{\psi}_0$ is a stream function of the incompressible fluid flow. Comparing two problem formulations (6.3.13) and (6.3.14), one may conclude that the solutions of these equations are identical, but only in different variables (ξ, η) and (x, y); that is,

$$\tilde{\psi}(\xi, \eta) = \tilde{\psi}_0(x, y), \quad \frac{\partial \tilde{\psi}}{\partial \xi} = \frac{\partial \tilde{\psi}_0}{\partial x}, \quad \frac{\partial \tilde{\psi}}{\partial \eta} = \frac{\partial \tilde{\psi}_0}{\partial y}.$$

Using these equalities, we now transform (6.3.6) with the aid of the Bernoulli integral and the relation $\tilde{P} = \tilde{\rho} c_\infty^2$ to the form

$$\begin{aligned} \tilde{u} &= \frac{1}{1 - M_\infty^2} \cdot \frac{\partial \tilde{\psi}}{\partial y} = \frac{1}{1 - M_\infty^2} \cdot \frac{\partial \tilde{\psi}}{\partial \eta} \cdot \frac{\partial \eta}{\partial y} \\ &= \frac{1}{\sqrt{1 - M_\infty^2}} \cdot \frac{\partial \tilde{\psi}_0}{\partial y} = \frac{\tilde{u}_0}{\sqrt{1 - M_\infty^2}}, \\ \tilde{v} &= -\frac{\partial \tilde{\psi}}{\partial x} = -\frac{\partial \tilde{\psi}}{\partial \xi} = -\frac{\partial \tilde{\psi}_0}{\partial x} = \tilde{v}_0. \end{aligned} \tag{6.3.15}$$

The obtained formulas enable one to construct the solution of the problem of a compressible flow around a slender profile on the basis of the solution of a problem of the incompressible fluid flow around the same profile. The pressure coefficient C_p is expressed with regard for formulas (6.3.15) by formula

$$C_p = \frac{C_p^0}{\sqrt{1 - M_\infty^2}},$$

where C_p^0 is the pressure coefficient computed from the problem of the incompressible fluid flow around a profile. The obtained formula expresses the following *Prandtl-Glauert rule*: the distribution of a pressure coefficient in a planar irrotational linearized subsonic gas flow at a given value $M_\infty < 1$ can be obtained from the corresponding distribution in the incompressible fluid flow if one increases all ordinates of this distribution by a factor of $1/\sqrt{1 - M_\infty^2}$.

Consider the case of *supersonic flow* around a slender profile; that is, $M_\infty > 1$. Present the governing equation as

$$\frac{\partial^2 \tilde{\psi}}{\partial x^2} - \left(\frac{1}{M_\infty^2 - 1}\right) \cdot \frac{\partial^2 \tilde{\psi}}{\partial y^2} = 0. \tag{6.3.16}$$

Since $M_\infty > 1$ and $1/(M_\infty^2 - 1) > 0$, equation (6.3.16) will coincide with the wave equation whose general solution is expressed by formula

$$\tilde{\psi} = \Omega_1(x - \omega y) + \Omega_2(x + \omega y), \tag{6.3.17}$$

where $\omega = \sqrt{M_\infty^2 - 1}$ and Ω_1 and Ω_2 are arbitrary functions of their arguments. By analogy with the subsonic flow around a slender profile, we present the boundary condition on the contour in the form

$$\tilde{\psi} = -u_\infty h_{1,2}(x) \quad \text{at} \quad y = \pm 0, \quad a \leq x \leq b. \tag{6.3.18}$$

Considering the particular solution

$$\tilde{\psi}_1 = \Omega_1(x - \omega y)$$

we see that a family of straight lines $x - \omega y = \text{const}$ exists along which the stream function $\tilde{\psi}_1$ will be constant and, consequently, all the gas-dynamic functions will be constant. Similarly, $\tilde{\psi}_2 = \Omega_2(x + \omega y)$ preserve the constant values along $x + y\,\omega = \text{const}$. The lines $x + y\omega = \text{const}$ and $x - \omega y = \text{const}$ are the characteristics of two different families C_2 and C_1, respectively. Let us fill the flow region with the characteristics of the first family C_1 and second family C_2 (see Fig. 6.23). Using the properties of the solution (6.3.17) and the boundary condition (6.3.18), we can present the final solution in the form

$$\tilde{\psi} = -u_\infty h_{1,2}(x \mp \omega y), \tag{6.3.19}$$

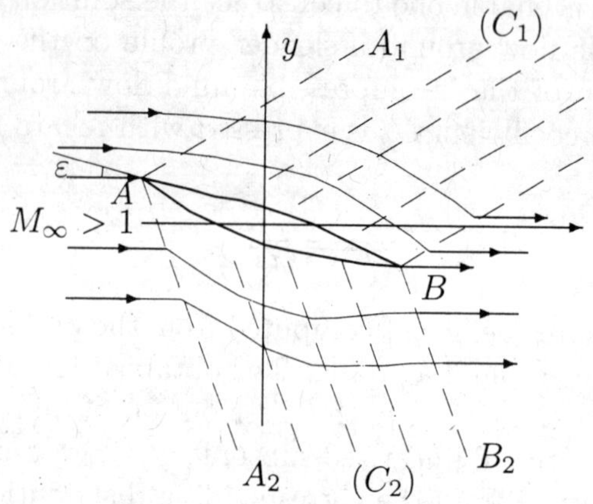

Figure 6.23: The qualitative gas flow pattern around a slender profile at $M_\infty > 1$.

where the upper contour part corresponds to index 1, and the lower contour part corresponds to index 2. It is seen from the constructed solution that the function $\tilde{\psi}(x, y)$ does not vanish at a large distance from the profile contour, but preserves the same distribution along x, as on the profile surface at $y \to \pm\infty$ inside the characteristics AA_1, BB_1 and AA_2, BB_2. Outside the above region, the flow remains homogeneous and its velocity is equal to u_∞. Thus, the complete disturbed flow will be presented by the stream function $\psi = u_\infty y - u_\infty h_{1,2}(x \mp \omega y)$. The streamlines $\psi = \text{const}$ represent the curves, which are obtained by a parallel translation of the upper and lower profile contours along the characteristics of the first and second families (Fig. 6.23). The streamlines will have a break at the characteristics AA_1, AA_2 and BB_1, BB_2. Let us compute the quantities $\tilde{u}$ and $\tilde{v}$ by formulas

$$\begin{aligned} \tilde{u} &= -\frac{1}{M_\infty^2 - 1}\frac{\partial\tilde{\psi}}{\partial y} = \mp\frac{u_\infty}{\sqrt{M_\infty^2 - 1}} \cdot h'_{1,2}(x \mp \omega y), \\ \tilde{v} &= -\frac{\partial\tilde{\psi}}{\partial x} = u_\infty h'_{1,2}(x \mp \omega y), \end{aligned}$$

where the primes denote the derivatives with respect to the complete argument, $z = x \mp \omega y$. Let us determine the angle $\Theta_{1,2}$ of the deviation of a tangent to the streamline in the disturbed region from the streamline of the nondisturbed flow:

$$\tan\Theta_{1,2} \approx \Theta_{1,2} \approx \frac{\tilde{v}}{u_\infty + \tilde{u}} \approx \frac{\tilde{v}}{u_\infty} \approx h'_{1,2}(x \mp \omega y),$$

and then

$$\tilde{u} = \mp\frac{u_\infty}{\sqrt{M_\infty^2 - 1}} \cdot \Theta_{1,2}, \quad \tilde{v} = u_\infty\Theta_{1,2}.$$

These formulas establish a relation between the streamwise and transverse components of the disturbance velocity at the given M_∞^2 and the local inclination angle $\Theta_{1,2}$. Computing the pressure coefficient $C_p = -\frac{2\tilde{u}}{u_\infty}$, we have:

$$C_p = \pm\frac{2\Theta_{1,2}(x \mp \omega y)}{\sqrt{M_\infty^2 - 1}}.$$

We find from here on the contour surface $y = \pm 0$:

$$C_p = \frac{\pm 2\Theta_{1,2}(x)}{\sqrt{M_\infty^2 - 1}}. \tag{6.3.20}$$

It should be noted that, at the subsonic flow, as in the incompressible fluid flow, the disturbance of the gasdynamic quantities at any flow point depends on the contour shape on the whole. In the case of supersonic flow, the variation of the contour shape near one of its points affects the magnitude of the disturbances of parameters only along that disturbance line which passes through this point. Such a local variation of the profile shape does not cause any distortions in the disturbances distribution in the remaining flow. This is related to different types of the differential equations governing the flows.

We can find the lift coefficient C_y by performing the integration from B to A by formula

$$C_y = \oint C_p d(\frac{x}{l}) = \frac{1}{l}\int_a^b (C_{p2} - C_{p1})\, dx,$$

where $l = b - a$, b and a are the abscissas of points B and A. Substituting the values C_{p2} and C_{p1}, we find:

$$\begin{aligned} C_y &= -\frac{2}{l\sqrt{M_\infty^2 - 1}}\int_a^b [h_2'(x) + h_1'(x)]\, dx \\ &= -\frac{2}{\sqrt{M_\infty^2 - 1}} \cdot \frac{2(y_B - y_A)}{(x_B - x_A)}. \end{aligned}$$

Introducing the angle of profile attack ε as the angle between the direction of the chord AB and u_∞, $\tan\varepsilon \approx (y_B - y_A)/(x_B - x_A)$, we find:

$$C_y = -\frac{4\varepsilon}{\sqrt{M_\infty^2 - 1}}.$$

This formula is called the *Ackeret's formula.* It is seen that the lift coefficient does not depend on the profile shape, but depends on the angle of attack ε and the freestream Mach number M_∞. In contrast with the subsonic flow, the quantity $C_y \neq 0$, and the D'Alembert's paradox does not take place in the given case. In the subsonic flow, the pressure in the trailing profile part is restored and creates a force counteracting the main vector of the pressure forces in the leading profile part. At a supersonic flow, however, there is no such equilibration. In the trailing expanding flow region, a phenomenon similar to the flow acceleration in the Laval nozzle takes place, where the flow continues to accelerate in the diverging part, and the pressure reduces, which leads to an additional suction force directed downstream. Thus, unlike the subsonic flow around the profile, the main vectors of the pressure forces on the leading and trailing parts of the profile surface do not cancel each other but on the contrary they are composed to form the integral wave drag force. The wave drag coefficient C_x is by definition equal to

$$C_x = \frac{R_x}{\frac{1}{2}\rho_\infty u_\infty^2 \cdot l}.$$

Computing

$$\begin{aligned} R_x &= -\oint P \cdot n_x \, dS = -\oint P \, dy = -\frac{1}{2}u_\infty^2 \rho_\infty \oint C_p \cdot \frac{dy}{dx} \, dx \\ &= \frac{1}{2}u_\infty^2 \rho_\infty \int_a^b \left(C_{p1}\frac{dy_1}{dx} - C_{p2}\frac{dy_2}{dx} \right) dx \end{aligned}$$

and taking (6.3.20) into account, we will have:

$$C_x = \frac{2}{l\sqrt{M_\infty^2 - 1}} \int_a^b \left\{ [h_1'(x)]^2 + [h_2'(x)]^2 \right\} dx.$$

Since

$$h_1'(x) = h_2'(x) = -\frac{y_A - y_B}{x_B - x_A} \approx \tan\varepsilon \approx \varepsilon,$$

we find the Ackeret formula for the wave drag coefficient in the form $C_x = 4\varepsilon^2/\sqrt{M_\infty^2 - 1}$. It should be noted that a detailed discussion of the question about the scope of applicability of the Ackeret's formulas as well as the experimental data may be found in the relevant literature[2].

6.4 Planar Irrotational Stationary Ideal Gas Flow (General Case)

We have already considered in Section 6.3 the planar irrotational flows of ideal gas in the case in which the disturbances introduced by the bodies in the flow are small. In the present section, we refuse this assumption and study the planar irrotational flows in a general case, in which the nonlinear terms in the governing equations play a significant role.

6.4.1 Characteristics of Stationary Irrotational Flows of Ideal Gas, Simple Wave: The Prandtl–Meyer Flow

The equations governing the planar stationary irrotational isentropic flows of ideal gas were derived in Section 6.3 [see equations (6.3.2) and (6.3.3)]. Retaining the same notations (u is the velocity along the x-axis, and v is the velocity along the y-axis), let us write the equations for the velocity:

$$\begin{aligned}&(c^2-u^2)\frac{\partial u}{\partial x}-uv\left(\frac{\partial u}{\partial y}+\frac{\partial v}{\partial x}\right)+(c^2-v^2)\frac{\partial v}{\partial y}=0,\\ &\frac{\partial u}{\partial y}-\frac{\partial v}{\partial x}=0,\end{aligned} \tag{6.4.1}$$

and the Bernoulli integral

$$\frac{u^2+v^2}{2}+\frac{c^2}{\gamma-1}=\frac{c_0^2}{\gamma-1},\quad c^2=\gamma\frac{P}{\rho},\quad P=A\rho^\gamma, \tag{6.4.2}$$

where c is the sound speed, A is a constant, and the subscript 0 refers to the stagnation point.

In order to construct a solution of the system of partial differential equations (6.4.1), it is necessary to find the characteristics and determine the system type. For this purpose, we use the general technique for finding the characteristics of a system of quasilinear equations presented above in Section 3.2. Let a characteristic of system (6.4.1) be given by equation $y=y(x)$. Then the following relations should hold along it:

$$\frac{\partial u}{\partial x}+\frac{dy}{dx}\frac{\partial u}{\partial y}=\frac{du}{dx},\quad \frac{\partial v}{\partial x}+\frac{dy}{dx}\frac{\partial v}{\partial y}=\frac{dv}{dx}. \tag{6.4.3}$$

Expressing from here the derivatives $\partial u/\partial x$, $\partial v/\partial x$ and substituting the obtained expressions in (6.4.1), we obtain:

$$\begin{aligned}&(y'(c^2-u^2)+uv)\frac{\partial u}{\partial y}-(y'uv+c^2-v^2)\frac{\partial v}{\partial y}\\ &=(c^2-u^2)\frac{du}{dx}-uv\frac{dv}{dx},\\ &\frac{\partial u}{\partial y}+y'\frac{\partial v}{\partial y}=\frac{dv}{dx},\end{aligned} \tag{6.4.4}$$

where $y'=\frac{dy}{dx}$. Considering (6.4.4) as a linear inhomogeneous system of equations for $\frac{\partial u}{\partial y}$ and $\frac{\partial v}{\partial y}$ and requiring that there is no uniqueness in the determination of these derivatives, we can find the equations for the determination of the characteristics:

$$\begin{vmatrix}(y'(c^2-u^2)+uv) & -(y'uv+c^2-v^2)\\ 1 & y'\end{vmatrix}=0, \tag{6.4.5}$$

$$\begin{vmatrix} (y'(c^2-u^2)+uv) & (\frac{dy}{dx}(c^2-u^2)-\frac{dv}{dx}uv \\ 1 & \frac{dv}{dx} \end{vmatrix} = 0. \qquad (6.4.6)$$

The calculation of determinant (6.4.5) yields the quadratic equation

$$(y')^2(u^2-c^2) - 2uvy' + v^2 - c^2 = 0, \qquad (6.4.7)$$

whose roots give the characteristics of system (6.4.1)[7]:

$$y'_{1,2} = \frac{uv \pm c\sqrt{u^2+v^2-c^2}}{u^2-c^2}. \qquad (6.4.8)$$

It follows from here that, for the subsonic flows, where $u^2+v^2<c^2$, the characteristics are complex and the system (6.4.1) belongs to the elliptic type. For a supersonic flow, $u^2+v^2>c^2$ the characteristics (6.4.8) are real, and the system (6.4.1) is hyperbolic.

Consider in more detail the supersonic flows for which there are two families of real characteristics:

$$\frac{dy_1}{dx} = \frac{uv + c\sqrt{u^2+v^2-c^2}}{u^2-c^2}, \quad C_+, \qquad (6.4.9)$$

$$\frac{dy_2}{dx} = \frac{uv - c\sqrt{u^2+v^2-c^2}}{u^2-c^2}, \quad C_-. \qquad (6.4.10)$$

Let us find the relations along the C_+ and C_- characteristics. Calculating the determinant (6.4.6):

$$(y'(c^2-u^2)+2uv)\frac{dv}{dx} - (c^2-u^2)\frac{du}{dx} = 0, \qquad (6.4.11)$$

we can find with regard for (6.4.7):

$$\frac{v^2-c^2}{y'}\frac{dv}{dx} + (u^2-c^2)\frac{du}{dx} = 0. \qquad (6.4.12)$$

Multiplying (6.4.9) and (6.4.10), we obtain the relation $y'_1y'_2 = (v^2-c^2)/(u^2-c^2)$, with regard for which, we can rewrite (6.4.12) as

$$\frac{dv}{dx} + \frac{y'}{y'_1y'_2}\frac{du}{dx} = 0. \qquad (6.4.13)$$

It follows from here that the relation

$$dv + \frac{1}{y'_2}du = 0 \qquad (6.4.14)$$

is satisfied along the C_+ characteristic, and along the C_- characteristic, the equation

$$dv + \frac{1}{y_1'}\, du = 0 \tag{6.4.15}$$

is satisfied.

The equations of characteristics (6.4.9) and (6.4.10) and the relations (6.4.14) and (6.4.15) on them will take a simpler form, if one uses instead of u and v the variables velocity module $q = \sqrt{u^2 + v^2}$, the angle between the velocity vector and the x-axis equal to $\theta = \arctan(v/u)$. The passage to the new variables is performed in accordance with formulas

$$u = q\cos\theta, \quad v = q\sin\theta. \tag{6.4.16}$$

Let us introduce the *Mach angle* α, determined by the expression

$$\sin\alpha = c/q \tag{6.4.17}$$

and the critical sound velocity $c_* = c = q$, which is related to the sound speed of the stagnation flow c_0 by formula

$$c_* = c_0\sqrt{\frac{2}{\gamma+1}}. \tag{6.4.18}$$

Using formulas (6.4.17) and (6.4.18), we can rewrite the Bernoulli integral (6.4.2) as follows:

$$\sin^2\alpha = \frac{\gamma+1}{2}\frac{c_*^2}{q^2} - \frac{\gamma-1}{2}. \tag{6.4.19}$$

Substituting the expressions for u and v from (6.4.16) in formula (6.4.8) and taking (6.4.17) and (6.4.19) into account, we can rewrite the equations for the characteristics in the form

$$y_{1,2}' = \tan(\theta \pm \alpha), \tag{6.4.20}$$

where the plus sign corresponds to the C_+ characteristic and the minus sign corresponds to the C_- characteristic. It follows from here that the velocity $\vec{v}$ with the components $q\cos\theta$ and $q\sin\theta$ is directed along a bisector of the angle formed by the tangents to the characteristics at some point O' (see Fig. 6.24). It follows from Fig. 6.24 and formula (6.4.17) that the velocity projection onto the normal to the characteristics is equal to the sound velocity $c = q\sin\alpha$.

Let us write the relations on the characteristics (6.4.14) and (6.4.15) in the variables q, θ. Computing the differentials with the aid of (6.4.16):

$$du = dq\cos\theta - q\sin\theta\, d\theta, \quad dv = dq\,\sin\theta + q\,\cos\theta\, d\theta \tag{6.4.21}$$

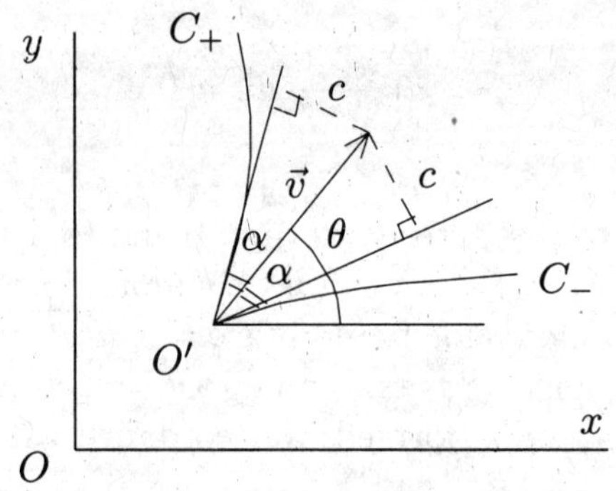

Figure 6.24: The mutual disposition of the C_+ and C_- characteristics and the velocity vector $\vec{v}$.

and substituting them in (6.4.14) and (6.4.15), we obtain with regard for the equation for the characteristics (6.4.20)[7,8]:

$$d\theta \mp \frac{\operatorname{ctg}\alpha}{q}\, dq = 0, \tag{6.4.22}$$

where the minus sign corresponds to the C_+ characteristic and the plus sign corresponds to the C_- characteristic. We can find from (6.4.19) the formula

$$\operatorname{ctg}\alpha = \sqrt{\frac{\lambda^2 - 1}{1 - (\frac{\gamma-1}{\gamma+1})\lambda^2}}, \qquad \lambda = \frac{q}{c_*} \tag{6.4.23}$$

the substitution of which in (6.4.22) yields, after integration, the *Riemann invariants*[7,8]:

$$\theta = \pm\zeta(\lambda) + \text{const},$$

$$\begin{aligned} \zeta(\lambda) &= \sqrt{\frac{\gamma+1}{\gamma-1}} \arctan\left(\sqrt{\left(\frac{\gamma-1}{\gamma+1}\right)\frac{(\lambda^2-1)}{\left(1-\left(\frac{\gamma-1}{\gamma+1}\right)\lambda^2\right)}}\right) \\ &- \arctan\sqrt{\frac{\lambda^2-1}{1-(\frac{\gamma-1}{\gamma+1})\lambda^2}}. \end{aligned} \tag{6.4.24}$$

Equations (6.4.24) describe a family of epicycloids (see Fig. 6.25) located in the u, v plane in the annulus

$$c_* < q < \sqrt{\frac{\gamma+1}{\gamma-1}}\, c_*. \tag{6.4.25}$$

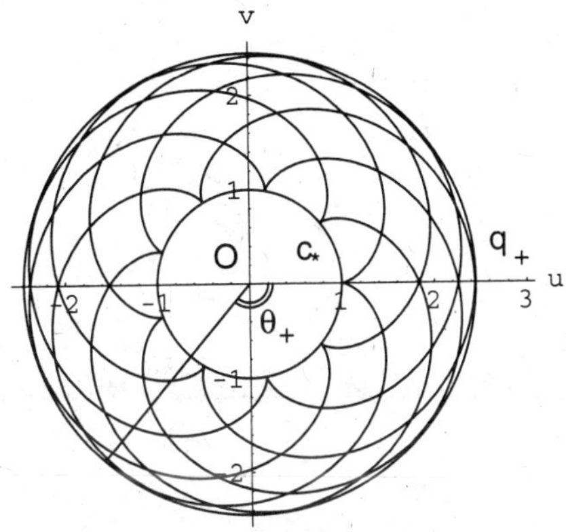

Figure 6.25: The family of epicycloids for $\gamma = 1.4$, which are produced by a point of a circle of diameter $q_+ - c_*$ rolling on the circle of the radius $q = c_*$.

It can be seen from formula (6.4.24) and Fig. 6.25 that all curves (6.4.24) may be obtained from a standard curve

$$\theta = -\zeta(\lambda) \tag{6.4.26}$$

by a rotation along the axis by a certain constant angle, or by a specular reflection with respect to the Ov-axis ($\theta \to -\theta$) and subsequent rotation. The limiting angle of the flow turning θ_+ (see Fig. 6.25) corresponds to $q_+ = \sqrt{\frac{\gamma+1}{\gamma-1}}c_*$, and it is equal in accordance with (6.4.24) to

$$\theta_+ = \frac{\pi}{2}\left(\sqrt{\frac{\gamma+1}{\gamma-1}} - 1\right). \tag{6.4.27}$$

Collecting the obtained results, one can write:

$$\begin{aligned} C_+ : \quad \frac{dy}{dx} &= \tan(\theta + \alpha), \quad \mathcal{I}_+ = \theta - \zeta(\lambda) = \text{const}, \\ C_- : \quad \frac{dy}{dx} &= \tan(\theta - \alpha), \quad \mathcal{I}_- = \theta + \zeta(\lambda) = \text{const}, \end{aligned} \tag{6.4.28}$$

where the $\mathcal{I}_+$ is conserved along the C_+ characteristic and $\mathcal{I}_-$ is conserved along the C_- characteristic.

Equations (6.4.28) are similar to equations (6.2.14) describing the one-dimensional nonstationary isentropic flows. There is consequently a full analogy between the solutions' behavior in these two cases. In particular, if one of the Riemann invariants remains identically constant, then such a flow is a *simple wave*. If the $\mathcal{I}_+$ invariant is identically constant in the flow region, then we can find from (6.4.28) that the θ and α are constant along the C_- characteristic; therefore, the C_-

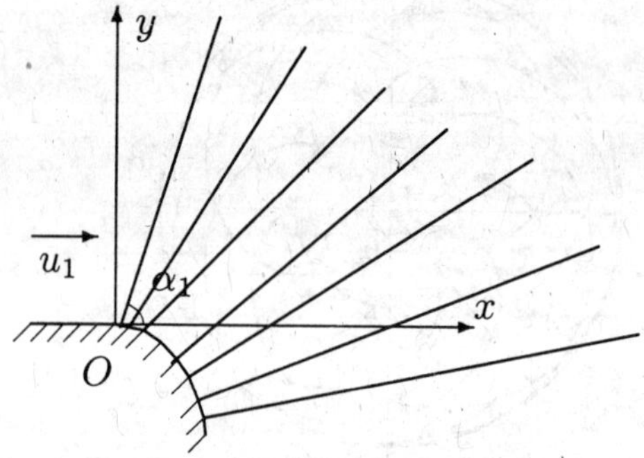

Figure 6.26: A simple rarefaction wave $\mathcal{I}_-$ arising in the flow around a convex surface $y_s = y_s(x_s)$.

characteristics will be straight lines, and the equation of the $\mathcal{I}_+$ wave will have the form

$$y = x \tan(\theta - \alpha) + f(\lambda), \quad \mathcal{I}_+ = \theta - \zeta(\lambda) = \mathcal{I}_+^0. \tag{6.4.29}$$

We can obtain in a similar way for the $\mathcal{I}_-$ wave that

$$y = x \tan(\theta + \alpha) + f(\lambda), \quad \mathcal{I}_- = \theta + \zeta(\lambda) = \mathcal{I}_-^0; \tag{6.4.30}$$

in this case, the C_+ characteristics will be the straight lines. A simple wave always adheres to a constant flow region. The joining of a constant flow to a simple $\mathcal{I}_+$ or $\mathcal{I}_-$ wave occurs along a C_- (or C_+) characteristic. In a simple compression (rarefaction) wave, the "fan handle" of the rectilinear characteristics is located ahead of (behind) the flow direction. The proof of this assertion is similar to that made in the one-dimensional nonstationary case in Section 6.2.2, and it is presented, for example, in[8].

We show in Fig. 6.26 a simple rarefaction wave $\mathcal{I}_-$ arising in the uniform supersonic flow $u = u_1|_{x=0}$, $u_1 > c_1$ around a convex surface $y_s = y_s(x_s)$. The solution in this wave is given by formulas (6.4.30):

$$y = x \tan(\theta + \alpha) + f(\lambda), \quad \theta + \zeta(\lambda) = \zeta(\lambda_1), \tag{6.4.31}$$

where $\lambda_1 = u_1/c_*$ is the velocity coefficient of the freestream and $\zeta(\lambda)$ is given by the second equation in (6.4.24). To determine the function $f(\lambda)$, we make use of the slip condition on the surface $y_s = y_s(x_s)$. By virtue of this condition, the gas velocity is directed along a tangent surface; therefore, at each surface point x_s, y_s, the following equations are satisfied:

$$\frac{dy_s}{dx_s} = \tan\theta, \qquad y_s = y_s(x_s), \tag{6.4.32}$$

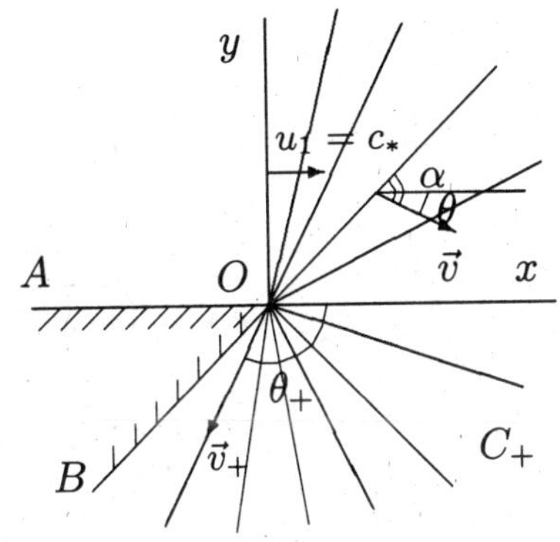

Figure 6.27: The picture of characteristics in the centered Prandtl–Meyer rarefaction wave.

the solution of which may be written parametrically as

$$y_s = y_s(\theta), \qquad x_s = x_s(\theta). \tag{6.4.33}$$

On the other hand, the equation for the C_+ characteristics [the first equation in (6.4.31)] is valid everywhere in the flow region, including the surface $y_s = y_s(x_s)$; therefore,

$$y - y_s = (x - x_s)\tan(\theta + \alpha), \tag{6.4.34}$$

from where it follows that

$$f(\lambda) = y_s - x_s \tan(\theta + \alpha). \tag{6.4.35}$$

Substituting here (6.4.33) and expressing $\theta = \theta(\lambda)$ from the second equation (6.4.31), we can find the expression for the arbitrary function:

$$f(\lambda) = y_s(\theta(\lambda)) - x_s(\theta(\lambda))\tan(\theta(\lambda) + \alpha(\lambda)), \tag{6.4.36}$$

where $\alpha(\lambda)$ is determined from (6.4.19).

In the particular case of the flow around a dihedral corner (Fig. 6.27), there is in the problem no quantity having the length dimension; therefore, $f(\lambda) = 0$. One can assume without loss of generality that a sonic flow with $u_1 = c_1 = c_*, \lambda = 1$ comes to the angle from the left; therefore, the substitution of the values $f(\lambda) = 0$, $\zeta(\lambda_1)|_{\lambda_1=1} = 0$ yields the solution of the type of a centered rarefaction wave, which is called the *Prandtl–Meyer wave* (see Fig. 6.27).

$$y = x\tan(\theta + \alpha), \qquad \theta + \zeta(\lambda) = 0. \tag{6.4.37}$$

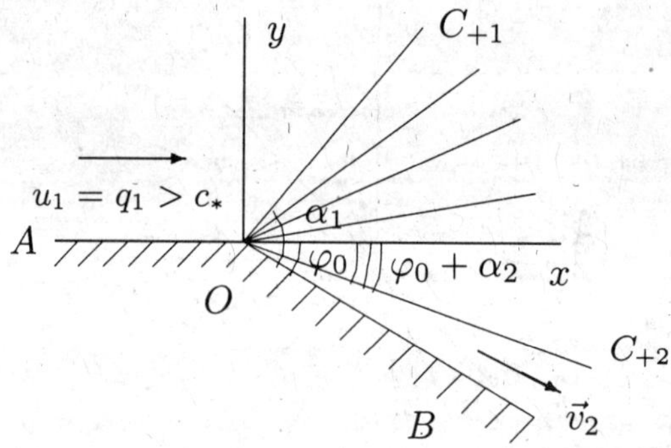

Figure 6.28: The picture of characteristics in the Prandtl-Meyer's wave in the flow around an arbitrary corner $\varphi_0 < 0$ at $|\varphi_0| < |\theta_+|$.

[The flow at an arbitrary freestream velocity $\lambda_1 > 1$ is obtained from (6.4.37) by a rotation by the angle $\zeta(\lambda_1)$; see (6.4.31) at $f = 0$.] The maximum angle θ_+ by which the flow turns is determined by formula (6.4.27). Here $\lambda_+ = \sqrt{\frac{\gamma+1}{\gamma-1}}$ and we obtain from (6.4.23) that $\alpha = 0$. It follows from here that the velocity $\vec{v}_+$ is directed along the left limiting characteristic $y = x \tan \theta_+$ and the velocity modulus $|\vec{v}_+| = q_+ = \sqrt{\frac{\gamma+1}{\gamma-1}}c_*$. There is a vacuum between this characteristic and the facet OB and $\rho = P = c = 0$ on the characteristic itself. If the dihedral angle $|\varphi_0| < |\theta_+|$ and the freestream has an arbitrary velocity $u_1 = \lambda_1 c_*$, $\lambda_1 > 1$, $v_1 = 0$, then the Prandtl–Meyer flow will be described by the equations

$$y = x \tan(\theta + \alpha), \quad \theta + \zeta(\lambda) = \zeta(\lambda_1). \tag{6.4.38}$$

This solution adheres to the constant flow $v_1 = 0$, $u_1 = q_1 > c_*$ along the characteristic C_{+1} and the constant flow $\vec{v}_2$ parallel with the OB line along the characteristic C_{+2} (see Fig. 6.28). It follows from here that the characteristics C_{+1}, C_{+2} will be described by the equations

$$y = x \tan \alpha_1, \quad y = x \tan(\varphi_0 + \alpha_2), \quad \varphi_0 < 0, \tag{6.4.39}$$

where $\sin \alpha_1 = 1/M_1$ and $\sin \alpha_2 = 1/M_2$. The Mach number $M = q/c$ is expressed in terms of the velocity coefficient λ by formula

$$M^2 - 1 = (\lambda^2 - 1)\Big/\left(1 - \left(\frac{\gamma - 1}{\gamma + 1}\right)\lambda^2\right). \tag{6.4.40}$$

In the freestream $\lambda_1 = u_1/c_*$, and in the flow turned by angle φ_0, the value λ_2 is found from the condition $\theta_2 = \varphi_0$ the substitution of which in (6.4.38) yields the equation for the determination of λ_2:

$$\varphi_0 + \zeta(\lambda_2) = \zeta(\lambda_1). \tag{6.4.41}$$

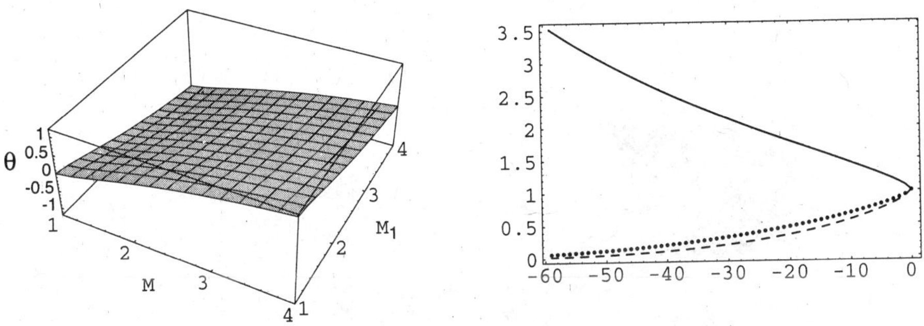

(a) (b)

Figure 6.29: The Prandtl–Meyer solution: (a) the surface $\theta = \theta(M_1, M)$ and (b) the distributions of the Mach number $M(\theta)$ (solid line), the pressure ratio $P(\theta)/P_1$ (dashed line), and the density ratio $\rho(\theta)/\rho_1$ (dotted line).

Collecting the formulas (6.4.24), (6.4.38), and (6.4.40), we can write the Prandtl–Meyer solution in the form convenient for applications:

$$\begin{aligned}\theta &= \sqrt{\frac{\gamma+1}{\gamma-1}}\left(\arctan\sqrt{\left(\frac{\gamma-1}{\gamma+1}\right)(M_1^2-1)}\right.\\ &\qquad\left.-\arctan\sqrt{\left(\frac{\gamma-1}{\gamma+1}\right)(M^2-1)}\right)\\ &\qquad-\left(\arctan\sqrt{M_1^2-1}-\arctan\sqrt{M^2-1}\right), \quad (6.4.42)\\ \alpha &= \arctan\frac{1}{\sqrt{M^2-1}}, \qquad y = x\tan(\theta+\alpha),\\ u &= q\cos\theta, \quad v = q\sin\theta, \quad M = q/c, \quad M_1 = q_1/c_1.\end{aligned}$$

The remaining parameters ρ, P are determined from the isentropic formulas (6.1.11).

We have implemented the Prandtl–Meyer solution (6.4.42) in the case $|\varphi_0| < |\theta_+|$ in our *Mathematica* Notebook `prog6-3.nb`. We present in Fig. 6.29 (a) the surface $\theta = \theta(M_1, M)$ determined by equation (6.4.42). Fig. 6.29 (b) shows the distribution of $M(\theta)$, $P(\theta)/P_1$, and $\rho(\theta)/\rho_1$. The values of the angle θ are given in degrees; these values are negative because the angle θ is measured in the negative (i.e., clockwise) direction.

Fig. 6.30 shows in different colors the local behavior of the pressure in the Prandtl–Meyer rarefaction wave in a spatial region. The local color in this picture (see also on the computer monitor the output

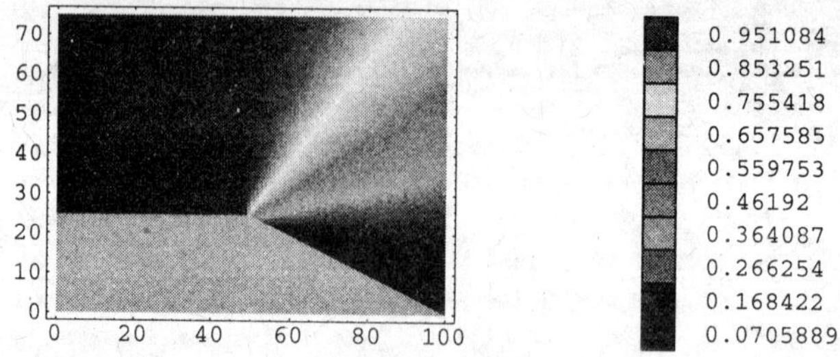

Figure 6.30: The color map of the pressure distribution in the Prandtl–Meyer gas flow around a dihedral corner.

of the *Mathematica* Notebook `prog4-3.nb`) depends on the local value $P(x, y)/P_1$. A colored column to the right of the flowfield picture in Fig. 6.30 gives the correspondence between the individual colors and the numerical values of the pressure ratio P/P_1. A more detailed description of the computational algorithm for the generation of color map of Fig. 6.30 may be found in[4].

Now consider the flow around a concave surface (see Fig. 6.31); the freestream is uniform and supersonic. As can be seen from Fig. 6.31, in this case a simple compression wave forms. The characteristics in the compression wave draw closer downstream, they intersect at some point A, and the solution becomes non-singlevalued. There arises indeed a shock wave AB at point A. The entropy increases behind the shock wave, and the flow becomes nonisentropic. The shock wave AB will affect the downstream flow with the aid of the C_- characteristics (the dashed lines in Fig. 6.31). Therefore, the solution of the simple compression wave type will exist only to the left of the line BAF. Note that the physical reason for the shock wave formation in this case is the same as in the piston problem (see Section 6.2.1) and is related to the nonlinearity of the gas dynamics equations.

Problem 6.5. Find the equation for the characteristics in the plane of the variables: the potential φ and the stream function ψ.

Solution: Find the formulas for the transformation from the variables x, y to $\varphi = \varphi(x, y)$, and $\psi = \psi(x, y)$. According to the definitions of the

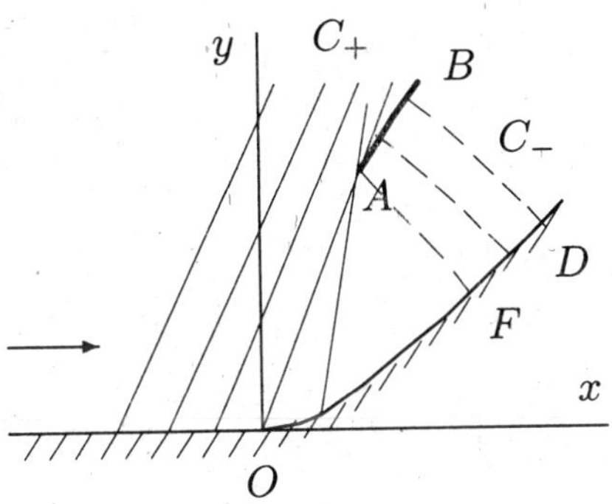

Figure 6.31: The picture of characteristics in the compression wave arising in a flow around a concave surface OD.

potential φ and the stream function ψ, we will have:

$$\begin{aligned} d\varphi &= \frac{\partial \varphi}{\partial x}dx + \frac{\partial \varphi}{\partial y}dy = u\,dx + v\,dy = q(\cos\theta\,dx + \sin\theta\,dy), \\ d\psi &= \frac{\partial \psi}{\partial x}dx + \frac{\partial \psi}{\partial y}dy = -\frac{\rho}{\rho_\infty}v\,dx + \frac{\rho}{\rho_\infty}u\,dy \\ &= \frac{\rho q}{\rho_\infty}(-\sin\theta\,dx + \cos\theta\,dy), \end{aligned}$$

Expressing from here dx and dy, we obtain:

$$dx = \frac{1}{q}(\cos\theta\,d\varphi - \frac{\rho_\infty}{\rho}\sin\theta\,d\psi), \quad dy = \frac{1}{q}(\sin\theta\,d\varphi + \frac{\rho_\infty}{\rho}\cos\theta\,d\psi).$$

Substituting these relations in the equation for characteristics (6.4.20) $dy = \tan(\theta \pm \alpha)\,dx$, we arrive at the expression

$$(\tan(\theta \pm \alpha)\cos\theta - \sin\theta)\,d\varphi = \frac{\rho_\infty}{\rho}(\tan(\theta \pm \alpha)\sin\theta + \cos\theta)d\psi,$$

the simplification of which yields the equation for the characteristics in terms of the variables φ and ψ:

$$d\varphi = \pm\frac{\rho_\infty}{\rho}\cot\alpha\,d\psi. \tag{6.4.43}$$

Problem 6.6. Generalize the program `prog6-3.nb` for the case of the Prandl–Meyer wave when $|\varphi_0| > |\theta_+|$. Generate also the color map of the density ratio ρ/ρ_1 in the flow field.

Hint: Compute the extreme left characteristic C_+, and assign the black color to the vacuum region between this characteristic and the neighboring facet of the dihedral corner (see also Fig. 6.27).

6.4.2 Chaplygin's Equations and Method

Consider a stationary potential inviscid gas flow in the plane $z = x + iy$. It follows from the definition of the potential φ and the stream function ψ (see Sections 6.3.1 and 6.3.2) that

$$\frac{\partial \varphi}{\partial x} = \frac{1}{\rho} \cdot \frac{\partial \psi}{\partial y}, \quad \frac{\partial \varphi}{\partial y} = -\frac{1}{\rho} \cdot \frac{\partial \psi}{\partial x}, \tag{6.4.44}$$

where ρ is the gas density related to the unit reference density $\rho_0 = 1$. We will study the barotropic motions; that is,

$$\rho = \rho(P). \tag{6.4.45}$$

As known from the previous results obtained, the relation between P and the gas motion velocity $q^2 = u^2 + v^2$ are established with the aid of the Bernoulli integral

$$\frac{q^2}{2} + \int_{P_0}^{P} \frac{dP}{\rho} = 0, \tag{6.4.46}$$

where P_0 is the pressure at the flow stagnation point. It should be noted that the constant in the Bernoulli integral will be the same in the overall flow domain.

The system of equations (6.4.44)–(6.4.46) is nonlinear, and its analysis is generally difficult. If we take the inclination angle θ of the velocity vector to the x-axis and the velocity modulus q as the independent variables, however, then the above system can be transformed to a linear one[2,11,12,14]. Let us indeed write on the basis of the definition:

$$\begin{aligned} d\varphi &= \frac{\partial \varphi}{\partial x} dx + \frac{\partial \varphi}{\partial y} dy = q(\cos\theta\, dx + \sin\theta\, dy), \\ d\psi &= \frac{\partial \psi}{\partial x} dx + \frac{\partial \psi}{\partial y} dy = -\rho q(\sin\theta\, dx - \cos\theta\, dy), \end{aligned}$$

where

$$\frac{\partial \varphi}{\partial x} = u = q\cos\theta, \quad \frac{\partial \varphi}{\partial y} = v = q\sin\theta, \quad \frac{\partial \psi}{\partial x} = -\rho v, \quad \frac{\partial \psi}{\partial y} = \rho u.$$

Multiplying the second relation by i/ρ and adding it to the first equation, we obtain:

$$dz = \frac{e^{i\theta}}{q} \cdot \left(d\varphi + i\frac{1}{\rho} \cdot d\psi\right). \tag{6.4.47}$$

Assume that the Jacobian

$$J = \frac{D(\theta, q)}{D(x, y)}$$

is different from zero, that is, the functions $\theta(x, y)$ and $q(x, y)$, are independent. Then we find from (6.4.47):

$$\frac{\partial z}{\partial q} = \frac{e^{i\theta}}{q}\left(\frac{\partial \varphi}{\partial q} + i\frac{1}{\rho} \cdot \frac{\partial \psi}{\partial q}\right), \quad \frac{\partial z}{\partial \theta} = \frac{e^{i\theta}}{q}\left(\frac{\partial \varphi}{\partial \theta} + i\frac{1}{\rho} \cdot \frac{\partial \psi}{\partial \theta}\right).$$

The differentiation of the first equation with respect to θ, the differentiation of the second equation with respect to q, and the subsequent subtraction of the obtained expressions from one another yields

$$\frac{ie^{i\theta}}{q}\left(\frac{\partial \varphi}{\partial q} + i\frac{1}{\rho}\frac{\partial \psi}{\partial q}\right) = e^{i\theta}\left(-\frac{1}{q^2} \cdot \frac{\partial \varphi}{\partial \theta} + i\frac{d}{dq}\left(\frac{1}{\rho q}\right)\frac{\partial \psi}{\partial \theta}\right).$$

Dividing both sides of this equation by $e^{i\theta}$ and separating the imaginary and real parts of the obtained equation, we find:

$$\frac{\partial \varphi}{\partial q} = q\frac{d}{dq}\left(\frac{1}{\rho q}\right)\frac{\partial \psi}{\partial \theta}; \qquad \frac{\partial \varphi}{\partial \theta} = \frac{q}{\rho} \cdot \frac{\partial \psi}{\partial q}. \tag{6.4.48}$$

The obtained system of equations (6.4.48) is already linear since its coefficients are the functions of the independent variable q only.

The flow in the physical plane (x, y) will be uniquely determined if the Jacobian

$$I = \frac{D(x, y)}{D(\theta, q)} = \frac{D(x, y)}{D(\varphi, \psi)} \cdot \frac{D(\varphi, \psi)}{D(\theta, q)} \neq 0.$$

The computation of this product yields

$$I = \frac{1}{\rho^2 q^3} \cdot \left[q^2\left(\frac{\partial \psi}{\partial q}\right)^2 + (1 - M^2)\left(\frac{\partial \psi}{\partial \theta}\right)^2\right],$$

where M is the Mach number. Since ρ and M are the known functions of q, the vanishing of the Jacobian I depends only on the given solution $\psi(\theta, q)$ in the flow region. For the subsonic flows, the Jacobian I can vanish only if the derivatives $\frac{\partial \psi}{\partial \theta}$ and $\frac{\partial \psi}{\partial q}$ are equal to zero, which is possible only at isolated points. At the supersonic speeds, in the hodograph plane (θ, q) the lines at the points of which $I(\theta, q) = 0$ may exist. It is conventional to call these lines the *critical lines*, and the corresponding lines in the physical plane (x, y) are called the *limiting lines*. There are two flows in the neighborhood of such lines, and these flows coincide on the line itself. Note that the above-constructed solution of the simple wave type (6.4.29), (6.4.30) possesses the property that $q = q(\theta)$ in this wave. Consequently, $1/I \equiv 0$ for this solution. Therefore, this solution is lost at a passage to the hodograph plane. Thus, the Chaplygin's equation derived below does not contain the solution of a simple wave type.

In the general case, it is reasonable to take as the independent variable certain arbitrary function $s = s(q)$ rather than the velocity modulus q. Then, we have instead of (6.4.48) the equations

$$\frac{\partial\varphi}{\partial\theta} = \left[\frac{q}{\rho\frac{dq}{ds}}\right]\frac{\partial\psi}{\partial s}; \quad \frac{\partial\varphi}{\partial s} = -\left[\frac{dq}{ds}\cdot\frac{1}{\rho q} - \frac{d}{ds}\left(\frac{1}{\rho}\right)\right]\frac{\partial\psi}{\partial\theta}.$$

Since the relation $s = s(q)$ is arbitrary, we choose it from the condition

$$\frac{q}{\rho\frac{dq}{ds}} = \frac{dq}{ds}\frac{1}{\rho q} - \frac{d}{ds}\left(\frac{1}{\rho}\right) = \sqrt{K(s)}, \tag{6.4.49}$$

where the function $K(s)$ is usually termed the *Chaplygin's function.* Owing to this condition, we obtain instead of (6.4.48) the canonical form of the system of equations:

$$\frac{\partial\varphi}{\partial\theta} = \sqrt{K}\frac{\partial\psi}{\partial s}, \quad \frac{\partial\varphi}{\partial s} = -\sqrt{K}\frac{\partial\psi}{\partial\theta}, \tag{6.4.50}$$

which is linear and has a convenient symmetric form obtained for the first time by Chaplygin for the adiabatic perfect gas flows and is called the *Chaplygin's equations*[12,13].

We can find from (6.4.49) and (6.4.46) after simple transformations:

$$K = \frac{1}{\rho^2}(1 - M^2), \tag{6.4.51}$$

where the sound velocity is determined by the dependence (6.4.45). The Chaplygin's function K vanishes at $M = 1$; $K > 0$ at $M < 1$; and $K < 0$ at $M > 1$. If $M = 0$, then the function $K = 1$ and the chosen value of the density ρ_0 corresponds to the density value at the flow stagnation point. It is easy to obtain from formulas (6.4.49) the equations for the determination of $\rho(s)$ and $q(s)$ at a given function $K(s)$; that is,

$$\frac{d\rho}{ds} = \frac{\rho^2K - 1}{\sqrt{K}}, \quad \frac{dq}{ds} = \frac{q}{\rho\sqrt{K}}. \tag{6.4.52}$$

It follows from (6.4.51) and the second relation (6.4.52) that, at $M > 1$, the quantity ds has purely imaginary values. In order to obtain for the flows at $M > 1$ similar equations in the real plane, we set $-K = H > 0$, $ds = i\,d\tau$ and take the θ and τ as the independent variables; that is, we introduce the variable $\tau(q)$ instead of s from the condition

$$\rho\frac{q}{\frac{dq}{d\tau}} = -\frac{dq}{d\tau}\cdot\frac{1}{\rho q} + \frac{d}{d\tau}\left(\frac{1}{\rho}\right) = \sqrt{H}, \tag{6.4.53}$$

and equations (6.4.50) will have the form

$$\frac{\partial \varphi}{\partial \theta} = \sqrt{H}\frac{\partial \psi}{\partial \tau}, \quad \frac{\partial \varphi}{\partial \tau} = \sqrt{H}\frac{\partial \psi}{\partial \theta}. \tag{6.4.54}$$

Solving system (6.4.53) with respect to $\rho(\tau)$ and $q(\tau)$, we find:

$$\frac{d\rho}{d\tau} = -\frac{\rho^2 H + 1}{\sqrt{H}}, \quad \frac{dq}{d\tau} = \frac{q}{\rho\sqrt{H}}. \tag{6.4.55}$$

Thus, as a result of the made transformations, the nonlinear system of equations (6.4.44)–(6.4.46) governing steady motions of a barotropic perfect gas may be transformed to a linear system for the Mach numbers $M < 1$ in the form (6.4.50), and at $M > 1$, it may be transformed to the form (6.4.54), where the Chaplygin's function is determined by formulas (6.4.45) and (6.4.51) and (6.4.52) and (6.4.55), respectively. The passage to a physical plane is performed by formulas (6.4.47). It should be noted that, at $M < 1$, the system (6.4.50) will have the elliptic type, and at $M > 1$, the system (6.4.54) will already have the hyperbolic type.

Some particular solutions in the hodograph variables. We will search for the solutions of (6.4.48) in the form

$$\varphi = -A_0\theta, \quad \psi = -\int_{q_{max}}^{q} \frac{\rho}{q}\, dq, \tag{6.4.56}$$

where A_0 is an arbitrary quantity. Using (6.4.47), we find with regard for (6.4.56):

$$dz = -\frac{A_0}{q}e^{i\theta}\, d\theta - i\frac{A_0}{q^2}e^{i\theta}\, dq.$$

Since the right-hand side is the total differential, we obtain as a result of integration:

$$z - z_0 = i\frac{A_0}{q}e^{i\theta},$$

where z_0 is some fixed point in the z-plane. Introduce the polar coordinates r and ϑ in the z-plane with a center at point $z = z_0$; that is, assume $z - z_0 = re^{i\vartheta}$. Then the velocity field of the given gas motion is determined by formula

$$q = \frac{A_0}{r}, \qquad \theta = \vartheta - \frac{\pi}{2}.$$

It is easy to see that the streamlines of this flow are the concentric circles with the common center at point $z = z_0$. The velocity on the streamlines decreases inversely proportionally to their radius, and the constant A_0 is related to the velocity circulation Γ along a closed streamline and

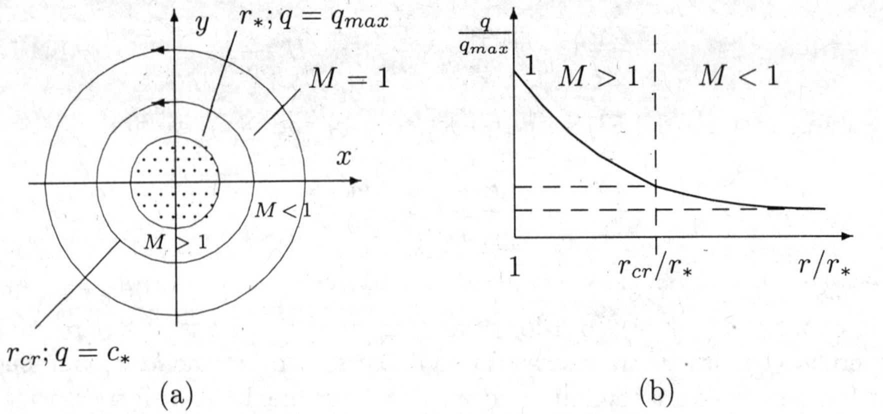

Figure 6.32: The particular solution of Chaplygin's equations: (a) the vortex streamlines and (b) the velocity dependence on radius in the vortex.

is determined from the relation $A_0 = \Gamma/(2\pi)$. The described flow represents a generalization for the case of compressible gas of a potential flow from a concentrated vortex in incompressible fluid. It should be noted, however, that for the adiabatic perfect gas flow such a flow will exist only outside the circle of radius r_* determined from the condition $r_* = \frac{\Gamma}{2\pi q_{max}}$. On the circle $r = r_*$, the gas velocity is equal to the maximum velocity of the adiabatic steady flow q_{max} and the pressure and density are equal to zero. At an unbounded growth of r, the gas velocity decreases down to zero and the pressure and density tend to their values in the stagnation state (see Fig. 6.32). On the circle $r = r_{cr} = \sqrt{\frac{\gamma+1}{\gamma-1}}r_*$ ($r_{cr} = 2.45r_*$ at $\gamma = 1.4$), the gas velocity is equal to the critical velocity ($v_{cr} = c_* = 0.41\ v_{max}$ at $\gamma = 1.4$), so that at the values $r > r_{cr}$ the gas flow is subsonic and, in the annular region $r_* < r < r_{cr}$, it is supersonic.

Consider the solution of type source (sink). We will search for a solution of (6.4.48) of the form

$$\psi = A_0\theta \ (A_0 > 0); \quad \varphi = A_0 \int_{q_{cr}}^{q} q\frac{d}{dq}\left(\frac{1}{\rho q}\right) dq. \tag{6.4.57}$$

Using (6.4.47) and (6.4.57), we find:

$$z - z_0 = A_0 \frac{1}{\rho q} e^{i\theta},$$

or in the polar coordinates:

$$r = A_0 \frac{1}{\rho q}, \qquad \vartheta = \theta. \tag{6.4.58}$$

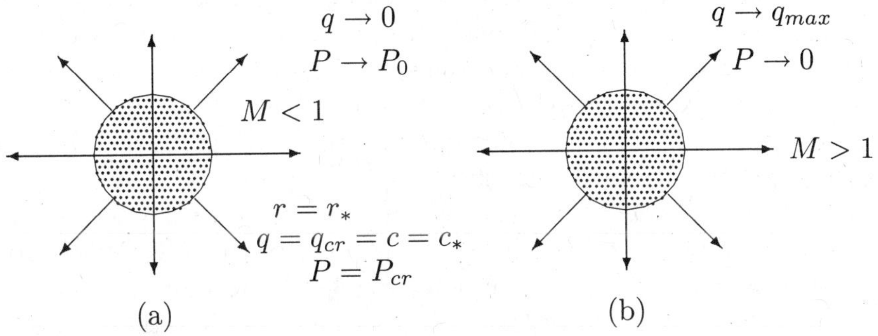

Figure 6.33: The solution of the source/sink type: (a) the subsonic source and (b) the supersonic source.

It follows from the obtained solution that the streamlines are directed along the beams emanating from the point $z = z_0$ (Fig. 6.33). The velocity magnitude on the streamlines is the same on each circle with a center at this point and varies with the radius r of the circle in accordance with the first relation (6.4.58), which expresses the conservation law for the gas mass. The constant A_0 is related to the gas flow rate across the circle by the relation $A_0 = Q/(2\pi)$. The presented flow represents a source (at $A_0 < 0$ it is a sink).

For the adiabatic flows of perfect gas, the dependence $q(r)$ has two solution branches [see Fig. 6.33 (a) and (b)]. Note that for the region $r < r_*$, where r_* corresponds to the critical gas velocity, there are no real values of q. Thus, we have two solutions (6.4.57), where the flow occurs on different leaves of the plane and is matched along the circle $r = r_*$ across which they cannot be prolonged towards $r < r_*$. The transformation Jacobian I on this circle is equal to zero, which is established directly, and the gas acceleration

$$\frac{dq}{dt} = q\frac{dq}{dr} = \frac{\rho q^3}{A_0(M^2 - 1)}$$

tends to infinity at $M \to 1$ on the line $r \to r_*$. Thus, the gas velocity varies on one solution branch from the critical velocity at $r = r_*$ to the zero velocity at $r \to \infty$, which corresponds to a subsonic outflow or a sink [see Fig. 6.33 (a)]. On the second solution branch at $r > r_*$, the flow velocity varies from the critical one to the maximum possible value at $r \to \infty$, and $P \to 0$ and $\rho \to 0$, which corresponds to a supersonic source or sink [see Fig. 6.33 (b)]. The line $r = r_*$ in this flow is one of the simple examples of the limiting line formations in the plane of the physical variables (x, y).

As a consequence of the linearity of the Chaplygin's system, a possibility of the existence of a solution in the form of a superposition of two above considered solutions follows. According to the above, two different flows exist one of which corresponds to the addition of the vorticity and the supersonic source (sink), and the second flow corresponds to the addition of a vortex and subsonic source (sink). It should be kept in mind that both of these flows will exist outside the limiting line on which the transformation Jacobian I vanishes.

Chaplygin's method[11,13]. When solving the problems of gas flow, such as the flow around a body or the gas flows in channels of various shapes, the boundary conditions for equations (6.4.44) are formulated in a natural way in the physical flow plane (x, y). While passing to the hodograph plane (θ, q), it is necessary to formulate the boundary conditions in this plane also. This is generally difficult because it is impossible to specify the velocity distribution on a given contour prior to the problem solution. This circumstance restricts significantly the classes of problems, which can be solved, and in some cases, it gives rise to intrinsically new formulations of the physical problems. The problem of the determination of a profile shape with a given velocity or pressure hodograph on its surface is one of such problems. It is conventional to call such problem classes the *inverse problems*, a natural generalization of which is related to the choice of a given value of the velocity or pressure hodograph from the solution of a variational problem. A variety of problems of the gas jet flows can be solved successfully in the variables of the hodograph plane, in which the pressure magnitude, and as follows from the Bernoulli integral, the velocity modulus q is given on an unknown jet boundary in a physical plane. Consider the application of the Chaplygin's method to the solution of gas jet problems for the adiabatic flows of a perfect gas. In this case, the dependence (6.4.3) takes the form

$$\frac{\rho}{\rho_0} = \left(1 - \frac{q^2}{q_{max}^2}\right)^{\frac{1}{\gamma-1}},$$

where ρ_0 is the stagnation density and q_{max} is the maximum gas velocity. If we introduce the variable $\tau = q^2/q_{max}^2$, then equations (6.4.48) take the form

$$\frac{\partial \varphi}{\partial \tau} = \frac{\frac{\gamma+1}{\gamma-1}\tau - 1}{2\tau(1-\tau)^{\frac{\gamma}{\gamma-1}}} \cdot \frac{\partial \psi}{\partial \theta}; \quad \frac{\partial \varphi}{\partial \theta} = \frac{2\tau}{(1-\tau)^{\frac{1}{\gamma-1}}} \frac{\partial \psi}{\partial \tau}.$$

This system of equations can easily be reduced by a cross differentiation to a single second-order equation for the stream function ψ:

$$\frac{\partial}{\partial \tau}\left(\frac{2\tau}{(1-\tau)^{\frac{1}{\gamma-1}}} \cdot \frac{\partial \psi}{\partial \tau}\right) + \frac{1 - \frac{\gamma+1}{\gamma-1}\tau}{2\tau(1-\tau)^{\frac{\gamma}{\gamma-1}}} \cdot \frac{\partial^2 \psi}{\partial \theta^2} = 0. \tag{6.4.59}$$

Since the coefficients of equation (6.4.59) depend only on one variable τ, one can use for the determination of its solutions the method of the variables separation in the form

$$\psi_n = \tau^{n/2} Y_n(\tau) e^{-in\theta},$$

where n is a constant. Substituting this solution in equation (6.4.59), we find that the function $Y_n(\tau)$ satisfies the equation:

$$\tau(1-\tau)\frac{d^2 Y_n}{d\tau^2} + \left[n+1+\left(\frac{1}{\gamma-1} - n - 1\right)\tau\right]\frac{dY_n}{d\tau} + \frac{n(n+1)}{2(\gamma-1)}Y_n = 0.$$

This equation determines the hypergeometric function that can be presented by a power series, which converges absolutely and uniformly at $0 \le \tau \le 1$, that is in the overall subsonic gas flow regime.

Chaplygin had applied successfully these solutions at the consideration of problems of the gas jets. In this way, he extended the known results of Helmholtz and Kirchhoff for incompressible fluid to the case of compressible gas flows. For a more detailed presentation of these results, we refer the reader to the original work[11] and the books[7,8].

The approximate Chaplygin's method and its extension as a method for the approximation of the equation of state. As was pointed out above, the systems (6.4.50) and (6.4.54) are linear; however, the Chaplygin's functions $\sqrt{K(s)}$ or $\sqrt{H(\tau)}$ have a sufficiently complex form and it is not always possible to obtain the exact integrals in closed form for the solution of boundary-value problems. In this connection, the approximate method for the solution of linear equations in the hodograph variables has gained a wide acceptance. The idea of this method was proposed by Chaplygin and it was developed and used actively in subsequent years by a number of the researchers.

Let us now briefly present the approximate Chaplygin's method. We will choose the functions $\sqrt{K(s)}$ or $\sqrt{H(\tau)}$ in such a form that the solution of systems of equations (6.4.50) and (6.4.54) can be obtained in a simple form. Then, after the integration of equations (6.4.52) or (6.4.55), the corresponding family of the functions $\rho(s)$ and $K(s(\rho))$ become known, which depend on a single arbitrary constant that will be assumed to be unknown for a while. The family of functions $P(\rho)$ corresponding to a certain function $K(\rho)$ can be determined in its turn from the relations

$$P = -\int \frac{q^2}{\rho^2 K - 1}\, d\rho + C_1, \quad q = \exp\left(\int \frac{d\rho}{\rho(\rho^2 K - 1)}\right) + C_2,$$

where C_1 and C_2 are the integration constants. These relations are obtained by integration of (6.4.52) and (6.4.46). Thus, we obtain for a fixed function $K(\rho)$ a family of the functions $P(\rho)$, depending on two arbitrary integration constants. The freedom in the choice of these constants can be used for the approximation of the equation of state, which is generally determined from experiments.

Similar results take place for the function $\sqrt{H}$. We now show how the above algorithm can be applied to the solution of the equation

$$\frac{\partial^2 \psi}{\partial s^2} + \frac{\partial^2 \psi}{\partial \theta^2} + T(s)\frac{\partial \psi}{\partial s} = 0, \tag{6.4.60}$$

where $T(s) = \frac{1}{2}\frac{d(\ln K)}{ds}$, which can be obtained easily by a cross differentiation of system (6.4.50) and elimination of the function φ. We will search for the solution of (6.4.60) in the form

$$\psi = \sum_{n=0}^{\infty} \Phi_n(s,\theta) \cdot f_n(s), \tag{6.4.61}$$

where all coefficients $\Phi_n(s,\theta)$ $(n = 0, 1, 2, \ldots)$ are the harmonic functions. It should be noted that the functions $\Phi_n(s,\theta)$ can generally be chosen in any other form. Substituting (6.4.61) in (6.4.60), we find:

$$\sum_{n=0}^{\infty}\left[\Phi_n\left(\frac{d^2 f_n}{ds^2} + T(s)\frac{df_n}{ds}\right) + \frac{\partial \Phi_n}{\partial s}\left(2\frac{df_n}{ds} + Tf_n\right)\right] = 0. \tag{6.4.62}$$

We now take such functions Φ_n and f_n, which turn this equation into identity. Let us assume that the functions Φ_n satisfy the equations

$$\Delta\Phi_0 = 0, \quad \Phi_{n-1} = \frac{\partial \Phi_n}{\partial s}, \quad n = 1, 2, \ldots.$$

If we then determine the coefficients f_n from the conditions

$$\begin{gathered} 2\frac{df_0}{ds} + Tf_0 = 0, \\ 2\frac{df_{n+1}}{ds} + Tf_{n+1} = -\left(\frac{d^2 f_n}{ds^2} + T\frac{df_n}{ds}\right), \\ n = 0, 1, 2, \ldots, \end{gathered} \tag{6.4.63}$$

it is easy to be sure of the fact that equation (6.4.62) turns into identity. It follows from the properties of the harmonic functions that all functions $\Phi_n(s,\theta) = \int \Phi_{n-1}\, ds$ will be harmonic if the function $\Phi_{n-1}(s,\theta)$ is harmonic. Let us prove this fact. Let Φ_{n-1} be a harmonic function. Let us apply the operation Δ to both sides of the relation $\frac{\partial \Phi_n}{\partial s} = \Phi_{n-1}$, and interchanging the operators $\frac{\partial}{\partial s}$ and Δ, we have:

$$\frac{\partial}{\partial s}(\Delta\Phi_n) = \Delta\Phi_{n-1},$$

and the integration of this relation with respect to s yields

$$\Delta\Phi_n = \int \Delta\Phi_{n-1} ds.$$

Since $\Delta\Phi_{n-1} = 0$, we have $\Delta\Phi_n = 0$.

Thus, if the function $\sqrt{K(s)}$ is known, then it follows from (6.4.63) that $f_0 = K^{-1/4}$ and the functions

$$f_n = -\frac{f'_{n-1}}{2} - \frac{1}{8}K^{-\frac{1}{4}} \int f'_{n-1} K' K^{-\frac{3}{4}} ds$$

can be computed in terms of the elementary functions, depending on the form of $K(s)$ or $P = P(\rho)$.

Restricting to a finite number of terms in the series (6.4.61), we will have from relations (6.4.62) an excessive equation of the form

$$\frac{d^2 f_n}{ds^2} + T f_n = 0.$$

In this case, the problem of the determination of the functions $f_n(s)$ will be overdetermined. Therefore, in order to close the problem of the determination of functions $f_n(s)$, we will assume the function $K(s)$ to be unknown and, as follows from the above presented-algorithm, this corresponds to a certain choice of the dependence $P(\rho)$.

Consider the following two examples.

1) Let $n = 0$. Then $\psi = \Phi_0(s,\theta) \cdot f_0(s)$, where $\Delta\Phi_0(s,\theta) = 0$ and $T = \frac{1}{2}\frac{d}{ds}(\ln K)$:

$$2f'_0 + T f_0 = 0, \quad f''_0 + T f_0 = 0.$$

Eliminating f_0 from the last two equations, we find:

$$K'' - \frac{3}{4}\frac{(K')^2}{K} = 0 \quad \text{or} \quad K = [A(s+b)]^4,$$

where A and B are the arbitrary constants[13]. Thus, we have obtained a well-known approximation of the Chaplygin's function proposed by Sauer. If we choose $b = \frac{1}{A}$ and $A \to 0$, then we obtain the Chaplygin's result, where $K = 1$. It follows from (6.4.49) that, in this case, the expressions for the velocity and density may be presented in the form

$$g = \frac{B}{\sinh\sigma}, \quad \sigma = s - s_0, \quad \rho = -\tanh\sigma, \tag{6.4.64}$$

where B and s_0 are the integration constants. Chaplygin and Busemann proposed to determine these constants by using the exact adiabat values at a point of zero velocity. Von Kármán, Tsien[12,13], and other authors replaced the adiabat with a tangent to it at a point corresponding to the gas parameters in the freestream while studying the flow around closed profiles. The general solution of the basic problem for system (6.4.50) has in this case the form

$$\varphi + i\psi = F(\zeta), \tag{6.4.65}$$

where $F(\zeta)$ is an arbitrary analytic function of complex variable $\zeta = s - i\,\theta$; that is, the problem of the gas motion in Chaplygin's approximation reduces to the boundary-value problems of the theory of functions of complex variable.

Let us determine the relation between the compressible gas flows obtained in the above way and the incompressible fluid flows. We will interpret the obtained formula (6.4.65) as a solution of the Cauchy–Riemann equations for the incompressible fluid flow with a complex potential $w = \varphi + i\,\psi$ and the complex velocity $\frac{dw}{dz} = q_1\zeta = q_{in}e^{-i\theta}$, where q_{in} is the velocity modulus of the incompressible flow. For such a flow, we have:

$$dz_{in} = \frac{dw}{q_1\zeta} = \frac{e^{i\theta}}{q_{in}}(d\varphi + i\,d\psi). \tag{6.4.66}$$

Present formula (6.4.47) in the form

$$dz = \frac{e^{i\theta}}{q}\Big[\frac{\rho+1}{\rho}(d\varphi + i\,d\psi) + \frac{\rho-1}{\rho}(d\varphi - i\,d\psi)\Big].$$

Substituting the values of ρ and q from (6.4.64), we obtain after simple transformations:

$$dz = \frac{q_0}{2}\Big(\frac{d\varphi + i\,d\psi}{q_{in}e^{-i\theta}} - \frac{q_1^2}{q_0^2}\cdot\frac{q_{in}}{q_1}e^{i\theta}\frac{d\varphi - i\,d\psi}{q_1}\Big).$$

Comparing the obtained formula with (6.4.66), we can write:

$$\frac{2}{q_0}dz = dz_{in} - \omega\overline{\zeta^2 dz_{in}}, \tag{6.4.67}$$

where $\omega = q_1^2/q_0^2$ and the constant q_0 is determined from the condition for the coincidence $q_{in} = q_1 = q_\infty$. It is conventional to call the formula (6.4.67) the *Tsien's formula*[12], which establishes the correspondence between the Chaplygin's gas flow plane and the incompressible fluid flow plane for the same function $F(\zeta)$ at different values of the parameter ω.

2) Let $n = 1$. Then the governing system of equations will have a more complex form:

$$\psi = f_0\Phi_0 + f_1\Phi_1, \quad 2f_0' + Tf_0 = 0, \quad 2f_1' + Tf_1 = -(f_0'' + Tf_0'),$$
$$f_1'' + Tf_1' = 0, \quad T = \frac{1}{2}\frac{d}{ds}(\ln K).$$

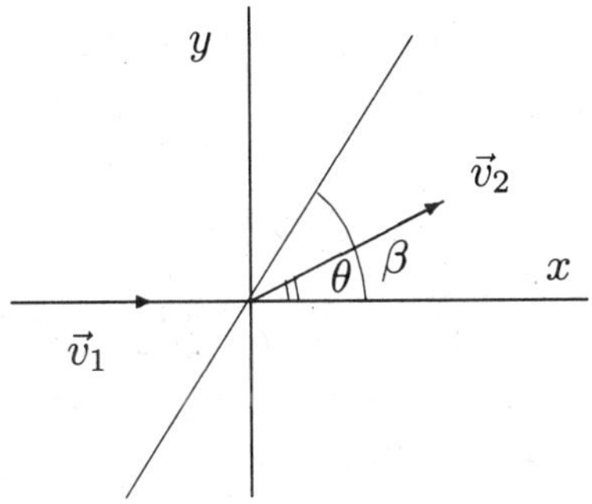

Figure 6.34: Oblique shock wave.

Problem 6.7. Construct the solution of the hyperbolic equation

$$\frac{\partial^2\psi}{\partial\theta^2} - \frac{\partial^2\psi}{\partial\tau^2} - \frac{1}{2}\frac{d(\ln H)}{d\tau}\frac{\partial\psi}{\partial\tau} = 0,$$

which can be obtained from system (6.4.54) by cross differentiation and elimination of the function φ.
Hint: Use the above-described technique.

6.4.3 Oblique Shock Waves

As was shown in Section 6.4.1, a planar, steady, continuous isentropic flow not always exists. In particular, an intersection of the characteristics of the same family occurs in compression wave by virtue of nonlinear effects, and a shock wave with curved front arises. The gas parameters undergo a jump across the shock wave whose magnitude is determined from the conservation laws for the mass, momentum, and energy (6.1.13). In contrast with the planar shock wave (6.1.30), the jump magnitudes depend not only on the freestream Mach number M_1, but also on the inclination of the flow velocity vector to the shock wave front.

One can choose in the neighborhood of each point of the shock wave front an infinitesimal planar area coinciding with the shock wave front and inclined at the angle β to the freestream velocity vector $\vec{v}_1$. Direct the velocity vector $\vec{v}_1$ along the x-axis and assume that it has the components $(u_1, 0)$. After the gas intersects the oblique shock wave its velocity vector changes its modulus and deviates by some angle θ (see Fig. 6.34). We will denote the components of gas velocity behind the shock wave front $\vec{v}_2$ by the letters u_2 and v_2, where u_2 is the velocity component along the x-axis, and v_2 is the velocity component along the y axis. According to (6.1.13), the tangent velocity component v_t does

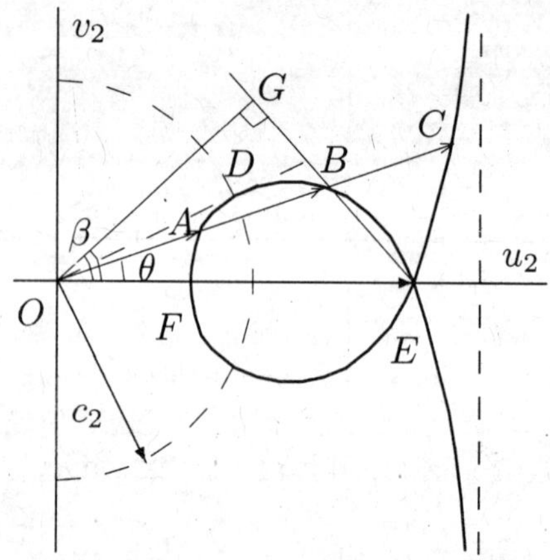

Figure 6.35: The shock polar in the hodograph plane (u_2, v_2), the segment OG is directed along the oblique shock wave; the segment OD is tangent to the polar at point D; $\vec{OB} = \vec{v}_2$, $\vec{OE} = \vec{v}_1$, $OG = v_{1t} = v_{2t}$, $GE = v_{1n}$, and $GB = v_{2n}$; and the dashed line is the sonic line $u_2^2 + v_2^2 = c_2^2$.

not change and the normal component v_n reduces spasmodically:

$$v_{1t} = v_{2t}, \quad v_{1n} > v_{2n}; \tag{6.4.68}$$

therefore, the velocity vector will always deviate, as shown in Fig. 6.34 (the angle between the tangent to the shock and the velocity vector will reduce). It follows from Fig. 6.34 that the tangent and normal velocity components before and behind the shock are given by formulas

$$\begin{aligned} v_{1t} &= u_1 \cos\beta, \qquad v_{1n} = u_1 \sin\beta, \\ v_{2t} &= q_2 \cos(\beta - \theta) = u_2 \cos\beta + v_2 \sin\beta, \\ v_{2n} &= q_2 \sin(\beta - \theta) = u_2 \sin\beta - v_2 \cos\beta, \end{aligned} \tag{6.4.69}$$

where $q_2 = \sqrt{u_2^2 + v_2^2}$. As was proved in Section 6.1, the normal velocity component of the gas ahead of the shock v_{1n} should exceed the sound speed c_1; that is, $v_{1n} > c_1$. Substituting here the second formula from (6.4.69), we obtain the inequality that should be satisfied by the inclination angle β of the oblique shock wave:

$$\alpha_1 < \beta < \pi/2, \quad \sin\alpha_1 = 1/M_1, \quad M_1 = u_1/c_1,$$

where α_1 is the Mach angle.

Making the substitutions $u_1 \to v_{1n}, u_2 \to v_{2n}, M_1 \to M_1 \sin\beta$ in the first formula in (6.1.30), we obtain:

$$\frac{u_2 \cos\beta + v_2 \sin\beta}{u_1 \sin\beta} = \frac{\gamma - 1}{\gamma + 1} + \frac{2}{\gamma + 1} \frac{1}{M_1^2 \sin^2\beta}. \tag{6.4.70}$$

We find from the condition for the continuity of the tangent velocity component:

$$u_1 \cos\beta = u_2 \cos\beta + v_2 \sin\beta. \tag{6.4.71}$$

Eliminating the angle β from equations (6.4.70) and (6.4.71), we obtain the equation for the *shock polar*[5–8]:

$$\left(\frac{v_2}{u_1}\right)^2 = \frac{\left(1-\frac{u_2}{u_1}\right)^2\left(\frac{2}{\gamma+1}\left(1-\frac{1}{M_1^2}\right)-\left(1-\frac{u_2}{u_1}\right)\right)}{1-\frac{u_2}{u_1}+\frac{2}{\gamma_1+1}\frac{1}{M_1^2}} \tag{6.4.72}$$

The equation of shock polar (6.4.72) describes in the hodograph plane u_2, v_2 a curve, which is a *strophoid* (see Fig. 6.35). It can be seen that three values of the velocity modulus at points A, B, and C correspond to a given value of θ. Let us study these solutions. Assuming $v_2 = 0$ in (6.4.72), we can find the points of intersection of the strophoid with the u_2-axis: $u_2 = u_1$ (point E) and

$$u_2 = u_1\left(1-\frac{2}{\gamma+1}\left(1-\frac{1}{M_1^2}\right)\right) = \frac{c_*^2}{u_1}$$

(point F), where $c_*^2 = \frac{\gamma-1}{\gamma+1}u_1^2 + \frac{2}{\gamma+1}c_1^2$ is the squared critical sound velocity [see formula (6.4.19)], in which one must set $q = u_1$, $c = c_1$. It can be seen from equation (6.4.71) that the inclination angle β of the shock wave is determined by a perpendicular drawn from point O to the line passing through point E and the point on the strophoid corresponding to the velocity behind the shock (point B is taken in Fig. 6.35). As point A approaches F, $\beta \to \pi/2$, and as B approaches E, the inclination angle of the chord BE will coincide with the inclination of a tangent to the strophoid at point E, equal to $\alpha_1 + \pi/2$. Since the triangle OGE is rectangular, then the angle β will coincide for a weak shock with the Mach angle α_1. Thus, point F corresponds to a normal shock in which $\beta = \pi/2$ and $v_2 = v_1 = 0$, and point E corresponds to an infinitesimally weak shock in which the velocity and, consequently, the angle θ tend to zero.

The shock corresponding to point C is nonphysical since $v_{2n} > v_{1n}$ on it, whereas the opposite inequality should be satisfied in the compression shock wave. For the same reason, the strophoid branches lying to the right of point E and tending to infinity should be eliminated. Points A and B determine a *strong* and *weak* shock wave, respectively. The sonic line intersects the strophoid very closely to point D lying on a beam tangent to the strophoid; therefore, the flow behind a strong shock wave is subsonic, and the flow behind a weak shock wave is supersonic. It can be seen from Fig. 6.35 that the angle of the flow turning θ does not exceed the value θ_{max}, corresponding to the tangent OD to the

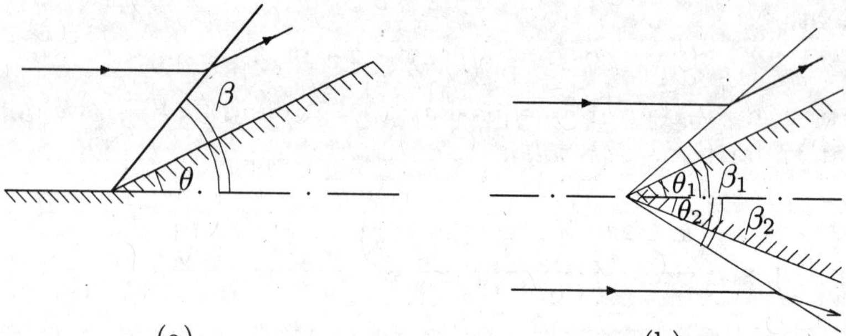

Figure 6.36: The supersonic flow ($M_1 > 1$) around (a) a corner and (b) a wedge.

strophoid. The quantity θ_{max} is a function of the Mach number M_1 and increases monotonously from $\theta_{max} = 0$ at $M_1 = 1$ to $\theta^*_{max} = \arcsin(1/\gamma)$ at $M_1 \to \infty$. It is easy to obtain the last formula if one notes that, at $M_1 \to \infty$, the strophoid equation (6.4.72) goes over to the circle equation:

$$\left(\frac{v_2}{u_1}\right)^2 = \left(1 - \frac{u_2}{u_1}\right)\left(\frac{u_2}{u_1} - \frac{\gamma - 1}{\gamma + 1}\right) \tag{6.4.73}$$

the tangent to which determines the value θ^*_{max}. The pressure P_2 and the density ρ_2 behind the oblique shock wave are found from the corresponding formulas (6.1.30) for the normal shock wave, if one repaces therein M_1 with $M_1 \sin\beta$:

$$P_2 = P_1\Big(1 + \frac{2\gamma}{\gamma+1}(M_1^2 \sin^2\beta - 1)\Big), \tag{6.4.74}$$

$$\rho_2 = \rho_1\left(1 + \frac{2\,(M_1^2 \sin^2\beta - 1)}{(\gamma - 1)M_1^2 \sin^2\beta + 2}\right). \tag{6.4.75}$$

The velocity turning angle is related to the freestream Mach number M_1 and the shock wave inclination angle by formula

$$\tan\theta \tan\beta \left(\frac{(\gamma+1)M_1^2}{2(M_1^2 \sin^2\beta - 1)} - 1\right) = 1, \tag{6.4.76}$$

and the Mach number behind the oblique shock wave $M_2 = q_2/c_2$ is equal to

$$M_2 = \sqrt{\frac{2 + (\gamma - 1)M_1^2}{2\gamma M_1^2 \sin^2\beta - (\gamma - 1)} + \frac{2M_1^2 \cos^2\beta}{2 + (\gamma - 1)M_1^2 \sin^2\beta}}. \tag{6.4.77}$$

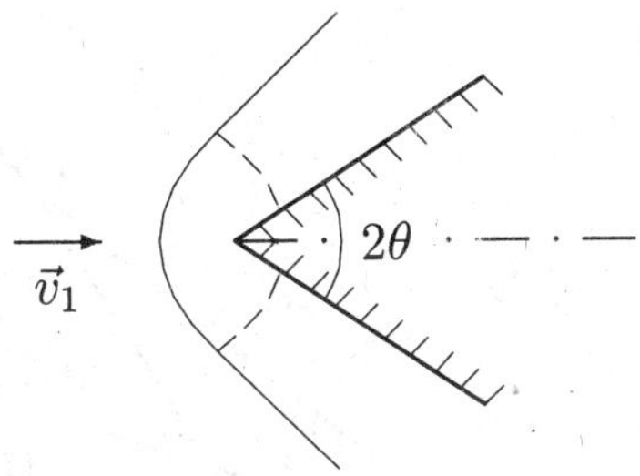

Figure 6.37: The detached shock wave arising in a supersonic gas flow around the wedge at $\theta > \theta_{max}$.

The above-obtained relations at oblique shocks enable one to construct the solution of the problem of the uniform supersonic flow around a concave corner and around the wedge (see Fig. 6.36). An oblique shock wave originates at the nose tip. The velocity direction behind the shock wave is parallel with the corner wall. Therefore, the velocity vector behind the oblique shock wave turns by angle θ. Using the known angle θ one can find the shock inclination angle β from the shock polar (see Fig. 6.35) and then all of the remaining gas parameters. As can be seen from Fig. 6.35, two values of β correspond to a given value of θ, which refer to a strong and weak shock wave. As the experience shows, at the flow around a corner bounding a finite body, a weak shock wave is realized, behind which the flow is supersonic. This choice is determined from the condition for the construction of a complete solution for the problem of the supersonic flow around a finite body. In particular, if one chooses the body with a sharp angle $\theta \ll 1$, then a weak oblique shock wave forms upstream of it, the state behind which is found in the neighborhood of point E (Fig. 6.35). As θ increases, this state is shifted along the supersonic branch of shock polar to the point D corresponding to the maximum angle θ_{max}. All of the above considerations are also valid for the wedge [Fig. 6.36 (b)]. In the upper half-plane, the solution is constructed, as in the problem of the flow around the angle θ_1, and in the lower half-plane, the solution is constructed, as in the problem of the flow around the angle θ_2.

If $\theta > \theta_{max}$, then the constructed solution does not exist. In this case, a detached curvilinear shock wave arises upstream of the corner or the wedge, which is shown for a wedge in Fig. 6.37. The flow in the neighborhood of a nose tip is subsonic, and then after the intersection of the dashed lines it becomes supersonic. As was noted above, the value θ_{max} is a monotonously increasing function of the Mach number M_1. Hence, at the flow around a concave fixed angle θ or the wedge with a given apex angle 2θ, a detached shock wave with increasing freestream

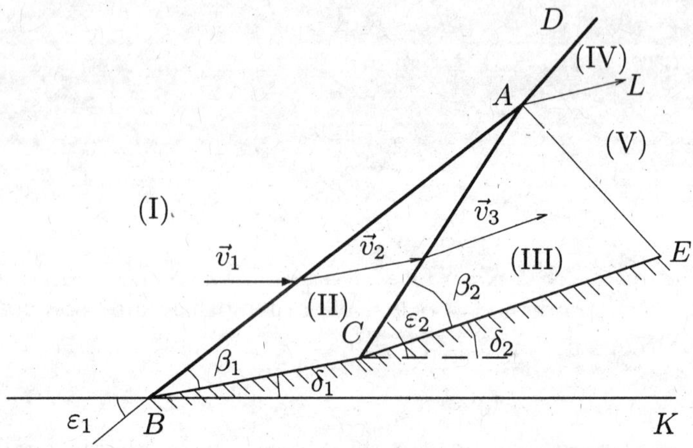

Figure 6.38: Interference of shocks of the same family.

Mach number at certain critical value M_1^* will arise (see Fig. 6.37). At the further increase in M_1, a distance between the detached shock wave and the body as well as the subsonic flow region will increase.

6.4.4 Interference of Stationary Shock Waves

There are two types of compression shocks interaction:

1) the interaction of two shocks belonging to the same family;
2) the interaction of shocks from different families.

Consider a qualitative picture of these two interactions[15–19]. The picture of the interaction process of two shocks belonging to the same family is shown in Fig. 6.38. Here AB and AC are the shock waves formed in a supersonic flow, impinging on the breaks of the wall BCE. The freestream supersonic flow velocity $\vec{v}_1$ is parallel with the line BK (see the notation of corresponding angles in Fig. 6.38). A passage from region (I) to region (II) and from region (II) to region (III), for the purpose of the determination of gasdynamic parameters, may easily be found with the use of the results of Section 6.4.3, if the angles δ_1 and δ_2 of the wall breaks are known. Assuming the flow to be stationary in all regions from (I) to (V), the pressure profile along the streamline AL should have no jump. The flow direction behind the jump AD will be the same as in region (III) and is determined by formulas of the oblique shock (Section 6.4.3), when the wall break angle is equal to $(\delta_1 + \delta_2)$. The pressure in region (IV) differs, however, from the pressure in region (III) because a compression in region (III) takes place as a result of two shocks AB and

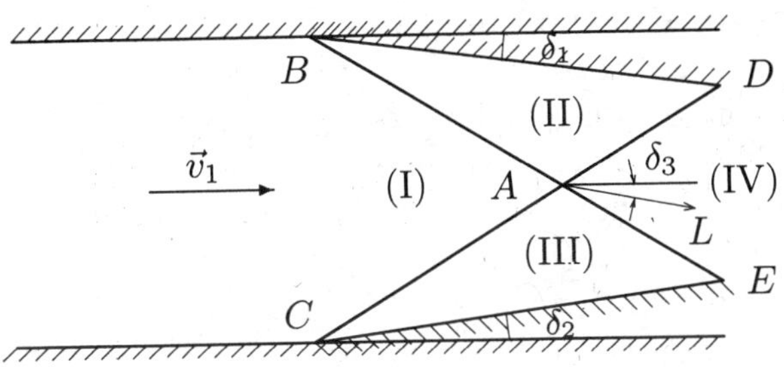

Figure 6.39: Interference of shocks of different families.

AC, and in region (IV), the gas is compressed only by a single shock AD, but with larger pressure jumps. In order to obtain along the streamline AL a single pressure and flow direction along $\vec{v}_2$ it is necessary to alter the pressure and flow direction in region (III). Therefore, the reflected wave AE should emanate from point A. The magnitude of this wave can easily be determined by using the results of Section 6.4.3. The reflected shock strength AE is usually very small, and it can be neglected in many cases, and the jump AD can be determined by using the expression for the deviation in region (IV) assuming that the flow direction is the same as in region (III), that is as at a passage across the shock AD. Then we have along the streamline AL a velocity jump at an equal pressure, which corresponds to a contact discontinuity.

Let us study the second interaction type, when there is an interference of shocks belonging to different families. Such an interaction kind occurs, for example, when the shocks arise on opposite duct walls, as shown in Fig. 6.39. Here BD and CE are the breaks of the duct walls under the corresponding angles δ_1 and δ_2 and, generally, $\delta_1 \neq \delta_2$. The jump BA forms at the expense of the deviation from the angle by angle δ_1, and the jump CA forms at the expense of the angle δ_2 having the direction opposite to the angle δ_1. The inclinations of the jumps BA and CA are generally different, and their strengths may be computed by using the results of Section 6.4.3. These jumps intersect. Depending on their strengths, that is, the initial parameters, two cases of the interaction are possible[15]: the regular and Mach interactions.

Let us at first consider the regular interaction. The flow conditions in regions (II) and (III) for a regular interaction of shocks can be determined from the values of the angles δ_1 and δ_2, depending on the flow conditions in region (I). At the same time, the flow conditions in region (IV) can be determined by taking into account the flow conditions at point A. The flow direction in region (II) differs from the flow direction in region (III) by angle $(\delta_1 + \delta_2)$. At a passage from region (II) to region (IV) across the shock AD, the deviation of the streamlines should be related to the deviation at a passage from region (III) to region (IV) across the jump AE, so that the flow direction along the streamline AL in region (IV) after two deviations is the same. As a consequence of the flow stationarity, the pressure along AL in regions DAL and EAL should be the same. These two conditions enable one to determine the changes of flow parameters along the lines AD and AE. Let δ_3 be the angle of the deviation in region (IV) with respect to the direction of $\vec{v}_1$ in region (I). At the same time, in region (II), this deviation angle is equal to $-\delta_1$, and in region (III), it is equal to $+\delta_3$. Then, at a passage across the shock AD, the deviation angle should be equal to $(\delta_3 + \delta_1)$, and at a passage across the shock AE, it is equal to $(\delta_2 - \delta_3)$. One can determine from the theory of the oblique shocks (cf. Section 6.4.3) the corresponding pressure value in region (IV) on the basis of the values of flow parameters in regions (II) and (III). Therefore, if we denote by $P_{3(II)}$ the pressure corresponding to the value δ_3, which is computed from the flow conditions in region (II), and denote by $P_{3(III)}$ the pressure value computed for the same values δ_2 from the flow conditions in region (III). Then in accordance with the condition of flow stationarity, the equality $P_{3(II)} = P_{3(III)}$ should take place. One can construct for each value of δ_3 the values $P_{3(II)}$ and $P_{3(III)}$ and determine the stationarity conditions at point A. By using the corresponding values $(\delta_2 - \delta_3)$ and $(\delta_3 + \delta_1)$, one can determine the inclination angles of shocks AE and AD.

Since the shock strengths differ significantly from each other, the entropy change in region EAL differs from the entropy change in the region LAD, and consequently, we have different velocities and densities at equal pressures in these regions along AL. In this case, a discontinuity along the line AL arises. It is conventional to call it the contact or tangential discontinuity. According to the experiment data[16,19], however, the regular reflection of shock waves is impossible in many cases and another kind of shock reflection takes place, which is usually termed the *Mach reflection of shock waves.* In this case, at the point of intersection of shock waves a configuration of three shock waves and a contact discontinuity arises. One can represent the Mach configuration qualitatively as follows: one part of the gas passes across two shock waves: the impinging shock and reflected shock. Another part of gas passes only through one shock. These two gas parts are separated from each other by a contact discontinuity line[15]. Let us study in more detail this physical phenomenon at the example of the reflection of a strong shock wave from a rigid wall or, as in the above-described example, in the case in which the angles $\delta_1 = \delta_2$.

Let a shock wave DC impinge on a rigid wall AB (see Fig. 6.40). At point C, a regular reflection of this shock from the wall AB takes place, consequently, the line CK is the reflected shock wave. A qualitative physical picture of such a phenomenon was already described above. We

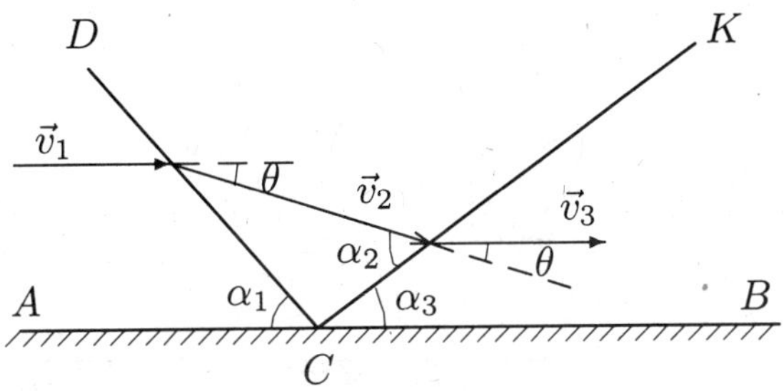

Figure 6.40: Regular shock wave reflection.

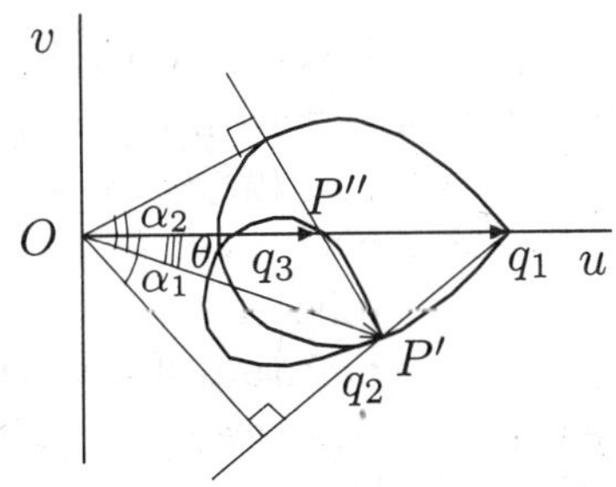

Figure 6.41: The shock polars in the hodograph plane.

now dwell only on some peculiarities. Knowing the values of $\vec{v}_1$ and P_1 for the freestream we can construct a shock polar (cf. Section 6.4.3) on the vector $\vec{v}_1$ (see Fig. 6.41). Knowing the shock strength, one can determine the flow velocity behind the shock, that is, OP' Using the found OP', we can construct a shock polar on the vector $\vec{v}_2$. Since the rigid wall AB is rectilinear, a finite gas flow OP'' should have the same direction as Ou, that is, $\vec{v}_3 \parallel \vec{v}_1$. Consequently the intersection of the shock polar corresponding to vector $\vec{v}_2$ with the Ou-axis determines the magnitude of the finite velocity q_3 and the angle of the reflected shock $\alpha_3 = \alpha_2 - \theta$. Usually $\alpha_1 \neq \alpha_3$, except for the case in which the impinging shock degenerates into the *Mach wave.* In this case, $\alpha_1 = \alpha_3$ and the strengths of the incident and reflected waves coincide.

If the incident shock DC possesses a very large strength, that is, the incidence angle of shock wave α_1 is large, then the shock polar $P'P''$ (see Fig. 6.41) will not intersect the Ou abscissa axis. It is impossible in this case to obtain a regular shock wave reflection, and the Mach reflection will take place (see Fig. 6.42). If the incident shock DC is rectilinear, then we will have a curved reflected shock wave CK and a Mach shock CM near the wall. The line CL is a tangential discontinuity, as this was shown above. In this case, however, as is known[15], the configuration of

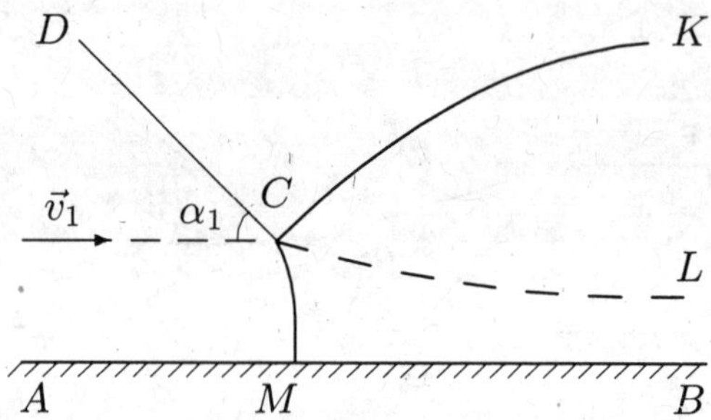

Figure 6.42: The Mach reflection of shock waves.

three shock waves will exist. This is confirmed by numerous experimental data[16,19]. It should be noted that the flow behind the Mach shock CM will be subsonic in contrast with the regular reflection, in which the supersonic gas flow will take place in all of the interaction regions.

The problem of the Mach reflection of shocks, however, is still far from its complete solution. Two criteria for a passage from a regular to Mach reflection[16,19] exist. What a criterion is followed by the nature, and what is the way in which the passage occurs at sufficiently large Mach numbers — these are the questions, which at present require final answers.

6.5 The Fundamentals of the Gasdynamic Design Technology

The computation of supersonic three-dimensional gas flows around the bodies represents a very complex problem of advanced aerodynamics. In some cases, however, it proves to be possible to construct the solution of a problem of the flow around three-dimensional bodies with the use of well-known exact solutions of a lesser dimension[20]. It is conventional to call such an approach the *method of gasdynamic (or aerodynamic) design.* In other words, the information about the particular solutions in one space is extended for a wider geometric space. By the space of aerodynamic design, we mean a particular solution or a set of particular solutions and algorithms, which are used in the specific modeling. The simplest aerodynamic design space consists of a single particular solution; for example, the solution of a problem of the supersonic gas flow around a wedge or of two supersonic uniform flows matched across an oblique shock wave.

A feature of the gasdynamic processes, which involves considerable difficulties for theoretic investigation, is their nonlinearity, which manifests itself in various and sometimes unexpected forms. This gives rise to a variety of the analysis techniques and specific peculiarities, which cannot be handled by a single scheme.

The system of equations for the streamlines governing the steady gas motion has the form

$$\frac{dy}{dx} = \frac{v_y}{v_x} = \varphi(x, y, z), \quad \frac{dz}{dx} = \frac{v_z}{v_x} = \psi(x, y, z),$$

whose solution may be presented in the form

$$y = y(x, C_1, C_2), \qquad z = z(x, C_1, C_2).$$

Such a representation of the streamlines and a goal-oriented coupling of the constants C_1 and C_2 is just a source and the basic idea of the aerodynamic design. The boundary condition is used here in the form of the slip condition $v_n = 0$.

6.5.1 The Basic Algorithm

Let us write the equations governing the flow of an ideal, non-heat-conducting gas in the divergence form for the stationary three-dimensional case. We will consider the Cartesian coordinates, for which the metric tensor $g_{ij} = \delta_{ij}$. Assuming the partial derivatives with respect to time to be equal to zero $\frac{\partial \varphi}{\partial t} = 0$ in (2.1.156), (2.1.154), and (2.1.161) and setting $\sigma^{ij} = -P\delta^{ij}$, $\nabla_j = \frac{\partial}{\partial x^j}$, we obtain:

the equations for the momentum

$$\begin{aligned} &\frac{\partial(\rho u^2 + P)}{\partial x} + \frac{\partial(\rho uv)}{\partial y} + \frac{\partial(\rho uw)}{\partial z} = 0, \\ &\frac{\partial(\rho uv)}{\partial x} + \frac{\partial(\rho v^2 + P)}{\partial y} + \frac{\partial(\rho vw)}{\partial z} = 0, \\ &\frac{\partial(\rho uw)}{\partial x} + \frac{\partial(\rho vw)}{\partial y} + \frac{\partial(\rho w^2 + P)}{\partial z} = 0; \end{aligned} \tag{6.5.1}$$

the equation for the mass conservation

$$\frac{\partial(\rho u)}{\partial x} + \frac{\partial(\rho v)}{\partial y} + \frac{\partial(\rho w)}{\partial z} = 0; \tag{6.5.2}$$

the energy equation

$$\frac{\partial(P + e)u}{\partial x} + \frac{\partial(P + e)v}{\partial y} + \frac{\partial(P + e)w}{\partial z} = 0. \tag{6.5.3}$$

Here ρ is the density, P is the pressure, u, v, and w are the projections of the velocity vector onto the corresponding axes, and $e = \rho[E + (u^2 +$

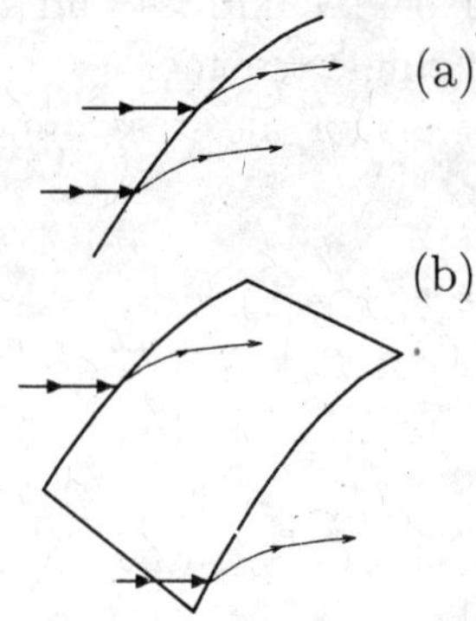

Figure 6.43: (a) the invariant solution and (b) the streamlines in the design space.

$v^2 + w^2)/2]$ is the total energy per unit volume. The system of equations (6.5.1)–(6.5.3) is completed by the equation of state $E = E(P, \rho)$, where $E(P, \rho)$ is the internal energy.

In the aerodynamic design, the two-dimensional solution

$$u_k = u_k(\alpha_1, \alpha_2), \quad u_i = P, \rho, \ldots, e$$

is considered as a particular solution of (6.5.1)–(6.5.3) in the three-dimensional space. We show in Fig. 6.43 (a) a two-dimensional flow, and (b) how this solution is represented in the space of aerodynamic design, that is, in the three-dimensional space, where the solution $u_k = u_k(\alpha_1, \alpha_2)$ is determined. The shock waves represent the surfaces of revolution if the particular solution was axisymmetric, $\alpha_1 = x, \alpha_2 = \sqrt{y^2 + z^2}$ and the ruled surfaces if the two-dimensional solution possesses a planar symmetry $\alpha_1 = x,\ \alpha_2 = y$.

Any streamline of the solution u_k can be taken as a rigid surface in an inviscid flow. Using only one particular solution $u_k(\alpha_1, \alpha_2)$, one can take as a surface around which the gas flows any streamline from the family $\alpha_1 = f_1(\alpha_2, c)$; there is a one-parameter arbitrariness here. On the other hand, there are no obstacles to translate the streamlines arbitrarily along the third coordinate (since we are in a three-dimensional space), and the arbitrariness increases and is now determined by a function $\varphi(\alpha_3)$. We have $\alpha_3 = z$ for the planar symmetry and $\alpha = \arctan(z/y)$ for the axial symmetry.

The aerodynamic design assumes the knowledge of a particular solution as a rule with a lesser dimension, which is considered in the three-dimensional space. The set of such solutions is unbounded at present because it is always possible to obtain a particular solution numerically in a planar or axisymmetrical case with a guaranteed accuracy. It will

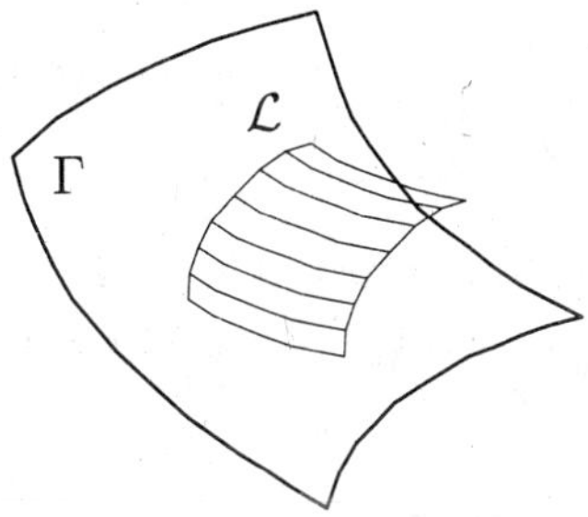

Figure 6.44: The surface construction.

be clear from the following that the presence of a planar or some other symmetry is not necessary, and it is desirable only for the purpose of the simplification of the formal operations.

The rigid surface, around which there is a gas flow, is constructed as follows: let us choose at the shock wave front a line, from each point of which we draw a streamline. As a result we obtain a liquid stream surface. This surface can be taken as a rigid wall in an inviscid gas (Fig. 6.44). The shock wave for this surface represents a part of the original shock of the used solution $u_k(\alpha_1, \alpha_2)$. The leading edge shape is not fixed in advance and determines the arbitrariness in the choice of the configuration or may serve as a subject for the formulation of a variational problem.

Thus, we will assume that we know the solution

$$u_k = u_k(\alpha_1, \alpha_2), \tag{6.5.4}$$

which depends on two variables, and the functions (6.5.4) themselves determine the distributions of gasdynamic parameters in the flow behind the bow shock, which is determined by the relation

$$\alpha_1 = \Gamma(\alpha_2). \tag{6.5.5}$$

The family of streamlines is determined by the ordinary differential equation $d\alpha_1/d\alpha_2 = F(\alpha_1, \alpha_2)$, where the right-hand side of the equation is a known function of the solution (6.5.4). The solution of the differential equation may be presented in the form

$$\alpha_1 = f_1(\alpha_2, c), \tag{6.5.6}$$

where c is the integration constant.

Let us direct a uniform freestream along the ordinate axis. The line determining the position and shape of the leading sharp edge lies at the front (6.5.5) and may be presented as an intersection of two cylindrical surfaces in the space of the variables α_1, α_2, and α_3:

$$\alpha_1 = \Gamma(\alpha_2), \qquad \alpha_2 = f_2(\alpha_3), \tag{6.5.7}$$

where $f_2(\alpha_3)$ is an arbitrary continuous single-valued function.

Present the equation of the flowstream surface in accordance with the solution (6.5.6) as

$$\alpha_1 = f_1(\alpha_2, c(\alpha_3)). \tag{6.5.8}$$

The function $c(\alpha_3)$ should be determined from the condition that (6.5.8) passes through the leading edge (6.5.7). As a result, we obtain the basic functional equation[21,22]:

$$\Gamma[f_2(\alpha_3)] = f_1[f_2(\alpha_3), c] \tag{6.5.9}$$

the solution of which determines in (6.5.8), for each value α_3, a streamline emanating from (6.5.7); that is, the streamlines are translated in space in a special way. which is determined by the form of f_2.

Summarizing the above, we see that the aerodynamic design of the solution assumes the knowledge of some particular flow $\Phi(u_1, \ldots, u_k)$, which has as a rule a lesser dimension $u_i = u_i(\alpha_1, \alpha_2)$. The set of such solutions is sufficiently vast and augmented continuously, and if we also add the numerical solutions of the planar and axisymmetric problems, which can now be obtained with a guaranteed accuracy on a computer, this class is practically unlimited. For a particular solution with planar symmetry, the solution does not depend on the third coordinate $\alpha_3 = z$, and for the axisymmetric flows, the solution does not depend on $\alpha_3 = y/z$.

The bow shock location is determined by the function $\alpha_1 = \Gamma(\alpha_2)$, which in the space of α_i, $i = 1, 2$, and 3, represents a certain cylindrical surface. The streamlines are determined as the intersections of the surfaces $\{\alpha_1 = f_1(\alpha_2, c)$ and $\alpha_3 = c_1\}$, where the constant c is an integration constant of the ODE $d\alpha_1/d\alpha_2 = F(\alpha_1, \alpha_2)$, determining the streamlines in the α_1, α_2 space.

The constant c_1 in the solution Φ is insignificant since this solution does not depend on the variable α_3; it is indeed not important in which section z the planar solution is considered, or in which meridional section the solution with axial symmetry is investigated. We point out that this "not important" refers only to the planar and axisymmetric problems. that is, to the particular solution, and this circumstance becomes extremely important if we include this particular solution in the space of the aerodynamic design.

The aerodynamic design consists of a goal-oriented coupling of the constants c and c_1: namely. one solution trajectory is considered in one section, and another solution trajectory is considered in another section. A condition for the choice is a certain goal. Certain constraints naturally arise here. For example, if we construct the solution of a problem of the flow around a configuration having a leading edge in the form $\{\alpha_1 = \Gamma(\alpha_2),\ \alpha_2 = f_2(\alpha_3)\}$, then the relation between the constants is determined by the equation[21,22]

$$\Gamma[f_2(c_1)] = f_1[f_2(c_1), c],$$

which determines the function $c = c(c_1) = c(\alpha_3)$ implicitly, and the surface $\alpha_1 = f_1(\alpha_2, c(\alpha_3))$ describes the desired spatial configuration. Note that, in a general case, a class of surfaces is obtained, which depends on a single arbitrary function $f_2(\alpha_3)$. The arising constraints are minimal: only the continuity is required.

Thus, one invariant solution in the space of two variables produces a family of the solutions of flow problems in the three-dimensional space.

6.5.2 The Superposition Procedure

We have presented above the equation of a surface $F(x, y, z) = 0$, in the neighborhood of which the chosen solution is realized. We will call the technique in which the solution is used for the flow description near the carrying system the *superposition operation.* We already have one surface: this is (6.5.8). The superposition operation enables us to describe the construction of a three-dimensional body.

The three-dimensional disturbing surface (6.5.8) is found under the conditions of the flow downstream from the bow shock, and the pressure can be different on another surface. The pressure jump on the one surface as well as the other surface can be a source for the creation of lift force. Such a body has been termed the waverider because of the specifics and the principle of the organization of its lifting properties.

In order to write the equation for the second stream surface to obtain some volume, we note that a ruled surface $\alpha_2 = f_2(\alpha_3)$ passes through the leading edge by definition (6.5.7) and does not introduce the disturbances in an uniform free stream. Hence, it will be the desired stream surface for the undisturbed flow.

We finally obtain a three-dimensional body (Fig. 6.45) bounded by the surfaces

$$\alpha_1 = f_1[\alpha_2, c(\alpha_3)], \tag{6.5.10}$$

$$\alpha_2 = f_2(\alpha_3). \tag{6.5.11}$$

The given set of the surfaces (6.5.10) and (6.5.11) is just the superposition operation. This is indeed the description of physical spaces in which the solutions Φ_1 and Φ_2 are determined, which produce the surfaces.

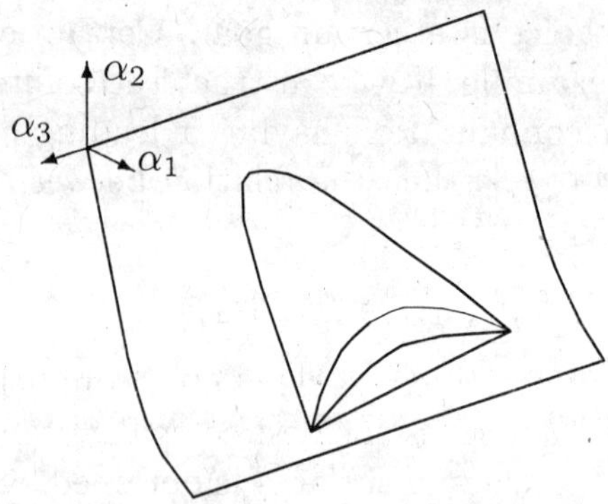

Figure 6.45: The implementation of the superposition procedure.

The bow shock attaches the body along a sharp leading edge (6.5.7); therefore, there is no interaction of the flows. A freestream condition is realized on (6.5.11), and (6.5.10) introduces the disturbances in a special way. The bow shock is a part of (6.5.5), and the flow downstream of it in the neighborhood of (6.5.10) is described by the solution (6.5.4).

We note once more that no errors are introduced in the process of the construction of the flow around a three-dimensional body, and the flow pattern is determined to the same accuracy as the initial (in the case considered, two-dimensional) solution. At a fixed particular solution (6.5.4), the arbitrariness in the configuration choice is obtained as a single function.

Let us apply the formal superposition procedure to the simplest planar flow around a sharp wedge with the attached planar shock wave. The function (6.5.5) determining the shock wave location is specified via the relation $x = y \operatorname{ctg} \beta$, where β is the inclination angle of the shock to the freestream velocity direction, which is parallel to the x-axis, upstream of the shock.

The differential equation for the streamlines of a constant supersonic flow downstream of the shock takes the form

$$\frac{dx}{dy} = \operatorname{ctg} \theta.$$

The turning angle of the flow at a shock wave θ is given by formula (6.4.76).

Since the bow shock is a plane, the flow turning angle θ on it is constant, and the family of the streamlines (6.5.6) can be written explicitly:

$$x = y \operatorname{ctg} \theta + c.$$

The solution of equation (6.5.9) for the given particular form of the flow

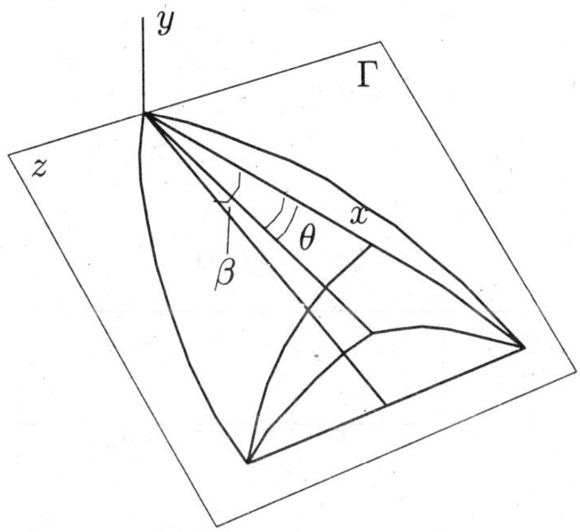

Figure 6.46: The complete view of the body (6.5.12).

can be obtained directly:

$$c = (\operatorname{ctg}\beta - \operatorname{ctg}\theta) f_2(z).$$

As a result, we obtain a solid body bounded by the surfaces composed of the stream line

$$x = y \operatorname{ctg}\theta + (\operatorname{ctg}\beta - \operatorname{ctg}\theta) f_2(z) = f_1(y, z), \quad y = f_2(z). \tag{6.5.12}$$

The superposition procedure gives only the head part of the bodies. In order to obtain a finite volume body, one can introduce the bottom section. If the bottom section edges are supersonic, then the flow in the wake does not change the upstream structure. The equation $x =$ const can serve as a bottom section with supersonic edges in the above-considered example. The general form of the three-dimensional body (6.5.12) is presented in Fig. 6.46. If we take a linear dependence

$$f_2 = \{z \tan\omega, \quad z \geq 0; \quad -z \tan\omega, \quad z \leq 0\}$$

as a function determining the shape of the leading edges and restrict the body (6.5.12) by the bottom section $x = l$, then we obtain a V-shaped wing (Fig. 6.47). The bibliography on the V-shaped wings includes many hundreds, if not thousands, of works only in Russia. The components of the velocity vector in the perturbed flow region downstream of the shock wave both for (6.5.12) and the V-shaped wing are equal to

$$v_x = U_\infty \left[1 - \frac{2(\sin^2\beta - \sin^2\mu_\infty)}{\gamma + 1}\right], \quad v_y = -\frac{2U_\infty \operatorname{ctg}\beta(\sin^2\beta - \sin^2\mu_\infty)}{\gamma + 1}.$$

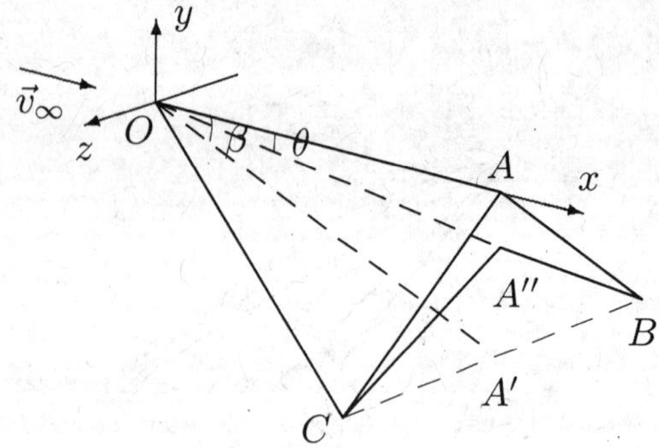

Figure 6.47: The V-shaped wing.

Here U_∞ is the modulus of the supersonic freestream velocity directed along the x-axis, $\mu_\infty = \arcsin(1/M_\infty)$ is the Mach angle corresponding to the Mach number M_∞.

Consider in more detail the superposition procedure at the example of the construction of the solution of the problem of stationary supersonic flow around a symmetric V-shaped wing (the waverider) as shown in Fig. 6.47. Assume that the freestream velocity vector $\vec{v}_\infty$ is directed along the rib OA of the wing. Let us direct the Ox-axis of a system of Cartesian coordinates $Oxyz$ along the rib OA. Then the upper wing facets are also parallel with the vector $\vec{v}_\infty$; therefore, they do not introduce any disturbances in the freestream.

A shock wave arises near the lower wing facets. Denote by θ the angle between the rib OA'' of the lower wing surface and the x-axis. We now require that the inclination angle β of the shock wave coincides with the angle between the plane OBC (see Fig. 6.47) and the x-axis. For this purpose, we make use of formula (6.4.76) at a given angle β and the freestream Mach number M_1 to find the angle θ. The found angle θ is the angle of gas flow turning behind the oblique shock wave. Since the rib OA'' belongs simultaneously to both lower facets, the gas velocity vectors behind the shock wave will be parallel with both of the lower wing facets, which ensures the satisfaction of the slip condition simultaneously on both lower facets.

On the basis of the given values of M_1 and β, we find from formulas (6.4.74) and (6.4.75) the pressure P_2 and the density ρ_2 behind the shock wave front. The pressure P_2 on the lower body surface is constant, and the pressure coefficient is equal to

$$c_p = \frac{P_2 - P_\infty}{\frac{1}{2}\rho_\infty U_\infty^2} = \frac{4}{\gamma+1}(\sin^2\beta - \sin^2\mu_\infty),$$

which is the source of the lifting capability of the waverider.

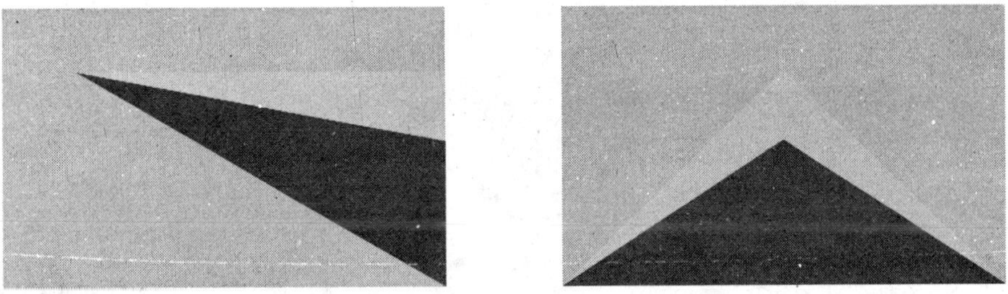

(a) (b)

Figure 6.48: The distribution of the gas pressure $P(x, y, z)$ in different sections: (a) in the wing symmetry plane and (b) in the plane $x = 1$.

We have implemented the above algorithm for the computation of the supersonic flow around the V-shaped wing in our *Mathematica* program `prog6-4.nb`. We present in Fig. 6.48 the distribution of the gas pressure in different wing sections. The blue color in these distributions corresponds to the freestream pressure P_1, and the red color corresponds to the pressure P_2 behind the oblique shock wave front.

We show in Fig. 6.49 the distribution of the gas velocity vectors in the flow around the same V-shaped wing, as in Fig. 6.48. To obtain the field of the velocity vectors, we have used the built-in function `PlotVectorField3D[_]` of the software system *Mathematica 3.0* (see a description of this function in Appendix A).

6.5.3 The Complement Procedure

The aerodynamic design of star-shaped configurations is an application of the general approach. Consider a V-shaped wing whose form was presented previously in the work by Maikapar. The shape of the type of a star-shaped penetrator is obtained if one turns this body n times by a certain angle around the longitudinal axis. A closed configuration is obtained if this angle is a multiple value of the complete revolution: $\alpha_n = 2\pi/n$, where n is an integer. This technique is called the complement procedure in the simple case, in which the generating function is linear.

In a general case, in which the leading sharp edge is arbitrary, the complement procedure is more complex. It can be seen from the above example for a linear function f that the peculiarities of the design of star shapes are related only to the external surface (6.5.11), in the neighborhood of which the freestream is realized, and do not depend on the chosen

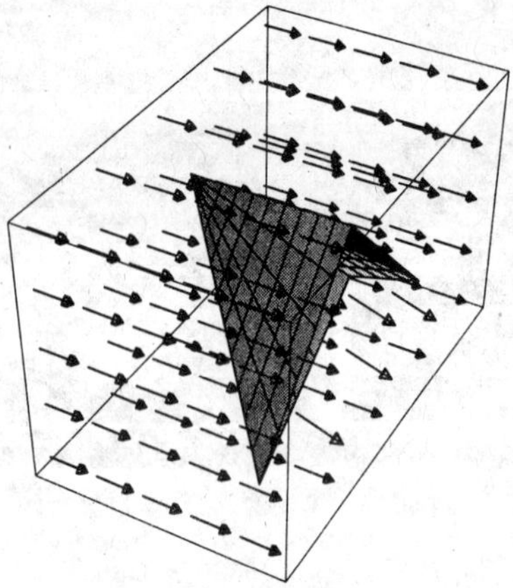

Figure 6.49: The distribution of the gas velocity vectors at a supersonic flow around the V-shaped wing with $M_\infty = 2.2,\ \beta = \pi/4$.

particular solution.

Let us subdivide the generating function into two parts:

$$f_2^{(1)}(z), \quad z \le 0; \qquad f_2^{(2)}(z), \quad z \ge 0.$$

It requires also the function continuity at point $z = 0$: $f_2^{(1)}(0) = f_2^{(2)}(0)$. In order to obtain a new shape with the aid of the complement procedure, it is necessary to ensure that the projections of the leading edges onto a plane perpendicular to the longitudinal axis coincide. Hence, the complement procedure reduces the class of possible curves and the arbitrariness in the choice of (6.5.11) reduces. If one part of the leading edge has been chosen, then the second one is the result of the rotation of the first leading edge by the angle α_n.

The complement procedure can be formalized easily. A rotation of a system of Cartesian coordinates by angle α_n around the x-axis is determined by the relations

$$x' = x, \quad y' = y\cos\alpha_n - z\sin\alpha_n, \quad z' = y\sin\alpha_n + z\cos\alpha_n.$$

The inverse transformation is

$$x = x', \quad y = y'\cos\alpha_n + z'\sin\alpha_n, \quad z = -y'\sin\alpha_n + z'\cos\alpha_n.$$

The determination of an additional leading edge branch by a given part of the basic leading edge is performed as follows. Let a single-valued function $f_2^{(1)}(z)$ be given. Then, at the rotation by angle α_n, this branch should coincide with the second branch, and this determines the second branch. Using the above formulas for coordinate transformation, we obtain the equation for the second branch in an implicit form:

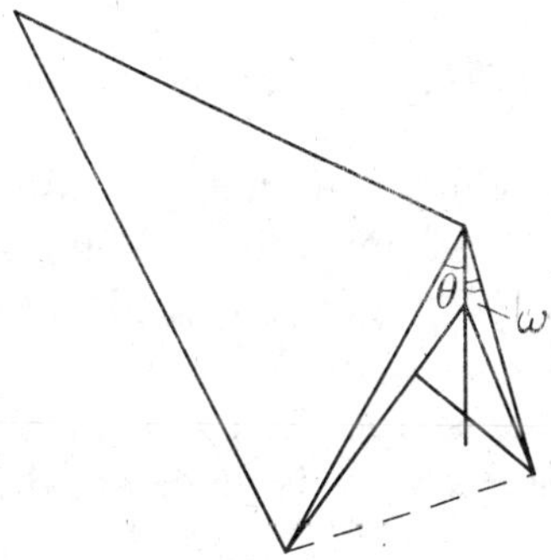

Figure 6.50: The asymmetric body: the initial stage for the design of a three-dimensional configuration with a moment, with respect to the longitudinal axis.

$$f_2^{(2)}: \quad y\cos\alpha_n - z\sin\alpha_n = f_2^{(1)}(y\sin\alpha_n + z\cos\alpha_n). \tag{6.5.13}$$

The peculiarities of the design of star shapes are related only to the external surface form (6.5.11) and do not affect the flow structure behind the shock wave. In particular, the same shape can have the parts designed on the basis of different solutions.

A V-shaped wing (the symmetric body) is shown in Fig. 6.47. The complement procedure gives both the symmetric and asymmetric shapes, with a central body of the fuselage type if one rotates the wings by the angle 2θ around the x-axis for a symmetric body and by the angle $\theta+\omega$ for an asymmetric body (Fig. 6.50). A closed shape of the star shape type is obtained if the angles 2θ and $\theta+\omega$ are the multiples of 2π. Repeating then the complement procedure n times (the multiplicity number), we obtain a shape with petal structure of the middle section. Note that the matching condition is always satisfied for the linear functions.

In the first case, we obtain a symmetric star-shaped configuration. In the second case, the shape will be asymmetric and will possess besides the wave drag a longitudinal moment with respect to the x-axis. Generally, if z_1 is a solution of equation $a - f_2^{(1)}(z) = 0$, where $f_2^{(1)}$ is a given branch of the leading edge, and it is equal to the solution z_2 of the equation $a - f_2^{(2)}(z) = 0$, where $f_2^{(2)}$ is determined by the relation (6.5.13), then the obtained star-shaped configuration will not have the moment; otherwise, the longitudinal moment will be different from zero. In particular, for the rectilinear leading edge $f_2^{(1)}(z) = z\,\mathrm{ctg}\,\theta$, we obtain from relation (6.5.13) the equation for the second branch:

$$y\cos\alpha_n - z\sin\alpha_n = (y\sin\alpha_n + z\cos\alpha_n)\,\mathrm{ctg}\,\theta.$$

Executing the elementary transformations, we obtain the equation:

$$y = f_2^{(2)}(z) = z\,\mathrm{ctg}\,(\alpha_n - \theta).$$

The condition $z_1 = z_2$ is equivalent for the case under study to the equality $\tan\theta = \tan(\alpha_n - \theta)$, from where it follows that $\theta = \alpha_n/2$. At $\theta = \alpha_n/2$, the star-shaped configuration will not have a longitudinal moment.

The wave drag of a star shape designed on a constant supersonic solution behind a planar shock wave is determined by the relation

$$c_w = c_p n \tan\sigma\, \tan\delta\, \tan(\alpha_n/2),$$

where σ is a user-specified parameter; it determines the inclination angle of the plane of the leading edges to the longitudinal axis. From the physical considerations, the shock wave inclination is always larger than the Mach angle $\sigma > \mu_\infty$, where μ_∞ is the freestream Mach angle. The quantity δ is the turning angle of the flow at a bow shock, and c_p is the pressure coefficient of the flow behind the shock. Since in our case this is a constant solution, the c_p does not depend on the procedures of aerodynamic design and is determined only by the parameters of the original solution:

$$c_p = 2\sin\sigma\, \sin\delta / \cos(\sigma - \delta) = c_p(\mu_\infty, \delta).$$

At the increase in the number of beams $n \to \infty$ or $n = 2\pi/n \to 0$, the wave drag tends to certain finite value $c_w^* = c_p\pi \tan\sigma \tan\delta$, depending only on the parameters of the particular solution: the freestream Mach angle μ_∞ and the inclination angle of the leading edges σ. Note that the transition to the limit considered in the exact formulation of the flow problem does not yield any paradoxical results for the drag. Whereas in an approximate formulation, for example, the c_p is determined from the hypersonic Newton's theory, we obtain $c_w \to 0$ at an unbounded growth of n.

The longitudinal moment will be maximal at $\omega = 0$ (ω is shown in Fig. 6.50) and is equal to $M_x = m(1 - \cos\alpha_n)$, where m depends on the determining parameters of the particular solution. With an increasing number of petals, it is easy to obtain for the aerodynamic moment the estimate $M_x \sim n^{-2}$; that is, the longitudinal moment vanishes as the number of beams of the star-shaped configuration increases. This is again an exact result. Consequently, the maximum longitudinal moment can be obtained at $\omega = 0$ and $n = 3$, the three-petal shape.

Thus, the presented algorithm enables one to construct a variety of model configurations of flying vehicles on the basis of the analytic solutions. This makes it possible to formulate and solve in a sufficiently simple way the optimization problems. This methodology is presented in sufficient detail in the relevant literature[21,22]. It should be noted, however, that an analysis of the possibilities of aerodynamic design reveals both its basic advantages and its shortcomings. First, the solution of a three-dimensional flow problem is obtained with the same accuracy as the original two-dimensional flow. A relative simplicity of the theoretical description of gasdynamic fields enables one to formulate and implement on a computer the optimization problems of aerodynamic design at a large number of conditions and constraints for the flight regimes and the flying vehicle geometry. The main shortcoming of the aerodynamic design is that the obtained solutions are valid only for a single regime in terms of the freestream Mach number. It should be noted here that, since the design flow regime is known sufficiently accurately, it can serve as certain *a priori* information for the construction of approximate flow models for the freestream Mach numbers different from the design values.

Problem 6.8. The asymmetric V-shaped wing may be an element of more complex aerodynamic shapes. Generalize the program `prog6-4.nb` for the case of an asymmetric V-shaped wing; that is, the wing whose upper and lower facets make different angles with the y-axis.

References

1. **Vallander, S.V.,** *Lectures in Hydroaeromechanics* (in Russian), Leningrad State University, Leningrad, 1978.
2. **Loitsyanskii, L.G.,** *Mechanics of Liquids and Gases*, Pergamon Press, Oxford, 1966.
3. **Laval, P.,** Time-dependent calculation method for transonic nozzle flows, *Proceedings of the Second International Conference on Numerical Methods in Fluid Dynamics, September 15-19, 1970, University of California, Berkeley* / Ed. M. Holt. Lecture Notes in Physics, Vol. 8. Springer-Verlag, Berlin, 1971, p. 187.
4. **Ganzha, V.G. and Vorozhtsov, E.V.,** *Numerical Solutions for Partial Differential Equations: Problem Solving Using Mathematica*, CRC Press, Boca Raton, 1996.
5. **Landau, L.D. and Lifschitz, E.M.,** *Hydrodynamics* (in Russian), Nauka, Moscow, 1986.

6. **Sedov, L.I.,** *Continuum Mechanics*, Vols. I and II (in Russian), Fifth Edition, Nauka, Moscow, 1994.
7. **Kochin, N.E., Kibel, I.A., and Rose, N.V.,** *Theoretical Hydromechanics* (in Russian), Vol. II, 4th Edition, Fizmatgiz, Moscow, 1963.
8. **Ovsyannikov, L.V.,** *Lectures on the Fundamentals of Gas Dynamics* (in Russian), Nauka, Moscow, 1981.
9. **Roždestvenskii, B.L. and Janenko, N.N.,** *Systems of Quasilinear Equations and Their Applications to Gas Dynamics.* Second Edition (in Russian), Nauka, Moscow, 1978. [English transl.: *Systems of Quasilinear Equations and Their Applications to Gas Dynamics*, Translations of Mathematical Monographs, Vol. 55 (American Mathematical Society, Providence, Rhode Island, 1983)].
10. **Stanyukovich, K.P.,** *Transient Motions of Continuum* (in Russian), Nauka, Moscow, 1971.
11. **Chaplygin, S.A.,** On gas jets, *Uchenye Zapiski MGU, otd. fisico-matematicheskih nauk* (in Russian), Vol. 24, Moscow, 1904.
12. **Chernyi, G.G.,** *Gas Dynamics* (in Russian), Nauka, Moscow, 1988.
13. **Dombrovskii, G.N.,** *Method for the Adiabat Approximation in the Theory of Planar Gas Motions* (in Russian), Nauka, Moscow, 1964.
14. **Mises von R.,** *Mathematical Theory of Compressible Fluid Flow*, Academic Press, New York, 1958.
15. **Courant, R., and Friedrichs, K.O.,** *Supersonic Flow and Shock Waves*, Interscience Publishers, Inc., New York, 1948.
16. **Ben-Dor, G.,** *Shock Wave Reflection Phenomena*, Springer-Verlag, New York, Berlin, 1992.
17. **Ferri, A.,** *Elements of Aerodynamics of Supersonic Flows*, MacMillan, New York, 1949.
18. **Liepmann, H.W., and Roshko, A.,** *Elements of Gas Dynamics*, John Wiley and Sons, Inc., New York, 1957.
19. **Fomin, V.M. et al.,** Experimental investigation of a transition to Mach reflection of stationary shock waves, *Doklady Rossiyskoi Akademii Nauk* (in Russian), Vol. 357:623, 1997.
20. **Maikapar, G.I.,** On the wave drag of asymmetric bodies in supersonic flow, *Prikladnaya Matematika i Mekhanika* (in Russian), Vol. 23:376, 1959.
21. **Shchepanovskii, V.A.,** *Gasdynamic Design* (in Russian), Nauka, Novosibirsk, 1991.
22. **Shchepanovskii, V.A. and Gutov, B.I.,** *Gasdynamic Design of Supersonic Inlets* (in Russian), Nauka, Novosibirsk, 1993.

7

Multiphase Media

The fundamentals of the mechanics of multiphase media are presented here for the first time within the framework of a course in fluid mechanics. This new branch of mechanics has appeared comparatively recently, about 40 years ago, in connection with the development of aerospace technology, nuclear power, and new technologies. At present, the general principles of the construction of the models of the mechanics of multiphase media have been formulated, and there are numerous applications. While presenting the material, we have aimed on the one hand to familiarize the reader with the mathematical models, which are applied for the description of various multiphase media, and on the other hand, to give an insight into the specific physical phenomena occurring in these media. Since the mechanics of multiphase media now enjoy a rapid development, the material of the present chapter does not have such a full-blown character as in the foregoing chapters. It will enable the reader to rapidly enter the details and become familiar with problems in the field of the mechanics of multiphase media.

The general equations of the mechanics of multiphase media are derived in Section 7.1 on the basis of the conservation laws of the principle of interpenetrating continua and the assumption on a local thermodynamic equilibrium within each phase. These equations contain the indeterminate terms related to the exchange by mass, momentum, and energy between the phases, and therefore they are not closed. Thus, the main problem of the mechanics of multiphase media is the determination of the exchange terms or, in other words, the closure relationships for each multiphase medium. In this section, we present a complete system of equations for two cases: the gas-particle mixture and the gas mixture. In the first case, the indeterminate exchange terms are determined from the experiment, and in the second case, they are determined with the aid of the methods of the kinetic theory of gases.

The mathematical properties of a model of gas-particle mixture are investigated in Section 7.2. In particular, the characteristics of the system of equations in the one-dimensional nonstationary case are determined and the well-posedness of the Cauchy problem for this system is discussed. The discontinuous solutions are investigated, for which the jump conditions are derived. It is shown that a new type of discontinuities arises here, which have a finite surface mass.

The quasi-one-dimensional flows of a gas-particle mixture in the Laval nozzle are considered in Section 7.3 at a small volume concentration of particles. This is one of the first problems, which was solved in the mechanics of multiphase media. Since the Laval nozzle is an integral part of the solid- and liquid-propellant propulsion units, this problem is also of practical importance today.

A continual-discrete model for gas-particle mixture is studied in Section 7.4, which enables us to describe the flows with intersecting particles trajectories. As a result of the intersection of particles trajectories, integrable singularities arise in the particles concentration, which are called the caustics. The caustics in a pseudogas of particles are investigated on the basis of the above model. The conditions are found for the formation of caustics, and the boundedness from above is proved for the total number of particles on a caustic.

The problem of the interaction of a shock wave with a cloud of solid spherical particles is studied in Section 7.5. The effects of the formation of a collective shock wave ahead of the particles cloud, the breakdown of the cloud under the development of an instability on its boundary, and the raising of the particles cloud behind the "sliding" shock wave are considered. Note that these effects are at present at the focus of the attention of the researchers. This is related, first of all, to insuring trouble-free work at coal mines whose drifts are dust-laden with small coal particles, at the mills, etc. Therefore, these effects are studied intensively both experimentally and theoretically.

The inertialess flows of multiphase media are considered in Section 7.6. The solution of a classical problem of the Brownian motion of small particles in fluid is presented. The filtration of a bicomponent mixture of incompressible fluids in a nondeformable porous skeleton is studied in detail. As an example, we consider the petroleum displacement by water, which is an important technique in the oil production.

The wave processes in a bubbly liquid are investigated in Section 7.7. The motion equations as well as the closing relationships are obtained for a bubbly liquid with the aid of a variational principle. They reduce for small disturbances to the Burgers–Korteweg–de Vries equation on the basis of which the solutions of the progressive wave type are obtained in the form of a soliton, monotonous, and oscillatory shock wave.

7.1 Mathematical Models of Multiphase Media

7.1.1 General Equations of the Mechanics of Multiphase Media

Consider a multiphase medium whose components differ significantly with respect to their physicomechanical properties. We present as examples a gas-particle mixture, mixture of liquid and gas bubbles, gas mixture, etc. It is hard to believe that the exact solution of a problem of the motion of such a medium would be obtained in principle. Therefore, one proceeds to a more rough description of the behavior of the mean quantities. By analogy with statistical mechanics, the averaging is performed in each phase. After that, each phase is represented as a continuum, which interacts with the remaining continua. One can in principle use the averagings over the ensemble, time, and volume. The averaging over the volume[1–8] is used most frequently in the mechanics of multiphase media, and we will use just this averaging in the following.

Let some microquantity φ_i' be defined for the ith phase (component) within certain volume V. Then the mean value over the volume will be determined by formula

$$\varphi_i = \frac{1}{V}\int_V \varphi_i'\, dV. \tag{7.1.1}$$

We will assume that there are in the volume V a sufficiently large number of particles and the inclusions of the ith phase, so that the averaging procedure is correct. In addition, the linear size of the averaging region $V^{1/3}$ should be substantially smaller than the reference length L of the variation of the mean parameters φ_i. We write both of these conditions in the form of the inequalities

$$d \ll V^{1/3} \ll L, \tag{7.1.2}$$

where d is a reference size of a particle, inclusion, etc. The functions of microquantities $\varphi_i'(x^k)$ may contain the discontinuities of the first kind at the phase boundaries; however, the functions of the mean quantities $\varphi_i(x^k)$ will be continuous at each point, and consequently, they determine the ith continuum. N continua $\varphi_i(x^k)$ are defined at each point x^k. The number N is equal to the number of phases (components) N.

Let us determine for each ith continuum, with the aid of the averaging procedure (7.1.1), the mean density ρ_i, velocity $\vec{v}_i$, tensor of stresses σ_i^{kl}, and other dynamic variables. Using then the conservation laws for the mass, momentum, and energy, we will derive the equations of continuity, motion, and energy for each continuum. We introduce for definiteness the Eulerian coordinate system x^k with the basis vectors $\vec{e}_k$ and the metric tensor $g_{kl} = \vec{e}_k \cdot \vec{e}_l = \delta_{kl}$, in which we define at time t a small

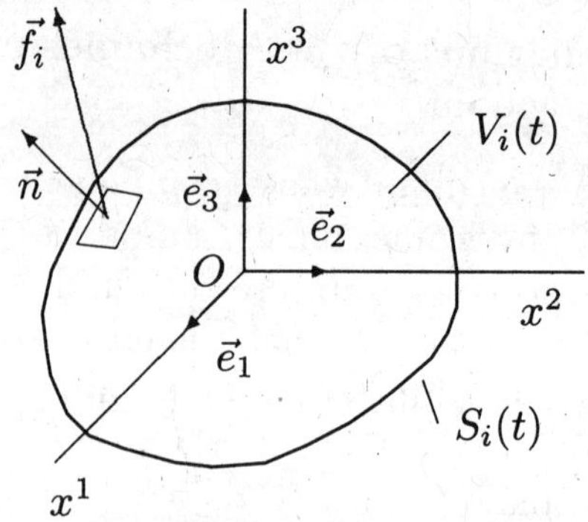

Figure 7.1: The individual volume of the ith phase.

volume V containing N phases. For each ith phase, we introduce an individual volume V_i which consists at all times of the same particles of the ith phase that coincides at time t with the volume V (see Fig. 7.1) and depends on time t. Let us write the conservation laws for the mass, momentum, and energy for each individual volume (see Section 2.1):

$$\frac{d}{dt}\int_{V_i}\rho_i\,dV = \int_{V_i}\sum{}'\kappa_{ji}\,dV,$$

$$\frac{d}{dt}\int_{V_i}\rho_i v_i^k\,dV = \int_{S_i} f_i^k\,dS + \int_{V_i}\rho_i F_i^k\,dV + \int_{V_i}\sum{}'P_{ji}^k\,dV,$$

$$\frac{d}{dt}\int_{V_i}\rho_i E_i\,dV = \int_{S_i}(f_i^k v_i^k)\,dS + \int_{V_i}\rho_i F_i^k v_i^k\,dV \tag{7.1.3}$$

$$+\int_{V_i}\sum{}'E_{ji}\,dV + \int_{S_i} Q_i^n\,dS,$$

where $\vec{v}_i = v_i^k\vec{e}_k$ is the velocity, $\vec{f_i} = f_i^k\vec{e}_k$ is the surface force, $\vec{F} = F_i^k\vec{e}_k$ is the body force per unit mass, $E_i = \frac{v_i^2}{2} + e_i$ is the specific total energy, e_i is the specific internal energy, Q_i^n is the heat flux to a unit surface S_i, the subscripts i and j denote the ith and jth phases, respectively, and the index k numbers in this section the components ($k = 1, 2, 3$). In contrast with the equations of monophase (one-component) continuum considered in Section 2.1, the terms stand on the right-hand sides of equations (7.1.3), which describe the exchange by the mass $\sum{}'\kappa_{ji}$, momentum $\sum{}'P_{ji}^k$, and energy $\sum{}'E_{ji}$ between the ith phase and all remaining phases (the symbol $\sum{}' = \sum_{j=1,(j\neq i)}^N$ denotes the summation over all phases except for the ith phase). It is obvious that the amount of mass κ_{ji}, momentum P_{ji}^k, and energy E_{ji} flowing out of the ith phase is equal to the amount of mass κ_{ji}, momentum P_{ij}^k, and energy E_{ij} flowing into the jth phase. For this reason, the tensors κ_{ji}, P^k_{ji}, and E_{ji} are antisymmetric, with respect to the subscripts i and j:

$$\kappa_{ji} = -\kappa_{ij}, \quad P^k_{ji} = -P^k_{ij}, \quad E_{ji} = -E_{ij}. \tag{7.1.4}$$

Using the formula (1.1.66)

$$\frac{d}{dt}\int_{V(t)} \Phi\, dV = \int_{V(t)} \left(\frac{\partial \Phi}{\partial t} + \nabla_k(\Phi v^k)\right), dV$$

let us rewrite the left-hand sides of equations (7.1.3) in the form

$$\begin{aligned}
\frac{d}{dt}\int_{V_i(t)} \rho_i dV &= \int_{V_i(t)} \left(\frac{\partial \rho_i}{\partial t} + \nabla_k(\rho_i v^k_i)\right) dV,\\
\frac{d}{dt}\int_{V_i(t)} \rho_i v^k_i dV &= \int_{V_i(t)} \left(\frac{\partial \rho_i v^k_i}{\partial t} + \nabla_l(\rho_i v^k_i v^l_i)\right) dV,\\
\frac{d}{dt}\int_{V_i(t)} \rho_i E^i dV &= \int_{V_i(t)} \left(\frac{\partial \rho_i E_i}{\partial t} + \nabla_l(\rho_i E_i v^l_i)\right) dV.
\end{aligned} \tag{7.1.5}$$

With the aid of the Cauchy formula (1.3.10) $f^k_i = \sigma^{kl}_i n_l$ and the Ostrogradsky–Gauss theorem (1.1.63), let us go over from the integration over the area S_i to the integration over the volume V_i in the right-hand side of (7.1.3):

$$\begin{aligned}
\int_{S_i} f^k_i\, dS &= \int_{S_i} \sigma^{kl}_i n_l\, dS = \int_{V_i} \nabla_l \sigma^{kl}_i\, dV,\\
\int_{S_i} f^k_i v^k_i\, dS &= \int_{S_i} \sigma^{kl}_i v^k_i n_l\, dS = \int_{V_i} \nabla_k(\sigma^{kl}_i v^l_i)\, dV,\\
\int_{S_i} Q^n_i\, dS &= \int_{S_i} \sigma^l_i n_l\, dS = \int_{V_i} \nabla_l Q^l_i\, dV.
\end{aligned} \tag{7.1.6}$$

Substituting (7.1.5) and (7.1.6) in (7.1.3), we obtain with regard for the arbitrariness of the integration volume V_i:

$$\frac{\partial \rho_i}{\partial t} + \nabla_l(\rho_i v^l_i) = \sum{}' \kappa_{ji}, \tag{7.1.7}$$

$$\frac{\partial \rho_i v^k_i}{\partial t} + \nabla_l(\rho_i v^k_i v^l_i) = \nabla_l \sigma^{kl}_i + \rho_i F^k_i + \sum{}' P^k_{ji}, \tag{7.1.8}$$

$$\frac{\partial \rho_i E_i}{\partial t} + \nabla_l(\rho_i E_i v^l_i) = \nabla_k(\sigma^{kl}_i v^l_i) + \rho_i F^k_i v^k_i + \sum{}' E_{ji} + \nabla_l Q^n_i. \tag{7.1.9}$$

Multiplying sequentially the continuity equation (7.1.7) by v^k_i and E_i and subtracting from (7.1.8) and (7.1.9), respectively, we will find a system of differential equations governing the multiphase medium motion:

$$\frac{d_i \rho_i}{dt} + \rho_i \nabla_l v^l_i = \sum{}' \kappa_{ji}, \tag{7.1.10}$$

$$\rho_i \frac{d_i v_i^k}{dt} = \nabla_l \sigma_i^{kl} + \rho_i F_i^k + \sum{}'(P_{ji}^k - \kappa_{ji} v_i^k), \tag{7.1.11}$$

$$\rho_i \frac{d_i E_i}{dt} = \nabla_k(\sigma_i^{kl} v_i^l) + \rho_i F_i^k v_i^k + \nabla_l Q_i^n + \sum{}'(E_{ji}^k - \kappa_{ji} E_i), \tag{7.1.12}$$

where

$$\frac{d_i}{dt} = \frac{\partial}{\partial t} + v_i^l \nabla_l.$$

The system (7.1.10)–(7.1.12) should be augmented by the condition for the symmetry of the stress tensor (1.3.20):

$$\sigma_i^{kl} = \sigma_i^{lk}, \tag{7.1.13}$$

which is valid for the media that do not have the internal moments.

We will assume that the hypothesis of a local thermodynamic equilibrium is satisfied for each ith phase at any point x^k. This means that one can introduce at point x^k the temperature T_i, entropy S_i, and other thermodynamic parameters. Then the equations (7.1.10)–(7.1.13) will be closed by the equations of state for each phase:

$$\sigma_i^{kl} = \sigma_i^{kl}(S_i, \varepsilon_i^{kl}, \dot{\varepsilon}_i^{kl}, \ldots), \quad \kappa_{ji}, \quad E_{ji}, \quad P_{ji}. \tag{7.1.14}$$

These functions are different for each multiphase medium, and their determination constitutes one of the main problems of the mechanics of multiphase media.

The above-developed approach is incomplete, because we have neglected the pulsations of the microparameters $\varphi_i'' = \varphi_i' - \varphi_i$, where $\langle \varphi_i'' \rangle = \frac{1}{V}\int_V \varphi_i''\, dV = 0$. In order to take into account the contribution of the pulsations to equations (7.1.10)–(7.1.12), it is necessary to average the differential equations of continuity, motion, and energy for the P_i', $\vec{v}_i'$ e_i' describing the micromotions of the ith phase. Such a program was realized sequentially in[3,9], and it was shown that the pulsations lead to the appearance of the pulsation stresses $\pi_i^{kl} = -\langle \rho_i' v_i''^k v_i''^l \rangle$, $(v_i''^k = v_i'^k - v_i^k)$, similar to the Reynolds stresses in a turbulent fluid. In the energy equation, a new term representing the energy of the pulsation motion $K_i'' = \frac{1}{2}\langle \rho_i' (v_i'')^2 \rangle$ arises, for which the corresponding equation was obtained in[3]. The determination of the closing relations of the form π_i^{kl} and K_i'' in terms of the mean values ρ_i, v_i^k, etc., represents a very complex problem, which has not been solved so far.

There exist two types of multiphase media: homogeneous and heterogeneous media. In a homogeneous medium, each phase (component) takes the overall volume and the corresponding components are mixed at a molecular level. The example of a homogeneous medium is the mixture of gases or the solution of one substance in another. The heterogeneous media are such media in which each phase occupies certain macroscopic volume, inside which the phases are not mixed. In the case of a heterogeneous gas-particle mixture, this volume coincides with the particle volume (for the bubbly liquid it coincides with the gas bubbles volume, etc.).

7.1.2 Equations of a Two-Phase Medium of the Type of Gas–Solid Particles

Consider the simplest example of a heterogeneous medium of the type of gas–solid spherical particles without phase transitions. Determine the fraction m_2 of the unit volume occupied by the particles and the fraction m_1 occupied by gas by formulas

$$m_2 = \frac{\pi d^2}{6} n, \quad m_1 = 1 - m_2, \tag{7.1.15}$$

where d is the particles diameter and n is the concentration of particles, that is, the number of particles in the unit volume. We will call the quantities m_2 and m_1, for brevity, the volume concentration of particles and gas. Let the viscosity and heat conduction be taken into account only in the force of the interphase interaction and heat exchange. Then

$$\kappa_{ij} = 0, \quad Q_i^k = 0, \quad \sigma_1'^{kl} = -P' g^{kl}, \tag{7.1.16}$$

where the $\sigma_1'^{kl}$ and P' are the true stress and pressure in gas, respectively, and $g^{kl} = (\vec{e}^{\,k} \cdot \vec{e}^{\,l})$ is the metric tensor. Determine the mean stress in the gas continuum by formula

$$\sigma_1^{kl} = \frac{1}{V} \int_{V_1} (-P') g^{kl}\, dV = \frac{V_1}{V} \left(\int_{V_1} (-P') g^{kl}\, dV \right) = -m_1 P g^{kl}, \tag{7.1.17}$$

where $m_1 = V_1/V$, $P = \frac{1}{V_1} \int_{V_1} P'\, dV$ is the mean true gas pressure. The mean stress in the continuum of particles is found by the formula

$$\sigma_2^{kl} = \frac{1}{V} \int_{V_2} \sigma_2'^{kl}\, dV = \frac{V_2}{V} \left(\frac{1}{V_2} \int_{V_2} \sigma_2'^{kl}\, dV \right) = m_2 \langle \sigma_2'^{kl} \rangle_2,$$

where

$$\langle \sigma_2'^{kl} \rangle_2 = \frac{1}{V_2} \int_{V_2} \sigma_2'^{kl}\, dV.$$

If the volume concentration of particles is small, i.e., $m_2 \ll 1$, one can neglect the pulsation stresses in the gas and the collisions between the particles. Then the stresses in particles will arise only at the expense of the interaction with gas. Strictly speaking, it is necessary for the

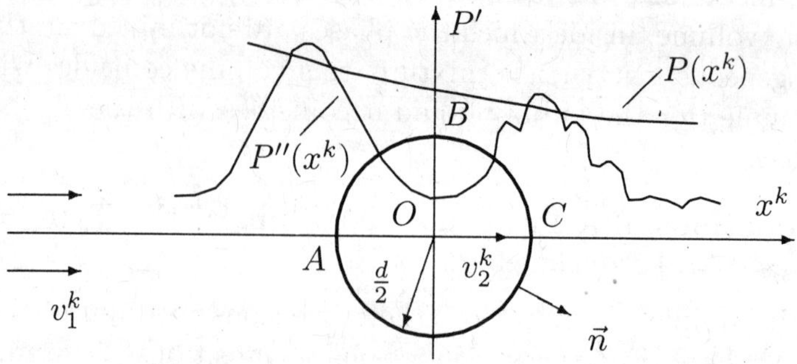

Figure 7.2: A qualitative picture of the variation of the mean pressure $P(x^k)$ and the pulsatile pressure $P''(x^k)$ in the particle neighborhood.

determination of ${\sigma'_2}^{kl}$ to solve a problem of the determination of the prestressed-deformed state of a spherical particle, to the boundary of which γ_{12} the gas pressure P' is applied. This problem is very complex. It nevertheless follows from the equilibrium conditions for solid particles that the mean true stress at particles $\langle {\sigma'_2}^{kl} \rangle_2$ is equal to the true pressure in gas $-Pg^{kl}$. (Note that, for the liquid drops, these quantities can differ by the magnitude of the Laplacian pressure $4\sigma/d$ conditioned by the surface tension σ and the curvature $2/d$ of the droplet surface.) With regard for the above, the mean stress in the continuum of particles without regard for the collisions between the particles is equal to

$$\sigma_2^{kl} = -m_2 P g^{kl}. \tag{7.1.18}$$

In the case of the Cartesian coordinates, it is necessary to assume in formulas (7.1.17) and (7.1.18) that $g^{kl} = \delta^{kl}$. Let us represent the force acting on a single particle from the gas as a sum of the two items:

$$R_p^k = R_1^k(\nabla P) + f_p^k(v_1^k - v_2^k). \tag{7.1.19}$$

The first item R_1^k is related to the variation of the mean pressure and depends on ∇P in the particle neighborhood (see Fig. 7.2). If the gas velocity v_i^k is larger than the particles velocity v_2^k, then, as a result of the deceleration in the neighborhood of point A, the pressure will locally increase, then decrease at point B, and again increase in the neighborhood of point C (see Fig. 7.2, where we show a rough qualitative picture of the pressure distribution). Thus, the pressure at point P' in the particle neighborhood will contain a pulsation component P'' with a reference pulsation size $L \sim d$. If the flow around the particle were ideal, then the pressure distribution $P''(x^k)$ would be a symmetric function

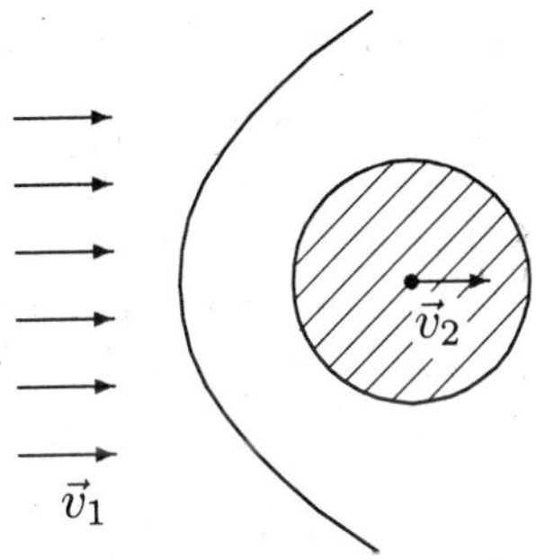

Figure 7.3: The shock wave arising upstream of the particle at a supersonic flow with $M_{12} = |\vec{v}_1 - \vec{v}_2|/c > 1$.

with respect to the plane $x^k = 0$. The distribution $P''(x^k)$ is indeed nonsymmetric at the expense of viscosity and flow turbulization, and the integral of $P''(x^k)$ over the particle surface will be equal to the particle drag force f_p^k, depending on the velocities difference $(v_1^k - v_2^k)$, which may be written in the form

$$f_p^k = C_d \frac{\pi d^2}{4} \rho_{11} \frac{|\vec{v}_1 - \vec{v}_2|(v_1^k - v_2^k)}{2}, \tag{7.1.20}$$

where ρ_{11} is the true gas density and C_d is the drag coefficient depending on the relative Reynolds number $\mathrm{Re} = |\vec{v}_1 - \vec{v}_2| d/\nu$, the relative Mach number $M_{12} = |\vec{v}_1 - \vec{v}_2|/c$, and the volume concentration of particles m_2. The determination of C_d is a complex problem, which has not been solved until now. For a small volume concentration of particles $m_2 \ll 1$, one can neglect in the first approximation the dependence of C_d on m_2. Then it is natural to expect that the C_d will be close to the drag coefficient of a single spherical particle in a homogeneous gas flow C_d. The value of C_d was determined by many experimentalists (cf.[10–11]), and there are several bulky empirical formulas. We present the simplest formula[12]:

$$C_d(\mathrm{Re}_{12}, M_{12}) \approx \left(1 + \exp\left(-\frac{0.43}{M_{12}^{4.67}}\right)\right)\left(0.38 + \frac{4}{\sqrt{\mathrm{Re}}} + \frac{24}{\mathrm{Re}}\right). \tag{7.1.21}$$

The right-hand side of expression (7.1.21) consists of two factors: the first of which depends on the Mach number M_{12}, and the second factor depends on the Reynolds number Re. A passage from the subsonic ($M_{12} < 1$) to the supersonic ($M_{12} > 1$) flow around a particle leads to the increase in the first factor, so that the C_d increases approximately by a factor of two. This is related to the fact that there arises a shock wave ahead of the particle at a supersonic flow (see Fig. 7.3), and the irreversible losses in it $[S_1] > 0$ lead to the increase in the force acting on the particle from the gas.

The second factor is equal to $24/\mathrm{Re}_{12}$ at small Reynolds numbers $\mathrm{Re} = |\vec{v}_1 - \vec{v}_2|d/\nu \ll 1$, which corresponds to the laminar Stokes flow regime considered in Section 5.3. At large Reynolds numbers $\mathrm{Re} \gg 1$, a turbulent flow takes place, and the second factor tends to the value 0.38.

Let us turn to the first item in formula (7.1.19) and present it in terms of an integral of the mean pressure P over the particle surface S_0:

$$R_1^k = -\int_{S_0} Pn^k\, dS = -\int_{V_0} \nabla^k P\, dV,$$

where we have used formula (1.1.63) at a passage to the integral over the particle volume V_0. Since the reference size of the mean pressure variation $L \gg d$, then one can assume $\nabla^k P$ to be constant and write:

$$R_1^k \approx -\frac{\pi d^3}{6}\nabla^k P.$$

Substituting this formula in (7.1.19), we will find the force acting on a particle:

$$R_p^k = -\frac{\pi d^3}{6}\nabla^k P + f_p^k, \tag{7.1.22}$$

where the f_p^k is determined by formulas (7.1.20) and (7.1.21). Using formula (7.1.22), let us write the equation for the motion of a particle moving at a mean velocity v_2^k:

$$(m_p + m_e)\frac{dv_2^k}{dt} = -\frac{\pi d^3}{6}\nabla^k P + f_p^k,$$

where $m_p = \frac{\pi d^3}{6}\rho_{22}$ is the particle mass and $m_e = \frac{\pi d^3}{12}\rho_{11}$ is the virtual mass. Multiplying this equation by the particles concentration n, we obtain with regard for $\rho_2 = nm_p = \rho_{22}m_2$ and $nm_e = \frac{1}{2}m_2\rho_{11}$:

$$\left(\rho_2 + \frac{1}{2}m_2\rho_{11}\right)\frac{dv_2^k}{dt} = -m_2\nabla^k P + nf_p^k.$$

The true gas density ρ_{11} is by three orders of magnitude smaller than the true particle density ρ_{22}; therefore, $m_2\rho_{11} \ll \rho_2$ and the last equation may be written as

$$\rho_2\frac{dv_2^k}{dt} = -m_2\nabla^k P + nf_p^k. \tag{7.1.23}$$

On the other hand, the equation for particles motion for the mean velocity v_2^k can be found by substituting the relations (7.1.18), $\kappa_{ji} = 0$, $j = 1$, $i = 2$ in equation (7.1.11), from which it follows that

$$\rho_2\frac{d_2 v_2^k}{dt} = -\nabla^k m_2 P + P_{12}^k. \tag{7.1.24}$$

The substantive derivative $\frac{d_2 v_2^k}{dt}$ on the left-hand side of (7.1.24) is equal to the acceleration of particles moving at a velocity v_2^k; therefore, the left-hand sides of equations (7.1.23) and (7.1.24) are equal to each other, and hence the right-hand sides are also equal to each other, from where it follows the expression for the rate of momentum exchange between the phases:

$$P_{12}^k = P\nabla^k m_2 + n f_p^k. \tag{7.1.25}$$

Using the relations $m_1 + m_2 = 1,\ P_{21}^k = -P_{12}^k$, we obtain:

$$P_{21}^k = P\nabla^k m_1 - n f_p^k. \tag{7.1.26}$$

The first item on the right-hand sides of (7.1.25) and (7.1.26) is related to the compression (expansion) of the gas stream tube $\nabla^k m_1 \neq 0$ and is called the Rakhmatulin's force[1,2,4]. In the formula for the energy exchange rate E_{12}, we identify by analogy with (7.1.25) the term related to the work made at a compression of a stream tube $-P\frac{\partial m_2}{\partial t}$, to the heat exchange between the gas and particles nq_p, and to the work of the drag force acting on the particles from the gas $nf_p^k v_2^k$. Taking formulas $m_1 + m_2 = 1,\ E_{12} = -E_{21}$ into account, we can write:

$$\begin{aligned} E_{12} &= -P\frac{\partial m_2}{\partial t} + n f_p^k v_{2k} + n q_p, \\ E_{21} &= P\frac{\partial m_1}{\partial t} - n f_p^k v_{2k} - n q_p. \end{aligned} \tag{7.1.27}$$

The heat flux from the gas to a single particle is found by formula[2–8]

$$q_p = \pi d^2 \alpha (T_1 - T_2), \tag{7.1.28}$$

where α is the heat transfer coefficient, which is often expressed in terms of the Nusselt number $\mathrm{Nu} = \alpha d/\lambda$ in the form

$$q_p = \pi d \lambda \,\mathrm{Nu}\,(T_1 - T_2). \tag{7.1.29}$$

When writing formulas (7.1.28) and (7.1.29), it was assumed that the heat conduction in the particle is substantially larger than the heat conduction in gas; therefore, the heat flux is determined by the rate of the heat supply (removal) in gas. It is clear that this process depends on the character of the flow around the particle. The transition from laminar to turbulent flow intensifies the heat exchange; therefore, the Nusselt number should increase with the Reynolds number. As a result of the processing of experimental data, a large number of formulas were obtained, which determine the dependence of Nu on Re[4,7,8]. The simplest of them has the form:

$$\mathrm{Nu} \approx 2 + 0.6\,\mathrm{Re}^{0.5}\mathrm{Pr}^{0.33}, \tag{7.1.30}$$

where $\Pr = C_p\mu/\lambda$ is the Prandtl number, C_p is the gas specific heat, μ is the gas viscosity, and λ is the thermal conductivity coefficient of gas.

Let in volume V the gas occupy the volume V_1, in which at each point the density $\rho_{11}(x^k)$ is defined, and let the particles occupy the volume V_2 and $V = V_1 + V_2$. Then we can determine the mean gas density ρ_1 by formula

$$\rho_1 = \frac{1}{V}\int_V \rho'_{11}\,dV = \frac{V_1}{V}\cdot\frac{1}{V_1}\int_{V_1}\rho'_{11}\,dV = m_1\rho_{11}, \tag{7.1.31}$$

where $m_1 = V_1/V$ is the volume fraction occupied by gas and

$$\rho_{11} = \frac{1}{V_1}\int_{V_1}\rho'_{11}\,dV$$

is the mean true gas density within the volume V_1. We will obtain in a similar way for solid (incompressible) particles:

$$\rho_2 = \rho_{22}m_2, \qquad \rho_{22} = \text{const}, \tag{7.1.32}$$

where ρ_{22} is the true particles density.

Let us find the gas equation of state for the mean quantities. Substituting in the equation of state $P' = \rho'_{11}RT'_1$ the values $\rho'_{11} = \rho_{11} + \rho''_{11}$, $T'_1 = T_1 + T''_1$, multiplying by dV, integrating over the volume V_1, and neglecting the mean value of the pulsation

$$\langle\rho''_{11}T''_1\rangle_1 = \frac{1}{V_1}\int_{V_1}\rho''_{11}T''_1\,dV \approx 0,$$

we obtain:

$$P = \rho_{11}RT_1, \tag{7.1.33}$$

where P was determined above [see (7.1.17)] and ρ''_{11}, T''_1 are the pulsations of the density and temperature satisfying the condition $\langle\rho''_{11}\rangle_1 = 0$, $\langle T''_1\rangle_1 = 0$.

Substituting (7.1.16)–(7.1.19), (7.1.25)–(7.1.27); (7.1.31), and (7.1.32) in (7.1.10)–(7.1.12), we obtain the system of equations for the two-phase medium gas–solid particles at a small volume concentration of particles $m_2 \ll 1$:

$$\frac{\partial\rho_1}{\partial t} + \nabla_k(\rho_1 v_1^k) = 0, \quad \rho_1 = \rho_{11}m_1, \quad m_1 + m_2 = 1,$$

$$\frac{\partial\rho_2}{\partial t} + \nabla_k(\rho_2 v_2^k) = 0, \quad \rho_2 = \rho_{22}m_2, \quad \rho_{22} = \text{const},$$

$$\rho_1\frac{d_1 v_1^k}{\partial t} = -m_1\nabla^k P - nf_p^k, \quad \frac{d_1}{dt} = \frac{\partial}{\partial t} + v_1^l\nabla_l,$$

$$\rho_2 \frac{d_2 v_2^k}{\partial t} = -m_2 \nabla^k P + n f_p^k, \quad \frac{d_2}{dt} = \frac{\partial}{\partial t} + v_2^l \nabla_l, \tag{7.1.34}$$
$$\rho_1 \frac{d_1 E_1}{\partial t} = -\nabla_k (m_1 v_1^k P) - P \frac{\partial m_1}{\partial t} - n f_p^k v_2^k - n q_p,$$
$$\rho_2 \frac{d_2 E_2}{\partial t} = -m_2 v_2^k \nabla_k P + n f_p^k v_2^k + n q_p, \quad P = \rho_{11} R T_1,$$
$$E_1 = \frac{v_1^2}{2} + e_1, \quad E_2 = \frac{v_2^2}{2} + e_2, \quad e_1 = C_V T_1, \quad e_2 = C_s T_2$$

where C_V is the gas specific heat at a constant volume and C_s is the specific heat of particles. Note that the system of equations (7.1.34) was obtained in such a form in[13] and for the barotropic case in[1].

Problem 7.1. Find the rate of the total entropy production in a unit volume for a gas-particle mixture, which is described by equations (7.1.34).

Solution: It follows from the Gibbs relation (2.1.64) that the rate of change of the specific entropy S_1 of gas is equal in the given notation to

$$\frac{d_1 S_1}{dt} = \frac{1}{T_1}\left(\frac{d_1 e_1}{dt} + P \frac{d_1}{dt}\left(\frac{1}{\rho_{11}}\right)\right), \tag{7.1.35}$$

and for the solid particles ($\rho_{22} = \text{const}$)

$$\frac{d_2 S_2}{dt} = \frac{1}{T_2} \frac{d_2 e_2}{dt}, \tag{7.1.36}$$

where S_2 is the specific entropy of particles. Multiplying the equation for the particles motion in (7.1.34) by v_2^k and summing over k, we find:

$$\rho_2 \frac{d_2}{dt}\left(\frac{v_2^2}{2}\right) = -v_2^k m_2 \nabla^k P + n f_p^k v_2^k, \quad v_2^2 = v_2^k v_2^k.$$

Subtracting this equation from the equation for the particles energy in (7.1.34), we obtain the relation:

$$\rho_2 \frac{d_2 e_2}{dt} = n q_p, \tag{7.1.37}$$

the substitution of which in (7.1.36) yields:

$$\frac{d_2 S_2}{dt} = \frac{n q_p}{\rho_2 T_2}. \tag{7.1.38}$$

Multiply the gas motion equation in (7.1.34) by v_1^k:

$$\rho_1 \frac{d_1}{dt}\left(\frac{v_1^2}{2}\right) = -m_1 v_1^k \nabla^k P - n f_p^k v_1^k,$$

and subtract this equation from the energy equation for gas in (7.1.34). With regard for the relation $v_{1k}\nabla^k = v_1^k \nabla_k$, we obtain:

$$\rho_1 \frac{d_1 e_1}{dt} = -P\Big(\frac{\partial m_1}{\partial t} + \nabla_k(m_1 v_1^k)\Big) + n f_p^k (v_1^k - v_2^k) - n q_p.$$

The expression

$$\frac{\partial m_1}{\partial t} + \nabla_k(m_1 v_1^k) = -\frac{m_1}{\rho_{11}} \frac{d_1 \rho_{11}}{dt}$$

follows from the first equation in (7.1.34). With regard for this, we can rewrite the equation for the specific internal energy of gas e_1 in the form

$$\frac{d_1 e_1}{dt} = -P \frac{d_1}{dt}\Big(\frac{1}{\rho_{11}}\Big) + \frac{n f_p^k}{\rho_1}(v_1^k - v_2^k) - \frac{n q_p}{\rho_1}. \tag{7.1.39}$$

Comparing this formula with (7.1.37), we see that the energy dissipation at the expense of the interphase interaction force $n f_p^k (v_1^k - v_2^k)$ occurs only in the gas phase. This fact is a consequence of the incompressibility of particles, in which the strain rate is equal to zero. If both phases, for example, i and j, are deformable, then the energy dissipation at the expense of the interphase interaction force f_{ji}^k will occur in both phases and is equal to

$$Q_{Di} = \alpha_{ji} f_{ji}^k (v_j^k - v_i^k), \quad \alpha_{ji} + \alpha_{ij} = 1. \tag{7.1.40}$$

The coefficients α_{ji} determine the fraction of energy dissipated in the ith phase and should be determined from the solution of a problem on the joint deformation of phases.

Returning to our problem, we substitute (7.1.39) in (7.1.35) and find:

$$\frac{d_1 S_1}{dt} = \frac{1}{\rho_1 T_1}(n f_p^k (v_1^k - v_2^k) - n q_e). \tag{7.1.41}$$

Writing the interphase interaction force (7.1.20) and the heat exchange term (7.1.29) in the form

$$f_p^k = \frac{v_1^k - v_2^k}{\tau_v}, \quad q_p = \frac{T_1 - T_2}{\tau_T}, \tag{7.1.42}$$

we find from (7.1.38), (7.1.41), and (7.1.42):

$$\rho_1 \frac{d_1 S_1}{dt} + \rho_2 \frac{d_2 S_2}{dt} = \frac{n}{T_1 \tau_v}\Big((\vec{v}_1 - \vec{v}_2)^2 - \frac{\tau_v (T_1 - T_2)^2}{\tau_T T_2}\Big), \tag{7.1.43}$$

where $(\vec{v}_1 - \vec{v}_2)^2 = (v_1^k - v_2^k)(v_1^k - v_2^k)$ and the summation is carried out over the index k. Multiplying the continuity equations for gas and particles in (7.1.34) by the corresponding entropies S_1 and S_2, we obtain:

$$S_i \frac{d_i \rho_i}{dt} + S_i \rho_i \nabla_k v_i^k = 0, \quad i = 1, 2.$$

Adding these formulas to (7.1.34), we arrive at the expression:

$$\frac{\partial}{\partial t}(\rho_1 S_1 + \rho_2 S_2) + \nabla_k(\rho_1 v_1^k S_1 + \rho_2 v_2^k S_2)$$
$$= \frac{n}{T_1 \tau_v}\left((\vec{v}_1 - \vec{v}_2)^2 + \frac{\tau_v}{\tau_T}\frac{(T_1 - T_2)^2}{T_2}\right). \tag{7.1.44}$$

It follows from here that the total entropy variation in a unit volume occurs at the expense of the inflow (outflow) of the entropy together with the mass flow, the entropy production at the expense of the dissipative processes, and the heat exchange between the phases. The entropy production rate is determined by the right-hand side of equation (7.1.44) and is always greater than or equal to zero. Thus, it has been proved that the *system of equations (7.1.34) satisfies the second law of thermodynamics.*

7.1.3 Equations of a Bicomponent Medium of Gas Mixture Type

Consider a bicomponent homogeneous medium of the gas mixture type. In this case, the molecules of gases are mixed at a molecular level, and each component occupies the overall volume enclosing the mixture. Taking into account the viscosity and heat conduction only in the interaction force and in the energy exchange between the components (the Eulerian approximation) and neglecting the body forces, let us write the right-hand sides of the equations for the ith component:

$$\begin{aligned}
&\sigma_i^{kl} = -P_i g^{kl}, \quad Q_i^k = 0, \quad F_i^k = 0, \quad \kappa_{ij} = 0,\\
&P_{ji}^k = \frac{(v_j^k - v_i^k)}{\tau_v},\\
&E_{ji} = \frac{T_j - T_i}{\tau_T} + \alpha_{ji}\frac{(\vec{v}_j - \vec{v}_i)^2}{\tau_v} + \frac{(v_j^k - v_i^k)v_i^k}{\tau_v},\\
&\alpha_{ji} + \alpha_{ij} = 1,
\end{aligned} \tag{7.1.45}$$

where the α_{ji} is the energy fraction dissipated in the ith component at the expense of the force of intercomponent (interphase) interaction. The characteristic times τ_T and τ_v can be determined by the methods of the molecular/kinetic theory by solving a system of Boltzmann's kinetic equations for the mixture of gases[8]:

$$\frac{1}{\tau_T} = \frac{3k}{\mu_i^* + \mu_j^*}\frac{1}{\tau_v} = 16k\frac{\rho_i \rho_j}{\mu_i^* + \mu_j^*}\Omega_{ij}^{(1,1)}, \quad \alpha_{ji} = \frac{v_{T,i}^2}{v_{T,ij}^2}, \tag{7.1.46}$$

where μ_i^* and μ_j^* are the masses of the molecules of the ith and jth kinds, k is the Boltzmann's constant, and $\Omega_{ij}^{(1,1)}$ is the Chapman–Enskog integral determined by the section of the molecules collision:

$$\Omega_{ij}^{(l,s)} = \pi v_{T,ij} \int e^{-x^2} x^{2s+3} Q^{(l)}\, dx, \quad Q^{(l)} = \int (1 - \cos^l \chi)\, b\, db,$$
$$v_{T,ij}^2 = v_{T,i}^2 + v_{T,j}^2, \quad v_{T,i}^2 = \frac{2kT_i}{\mu_i},$$

where χ is the angle of the molecules scatter.

Substituting (7.1.45) into (7.1.10)–(7.1.12), we obtain a system of equations for a bicomponent medium ($i = 1, 2,\ j = 2, 1$):

$$\frac{\partial \rho_i}{\partial t} + \nabla_k \rho_i v_i^k = 0,$$
$$\rho_i \frac{d_i v_i^k}{dt} = -\nabla^k P_i + \frac{(v_j^k - v_i^k)}{\tau_v}, \quad \frac{d_i}{dt} = \frac{\partial}{\partial t} + v_i^l \nabla_l, \tag{7.1.47}$$
$$\rho_i \frac{d_i E_i}{dt} = -\nabla^k v_i^k P_i + \frac{(T_j - T_i)}{\tau_T} + \alpha_{ji} \frac{(\vec{v}_j - \vec{v}_i)^2}{\tau_v} + \frac{(v_j^k - v_i^k) v_i^k}{\tau_v},$$
$$E_i = e_i + \frac{v_i^2}{2},$$

which should be augmented by the equation of state and by the Gibbs equation:

$$e_i = C_{vi} T_i, \quad P_i = P_i(\rho_i, S_i),$$
$$\frac{d_i e_i}{dt} = T_i \frac{d_i S_i}{dt} - P_i \frac{d_i}{dt}\left(\frac{1}{\rho_i}\right), \quad \alpha_{ji} = \frac{\frac{T_i}{\mu_i^*}}{\frac{T_i}{\mu_i^*} + \frac{T_j}{\mu_j^*}}, \tag{7.1.48}$$
$$\alpha_{ji} + \alpha_{ij} = 1.$$

For the rarefied gases, the equation of state will be written in the form $P_i = \rho_i R_i T_i$.

Problem 7.2. Find the entropy production rate for a mixture of gases described by the equations (7.1.47) and (7.1.48).

Solution: Multiplying the second equation (7.1.47) by v_i^k, summing over k and subtracting from the third energy equation, we obtain:

$$\rho_i \frac{d_i e_i}{dt} = -P_i \nabla_k v_i^k + \frac{T_j - T_i}{\tau_T} + \alpha_{ji} \frac{(\vec{v}_j - \vec{v}_i)^2}{\tau_v}. \tag{7.1.49}$$

Determining from the continuity equation the derivative

$$\nabla_k v_i^k = -\frac{1}{\rho_i} \frac{d_i \rho_i}{dt} = \rho_i \frac{d_i}{dt}\left(\frac{1}{\rho_i}\right),$$

rewrite (7.1.49) in the form

$$\frac{d_i e_i}{dt} = -P_i \frac{d_i}{dt}\left(\frac{1}{\rho_i}\right) + \frac{T_j - T_i}{\rho_i \tau_T} + \alpha_{ji}\frac{(\vec{v}_j - \vec{v}_i)^2}{\rho_i \tau_v}.$$

Substituting this expression in the Gibbs relation (7.1.48), we find:

$$\rho_i \frac{d_i S_i}{dt} = \frac{T_j - T_i}{T_i \tau_T} + \alpha_{ji}\frac{(\vec{v}_j - \vec{v}_i)^2}{T_i \tau_v}. \tag{7.1.50}$$

Assuming here $i = 1,\ j = 2$ and $i = 2,\ j = 1$, we obtain the entropy production rate in the first and second components. Summing them, we find:

$$\rho_1 \frac{d_1 S_1}{dt} + \rho_2 \frac{d_2 S_2}{dt} = \frac{(T_2 - T_1)^2}{T_1 T_2 \tau_T} + \left(\frac{\alpha_{21}}{T_1} + \frac{\alpha_{12}}{T_2}\right)\frac{(\vec{v}_1 - \vec{v}_2)^2}{\tau_v}. \tag{7.1.51}$$

Multiplying the continuity equations by S_i and adding to (7.1.51), we can rewrite (7.1.51) as

$$\frac{\partial}{\partial t}(\rho_1 S_1 + \rho_2 S_2) + \nabla_k(\rho_1 v_1^k S_1 + \rho_2 v_2^k S_2)$$
$$= \frac{(T_2 - T_1)^2}{T_1 T_2 \tau_T} + \frac{(\mu_1^* + \mu_2^*)}{(\mu_1^* T_2 + \mu_2^* T_1)}\frac{(\vec{v}_1 - \vec{v}_2)^2}{\tau_v}. \tag{7.1.52}$$

The nonnegative right-hand side of equation (7.1.52) determines the entropy production as a consequence of the equalization of the velocities and temperatures.

7.2 Correctness of the Cauchy Problem: Relations at Discontinuities in Multiphase Media

7.2.1 The Characteristics of a System of Equations for Gas-Particle Mixtures and Correctness of the Cauchy Problem

As was noted in Section 3.2, it is necessary to know the type of system of equations, which is determined by its characteristics, for the formulation and solution of initial- and boundary-value problems. Let us find the characteristics of the system of equations for gas-particle mixture (7.1.34) in the one-dimensional nonstationary case for the flows with plane ($\nu = 0$), cylindrical ($\nu = 1$), and spherical ($\nu = 2$) symmetry. In this case, equations (7.1.34) can be written as

$$\frac{\partial \rho_1}{\partial t} + \rho_1 \frac{\partial v_1}{\partial x} + v_1 \frac{\partial \rho_1}{\partial x} + \frac{\nu \rho_1 v_1}{x} = 0, \quad \rho_1 = \rho_{11} m_1,$$

$$
\begin{aligned}
&\frac{\partial \rho_2}{\partial t} + \rho_2 \frac{\partial v_2}{\partial x} + v_2 \frac{\partial \rho_2}{\partial x} + \frac{\nu \rho_2 v_2}{x} = 0, \quad \rho_2 = \rho_{22} m_2, \\
&\frac{\partial v_1}{\partial t} + v_1 \frac{\partial v_1}{\partial x} + \frac{1}{\rho_{11}} \frac{\partial P}{\partial x} + \frac{n f_p}{\rho_1} = 0, \\
&\frac{\partial v_2}{\partial t} + v_2 \frac{\partial v_2}{\partial x} + \frac{1}{\rho_{22}} \frac{\partial P}{\partial x} - \frac{n f_p}{\rho_2} = 0, \\
&\frac{\partial H}{\partial t} + v_1 \frac{\partial H}{\partial x} - \frac{1}{\rho_{11}} \left(\frac{\partial P}{\partial t} + v_1 \frac{\partial P}{\partial x} \right) + \dot{Q} = 0, \\
&\frac{\partial e_2}{\partial t} + v_2 \frac{\partial e_2}{\partial x} - \frac{n q_p}{\rho_2} = 0, \\
&m_1 + m_2 = 1, \quad \rho_{22} = \text{const}, \quad H = e_1 + \frac{P}{\rho_{11}}, \\
&\dot{Q} = \frac{n q_p - n f_p (v_1 - v_2)}{\rho_1}, \quad e_1 = C_V T_1, \quad e_2 = C_s T_2,
\end{aligned}
\tag{7.2.1}
$$

where H is the gas enthalpy and the remaining notations coincide with the notations adopted in Section 7.1.2. The gas equation of state is specified by the function $H = H(P, \rho_{11})$, and one can assume for ideal gas that $H = C_p T_1$. It is easy to see that the last two equations in (7.2.1) are written in characteristic form (3.2.12). Introducing indeed the function $\mathcal{I}_0 = H - \int_{P_0}^{P} \frac{dP}{\rho_{11}}$, we can rewrite the two last equations of (7.2.1) in the characteristic form (3.2.12):

$$
\frac{\partial \mathcal{I}_0}{\partial t} + v_1 \frac{\partial \mathcal{I}_0}{\partial x} + \dot{Q} = 0, \quad \frac{\partial e_2}{\partial t} + v_2 \frac{\partial e_2}{\partial t} - \frac{n q_p}{\rho_2} = 0, \tag{7.2.2}
$$

along the characteristics

$$
\frac{dx}{dt} = v_1, \quad \frac{dx}{dt} = v_2 \tag{7.2.3}
$$

coinciding with the trajectories of gas and particles. Let us find the characteristics of the four remaining first equations in (7.2.1). Using the equation of state, we eliminate ρ_1 from the set of dependent variables. For this purpose, we express that $\rho_{11} = \rho_{11}(H, P)$, and with regard for the relations $H = C_p T_1$, $m_2 = 1 - m_1$ we can write the substantive derivative as

$$
\begin{aligned}
\frac{d_1 \rho_1}{dt} &= m_1 \frac{d_1 \rho_{11}}{dt} + \rho_{11} \frac{d_1 m_1}{dt} \\
&= m_1 \left(\rho_{11,p} \frac{d_1 P}{dt} + \frac{\rho_{11,T}}{C_p} \frac{d_1 H}{dt} \right) - \rho_{11} \frac{d_1 m_2}{dt},
\end{aligned}
$$

where

$$
\rho_{11,p} = \frac{\partial \rho_{11}}{\partial P}, \quad \rho_{11,T} = \frac{\partial \rho_{11}}{\partial T_1}, \quad \frac{d_1}{dt} = \frac{\partial}{\partial t} + v_1 \frac{\partial}{\partial x}.
$$

Using the energy equation for gas [the last but one equation in (7.2.1)] and a formula for the sound velocity $c^2 = 1/(\rho_{11,p} + \rho_{11,T}/(C_p\rho_{11}))$, we can rewrite this expression in the form

$$\frac{d_1\rho_1}{dt} = m_1\left(\rho_{11,p} + \frac{\rho_{11,T}}{C_p\rho_{11}}\right)\frac{d_1P}{dt} - \frac{\rho_{11}}{\rho_{22}}\frac{d_1\rho_2}{dt} - \frac{m_1}{C_p}\rho_{11,T}\dot{Q}$$
$$= \frac{m_1}{c^2}\frac{d_1P}{dt} - \frac{\rho_{11}}{\rho_{22}}\frac{d_1\rho_2}{dt} - \frac{m_1}{C_p}\rho_{11,T}\dot{Q}.$$

Substituting this formula in the first equation of (7.2.1) and multiplying by c^2/m_1, we obtain:

$$\frac{\partial P}{\partial t} + v_1\frac{\partial P}{\partial x} + \rho_{11}c^2\frac{\partial v_1}{\partial x} - \frac{\rho_{11}c^2}{\rho_{22}m_1}\left(\frac{\partial\rho_2}{\partial t} + v_1\frac{\partial\rho_2}{\partial x}\right) = \frac{c^2\rho_{11,T}}{C_p}\dot{Q} - \nu\rho_{11}c^2\frac{v_1}{x}.$$

Expressing the derivative $\frac{\partial\rho_2}{\partial t}$ from the second equation of (7.2.1) and substituting here, we can write the given equation as well as the second, third, and fourth equations in (7.2.1) in matrix form with respect to the unknown variables P, ρ_1, v_1, and v_2:

$$\frac{\partial}{\partial t}\begin{pmatrix} P \\ \rho_2 \\ v_1 \\ v_2 \end{pmatrix} + \begin{pmatrix} v_1 & \frac{\rho_{11}c^2}{\rho_{22}m_1}(v_2 - v_1) & \rho_{11}c^2 & \rho_{11}c^2\frac{m_2}{m_1} \\ 0 & v_2 & 0 & \rho_2 \\ \frac{1}{\rho_{11}} & 0 & v_1 & 0 \\ \frac{1}{\rho_{22}} & 0 & 0 & v_2 \end{pmatrix}\frac{\partial}{\partial x}\begin{pmatrix} P \\ \rho_2 \\ v_1 \\ v_2 \end{pmatrix}$$
$$= \begin{pmatrix} f_1 \\ f_2 \\ f_3 \\ f_4 \end{pmatrix}, \tag{7.2.4}$$

where

$$f_1 = \frac{c^2\rho_{11,T}}{C_p}\dot{Q} - \frac{\nu c^2}{x}\left(\rho_{11}v_1 + \rho_{11}v_2\frac{m_2}{m_1}\right), \quad f_2 = -\frac{\nu\rho_2 v_2}{x},$$
$$f_3 = -\frac{nf_p}{\rho_1}, \quad f_4 = \frac{nf_p}{\rho_2}.$$

According to (3.2.2), (3.2.7), and (3.2.8), the characteristics $\frac{dx}{dt} = \lambda_i$ are determined by the eigenvalues λ_i of the matrix C standing on the left-hand side of (7.2.4):

$$\begin{vmatrix} v_1 - \lambda & \frac{\rho_{11}c^2}{\rho_{22}m_1}(v_2 - v_1) & \rho_{11}c^2 & \rho_{11}c^2\frac{m_2}{m_1} \\ 0 & v_2 - \lambda & 0 & \rho_2 \\ \frac{1}{\rho_{11}} & 0 & v_1 - \lambda & 0 \\ \frac{1}{\rho_{22}} & 0 & 0 & v_2 - \lambda \end{vmatrix} = 0.$$

Calculating the determinant, we arrive at the fourth-order equation

$$(v_2-\lambda)^2(c^2-(v_1-\lambda)^2)+\frac{m_2\rho_{11}c^2}{m_1\rho_{22}}(v_1-\lambda)^2=0. \quad (7.2.5)$$

It is impossible to solve this equation with respect to λ in a general case; therefore, the solution of (7.2.5) is sought for in the form of a series expansion with respect to the small parameter $\rho_{11}/\rho_{22} \ll 1$[7,13]:

$$\begin{aligned}\lambda_{1,2} &= v_1 \pm c\left(1+\frac{\rho_{11}m_2c^2}{2\rho_{22}m_1(v_2-v_1\pm c)^2}\right)+O\Big(\Big(\frac{\rho_{11}}{\rho_{22}}\Big)^4\Big),\\ \lambda_{3,4} &= v_2 \pm \frac{c\rho_{11}(v_1-v_2)}{\rho_{22}\sqrt{\frac{\rho_1}{\rho_2}((v_1-v_2)^2-c^2)}}+O\Big(\Big(\frac{\rho_{11}}{\rho_{22}}\Big)^2\Big). \end{aligned} \quad (7.2.6)$$

It can be seen from here that the first two zeroes $\lambda_{1,2}$ are always real, and the third and fourth zeroes $\lambda_{3,4}$ are real only at a supersonic flow around the particles: $M_{12}=\frac{|v_1-v_2|}{c}>1$. At a subsonic flow around the particles when $M_{12}<1$ the zeroes $\lambda_{3,4}$ are complex. It follows from here that the system (7.2.1) *will be hyperbolic* at $M_{12}>1$, and at $M_{12}<1$, *it will have a composite type.*

The presence of complex characteristics λ_i leads to the growth of small disturbances and to the ill-posedness of the Cauchy problem for system (7.2.1). Let at time $t=0$ small disturbances be given:

$$\varphi_i'\Big|_{t=0}=\delta\varphi_i\exp(ikx), \quad \varphi_i'=(\rho_1,\rho_2,v_1,v_2,H,e_2).$$

As was shown in Section 3.2, small disturbances propagate along the characteristics; therefore, the solution of (7.2.1) under the initial conditions $\varphi_i'|_{t=0}$ may be written in the form

$$\varphi_i'=\sum_j S_{ij}\,\delta\varphi_j e^{ik(x-\lambda_j t)}, \quad (7.2.7)$$

where S_{ij} is the matrix determined by formula (3.2.10). Substituting (7.2.6) in (7.2.7), we find that the small disturbances will increase by the exponential law[14,15]:

$$\begin{aligned}&\varphi_i'\sim\delta\varphi_i\,e^{k\beta t}, \quad \beta=cM_{12}\sqrt{\frac{m_2\rho_{11}}{m_1\rho_{22}(1-M_{12}^2)}},\\ &M_{12}=\frac{|v_1-v_2|}{c}<1. \end{aligned} \quad (7.2.8)$$

A physical reason for the nonhyperbolicity of system (7.2.1) is the absence of the eigenpressure in the pseudogas of particles. Let us briefly

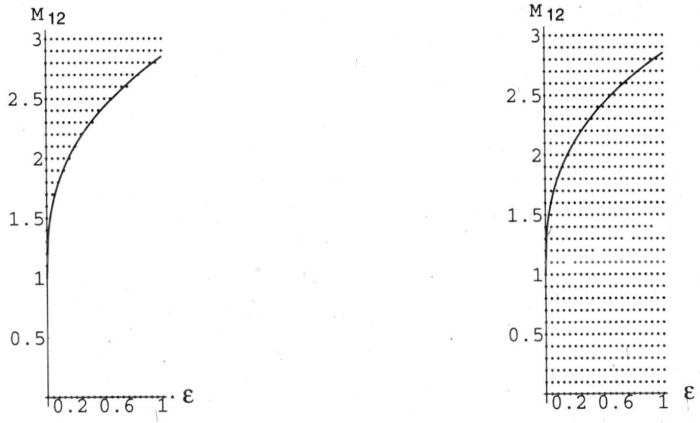

(a) (b)

Figure 7.4: The region of real roots of equation (7.2.4) in the (ε, M_{12}) plane: (a) the roots λ_2, λ_3 and (b) the roots λ_1, λ_4.

describe the mechanism of the development of the given instability[14]. Assume that at some point x the mean density of particles has increased in a random way by an infinitesimal quantity $\delta\rho_2$. This will lead to the narrowing of the effective gas stream tube whose "walls" are formed by the particles of the quantity δm_1. According to the quasi-one-dimensional flow theory of gas flow in a variable section duct, developed in Section 6.1, at this point will occur at $M_{12} < 1$ the increase in the gas velocity and the pressure decrease. As a result, a pressure gradient arises under which the particles begin to move toward this point, which will lead to a further increase in ρ_2 and the narrowing of the effective gas stream tube.

We recall that it was assumed above at the solution of equation (7.2.4) that $\rho_{11}/\rho_{22} \ll 1$. Since $m_2 \ll 1$ and $m_1 \sim 1$, all of these results are valid only asymptotically at $\varepsilon = \frac{m_2\rho_{11}}{m_1\rho_{22}} \to 0$. With the aid of the *Mathematica* Notebook `prog7-1.nb`, one can find the exact zeroes of equation (7.2.4). Two obtained zeroes are real, and the two remaining zeroes can be either real or complex, depending on the values of the nondimensional parameters M_{12} and ε. In Fig. 7.4, we show by the dots a region of the existence of real zeroes in the plane of the variables M_{12}, ε, and the zeroes are complex in the remaining part of the rectangular region. The boundary separating the regions of real and complex zeroes can be approximated by the formula

$$M_{12} = 1 + 1.85\varepsilon^{0.37}$$

(see also the solid line in Fig. 7.4). It follows from this formula that, at finite ε, the nonhyperbolicity region shifts into the region of supersonic flow, where $M_{12} > 1$ and approaches the value $M_{12} \approx 2.85$ as $\varepsilon \to 1$.

It should be noted that the above physical mechanism is valid only in the region of short wavelengths $k \sim 1/d$. For long wavelengths $k \sim 1/L$, one can neglect the term $\frac{\pi d^3}{6}\nabla^k P$, as compared to f_p^k in the force R_p^k (7.1.22) acting on a particle from gas. This assertion follows from the estimate

$$\left|\frac{\pi d^3}{6}\nabla^k P/f_p^k\right| \sim \frac{\Delta P}{C_d\rho_{11}(\vec{v}_1-\vec{v}_2)^2}\frac{d}{L} \sim \frac{d}{L} \ll 1,$$

where we have used formulas (7.1.20) and (7.1.21) and $C_d > 1/2$, $\Delta P \sim \rho_{11}(\vec{v}_1 - \vec{v}_2)^2$. Consequently, for sufficiently smooth flows ($L \gg d$) in the complete system of equations for the gas-particle mixtures (7.1.34), one can neglect the term $m_2\nabla^k P$, as compared to nf_p^k as well as the difference of the mean gas density from the true gas density, i.e., $\rho_1 \approx \rho_{11}$. (We recall that small volume concentrations of particles are considered, i.e., $m_2 \ll 1$.) As a result, we obtain the Kliegel–Nickerson system of equations, which was applied for the first time for the description of a flow of a gas-particle mixture in nozzles[16]:

$$\begin{aligned}
&\frac{\partial\rho_1}{\partial t} + \nabla_k(\rho_1 v_1^k) = 0, \quad \rho_1 = \rho_{11},\\
&\frac{\partial\rho_2}{\partial t} + \nabla_k(\rho_2 v_2^k) = 0, \quad \rho_2 = \rho_{22}m_2, \quad \rho_{22} = \text{const},\\
&\rho_1\frac{d_1 v_1^k}{dt} = -\nabla^k P - nf_p^k, \quad \frac{d_1}{dt} = \frac{\partial}{\partial t} + v_1^l\nabla_l,\\
&\rho_2\frac{d_2 v_2^k}{dt} = nf_p^k, \quad \frac{d_2}{dt} = \frac{\partial}{\partial t} + v_2^l\nabla_l,\\
&\frac{d_1 H}{dt} - \frac{1}{\rho_1}\frac{d_1 P}{dt} + \dot{Q} = 0, \quad \dot{Q} = nq_p - nf_p^k(v_1^k - v_2^k),\\
&\frac{d_2 e_2}{dt} - \frac{nq_p}{\rho_2} = 0, \quad e_1 = C_V T_1, \quad H = e_1 + \frac{P}{\rho_1}, \quad e_2 = C_s T_2.
\end{aligned} \tag{7.2.9}$$

The system (7.2.9) is closed by the equation of state $H = H(\rho_1, P)$, in particular, for the ideal gas $P = \rho_1 RT$, where R is the universal gas constant.

Let us find the characteristics of the Kliegel–Nickerson system (7.2.9) in the one-dimensional planar ($\nu = 0$) nonstationary case. The last two equations in (7.2.9) are written along the characteristics $\frac{dx}{dt} = v_1$, $\frac{dx}{dt} = v_2$; therefore, it is sufficient to consider the first four equations

$$\frac{\partial\rho_1}{\partial t} + \rho_1\frac{\partial v_1}{\partial x} + v_1\frac{\partial\rho_1}{\partial x} = 0,$$

$$\begin{aligned}
&\frac{\partial \rho_2}{\partial t} + \rho_2 \frac{\partial v_2}{\partial x} + v_2 \frac{\partial \rho_2}{\partial x} = 0, \\
&\frac{\partial v_1}{\partial t} + v_1 \frac{\partial v_1}{\partial x} + \frac{1}{\rho_1}\frac{\partial P}{\partial x} = -\frac{1}{\rho_1} n f_p, \\
&\frac{\partial v_2}{\partial t} + v_2 \frac{\partial v_2}{\partial x} = \frac{1}{\rho_2} n f_p.
\end{aligned} \tag{7.2.10}$$

Let us eliminate the variable ρ_1 with the aid of the equation of state $\rho_1 = \rho_1(H, P)$, $H = C_p T_1$, the energy equation [the second to last equation in (7.2.9)], and the formula for the sound velocity

$$c^2 = 1/\left(\frac{\partial \rho_1}{\partial P} + \frac{1}{C_p \rho_1}\frac{\partial \rho_1}{\partial T}\right), \tag{7.2.11}$$

$$\begin{aligned}
\frac{d_1 \rho_1}{dt} &= \frac{\partial \rho_1}{\partial H}\frac{d_1 H}{dt} + \frac{\partial \rho_1}{\partial P}\frac{d_1 P}{dt} \\
&= \left(\frac{\partial \rho_1}{\partial P} + \frac{1}{\rho_1}\frac{\partial \rho_1}{\partial H}\right)\frac{d_1 P}{dt} - \frac{\partial \rho_1}{\partial H}\dot{Q} = \frac{1}{c^2}\frac{d_1 P}{dt} - \frac{\rho_{1,T}}{C_p}\dot{Q},
\end{aligned}$$

where $\rho_{1,T} = \frac{\partial \rho_1}{\partial T}$. Substituting this formula into the first equation of (7.2.10), we can rewrite (7.2.10) as follows:

$$\begin{aligned}
&\frac{\partial P}{\partial t} + v_1 \frac{\partial P}{\partial x} + \rho_1 c^2 \frac{\partial v_1}{\partial x} = \frac{c^2 \rho_{1,T}}{C_p}\dot{Q}, \\
&\frac{\partial \rho_2}{\partial t} + \rho_2 \frac{\partial v_2}{\partial x} + v_2 \frac{\partial \rho_2}{\partial x} = 0, \\
&\frac{\partial v_1}{\partial t} + v_1 \frac{\partial v_1}{\partial x} + \frac{1}{\rho_1}\frac{\partial P}{\partial x} = -\frac{n f_p}{\rho_1}, \\
&\frac{\partial v_2}{\partial t} + v_2 \frac{\partial v_2}{\partial x} = \frac{n f_p}{\rho_2}.
\end{aligned} \tag{7.2.12}$$

Writing this system of equations in the matrix form $\frac{\partial u}{\partial t} + C\frac{\partial u}{\partial x} = f$, $u^T = (P, \rho_2, v_1, v_2)$, we can find the equation for the characteristics (3.2.8):

$$\begin{vmatrix}
v_1 - \lambda & 0 & \rho_1 c^2 & 0 \\
0 & v_2 - \lambda & 0 & \rho_2 \\
\frac{1}{\rho_1} & 0 & v_1 - \lambda & 0 \\
0 & 0 & 0 & v_2 - \lambda
\end{vmatrix} = 0.$$

Calculating this determinant, we obtain the equation:

$$(v_2 - \lambda)^2((v_1 - \lambda)^2 - c^2) = 0,$$

whose solution is

$$\lambda_{1,2} = v_1 \pm c, \quad \lambda_{3,4} = v_2. \tag{7.2.13}$$

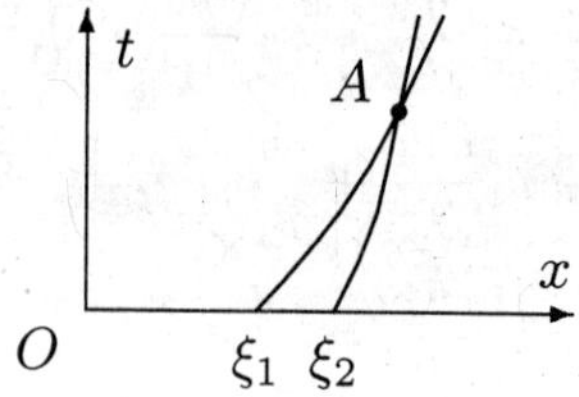

Figure 7.5: To the explanation of the growth mechanism of the disturbances of a particle's density.

It can be seen from here that all characteristics are real; however, two of them are multiple, that is, coincide with one another. In this case, for the elucidation of the type of system (7.2.12), it is necessary to know whether this system can be reduced to a canonical form (3.2.11). Dividing the first equation in (7.2.12) by $\rho_1 c$ and adding to and subtracting from the third equation in (7.2.12), we obtain instead of (7.2.12):

$$\begin{aligned}
&\frac{\partial \mathcal{I}_+}{\partial t} + (v_1 + c)\frac{\partial \mathcal{I}_+}{\partial x} = \frac{c\rho_{1,T}}{C_p\rho_1}\dot{Q} - \frac{nf_p}{\rho_1}, \quad \mathcal{I}_+ = v_1 + \int \frac{dP}{\rho_1 c},\\
&\frac{\partial \mathcal{I}_-}{\partial t} + (v_1 - c)\frac{\partial \mathcal{I}_-}{\partial x} = -\frac{c\rho_{1,T}}{C_p\rho_1}\dot{Q} - \frac{nf_p}{\rho_1}, \quad \mathcal{I}_- = v_1 - \int \frac{dP}{\rho_1 c},\\
&\frac{\partial \rho_2}{\partial t} + v_2\frac{\partial \rho_2}{\partial x} + \rho_2\frac{\partial v_2}{\partial x} = 0, \qquad\qquad (7.2.14)\\
&\frac{\partial v_2}{\partial t} + v_2\frac{\partial v_2}{\partial x} = \frac{nf_p}{\rho_2},
\end{aligned}$$

where $\mathcal{I}_+$ and $\mathcal{I}_-$ are the Riemann invariants. It is seen from here that the third equation and, consequently, the system of equations cannot be reduced to the canonical form $\frac{\partial\varphi}{\partial t} + \lambda\frac{\partial\varphi}{\partial x} + g\varphi = h$, and therefore, the Kliegel–Nickerson system (7.2.9), (7.2.10) is not hyperbolic.

The development of small short wavelength disturbances in the Kliegel–Nickerson model (7.2.10) was studied in[15]. It was shown that all disturbances, except for the particle's density ρ_2' remain finite, and the last one "slowly" increases by the linear law

$$\rho_2' = \delta\rho_2 \sin k\xi - \rho_2^0 \delta v_2\, kt \cos k\xi, \quad \xi = x - \int_0^t v_2\, dt'. \qquad (7.2.15)$$

The physical reason for the nonhyperbolicity and the growth of the density disturbances is also the absence of the eigenpressure in a pseudogas of particles. The mechanism of the disturbances growth corresponding to formula (7.2.15) is shown in Fig. 7.5. Assume that a disturbance of the velocity v_2 at time $t = 0$ arises, at which the velocity of particles with the Lagrangian coordinate ξ_1 is larger than the velocity of particles with the Lagrangian coordinate ξ_2: $v_2(\xi_1) > v_2(\xi_2) > 0$ and $\xi_1 < \xi_2$. As a result, the "back" particles ($x|_{t=0} = \xi_1$) will overtake the "forward" particles ($x|_{t=0} = \xi_2$), and from the mass conservation law for the two Lagrangian trajectories ξ_1 and ξ_2, we obtain an increase in the particle's density in accordance with the law $\rho_2 = \rho_2^0(\xi_2 - \xi_1)/(x_2 - x_1)$, where $x_2 = x(\xi_2, t)$ and $x_1 = x(\xi_1, t)$ are the particles trajectories. At the point of the intersection of particles trajectories A, we have $x_2 \to x_1$; therefore, $\rho_2 \to \infty$. As was shown in[8,17], only the short-wavelength disturbances whose wave vector $k = \frac{2\pi}{\lambda}$ satisfies the inequality

$$\delta v_2 \, k\tau_v > 1 \tag{7.2.16}$$

lead to the intersection of the particles, where τ_v is the time of the velocity relaxation (7.1.42). If the inequality (7.2.16) is not satisfied, then, at the expense of the interaction with gas the particles velocities $v_2(\xi_1)$ and $v_2(\xi_2)$ are equalized earlier than occurs an intersection of their trajectories.

Thus, the Cauchy problem for the system of equations (7.2.9) and (7.2.10) will be well-posed if only such flows in which no intersection of particles trajectories occurs are considered. The admissible disturbances should satisfy the inequality $\delta v_2 \, k\tau_v < 1$. The same condition should also be satisfied for system (7.2.1); however, in this case, the disturbances must satisfy the inequality $\delta\varphi_i < \exp(-k\beta t_m)$, where β is determined by formula (7.2.8) and t_m is the maximum time of the process under study. The given problems are called "conditionally correct"[18].

Consider the flows of two-phase gas-particle media in which the intersections of particles trajectories take place. In connection with the above-presented analysis, a question on the description of such flows arises. Three cases are possible here when the Knudsen number of the particles pseudogas $\mathrm{Kn} \ll 1$, $\mathrm{Kn} \sim 1$, and $\mathrm{Kn} \gg 1$, where the number

$$\mathrm{Kn} = \frac{d}{6m_2 L} \tag{7.2.17}$$

is defined as a ratio of the free path of particles of diameter d between two collisions $\lambda_* = 1/(\pi d^2 n)$ to the reference flow size L.

In the first case $\mathrm{Kn} \ll 1$, the collisions between the particles play a significant role, and in the mean tensor of the particles stresses σ_2^{kl}, it is necessary to take into account the eigenpressure of the pseudogas of particles P_2. As a result, we will have instead (7.1.18) the formula

$$\sigma_2^{kl} = -(m_2 P + P_2) g^{kl}. \tag{7.2.18}$$

The pressure P_2 can be found from the kinetic equation for the pseudogas

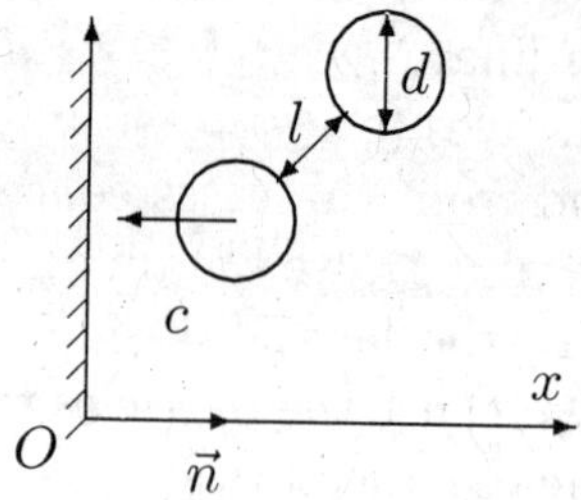

Figure 7.6: To the computation of the pressure P_2 in the pseudogas of particles.

of particles[19]

$$\frac{\partial f}{\partial t}+v_2^k\nabla_k f+\frac{\partial}{\partial v_2^k}\left(\left(f_p^k-\frac{1}{\rho_{22}}\nabla^k P\right)f\right)=S(f,f_1)+D_p\frac{\partial f}{\partial v_2^k\partial v_2^k}, \quad (7.2.19)$$

where $f=f(t,x^k,v_2^k)$ is the one-particle distribution function of particles, $S(f,f_1)$ is the integral of collisions, and D_p is the diffusion coefficient in the space of the velocities v_2^k conditioned by the interaction of particles with the gas pulsations. In this case $\mathrm{Kn}\ll 1$, the kinetic equation is solved by the Chapman–Enskog method[19]; as a result, one obtains the continuum equations for a pseudogas of particles with the tensor of stresses (7.2.18). The presentation of this method does not belong to the scope of this book; therefore, we will find a formula for P_2 by the methods of the elementary kinetic theory[20].

Consider a pseudogas of particles consisting of the spherical particles of diameter d moving chaotically and colliding with each other. Let the distance between the particles $l<d$; therefore, a given particle collides only with its nearest neighbors, which means that the screening effect takes place (see Fig. 7.6). Then the free path of a particle is of order $2l$, and the time of free path $\tau\approx 2l/(\sqrt{3}c)$, where c is the velocity of chaotic motion along some axis. It is assumed that the system is in the thermodynamic equilibrium state; therefore, there is the energy $\frac{1}{2}mc^2$ per each degree of freedom. If the particles are smooth and do not rotate, then the energy per one particle is equal to $\frac{3}{2}mc^2$. For the rough particles, one must also take into account three rotational degrees of freedom; therefore, the energy of a single particle is $3mc^2$. Denote by m_2^* the volume concentration of particles for a close packing. Then the distance between the particles $l=0$. The distance l between the particles at an arbitrary $m_2<m_2^*$ can be approximated from geometric

considerations by formula[20]

$$l = d\left(\left(\frac{m_2^*}{m_2}\right)^{1/3} - 1\right). \tag{7.2.20}$$

Let us choose a unit area with the normal $\vec{n}$ parallel with the x-axis and compute the momentum transferred by the particles at elastic collisions, which is equal to P_2. As a consequence of the screening effect, the particles located in the layer of thickness $l + d \approx d$ will collide with the area. Each particle makes per unit time $c/(2l)$ collisions with the area, and at each collision it transfers the momentum $2mc$. Consequently, the total transferred momentum per unit time is equal to the pressure

$$P_2 = (2m_pc)\left(\frac{c}{2l}\right)nd,$$

and with regard for (7.2.20) and the relation $\rho_2 = nm_p$, we obtain the formula for the eigenpressure in the pseudogas of particles:

$$P_2 = \frac{\rho_2 c^2}{\left(\frac{m_2^*}{m_2}\right)^{1/3} - 1}. \tag{7.2.21}$$

It follows from (7.2.21) that, at $m_2 < m_2^*$

$$\frac{\partial P_2}{\partial \rho_2} = \frac{c^2}{\left(\left(\frac{m_2^*}{m_2}\right)^{1/3} - 1\right)^2}\left(\frac{1}{3}\left(\frac{m_2^*}{m_2}\right)^{1/3} - 1\right) > 0;$$

therefore, the local increase in the particle's density ρ_2 will lead to the increase of pressure P_2, which will hinder a further growth of the density ρ_2. Note that formula (7.2.21) contains, besides the density ρ_2, the quantity c^2 proportional to the chaotic motion energy $e_{2c} = \beta c^2$, where $\beta = \frac{3}{2}$ in the presence of the translational degrees of freedom and $\beta = 3$ in the presence of the translational and rotational degrees of freedom. Substituting in (7.1.11) and (7.1.12) the expressions $E_2 = e_2 + e_{2c} + \frac{1}{2}v_2^2$, $\sigma_2^{kl} = -(m_2P + P_2)g^{kl}$, we obtain with regard for (7.1.34) the momentum and energy equations for particles:

$$\begin{aligned} \rho_2\frac{d_2v_2^k}{dt} &= -m_2\nabla^kP - \nabla^kP_2 + nf_p^k, \\ \frac{d_2e_{2c}}{dt} &= -P_2\frac{d_2}{dt}\left(\frac{1}{\rho_2}\right) + (\dot{Q}_c - \dot{Q}_g)/\rho_2, \\ \frac{\rho_2 d_2e_2}{dt} &= \dot{Q}_g + nq_p, \end{aligned} \tag{7.2.22}$$

where $\dot{Q}_c$ and $\dot{Q}_g$ are the rate of the inflow of energy and of the dissipation of the energy of chaotic motion of particles in a unit volume,

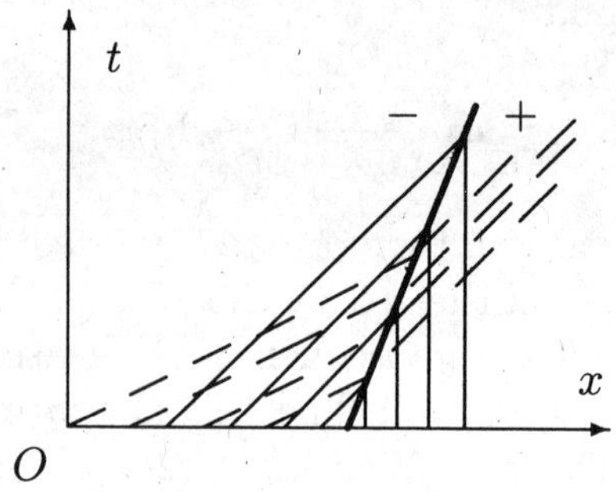

Figure 7.7: The particle's sheet. The solid lines are the particle trajectories, and the dashed lines are the gas trajectories.

respectively. At the expense of a velocity nonequilibrium $v_1^k \neq v_2^k$ and the action of the drag force f_p^k, a part of the kinetic energy $\beta_1 n f_p^k (v_1^k - v_2^k)$ immediately transforms into heat and a part $\beta_2 n f_p^k (v_{1k} - v_{2k})$ transforms into the energy of the gas pulsations, where $\beta_1 + \beta_2 = 1$. Some part α_{22} of the pulsation energy is spent for increasing the energy of the chaotic motion of particles, and the fraction α_{21} dissipates into the heat, where $\alpha_{21} + \alpha_{22} = 1$. Consequently, the rate of the inflow of the energy of chaotic particles motion will be given by the formula

$$\dot{Q}_c = \alpha_{22} \beta_2 n f_p^k (v_1^k - v_2^k). \tag{7.2.23}$$

The calculation of the coefficients β_1, β_2, α_{21}, and α_{22} is impossible without the description of the oscillatory gas motion. Since this problem has not been solved until now, the above coefficients should be determined from the experiment. Neglecting the energy of pulsations, let us rewrite the equation for the gas internal energy (7.1.39) as

$$\frac{d_1 e_1}{dt} = -P \frac{d_1}{dt}\left(\frac{1}{\rho_{11}}\right) + \frac{(\beta_1 + \alpha_{21}\beta_2) n f_p^k (v_1^k - v_2^k)}{\rho_1} - \frac{n q_p}{\rho_1}. \tag{7.2.24}$$

The equations for the gas motion and continuity in (7.1.34) do not change their forms when the effects of particle's collisions are taken into account. The energy dissipation $\dot{Q}_g$ occurs mainly at the particle's collisions at the expense of inelastic deformations of particles. If the particle velocity before the collision was c, and after the collision it became equal to kc, then the energy $(1 - k^2) mc^2$ transforms into heat at a single collision; hence, the rate of the energy dissipation within a unit volume is[20]

$$\dot{Q}_g = \frac{\sqrt{3}}{4}(1 - k^2)\rho_{22}\frac{c^3}{d}\frac{m_2}{\left(\frac{m_2^*}{m_2}\right)^{1/3} - 1}, \tag{7.2.25}$$

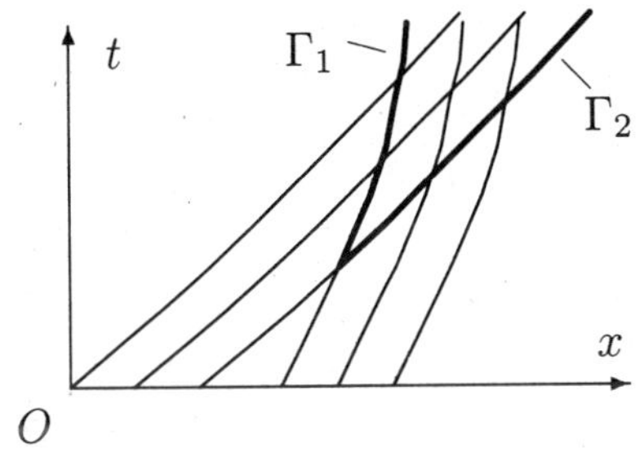

Figure 7.8: The intersection of the particle's trajectories in the region bounded by lines Γ_1 and Γ_2.

where $0 < k < 1$ is the coefficient of restoration at a particle's collision.

If the region of increased density ρ_2, which has appeared as a result of the intersection of trajectories and the particle's collisions, has a small spatial dimension $\Delta x \ll L$, then it can be replaced with a sheet carrying the surface mass[15] (see Fig. 7.7). It is assumed that the particle's collisions are absolutely inelastic; therefore, the particles are packed on the sheet up to a close packing state m_2^*. The gas parameters φ_1 at a flow through the sheet experience a jump $[\varphi_1] = \varphi_+ - \varphi_-$ whose magnitude is computed below [see (7.2.37)]. The sheet is governed by ordinary differential equations (ODEs) (7.2.38); therefore, the Cauchy problem for such a model proves to be correct.

In the case in which Kn ~ 1, it is necessary to describe the behavior of the pseudogas of particles in the region of the intersection of particle's trajectories by the kinetic equation (7.2.19). The solution of this integrodifferential equation represents a very complex problem. Therefore, one uses for the computation of such flows the method of direct statistical modeling of the particles motion[21]. The computation is subdivided into two stages: at the first stage, the particles collisions are modeled in a random way, and at the second stage, the motions of particles are computed until the next collision. As the numerical computations show, the particles collisions bound the growth of the mean particle's density in the places of the intersections of their trajectories[22].

In the case in which Kn $\gg 1$ the particles do not collide with one another, the particle's trajectories can intersect with one another. Then, in the region of intersection, several velocities of particles are determined at each t, x point (see Fig. 7.8); therefore, the representation of a pseudogas of particles in the form of a continuum is impossible in principle. It is necessary to use, in this case, for a description of particles, the collisionless kinetic equation[8,17]. The corresponding equations and the studies of flows with the intersections of particles trajectories are presented in

Section 7.4.

Problem 7.3. Find the characteristics of the system of equations of a bicomponent gas mixture (7.1.47), (7.1.48) in the one-dimensional nonstationary case.

Solution: Note that the energy equations in (7.1.47) are reduced to equations (7.1.50) in which the derivatives along the trajectories $\frac{dx}{dt} = v_i$ coinciding with the components trajectories stand on the left-hand side. Therefore, it is sufficient to consider the continuity and motion equations:

$$\frac{d_i\rho_i}{dt} + \rho_i\frac{\partial v_i}{\partial x} = 0, \quad \frac{d_i}{dt} = \frac{\partial}{\partial t} + v_i\frac{\partial}{\partial x}, \quad i = 1, 2,$$
$$\rho_i\frac{d_iv_i}{dt} + \frac{\partial P_i}{\partial x} = \frac{v_j - v_i}{\tau_v}, \quad P_i = P_i(\rho_i, S_i). \tag{7.2.26}$$

Expressing the density $\rho_i = \rho_i(P_i, S_i)$ from the last equation, we find with regard for (7.1.50):

$$\frac{d_i\rho_i}{dt} = \left(\frac{\partial\rho_i}{\partial P_i}\right)_{S_i}\frac{d_iP_i}{dt} + \left(\frac{\partial\rho_i}{\partial S_i}\right)_{P_i}\frac{d_iS_i}{dt} = \frac{1}{c_i^2}\frac{d_iP_i}{dt} + g_i\dot{Q}_{ji},$$
$$g_i = \frac{1}{\rho_i}\left(\frac{\partial\rho_i}{\partial S_i}\right)_{P_i}, \quad \dot{Q}_{ji} = \frac{T_j - T_i}{T_i\tau_T} + \alpha_{ji}\frac{(\vec{v}_j - \vec{v}_i)^2}{T_i\tau_v}, \tag{7.2.27}$$

where no summation is carried out over the repeating indices. Substituting the derivative $\frac{d_i\rho_i}{dt}$ from (7.2.27) in the first equation (7.2.26) and assuming at first that $i = 1$ and $j = 2$ and then $i = 2$ and $j = 1$, we obtain the system of equations:

$$\frac{\partial P_1}{\partial t} + v_1\frac{\partial P_1}{\partial x} + \rho_1c_1^2\frac{\partial v_1}{\partial x} = -c_1^2g_1\dot{Q}_{21},$$
$$\frac{\partial v_1}{\partial t} + \frac{1}{\rho_1}\frac{\partial P_1}{\partial x} + v_1\frac{\partial v_1}{\partial x} = \frac{v_2 - v_1}{\rho_1\tau_v},$$
$$\frac{\partial P_2}{\partial t} + v_2\frac{\partial P_2}{\partial x} + \rho_2c_2^2\frac{\partial v_2}{\partial x} = -c_2^2g_2\dot{Q}_{12}, \tag{7.2.28}$$
$$\frac{\partial v_2}{\partial t} + \frac{1}{\rho_2}\frac{\partial P_2}{\partial x} + v_2\frac{\partial v_2}{\partial x} = \frac{v_1 - v_2}{\rho_2\tau_v}.$$

Introducing the transposed vector $u^T = (P_1, v_1, P_2, v_2)$ and writing the system (7.2.28) in the matrix form $\frac{\partial u}{\partial t} + C\frac{\partial u}{\partial x} = f$, we can find the equation for the characteristics $|C - \lambda E| = 0$ in the form:

$$\begin{vmatrix} v_1 - \lambda & \rho_1c_1^2 & 0 & 0 \\ \frac{1}{\rho_1} & v_1 - \lambda & 0 & 0 \\ 0 & 0 & v_2 - \lambda & \rho_2c_2^2 \\ 0 & 0 & \frac{1}{\rho_2} & v_2 - \lambda \end{vmatrix} = 0.$$

Calculating the determinant, we obtain the equation:

$$((v_1-\lambda)^2-c_1^2)((v_2-\lambda)^2-c_2^2)=0$$

whose solution is

$$\lambda_{1,2}=v_1\pm c_1,\quad \lambda_{3,4}=v_2\pm c_2, \tag{7.2.29}$$

where $c_i=(\frac{\partial P}{\partial \rho_i})_{S_i}$ is the sound velocity of the ith component ($i=1,2$). Thus, the system of equations (7.2.26) for a mixture of gases has the real characteristics $\frac{dx}{dt}=\lambda_i$ (7.2.29) and belongs to the hyperbolic type.

7.2.2 Jump Relations

The above-investigated systems of the equations of the mechanics of multiphase media are valid for smooth flows in which the characteristic size of the variation of the mean parameters L is substantially larger than the size of the inclusions (inhomogeneity) d. If this condition is not satisfied in some transitional region, then the transitional region is replaced with a discontinuity surface, the relations across which are obtained from the conservation laws for mass, momentum, and energy. The procedure of obtaining the jump relations is in principle similar to the determination of the relations at the contact discontinuities and shock waves in gas dynamics. Certain peculiarities can arise here, however, which are related to the nondivergence form of the equations of the mechanics of multiphase media and the finiteness of the surface mass of particles at a discontinuity[8,15,23].

Consider for definiteness a gas-particle mixture at a large volume concentration of particles, which is described in the one-dimensional nonstationary case by the equations (7.2.1) and (7.2.22)–(7.2.24):

$$\begin{aligned}
&\frac{\partial \rho_1}{\partial t}+\frac{\partial}{\partial x}(\rho_1 v_1)=0,\quad \rho_1=\rho_{11}m_1,\\
&\frac{\partial \rho_2}{\partial t}+\frac{\partial}{\partial x}(\rho_2 v_2)=0,\quad \rho_2=\rho_{22}m_2,\quad \rho_{22}=\text{const},\\
&\rho_1\frac{d_1v_1}{dt}=-m_1\frac{\partial P}{\partial x}-nf_p,\quad \frac{d_1}{dt}=\frac{\partial}{\partial t}+v_1\frac{\partial}{\partial x},\\
&\rho_2\frac{d_2v_2}{dt}=-m_2\frac{\partial P}{\partial x}-\frac{\partial P_2}{\partial x}+nf_p,\quad \frac{d_2}{dt}=\frac{\partial}{\partial t}+v_2\frac{\partial}{\partial x},\\
&\frac{d_1e_1}{dt}=-P\frac{d_1}{dt}\Big(\frac{1}{\rho_{11}}\Big)+\frac{1}{\rho_1}r_1nf_p(v_1-v_2)-\frac{1}{\rho_1}nq_p,\\
&\frac{d_2e_{2c}}{dt}=-P_2\frac{d_2}{dt}\Big(\frac{1}{\rho_2}\Big)+\frac{1}{\rho_2}\dot{Q}_c-\frac{1}{\rho_2}\dot{Q}_g,\\
&\frac{d_2e_2}{dt}=\frac{1}{\rho_2}\dot{Q}_g+\frac{nq_p}{\rho_2},
\end{aligned} \tag{7.2.30}$$

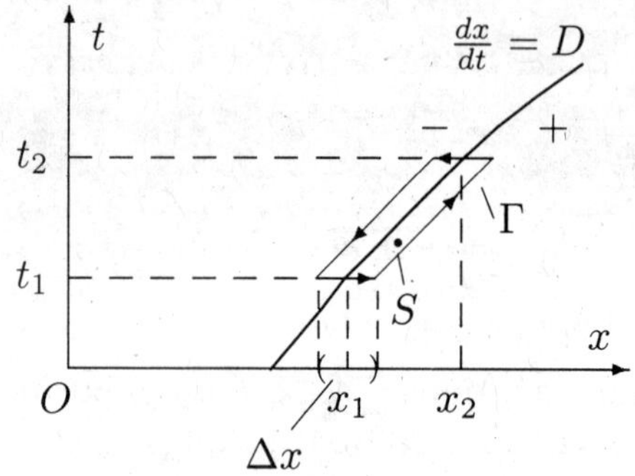

Figure 7.9: The discontinuity trajectory in the x, t plane.

$$\dot{Q}_c = r_2 n f_p(v_1 - v_2), \quad r_1 = \beta_1 + \alpha_{21}\beta_2, \ r_2 = \alpha_{22}\beta_2, \ r_1 + r_2 = 1,$$
$$P_2 = P_2(\rho_2, e_{2c}), \ e_1 = C_V T_1, \ e_2 = C_s T_2,$$

where the pressure P_2 is determined by equation (7.2.21). Combining these equations, we obtain:

$$\begin{aligned}
&\frac{\partial \rho_1}{\partial t} + \frac{\partial}{\partial x}(\rho_1 v_1) = 0, \quad \rho_1 = \rho_{11} m_1,\\
&\frac{\partial \rho_2}{\partial t} + \frac{\partial}{\partial x}(\rho_2 v_2) = 0, \quad \rho_2 = \rho_{22} m_2, \quad \rho_{22} = \text{const},\\
&\frac{\partial}{\partial t}(\rho_1 v_1) + \frac{\partial}{\partial x}(\rho_1 v_1^2 + m_1 P) = P\frac{\partial m_1}{\partial x} - n f_p,\\
&\frac{\partial}{\partial t}(\rho_2 v_2) + \frac{\partial}{\partial x}(\rho_2 v_2^2 + m_2 P + P_2) = -P\frac{\partial m_1}{\partial x} + n f_p,\\
&\frac{\partial}{\partial t}\left(\rho_1\left(e_1 + \frac{v_1^2}{2}\right)\right) + \frac{\partial}{\partial x}\left(\rho_1 v_1\left(H_1 + \frac{v_1^2}{2}\right)\right) \qquad (7.2.31)\\
&= -P\frac{\partial}{\partial x}(m_2 v_2) + r_1(v_1 - v_2) n f_p - v_1 n f_p - n q_p,\\
&\frac{\partial}{\partial t}\left(\rho_2\left(\tilde{e}_2 + \frac{v_2^2}{2}\right)\right) + \frac{\partial}{\partial x}\left(\rho_2 v_2\left(\tilde{H}_2 + \frac{v_2^2}{2}\right)\right)\\
&= P\frac{\partial}{\partial x}(m_2 v_2) + r_2(v_1 - v_2) n f_p + v_2 n f_p + n q_p,\\
&\tilde{e}_2 = e_{2c} + e_2, \quad \tilde{H}_2 = \tilde{e}_2 + \frac{P_2}{\rho_2} + \frac{P}{\rho_{22}}, \quad H_1 = e_1 + \frac{P}{\rho_{11}}.
\end{aligned}$$

It follows from here that the system of equations (7.2.31) cannot be reduced to a divergence form $\frac{\partial \varphi}{\partial t} + \frac{\partial \psi}{\partial x} = g(\varphi, \psi)$ because the derivatives $P\frac{\partial m_1}{\partial x}$ and $P\frac{\partial}{\partial x}(m_2 v_2)$ stand on the right-hand side.

Consider a discontinuity surface, the normal to which is parallel with the x-axis, and moves along the x-axis at a velocity $\frac{dx}{dt} = D$ (see Fig.

7.9). If the equations $\frac{\partial\varphi}{\partial t} + \frac{\partial\psi}{\partial x} = g$ are valid in the smooth flow region, then in the presence of a discontinuity surface we will be basing on the integrodifferential equation

$$\iint \left(\frac{\partial\varphi}{\partial t} + \frac{\partial\psi}{\partial x}\right) dx\, dt = \iint g\, dx\, dt. \tag{7.2.32}$$

Let us choose the surface S bounded by contour Γ (see Fig. 7.9) as an integration region. The lower and upper pieces of the boundary Γ are parallel with the Ox-axis, and the lateral sides of Γ are parallel with the discontinuity trajectory $\frac{dx}{dt} - D$. Thus, S represents a small strip of width Δx containing the discontinuity trajectory. Using the Stokes theorem (it is called the Green's formula in the two-dimensional case), we go over in the left-hand side of (7.2.32) from the integral over the area S to the integral along the contour Γ:

$$\oint_\Gamma -\varphi\, dx + \psi\, dt = \iint_S g\, dx\, dt. \tag{7.2.33}$$

Denoting the parameters to the left and right of the discontinuity by indices minus and plus, respectively (see Fig. 7.9), we write:

$$\oint_\Gamma \varphi\, dx = \int \varphi\, dx\Big|_{t_1} - \int_{x_1}^{x_2} \varphi_-\, dx - \int \varphi\, dx\Big|_{t_2} + \int_{x_1}^{x_2} \varphi_+\, dx,$$
$$\oint_\Gamma \psi\, dt = -\int_{t_1}^{t_2} \psi_-\, dt + \int_{t_1}^{t_2} \psi_+ dt. \tag{7.2.34}$$

Substituting (7.2.34) into (7.2.33) and using $dx = D\, dt$, we find:

$$\int \varphi\, dx\Big|_{t_2} - \int \varphi\, dx\Big|_{t_1} + \int_{t_1}^{t_2} ((\psi_+ - \varphi_+ D) - (\psi_- - \varphi_- D))dt$$
$$= \iint_S g\, dx\, dt. \tag{7.2.35}$$

Assuming the interval $\Delta t = t_2 - t_1$ to be small, we obtain from (7.2.35):

$$\frac{d\Theta}{dt} + [\psi - \varphi D] = \frac{1}{\Delta t}\iint_S g\, dx\, dt,$$
$$\Theta = \int_{x_p - \Delta x/2}^{x_p + \Delta x/2} \varphi\, dx, \tag{7.2.36}$$

where x_p is the discontinuity coordinate at time t satisfying the inequalities $t_1 < t < t_2$ and Δx is the width of a transitional region, which is replaced with the discontinuity surface. Applying formula (7.2.36) to each of the equations (7.2.31), we find, with regard for the relations

$m_1 \sim m_2$, $\rho_{11} \ll \rho_{22}$, $\rho_1 \ll \rho_2$, the jump relations in the one-dimensional case for the gas:

$$[\rho_1 u_1] = 0, \quad [j_1 u_1 + p m_1] = F^\sigma,$$
$$j_1[H_1 + \frac{u_1^2}{2}] = -Q^\sigma, \quad j_1 = \rho_1 u_1, \quad u_1 = v_1 - D, \quad (7.2.37)$$

and for particles:

$$\frac{dM_p}{dt} + [\rho_2 u_2] = 0, \quad \frac{d}{dt}(M_p D) + [j_2 u_2 + P m_2 + P_2] = -F^\sigma,$$
$$\frac{d}{dt}\left(M_p\left(\tilde{e}_2 + \frac{D^2}{2}\right)\right) + \left[j_2\left(\tilde{H}_2 + \frac{u_2^2}{2}\right)\right] = Q^\sigma, \quad (7.2.38)$$
$$j_2 = \rho_2 u_2, \quad u_2 = v_2 - D,$$

where $M_p = \int \rho_2 dx \approx \rho_2 \Delta x$ is the particles mass at the discontinuity. If we introduce the mean gas pressure at the discontinuity P^σ, then we can find the following approximate expressions for the surface force F^σ and the energy flux Q^σ:

$$F^\sigma = \frac{1}{\Delta t}\int\int_S \left(P\frac{\partial m_1}{\partial x} - n f_p\right)dx\, dt \approx P^\sigma[m_1] - N f_p,$$
$$Q^\sigma = \frac{1}{\Delta t}\int\int_S \left(P\frac{\partial}{\partial x}(m_2 v_2) + (r_1 v_1 + r_2 v_2) n f_p + n q_p\right)dx\, dt \quad (7.2.39)$$
$$\approx P^\sigma[m_2 v_2] + N(r_1 v_1 + r_2 v_2) n f_p + n q_p,$$

where $N \approx n\,\Delta x$ is the particle's number at a dscontinuity. The quantity P^σ depends on the discontinuity nature and is to be determined from the experiment. In the general case in which the tangential velocity components of gas v_{1t} and particles v_{2t} are different from zero at a discontinuity surface, it is necessary to use instead (7.2.32) the formula (3.3.7). If we carry out the integration in formula (3.3.7) and take into account the nonzero right-hand side in (7.2.31) and the surface mass M_p, then we obtain the relationships, which coincide completely with (7.2.37)–(7.2.39), in which $u_1 = v_{1n} - D$ and $u_2 = v_{2n} - D$ and are augmented by the conditions

$$[v_{1t}] = 0, \quad [v_{2t}] = 0. \quad (7.2.40)$$

Depending on the transitional region nature, the quantity Δx will be of the order of magnitude of the molecules free path in gas λ_* or it will be of the order of the particle diameter d. In the first case, $M_p = 0$, and in the second case, the particle's mass at a discontinuity surface is finite and has the order of magnitude $M_p \approx \rho_2 d$.

Basing on the system of equations (7.2.37)–(7.2.40), we now carry out the classification of possible discontinuities as follows:

1) *A stationary tangential discontinuity in gas and particles* at which the following quantities change spasmodically:

$$\rho_1, \quad \rho_2, \quad v_{1t}, \quad v_{2t}, \quad e_1, \quad e_2;$$

however, $j_1 = 0$, $j_2 = 0$, $\frac{dM_p}{dt} = 0$, and $\frac{dD}{dt} = 0$.
Substituting the last equalities in (7.2.37) and (7.2.38), we can find:

$$[P] = 0, \quad [P_2] = 0, \quad u_1 = 0, \quad Q^\sigma = 0, \quad P^\sigma = P, \quad F^\sigma = P[m_1].$$

The gas and particles move in this case along the tangential discontinuity surface, which is common for both phases. Note that the above jumps of density, tangential velocity, and internal energy are arbitrary.

2) *Tangential discontinuity in gas* at which the following quantities experience an arbitrary jump, ρ_1, v_{1t}, e_1, and the mass flow rate of gas $j_1 = 0$. The transitional region thickness $\Delta x \approx \lambda_*$; therefore, $M_p = 0$. The particles can intersect the discontinuity surface; therefore, $j_2 \neq 0$. We can find from equations (7.2.37) and (7.2.38) that $u_1 = 0$, $[P] = 0$, $[m_1] = 0$, $F^\sigma = 0$, $Q^\sigma = 0$, $[u_2] = 0$, $[P_2] = 0$, $[\tilde{e}_2] = 0$, $[\rho_2] = 0$, and $[v_{2t}] = 0$. It follows from here that the particle's parameters remain continuous when crossing the discontinuity, only its derivatives will change spasmodically.

3) *Shock wave propagating in gas*:

$$j_1 \neq 0, \quad j_2 \neq 0, \quad \Delta x \approx \lambda_*, \quad M_p = 0.$$

Since the gas density is substantially smaller than the particle's density, i.e., $\rho_{11} \ll \rho_{22}$, the particle will intersect the shock wave front without a change in the velocity and temperature; consequently,

$$[u_2] = 0, \quad [\tilde{e}_2] = 0, \quad [\rho_2] = 0, \quad [v_{2t}] = 0, \quad F^\sigma = 0, \quad Q^\sigma = 0.$$

Substituting these expressions in (7.2.37), we obtain the formulas:

$$[\rho_{11} u_1] = 0, \quad [\rho_{11} u_1^2 + P] = 0, \quad \left[H_1 + \frac{u_1^2}{2}\right] = 0, \quad [v_{1t}] = 0, \qquad (7.2.41)$$

which coincide with the relations at a shock wave in ideal gas. The equalization of the velocities and temperatures of gas and particles will take place behind the shock wave front. This region is called the *relaxation zone*.

4) *A nonstationary combined discontinuity* without the eigenpressure of the pseudogas of particles:

$$j_1 \neq 0, \quad j_2 \neq 0, \quad P_2 = 0, \quad \Delta x \approx d, \quad M_p \neq 0.$$

Assuming that $P_2 = 0$ and $e_{2c} = 0$ in (7.2.38), we can write:

$$[\rho_1 u_1] = 0, \quad [j_1 u_1 + P m_1] = F^\sigma, \quad j_1\left[H_1 + \frac{u_1^2}{2}\right] = -Q^\sigma,$$
$$\frac{dM_p}{dt} + [\rho_2 u_2] = 0, \quad \frac{d}{dt}(M_p D) + [j_2 u_2 + P m_2] = -F^\sigma,$$
$$\frac{d}{dt}\left(M_p\left(e_2 + \frac{D^2}{2}\right)\right) + j_2\left[H_2 + \frac{u_2^2}{2}\right] = Q^\sigma \tag{7.2.42}$$
$$[v_{1t}] = 0, \quad [v_{2t}] = 0, \quad H_2 = e_2 + \frac{P}{\rho_{22}}.$$

It follows from here that the nonstationary effects $\frac{dM_p}{dt} \neq 0$, $\frac{dD}{dt} \neq 0$, and $\frac{de_2}{dt} \neq 0$ arise at the expense of the nonzero mass, momentum, and energy flow rates of particles at the discontinuity surface. In the particular case in which

$$[j_2] = 0, \quad j_2[u_2] + [P m_2] = -F^\sigma, \quad j_2\left[H_2 + \frac{u_2^2}{2}\right] = Q^\sigma \tag{7.2.43}$$

the combined discontinuity will be stationary. The approximate formulas for F^σ and Q^σ were presented above in (7.2.39). To determine P^σ, we make use of the fact that a subsonic inflow into a region of an increased volume concentration of particles occurs adiabatically:

$$[S_1] = 0 \text{ at } M_{12} < 1, \quad [m_2] > 0. \tag{7.2.44}$$

At a subsonic outflow from the region of increased volume concentration of particles, an expansion of the gas stream tube and the pressure growth take place, which leads to the flow separation. In this case, the flow scheme of the Bird's type is realized:

$$P^\sigma = P^- \text{ at } M_{12} < 1, \quad [m_2] < 0. \tag{7.2.45}$$

At a supersonic outflow from the region of an increased concentration of particles, two regimes can be realized, with the choked separated zones and the nonchoked separated zones. We will neglect this difference and assume that

$$P^\sigma = P^+ \text{ at } M_{12} > 1, \quad [m_2] < 0. \tag{7.2.46}$$

Equations (7.2.44)–(7.2.46) are similar to the closure relations used for the computation of the gas flow in a channel with a cross-section jump and of the flow through porous membrane.

5) *A stationary combined discontinuity:*

$$j_1 \neq 0, \; j_2 = 0, \; P_2 \neq 0, \; \Delta x \geq d, \; M_p \neq 0, \; \frac{dM_p}{dt} = 0, \; \frac{dD}{dt} = 0, \; \frac{d\tilde{e}_2}{dt} = 0.$$

The jump conditions follow from equations (7.2.37) and (7.2.38):

$$[\rho_1 u_1] = 0, \quad [j_1 u_1 + P m_1] = F^\sigma, \quad \left[H_1 + \frac{u_1^2}{2}\right] = 0,$$
$$u_2 = 0, \quad [P_1 m_2 + P_2] = -F^\sigma, \quad Q^\sigma = 0, \tag{7.2.47}$$
$$[v_{1t}] = 0, \quad [v_{2t}], \ [\tilde{H}_2] \ \text{are arbitrary.}$$

6) *Shock wave in the pseudogas of particles*:

$$j_1 \neq 0, \quad j_2 \neq 0, \quad P_2 \neq 0, \quad \Delta x \geq d.$$

Neglecting the nonstationary items in (7.2.37) and (7.2.38), we can write the jump relations for the given case as

$$[\rho_1 u_1] = 0, \quad [j_1 u_1 + P m_1] = F^\sigma, \quad j_1\left[H_1 + \frac{u_1^2}{2}\right] = -Q^\sigma,$$
$$[\rho_2 u_2] = 0, \quad [j_2 u_2 + P m_2 + P_2] = -F^\sigma, \tag{7.2.48}$$
$$j_2\left[\tilde{H}_2 + \frac{u_2^2}{2}\right] = Q^\sigma, \quad [e_2] = 0, \quad [v_{1t}] = 0, \ [v_{2t}] = 0.$$

It follows from here that not only the particle's parameters, but also the gas parameters change spasmodically at a shock wave in a pseudogas of particles. The variation of the gas parameters is related to a jump of the volume concentration of particles $[m_1] \neq 0$ and to a nonzero gas flow rate $j_1 \neq 0$.

Using the above-obtained relations, let us find the equation for the sheet motion and the relations involving the gas parameters to the left of the sheet φ^- and to the right of the sheet φ^+. We will assume that the particles are closely packed on the sheet up to the volume concentration $m_2^* < 1$ and the gas velocity $v_1 > 0$ exceeds the sheet velocity $D > 0$, $u_1 = v_1 - D > 0$ (see Fig. 7.7). At the sheet inlet, the volume concentration of particles changes spasmodically from m_2^- to m_2^* and the jumps of the gas parameters are related to each other by the first three equations in (7.2.42):

$$[j_1] = 0, \quad j_1 = \rho_1^- u_1^- = \rho_1^* u_1^*,$$
$$j_1 u_1^* + P^* m_1^* - j_1 u_1^- - P^- m_1^- = F_-^\sigma,$$
$$j_1\left(H_1^* + \frac{(u_1^*)^2}{2} - H_1^- - \frac{(u_1^-)^2}{2}\right) = -Q_-^\sigma.$$

At the sheet exit, $[m_2] = m_2^+ - m_2^*$, and it follows from (7.2.42) that

$$[j_1] = 0, \quad j_1 = \rho_1^+ u_1^+ = \rho_1^* u_1^*,$$
$$j_1 u_1^+ + P^+ m_1^+ - j_1 u_1^* - P^* m_1^* = F_+^\sigma,$$
$$j_1\left(H_1^+ + \frac{(u_1^+)^2}{2} - H_1^* - \frac{(u_1^*)^2}{2}\right) = -Q_+^\sigma.$$

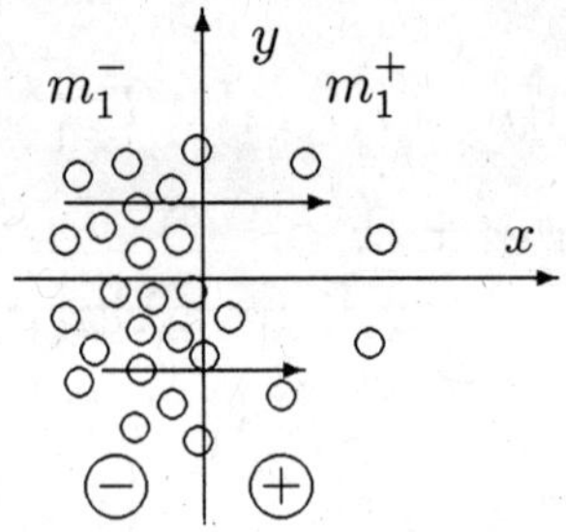

Figure 7.10: A combined discontinuity in the gas-particle mixture.

Eliminating the parameters with asterisk from these equations, we can find the jump relations for gas on the sheet (Fig. 7.7):

$$[j_1] = 0, \quad [j_1 u_1 + P m_1] = F_{12}, \quad j_1 \left[H_1 + \frac{u_1^2}{2} \right] = -Q_{12},$$
$$F_{12} = F_+^\sigma + F_-^\sigma, \quad Q_{12} = Q_1^\sigma + Q_-^\sigma, \quad [\varphi] = \varphi_+ - \varphi_-. \quad (7.2.49)$$

The equations for the sheet motion are given by the fourth, fifth, and sixth equations in (7.2.42) in which one must substitute F_{12} instead of F^σ and replace Q^σ with Q_{12}:

$$\frac{dM_p}{dt} + [j_2] = 0, \quad \frac{d}{dt}(M_p D) + [j_2 u_2 + P m_2] = -F_{12},$$
$$\frac{d}{dt}\left(M_p \left(e_2 + \frac{D^2}{2} \right) \right) + j_2 \left[H_2 + \frac{u_2^2}{2} \right] = Q_{12}, \quad j_2 = \rho_2 u_2.$$

Expressing from (7.2.49) the quantities F_{12} and Q_{12} and substituting in the given expressions, we obtain the equations for the sheet motion[15]:

$$\frac{dM_p}{dt} + [j_2] = 0, \quad \frac{d}{dt}(M_p D) + [j_2 u_2 + j_1 u_1 + P] = 0,$$
$$\frac{d}{dt}\left(M_p \left(e_2 + \frac{D^2}{2} \right) \right) + j_1 \left[H_1 + \frac{u_1^2}{2} \right] + j_2 \left[H_2 + \frac{u_2^2}{2} \right] = 0. \quad (7.2.50)$$

Problem 7.4. Neglecting the nonstationary effects, find the gas entropy jump $[S_1]$ across a combined discontinuity at $M_{12} \ll 1$, $[m_1] > 0$, and $\mathrm{Re} \gg 1$, $[m_1] \ll 1$.

Solution: Direct the x-axis along a normal to the discontinuity surface, and position the discontinuity in the plane $x = 0$ (see Fig. 7.10). We will assume that the gas flows from the left to the right, i.e., $v_1 > 0$, and we will mark the parameters to the left of the discontinuity by the minus sign and to the right of discontinuity by the plus sign. The particles velocity at a combined discontinuity is equal to zero: $v_2 = 0$. Neglecting the heat exchange between the gas and particles as well as the nonstationary terms at the discontinuity, we can rewrite the jump relations for gas (7.2.42) as

$$\begin{gathered}[j_1] = 0, \quad [j_1 v_1 + P m_1] = F^\sigma, \quad \left[H_1 + \frac{v_1^2}{2}\right] = 0,\\ j_1 = \rho_1 v_1 = \rho_{11} m_1 v_1, \quad F^\sigma = P^\sigma [m_1] - N f_p,\\ N \approx nd, \qquad D = 0, \qquad v_1 = u_1.\end{gathered} \tag{7.2.51}$$

In this case, the flow scheme of the Bird's type (7.2.45) is realized across the discontinuity: $P^\sigma = P^-$. Using the formula for a force acting on the particle (7.1.20) $f_p = C_d \frac{\pi d^2}{4} \frac{\rho_{11} v_1^2}{2}$, we obtain the estimate:

$$\frac{N f_p}{P^\sigma [m_1]} \approx \frac{n d f_p}{P^- m_2} \approx \frac{3}{4} C_d M_{12}^2 \approx \frac{3}{8} M_{12}^2 \ll 1,$$

from which it follows that

$$F^\sigma \approx P^- [m_1]. \tag{7.2.52}$$

Substituting (7.2.52) into (7.2.51), we find:

$$\begin{gathered}j_1 = \rho_{11}^- v_1^- m_1^- = \rho_{11}^+ v_1^+ m_1^+, \quad j_1 (v_1^- - v_1^+) = (P^+ - P^-) m_1^+,\\ H_1^- + \frac{(v_1^-)^2}{2} = H_1^+ + \frac{(v_1^+)^2}{2}.\end{gathered} \tag{7.2.53}$$

Introduce the gas specific volume $V = 1/\rho_{11}$, and rewrite (7.2.53) as follows:

$$\begin{gathered}j_1^2 \left(\frac{V^-}{m_1^-} - \frac{V^+}{m_1^+}\right) = (P^+ - P^-) m_1^+,\\ H_1^- - H_1^+ + \frac{j_1^2}{2} \left(\frac{V^-}{m_1^-} - \frac{V^+}{m_1^+}\right)\left(\frac{V^-}{m_1^-} + \frac{V^+}{m_1^+}\right) = 0,\\ v_1^- = \frac{j_1 V^-}{m_1^-}, \qquad v_1^+ = \frac{j_1 V^+}{m_1^+}.\end{gathered} \tag{7.2.54}$$

Eliminating the j_1^2 from the first two equations, we arrive at the equation

$$H_1^- - H_1^+ + \frac{1}{2}\left(\frac{V^-}{m_1^-} + \frac{V^+}{m_1^+}\right)(P^+ - P^-) m_1^+ = 0. \tag{7.2.55}$$

Choosing P and S_1 as the independent variables, we can write: $H_1 = H_1(P, S_1)$, $V = V(P, S_1)$. Assuming all jumps to be small, we expand $[H_1]$ into a series

$$H_1^+ - H_1^- \approx T_1^- (S_1^+ - S_1^-) + V^- (P^+ - P^-),$$

where we have taken into account the fact that $T_1 = \frac{\partial H}{\partial S_1}$, $V = \frac{\partial H_1}{\partial P}$. Substituting this expression in (7.2.55), we obtain:

$$S_1^+ - S_1^- = \frac{1}{2T_1^-}\left(V^+ + \frac{m_1^+}{m_1^-}V^- - 2V^-\right)(P^+ - P^-).$$

Assuming in this formula $V^+ \approx V^-$, we finally find:

$$[S_1] = \frac{V^-}{2T^- m_1^-}[m_1][P], \tag{7.2.56}$$

where $[\varphi] = \varphi^+ - \varphi^-$. For a subsonic flow $M_{12} < 1$, the expansion of the stream tube $[m_1] > 0$ is accompanied by the pressure growth: $[P] > 0$; therefore, $[S_1] > 0$. It can be seen that at a combined discontinuity a more intense entropy growth takes place than at a shock wave [cf. (3.3.50), where $[S] \sim [P^3]$, the $[P]$ is small]. This is related to the flow separation and the dissipation of the energy of vortices at the expense of viscosity. This effect is described integrally by the Bird's condition (7.2.45).

Problem 7.5. Obtain the jump relations, and give their classification in a mixture of two gases[8].

Solution: Let us transform the equations for a gas mixture (7.1.47) to the divergence form in the one-dimensional nonstationary case:

$$\begin{aligned}
&\frac{\partial \rho_i}{\partial t} + \frac{\partial}{\partial x}(\rho_i v_i) = 0,\\
&\frac{\partial}{\partial t}(\rho_i v_i) + \frac{\partial}{\partial x}(\rho_i v_i^2 + P_i) = f_{ji},\ i = 1,2,\ j = 2,1,\\
&\frac{\partial}{\partial t}\left(\rho_i\left(e_i + \frac{v_i^2}{2}\right)\right) + \frac{\partial}{\partial x}\left(\rho_i v_i\left(H_i + \frac{v_i^2}{2}\right)\right) = Q_{ji},\\
&H_i = e_i + \frac{P_i}{\rho_i},\quad f_{ji} = \frac{v_j - v_i}{\tau_v},\quad P_i = \rho_i R_i T_i,\\
&Q_{ji} = \frac{T_j - T_i}{\tau_T} + \alpha_{ji}\frac{(v_j - v_i)^2}{\tau_v} + \frac{(v_j - v_i)v_i}{\tau_v},\quad e_i = C_{Vi}T_i.
\end{aligned} \tag{7.2.57}$$

The thickness of the discontinuities in a mixture of gases $\Delta x \approx \lambda_*$; therefore, the gas mass at a discontinuity tends to zero, and it is necessary to set $\Theta = 0$ in (7.2.36). Directing the x-axis along a normal to the discontinuity surface, we obtain from (7.2.36) and (7.2.57) the jump relations in a mixture of gases:

$$\begin{aligned}
&[j_i] = 0,\qquad [j_i u_i + P_i] = 0,\\
&\left[H_i + \frac{u_i^2}{2}\right] = 0,\quad j_i = \rho_i u_i,\quad u_i = v_i - D,
\end{aligned} \tag{7.2.58}$$

where

$$P_i = \rho_i R_i T_i, \quad e_i = C_{vi} T_i. \tag{7.2.59}$$

In a general case in which the gas velocities are not perpendicular to the discontinuity surface, it is necessary to augment the relations (7.2.58) by the condition for the continuity of the tangential velocity components:

$$j_i [v_{it}] = 0. \tag{7.2.60}$$

The relations (7.2.58)–(7.2.60) enable us to give the following classification of discontinuities.

1) *Contact discontinuity*: $j_i = 0$, at which $[P_i] = 0$, $[u_i] = 0$ and the jumps $[\rho_i] \neq 0$, $[v_{it}] \neq 0$, and $[T_i] \neq 0$ are arbitrary.

2) *Combined discontinuity*: $j_1 = 0$, $j_2 \neq 0$, at which $[P_1] = 0$, $[u_1] = 0$, $[v_{2t}] = 0$; however, $[v_{1t}] \neq 0$, $[P_2] \neq 0$, $[u_2] \neq 0$, $[\rho_i] \neq 0$, and $[T_i] \neq 0$. Thus, a contact discontinuity in gas 1 and a shock wave in gas 2 take place. The opposite case $j_2 = 0$, $j_1 \neq 0$ is obtained by the permutation of indices 1 and 2.

3) *Shock wave in both gases*: $j_1 \neq 0$, $j_2 \neq 0$, at which the quantities v_{it} are continuous and the remaining parameters undergo the jumps. If we assume that the gas mixture is in an equilibrium with respect to velocity and temperature ahead of the shock wave front:

$$u_1^- = u_2^- = u^-, \quad T_1^- = T_2^- = T^-,$$

then we obtain from (7.2.58) and (7.2.59) the relations at the shock waves:

$$\rho_i^+ u_i^+ = \rho_i^- u^-, \quad \rho_i^+ (u_i^+)^2 + P_i^+ = \rho_i^- (u^-)^2 + P_i^-,$$
$$\frac{\gamma P_i^+}{(\gamma - 1)\rho_i^+} + \frac{(u_i^+)^2}{2} = \frac{\gamma P_i^-}{(\gamma - 1)\rho_i^-} + \frac{(u^-)^2}{2}, \quad i = 1, 2. \tag{7.2.61}$$

Eliminating the ρ_i^+, u^-, u_i^+ from (7.2.61), we obtain the equation:

$$(P_1^+ - P_1^-)(P_2^+ - P_2^-)((\gamma + 1)(P_1^+ \rho_2^- - P_2^+ \rho_1^-)$$
$$+(\gamma - 1) \cdot (P_1^- \rho_2^- - P_2^- \rho_1^-)) = 0,$$

which has the following three solutions corresponding to three types of shock waves:

1) $P_1^+ = P_1^-$;
2) $P_2^+ = P_2^-$;
3) $P_1^+ = \rho_1^- P_2^+ / \rho_2^- + (\gamma - 1)(P_2^- \rho_1^- / \rho_2^- - P_1^-)/(\gamma + 1)$.

A detailed investigation of shock waves in the mixtures of gases may be found in[8].

7.3 Quasi-One-Dimensional Flows of a Gas-Particle Mixture in Laval Nozzles

7.3.1 The Equations of the Quasi-One-Dimensional Flow of a Gas-Particle Mixture

Consider a flow of a gas-particle mixture in the Laval nozzle in the quasi-one-dimensional approximation at a small volume concentration of particles $m_2 \ll 1$ [24]. Neglecting the collisions between the particles, we will describe the flow of a gas-particle mixture with the aid of the Kliegel–Nickerson system of equations (7.2.9):

$$
\begin{aligned}
&\frac{\partial \rho_1}{\partial t} + \nabla_k \rho_1 v_1^k = 0, \quad \frac{\partial \rho_2}{\partial t} + \nabla_k \rho_2 v_2^k = 0, \quad \rho_2 = \rho_{22} m_2, \\
&\frac{d_1 \vec{v}_1}{dt} = -\frac{1}{\rho_1}\nabla P - \frac{\rho_2}{\rho_1}\frac{(\vec{v}_1 - \vec{v}_2)}{\tau_v}, \quad \frac{d_1}{dt} = \frac{\partial}{\partial t} + v_1^k \nabla_k, \\
&\frac{d_2 \vec{v}_2}{dt} = \frac{(\vec{v}_1 - \vec{v}_2)}{\tau_v}, \quad \frac{d_2}{dt} = \frac{\partial}{\partial t} + v_2^k \nabla_k, \\
&\frac{\partial}{\partial t}(\rho_1 E_1 + \rho_2 E_2) + \nabla_k \left(\rho_1 v_1^k E_1 + \rho_2 v_2^k E_2 + \rho_1 v_1^k \frac{P}{\rho_1}\right) = 0, \quad (7.3.1) \\
&E_1 = e_1 + \frac{1}{2}v_1^2, \quad E_2 = e_2 + \frac{1}{2}v_2^2, \\
&\frac{d_2 T_2}{dt} = \frac{T_1 - T_2}{\tau_T}, \quad e_1 = C_V T_1, \quad E_2 = C_s T_2, \\
&P = \rho_1 R T_1, \qquad R = C_p - C_V,
\end{aligned}
$$

where R is the universal gas constant. When writing (7.3.1), it was assumed that the regime of Stokes flow around a particle takes place: $M_{12} \ll 1$, $\mathrm{Re}_{12} = |v_1 - v_2| d/\nu \ll 1$, at which we obtain from (7.1.21) and (7.1.30):

$$
C_d = \frac{24}{\mathrm{Re}_{12}}, \qquad \mathrm{Nu} = 2. \qquad (7.3.2)
$$

Substituting (7.3.2) in the expression for the force (7.1.20) acting on a particle

$$
\vec{f}_p = \frac{3}{4} m_p \frac{\mathrm{Re}_{12}\,\mu}{\rho_{22} d^2} C_d (\vec{v}_1 - \vec{v}_2), \quad m_p = \frac{\pi d^3}{6}\rho_{22}, \qquad (7.3.3)
$$

and in the expression for a heat flux to a particle (7.1.29)

$$
q_p = \pi d \,\mathrm{Nu}\, \lambda (T_1 - T_2), \qquad (7.3.4)
$$

we write:

$$
\begin{aligned}
\frac{d_2 \vec{v}_2}{dt} &= \frac{\vec{v}_1 - \vec{v}_2}{\tau_v} = \frac{n \vec{f}_p}{\rho_2} = \frac{18\mu}{\rho_{22} d^2}(\vec{v}_1 - \vec{v}_2), \\
\frac{d_2 T_2}{dt} &= \frac{T_1 - T_2}{\tau_T} = \frac{n q_p}{C_s \rho_2} = \frac{12\lambda}{C_s \rho_{22} d^2}(T_1 - T_2).
\end{aligned}
$$

From these formulas, we obtain the expressions for the reference times of the velocity relaxation τ_v and the thermal relaxation τ_T:

$$\tau_v = \frac{\rho_{22} d^2}{18\mu}, \quad \tau_T = \frac{3}{2}\tau_v \mathrm{Pr}\,\frac{C_s}{C_p}, \quad \mathrm{Pr} = \frac{\mu C_p}{\lambda}, \tag{7.3.5}$$

where Pr is the Prandtl number.

Let a two-phase gas-particle mixture move in a variable section duct as shown in Fig. 6.1 in Section 6.1. Introducing the Cartesian coordinate system (see Fig. 6.1) we will assume that the duct crosssection varies slowly along the Ox-axis. In this case, it is natural to assume that the following inequalities are valid:

$$\begin{aligned} &v_{1y} \sim v_{1z} \ll v_{1x}, \quad v_{2y} \sim v_{2z} \ll v_{2x}, \\ &\frac{dv_{1y}}{dt} \sim \frac{dv_{1z}}{dt} \ll \frac{dv_{1x}}{dt}, \quad \frac{dv_{2y}}{dt} \sim \frac{dv_{2z}}{dt} \ll \frac{dv_{2x}}{dt}. \end{aligned} \tag{7.3.6}$$

Let us simplify the system of equations (7.3.1) by using (7.3.6):

$$\begin{aligned} &\frac{\partial \rho_1}{\partial t} + \nabla_k \rho_1 v_1^k = 0, \quad \frac{\partial \rho_2}{\partial t} + \nabla_k \rho_2 v_2^k = 0, \\ &\frac{\partial v_{1x}}{\partial t} + v_{1x}\frac{\partial v_{1x}}{\partial x} = -\frac{1}{\rho_1}\frac{\partial P}{\partial x} - \frac{\rho_2}{\rho_1}\frac{(v_{1x} - v_{2x})}{\tau_v}, \\ &\frac{\partial P}{\partial y} = 0, \quad \frac{\partial P}{\partial z} = 0, \\ &\frac{\partial v_{2x}}{\partial t} + v_{2x}\frac{\partial v_{2x}}{\partial x} = \frac{v_{1x} - v_{2x}}{\tau_v}, \\ &\frac{\partial}{\partial t}(\rho_1 E_1 + \rho_2 E_2) + \nabla_k\left(\rho_1 v_1^k\left(H_1 + \frac{v_1^2}{2}\right) + \rho_2 v_2^k\left(e_2 + \frac{v_2^2}{2}\right)\right) = 0, \\ &\frac{\partial T_2}{\partial t} + v_{2x}\frac{\partial T_2}{\partial x} = \frac{T_1 - T_2}{\tau_T}, \quad H_1 = C_p T_1, \quad e_2 = C_s T_2, \\ &P = \rho_1 R T_1, \qquad R = C_p - C_V, \end{aligned} \tag{7.3.7}$$

where

$$\nabla_k(\psi_i v_i^k) = \frac{\partial}{\partial x}(\psi_i v_{ix}) + \frac{\partial}{\partial y}(\psi_i v_{iy}) + \frac{\partial}{\partial z}(\psi_i v_{iz}),$$

$$\psi_i = \left\{\rho_1,\ \rho_2,\ \rho_1\left(H_1 + \frac{v_1^2}{2}\right),\ \rho_2\left(e_2 + \frac{v_2^2}{2}\right)\right\}, \qquad i = 1, 2.$$

It follows from the fourth and fifth equations (7.3.7) that $P = P(x,t)$. Assume that $\rho_i = \rho_i(x,t)$, $T_i = T_i(x,t)$, $v_{ix} = v_{ix}(x,t)$, $i = 1, 2$, and average the equations (7.3.7) over a duct crosssection. For this purpose, we write the equations of continuity and total energy in an unified form:

$$\frac{\partial \varphi_i}{\partial t} + \nabla_k(\psi v_i^k) = 0, \tag{7.3.8}$$

where the ψ_i was defined above and the φ_i has different forms depending on the equation: $\varphi_i = \{\rho_1,\ \rho_2,\ \rho_1 E_1 + \rho_2 E_2\}$. Multiply equation (7.3.8) by dV and integrate over a fixed volume V, then we obtain with regard for the Ostrogradsky–Gauss theorem:

$$\frac{\partial}{\partial t}\int_V \varphi_i dV + \int_S \psi_i v_{in}\, dS = 0, \tag{7.3.9}$$

where $v_{in} = \vec{v}_i \cdot \vec{n} = v_i^k n^k$, $\vec{n}$ is a vector of a unit normal to the surface S. Let us choose as V a part of the duct volume bounded by the planar crosssections ab and $a'b'$, and by the lateral duct surfaces aa' and bb'. We will assume that the ab plane is located at point x and that the plane $a'b'$ is located at an infinitesimally close point $x+\Delta x$. Then we can find under the assumption that the functions φ_i are smooth:

$$\int_V \varphi_i\, dV \approx \int_F \varphi_i\, dS\, \Delta x \approx \varphi_i\, F\, \Delta x. \tag{7.3.10}$$

Assume that the slip condition $v_{1n} = v_{2n} = 0$ is satisfied both for gas and for particles. Then

$$\begin{aligned}\int_S \psi_i\, v_{in}\, dS &= \int_{F(x+\Delta x)} \psi_i v_{ix} dS - \int_{F(x)} \psi_i v_{ix}\, dS \\ &\approx F\, \psi_i v_{ix}\Big|_{x+\Delta x} - F\, \psi_i v_{ix}\Big|_x \approx \frac{\partial}{\partial x}(F \psi_i\, v_{ix}\, \Delta x).\end{aligned} \tag{7.3.11}$$

Substituting (7.3.10) and (7.3.11) into (7.3.9), we obtain the equation for the quasi-one-dimensional flow:

$$\frac{\partial}{\partial t}(\varphi_i F) + \frac{\partial}{\partial x}(F\psi_i v_{ix}) = 0. \tag{7.3.12}$$

It follows from here that the averaged equations of continuity and total energy in (7.3.7) may be written as

$$\begin{aligned}&\frac{\partial}{\partial t}(\rho_1 F) + \frac{\partial}{\partial x}(\rho_1 F v_{1x}) = 0, \quad \frac{\partial}{\partial t}(\rho_2 F) + \frac{\partial}{\partial x}(\rho_2 F v_{2x}) = 0, \\ &\frac{\partial}{\partial t}(\rho_1 F E_1 + \rho_2 F E_2) + \frac{\partial}{\partial x}\Big(\rho_1 F v_{1x}\Big(C_p T_1 + \frac{v_1^2}{2}\Big) \\ &+ \rho_2 F v_{2x}\Big(C_s T_2 + \frac{v_2^2}{2}\Big)\Big) = 0,\end{aligned} \tag{7.3.13}$$

and the equations of particles motion and energy do not depend on y and z; therefore, they do not change at the averaging. Omitting the subscript x by the velocities v_{1x} and v_{2x}, we write the system of equations (7.3.7) and (7.3.13) in the stationary case:

$$
\begin{aligned}
&\frac{d}{dx}(\rho_1 v_1 F) = 0, \quad \frac{d}{dx}(\rho_2 v_2 F) = 0,\\
&\rho_1 v_1 \frac{dv_1}{dx} + \rho_2 v_2 \frac{dv_2}{dx} + \frac{dP}{dx} = 0,\\
&v_2 \frac{dv_2}{dx} = \frac{v_1 - v_2}{\tau_v},\\
&\frac{d}{dx}\Big(\rho_1 F v_1 \Big(C_p T_1 + \frac{v_1^2}{2}\Big) + \rho_2 F v_2 \Big(C_s T_2 + \frac{v_2^2}{2}\Big)\Big) = 0,\\
&v_2 \frac{dT_2}{dx} = \frac{T_1 - T_2}{\tau_T},\\
&P = \rho_1 R T_1, \qquad C_p - C_V = R.
\end{aligned}
\tag{7.3.14}
$$

Integrating the equations of continuity and total energy, let us rewrite (7.3.14) as

$$
\begin{aligned}
&\rho_1 v_1 F = Q_1, \quad \rho_2 v_2 F = Q_2 = \kappa Q_1,\\
&C_p T_1 + \frac{v_1^2}{2} + \kappa\Big(C_s T_2 + \frac{v_2^2}{2}\Big) = B,\\
&\rho_1 v_1 \frac{dv_1}{dx} + \rho_2 v_2 \frac{dv_2}{dx} + \frac{dP}{dx} = 0, \quad P = \rho_1 R T_1,\\
&v_2 \frac{dv_2}{dx} = \frac{v_1 - v_2}{\tau_v}, \qquad v_2 \frac{dT_2}{dx} = \frac{T_1 - T_2}{\tau_T},
\end{aligned}
\tag{7.3.15}
$$

where Q_1, Q_2, B, and κ are the integration constants, which do not depend on x. The quantity Q_1 specifies the gas flow rate in the duct, and Q_2 is the flow rate of particles, in which $\kappa = Q_2/Q_1$ is their ratio. Eliminating the quantities ρ_1, ρ_2, P, T_1 from the set of variables, we will arrive at a system of three equations:

$$
\begin{aligned}
\frac{d\ln v_1}{dx} &= \frac{1}{M^2 - 1}\frac{d\ln F}{dx}\\
&+ \frac{\kappa}{M^2 - 1}\Big(\frac{M^2}{v_1}\Big((\gamma - 1)\frac{v_2}{v_1} - \gamma\Big)\frac{dv_2}{dx} + \frac{C_s}{C_p}\frac{1}{T_1}\frac{dT_2}{dx}\Big),
\end{aligned}
\tag{7.3.16}
$$

$$
\begin{aligned}
\gamma &= \frac{C_p}{C_V}, \quad M^2 = \frac{v_1^2}{c_1^2} = \frac{v_1^2}{\gamma R T_1},\\
v_2 \frac{dv_2}{dx} &= \frac{v_1 - v_2}{\tau_v}, \quad v_2 \frac{dT_2}{dx} = \frac{T_1 - T_2}{\tau_T},
\end{aligned}
$$

where T_1 in the last equation is determined from the third equation in (7.3.15).

Let us apply the system of equations (7.3.16) to the description of the flow of a gas-particle mixture in the Laval nozzle consisting of a

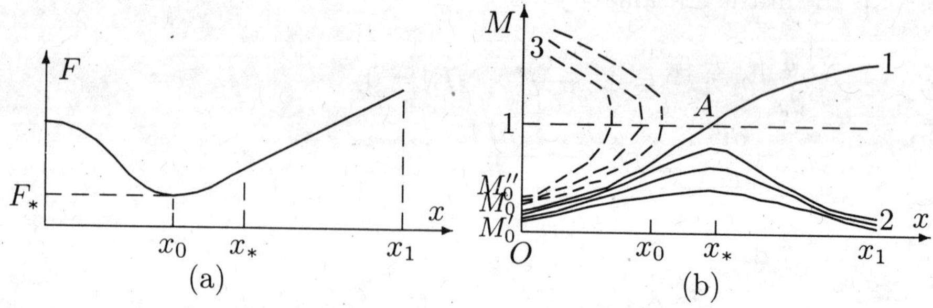

Figure 7.11: (a) the Laval nozzle and (b) the dependence of the Mach number on the x-coordinate in the Laval nozzle.

converging/diverging tube [see Fig. 7.11 (a)]. As in Section 6.1, it is convenient to introduce instead of the velocity v_1 the Mach number $M = v_1/\sqrt{\gamma RT_1}$ as the variable. Then we obtain instead of (7.3.16):

$$\frac{d\ln M}{dx} = \frac{\left(1+\frac{\gamma-1}{2}M^2\right)}{(M^2-1)}\left\{\frac{1}{F}\frac{dF}{dx} + \frac{(\gamma-1)\kappa}{\gamma RT_1}\cdot\left(\frac{1+\gamma M^2}{2+(\gamma-1)M^2}\right)\cdot\left(C_s\frac{dT_2}{dx} + v_2\frac{dv_2}{dx}\right) - \frac{\kappa v_1}{RT_1}\frac{dv_2}{dx}\right\}, \quad (7.3.17)$$

$$v_2\frac{dv_2}{dx} = \frac{v_1 - v_2}{\tau_v}, \qquad v_2\frac{dT_2}{dx} = \frac{T_1 - T_2}{\tau_T}.$$

In order to determine from (7.3.17) the functions $M(x)$, $v_2(x)$, $T_2(x)$, it is necessary to specify the initial conditions

$$M|_{x=0} = M^0, \quad v_2|_{x=0} = v_2^0, \quad T_2|_{x=0} = T_2^0,$$

and $v_1|_{x=0} \geq v_2|_{x=0}$ and $T_1|_{x=0} \geq T_2|_{x=0}$. It is easy to see that, if there are no particles, i.e., $\kappa = 0$, then the first equation in (7.3.17) can be integrated. As a result, formula (6.1.12) is obtained. It follows from (6.1.12) that for the ideal gas (without particles) the condition $M = 1$ is satisfied in the minimal section $F_* = F(x_0)$.

In the case of a gas-particle mixture, $\kappa \neq 0$ and the transition point A at which $M = 1$ is shifted to the diverging nozzle part $x_* > x_0$. Let us prove this assertion. Turn to the first equation in (7.3.17), from which it follows that at $M = 1$ the denominator on the right-hand side vanishes; therefore, the expression in the curly brackets should also vanish. We at first note that the particles at their motion always lag behind the gas $v_1 > v_2$; therefore, we obtain from the second equation of (7.3.17) that $\frac{dv_2}{dx} > 0$. Thus, the third equation in the curly brackets will always be negative; in addition, its modulus is larger than the second item modulus. Consequently, in order for the expression in curly brackets to vanish, it is necessary to require that the first item be positive, which is possible only in the diverging nozzle part $\frac{dF}{dx} > 0$. The point A, where $M = 1$, $x = x_*$ is a singular saddle-like point of the differential equation (7.3.17), and this conditions the complexity of the integration of this equation at a given nozzle shape $F(x)$. (It is assumed that the $v_2(x)$ and $T_2(x)$ are determined as the functions of the x-coordinate.) In order to get to the supersonic branch of the solution 1 from the initial state, one must know exactly the initial value of the Mach number M_0. A small deviation of the Mach number toward the lesser values M_0' will lead to the fact that the solution proceeds along the subsonic branch 2, and the deviation towards a larger value M_0'' leads to a nonphysical branch 3 [see Fig. 7.11 (b)]. Note that a similar problem also takes place at the ideal gas flow in a nozzle. To surmount this problem, one usually solves in practice an inverse problem: on the basis of a given law of the variation of some function of the coordinate, for example, the velocity $v_1(x)$, one determines the corresponding nozzle shape $F(x)$. If the nozzle was specified initially, then one can fit the nozzle to a given one by solving a number of such problems.

7.3.2 The Flow of a Gas-Particle Mixture in the Laval Nozzle with Small Velocity and Temperature Lags of Particles

Consider the flows of gas-particle mixtures in the Laval nozzle when only small lags of the velocity v_2 and temperature T_2 of particles from the velocity v_1 and temperature T_1 of gas take place. Similarly to[24], we introduce a small parameter $\varepsilon = c_{*e}\tau_v$ and will assume that

$$v_1 - v_2 = \varepsilon v' \ll v_1, \quad T_1 - T_2 = \varepsilon T' \ll T_1. \tag{7.3.18}$$

Expand the functions sought for in a series in powers of ε near the equilibrium solution, which will be marked below by subscript "e." The quantity c_{*e} entering the definition of parameter ε is the critical sound velocity of equilibrium flow $v_{*e} = c_{*e}$. Restricting to the first-order terms in ε, we write:

$$v_1 = v_e + \varepsilon v_1', \quad T_1 = T_e + \varepsilon T_1', \quad \rho_1 = \rho_{1e} + \varepsilon \rho_1', \quad P = P_e + \varepsilon P', \tag{7.3.19}$$

where the functions marked by subscript "e" and the prime $\varphi' = \frac{\partial \varphi}{\partial \varepsilon}|_{\varepsilon=0}$ are to be determined. Note that in the equilibrium state

$$v_1 = v_2 = v_e, \qquad T_1 = T_2 = T_e. \tag{7.3.20}$$

Using (7.3.18) and (7.3.19), let us present the velocity and temperature of particles in the form

$$v_2 = v_e + \varepsilon(v_1' - v'), \quad T_2 = T_e + \varepsilon(T_1' - T'). \tag{7.3.21}$$

Since $\tau_v \sim d^2$ in accordance with (7.3.5), the flows close to the equilibrium ones will take place for sufficiently small particles. Substituting (7.3.18), (7.3.19), and (7.3.21) in (7.3.15), deleting the terms of order ε^2, and equalling separately the coefficients affecting ε^0 and ε^1, we obtain the *equations for the equilibrium variables*:

$$\rho_e v_e \frac{dv_e}{dx} + \frac{dP_e}{dx} = 0, \quad P_e = \rho_e R_e T_e,$$
$$C_{pe} T_e + \frac{v_e^2}{2} + B_e, \quad \rho_e v_e F = Q_e, \tag{7.3.22}$$

where

$$\rho_e = \rho_{1e}(1+\kappa), \; R_e = \frac{R}{1+\kappa}, \; C_{pe} = \frac{C_p + \kappa C_s}{1+\kappa}, \quad B_e = \frac{B}{1+\kappa},$$
$$C_{ve} = C_{pe} - R_e, \quad \gamma_e = \frac{C_{pe}}{C_{Ve}} = \gamma\left(\frac{1+\kappa C_s/C_p}{1+\gamma\kappa C_s/C_p}\right), \tag{7.3.23}$$

as well as the *equations for perturbations*:

$$\frac{\rho_1'}{\rho_{1e}} + \frac{v_1'}{v_e} = 0, \quad \frac{P'}{P_e} = \frac{T_1'}{T_e} + \frac{\rho_1'}{\rho_{1e}},$$
$$C_{pe} T_1' + v_e v_1' = \frac{\kappa}{(1+\kappa)}(v_e v' + C_s T'),$$
$$\rho_e v_e \left(\frac{dv_1'}{dx} - \frac{\kappa}{(1+\kappa)}\frac{dv'}{dx}\right) + \frac{dP'}{dx} = 0, \tag{7.3.24}$$
$$v_e \frac{dv_e}{dx} = c_{*e} v', \quad v_e \frac{dT_e}{dx} = \frac{\tau_v}{\tau_T} c_{*e} T'.$$

Equations (7.3.22) coincide with the corresponding equations for ideal gas in Section 6.1.2, if we replace in the latter equations $C_p \to C_{pe}$, $\gamma \to \gamma_e$; therefore, the solution of (7.3.22) will be given by formulas (6.1.11) and (6.1.12), if one makes in the latter equations the above substitution. In particular, for the equilibrium pseudogas, the Mach number is equal to $M_e = v/\sqrt{\gamma_e R_e T}$, and for the ideal gas, $M = v/\sqrt{\gamma R T}$, from where we find with regard for (7.3.23) the relation

$$\frac{M_e}{M} = \sqrt{(1+\kappa)\left(\frac{1+\gamma\kappa C_s/C_p}{1+\kappa C_s/C_p}\right)} > 1. \tag{7.3.25}$$

According to (6.1.12), in the minimal section $F_* = F(x_0)$ [see Fig. 7.11 (a)], the Mach number of pseudogas $M_e = 1$. We obtain from formula (7.3.25) that the gas flow in this section $x = x_0$ is subsonic, $M < 1$, and a passage through the sound velocity in gas takes place in the supersonic nozzle part $M(x_*) = 1$, $x_* > x_0$.

It is convenient to go in this section from the Mach number $M = \frac{v}{c}$ to the velocity coefficient $\lambda = v/c_*$, where c_* is the critical sound velocity determined from the condition $M = 1$; that is, $v = c = c_*$. Assuming that $M = 1$ in (6.1.8), we obtain the formula:

$$c_* = \sqrt{\frac{2}{\gamma + 1}}c_0, \tag{7.3.26}$$

with regard for which

$$M_0^2 = \frac{v^2}{c_0^2} = \frac{v^2}{c_*^2}\frac{c_*^2}{c_0^2} = \lambda^2\frac{2}{\gamma + 1}.$$

On the other hand, we can write with the use of (6.1.8):

$$M^2 = \frac{v^2}{c^2} = \frac{v^2}{c_0^2} \cdot \frac{c_0^2}{c^2} = M_0^2\left(1 + \frac{\gamma - 1}{2}M^2\right).$$

Expressing from here M_0^2 and substituting in the foregoing formula, we find a relation between λ and M:

$$M^2 = \frac{2\lambda^2}{\gamma + 1}/\left(1 - \left(\frac{\gamma - 1}{\gamma + 1}\right)\lambda^2\right). \tag{7.3.27}$$

Combining this formula with (6.1.12), we can find the dependence of F/F_* on λ:

$$\frac{F}{F_*} = \frac{\left(\frac{2}{\gamma+1}\right)^{\frac{1}{\gamma-1}}\frac{1}{\lambda}}{\left(1 - \left(\frac{\gamma-1}{\gamma+1}\right)\lambda^2\right)^{\frac{1}{\gamma-1}}}. \tag{7.3.28}$$

Note that a qualitative dependence of F/F_* on λ is similar to the dependence of F/F_* on M as shown in Fig. 6.4. The only difference is that the Mach number varies in the interval $0 \le M \le \infty$, whereas the velocity coefficient varies in the interval $0 \le \lambda \le \sqrt{\frac{\gamma+1}{\gamma-1}}$. Proceeding to the equilibrium pseudogas, we must substitute γ in formulas (7.3.26) for γ_e, so that

$$\lambda = \frac{v_e}{c_{*e}}, \quad c_{*e} = \sqrt{\frac{2}{\gamma_e + 1}}c_0, \quad c_0 = \sqrt{\gamma_e R_e T_0}, \tag{7.3.29}$$

where T_0 is the stagnation temperature for the pseudogas ($v_e = 0$).

An important characteristic of the flow escaping through the Laval nozzle from a high-pressure chamber is its specific impulse

$$I = F(P + \rho_1 v_1^2 + \rho_2 v_2^2)/(Q_1 + Q_2). \tag{7.3.30}$$

Assuming that here $v_1 = v_2 = v_e$ and $I = I_e$ and using the formulas

$$Q_1 + Q_2 = \rho_1 v_e F(1+\kappa), \quad v_e/c_{*e} = \lambda, \quad P = \rho_e R_e T = \rho_1 RT$$

rewrite (5.3.30) in the form

$$I_e = \sqrt{\frac{(\gamma_e+1)RT_0}{2\gamma_e(1+\kappa)}}\Big(\lambda + \frac{1}{\lambda}\Big). \tag{7.3.31}$$

It follows from here that the presence of particles leads to the diminution, approximately by a factor of $\sqrt{1+\kappa}$, of the specific impulse of a two-phase equilibrium jet as compared to the ideal gas.

Let us now go over to the calculation of the *first-order terms* in formulas (7.3.19) and (7.3.21), which are related to the velocity and temperature lag of particles. The corresponding coefficients in these terms are found from the system of equations (7.3.24). Substituting $v_e = \lambda c_{*e}$ in the last two equations of (7.3.24), we find:

$$v' = c_{*e}\lambda\frac{d\lambda}{dx}, \quad T' = \lambda\frac{\tau_T}{\tau_v}\frac{dT_e}{dx}. \tag{7.3.32}$$

Let us express $\frac{dT_e}{dx}$ in terms of $\frac{d\lambda}{dx}$. Substituting (7.3.27) in (6.1.7) and replacing $T \to T_e$, $\gamma \to \gamma_e$, we obtain:

$$\frac{T_e}{T_0} = 1 - \Big(\frac{\gamma_e - 1}{\gamma_e + 1}\Big)\lambda^2. \tag{7.3.33}$$

Differentiating this equation with respect to x and substituting in the second equation of (7.3.32), we will have with regard for (7.3.29):

$$v' = c_{*e}\lambda\frac{d\lambda}{dx}, \quad T' = -\frac{\tau_T}{\tau_v}\cdot\frac{c_{*e}^2}{C_{pe}}\lambda^2\frac{d\lambda}{dx}, \tag{7.3.34}$$

where $\lambda(x)$ is a known function, which is determined from equations (7.3.28). Differentiating (7.3.28), we find the derivative:

$$\frac{d\lambda}{dx} = \frac{d\lambda}{dF}\cdot\frac{dF}{dx} = \frac{\lambda\Big(1 - \frac{\gamma_e - 1}{\gamma_e + 1}\lambda^2\Big)}{F(\lambda^2 - 1)}\frac{dF}{dx}, \tag{7.3.35}$$

which enters the right-hand sides of (7.3.34).

Let us write the third and fourth equations of (7.3.24) in the differential form:

$$v_e\, dv_1' = \frac{\kappa}{1+\kappa}d(v_e v' + C_s T') - v_1' dv_e - C_{pe} dT_1',$$

$$v_e\, dv_1' = \frac{\kappa v_e}{1+\kappa}dv' - \frac{dP'}{\rho_e}.$$

Equalling the right-hand sides of these equations, we obtain:

$$C_{pe}dT_1' = \frac{dP'}{\rho_e} + \frac{\kappa}{1+\kappa}(v'dv_e + C_s dT') - v_1'dv_e. \tag{7.3.36}$$

Using the first two equations in (7.3.24) and formulas (7.3.22)

$$C_{pe}dT + d\Big(\frac{v_e^2}{2}\Big) = 0, \quad \frac{dT_e}{T_e} = \frac{\gamma_e - 1}{\gamma_e}\frac{dP_e}{P_e},$$

we find:

$$v_1'dv_e = C_{pe}T_e\Big(T_1'd\Big(\frac{1}{T_e}\Big) - \frac{(\gamma_e-1)}{\gamma_e}P'd\Big(\frac{1}{P_e}\Big)\Big).$$

Substituting this expression in (7.3.36), we obtain with regard for formula $P_e = \frac{(\gamma_e-1)}{\gamma_e}C_p\rho_e T_e$:

$$\frac{dT_1'}{T_e} + T_1'd\Big(\frac{1}{T_e}\Big) = \Big(\frac{dP'}{P_e} + P'd\Big(\frac{1}{P_e}\Big)\Big)\Big(\frac{\gamma_e-1}{\gamma_e}\Big) + \frac{\kappa}{1+\kappa}\Big(\frac{v'dv_e + C_s dT'}{C_{pe}T_e}\Big).$$

Integrating this expression, we find:

$$\frac{T_1'}{T_e} = \frac{(\gamma_e-1)}{\gamma_e}\frac{P'}{P_e} + \frac{\kappa}{1+\kappa}\int \frac{v'dv_e + C_s dT'}{C_{pe}T_e} + C_1,$$

where C_1 is the integration constant. Substituting here the expressions (7.3.34) instead of v' and T' and integrating by parts, we arrive at the equation[24]

$$\frac{T_1'}{T_e} = \frac{(\gamma_e-1)}{\gamma_e}\frac{P'}{P_e} + \frac{\kappa}{1+\kappa}\Big(\zeta(\lambda) - \frac{2\alpha\omega\lambda^2}{T(\lambda)}\frac{d\lambda}{dx} - \xi\Big), \tag{7.3.37}$$

where

$$\zeta(\lambda) = 2\omega\int_{\lambda(0)}^{\lambda} \frac{\Phi(\lambda)\lambda\, d\lambda}{\big(\frac{dx}{d\lambda}\big)T^2(\lambda)}, \quad \omega = \frac{\gamma_e-1}{\gamma_e+1}, \quad \Phi(\lambda) = 1 - \omega(1-2\alpha^2)\lambda^2,$$

$$\alpha = \frac{C_s}{C_{pe}}\frac{\tau_T}{\tau_v}, \quad T(\lambda) = 1 - \omega\lambda^2, \quad \lambda(0) = \lambda|_{x=0},$$

and ξ is the integration constant, which will be determined below. Substituting (7.3.34) in the right-hand side of the third equation in (7.3.24), we can rewrite the first three equations (7.3.24) as

$$\frac{v_1'}{v_e} + \frac{\rho_1'}{\rho_{1e}} = 0, \quad \frac{P'}{P_e} - \frac{T_1'}{T_e} - \frac{\rho_1'}{\rho_{1e}} = 0,$$
$$\frac{C_{pe}T_1'}{v_e^2} + \frac{v_1'}{v_e} = \frac{\kappa}{1+\kappa}(1-\alpha)\frac{d\lambda}{dx}. \tag{7.3.38}$$

Thus, we have obtained a system of four linear equations (7.3.37) and (7.3.38) for four unknowns, the solution of which gives

$$\begin{aligned}
\frac{P'}{P_e} &= \frac{\kappa}{(1-\lambda^2)(1+\kappa)}\left(\frac{2\gamma_e\lambda^2}{(\gamma_e+1)}\left(\frac{1+\omega(2\alpha-1)\lambda^2}{1-\omega\lambda^2}\right)\frac{d\lambda}{dx}\right.\\
&\qquad\left.-\frac{\gamma_e}{(\gamma_e-1)}(\zeta-\xi)(1+\omega\lambda^2)\right),\\
\frac{T_1'}{T_e} &= \frac{2\omega\lambda^2\kappa}{(1-\lambda^2)(1+\kappa)}\left((1-\alpha+\omega(2\alpha-1)\lambda^2+\alpha\lambda^2)\frac{1}{T(\lambda)}\frac{d\lambda}{dx}\right.\\
&\qquad\left.-\frac{\gamma_e}{(\gamma_e-1)}(\zeta-\xi)\right), \qquad (7.3.39)\\
\frac{v_1'}{v_e} &= \frac{\lambda^2\kappa}{(1-\lambda^2)(1+\kappa)}\left((\gamma-2\alpha\omega-1)\frac{d\lambda}{dx}+\frac{T(\lambda)}{\lambda^2}\frac{\gamma_e}{(\gamma_e-1)}(\zeta-\xi)\right),\\
\frac{\rho_1'}{\rho_{1e}} &= -\frac{v_1'}{v_e},
\end{aligned}$$

where $\lambda(x)$ and $\frac{d\lambda}{dx}(x)$ are the known functions of x, which are determined from (7.3.28) and (7.3.35). Note that the factor $1-\lambda^2$ stands in the denominator of formulas (7.3.39), which vanishes at a critical point $\lambda = 1$. Since all quantities with primes should be finite at this point $\lambda = 1$ the numerator should also vanish. Equalling the expressions in the parentheses in these numerators to zero and assuming $\lambda = 1$, we find the constant:

$$\xi = \zeta_* - \frac{(\gamma_e-1)}{\gamma_e}\Big(1+\alpha(\gamma_e-1)\Big)\left(\frac{d\lambda}{dx}\right)_*. \qquad (7.3.40)$$

The formulas (7.3.18), (7.3.19), (7.3.28), (7.3.29), (7.3.34), (7.3.39), and (7.3.40) determine the flow of a gas-particle mixture in the Laval nozzle with a small lag of particles in the quasi-one-dimensional approximation. It follows from these formulas that the initial values of the flow parameters at the nozzle inlet are different from their equilibrium values. If there is a break in the nozzle contour, then the derivative $\frac{d\lambda}{dx}$ and all flow parameters experience a discontinuity. In the nonlinear theory, all parameters will be continuous, and their derivatives will undergo the jumps.

Note that it was assumed at the derivation of the equations (7.3.14) governing the quasi-one-dimensional flow of a gas-particle mixture that the normal component of the particles velocity at the nozzle contour is equal to zero: $v_{2n} = 0$. In real flows, this condition can be violated. We show in Fig. 7.12 a supersonic nozzle part in which a two-phase gas-particle mixture flows from the left to the right. At point A, a separation of particles from the nozzle wall takes place, $v_{2n} > 0$ at this point, and to the right of point B the particles form a sediment on the nozzle wall. Therefore, it is necessary to formulate on the wall the condition $v_{2n} < 0$.

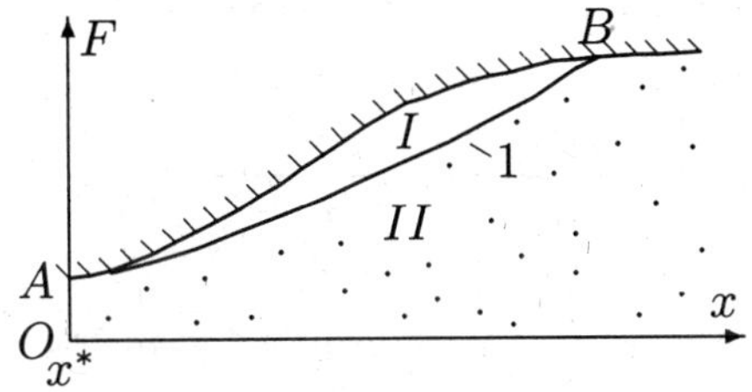

Figure 7.12: The two-phase jet in supersonic nozzle part.

A pure gas flows in region I, and a gas-particle mixture flows in region II. The boundary 1 separating these regions coincides with the streamline of particles beginning at point A and ending at point B.

Problem 7.6. Find the equation for the nozzle contours, which correspond to the two-phase flows with a constant particle's lag[24,25]. Determine the ratio I_k/I_e.

Solution: The constant particles lag means that

$$v_2 = kv_1, \quad T_2 = T_0 + l(T_1 - T_0), \tag{7.3.41}$$

where T_0 is the stagnation temperature of the flow ($\lambda = M = 0$) and k and l are the constant coefficients characterizing the velocity and temperature lag of particles. The quantities $k = 1$ and $l = 1$ correspond to the equilibrium flow, $k = 0$ and $l = 0$ correspond to the frozen flow, and in the general case, $0 < k < 1$, $0 < l < 1$. Differentiating the third equation in (7.3.15) for the total energy, we write:

$$C_p dT_1 + v_1 dv_1 + \kappa C_s dT_2 + v_2 dv_2 = 0.$$

Substituting here the expressions (7.3.41), we obtain at a constant particles lag:

$$v_1 dv_1 = -\frac{C_p + \kappa C_s l}{1 + \kappa k^2} dT_1.$$

From the total momentum equation (7.3.15), we find at $v_2 = kv_1$:

$$v_1 dv_1 = -\frac{dP}{\rho_1(1 + \kappa k)}.$$

Equalling the right-hand sides of these equations, we obtain:

$$\frac{dP}{\rho_1(1 + \kappa k)} = \frac{C_p + \kappa C_s l}{1 + \kappa k^2} dT_1.$$

Substituting and integrating here the relation $dP = \rho_1 R\, dT_1 + RT_1\, d\rho_1$, we find:

$$P = A\rho_1^{\gamma_k}, \quad \gamma_k = \frac{\gamma}{\left(\gamma - (\gamma-1)\left(\frac{1+\kappa k^2}{1+\kappa k}\right)\frac{1}{1+\kappa C_s C_p^{-1} l}\right)}, \tag{7.3.42}$$

where A is the integration constant. Assuming that $v_2 = kv_1$, $v_k = v_1$, $P_k = P$, $dT_2 = e\, dT_1$, and $T_k = T_1$ in equations (7.3.15) for the total momentum and energy, we can rewrite these equations in the form

$$\begin{gathered} \rho_k v_k \frac{dv_k}{dx} + \frac{dP_k}{dx} = 0, \quad \rho_k = \rho_1(1+\kappa k), \\ C_{pk}\frac{dT_k}{dx} + v_k\frac{dv_k}{dx} = 0, \quad C_{pk} = \frac{C_p + \kappa C_s l}{1+\kappa k^2}. \end{gathered} \tag{7.3.43}$$

By virtue of equality $P_k = P$, we have $\rho_k R_k T_k = \rho_1 R T_1$, from where we obtain with regard for $T_k = T_1$:

$$R_k = \frac{R}{1+\kappa k}. \tag{7.3.44}$$

It follows from equations (7.3.42)–(7.3.44) that the *flow of pseudogas (of two-phase flow with constant particles lag) obeys the same equations as the flow of usual ideal gas.*

Let us find the pseudogas flow rate $Q_k = Q_1 + Q_2 = (\rho_1 v_1 + \rho_2 v_2)F = \rho_1 v_1(1+\kappa k)F = \rho_k v_k F$; thus

$$Q_k = (1+\kappa k)Q_1. \tag{7.3.45}$$

Substituting $v_2 = kv_1$ in equation for particles motion [the second to last equation in (7.3.15)], we arrive at the equation

$$\frac{dv_1}{dx} = \frac{1}{k\tau_v}\left(\frac{1}{k} - 1\right),$$

the integral of which under the initial condition $v_1|_{x=0} = 0$ is

$$v_1 = \frac{x}{k\tau_v}\left(\frac{1}{k} - 1\right). \tag{7.3.46}$$

Substituting this expression in the equation for the pseudogas energy [the third equation in (7.3.43)] at $v_k = v_1$, $T_k = T_1$) and integrating it under the initial condition $T_1|_{x=0} = T_0$, we obtain the distribution of gas temperature:

$$T_1 = T_0 - \frac{x^2}{2\tau_v C_{pk}}\left(\frac{1}{k} - 1\right)^2 \tag{7.3.47}$$

and of particles

$$T_2 = T_0 + l(T_1 - T_0).$$

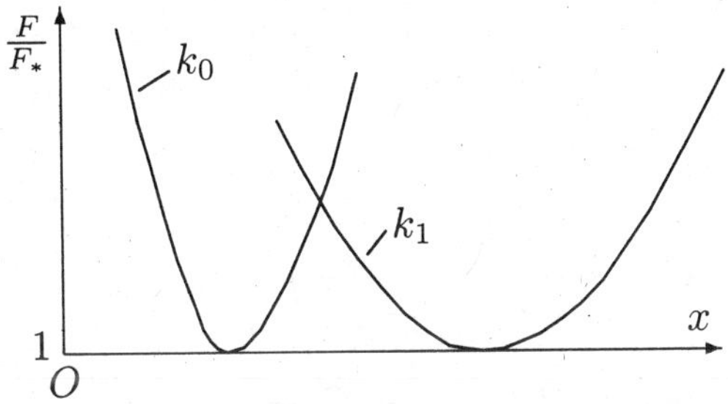

Figure 7.13: The nozzle contours at two values of the lag constants k_0 and k_1, where $k_0 < k_1$.

On the other hand, the temperatures of gas and particles are related to one another by the equation for particles energy [the last equation in (7.3.15)]. Substituting the found temperatures $T_1(x)$ (7.3.47) and $T_2(x)$ in the last equation of (7.3.15), we find that it will be satisfied only under the condition:

$$\frac{1}{l} = 1 + 2\frac{\tau_T}{\tau_v}\left(\frac{1}{k} - 1\right). \qquad (7.3.48)$$

Thus, the coefficient of the temperature lag l is uniquely related to the coefficient of the velocity lag k, and the constructed solution (7.3.42)–(7.3.48) is oneparametric and depending on the only parameter k. The solution of equations (7.3.42) and (7.3.43) coincides with the solution (6.1.12) for ideal gas, which was obtained in Section 6.1. It is more convenient, however, to introduce instead of the Mach number the velocity coefficient $\lambda_k = v_1/c_{*k}$ and to write the solution in the form (7.3.28):

$$\frac{F_*}{F(x)} = \left(\frac{\gamma_k + 1}{2}\right)^{\frac{1}{(\gamma_k - 1)}} \lambda_k \left(1 - \left(\frac{\gamma_k - 1}{\gamma_k + 1}\right)\lambda_k^2\right)^{\frac{1}{\gamma_k - 1}}. \qquad (7.3.49)$$

Using formula (7.3.46), we can find the relation between the velocity coefficient λ_k and the coordinate x:

$$\lambda_k = \sqrt{\frac{(\gamma_k + 1)}{2\gamma_k}\frac{(1 + xk)}{RT_0}\frac{x}{k\tau_v}\left(\frac{1}{k} - 1\right)}, \qquad (7.3.50)$$

where we have used a formula for the sound velocity of pseudogas $c_{*k} = \sqrt{\frac{2\gamma_k}{\gamma_k+1}R_k T_0}$ as well as formula (7.3.44). We present in Fig. 7.13 a qualitative picture of the behavior of the solution $F(x)/F_*$ depending on the x-coordinate at two values of $k_0 < k_1$, which corresponds to equations (7.3.49) and (7.3.50). It is seen that with increasing lag (reducing k) the nozzle contour becomes steeper.

Let us determine the specific impulse of pseudogas I_k with a constant lag k. Substituting the relations $v_1 = v_k$, $v_2 = kv_1 = kv_k$, $P = P_k = \rho_k R_k T_1$, $\lambda_k = v_k/c_{*k}$, and formulas (7.3.44) and (7.3.45) in formula (7.3.30), we obtain:

$$I_k = \sqrt{\frac{(\gamma_k+1)RT_0}{2\gamma_k}\frac{\sqrt{1+\kappa k}}{1+\kappa}}\left(\lambda_k + \frac{1}{\lambda_k}\right). \qquad (7.3.51)$$

Dividing this expression by (7.3.31), we find:

$$\frac{I_k}{I_e} = \sqrt{\frac{(1+\kappa k)(\gamma_k+1)\gamma_e}{(1+\kappa)(\gamma_e+1)\gamma_k}}\left(\lambda_k + \frac{1}{\lambda_k}\right)/\left(\lambda + \frac{1}{\lambda}\right). \qquad (7.3.52)$$

It can be seen from here that, with increasing the velocity nonequilibrium (reducing k), the ratio of the specific impulse I_k to the equilibrium specific impulse I_e reduces.

7.4 The Continual-Discrete Model and Caustics in the Pseudogas of Particles

7.4.1 The Equations of the Continual-Discrete Model of a Gas-Particle Mixture at a Small Volume Concentration of Particles

At a small volume concentration of particles $m_2 \ll 1$, when there are no collisions between the particles (Kn $\gg 1$), it is necessary to describe a pseudogas of particles by the collisionless kinetic equation[8,17]. The particles are characterized by a one-particle distribution function $f(t, x^k, v_2^k, T_2, r_p)$ in the phase space t, x^k, v_2^k, T_2, r_p, $k = 1, 2, 3$. (Note that the particle's temperature T_2 here is a dynamic variable.) The particle's concentration n, volume concentration m_2, mean velocity of particles $\langle\vec{v}_2\rangle$, and temperature $\langle T_2\rangle$ are found by the formulas

$$\begin{aligned} n &= \int f\,dV_p, \qquad m_2 = \int \frac{4}{3}\pi r_p^3 f\,dV_p, \\ \langle\vec{v}_2\rangle &= \frac{1}{n}\int \vec{v}_2 f\,dV_p, \quad \langle T_2\rangle = \frac{1}{n}\int T_2\,dV_p, \\ \vec{v}_2 &= v_2^k\vec{e}_k, \qquad dV_p = dv_2^1\,dv_2^2\,dv_2^3\,dT_2\,dr_p, \end{aligned} \qquad (7.4.1)$$

where $r_p = d/2$ is the particles radius and dV_p is an infinitesimal element of the phase volume. The integration is carried out in (7.4.1) over the total phase volume. The distribution function f satisfies the continuity equation in the phase space, which is similar to a corresponding equation in the rarefied gas dynamics[26]

$$\frac{\partial f}{\partial t} + v_2^k \frac{\partial f}{\partial x^k} + \frac{\partial}{\partial v_{2k}}(F^k f) + \frac{\partial}{\partial T_2}(qf) = 0,$$
$$F^k = \frac{v_1^k - v_2^k}{\tau_v} + g^k, \qquad q = \frac{T_1 - T_2}{\tau_T}, \tag{7.4.2}$$

where $\vec{g} = g^k \vec{e}_k$ is a body force per unit mass and τ_v and τ_T are the velocity and temperature relaxation times (7.3.5). Note that equation (7.4.2) may be obtained from equation (7.2.19) if one neglects in the latter equation the integral of collisions and the diffusion in the velocities space. It is necessary to augment equations (7.4.1) and (7.4.2) by the equations for gas, which we will take from the system (7.2.9):

$$\frac{\partial \rho_1}{\partial t} + \nabla_k(\rho_1 v_1^k) = 0, \quad \rho_1 = \rho_{11},$$
$$\rho_1 \frac{d_1 v_1^k}{dt} = -\nabla^k P - \rho_2 \frac{(v_1^k - \langle v_2^k \rangle)}{\tau_v} + \rho_1 g^k, \quad \frac{d_1}{dt} = \frac{\partial}{\partial t} + v_1^k \nabla_k,$$
$$\frac{d_1 e_1}{dt} = -P \frac{d_1}{dt}\left(\frac{1}{\rho_1}\right) + \frac{1}{\rho_1}\dot{Q}_{12}, \tag{7.4.3}$$
$$\dot{Q}_{12} = \int m_p \left(\frac{(\vec{v}_1 - \vec{v}_2)^2}{\tau_v} - \frac{(T_1 - T_2)}{\tau_T} \right) f \, dV_p,$$
$$m_p = \frac{\pi d^3}{6}\rho_{22}, \quad P = P(\rho_1, T_1), \quad e_1 = C_V T_1.$$

The system of equations (7.4.1)–(7.4.3) is applicable at $\mathrm{Kn} \approx \frac{d}{6 m_2 L} \gg 1$ and enables one to compute the flows with the particle's intersections.

We will search for the solution of kinetic equation (7.4.2) by the method of characteristics, which are determined from the equations

$$\frac{dx^k}{dt} = v_2^k, \quad \frac{dv_2^k}{dt} = F^k, \quad \frac{dT_2}{dt} = q. \tag{7.4.4}$$

Let us identify in (7.4.2) a derivative along the characteristics (7.4.4):

$$\frac{d'}{dt} = \frac{\partial}{\partial t} + v_2^k \frac{\partial}{\partial x^k} + F^k \frac{\partial}{\partial v_2^k} + q \frac{\partial}{\partial T_2}.$$

Then we can rewrite (7.4.2) in the form

$$\frac{d' f}{dt} = -\left(\frac{\partial F^k}{\partial v_2^k} + \frac{\partial q}{\partial T_2} \right) f.$$

This equation has the following solution:

$$f(t, x^k, v_2^k, T_2) = f^0 \exp\left(- \int_0^t \left(\frac{\partial F^k}{\partial v_2^k} + \frac{\partial q}{\partial T_2} \right) dt' \right), \tag{7.4.5}$$

where the integral is taken along the characteristics (7.4.4) and

$$\begin{aligned} f^0 &= f\Big|_{t=0} = f(x^{*k},\ v_2^{*k},\ T_2^*), \\ x^{*k} &= x^k\Big|_{t=0}, \quad v_2^{*k} = v_2^k\Big|_{t=0}, \quad T_2^* = T_2\Big|_{t=0}. \end{aligned} \tag{7.4.6}$$

Thus, in order to find the solution of the kinetic equation, one must know the characteristics (7.4.4). To find the latter, one must also solve the system of equations (7.4.3). This is a very complex problem, which is generally solved with the use of numerical methods.

One can neglect the influence of particles on gas in the one-dimensional nonstationary case at $\rho_2/\rho_1 \ll 1$ and construct the solution of equations (7.4.1) and (7.4.2) for the constant gas parameters $(\rho_1, \vec{v}_1, P, T_1) =$ const and $\vec{g} = 0$. Let us direct the x^1-axis along the gas velocity:

$$v_1^2 = v_1^3 = \langle v_2^2 \rangle = \langle v_2^3 \rangle = 0.$$

We will assume for simplicity that all particles have the same radius $r_p = d/2$ and temperature $T_2 = T_1$. We will search for the solution of equation (7.4.2) in the form

$$f = f_1(t, x^1, v_2^1)\, f_2(t, v_2^2)\, f_3(t, v_2^3)\, \delta(r_p - d/2)\, \delta(T_2 - T_1). \tag{7.4.7}$$

Substituting (7.4.7) into (7.4.2), we obtain three equations:

$$\begin{aligned} &\frac{\partial f_1}{\partial t} + v_2^1 \frac{\partial f_1}{\partial x^1} + \frac{(v_1^1 - v_2^1)}{\tau_v} \frac{\partial f_1}{\partial v_2^1} - \frac{f_1}{\tau_v} = 0, \\ &\frac{\partial f_2}{\partial t} - \frac{v_2^2}{\tau_v} \frac{\partial f_2}{\partial v_2^2} - \frac{f_2}{\tau_v} = 0, \\ &\frac{\partial f_3}{\partial t} = -\frac{v_2^3}{\tau_v} \frac{\partial f_3}{\partial v_2^3} - \frac{f_3}{\tau_v} = 0. \end{aligned} \tag{7.4.8}$$

The last two equations of (7.4.8) describe the relaxation of the velocities v_2^2 and v_2^3 and have the solutions

$$f_2 = f_2^0 e^{t/\tau_v}, \quad f_3 = f_3^0 e^{t/\tau_v}.$$

Integrating the equation $\frac{dv_2^2}{dt} = -\frac{v_2^2}{\tau_v}$, we obtain the equation for the characteristics

$$v_2^2 = v_2^{*2} e^{-t/\tau_v},$$

with the aid of which we can find the integral

$$\int_{-\infty}^{\infty} f_2\, dv_2^2 = \int_{-\infty}^{\infty} f_2^0\, e^{t/\tau_v} \frac{\partial v_2^2}{\partial v_2^{*2}}\, dv_2^{*2} = \int_{-\infty}^{\infty} f_2^0\,(v_2^{*2})\, dv_2^{*2} = C_2. \tag{7.4.9}$$

We obtain in a similar way:

$$\int_{-\infty}^{\infty} f_3\, dv_2^3 = \int_{-\infty}^{\infty} f_3^0(v_2^{*3})\, dv_2^{*3} = C_3. \tag{7.4.10}$$

Let us choose the f_2^0 and f_3^0 in such a way that $C_2 = C_3 = 1$. Let some quantity $A(t, x^1, v_2^1)$ be given, and then find with the aid of (7.4.7), (7.4.9), and (7.4.10):

$$\begin{aligned} \langle A \rangle &= \frac{1}{n} \cdot \int A(t, x^1, v_2^1)\, f_1 f_2 f_3\, \delta(r_p - d/2)\, \delta(T_2 - T_1)\, dv_2^1\, dv_2^2\, dv_2^3\, dr_p\, dT_2 \\ &= \frac{1}{n} \int A(t, x^1, v_2^1)\, f_1 f_2 f_3\, dv_2^1\, dv_2^2\, dv_2^3 \\ &= \frac{1}{n} \int_{-\infty}^{\infty} A(t, x^1, v_2^1)\, v_1\, dv_2^1 \int_{-\infty}^{\infty} f_2\, dv_2^2 \int_{-\infty}^{\infty} f_3\, dv_2^3 \\ &= \frac{1}{n} \int_{-\infty}^{\infty} A(t, x^1, v_2^1)\, f_1\, dv_2^1. \end{aligned} \tag{7.4.11}$$

If one specifies the initial conditions $f_1|_{t=0} = f_1^0(x^{*1}, v_2^{*1})$, then the solution of the first equation of (7.4.8) can be written similarly to (7.4.5):

$$f_1 = f_1^0(x^1, v_2^{*1}) e^{t/\tau_v}. \tag{7.4.12}$$

From the equation $\frac{dv_2^1}{dt} = \frac{v_1^1 - v_2^2}{\tau_v}$, we obtain at $v_1^1 = \text{const}$, $v_2^1|_{t=0} = v_2^{*1}$:

$$v_2^1 = v_1^1 + (v_2^{*1} - v_1^1)\, e^{-t/\tau_v}, \qquad \frac{\partial v_2^1}{\partial v_2^{*1}} = e^{-t/\tau_v}. \tag{7.4.13}$$

Substituting (7.4.12) and (7.4.13) into (7.4.11), we find:

$$\langle A \rangle = \int_{-\infty}^{\infty} A(t, x^1, v_2^1)\, f_1^0(x^{*1}, v_2^{*1})\, dv_2^{*1}. \tag{7.4.14}$$

Substituting instead of A the values $A = 1$ and $A = v_2$ and omitting the index "1," we find the mean concentration and velocity of particles by the formulas:

$$n = \int_{-\infty}^{\infty} f^0(x^*, v_2^*)\, dv_2^*, \quad \langle v_2 \rangle = \frac{1}{n} \int_{-\infty}^{\infty} v_2\, f^0(x^*, v_2^*)\, dv_2^*. \tag{7.4.15}$$

The expressions (7.4.15) enable one to find $n = n(t, x)$ and $\langle v_2 \rangle = \langle v_2 \rangle(t, x)$ at point x, which is related to the initial coordinate x^* with the aid of the trajectory equation

$$x = x^* + \Psi(t, v_2^*), \quad v_2 = \Phi(v_2^*, t). \tag{7.4.16}$$

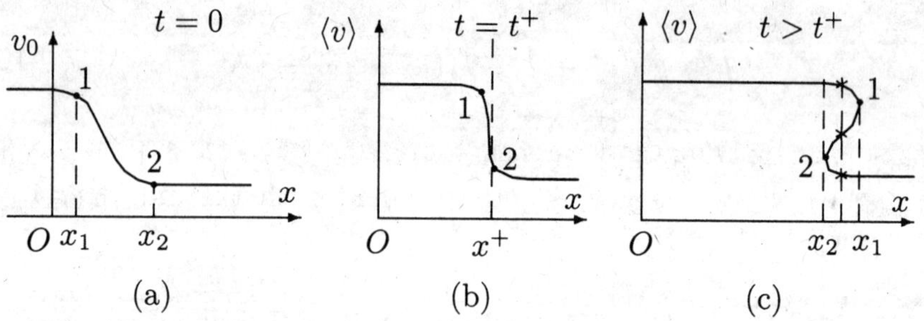

Figure 7.14: The mean velocity profiles of particles at three moments of time: (a) $t = 0$ and (b) $t = t^+$; (c) $t > t^+$.

The functions Φ and Ψ are found as a result of the integration of the equations for the characteristics (7.4.4) in the one-dimensional case:

$$\frac{dx}{dt} = v_2, \qquad \frac{dv_2}{dt} = F \tag{7.4.17}$$

at $F = (v_1 - v_2)/\tau_v$ under the initial conditions

$$x|_{t=0} = x^*, \qquad v_2|_{t=0} = v_2^*. \tag{7.4.18}$$

7.4.2 Investigation of Caustics in the Pseudogas of Particles

The obtained solution (7.4.15)–(7.4.18) of the collisionless kinetic equation (7.4.8) enables us to investigate the caustics in the one-dimensional nonstationary case at a constant gas velocity $v_1 = \text{const}$ [8,17].

Assume that the distribution function of particles at time $t = 0$ is determined by the formula

$$f^0 = \frac{n_0}{\sqrt{2\pi\sigma}} \exp\left(-\frac{(v^* - v^0(x^*))^2}{2\sigma}\right) \tag{7.4.19}$$

(the index "2" by the particles velocity will be omitted below in this Subsection). Let us specify the velocity $v^0(x^*)$ in the form of a "smeared" step as shown in Fig. 7.14 (a). Since $\langle v\rangle(x_1) > \langle v\rangle(x_2)$, particle "1" will catch up with particle "2" until an overlapping at some time t^+ arises [Fig. 7.14 (b)]. After that, the first particle will overtake the second particle and three velocities marked by the asterisks in Fig. 7.14 (c) will be determined at each point of the interval $x_2 < x < x_1$. The corresponding picture of the trajectories in the t, x plane is shown in Fig. 7.15. The solid lines Γ_1 and Γ_2 are the envelopes of the family of trajectories and called the *caustics*. The caustic arises at point t^+, x^+, and three trajectories pass inside it through each point. If

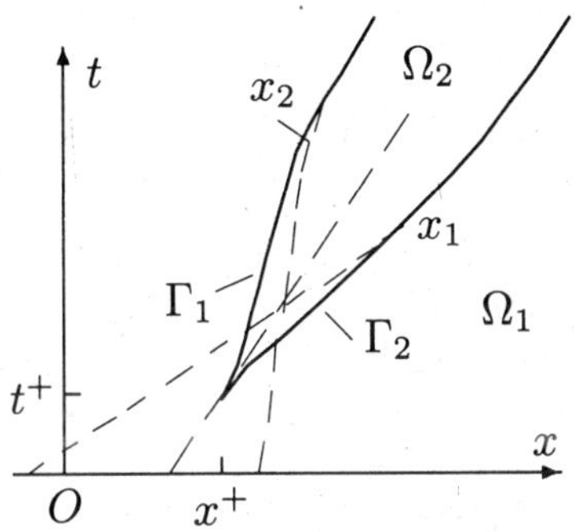

Figure 7.15: The picture of trajectories (dashed lines) and caustics (solid lines Γ_1 and Γ_2) in the t, x plane.

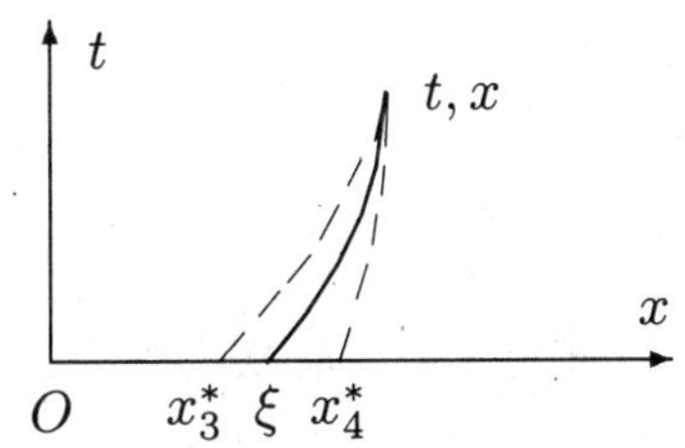

Figure 7.16: To the computation of integral (7.4.20).

all particles had the same velocity, $f^0 = n\,\delta(v^* - v_0(x^*))$ (δ is the Dirac delta function), then the particle's concentration would turn into infinity on the caustic. The presence of dispersion in the velocities $\sigma \neq 0$ leads to the fact that the concentration becomes finite on the caustic. Let us find its magnitude.

Substituting (7.4.19) into (7.4.15), we obtain with regard for (7.4.16) the expression for the concentration $n(t, x)$:

$$\frac{n}{n_0} = \frac{1}{\sqrt{2\pi\sigma}} \int_{-\infty}^{\infty} \exp\left(-(v^* - v^0(x^*))^2/(2\sigma)\right) dv^*, \;\; x = x^* + \Psi(t, v^*), \tag{7.4.20}$$

where t and x are fixed. Let us take some arbitrary point t, x (see Fig. 7.16). The points x^* lying in the neighborhood of point ξ, at which the minimum of the function $\varphi^2 = (v^* - v^0(x^*))^2$ is achieved give the main contribution to the integral (7.4.20). It is clear that φ^2 is minimal at $v^* = v^0(\xi)$, from where we obtain with the aid of the second equation (7.4.20) a relation for determination of ξ:

$$x = \xi + \Psi(t, v^0(\xi)). \tag{7.4.21}$$

Expanding φ into the Taylor series near the minimum, we can write:

$$\begin{aligned}\varphi &\approx \frac{\partial \varphi}{\partial v^*}\bigg|_{v^0(\xi)}(v^* - v^0(\xi)) + \frac{1}{2}\frac{\partial^2 \varphi}{\partial (v^*)^2}\bigg|_{v^0(\xi)}(v^* - v^0(\xi))^2 \\ &+ \frac{1}{6}\frac{\partial^2 \varphi}{\partial (v^*)^3}\bigg|_{v^0(\xi)}(v^* - v^0(\xi))^3. \end{aligned} \tag{7.4.22}$$

We find from formulas $\varphi = v^* - v^0(x^*)$, $x = x^* + \psi(t, v^*)$ for a fixed x:

$$\frac{\partial \varphi}{\partial v^*} = 1 - \frac{\partial v^0}{\partial x^*}\frac{\partial x^*}{\partial v^*}, \quad \frac{\partial x}{\partial x^*} = 0 = 1 + \frac{\partial \psi}{\partial v^*}\frac{\partial v^*}{\partial x^*},$$

from where it follows that $\frac{\partial \varphi}{\partial v^*} = 1 + \frac{\partial v^0}{\partial x^*}\frac{\partial \psi}{\partial v^*}$. Substituting here $v^* = v^0(\xi)$, we obtain:

$$\frac{\partial \varphi}{\partial v^*}\bigg|_{v^0(\xi)} = 1 + \frac{\partial v^0}{\partial \xi} \cdot \frac{\partial \psi}{\partial v^0}.$$

On the other hand, it follows from equation (7.4.21) that

$$\frac{\partial x}{\partial \xi} = 1 + \frac{\partial \psi}{\partial v^0}\frac{\partial v^0}{\partial \xi}.$$

Comparing this equation with the foregoing one, we obtain:

$$\frac{\partial \varphi}{\partial v^*}\bigg|_{v^0(\xi)} = \frac{\partial x}{\partial \xi}.$$

We can find similarly the remaining derivatives:

$$\begin{aligned}\frac{\partial^2 \varphi}{\partial (v^*)^2}\bigg|_{v^0(\xi)} &= \frac{\partial^2 \varphi}{\partial (v^0)^2} = \frac{\partial}{\partial v^0}\left(\frac{\partial x}{\partial \xi}\right) = \frac{\partial^2 x}{\partial \xi^2}\frac{\partial \xi}{\partial v^0}, \\ \frac{\partial^3 \varphi}{\partial (v^*)^3}\bigg|_{v^0(\xi)} &= \frac{\partial \varphi}{\partial (v^0)}\left(\frac{\partial^2 \varphi}{\partial (v^0)^2}\right) = \frac{\partial}{\partial v^0}\left(\frac{\partial^2 x}{\partial \xi^2}\frac{\partial \xi}{\partial v^0}\right) \\ &= \frac{\partial^3 x}{\partial \xi^3}\left(\frac{\partial \xi}{\partial v^0}\right)^2 + \frac{\partial^2 x}{\partial \xi^2}\frac{\partial^2 \xi}{\partial (v^0)^2}.\end{aligned}$$

Substituting the found derivatives in (7.4.22), we write the expansion $\varphi = v^* - v^0(x^*)$ into a series in powers of $(v^* - v^0(\xi))$:

$$\begin{aligned}\varphi &\approx \frac{\partial x}{\partial \xi}(v^* - v^0(\xi)) + \frac{1}{2}\frac{\partial^2 x}{\partial \xi^2}\frac{\partial \xi}{\partial v^0}(v^* - v^0(\xi))^2 + \frac{1}{6}\left(\frac{\partial^3 x}{\partial \xi^3}\left(\frac{\partial \xi}{\partial v^0}\right)^2\right. \\ &+ \left.\frac{\partial^2 x}{\partial \xi^2}\frac{\partial^2 \xi}{\partial (v^0)^2}\right)(v^* - v^0(\xi))^3, \end{aligned} \tag{7.4.23}$$

where ξ is found from (7.4.21) at fixed t and x. Retaining the nonlinear term in (7.4.23) and substituting it in the integral (7.4.20), we obtain:

$$\begin{aligned}\frac{n}{n_0} &= \frac{1}{\sqrt{2\pi\sigma}}\int_{-\infty}^{\infty}\exp\left(-\left(\frac{\partial x}{\partial \xi}(v^* - v^0(\xi))^2/(2\sigma)\right)dv^*\right. \\ &= \sum_{i=1}^{n}\frac{1}{|x'|\sqrt{2\pi}}\int_{-\infty}^{\infty}e^{-y^2/2}dy = \sum_i \frac{1}{|x'|_i}, \quad x' = \frac{\partial x}{\partial \xi},\end{aligned}$$

where the summation is carried out over all n trajectories (7.4.21) passing through the given point t, x. Consequently, the dependence $n(t,x)$ is given by formulas

$$n = \sum_{i=1}^{n} n_0/|\frac{\partial x}{\partial \xi}|_i, \quad x = \xi + \psi(t, v^0(\xi)). \tag{7.4.24}$$

As can be seen from Fig. 7.15, $n = 1$ in the region Ω_1, and in the region Ω_2 bounded by the caustics Γ_1 and Γ_2, we have $n = 3$. The first formula in (7.4.24) represents the continuity equation written in the Lagrangian coordinates. It is valid everywhere, except for the lines

$$\frac{\partial x}{\partial \xi} = 0, \tag{7.4.25}$$

which are the *envelopes of the family of the trajectories (7.4.21) and are called the caustics* (they are denoted by Γ_1 and Γ_2 in Fig. 7.15). Differentiating (7.4.21), let us rewrite the caustic equation (7.4.25) as follows:

$$1 + \frac{\partial \psi}{\partial v^0}\cdot\frac{\partial v^0}{\partial \xi} = 0, \quad x = \xi + \psi(t, v^0(\xi)). \tag{7.4.26}$$

The first equation determines the dependence $t = t(\xi)$, which is the caustic equation in the Lagrangian coordinates t, ξ. Substituting this dependence in the second equation (7.4.26), we obtain the caustic equation in the Eulerian coordinates $x = x(\xi, t(\xi))$. Let us differentiate equation (7.4.25) with respect to ξ:

$$\frac{\partial^2 x}{\partial \xi^2} + \frac{\partial^2 x}{\partial t \partial \xi}\frac{\partial t}{\partial \xi} = 0.$$

The curve $t = t(\xi)$ determining the caustic in the Lagrangian coordinates has a minimum $\frac{\partial t}{\partial \xi} = 0$ at $t = t^+$ (the time of the caustic formation). Substituting this condition in the foregoing equation, we find the equations:

$$\frac{\partial^2 x}{\partial \xi^2} = 0, \quad \frac{\partial x}{\partial \xi} = 0, \tag{7.4.27}$$

from which one can determine the point of the caustic formation t^+, x^+.

Let us find the particle's concentration on a caustic. In this case, the linear term in the expansion (7.4.23) is equal to zero, and retaining the quadratic term, we obtain:

$$\varphi \approx \frac{1}{2}\frac{x''}{v^{0\,\prime}}(v^* - v^0)^2, \quad v^0 = v^0(\xi), \quad x'' = \frac{\partial^2 x}{\partial \xi^2}, \; v^{0\,\prime} = \frac{dv^0}{d\xi}.$$

Substituting this expression into (7.4.20), we find:

$$\begin{aligned}\frac{n}{n_0} &= \frac{2}{\sqrt{2\pi\sigma}}\int_{-\infty}^{\infty} \exp\left(-\frac{1}{8}\left(\frac{x''}{v^{0\,\prime}}\right)^2 \frac{(v^* - v^0)^4}{\sigma}\right) dv^* \\ &= \frac{2}{8^{1/4}\sqrt{\pi}\sigma^{1/4}|\alpha|^{1/2}}\int_0^{\infty} t^{-3/4}e^{-t}\, dt, \qquad (7.4.28)\end{aligned}$$

where

$$t = \frac{1}{8}\frac{\alpha^2 u^4}{\sigma}, \quad \alpha^2 = \left(\frac{x''}{v^{0\,\prime}}\right)^2, \quad u = v^* - v^0.$$

The coefficient 2 takes into account the fact that two trajectories, each of which gives a contribution to n, come to a caustic at each moment of time [see the neighborhood of points 1 and 2 in Fig. 7.14 (c)]. Using the definition of the Gamma function

$$\Gamma(z) = \int_0^{\infty} t^{z-1}e^{-t}\, dt,$$

we can rewrite (7.4.28) in the form

$$\frac{n}{n_0} = 2\Gamma(\tfrac{1}{4})/\sqrt{\pi 8^{1/2}\sigma^{1/2}\left|\frac{\partial^2 x}{\partial \xi^2}\right|/\left|\frac{dv^0}{d\xi}\right|}. \qquad (7.4.29)$$

Thus, the concentration on a caustic is $n/n_0 \sim 1/\sigma^{1/4}$, and if the velocity dispersion is equal to zero $\sigma = 0$, then the particle's concentration on a caustic becomes infinite: $n/n_0 \to \infty$.

Let us find the particle's concentration in the neighborhood of the caustic $\frac{\partial x}{\partial \xi} = 0$. Let us fix the moment of time t and expand the function (7.4.21) $x = x(t, \xi)$ into a series in powers of $(\xi - \xi_k)$ in the caustic neighborhood $x_k = x(t, \xi_k(t))$:

$$x \approx x_k + \left(\frac{\partial x}{\partial \xi}\right)_{\xi_k}(\xi - \xi_k) + \frac{1}{2}\left(\frac{\partial^2 x}{\partial \xi^2}\right)_{\xi_k}(\xi - \xi_k)^2.$$

Since $(\frac{\partial x}{\partial \xi})_{\xi_k} = 0$ on a caustic, then

$$x = x_k + \frac{1}{2}\left(\frac{\partial^2 x}{\partial \xi^2}\right)_{\xi^k}(\xi - \xi_k)^2, \qquad (7.4.30)$$

from where

$$\xi - \xi_k = \pm\sqrt{2|x - x_k|/|x''|}, \quad x'' = \frac{\partial^2 x}{\partial \xi^2}|_{\xi^k}. \tag{7.4.31}$$

Differentiating (7.4.30) and using (7.4.31), we find:

$$\left|\frac{\partial x}{\partial \xi}\right| = |x''(\xi - \xi_k)| = \sqrt{2|x''|\,|x - x_k|}.$$

Substituting this formula into the first equation in (7.4.24) and multiplying by two, we obtain $n(t, x)$ in the caustic neighborhood:

$$\frac{n}{n_0} = \sqrt{\frac{2}{|x''|\,|x - x_k|}}, \tag{7.4.32}$$

where $x'' = \frac{\partial^2 x}{\partial \xi^2}|_{\xi^k}$, $x_k = x(t, \xi_k(t))$ is the caustic equation. The singularity in (7.4.32) is integrable since the total number of particles N_k on a caustic is finite:

$$N_k = \int_0^{\Delta x} n\,dx = n_0\sqrt{\frac{2}{|x''|}}\int_0^{\Delta x}\frac{dy}{\sqrt{y}} = 2n_0\sqrt{\frac{2\Delta x}{|x''|}},$$

where $\Delta x = x - x_k$ is the caustic "thickness." Assuming that $n \approx n_0$ in (7.4.32), we obtain the estimate $\Delta x \approx 2/|x''|$, with regard for which:

$$N_k \approx \frac{4n_0}{|x''|}. \tag{7.4.33}$$

The obtained formula is applicable at all caustic points, except for t^+, x^+, where $x'' = 0$ and $x' = 0$. At this point, only the third item will be different from zero in expansion (7.4.23):

$$\varphi \approx \frac{1}{6}\frac{x'''}{(v^{0\,\prime})^2}(v^* - v^0)^3, \quad x''' = \frac{\partial^3 x}{\partial \xi^3}, \quad v^{0\,\prime} = \frac{dv^0}{d\xi}.$$

Substituting these expressions into (7.4.20), we find at point t^+, x^+:

$$\begin{aligned} \frac{n}{n_0} &= \frac{1}{\sqrt{2\pi\sigma}}\int_{-\infty}^{\infty}\exp\left(-\left(\frac{x'''}{6(v^{0\,\prime})^2}\right)^2 u^6\right)du \\ &= \frac{\Gamma\left(\frac{1}{6}\right)}{3^{2/3}\sqrt{\pi}}\left(\frac{(v^{0\,\prime})^2}{\sigma x'''}\right)^{1/3}. \end{aligned} \tag{7.4.34}$$

Thus, the estimate $\frac{n}{n_0}|_{x^+} \sim \frac{1}{\sigma^{1/3}}$ holds at point t^+, x^+. At a given t, we have in the neighborhood of the x^+ point:

$$x = x^+ + \frac{1}{6}x'''(\xi - \xi^+)^3, \quad x' = \frac{1}{2}x'''(\xi - \xi^+)^2.$$

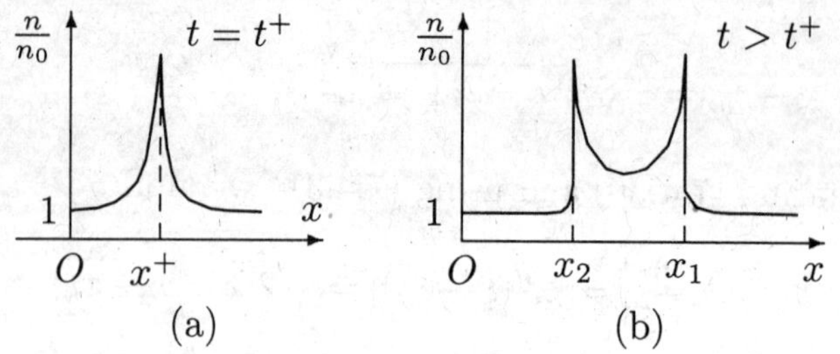

Figure 7.17: The dependence $n(x)$ (a) at the time of caustic formation t^+ and (b) at time $t > t^+$.

Eliminating the ξ^+ from the second formula and substituting in (7.4.24), we obtain the particle's concentration in the neighborhood of x^+:

$$\frac{n}{n_0} = \left(\frac{\sqrt{2}}{3|x'''|^{1/2}|x - x^+|}\right)^{2/3}. \tag{7.4.35}$$

This singularity is also integrable, and the total particles number on a caustic at the time of its formation $t = t^+$ can be estimated by formula

$$N_k^+ = 2\int_0^{\Delta x} n\,dx = 2\Big(\frac{\sqrt{2}}{3}^{\frac{2}{3}} \frac{n_0}{|x'''|^{1/3}} \int_0^{\Delta x} y^{-2/3} dy = 48^{1/3} n_0 \frac{\Delta x^{\frac{1}{3}}}{|x'''|^{1/3}}.$$

We can find the caustic "thickness" from formula (7.4.35) by setting $n \sim n_0$ therein. As a result, we obtain:

$$\Delta x \approx \frac{\sqrt{2}}{3|x'''|^{1/2}}, \quad N_k^+ \approx 2^{3/2} 3^{1/3} n_0/|x'''|^{1/2}. \tag{7.4.36}$$

Let us construct a qualitative dependence $n(x)$ at the time of the caustic formation $t = t^+$ [Fig. 7.17 (a)] and at a later time $t > t^+$ [Fig. 7.17 (b)] by using formulas (7.4.24), (7.4.29), (7.4.32), (7.4.34), and (7.4.36). It can be seen that the caustic forming at time t^+ is unstable, and it splits into two branches with coordinates $x_1(t)$ and $x_2(t)$.

Note that the caustics not always arise, but only under the satisfaction of a certain condition, which we will obtain below. Integrating the trajectories' equations $\frac{dx}{dt} = v$, $\frac{dv}{dt} = \frac{v_1 - v}{\tau_v}$ at a constant velocity $v_1 = \text{const}$ and $v|_{t=0} = v^0(\xi)$, $x|_{t=0} = \xi$, we obtain:

$$\begin{aligned} v &= v_1 + (v^0(\xi) - v_1)\, e^{-t/\tau_v}, \\ x &= \xi + v_1 t + (v^0(\xi) - v_1)K(t), \\ K(t) &= \tau_v(1 - e^{-t/\tau_v}). \end{aligned} \tag{7.4.37}$$

Differentiating the second equation with respect to ξ and substituting in the caustic equation (7.4.25), we find:

$$1 + \frac{dv^0}{d\xi} K(t) = 0. \tag{7.4.38}$$

As seen from Fig. 7.14, the caustic arises when the "rear" particles move faster than the forward particles; therefore, the following estimate holds for the derivative $\frac{dv^0}{d\xi}$: $\frac{dv^0}{d\xi} \approx -|\Delta v^0|/L$, where L is a reference length of the variation of the mean velocity Δv^0 at $t = 0$. Substituting this estimate into (7.4.38), we obtain:

$$1 - |\Delta v^0| K(t)/L = 0.$$

It follows from the last equation (7.4.37) that $K(t) \leq \tau_v$; therefore, the condition for the caustic formation may be written in the form

$$\mathrm{St} \geq 1, \quad \mathrm{St} = \frac{l_p}{L}, \quad l_p = |\Delta v^0| \tau_v, \tag{7.4.39}$$

where St is the Stokes number. This condition is related to the non-Hamiltonian form of a system for gas and particles. If the inverse inequality $\mathrm{St} < 1$ is satisfied, then $L > l_p$ and the particles velocity will become equal to the gas velocity earlier than a caustic forms. For the Hamiltonian systems of particles without the eigenpressure[27], the caustics arise on the contrary at any irregular velocity field containing at $t = 0$ a region where $\frac{dv^0}{dx} < 0$. As a consequence of the singularity integrability, the number of particle's collisions on a caustic is small, and therefore the collisionless model (7.4.1)–(7.4.3) is also applicable for them. Let us estimate the number of the collisions ν_k when a particle intersects a caustic at time $t > t^+$. Using (7.4.33), we find:

$$\nu_k = \pi d^2 \int_{x_k}^{x_k + \Delta x} n\, dx = \pi d^2 N_k \approx \frac{4\pi d^2 n_0}{|x''|}. \tag{7.4.40}$$

Differentiating the second equation in (7.4.37) twice with respect to ξ, we can estimate the second derivative:

$$|x''| = \left| \frac{d^2 v^0}{d\xi^2} \right| K(t) \approx \frac{|\Delta v^0| \tau_v}{L^2} = \frac{\mathrm{St}}{L}.$$

We obtain with regard for (7.4.39) and the given estimate:

$$\nu_k \approx \frac{4\pi d^2 n_0}{\mathrm{St}} L < \frac{24 m_2^0 L}{d} = \frac{4}{\mathrm{Kn}^0} \ll 1, \tag{7.4.41}$$

where $\mathrm{Kn}^0 = \frac{d}{6 m_2^0 l} \gg 1$ is the initial Knudsen number (at $t = 0$) of the pseudogas of particles. Since the quantity $L < l_p$, the number of

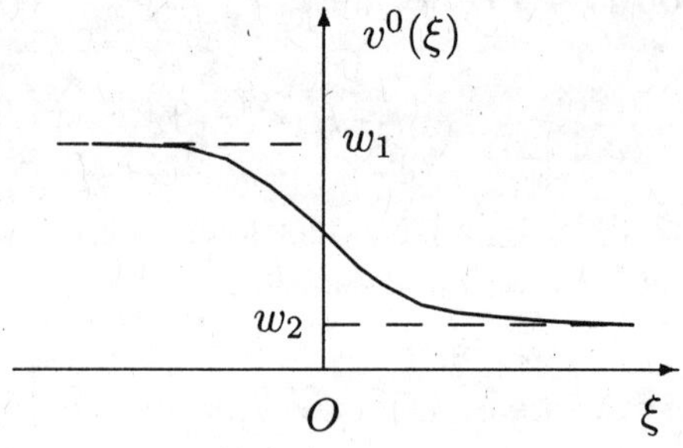

Figure 7.18: The mean velocity dependence on the coordinate at the initial moment of time.

particles on a caustic N_k and the number of collisions ν_k prove to be bounded from above:

$$N_k < 4n_0\, l_p, \quad \nu_k < \frac{4}{\mathrm{Kn}^l} \ll 1, \quad \mathrm{Kn}^l = \frac{d}{6m_2^0 l_p}. \qquad (7.4.42)$$

Similar inequalities are valid at point t^+, x^+. It is clear that such limitations are absent for the Hamiltonian systems of particles.

The theory of caustics was constructed initially for a dusty medium moving in a gravitational field[28]. It was extended subsequently for arbitrary Hamiltonian systems and is now called the *catastrophe theory*[29]. We would like to stress here that the non-Hamiltonian character of the system gives rise to a number of substantially new properties of the caustics: the limitation from above of the particles mass on a caustic and the presence of the criterion (7.4.39) for the caustic formation.

Problem 7.7. Neglecting the influence of particles on gas, find the caustic equation and the particles concentration on a caustic in the one-dimensional nonstationary case at $v_1 = \text{const}$, $v^0(\xi) = w_s - w \arctan \alpha\xi$,

$$w_s = (w_1 + w_2)/2, \quad w = (w_1 - w_2)/\pi, \quad w_1 > w_2 > v_1 > 0$$

(see Fig. 7.18).

Solution: The equation for the particle's trajectories is given by formula (7.4.37), in which one must substitute

$$v^0(\xi) = w_s - w \arctan \alpha\xi. \qquad (7.4.43)$$

Differentiating (7.4.37) and (7.4.43) with respect to ξ, we obtain:

$$\frac{\partial x}{\partial \xi} = 1 - \frac{w\alpha K}{1 + (\alpha\xi)^2}. \qquad (7.4.44)$$

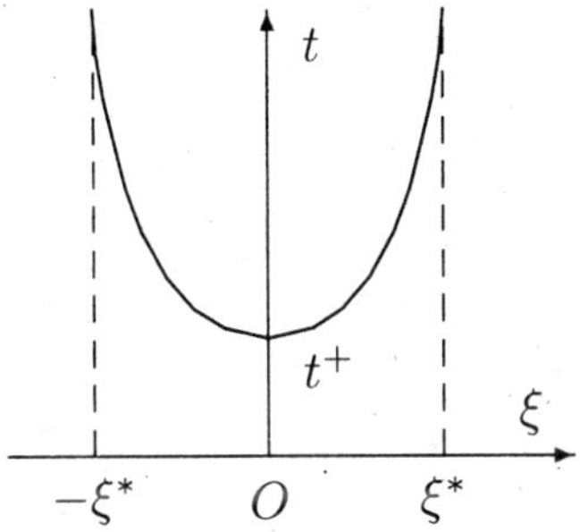

Figure 7.19: The caustic in Lagrangian coordinates.

Substituting this formula into (7.4.24), we can determine the particles concentration outside the caustic:

$$n = n_0 \sum_{i=1}^{n} 1/\left|1 - \frac{w\alpha K}{1+(\alpha\xi)^2}\right|. \tag{7.4.45}$$

From the condition (7.4.25) $\frac{\partial x}{\partial \xi} = 0$ and (7.4.44), we find the caustic equation in the Lagrangian coordinates:

$$\xi_{1,2} = \pm\frac{1}{\alpha}\sqrt{w\alpha K - 1}, \quad K = \tau_v(1 - e^{-t/\tau_v}) \tag{7.4.46}$$

(see Fig. 7.19), where $\xi^* = \frac{1}{\alpha}\sqrt{w\alpha\tau - 1}$. The subradical expression on the right-hand sides of (7.4.46) should be greater than zero; therefore, $w\alpha K > 1$. The time of the caustic formation t^+ is found from the condition $w\alpha K(t^+) = 1$, from where it follows that

$$t^+ = -\tau_v \ln\left(1 - \frac{1}{\alpha w \tau_v}\right). \tag{7.4.47}$$

The logarithm exists at $\alpha w \tau_v > 1$, from where the condition for the caustic formation (7.4.39) follows:

$$\mathrm{St} = \frac{l_p}{L} > 1, \quad l_p = w\tau_v, \quad L = 1/\alpha. \tag{7.4.48}$$

The caustic equation in the Eulerian coordinates follows from (7.4.37) and (7.4.46) (see Fig. 7.20):

$$x_{1,2} = v_1(t-K) + wK \pm \left(\frac{1}{\alpha}\sqrt{w\alpha K - 1} - wK \arctan\sqrt{w\alpha K - 1}\right). \tag{7.4.49}$$

Substituting (7.4.47) into (7.4.49), we find the coordinate of the caustic formation:

$$x^+ = -v_1\tau_v \ln(1 - 1/(\alpha w \tau_v)) + (w - v_1)/(\alpha w). \tag{7.4.50}$$

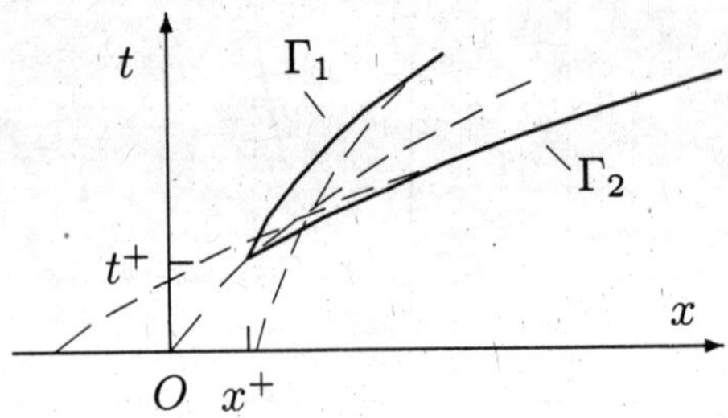

Figure 7.20: The caustic in Eulerian coordinates.

Inside a region bounded by the caustics Γ_1 and Γ_2, three trajectories pass through each point; therefore, it is necessary to assume $n = 3$ in formula (7.4.45) and $n = 1$ outside this region. Let us use the formulas (7.4.29), (7.4.32), (7.4.34), and (7.4.35) on the caustic and in its neighborhood. Let us calculate the derivatives of (7.4.44) on a caustic:

$$\frac{\partial^2 x}{\partial \xi^2}\bigg|_{\xi_{1,2}} = \frac{2w\alpha^3 K\xi}{(1+(\alpha\xi)^2)^2}\bigg|_{\xi_{1,2}} = \pm\frac{2\alpha\sqrt{w\alpha K - 1}}{w\alpha K},$$

$$\frac{\partial^3 x}{\partial \xi^3}\bigg|_{x^+} = \left(\frac{2w\alpha^3 K}{(1+(\alpha\xi)^2)^2} - \frac{4w\alpha^5 K\xi^2}{(1+(\alpha\xi)^2)^3}\right)\bigg|_{\xi=0} = 2\alpha^2, \quad \frac{dv^0}{d\xi} = -\frac{1}{K}.$$

Substituting these expressions into the above-listed formulas, we find: the concentration on a caustic

$$\frac{n}{n_0}\bigg|_{x_{1,2}} = \frac{\Gamma(1/4)}{2^{1/4}\sqrt{\pi}}\left(\frac{w^2}{\sigma}\right)^{1/4}\frac{1}{(w\alpha K - 1)^{1/4}};$$

in the neighborhood of the caustics $x_{1,2}$

$$\frac{n}{n_0}\bigg|_{x\sim x_{1,2}} = \left(\frac{w\alpha K}{\alpha\sqrt{w\alpha K - 1}|x - x_{1,2}|}\right)^{1/2};$$

at the point of the caustic formation x^+

$$\frac{n}{n_0}\bigg|_{x^+} = \frac{\Gamma(1/6)}{18^{1/3}\sqrt{\pi}}\left(\frac{w^2}{\sigma}\right)^{1/3};$$

in the neighborhood of x^+

$$\frac{n}{n_0}\bigg|_{x\sim x^+} = \frac{1}{3^{2/3}}\frac{1}{(\alpha|x - x^+|)^{2/3}}.$$

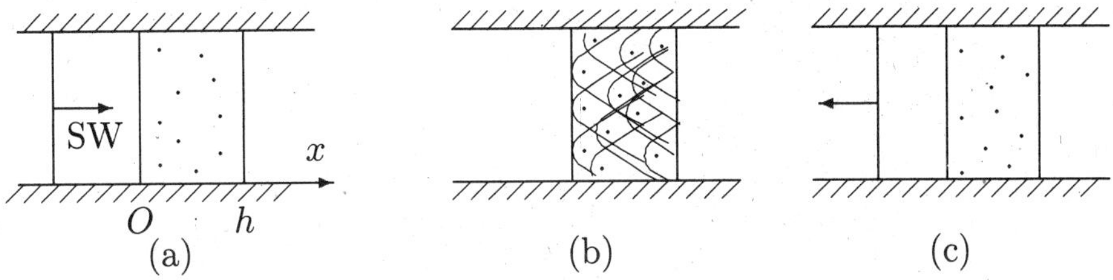

Figure 7.21: The stages of shock wave interaction with a cloud of particles: (a) the shock wave impinges on a cloud of particles, (b) a Mach shock wave is formed on each particle, and (c) a collective shock wave is formed upstream of the cloud.

7.5 Nonstationary Processes in Gas-Particle Mixtures

7.5.1 Interaction of a Shock Wave with a Cloud of Particles

Consider a cloud of solid spherical particles in a rectangular channel, and a shock wave impinges on the cloud from the left [see Fig. 7.21 (a)]. The shock wave is sufficiently strong, so that the gas flow behind it is supersonic: $M_1 = v_1/c > 1$. In Fig. 7.21, we show the gas flow pattern arising in the particle's cloud as a result of the shock wave interaction with the particle's cloud obtained in experiment[12]. At a volume concentration of particles in the cloud $m_2 \sim 10^{-3}$, the gas flow is supersonic; therefore, a Mach shock wave forms in the neighborhood of each particle [see Fig. 7.21 (b)]. At $m_2 \geq 10^{-2}$, the flow character changes drastically, because a collective shock wave forms before the cloud [see Fig. 7.21 (c)], and the gas flow in the cloud becomes subsonic. The experiments were conducted with the particles of bronze ($\rho_{22} = 8.6\ g/cm^3$, $80\ \mu m < d < 130\ \mu m$) and acrylic plastic ($\rho_{22} = 1.2\ g/cm^3$, $80\ \mu m < d < 300\ \mu m$). The collective shock wave forms not at once, but in a time of the order of $50\ \mu s$.

Consider a mechanism of the formation of a collective shock wave. The simplest hypothesis consists of the fact that the collective shock wave forms at the joining of the transonic zones, which form behind the individual particles. The distance l between the particles should be here of the order of the particle diameter d. In the experiment, however, the shock wave arises when $m_2 \sim 10^{-2}$, $l/d \sim 4 \div 5$. In addition, the time of its formation in accordance with the given mechanism should be of the order of the time for the formation of separated zones $t' \sim d/c \sim 1\ \mu s$ at $d \sim 300\ \mu m$ and $c = 300\ m/s$, which is nearly by two orders of magnitude smaller than the time observed in experiment. Thus, another mechanism of the formation of a collective shock wave should take place

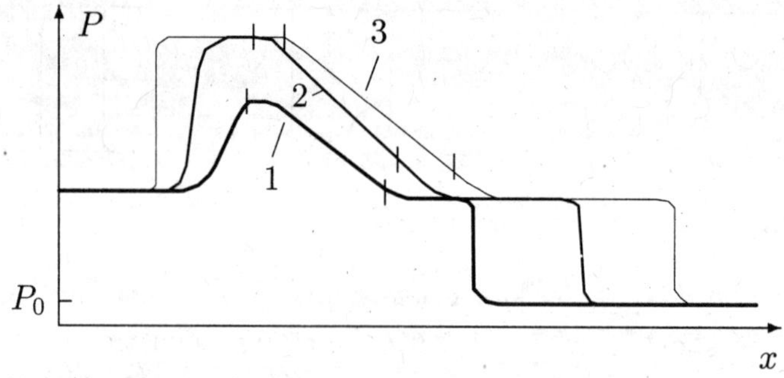

Figure 7.22: The profiles of the pressure $P(x)$ at the moments of time $t_1 < t_2 < t_3$ for the cloud of acrylic plastic particles at $m_2 = 3 \cdot 10^{-2}$.

here.

For its elucidation, the complete system of equations (7.4.1)–(7.4.3) of a continual-discrete model in the two-dimensional case was solved numerically in[12]. The characteristic times of the velocity relaxation τ_v and the thermal relaxation τ_T in (7.4.2) were found by formulas

$$\frac{1}{\tau_v} = \frac{3}{4}\left(\frac{\mathrm{Re}\,\mu}{\rho_{22}d^2}\right)C_d(\mathrm{Re}, M_{12}), \quad \frac{1}{\tau_T} = \frac{6\lambda\,\mathrm{Nu}}{C_s\rho_{22}\,d^2}, \tag{7.5.1}$$

where the functions $C_d(\mathrm{Re}, M_{12})$ and $\mathrm{Nu}(\mathrm{Re}, \mathrm{Pr})$ were determined above [see (7.1.21) and (7.1.30)]. The numerical algorithm was as follows. A rectangular Eulerian grid was constructed in the x, y plane. The finite-difference equations were written on this grid in accordance with a third-order scheme approximating the differential equations for gas (7.4.3). The collisionless kinetic equation (7.4.2) was solved in the Lagrangian variables. The region occupied by particles at time $t = 0$ was subdivided into rectangular Lagrangian cells. Within each cell, all particles had the same velocities, temperature, and radius. The alteration of the cell state was determined from the solution of the equations for the characteristics (7.4.4). The computation of the grid parameters on the gas grid and of the gas parameters in the particle cells was carried out with the aid of linear interpolation.

We present in Fig. 7.22 the dependencies of the pressure in gas on the x-coordinate for three moments of time $t_3 > t_2 > t_1$. The vertical lines show the left- and right-hand boundaries of the cloud of acrylic plastic particles with initial volume concentration $m_2 = 3 \cdot 10^{-2}$. It can be seen that, as a result of gas deceleration, a compression wave forms in the cloud, which at the expense of nonlinear effects transforms into a shock wave propagating towards the flow (see Fig. 7.22). The time of the formation of a collective, reflected shock wave t^* is of the same

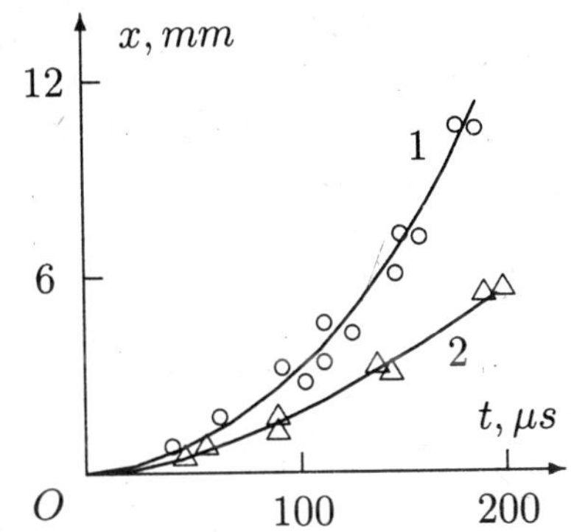

Figure 7.23: The trajectories of bronze particles at the left cloud boundary at $P_0 = 0.1$ MPa, and the Mach number of the incident shock wave $M_0 = D/c_0 = 2.8$ [12]. At $m_2 = 10^{-3}$ 1 is the numerical computation, $\circ$ is experiment. At $m_2 = 10^{-2}$ 2 is numerical computation and $\triangle$ is experiment.

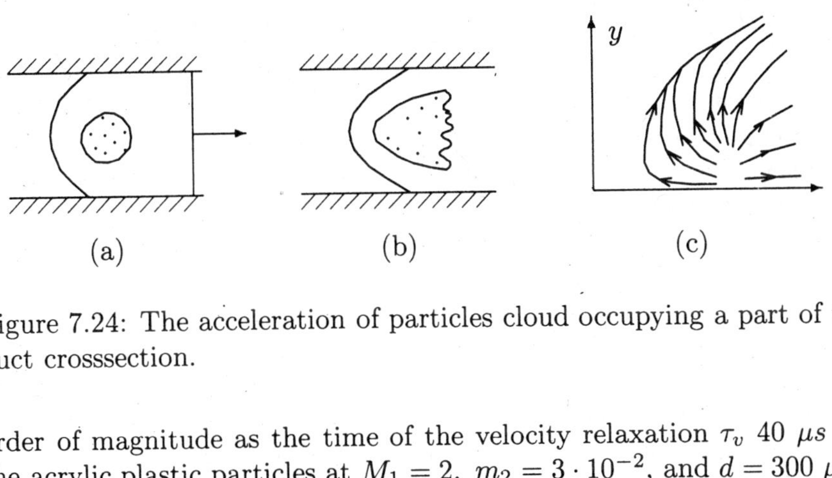

Figure 7.24: The acceleration of particles cloud occupying a part of the duct crosssection.

order of magnitude as the time of the velocity relaxation τ_v 40 μs for the acrylic plastic particles at $M_1 = 2$, $m_2 = 3 \cdot 10^{-2}$, and $d = 300$ μm. At a smaller volume concentration $m_2 \leq 10^{-3}$, the amplitude of the compression wave in the cloud is small, and during the time τ_v it is unable to go over to a shock wave because of a weak nonlinearity. The flow remains supersonic in the overall region.

The numerical results agree well with experimental data (see Fig. 7.23). An analysis of the numerical results and experiments (Fig. 7.23) shows that the rarefied cloud accelerates more strongly behind the passing shock wave. A force acting on a particle from the gas at $v_2 \ll v_1$ (the initial acceleration stage) can be estimated by the formula $f_p \approx C_d \frac{\pi d^2}{4} \rho_{11} v_1^2/2$, from which we can find the particle acceleration $a \sim C_d \rho_{11} v_1^2/d$. The gas flow rate $j = \rho_{11} v_1$ for a dense and rarefied gas is approximately the same; thus, the difference reduces to $a \sim C_d v_1$. Upstream of the dense cloud with $m_2 \sim 10^{-2}$, a collective shock wave forms, in which the gas decelerates to the subsonic velocities $M_{12} < 1$. According to (7.1.21), this leads to a diminution of C_d by a factor of about two, as compared to the supersonic flow. Since the velocity of a flow around the particle v_1 in a dense cloud is also smaller than in the rarefied flow, the particle acceleration will also be smaller: $a \sim C_d v_1$. Note that, if we consider an inverse problem on the deceleration of a particle's cloud in gas, then the dense cloud will decelerate for the same reason much less than in the case of a rarefied cloud.

If a dense cloud with $m_2 \sim 10^{-2}$ occupies a part of the channel crosssection, then after the shock wave passage a collective shock wave forms upstream of it with a curvilinear front [see Fig. 7.24 (a)]. The cloud itself with time takes a comet-shaped form with a dense kernel $m_2 \sim 10^{-2}$ and a low-density tail $m_2 \sim 10^{-4}$ [see Fig. 7.24 (b)]. In Fig. 7.24 (c) we show the particle's trajectories in the upper cloud half moving in the system of the mass center of particles cloud. It is seen that on the forward cloud boundary a boundary caustic arises, where the particle's concentration increases. It is seen from Fig. 7.24 (b) that, at a shift upwards or downwards from the channel axis, the shock wave front turns monotonously with respect to the oncoming flow. This leads to the increase in the gas velocity v_1 behind the collective shock wave. The particles ascend and get into a region of a large pressure head [see Fig. 7.24 (c)], which leads to their more intense acceleration and the cloud stretching in the downstream direction, that is, to the tail formation. Note that a negative pressure gradient $\frac{\partial P}{\partial x}$ forms in the cloud itself (see Fig. 7.22), which compensates for the action of the drag force from the particles on gas nf_p and ensures the gas flow through the cloud.

If at the initial moment of time the particle's cloud boundary was subjected to a perturbation, then this perturbation increases with time. As a result, an instability of the boundary develops, which leads to the destruction of the cloud into clusters. We show in Fig. 7.25 the instability development of a boundary of a acrylic plastic particle's cloud, which was computed in[30] at $m_2 = 3 \cdot 10^{-2}$ and $d = 20\ \mu m$ at a normal impingement of shock wave. The initial perturbations are presented in Fig. 7.25 (a). These perturbations increase with time, and under the action of secondary vortex flows, the cloud decomposes into separate clusters [see Fig. 7.25 (b)].

We show in Fig. 7.26 the instability development of a boundary of the same cloud, but at the interaction with a sliding shock wave[30]. The forming secondary vortex flows lead to the disturbances growth and cloud breakdown. It has been shown in the specially conducted experiments that the boundary separating the pure gas region from the two-phase flow rapidly ascends with time. It reaches an upper channel wall, and

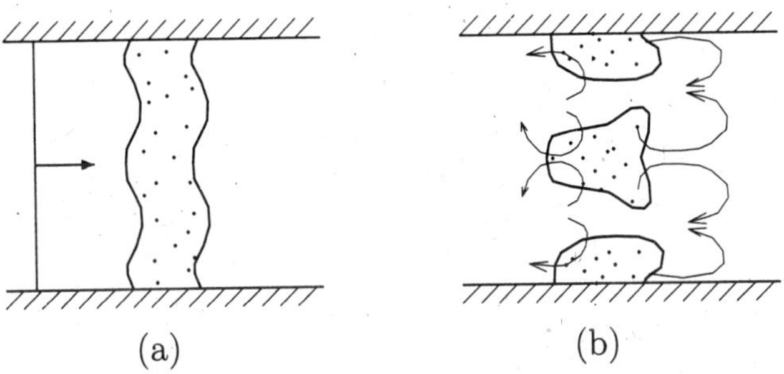

Figure 7.25: (a) The initial perturbation of the particles cloud (b) the disintegration of the particle's cloud into separate clusters. The arrows show the secondary vortex flows in a reference frame comoving at a mean gas velocity.

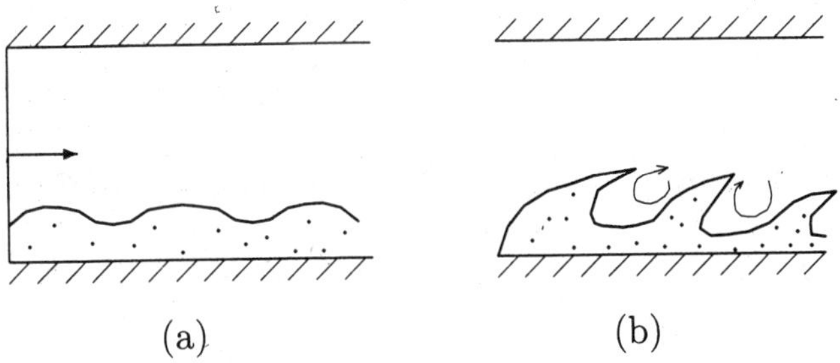

Figure 7.26: (a) The initial perturbation of the particles cloud, and (b) the instability development at the cloud boundary.

the two-phase flow takes the overall channel. In the case of small particles $d \sim 10\ \mu m$, the reason for particle's ascent is the turbulent pulsations in channel[31]. As was shown in[32], the large particles with $d > 100\ \mu m$ ascend under the action of the Magnus force. Let the particles be poured in a thin layer on the channel bottom. Under the action of a pressure head, the particles begin to move over the surface; they collide with other particles (see Fig. 7.27). After the collisions, the particles acquire an angular velocity ω_0 and the Magnus lift force begins to act on them. The magnitude of the Magnus force at $\mathrm{Re} \gg 1$ is given by formula (7.5.7) (see Problem 7.8 at the end of this section):

$$\vec{F}_M = \frac{8\pi}{3}\rho_{11} r_p^3 [(\vec{v}_1 - \vec{v}_2) \times \omega_0]. \tag{7.5.2}$$

Neglecting the particles velocity $\vec{v}_2 \ll \vec{v}_1$ in (7.5.2) and assuming that $\vec{\omega}_0$ is perpendicular to $\vec{v}_1$, we write the motion equation for a particle of mass $m_p = \frac{4}{3}\pi r_p^3 \rho_{22}$ along the y-axis:

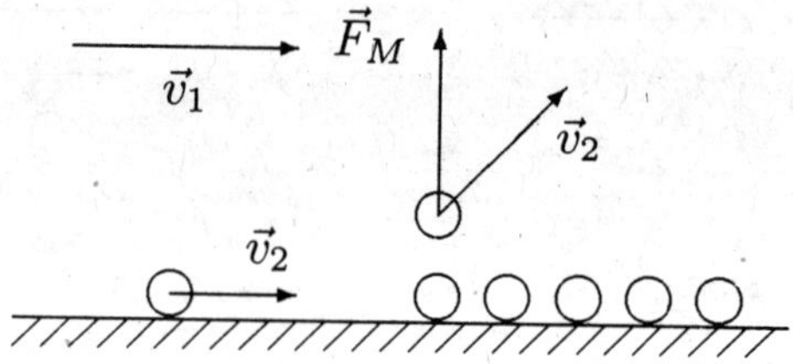

Figure 7.27: The picture of particles motion near the lower channel wall.

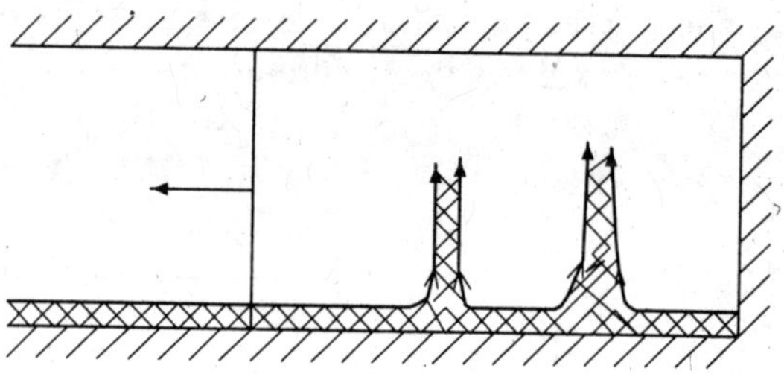

Figure 7.28: The dust ascent behind the shock wave reflected from a wall.

$$\frac{d^2y}{dt^2} = \frac{2\rho_{11}}{\rho_{22}}\omega_0 v_1. \tag{7.5.3}$$

Assuming that $v_1 = \text{const}$, $y|_{t=0} = 0$ and $\frac{dy}{dt}|_{t=0} = 0$, we easily find the integral of (7.5.3):

$$y = \frac{\rho_{11}}{\rho_{22}}\omega_0 v_1 t^2. \tag{7.5.4}$$

The parabolic law (7.5.4) agrees well with the dependence of the coordinate of the interface y_γ between the pure gas and two-phase flow on time t measured in the experiments for large particles[32].

Boiko [32] has also revealed the following interesting effect. If one takes a channel with a butt-end wall and lets a shock wave move toward the butt-end wall, then the particles will ascend behind the shock wave reflected from the wall, as shown in Fig. 7.28. The mechanism for the dust ascent in this case has still not been elucidated. As a matter of fact, the gas behind the reflected shock wave is at rest on the average; therefore, it is not clear from where the particles take the energy for their ascent. One of the hypotheses consists of the fact that[33] as a consequence of the friction on particles the front of an incident shock wave is distorted. As a result, the vortices behind the reflected shock wave arise, which lift the particles upwards.

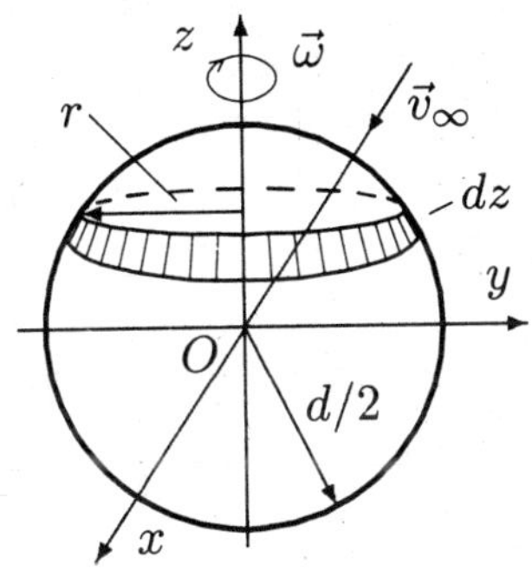

Figure 7.29: The ideal fluid flow around the rotating particle.

Problem 7.8. Find the Magnus force acting on a rotating particle placed in a homogeneous fluid flow in the case in which $\mathrm{Re} = \rho v_\infty d/\mu \gg 1$ (see Fig. 7.29).

Hint: Use the hypothesis of planar flows[20] by subdividing a particle into the cylindrical strips and assuming that they are flowed past by a rotating ideal fluid flow.

Solution: Let a particle of radius $r_p = d/2$ rotate around the z-axis at an angular velocity ω, and let a fluid flow having at infinity the velocity along the x-axis equal to v_∞ and the density ρ take place around the particle. If the Reynolds number is large, $\mathrm{Re} \gg 1$, then the viscosity plays an important role only in a thin boundary layer near the particle surface. Under the viscosity action, the particle rotation is transferred via this layer to the fluid surrounding the particle. Consequently, one can assume in the first approximation that an ideal fluid flow around the particle with certain circulation takes place. Following[20], let us partition the sphere surface into the strips of width dz (see Fig. 7.29), and we will consider the flow around each strip independently of other strips. Then a strip will be subject to the action of a force equal to the Joukowsky lift force for a cylinder of radius r and width dz:

$$dR = \rho v_\infty \Gamma \, dz. \tag{7.5.5}$$

The slip condition is satisfied on the cylinder surface; therefore,

$$\Gamma = \oint \vec{v} \, d\vec{l} = 2\pi r \, (\omega r) = 2\pi r^2 \omega.$$

Substituting and integrating this formula into (7.5.5) we find:

$$R = \rho v_\infty \int_{-r_p}^{r_p} \Gamma(z) \, dz = 2\pi \, \rho v_\infty \omega \int_{-r_p}^{r_p} (r_p^2 - z^2) dz = \frac{8\pi}{3} r_p^3 \rho \, v_\infty \omega. \tag{7.5.6}$$

The Magnus force R acts along the y-axis in the negative direction. In the case of an arbitrary orientation of the vectors $\vec{\omega}$ and $\vec{v}_\infty$, the expression (7.5.6) for the Magnus force will have the form

$$\vec{R} = \frac{\pi d^3}{3} \rho \left(\vec{v}_\infty \times \vec{\omega}\right). \tag{7.5.7}$$

7.5.2 Acoustic Approximation in the Problem of Shock Wave Interaction with a Particle's Cloud at a Small Volume Concentration

Let a shock wave impinge from the left on a cloud of spherical particles [see Fig. 7.21 (a)]. The volume concentration of particles in the cloud is assumed to be small: $m_2 \leq 10^{-3}$, and the cloud thickness h is less than the length of the relaxation zone l_p. In this case the shock wave passes through the cloud with practically no change in its parameters. After it escapes from the channel, a constant gas flow forms with the velocity v_1^0, pressure P^0, and density ρ_1^0. The particles begin to accelerate in this flow, and the gas decelerates. The compression waves arise in the particle's cloud, which propagate beyond the cloud limits. The wave's amplitude will be small at a small volume concentration of particles $m_2 \leq 10^{-3}$, and one can use the acoustic approximation[34] for their computation. The acceleration and velocity of particles will be by a factor of $\rho_{11}/\rho_{22} \ll 1$ smaller than the acceleration and velocity of gas. Therefore, if we restrict ourselves to the times $t \ll \tau_v$, then we can neglect the particles motion; that is, we can take $v_2 = 0$ and assume all particle's parameters to be constant. In this approximation, it is sufficient to solve only the equations for gas (7.4.3) instead of the complete system of equations (7.4.1)–(7.4.3). Assuming that $v_2^k = 0$ in (7.4.3) and neglecting the heat exchange between the gas and particles, we can write the system of equations for gas in the one-dimensional nonstationary case as

$$\begin{gathered}
\frac{\partial \rho_1}{\partial t} + \rho_1 \frac{\partial v_1}{\partial x} + v_1 \frac{\partial \rho_1}{\partial x} = 0, \quad \rho_1 \approx \rho_{11}, \\
\rho_1 \left(\frac{\partial v_1}{\partial t} + v_1 \frac{\partial v_1}{\partial x} \right) = -\frac{\partial P}{\partial x} - \frac{m_2}{\tau} \rho_1 v_1, \quad \tau = \frac{\rho_1}{\rho_{22}} \tau_v, \\
T_1 \left(\frac{\partial S_1}{\partial t} + v_1 \frac{\partial S_1}{\partial x} \right) = \frac{m_2}{\tau} v_1^2, \quad P = \rho_1 R T_1, \\
m_2 = \begin{cases} m_2^0, & 0 < x < h, \\ 0, & x < 0, \quad x > h. \end{cases}
\end{gathered} \tag{7.5.8}$$

Assuming that $\mathrm{Re} = \rho_1 v_1^0 d / \mu$ in the first formula (7.5.1) we obtain the time of the velocity relaxation:

$$\tau_v = \frac{4}{3} \frac{d}{C_d v_1^0} \frac{\rho_{22}}{\rho_1}. \tag{7.5.9}$$

The gas entropy S_1 is expressed in terms of the pressure P and density ρ_1 by formula

$$S_1 = C_V \ln \frac{P}{\rho_1^\gamma} + S_0. \tag{7.5.10}$$

Let us make use of the acoustic approximation and search for the solution of (7.5.8)–(7.5.10) in the form

$$\varphi = \varphi^0 + \varphi', \quad \varphi = (v_1, P, \rho_1, T_1, S_1), \tag{7.5.11}$$

where φ^0 are the gas's parameters behind the incident shock wave and $\varphi' \sim m_2^0$ are the disturbances in gas, which are created by the cloud. Substituting (7.5.11) in the system of equations (7.5.8)–(7.5.10) and retaining the terms of the order φ' and m_2^0, we obtain the equations for disturbances:

$$\begin{gathered}\frac{\partial \rho_1'}{\partial t} + \rho_1^0 \frac{\partial v_1'}{\partial x} + v_1^0 \frac{\partial \rho_1'}{\partial x} = 0,\\ \frac{\partial v_1'}{\partial t} + v_1^0 \frac{\partial v_1'}{\partial x} + \frac{(c^0)^2}{\rho_1^0}\frac{\partial \rho_1'}{\partial x} + \frac{P^0}{C_V \rho_1^0}\frac{\partial S_1'}{\partial x} = -\frac{m_2^0 v_1^0}{\tau}\Theta(x), \\ \frac{\partial S_1'}{\partial t} + v_1^0 \frac{\partial S_1'}{\partial x} = \frac{m_2^0 (v_1^0)^2}{\tau T_1^0}\cdot \Theta(x), \ \Theta(x) = \begin{cases} 1, & 0 < x < h, \\ 0, & x < 0, \ x > h. \end{cases}\end{gathered} \tag{7.5.12}$$

Introducing the nondimensional variables $\eta = \rho_1'/\rho_1^0$, $v = v_1'/v_1^0$, and $s = S'/C_V$ and omitting the subscript "1" by the gas velocity $v^0 \equiv v_1^0$, let us rewrite (7.5.12) as follows:

$$\begin{gathered}\frac{1}{v^0}\frac{\partial \eta}{\partial t} + \frac{\partial \eta}{\partial x} + \frac{\partial v}{\partial x} = 0,\\ \frac{\partial v}{\partial t} + v^0 \frac{\partial v}{\partial x} + \frac{(c^0)^2}{v^0}\frac{\partial \eta}{\partial x} = -\frac{(c^0)^2}{\gamma v^0}\frac{\partial s}{\partial x} - \frac{m_2^0}{\tau}\Theta(x),\\ \frac{\partial s}{\partial t} + v^0 \frac{\partial s}{\partial x} = \gamma(\gamma - 1)M^2 \frac{m_2^0}{\tau}\Theta(x), \quad M = \frac{v^0}{c^0},\\ \Theta(x) = \begin{cases} 1, & 0 < x < h, \\ 0, & x < 0, \quad x > h, \end{cases}\end{gathered} \tag{7.5.13}$$

where M is the Mach number of the flow behind the incident shock wave.

The system of equations (7.5.13) is solved in the region $-\infty < x < +\infty$, $t > 0$ under the initial conditions at $t = 0$ (the Cauchy problem):

$$\eta = 0, \quad v = 0, \quad s = 0. \tag{7.5.14}$$

Multiplying the first equation in (7.5.13) by c_0 and adding to and subtracting from the second equation, we can reduce the system of equations (7.5.13) to the characteristic form

$$\frac{d^\pm I^\pm}{dt} = \varphi, \quad \frac{ds}{dt} = \psi, \quad I^\pm = v \pm \frac{c^0}{v^0}\eta, \tag{7.5.15}$$

where $\mathcal{I}^{\pm}$ are the Riemann invariants, and $\frac{d^{\pm}}{dt}$ and $\frac{d}{dt}$ are the derivatives along the characteristics $C_{\pm}$ and C_0:

$$C_{\pm}: \quad x = (v^0 \pm c^0)t + \xi_{\pm}, \quad C_0: \quad x = v_0 t + \xi_0. \tag{7.5.16}$$

The quantities φ and ψ are determined according to (7.5.13) by formulas

$$\begin{aligned} \varphi &= \varphi_1 + \varphi_2, \\ \varphi_1 &= -\frac{(c^0)^2}{\gamma v^0}\frac{\partial s}{\partial x}, \quad \varphi_2 = -\frac{m_2^0}{\tau}\Theta(x), \\ \psi &= \gamma(\gamma - 1)M^2\frac{m_2^0}{\tau}\Theta(x). \end{aligned} \tag{7.5.17}$$

Integrating (7.5.15), we obtain with regard for (7.5.14):

$$\begin{aligned} v &= \frac{1}{2}\left(\int_{C_+}\varphi\, dt + \int_{C_-}\varphi\, dt\right), \quad \eta = \frac{1}{2}\frac{v^0}{c^0}\left(\int_{C_+}\varphi\, dt - \int_{C_-}\varphi\, dt\right), \\ s &= \int_{C_0}\psi\, dt, \end{aligned} \tag{7.5.18}$$

where $\varphi = \varphi_1 + \varphi_2$. We have along the C_+ characteristic:

$$\begin{aligned} \int_{C_+}\frac{\partial s}{\partial x}\, dt &= \frac{1}{(v^0 + c^0)}\int_{C_+}\left(\frac{\partial s}{\partial t} + (v^0 + c^0)\frac{\partial s}{\partial x}\right) dt \\ &- \frac{1}{(v^0 + c^0)}\int_{C_+}\frac{\partial s}{\partial t}\, dt = \frac{s}{v^0 + c^0} - \frac{1}{(v^0 + c^0)}\int_{C_+}\frac{\partial s}{\partial t}\, dt. \end{aligned}$$

We can find in a similar way along the C_- characteristic:

$$\int_{C_-}\frac{\partial s}{\partial x}\, dt = \frac{s}{v^0 - c^0} - \frac{1}{(v^0 - c^0)}\int_{C_-}\frac{\partial s}{\partial t}\, dt.$$

Adding this formula to the foregoing one and taking into account the second equation in (7.5.17), we obtain:

$$\begin{aligned} \int_{C_+}\varphi_1 dt + \int_{C_-}\varphi_1 dt &= -\frac{s}{\gamma(M^2 - 1)} \\ &+ \frac{1}{2\gamma M}\left(\frac{1}{(M + 1)}\int_{C_+}\frac{\partial s}{\partial t}\, dt + \frac{1}{(M - 1)}\int_{C_-}\frac{\partial s}{\partial t}\, dt\right). \end{aligned} \tag{7.5.19}$$

Substituting (7.5.17) and (7.5.19) in equations (7.5.18), we can write the general solution of (7.5.13) with the initial conditions (7.5.14):

$$v = -\frac{s}{\gamma(M^2 - 1)} + \frac{1}{2\gamma M}\left(\frac{1}{(M + 1)}\int_{C_+}\frac{\partial s}{\partial t}\, dt\right.$$

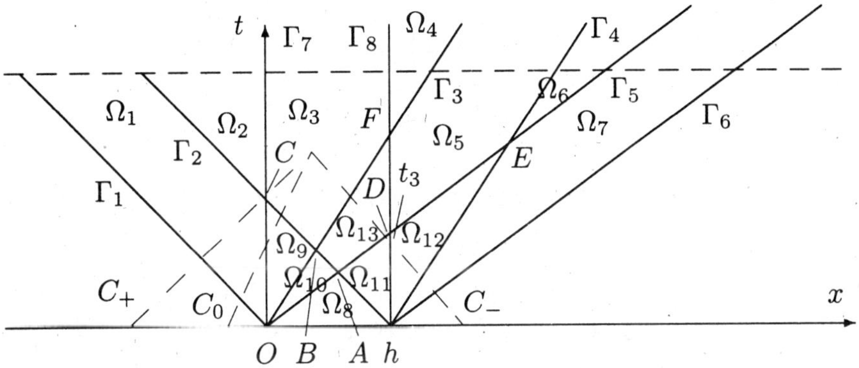

Figure 7.30: The picture of characteristics in the case of a subsonic gas flow ($M < 1$).

$$
\begin{aligned}
&+ \frac{1}{(M-1)}\int_{C_-}\frac{\partial s}{\partial t}\,dt\Bigg) - \frac{m_2^0}{2\tau}\left(\int_{C_+}\Theta\,dt + \int_{C_-}\Theta\,dt\right), \\
\eta &= \frac{s}{\gamma(M^2-1)} + \frac{1}{2\gamma}\left(\frac{1}{(M+1)}\int_{C_+}\frac{\partial s}{\partial t}\,dt\right. \qquad (7.5.20)\\
&\qquad\qquad \left. - \frac{1}{(M-1)}\int_{C_-}\frac{\partial s}{\partial t}\,dt\right) \\
&\qquad\qquad - \frac{m_2^0}{2\tau}\left(\int_{C_+}\Theta\,dt - \int_{C_-}\Theta\,dt\right), \\
s &= \gamma(\gamma-1)\frac{m_2^0}{\tau}M^2\int_{C_0}\Theta\,dt.
\end{aligned}
$$

We will calculate the integrals in (7.5.20) separately for a subsonic gas flow $M < 1$ and supersonic gas flow $M > 1$. At $M < 1$, the picture of characteristics is shown in Fig. 7.30. The half-plane $-\infty < x < \infty, t > 0$ is partitioned by the lines Γ_i into 13 regions Ω_i. The lines Γ_i coincide with the characteristics and the cloud boundaries and are described by the equations

$$
\begin{aligned}
&x = (v^0 - c^0)t, \quad \Gamma_1; \quad x = (v^0 - c^0)t + h, \quad \Gamma_2; \\
&x = v^0 t, \quad \Gamma_3; \quad x = v^0 t + h, \quad \Gamma_4; \\
&x = (v^0 + c^0)t, \quad \Gamma_5; \quad x = (v^0 + c^0)t + h, \quad \Gamma_6; \\
&x = 0, \quad \Gamma_7; \quad x = h, \quad \Gamma_8.
\end{aligned}
$$

The lines Γ_i intersect at points A–F with coordinates

$$\left(x_A = \frac{h}{2}(1+M),\ t_A = \frac{h}{2c^0}\right), \quad \left(x_B = hM,\ t_B = \frac{h}{c^0}\right),$$
$$\left(x_C = 0,\ t_C = \frac{h}{c^0 - v^0}\right), \quad \left(x_D = h,\ t_D = \frac{h}{c^0 + v^0}\right),$$
$$\left(x_E = h(1+M),\ t_E = \frac{h}{c^0}\right), \quad \left(x_F = h,\ t_F = \frac{h}{v^0}\right),$$
$$t_A < t_D < t_B = t_E < t_F < t_C.$$

Let us construct, for example, the solution in the region Ω_3 by using Fig. 7.30. Choosing point G with coordinates t, x, we calculate the integrals in (7.5.20):

$$\int_{C_0} \Theta \frac{dt}{\tau} = \frac{t - t_1}{\tau} = \frac{x}{v^0 \tau},$$

where t_1 is the time of the intersection of the C_0 characteristic with the line $x = 0$. Substituting this value in the last equation of (7.5.20), we obtain:

$$s = \gamma(\gamma - 1)M^2 \frac{m_2^0 x}{\tau v^0}. \tag{7.5.21}$$

The integrals along the C_+ and C_- are computed in terms of elementary functions:

$$\int_{C_+} \Theta \frac{dt}{\tau} = \frac{t - t_2}{\tau} = \frac{x}{(v^0 + c^0)\tau}, \quad \int_{C_-} \Theta \frac{dt}{\tau} = \frac{t - t_3}{\tau} = \frac{h - x}{(c^0 - v^0)\tau}. \tag{7.5.22}$$

The entropy is determined in Ω_3 and Ω_9 by formula (7.5.21); therefore, in these regions $\frac{\partial s}{\partial t} = 0$, and the integral $\int_{C_+} \frac{\partial s}{\partial t}\, dt = 0$. In the region Ω_{13}, we have

$$s = \gamma(\gamma - 1)M^2 \frac{m_2^0 t}{\tau}, \quad \frac{\partial s}{\partial t} = \gamma(\gamma - 1)\frac{M^2 m_2^0}{\tau}.$$

When calculating the entropy in Ω_{12}, a contribution to the integral is given by the interval $0 < t < t_4$, where the time t_4 is found from the equations $x = v^0 t + \xi$, $h = v^0 t_4 + \xi$ and equal to $t_4 = t + (h - x)/v^0$. Consequently, in Ω_{12}, the partial derivative of the entropy s with respect to time is different from zero: $\frac{\partial s}{\partial t} = \gamma(\gamma-1)M^2 m_2^0/\tau$. Thus, the integral

$$\int_{C_-} \frac{\partial s}{\partial t}\, dt = \gamma(\gamma - 1)M^2 m_2^0 \frac{(t_6 - t_5)}{\tau} = \gamma(\gamma - 1)M^2 m_2^0 \frac{h}{c^0 \tau}, \tag{7.5.23}$$

where the difference $t_6 - t_5 = h/c^0$ is determined from the equations of characteristics [see Fig. 7.30)]:

$$x = (v^0 - c^0)t + \xi_-, \quad x_5 = (v^0 - c^0)t_5 + \xi_-,$$
$$x_5 = v^0 t_5 + h, \quad x_6 = (v^0 - c^0)t_6 + \xi_-, \quad x_6 = v^0 t_6.$$

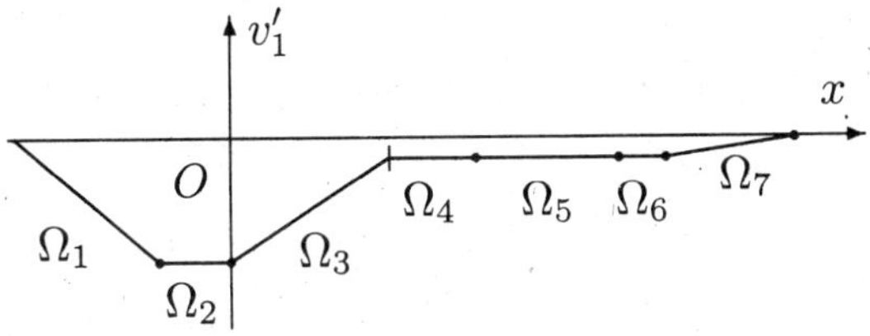

Figure 7.31: The dependence of the velocity disturbance $v_1'(x)$ at a given moment of time.

Substituting (7.5.21)–(7.5.23) into the first equation of (7.5.20), we obtain with regard for equation $v_1' = vv^0$:

$$v_1' = -\frac{m_2^0 h}{2\tau}(1+(\gamma-1)M)\frac{M}{(1-M)} + \frac{\gamma m_2^0}{\tau}\frac{M^2}{(1-M^2)}x. \tag{7.5.24}$$

We can calculate in a similar way the integrals in the remaining regions Ω_i. We present here the results of the calculation of the velocity v_1' only for the first seven regions:

$$\begin{aligned} v_1' &= -\frac{m_2^0 v^0}{2\tau}(1+(\gamma-1)M)\Big(t+\frac{x}{c^0-v^0}\Big), \quad \Omega_1; \\ v_1' &= -\frac{m_2^0 h}{2\tau}(1+(\gamma-1)M)\frac{M}{(1-M)}, \quad \Omega_2; \\ v_1' &= -\frac{m_2^0 h}{2\tau}(1-(\gamma-1)M)\frac{M}{(1+M)}, \quad \Omega_4, \Omega_5, \Omega_6; \\ v_1' &= -\frac{m_2^0 v^0}{2\tau}(1-(\gamma-1)M)\Big(t+\frac{h-x}{v^0+c^0}\Big), \quad \Omega_7. \end{aligned} \tag{7.5.25}$$

We show in Fig. 7.31 a qualitative dependence $v_1'(x)$ constructed by formulas (7.5.24) and (7.5.25) for a given moment of time t as shown by a dashed line in Fig. 7.30. The regions Ω_i are determined from the conditions of the intersection of a horizontal dashed line and the characteristics Γ_i in Fig. 7.30. It can be seen from Fig. 7.31 that a compression wave, in which the gas deceleration takes place, propagates towards the gas flow. In the cloud itself (region Ω_3), the gas accelerates in the rarefaction wave to the final state (region Ω_4), which is determined by the irreversible losses of the gas velocity at the expense of a friction on the particles.

In the case of a supersonic flow $M > 1$, the picture of the characteristics has a qualitatively different form (see Fig. 7.32). The C_- characteristics do not penetrate a region to the left of the particle's cloud $x < 0$; therefore, the flow remains undisturbed there.

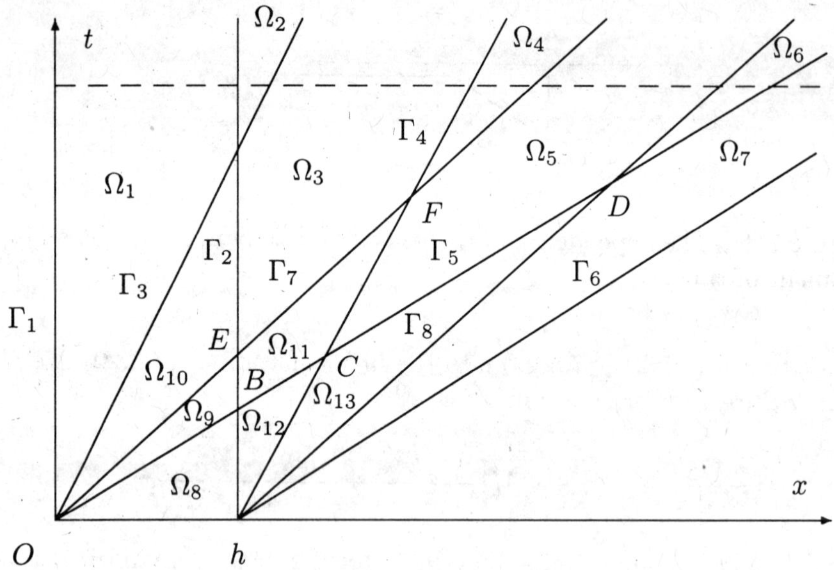

Figure 7.32: The picture of characteristics in the case of a supersonic gas flow ($M > 1$).

The half-plane $-\infty < x < \infty$, $t > 0$ is subdivided into 13 regions Ω_i separated by the line Γ_i:

$$x = 0,\ \ \Gamma_1; \quad x = h,\ \ \Gamma_2; \quad x = (v^0 - c^0)t,\ \ \Gamma_3;$$
$$x = (v^0 - c^0)t + h,\ \ \Gamma_4; \quad x = (v^0 + c^0)t,\ \ \Gamma_5;$$
$$x = (v^0 + c^0)t + h,\ \ \Gamma_6; \ \ x = v^0 t,\ \ \Gamma_7; \quad x = v^0 t + h,\ \ \Gamma_8.$$

The points of intersection of Γ_i determine the moments of time

$$t_A = \frac{h}{v^0 - c^0},\ t_B = \frac{h}{v^0 + c^0},\ t_C = \frac{h}{2c^0},\ t_D = \frac{h}{c^0},\ t_E = \frac{h}{c^0},\ t_F = \frac{h}{c^0},$$

which at $M = v^0/c^0 > 1$ obey the inequalities $t_B < t_C < t_E = t_D = t_F < t_A$. The solution in the regions Ω_i is determined from equations (7.5.20) and constructed similarly to (7.5.24). Let us present the formulas for the velocity disturbance v_1' in the first seven regions:

$$v_1' \ = \ -\frac{\gamma m_2^0 x}{\tau}\frac{M^2}{(M^2-1)}, \quad \Omega_1;$$

$$v_1' \ = \ -\frac{\gamma m_2^0 h}{\tau}\frac{M^2}{(M^2-1)}, \quad \Omega_2;$$

$$v_1' \ = \ -\frac{m_2^0}{\tau}\Big(\gamma h \frac{M^2}{(M^2-1)}$$

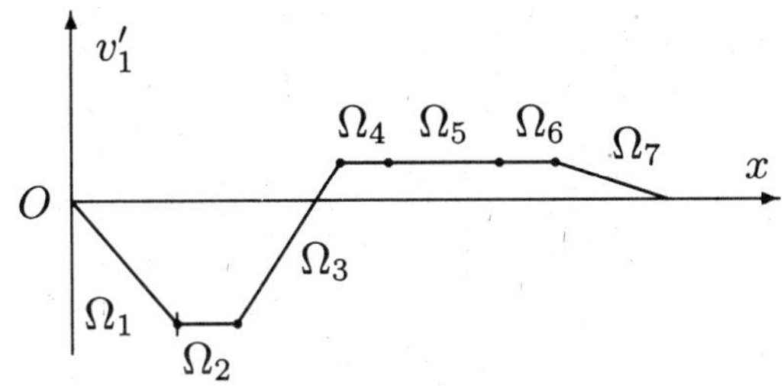

Figure 7.33: The dependence $v_1'(x)$ at a given moment of time.

$$+\quad \frac{1}{2}(1+(\gamma-1)M)\Big(v^0 t-\frac{xM}{(M-1)}\Big)\Big),\quad \Omega_3; \qquad (7.5.26)$$

$$v_1' = -\frac{m_2^0 h}{2\tau}(1-(\gamma-1)M)\frac{M}{(M+1)},\quad \Omega_4,\Omega_5,\Omega_6;$$

$$v_1' = -\frac{m_2^0 v^0}{2\tau}(1-(\gamma-1)M)\Big(t+\frac{h-x}{v^0+c^0}\Big),\quad \Omega_7.$$

We present in Fig. 7.33 a qualitative dependence $v_1'(x)$, which is described by formulas (7.5.26). The moment of time is fixed (it is shown by a dashed line in Fig. 7.32), and the sizes of the regions Ω_i are determined by the point of intersection of the horizontal dashed line in Fig. 7.32 with the characteristics Γ_i. It is seen that, within the cloud, the gas decelerates in the compression wave, and behind the cloud, it accelerates in the rarefaction wave to the velocity exceeding the gas velocity at the cloud inlet. As follows from Fig. 7.32 and formulas (7.5.26), a stationary flow is established in the cloud at $t > t_A$. Let us find the variation of the Mach number of gas flow $M_1 = v_1/c$ in the cloud for this case. Assuming that $v_1 = v^0 + v_1'$, $c = c^0 + c'$ and using the formula $M = v^0/c^0$, we obtain:

$$\frac{M_1}{M} \approx 1+\frac{v_1'}{v^0}-\frac{c'}{c^0}. \qquad (7.5.27)$$

From the equations

$$P = B\rho_1^\gamma \exp{(S_1/C_V)},\quad c = c^0 + c',\quad c^2 = \left(\frac{\partial P}{\partial \rho_1}\right)_{S_1},$$

we can find the expression for the perturbation of the sound velocity:

$$\frac{c'}{c^0} = \frac{s}{2}+\frac{(\gamma-1)}{2}\eta. \qquad (7.5.28)$$

The dependences $\eta(x)$ and $s(x)$ are determined from (7.5.20):

$$\eta = \gamma\frac{m_2^0}{\tau v^0}\frac{M^2}{(1-M^2)}x,\quad s = \gamma(\gamma-1)\frac{m_2^0 M^2}{v^0\tau}x. \qquad (7.5.29)$$

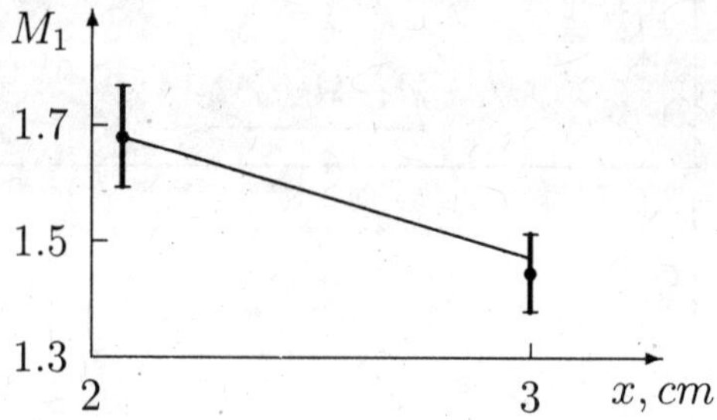

Figure 7.34: The dependence of the Mach number of the gas flow in the cloud of particles. The line is the result of computation by (7.5.30); the points are obtained from the experiment.

Substituting (7.5.28) and (7.5.29) and the first equation from (7.5.26) in (7.5.27), we obtain:

$$\frac{M_1}{M} = 1 - \frac{\gamma m_2^0}{\tau v^0} \frac{M^2}{(M^2-1)} \left(1 + \frac{\gamma - 1}{2} M^2\right) x, \quad M = \frac{v^0}{c^0}. \qquad (7.5.30)$$

We show in Fig. 7.34 the experimentally measured values of the flow Mach number M_1 at the cloud inlet and outlet[35]. The vertical lines correspond to the experiment error. The solid line is the result of computation by formula (7.3.30) for the given experiment: $M = 1.68$, $m_2^0 = 2 \cdot 10^{-3}$, $d = 400\ \mu m$, and $\rho_{22} = 1.2\ g/cm^3$ (the acrylic plastic). It can be seen that there is the good agreement between the computation and the experiment. As is seen from (7.5.30), the flow Mach number M behind the incident shock wave has a significant effect on the line inclination $\frac{dM_1}{dx}$.

Problem 7.9. Find the dependences of the nondimensional density $\eta(t, x)$ and entropy $s(t, x)$ in the regions $\Omega_1 \div \Omega_7$ at $M < 1$ and $M > 1$ (see Figs. 7.29 and 7.31) in the acoustic approximation.

Hint: Use the formulas (7.5.20).

7.6 The Flows of Heterogeneous Media without Regard for Inertial Effects

7.6.1 The Brownian Motion of Particles in a Fluid

Consider a fluid at rest, in which a large number of spherical particles are suspended. We will assume that the mixture is in the field of the gravity force with the gravity acceleration $\vec{g}$. The particles will descend under the gravity force, and, if there were no Brownian motion, all of them would form sediment on the bottom. The Brownian motion of particles, however, hinders their sedimentation. As a result, a stationary distribution of particles in the fluid is established.

Let us derive the equations determining the particle's concentration n in the fluid. The Brownian motion of particles arises under the influence of the fluid molecules, which perform a thermal motion, collide with a particle, and transfer some impulse to it. Under the action of these random impulses, the particles begin to chaotically move in space, and such a motion is called the *Brownian* motion. Since the collisions of molecules occur uniformly over the total particle surface, the mean force acting on the particle will be equal to zero. Consequently, the motion equation of particles in the system (7.1.34) remains unchanged (the motion equations in the liquid and gas coincide). Assuming the pressure in fluid to be constant, $\nabla^k P = 0$ and neglecting the inertial term $\rho_2 \frac{d_2 v_2^k}{dt} \approx 0$, we write the equation for particles motion with regard for the gravity force:

$$n\vec{f}_p + \rho_2 \vec{g} = 0, \tag{7.6.1}$$

where $\rho_2 = n\, m_p$ is the mean particle's density and $m_p = \frac{\pi d^3}{6}\rho_{22}$ is the particle mass. The force f_p acting on a particle moving at a velocity $\vec{v}$ in the fluid at rest is given at small Reynolds numbers $\mathrm{Re} \ll 1$ by the Stokes formula

$$\vec{f}_p = -3\pi\eta d\, \vec{v}, \tag{7.6.2}$$

where η is the fluid viscosity. The formula (7.6.2) follows from equations (7.1.20) and (7.1.21) if one assumes therein $\vec{v}_1 = 0$, $\vec{v}_2 = \vec{v}$, $\eta = \rho_{11}\nu$, and $M_{12} \to 0$. Substituting (7.6.2) into (7.6.1), we obtain the velocity of particle motion in the gravity field:

$$\vec{v} = b\vec{F}, \quad b = \frac{1}{3\pi\eta d}, \quad \vec{F} = m_p \vec{g}. \tag{7.6.3}$$

The constant b is the mobility, and $\vec{F}$ is a force acting on a single particle.

The presence of the Brownian motion will lead to the alteration of the continuity equation for particles:

$$\frac{\partial n}{\partial t} + \operatorname{div} \vec{j} = 0, \quad \vec{j} = n\vec{v}, \quad n = \frac{\rho_2}{m_p}, \tag{7.6.4}$$

where $\vec{j} = n\vec{v}$ is the particle's flow related to the motion at a velocity $\vec{v}$. As a result of a Brownian motion, an additional diffusive flux of particles[36] arises:

$$\vec{\imath} = -D\,\nabla n, \tag{7.6.5}$$

so that the total particle's flux will be equal to

$$\vec{J} = n\vec{v} - D\,\nabla n, \tag{7.6.6}$$

where D is the diffusion coefficient of particles in the fluid. Substituting (7.6.6) into the first equation of (7.6.4) instead of $\vec{j}$, we obtain:

$$\frac{\partial n}{\partial t} + \operatorname{div} n\vec{v} = D\Delta n. \tag{7.6.7}$$

Note that formula (7.6.5) is similar to formula (2.1.108) determining the heat flux. This is not accidental and is related to the fact that both of these fluxes are conditioned by the molecular nature of substance. They arise in the presence of the gradients of concentration n and temperature T and lead to an irreversible entropy growth.

Let us find the diffusion coefficient D of particles in the fluid. For this purpose, we will consider the particles in the fluid as a weak solution[37], in which the particles play the role of a dissolved substance and the fluid plays the role of solvent. In the theory of solvents, the concentration $c = n/N \ll 1$ is a solvent characteristic, where N is the concentration of solvent particles and n is the concentration of the dissolved substance. If the chemical potential (the Gibbs potential) of a pure solvent is $G_0 = N\mu_0(P,T)$, then we will assume that at the addition of one particle of a substance to be dissolved it changes by the amount $\varepsilon(P,T,N)$. Assuming the solution to be weak and neglecting the interaction between the particles of the substance to be dissolved, we find the expressions for the solution chemical potential in which n particles are dissolved:

$$G = N\,\mu_0(P,T) + n\varepsilon(P,T,N) + T\ln n!. \tag{7.6.8}$$

The appearance of the last item is related to the indistinguishability of particles of the dissolved substance. In this case, one must subdivide the statistical sum Z into a number of possible permutations of particles $n!$, and the expression for free energy will have the form[37]

$$F = -T\ln\frac{Z}{n!} = -T\ln Z + T\ln n!.$$

One must do the same correction in the chemical potential G. For $n \gg 1$, one can use the Stirling's formula $\ln n! = n\ln(n/e)$, with regard for which,

$$G = N\mu_0 + n(\varepsilon + T\ln\frac{n}{e}) = N\mu_0 + nT\ln\left(\frac{n}{e}e^{\varepsilon/T}\right).$$

The chemical potential G should be a homogeneous function of the first order of n and N. This is possible when $\exp(\varepsilon/T) = \varphi'(P,T)/N$; then

$$G = N\mu_0 + nT\ln\left(\frac{n}{N}\frac{\varphi'}{e}\right). \tag{7.6.9}$$

The chemical potential differential of the solution is given by formula

$$dG = -S\,dT + V\,dP + \mu_e\,dN + \mu_p\,dn. \tag{7.6.10}$$

We can find from formulas (7.6.9) and (7.6.10) the chemical potential of the solvent:

$$\mu_e = \frac{\partial G}{\partial N} = \mu_0 - \frac{n}{N}T \tag{7.6.11}$$

and of the dissolved substance:

$$\mu_p = \frac{\partial G}{\partial n} = T \ln \frac{n}{N} + \varphi(P, T), \tag{7.6.12}$$

where $\varphi(P, T) = T \ln \varphi'(P, T)$.

Let us apply formula (7.6.12) to our case of a weak solution of particles in the fluid. Assuming the pressure, temperature, and density of fluid to be constant, we can find from equation (7.6.12) the expression

$$\nabla \mu_p = \frac{T}{n} \nabla n \tag{7.6.13}$$

with regard for which the total particle's flux (7.6.6) may be rewritten as

$$\vec{J} = n\vec{v} - \frac{Dn}{T} \nabla \mu_p, \tag{7.6.14}$$

where μ_p is a chemical potential of the suspended particles. Consider an equilibrium state of the particles and fluid in the gravity field. In this case, the additional items will enter the chemical potential:

$$dG = -S\,dT + v\,dP + (\mu_e + U_e)\,dN + (\mu_p + U_p)\,dn, \tag{7.6.15}$$

where $U_e(\vec{r})$ is the potential energy per one liquid particle and $U_p(\vec{r})$ is the potential energy per one solid particle, so that $\vec{F} = -\nabla U_p$. It follows from (7.6.15) that

$$\left(\frac{\partial G}{\partial n}\right)_{T,P,N} = \mu_p + U_p. \tag{7.6.16}$$

If the pressure, temperature, and density of fluid are constant, then we obtain from (7.6.16):

$$G = \int (\mu_p + U_p)\,dn. \tag{7.6.17}$$

In the equilibrium, the total particles flux is equal to zero: $\vec{J} = 0$ and the chemical potential G achieves its minimum. The condition for the minimality of G leads to equation[37]

$$\mu_p + U_p = \text{const}, \tag{7.6.18}$$

from where we find:

$$\nabla \mu_p = -\nabla U_p = +\vec{F}. \tag{7.6.19}$$

Substituting (7.6.3) and (7.6.19) in equation (7.6.14) and assuming $\vec{J} = 0$ therein, we obtain:

$$D = bT = \frac{T}{3\pi\eta d}. \tag{7.6.20}$$

Equation (7.6.20) is known in the literature as the Einstein relation. The viscosity η entering formula (7.6.20) depends on the volume concentration of particles m_2. At a small concentration $m_2 \ll 1$, the Einstein formula[36]

$$\eta = \eta_0\left(1 + \frac{5}{2}m_2\right) \tag{7.6.21}$$

is valid for spherical particles, where η_0 is the pure liquid viscosity.

Problem 7.10. The particles in the fluid perform a Brownian motion. At the initial moment of time, one particle is located at a distance x_0 from the wall. Assuming that the particle sticks to the wall after it reaches the wall, determine the probability of a sticking of the particle during the time t [36].

Solution: The particle motion in the fluid can be considered as diffusion. The probability $\varphi(\vec{r}, t)$ plays the role of concentration n in equation (7.6.7), for which at $\vec{v} = 0$, we have the equation

$$\frac{\partial\varphi}{\partial t} = D\,\Delta\varphi. \tag{7.6.22}$$

The initial and boundary conditions for equation (7.6.22) have the form

$$\varphi\Big|_{t=0} = \delta(x - x_0), \quad \varphi\Big|_{x=0} = 0, \quad x > 0. \tag{7.6.23}$$

One can replace the initial- and boundary-value problem (7.6.22),(7.6.23) with the Cauchy problem $t > 0$, $-\infty < x < \infty$ with the initial conditions

$$\varphi|_{t=0} = \delta(x - x_0) - \delta(x + x_0). \tag{7.6.24}$$

At first consider the Cauchy problem for equation (7.6.22) at an arbitrary dependence $\varphi|_{t=0} = \varphi^0(\vec{r})$. Let us present the function $\varphi(\vec{r}, t)$ sought for in terms of the Fourier integral:

$$\varphi(\vec{r}, t) = \int \varphi_k(t) e^{i\vec{k}\cdot\vec{r}} \frac{d^3k}{(2\pi)^3}, \quad \varphi_k(t) = \int \varphi(\vec{r}, t) e^{-i\vec{k}\cdot\vec{r}} d^3x, \tag{7.6.25}$$

where

$$\vec{k}\cdot\vec{r} = k_x x + k_y y + k_z z, \quad d^3k = dk_x\, dk_y\, dk_z, \quad d^3x = dx\, dy\, dz.$$

Differentiating $\varphi(\vec{r},t)$ with respect to x and t, we find with the aid of (7.6.25):

$$\frac{\partial\varphi}{\partial t}=\int\frac{\partial\varphi_k}{\partial t}e^{i\vec{k}\cdot\vec{r}}\frac{d^3k}{(2\pi)^3},\ \Delta\varphi=-\int k^2\varphi_k(t)e^{i\vec{k}\cdot\vec{r}}\frac{d^3k}{(2\pi)^3},\ k^2=k_x^2+k_y^2+k_z^2.$$

Substituting these formulas in (7.6.22), we obtain the ordinary differential equation:

$$\frac{d\varphi_k}{dt}+Dk^2\varphi_k=0$$

with the initial condition

$$\varphi_k|_{t=0}=\varphi_k^0=\int\varphi^0(\vec{r}')\,e^{-i\vec{k}\cdot\vec{r}}d^3x',$$

which has the solution

$$\varphi_k=\varphi_k^0e^{-k^2Dt}.$$

Substituting this solution into the first equation of (7.6.25) we obtain the solution of the Cauchy problem for an arbitrary function $\varphi^0(\vec{r}')$:

$$\varphi(\vec{r},t)=\iint_{-\infty}^{\infty}\varphi^0(\vec{r}')e^{i\vec{k}\cdot(\vec{r}-\vec{r}')-k^2Dt}\frac{d^3k}{(2\pi)^3}\,d^3x'. \tag{7.6.26}$$

The integral over d^3k is equal to the product of three equal integrals of the form

$$\int_{-\infty}^{\infty}e^{-k_x^2Dt}\cos k_x(x-x')\,dk_x=\left(\frac{\pi}{Dt}\right)^{1/2}\exp\left(-\frac{(x-x')^2}{4Dt}\right),$$

where we have taken into account the fact that by virtue of the oddness of the function $\sin k_x(x-x')$ the integral

$$i\int_{-\infty}^{\infty}e^{-k_x^2Dt}\sin k_x(x-x')\,dk_x=0.$$

Multiplying the integrals over dk_x, dk_y, dk_z and substituting in (7.6.26), we can write:

$$\varphi(\vec{r},t)=\int_{-\infty}^{\infty}\varphi^0(\vec{r}')\exp\left(-\frac{(x-x')^2+(y-y')^2+(z-z')^2}{4Dt}\right)\frac{d^3x'}{(4\pi Dt)^{\frac{3}{2}}}. \tag{7.6.27}$$

The function (7.6.24) is in our case antisymmetric in x; therefore, assuming that $\varphi^0(x,y,z)=-\varphi^0(-x,y,z)$, we can rewrite (7.6.27) as

$$\varphi(\vec{r},t)=\int_{-\infty}^{\infty}\int_{-\infty}^{\infty}\int_0^{\infty}\varphi^0(x',y',z')\exp\left(-\frac{(y-y')^2+(z-z')^2}{4Dt}\right)\cdot$$

$$\left\{\exp\left(-\frac{(x-x')^2}{4Dt}\right)-\exp\left(-\frac{(x+x')^2}{4Dt}\right)\right\}\frac{dy'dz'dx'}{(4\pi Dt)^{3/2}}. \tag{7.6.28}$$

If φ^0 does not depend on y and z, then, by computing in (7.6.28) the integrals

$$\int_{-\infty}^{\infty} \exp\left(-\frac{(y'-y)^2}{4Dt}\right)\frac{d(y'-y)}{\sqrt{4Dt}} = \int_{-\infty}^{\infty} \exp\left(-\frac{(z'-z)^2}{4Dt}\right)\frac{d(z'-z)}{\sqrt{4Dt}} = \int_{-\infty}^{\infty} e^{-\xi^2}\, d\xi = \sqrt{\pi}$$

we obtain:

$$\varphi(\vec{r},t) = \int_0^{\infty} \varphi^0(x')\left\{\exp\left(-\frac{(x-x')^2}{4Dt}\right) - \exp\left(-\frac{(x+x')^2}{4Dt}\right)\right\}\frac{dx'}{(4\pi Dt)^{1/2}}. \tag{7.6.29}$$

Substituting (7.6.24) into (7.6.29), we find, with regard for the equality to zero of the integrals of $\delta(x'+x_0)$ and

$$\int_0^{\infty} \delta(x'-x_0)\exp\left(-\frac{(x\pm x')^2}{4Dt}\right) dx' = \exp\left(-\frac{(x\pm x_0)^2}{4Dt}\right),$$

the probability of a particle stay in a plane standing at a distance x from the coordinate origin at a given moment of time t:

$$\varphi(x,t) = \frac{1}{2\sqrt{\pi Dt}}\left(\exp\left(-\frac{(x-x_0)^2}{4Dt}\right) - \exp\left(-\frac{(x+x_0)^2}{4Dt}\right)\right). \tag{7.6.30}$$

The probability of sticking to the wall located at the coordinate origin $x=0$ per unit time is equal to the diffusive flux $D\frac{\partial\varphi}{\partial x}|_{x=0}$. Using the constructed solution (7.6.30), we can find the probability of a sticking during the time t:

$$w(t) = D\int_0^t \frac{\partial\varphi}{\partial x}\bigg|_{x=0} dt = \int_0^t \frac{x_0}{2\sqrt{\pi Dt^3}}\exp\left(-\frac{x_0^2}{4Dt}\right) dt = \frac{2}{\sqrt{\pi}}\int_{\frac{x_0}{\sqrt{4Dt}}}^{\infty} e^{-\xi^2}\, d\xi = 1-\operatorname{erf}\left(\frac{x_0}{\sqrt{4Dt}}\right), \tag{7.6.31}$$

where

$$\operatorname{erf} x = \frac{2}{\sqrt{\pi}}\int_0^x e^{-\xi^2}\, d\xi, \quad \operatorname{erf}(\infty) = 1,$$

the error function. At $t\to\infty$ $\operatorname{erf}(x/\sqrt{4Dt})\to 0$, $w\to 1$ and the particle will necessarily stick to the surface as the time increases. We show in Fig. 7.35 a qualitative dependence of the sticking probability w on time t (7.6.31). It can be seen that a characteristic time for the sticking of a particle at the expense of diffusion in the fluid is $t^*\sim x_0^2/(4D)$.

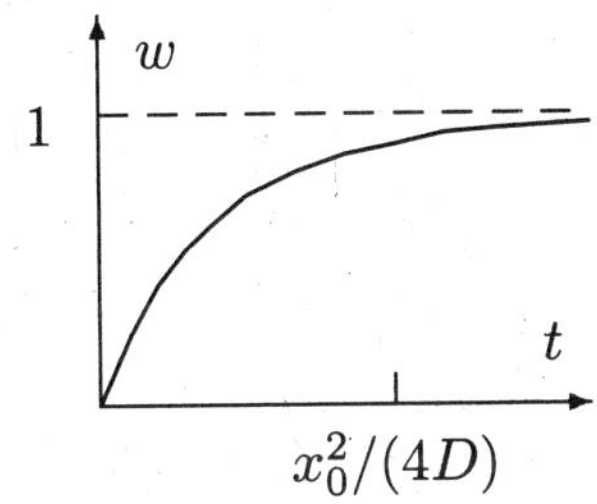

Figure 7.35: A qualitative dependence of the probability of particle sticking w on time t.

Problem 7.11. A suspension of fluid and solid particles is in a cylindrical glass, which rotates around its axis at a frequency ω. Find the dependence of the particles concentration n on the distance from the r axis.

7.6.2 Fluid Filtration in a Porous Medium

Following[5], let us consider the fluid filtration in a porous medium, which we will consider to be nondeformable, i.e., $\vec{v}_2 = 0$, with a constant porosity $m_1 =$ const. The fluid filtration is a slow process; therefore, one can neglect in the fluid motion equations [$i = 1$ in equation (7.1.11)] the inertial terms $\rho_1 \frac{d_1 v_1^k}{dt}$. We will assume that

1) the porous skeleton possesses a large heat conductivity, and the temperatures of fluid and porous skeleton are equal to $T_1 = T_2 =$ const;

2) a mass exchange between the porous skeleton and fluid is absent, $\kappa_{12} = \kappa_{21} = 0$;

3) the viscosity is taken into account only in the forces of interphase interaction, $\sigma_1^{kl} = -P\delta^{kl}$, $P_{21}^k = -m_1 \eta v_1^k / k$, where η is the fluid viscosity, and k is the absolute permeability of a porous medium.

Substituting these relations in equations (7.1.10)–(7.1.12), we obtain the equations for fluid filtration in an inert porous medium:

$$\frac{\partial \rho_1}{\partial t} + \nabla_k \rho_1 v_1^k = 0, \quad \vec{v}_1 = -\frac{k}{m_1 \eta} \nabla P,$$
$$\vec{v}_1 = v_1^k \vec{e}_k, \quad \rho_1 = \rho_{11} m_1, \quad \rho_{11} = \text{const}. \tag{7.6.32}$$

The second equation in (7.6.32) is called the *Darcy law*. The absolute permeability of a porous medium k is determined from experiment and depends on the physicomechanical properties of the porous skeleton. The

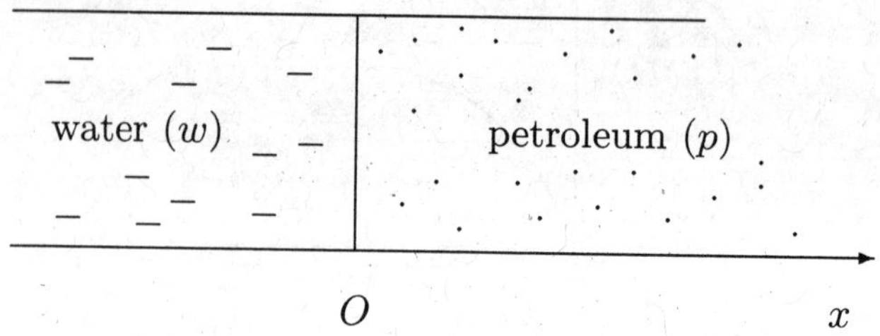

Figure 7.36: The problem of the petroleum displacement by water.

fluid filtration in porous skeleton is similar to a viscous fluid flow in a tube at small Reynolds numbers. Therefore, it follows from the comparison of (7.6.32) with the Poiseuille formula that the absolute permeability $k \sim r^2$, where r is the mean pore radius.

The filtering fluid consists as a rule of several components. Consider an important particular case of the filtration of a two-component fluid at the example of the displacement of petroleum by water (see Fig. 7.36). We will mark the parameters referring to petroleum and water by indices p and w, respectively, and we will omit index 1 for brevity. Introduce the volume concentrations of water (the water saturation) S_w and petroleum (petroleum saturation) S_p. Then we obtain from (7.6.32) the continuity equations:

$$\frac{\partial \rho_w}{\partial t} + \nabla_k \rho_w v_w^k = 0, \quad \rho_w = m S_w \rho_w^0,$$
$$\frac{\partial \rho_p}{\partial t} + \nabla_k \rho_p v_p^k = 0, \ \rho_p = m S_p \rho_p^0, \ S_p + S_w = 1, \quad (7.6.33)$$

and the motion equations:

$$m S_w \vec{v}_w = -\frac{k K_w}{\eta_w} \nabla P, \quad m S_p \vec{v}_p = -\frac{k K_p}{\eta_p} \nabla P, \qquad (7.6.34)$$

where ρ_w^0 is the true water density, ρ_p^0 is the true petroleum density, m is the porosity, and $K_p = K_p(S_p)$, $K_w = K_w(S_w)$ are the phase permeabilities. A qualitative dependence of phase permeabilities on the volume concentrations is determined from the experiment and shown in Fig. 7.37 [5,38]. It can be seen that, at some value of the water saturation S_w^*, already no petroleum displacement occurs, $\vec{v}_p = 0$ because $K_p(S_w^*) = 0$. This is related to the fact that a part of petroleum $S_p^* = 1 - S_w^*$ is restrained in narrow capillaries by the water plugs at the expense of the capillary effects (see Fig. 7.38). The dependence of the residual petroleum S_p^* on the nondimensional capillary number

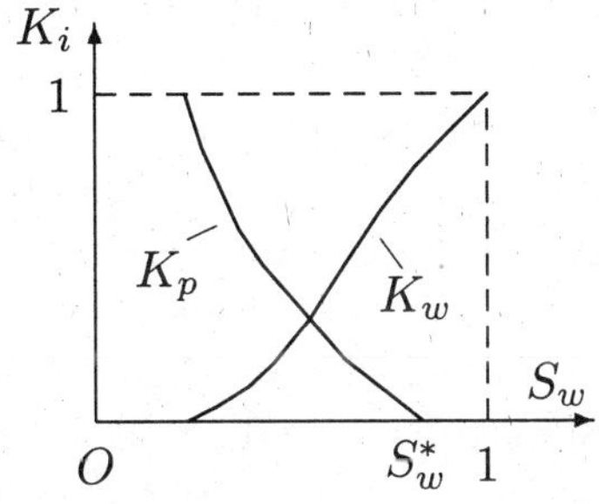

Figure 7.37: A qualitative dependence of the permeability on the volume concentration S_w ($S_p + S_w = 1$).

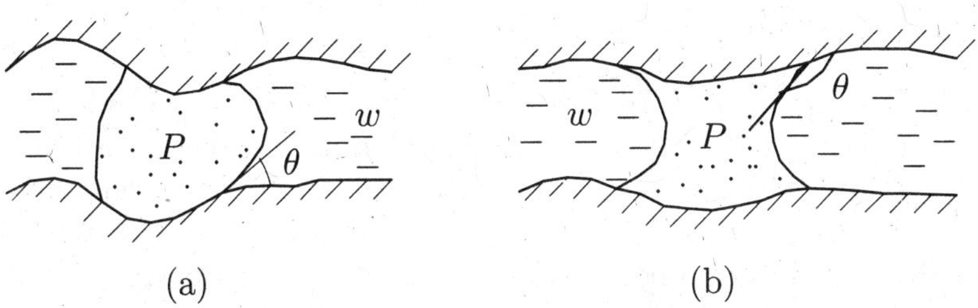

Figure 7.38: The pinch of petroleum (a) moistened ($\theta < 90^o$) and (b) not moistened ($\theta > 90^o$) by water.

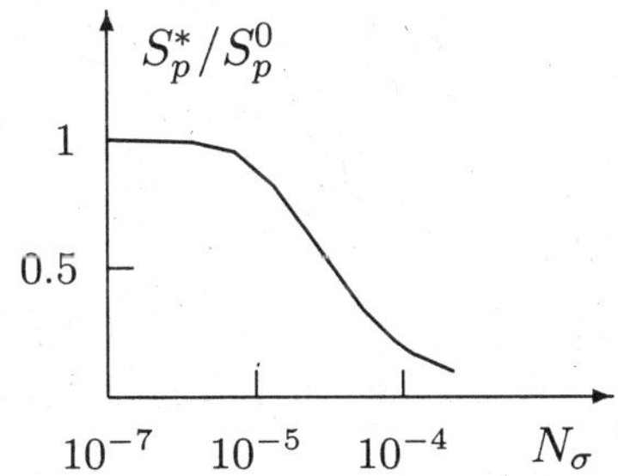

Figure 7.39: The dependence of the residual petroleum saturation on the capillary number in the sand, where $S_p^0 = S_p^*$ at $N_\sigma = 10^{-7}$.

$N_\sigma = k\nabla P/\sigma \sim r\nabla P/(\sigma/r)$, which is shown in Fig. 7.39, has been obtained from experiment[39]. Here σ is the surface tension coefficient, and $r \sim \sqrt{k}$ is the radius of pores. At the displacement of petroleum by water, usually $N_\sigma \sim 10^{-7}$ and $S_p^0 \approx 0.4$ for $\sigma = 30\ g/s^2$ [39]. An efficient displacement of the petroleum restrained in the capillaries is possible at $N_\sigma \approx 10^{-4}$. The increase in N_σ from 10^{-7} to 10^{-3} is possible at the expense of the reduction of the surface tension coefficient of fluid σ at the interphase boundaries. To reduce σ one uses the surface-active substances, under the action of which the petroleum and water form a micellar solution in the form of petroleum and small water particles. Consider a one-dimensional nonstationary problem of the petroleum displacement by water[5] (see Fig. 7.36). Assuming the porosity constant, $m =$ const, we rewrite equations (7.6.33) and (7.6.34) in the one-dimensional nonstationary case as

$$\frac{\partial S_w}{\partial t} + \frac{\partial}{\partial x}(S_w v_w) = 0, \quad \frac{\partial S_p}{\partial t} + \frac{\partial}{\partial x}(S_p v_p) = 0,$$
$$mS_w v_w = -\frac{kK_w}{\eta_w}\frac{\partial P}{\partial x}, \quad mS_p v_p = -\frac{kK_p}{\eta_p}\frac{\partial P}{\partial x}, \quad S_w + S_p = 1. \quad (7.6.35)$$

When writing equations (7.6.33) and (7.6.35), we have neglected a diffusive transport of the components w and p as compared to the convective transport. This approximation is valid at large Péclet numbers $\mathrm{Pe} = v_p L/D \gg 1$, when $D\nabla S_p \ll S_p v_p$. Adding the first two equations in (7.6.35) and multiplying the sum by $m =$ const, we obtain the following equation with regard for the relations $S_w + S_p = 1$, $\frac{\partial}{\partial t}(S_w + S_p) = 0$:

$$\frac{\partial}{\partial x}(mS_w v_w + mS_p v_p) = 0,$$

the integral of which determines the volume flow rate of the mixture:

$$W(t) = mS_w w_w + mS_p v_p. \quad (7.6.36)$$

Introduce the functions F_w and F_p determining the fractions of water and petroleum in the total flow rate W:

$$F_w = \frac{mS_w v_w}{W}, \quad F_p = \frac{mS_p v_p}{W}, \quad F_w + F_p = 1, \quad (7.6.37)$$

which are called the *Buckley–Leverett functions* in the literature. With regard for (7.6.36) and (7.6.37), the system of equations (7.6.35) may be rewritten as

$$m\frac{\partial S_w}{\partial t} + W(t)\frac{\partial F_w}{\partial x} = 0,$$
$$\frac{\partial P}{\partial x} = \frac{W(t)}{k(K_w/\eta_w + K_p/\eta_p)},$$

$$F_w = \frac{K_w/\eta_w}{K_w/\eta_w + K_p/\eta_p}, \quad F_p = \frac{K_p/\eta_p}{K_w/\eta_w + K_p/\eta_p}, \tag{7.6.38}$$
$$K_w = K_w(S_w), \qquad K_p = K_p(S_p).$$

Introducing the variable

$$\theta = \frac{1}{m}\int_0^t W(t)\,dt,$$

computing the derivative $\frac{\partial S_w}{\partial t} = \frac{\partial S_w}{\partial \theta}\frac{d\theta}{dt} = \frac{\partial S_w}{\partial \theta}\frac{W}{m}$, and substituting into the first equation of (7.6.38), we obtain the Buckley–Leverett equation:

$$\frac{\partial S_w}{\partial \theta} + F'_w \frac{\partial S_w}{\partial x} = 0, \quad F'_w = \frac{dF_w}{dS_w} \geq 0. \tag{7.6.39}$$

This equation enables one to find the function $S_w(x,t)$. Then $S_p(x,t) = 1 - S_w(x,t)$, and the phase permeability is restored on the basis of these functions:

$$K_w = K_w(S_w(x,t)), \quad K_p = K_p(S_w(x,t)),$$

and the pressure distribution $P(x,t)$ is finally determined from the second equation (7.6.38).

The characteristics of the quasilinear hyperbolic equation (7.6.39) are determined from equation $\frac{dx}{d\theta} = F'_w$ and have the form

$$x - \xi = \int_0^\theta F'_w(S_w(x,\theta))\,d\theta. \tag{7.6.40}$$

Thus, F'_w is the velocity of the propagation of the water saturation wave. We show in Figs. 7.40–7.42 (see[5]) the qualitative dependence of F_w on S_w as well as the picture of the propagation of the water saturation waves. It follows from Fig. 7.40 that the function F'_w is nonmonotone. In interval 1, the second derivative $\frac{d^2F_w}{dS_w^2} > 0$, and in interval 2, we have $\frac{d^2F_w}{dS_w^2} < 0$. Consequently, in interval 1, a wave with a larger amplitude S_w propagates with a larger velocity F'_w. It catches up with a wave with a smaller amplitude, and the effect of the wave turning over arises. To eliminate the solution ambiguity, one introduces the discontinuity surface (see Fig. 7.42). The jump relations follow from the conservation law for the mass of phases:

$$S_w^+(D - v_w^+) = S_w^-(D - v_w^-), \quad S_p^+(D - v_p^+) = S_p^-(D - v_p^-). \tag{7.6.41}$$

The discontinuity velocity D_θ in terms of the variables x, θ is

$$D_\theta = (F_w^+ - F_w^-)/(S_w^+ - S_w^-). \tag{7.6.42}$$

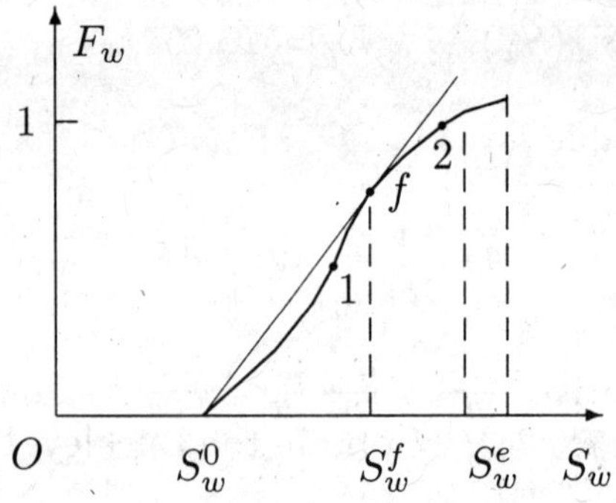

Figure 7.40: The Buckley–Leverett function.

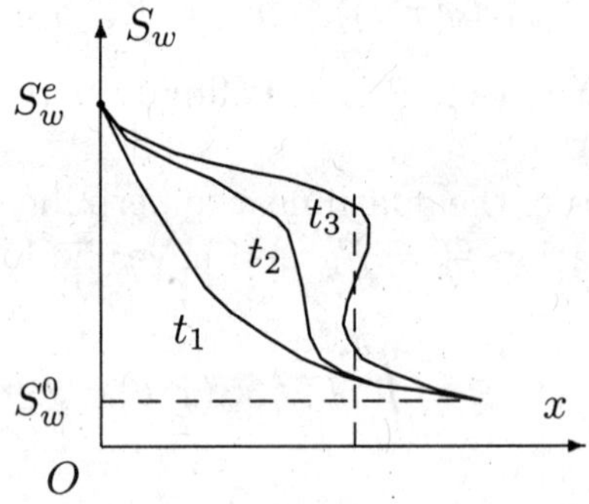

Figure 7.41: The tilting of the water saturation wave ($t_1 < t_2 < t_3$).

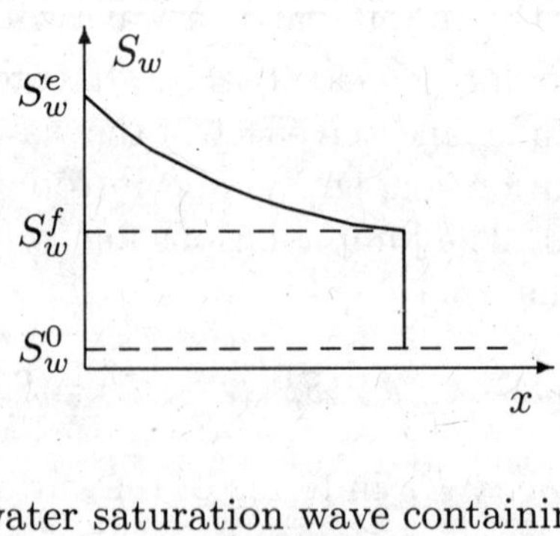

Figure 7.42: The water saturation wave containing a discontinuity.

Using the relations $\frac{dx}{dt} = \frac{dx}{d\theta} \cdot \frac{d\theta}{dt} = \frac{dx}{d\theta}\frac{W}{m}$, $D_\theta = \frac{dx}{d\theta}$, and formula (7.6.41), we can find the discontinuity velocity in terms of the variables x and t:

$$D = \frac{W}{m} \frac{(F_w^+ - F_w^-)}{(S_w^+ - S_w^-)}. \tag{7.6.43}$$

Three equations, (7.6.41) and (7.6.43), are insufficient for the determination of the four quantities S_w^+, v_w^+, v_p^+, D. An additional equation is found from the condition for the tangent of a straight line joining the points ahead of and behind the discontinuity and the curve $F_w(S_w)$. It is equivalent to the Chapman–Jouguet condition in the theory of detonation and gives the equation

$$D_\theta = c, \qquad c = \frac{dF_w}{dS_w}. \tag{7.6.44}$$

The tangent point is denoted in Fig. 7.40 by the letter f, and the jump magnitude is equal to $S_w^f - S_w^0$. If the jump ended at point 1, then $D_\theta < c$ and the compression waves would overtake the jump, thus strengthening it until the jump shifts to the point f. The jump cannot also end at point 2 since the straight line will intersect the curve $F_w(S_w)$ at more than one point, and the solution will be ambiguous. Such a jump will split into a faster jump corresponding to point f and a continuous water saturation wave.

In order to construct a continuous solution behind the jump, let us formulate the initial and boundary conditions in the problem of the petroleum displacement by water (see Fig. 7.36):

$$S_w(x) = S_w^0 \text{ at } t = 0, \quad S_w(t) = S_w^e \text{ at } x = 0. \tag{7.6.45}$$

Since no reference length and time are in the problem (7.6.39), (7.6.45), we will search for the solution of (7.6.39) in the self-similar variables

$$S_w = S_w(\zeta), \quad \zeta = \frac{x}{\theta}. \tag{7.6.46}$$

The substitution of (7.6.46) into (7.6.39) leads to the ODE

$$(\zeta - F_w'(S_w))\frac{dS_w}{d\zeta} = 0 \tag{7.6.47}$$

with the boundary conditions

$$S_w = S_w^e \text{ at } \zeta = 0, \quad S_w = S_w^0 \text{ at } \zeta = \infty. \tag{7.6.48}$$

Equation (7.6.47) has the solution of two types: in the form of a centered wave

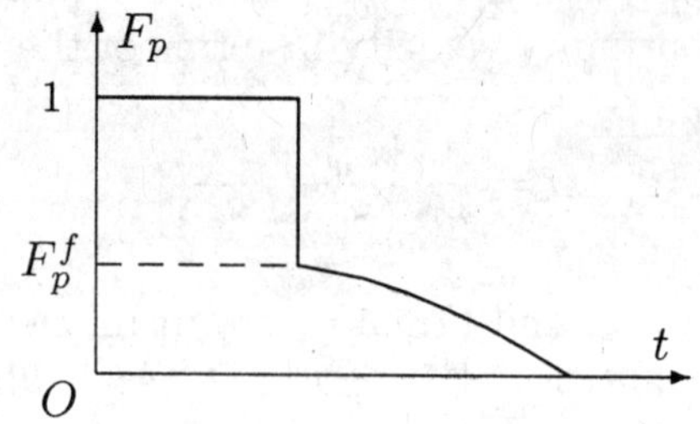

Figure 7.43: The petroleum flow rate at the seam exit.

$$\zeta = F'_w(S_w), \tag{7.6.49}$$

and a constant solution S_w = const. Since in our case $S_w^f < S_w^e$, then a centered wave will adjoin the discontinuous solution at point S_w^f, in which the water saturation will increase continuously from S_w^f to S_w^e. The corresponding solution is shown in Fig. 7.42. It follows from this solution that the petroleum will at first reach the outlet of the oil pool: $F_p = 1$, $F_w = 0$. Then the jump will come, in which F_p reduces spasmodically to $F_p^f = 1 - F_w(S_w^f)$, and the petroleum fraction in the total flow rate will subsequently reduce to zero monotonously (see Fig. 7.43). Note that more complex schemes of the petroleum displacement with the use of surface-active substances and the thickening polymer have been investigated in the literature[5,40]. In addition, the used model (7.6.35) is the simplest one in the series of other models, which take into account the fluid separation into mobile and immobile fluids as well as a difference in pressures in the phases at the expense of surface tension, etc. Their presentation goes beyond the scope of this book, and we refer the interested reader to the relevant literature[41,42].

7.7 Wave Processes in Bubbly Liquids

7.7.1 Equations of the Motion of a Bubbly Liquid

Equations of the mechanics of multiphase media can be obtained not only from the integral conservation laws (7.1.10)–(7.1.12), but also from Hamilton–Ostrogradsky's variational principle[43].

Consider a liquid with the bubbles of radius a whose concentration n is sufficiently large. We will assume that there are no body forces and the velocities of the liquid and bubbles coincide. Let us derive the equations for the motion of a mixture of a liquid and gas bubbles from Hamilton's variational principle. Let us carry out an averaging over the volume $\varphi_i = \frac{1}{V}\int \varphi'_i \, dV'$ and go over to a one-velocity continuum containing the

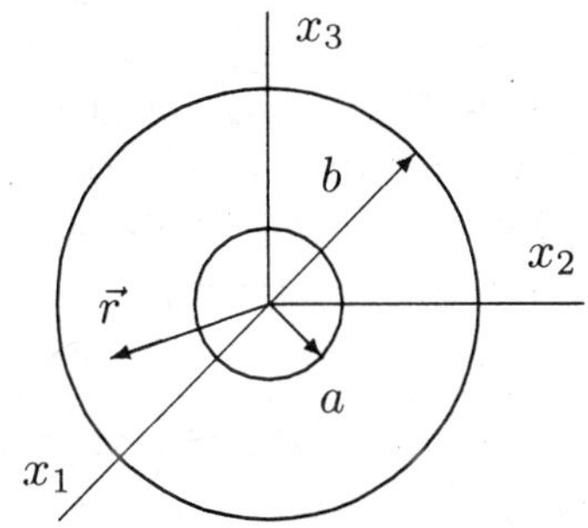

Figure 7.44: A spherical liquid cell with a gas bubble.

liquid and bubbles. We will refer index 1 to the bubbles and index 2 to the liquid. The volume concentration of bubbles is $m_1 = \frac{4}{3}\pi a^3 n$, and the volume concentration of liquid is $m_2 = 1 - m_1$. The mean gas density ρ_1 is expressed in terms of m_1 by formula $\rho_1 = \rho_{11} m_1$, and the mean liquid density is $\rho_2 = \rho_{22} m_2$. Consider the mixture motion within a given volume V in the Lagrangian coordinates ξ^i. Introducing a Cartesian coordinate system at $t = 0$, let us write the expression for the action S in a fixed Cartesian basis:

$$\begin{aligned} S &= \int_{t_0}^{t_1} \int_{V_0} L \, d\xi^1 d\xi^2 d\xi^3 \, dt, \\ L &= \rho_1^0 (K_1 - \langle U_1 \rangle) + \rho_2^0 (K_2 - \langle U_2 \rangle) - n_0 U_s, \end{aligned} \tag{7.7.1}$$

where L is the Lagrange function density and K, U are the specific (per unit mass) mean kinetic and internal energies, U_s is the surface energy of a single bubble, index 0 refers to the time $t = 0$, and the moments of time t_0 and t_1 are fixed. Let us present the fluid velocity $\vec{v}_t$ at some point $\vec{x}$ in the form $\vec{v}_t = \vec{v} + \vec{v}\,'$, where $\vec{v}$ is the mean velocity and $\vec{v}\,'$ is the pulsation velocity satisfying the condition $\langle \vec{v}\,' \rangle = 0$ (the parentheses denote the averaging over the volume). Thus, the oscillatory motion with the velocity $\vec{v}\,'$ is imposed on the mean flow at the velocity $\vec{v}$, which leads to the appearance of additional internal variables in the Lagrangian $L = K - U$. The expression for K_2 has the form

$$K_2 = \left\langle \frac{v_t^2}{2} \right\rangle = \left\langle \frac{v^2 + 2\vec{v} \cdot \vec{v}\,' + v'^2}{2} \right\rangle = \frac{v^2}{2} + \left\langle \frac{v'^2}{2} \right\rangle, \tag{7.7.2}$$

where we have taken into account the fact that $\langle \vec{v}\,' \cdot \vec{v} \rangle = \vec{v} \cdot \langle \vec{v}\,' \rangle = 0$. Introduce a Cartesian Eulerian coordinate system, then $\vec{v} = v^i \vec{e}_i$, $v^2 = v_i v_i$, and the following equations are valid:

$$\frac{dx_i}{dt} = v_i, \quad x_i\Big|_{t=0} = \xi_i. \tag{7.7.3}$$

To compute L, we subdivide the overall mixture volume into spherical cells, so that an air bubble of radius a is located at the center of each cell and the liquid occupies the layer $a < r < b$ (see Fig. 7.44). The cell radius b is related to m_1 and m_2 by formulas

$$m_1 = \left(\frac{a}{b}\right)^3, \quad m_2 = \frac{b^3 - a^3}{b^3}. \tag{7.7.4}$$

We will consider small volume concentrations of gas $m_1 = (a/b)^3 \ll 1$ in what follows. Since m_1 and m_2 are related to one another by the relation $m_1 + m_2 = 1$, one can introduce a single internal variable $\alpha = \frac{1}{m_2}$. Then by virtue of (7.7.4)

$$m_1 = \frac{\alpha - 1}{\alpha}, \quad m_2 = \frac{1}{\alpha}, \quad \alpha > 1. \tag{7.7.5}$$

In our case, the pulsation velocity is related to the radial oscillations of the bubble. In a spherical coordinate system related to the bubble center, this velocity is equal to $v'_r = v'_r(r)$, and in the Cartesian system, $v'_i = v'_r n_i$, where n_i are the cosines of the angles between the vector $\vec{r}$ and the coordinate axes x_i. It is known that $\langle n_i \rangle = \frac{1}{4\pi} \int n_i \sin\theta \, d\theta \, d\varphi = 0$; therefore, $\langle v'_i \rangle = 0$. To determine v'_r, we will assume the liquid to be incompressible, and from the conservation law for a liquid mass bounded by the spheres of radii r and a, we obtain:

$$r^3 - a^3 = r_0^3 - a_0^3,$$

where the subscript 0 refers to the time $t = 0$. Differentiating this equation with respect to t, we obtain:

$$v'_r = \frac{a^2}{r^2}\dot{a}, \quad v'_r = \frac{dr}{dt}. \tag{7.7.6}$$

Using the relations

$$b^3 - a^3 = b_0^3 - a_0^3 = \frac{a_0^3}{\alpha_0 - 1}, \quad \alpha - 1 = \frac{a^3}{b^3 - a^3} = \frac{a^3}{b_0^3 - a_0^3},$$

$$\dot{\alpha} = \frac{3a^2\dot{a}}{b_0^3 - a_0^3} = \frac{3a^2\dot{a}}{a_0^3}(\alpha_0 - 1),$$

we rewrite (7.7.6) in the form

$$v'_r = \frac{a_0^3\dot{\alpha}}{3(\alpha_0 - 1)}\frac{1}{r^2}. \tag{7.7.7}$$

It follows from here that

$$\langle v'^2\rangle = \langle v_r'^2\rangle = \frac{3}{b^3 - a^3}\int_a^b \Big(\frac{a_0^3\dot{\alpha}}{3(\alpha_0 - 1)}\Big)^2 \frac{1}{r^4} r^2\, dr = \frac{a_0^3\dot{\alpha}^2}{3(\alpha_0 - 1)}\Big(\frac{1}{a} - \frac{1}{b}\Big),$$

and with regard for the relation

$$\frac{1}{a} - \frac{1}{b} = \Big(\frac{1}{(\alpha - 1)^{1/3}} - \frac{1}{\alpha^{1/3}}\Big)\frac{(\alpha_0 - 1)^{1/3}}{a_0}$$

we have that

$$\langle v'^2\rangle = \frac{a_0^2\dot{\alpha}^2}{3(\alpha_0 - 1)^{2/3}}\Big(\frac{1}{(\alpha - 1)^{1/3}} - \frac{1}{\alpha^{1/3}}\Big). \tag{7.7.8}$$

We will restrict ourselves to the adiabatic flows for a while; therefore, we will assume the fluid entropy to be constant: $S_2 = \text{const}$. The specific internal energy of fluid at point $\vec{x}$ is equal to $U_2 = U_2(S_2, \tilde{\rho}_{22})$. Assuming that $\tilde{\rho}_{22} = \rho_{22} + \rho_{22}'$, $\langle\rho_{22}'\rangle = 0$, we expand U_2 into a series in ρ_{22}':

$$U_2 = U_2(\rho_{22}) + \frac{\partial U_2}{\partial\rho_{22}}\rho_{22}' + \frac{1}{2}\frac{\partial^2 U_2}{\partial\rho_{22}^2}\rho_{22}'^2$$

and after the averaging over the cell volume, we obtain:

$$\langle U_2\rangle = \frac{1}{2}\int U_2 dV = U_2(\rho_{22}) + \frac{1}{2}\frac{\partial^2 U_2}{\partial\rho_{22}^2}\langle\rho_{22}'^2\rangle.$$

Neglecting the terms of the second order of smallness of the order $\langle\rho_{22}'^2\rangle$, we can find that $\langle U_2\rangle = U_2(S_2, \rho_{22})$. Since $\rho_{22} = \rho_2/m_2 = \rho_2\alpha$, we finally obtain:

$$\langle U_2\rangle = U_2(S_2, \rho_2\alpha). \tag{7.7.9}$$

We have similarly for gas that $\langle U_1\rangle = U_1(S_1, \rho_{11})$. It follows from the relations $m_1 = \frac{\alpha - 1}{\alpha}$ and $\rho_{11} = \rho_1\frac{\alpha}{\alpha - 1}$ that

$$\langle U_1\rangle = U_1\Big(S_1, \rho_1\frac{\alpha}{\alpha - 1}\Big). \tag{7.7.10}$$

A surface energy per one bubble is equal to $U_s = 4\pi a^2\sigma$, where σ is the surface tension coefficient. Using the relations $\frac{4}{3}\pi b_0^3 n_0 = 1$ and $(a_0/b_0)^3 - (\alpha_0 - 1)/\alpha_0$, we can find the total surface energy of all bubbles per unit volume:

$$n_0 U_s = \frac{3\sigma}{a_0}\Big(\frac{\alpha - 1}{\alpha_0 - 1}\Big)^{2/3}\Big(\frac{\alpha_0 - 1}{\alpha_0}\Big). \tag{7.7.11}$$

Substituting formula (7.7.8) in (7.7.2), we write the expression for K_2:

$$K_2 = \frac{v^2}{2} + \frac{a_0^2\dot{\alpha}^2}{3(\alpha_0 - 1)^{2/3}}\Big(\frac{1}{(\alpha - 1)^{1/3}} - \frac{1}{\alpha^{1/3}}\Big), \tag{7.7.12}$$

where $v^2 = v_i v_i$, $v_i = \frac{\partial x_i}{\partial t}$. Neglecting the pulsation energy of gas in the bubble, we can write the specific kinetic energy of gas as

$$K_1 = \frac{v^2}{2}. \tag{7.7.13}$$

Substituting (7.7.9)–(7.7.13) into (7.7.1), we find the density of the Lagrange function:

$$\begin{aligned} L &= \rho_2^0\Big(\frac{v^2}{2} + \frac{a_0^2\dot{\alpha}^2}{6(\alpha_0-1)^{2/3}}\Big(\frac{1}{(\alpha-1)^{1/3}} - \frac{1}{\alpha^{1/3}}\Big)\Big) - \rho_2^0 U_2(S_2, \rho_2\alpha) \\ &+ \rho_1^0\frac{v^2}{2} - \rho_1^0 U_1\Big(S_1, \frac{\rho_1\alpha}{(\alpha-1)}\Big) - \frac{3\sigma}{a_0}\Big(\frac{\alpha-1}{\alpha_0-1}\Big)^{2/3}\frac{(\alpha_0-1)}{\alpha_0}. \end{aligned} \tag{7.7.14}$$

Thus, the density of the Lagrange function L depends on the independent variables x_i and α and on their derivatives:

$$L = L(\dot{x}_i, \partial x_i/\partial\xi_j, \alpha, \dot{\alpha}).$$

On the real motions, the variation of action is equal to zero:

$$\delta S = \int_0^{t_1}\int_{V_0} \delta L\, d\xi_1\, d\xi_2\, d\xi_3\, dt = 0, \tag{7.7.15}$$

where

$$\delta L = \frac{\partial L}{\partial \dot{x}_i}\delta\dot{x}_i + \frac{\partial L}{\partial(\partial x_i/\partial\xi_j)}\delta\Big(\frac{\partial x_i}{\partial\xi_j}\Big) + \frac{\partial L}{\partial\dot{\alpha}}\delta\dot{\alpha} + \frac{\partial L}{\partial\alpha}\delta\alpha$$

and $t_0 = 0$. Integrating (7.7.15) by parts and equalling to zero the expressions affecting the independent variations δx_i and $\delta\alpha$, we obtain the Lagrange equations

$$\begin{aligned} &\frac{\partial}{\partial t}\Big(\frac{\partial L}{\partial v_i}\Big) + \frac{\partial}{\partial\xi_j}\Big(\frac{\partial L}{\partial(\partial x_i/\partial\xi_j)}\Big) = 0, \\ &\frac{\partial}{\partial t}\Big(\frac{\partial L}{\partial\dot{\alpha}}\Big) + \frac{\partial L}{\partial\alpha} = 0. \end{aligned} \tag{7.7.16}$$

For the computation of $\frac{\partial L}{\partial x_{i,j}}$, we derive certain auxiliary formulas for $\frac{\partial\rho_1}{\partial x_{i,j}}$ and $\frac{\partial\rho_2}{\partial x_{i,j}}$ (we have introduced here the notation $x_{i,j} = \frac{\partial x_i}{\partial\xi_j}$). The densities ρ_1 and ρ_2 are determined from the continuity equation $\rho_1/\rho_1^0 = \rho_2/\rho_2^0 = 1/\sqrt{\bar{g}}$, where $\sqrt{\bar{g}} = \det\parallel \frac{\partial x_i}{\partial\xi_j}\parallel$. If the particle trajectory is given by equation $x_i = x_i(\xi_j, t)$, then at fixed t it determines a relation between the Eulerian and Lagrangian coordinates. We recall that the Eulerian and Lagrangian (at time $t = 0$) coordinate systems are

Cartesian; therefore, no difference is made between the contravariant and covariant quantities. Using the formula

$$\xi_{j,i} = \frac{1}{\sqrt{\hat{g}}} \frac{\partial \sqrt{\hat{g}}}{\partial x_{i,j}},$$

we obtain:

$$\frac{\partial \rho_k}{\partial x_{i,j}} = -\frac{\rho_k^0}{\hat{g}} \frac{\partial \sqrt{\hat{g}}}{\partial x_{i,j}} = -\frac{\rho_k^0}{\hat{g}} \sqrt{\hat{g}} \xi_{j,i} = -\rho_k \xi_{j,i}, \qquad (7.7.17)$$

where $k = 1, 2$ and $\xi_{j,i} = \frac{\partial \xi_j}{\partial x_i}$. Differentiating L with respect to v_i and $x_{i,j}$, we obtain:

$$\begin{aligned} \frac{\partial L}{\partial v_i} &= \rho_2^0 v_i + \rho_1^0 v_i, \\ \frac{\partial L}{\partial x_{i,j}} &= -\rho_2^0 \frac{\partial U_2}{\partial \rho_2} \frac{\partial \rho_2}{\partial x_{i,j}} - \rho_1^0 \frac{\partial U_1}{\partial \rho_1} \frac{\partial \rho_1}{\partial x_{i,j}} \\ &= \rho_2^0 \rho_2 \xi_{j,i} \frac{\partial U_2}{\partial \rho_2} - \rho_1^0 \rho_1 \xi_{j,i} \frac{\partial U_1}{\partial \rho_1}. \end{aligned} \qquad (7.7.18)$$

Let us express the right-hand sides of (7.7.18) in terms of the pressure:

$$\rho_1 \frac{\partial U_1}{\partial \rho_1} = \rho_1 \frac{\partial U_1}{\partial \rho_{11}} \frac{\partial \rho_{11}}{\partial \rho_1} = \rho_{11} \frac{\partial U_1}{\partial \rho_{11}} = \frac{m_1 P_1}{\rho_1},$$

where $P_1 = \rho_{11}^2 \frac{\partial U_1}{\partial \rho_{11}}$. Similarly, $\rho_2 \frac{\partial U_2}{\partial \rho_2} = \frac{m_2 P_2}{\rho_2}$, where $P_2 = \rho_{22}^2 \frac{\partial U_2}{\partial \rho_{22}}$. As a result, we obtain:

$$\frac{\partial L}{\partial x_{i,j}} = \xi_{j,i} \Big(\frac{\rho_2^0}{\rho_2} m_2 P_2 + \frac{\rho_1^0}{\rho_1} m_1 P_1 \Big) = \xi_{j,i} \frac{\rho_2^0}{\rho_2} (m_1 P_1 + m_2 P_2). \qquad (7.7.19)$$

Substituting (7.7.18) and (7.7.19) into the first equation of (7.7.16), we find:

$$(\rho_1^0 + \rho_2^0) \frac{\partial^2 x_i}{\partial t^2} + \frac{\partial}{\partial \xi_j} \Big(\frac{\partial \xi_j}{\partial x_i} \frac{\rho_2^0}{\rho_2} (m_1 P_1 + m_2 P_2) \Big) = 0. \qquad (7.7.20)$$

One can reduce this equation to the form

$$\rho^0 \frac{\partial^2 x_i}{\partial t^2} + \frac{\partial}{\partial \xi_j} \Big(\frac{\partial \xi_j}{\partial x_i} \frac{\rho^0}{\rho} P \Big) = 0, \qquad (7.7.21)$$

where $\rho = \rho_1 + \rho_2$, $\rho^0 = \rho_1^0 + \rho_2^0$, and $P = m_1 P_1 + m_2 P_2$ is the total mean pressure in a bubbly liquid.

Consider the second equation in the system (7.7.16). Using (7.7.14), we can find the derivatives:

$$\begin{aligned}\frac{1}{\rho_2^0}\frac{\partial L}{\partial \dot\alpha} &= \frac{a_0^2\dot\alpha}{3(\alpha_0-1)^{2/3}}\left(\frac{1}{(\alpha-1)^{1/3}}-\frac{1}{\alpha^{1/3}}\right),\\ \frac{1}{\rho_2^0}\frac{\partial L}{\partial \alpha} &= -\frac{a_0^2\dot\alpha^2}{18(\alpha_0-1)^{2/3}}\left(\frac{1}{(\alpha-1)^{4/3}}-\frac{1}{\alpha^{4/3}}\right)-\frac{\partial U_2}{\partial\alpha}\\ &- \frac{\rho_1^0}{\rho_2^0}\frac{\partial U_1}{\partial\alpha}-\frac{2\sigma}{a_0\rho_2^0\alpha_0}\left(\frac{\alpha_0-1}{\alpha-1}\right)^{1/3}\end{aligned} \quad (7.7.22)$$

Taking into account the relations $m_2=\frac{1}{\alpha}$, $\rho_{22}=\rho_2\alpha$, $m_1=\frac{\alpha-1}{\alpha}$, $\rho_{11}=\rho_1\frac{\alpha}{\alpha-1}$, $P_1=\rho_{11}^2\frac{\partial U_1}{\partial\rho_{11}}$, and $P_2=\rho_{22}^2\frac{\partial U_2}{\partial\rho_{22}}$ we now express the derivatives of the internal energy in terms of the pressure:

$$\begin{aligned}\frac{\partial U_2}{\partial\alpha} &= \frac{\partial U_2}{\partial\rho_{22}}\frac{\partial\rho_{22}}{\partial\alpha}=\frac{\partial U_2}{\partial\rho_{22}}\rho_2=\frac{P_2m_2}{\rho_{22}},\\ \frac{\rho_1^0}{\rho_2^0}\frac{\partial U_1}{\partial\alpha} &= \frac{\rho_1}{\rho_2}\frac{\partial U_1}{\partial\rho_{11}}\frac{\partial\rho_{11}}{\partial\alpha}=-\frac{\partial U_1}{\partial\rho_{11}}\frac{\rho_1^2}{\rho_2(\alpha-1)^2}\\ &= -\frac{P_1m_1^2}{\rho_2(\alpha-1)^2}=-\frac{P_1}{\rho_2\alpha^2}=-\frac{P_1m_2}{\rho_{22}}.\end{aligned} \quad (7.7.23)$$

Substituting (7.7.22) and (7.7.23) into (7.7.16), we obtain:

$$\frac{a_0^2\ddot\alpha}{3(\alpha_0-1)^{2/3}}\left(\frac{1}{(\alpha-1)^{1/3}}-\frac{1}{\alpha^{1/3}}\right)-\frac{a_0^2\dot\alpha^2}{18(\alpha_0-1)^{2/3}}\left(\frac{1}{(\alpha-1)^{4/3}}-\frac{1}{\alpha^{4/3}}\right)+(P_2-P_1)\frac{m_2}{\rho_{22}}+\frac{2\sigma}{a_0\rho_{22}}\left(\frac{\alpha_0-1}{\alpha-1}\right)^{1/3}=0. \quad (7.7.24)$$

We have approximately replaced ρ_{22}^0 in the left-hand side of (7.7.24) with ρ_{22}. Equations (7.7.21) and (7.7.24) determine together with the equations

$$\frac{\rho_1}{\rho_1^0}=\frac{\rho_2}{\rho_2^0}=\frac{1}{|\partial x_i/\partial\xi_j|},\quad \frac{dx_i}{dt}=v_i,\quad x_i\Big|_{t=0}=\xi_i \quad (7.7.25)$$

a complete system of equations.

At the derivation of this system of equations, we have neglected the viscous stresses in the liquid. The consideration of the liquid viscosity η at the radial bubble pulsations leads to the appearance of the term $\frac{4}{3}\frac{\eta\dot\alpha}{\alpha(\alpha-1)}$ on the left-hand side of equation (7.7.24) [see equations (7.7.28) in the problem presented below].

Problem 7.12. Find the equation for determination of $\alpha(t)$ of type (7.7.24) at a radial motion of viscous incompressible fluid in a spherical cell (see Fig. 7.44).

Solution: The radial motion equation for a viscous incompressible fluid in Eulerian coordinates has the form

$$\rho_{22}\Big(\frac{\partial v_r'}{\partial t}+v_r'\frac{\partial v_r'}{\partial r}\Big)=\frac{\partial \sigma_r'}{\partial r}+2\frac{(\sigma_r'-\sigma_\varphi')}{r},$$
$$\sigma_r'-\sigma_\varphi'=2\eta\Big(\frac{\partial v_r'}{\partial r}-\frac{v_r'}{r}\Big), \tag{7.7.26}$$

where η is the fluid viscosity. Substituting (7.7.7) into (7.7.26) and integrating over r from $r=a$ to certain r, we obtain:

$$\sigma_r'(r)\;-\;\sigma_r'(a)=\frac{\rho_{22}a_0^3}{3(\alpha_0-1)}\Big(\ddot\alpha\Big(\frac{1}{a}-\frac{1}{r}\Big)+\frac{\dot\alpha^2a_0^3}{6(\alpha_0-1)}\Big(\frac{1}{r^4}-\frac{1}{a^4}\Big)\Big)$$
$$+\;\frac{4}{3}\frac{\eta a_0^3\dot\alpha}{(\alpha_0-1)}\Big(\frac{1}{a^3}-\frac{1}{r^3}\Big). \tag{7.7.27}$$

Assuming that $r=b$ in (7.7.27), we obtain with regard for formulas $a=a_0(\frac{\alpha-1}{\alpha_0-1})^{1/3}$ and $b=\frac{a_0\alpha^{1/3}}{(\alpha_0-1)^{1/3}}$:

$$\frac{a_0^2\ddot\alpha}{3(\alpha_0-1)}\Big(\frac{1}{(\alpha-1)^{1/3}}-\frac{1}{\alpha^{1/3}}\Big)-\frac{a_0^2\dot\alpha^2}{18(\alpha_0-1)^{2/3}}\Big(\frac{1}{(\alpha-1)^{4/3}}$$
$$-\frac{1}{\alpha^{4/3}}\Big)+\frac{4\eta\dot\alpha}{3\alpha(\alpha-1)\rho_{22}}+\frac{(\sigma_r'(a)-\sigma_r'(b))}{\rho_{22}}=0. \tag{7.7.28}$$

Comparing (7.7.28) with equation (7.7.24), we find that we should take as the stresses on the internal radius $r=a$ and the external radius $r=b$ of the cell:

$$\sigma_r'(a)=-(P_1-2\sigma/a),\quad \sigma_r'(b)=-P, \tag{7.7.29}$$

where $P=m_1P_1+m_2P_2$. The consideration of fluid viscosity has resulted in the appearance of the term $\frac{4}{3}\frac{\eta\dot\alpha}{\alpha(\alpha-1)\rho_{22}}$ on the left-hand side of (7.7.28).

7.7.2 Equations for Weak Nonlinear Disturbances in Bubbly Liquids

Let us derive the equations governing the development of weak nonlinear disturbances in the one-dimensional nonstationary case at a small volume concentration of bubbles $m_1 \ll 1$. We will assume that the medium in its undisturbed state is at rest: $v^0=0$ and homogeneous: $(\rho^0,P^0,\alpha^0,a_0,n^0,\rho_1^0)=\text{const}$. We will search for the solution in the presence of disturbances in the form $\varphi=\varphi^0+\varphi'$, where $|\varphi'|\ll|\varphi^0|$. Substituting into the motion equation $\rho=\rho^0+\rho'$, $x_i=\xi_i+u_i'$, and $P=P^0+P'$ and neglecting the terms of the second-order smallness$\sim(\varphi')^2$, we find: $\rho_0 \frac{\partial v_i'}{\partial t} + \frac{\partial P'}{\partial x^i} = 0$, $v_i' = \frac{\partial u_i'}{\partial t}$, where u_i' is the displacement along the ith axis. Assuming here the index $i = 1$ and omitting it for brevity, we obtain:

$$\rho^0 \frac{\partial v'}{\partial t} + \frac{\partial P'}{\partial x} = 0. \tag{7.7.30}$$

In the derivation of (7.7.30), we have used the continuity equation in the Lagrangian coordinates for the one-dimensional case $\rho^0 d\xi = \rho\, dx$. Substituting into it $d\xi = (1 - \frac{\partial u}{\partial x})\, dx$ and differentiating with respect to time, we can write the continuity equation as

$$\frac{\partial \rho'}{\partial t} + \rho^0 \frac{\partial v'}{\partial x} = 0. \tag{7.7.31}$$

Let us turn to equation (7.7.24) for the volume concentration of bubbles. Adding the viscous term $\frac{4}{3}\frac{\eta\dot{\alpha}}{\rho_{22}\alpha(\alpha-1)}$ and neglecting the surface tension [the item $\frac{2\sigma}{a_0\rho_{22}}(\frac{\alpha_0-1}{\alpha-1})^{1/3}$], we obtain with regard for formulas $m_2P_2 - m_2P_1 = P - P_1$, $P = m_1P_1 + m_2P_2$, $P = P^0 + P'$, $P_1 = P^0 + P_1'$, and $\alpha = \alpha_0 + \alpha'$:

$$\frac{a_0^2\ddot{\alpha}'}{3(\alpha_0-1)^{2/3}}\left(\frac{1}{(\alpha_0-1)^{1/3}} - \frac{1}{\alpha_0^{1/3}}\right) - \frac{a_0^2\dot{\alpha}'^2}{18(\alpha_0-1)^{2/3}}\left(\frac{1}{(\alpha_0-1)^{4/3}} - \frac{1}{\alpha_0^{4/3}}\right) + \frac{P'-P_1'}{\rho_{22}} + \frac{4\eta\dot{\alpha}'}{3\alpha_0(\alpha_0-1)\rho_{22}} = 0, \tag{7.7.32}$$

where $\dot{\alpha}' = \frac{\partial\alpha'}{\partial t}$, $\ddot{\alpha}' = \frac{\partial^2\alpha'}{\partial t^2}$. It follows from the continuity equation (7.7.25) and the definition $\alpha = 1/m_2$ that $\rho/\rho_0 = n/n_0 = m_2/m_2^0 = \alpha_0/\alpha$. Differentiating these equations with respect to time, $\dot{\alpha} = -\dot{\rho}\alpha/\rho$, $\ddot{\alpha} = -\ddot{\rho}\alpha/\rho$, we find with the accuracy up to the squared disturbances:

$$\dot{\alpha}' = -\dot{\rho}'\alpha_0/\rho^0, \quad \ddot{\alpha}' = -\ddot{\rho}'\alpha_0/\rho^0. \tag{7.7.33}$$

Substituting these formulas into (7.7.32), we find with regard for formulas $\rho^0 = \rho_1^0 + \rho_2^0 \sim \rho_2^0 = \rho_{22}m_2^0$:

$$\frac{a_0^2\ddot{\rho}'}{3m_1^0m_2^0}(1-(m_1^0)^{1/3}) + \frac{a_0^2(\dot{\rho}')^2}{18(m_1^0)^2m_2^0\rho^0}(1-(m_1^0)^{4/3}) + \frac{4}{3}\frac{\eta}{\rho_{22}}\frac{\dot{\rho}'}{m_1^0} + P_1' - P' = 0, \tag{7.7.34}$$

where

$$\dot{\rho}' = \frac{\partial\rho'}{\partial t}, \quad \ddot{\rho}' = \frac{\partial^2\rho'}{\partial t^2}, \quad m_2^0 = \frac{1}{\alpha_0}, \quad m_1^0 = \frac{\alpha_0 - 1}{\alpha_0}.$$

Assume that the gas compression and expansion in a bubble occurs adiabatically; that is, $P_1 a^{3\gamma} = P^0 a_0^{3\gamma}$. Expressing the bubble radius in terms of α we can write:

$$P_1 = P^0 \left(\frac{a}{a_0}\right)^{-3\gamma} = P^0 \left(\frac{\alpha - 1}{\alpha_0 - 1}\right)^{-\gamma}.$$

Assuming that $P_1 = P^0 + P_1'$, $\alpha = \alpha_0 + \alpha'$ and expanding the right-hand side into the Taylor series up to quadratic terms

$$P_1' \approx -\gamma P^0 \frac{\alpha'}{(\alpha_0 - 1)} + \frac{\gamma(\gamma+1)}{2} P^0 \left(\frac{\alpha'}{\alpha_0 - 1}\right)^2,$$

we obtain with regard for the first formula (7.7.33) $\alpha' = -\rho' \alpha_0 / \rho^0$:

$$P_1' \approx c_0^2 \rho' + \frac{\gamma(\gamma+1)}{2} \frac{P^0}{(m_1^0)^2} \left(\frac{\rho'}{\rho^0}\right)^2, \tag{7.7.35}$$

where

$$c_0 = \sqrt{\frac{\gamma P^0}{\rho^0 m_1^0}} \tag{7.7.36}$$

is the low-frequency sound velocity in a bubbly liquid. Substituting (7.7.35) into (7.7.34), we can write:

$$\frac{a_0^2 \ddot{\rho}'}{3 m_1^0 m_2^0} (1 - (m_1^0)^{1/3}) + \frac{a_0^2 (\dot{\rho}')^2}{18 (m_1^0)^2 m_2^0 \rho^0} (1 - (m_1^0)^{4/3})$$
$$+ \frac{(\gamma+1)}{2} \frac{c_0^2 (\rho')^2}{m_1^0 \rho^0} + \frac{4}{3} \frac{\eta}{\rho_{22}} \frac{\dot{\rho}'}{m_1^0} + c_0^2 \rho' = P'. \tag{7.7.37}$$

Let us estimate the second and third items in this equation. Since $\dot{\rho}' \sim \rho'/\Delta t \sim \rho' v_r'/a_0$, we have with regard for relations $m_1^0 \ll 1$ and $m_2^0 \approx 1$:

$$\frac{a_0^2 (\dot{\rho}')^2}{18 (m_1^0)^2 m_2^0 \rho^0} (1 - (m_1^0)^{4/3}) \sim \frac{(\rho')^2}{18 \rho^0} \frac{v_r'^2}{(m_1^0)^2},$$
$$\frac{(\gamma+1)}{2} \frac{c_0^2 (\rho')^2}{m_1^0 \rho^0} \sim \frac{(\rho')^2}{2 \rho^0} \frac{c^2}{(m_1^0)^2}, \quad c^2 = \gamma \frac{P^0}{\rho^0}.$$

The velocity of the bubble oscillations v_r' is small and satisfies the inequality $v_r' \ll c$; therefore, one can neglect the second item in (7.7.37). As a result, we obtain:

$$\varphi_1 \frac{\partial^2 \rho'}{\partial t^2} + \varphi_2 \frac{\partial \rho'}{\partial t} + \varphi_3 c_0^2 (\rho')^2 + c_0^2 \rho' = P', \tag{7.7.38}$$

where

$$\varphi_1 = \frac{a_0^2}{3 m_1^0 m_2^0} (1 - (m_1^0)^{1/3}), \quad \varphi_2 = \frac{4}{3} \frac{\eta}{\rho_{22} m_1^0}, \quad \varphi_3 = \frac{\gamma+1}{2} \frac{1}{\rho^0 m_1^0}.$$

Differentiating equation (7.7.30) with respect to x and equation (7.7.31) with respect to t and subtracting one equation from another, we find:

$$\frac{\partial^2 \rho'}{\partial t^2} = \frac{\partial^2 P'}{\partial x^2}. \tag{7.7.39}$$

Differentiating (7.7.38) twice with respect to t and using (7.7.39), we obtain:

$$\varphi_1 \frac{\partial^4 P'}{\partial t^2 \partial x^2} + \varphi_2 \frac{\partial^3 P'}{\partial x^2 \partial t} + \varphi_3 c_0^2 \frac{\partial^2}{\partial t^2}(\rho')^2 + c_0^2 \frac{\partial^2 P'}{\partial x^2} - \frac{\partial^2 P'}{\partial t^2} = 0. \tag{7.7.40}$$

Here the first item describes the dispersion, the second item describes the viscous effects, and the third item describes the nonlinearity of a bubbly medium. The last two items represent the D'Alembert's wave operator. The first three items have the following orders of magnitude:

$$\begin{aligned} &\varphi_1 \frac{\partial^2 P'}{\partial t^2 \partial x^2} \sim \varepsilon_1 \frac{\partial^2 P'}{\partial t^2}, \quad \varepsilon_1 = \frac{a_0^2}{L^2 m_1^0} \ll 1, \\ &\varphi_2 \frac{\partial^3 P'}{\partial x^2 \partial t} \sim \varepsilon_2 \frac{\partial^2 P'}{\partial t^2}, \quad \varepsilon_2 = \frac{\eta}{\rho^0 m_1^0 c_0 L} \ll 1, \\ &\varphi_3 c_0^2 \frac{\partial^2}{\partial t^2}(\rho')^2 \sim \varepsilon_3 \frac{\partial^2 P'}{\partial t^2}, \quad \varepsilon_3 = \frac{\rho'}{\rho^0 m_1^0} \ll 1, \end{aligned} \tag{7.7.41}$$

where L is the reference length of the variation of the mean mixture parameters $\frac{\partial P'}{\partial x} \sim \frac{P'}{L}$. When the inequalities (7.7.41) $\varepsilon_1 \ll 1$, $\varepsilon_2 \ll 1$, $\varepsilon_3 \ll 1$ are satisfied, then the dispersive,viscous, and nonlinear terms in (7.7.40) will give small corrections to the wave equation $c_0^2 \frac{\partial^2 P'}{\partial x^2} - \frac{\partial^2 P'}{\partial t^2} = 0$, which has the solution in the form of the waves propagating to the left and to the right at the velocity c_0:

$$P' = F_1(x - c_0 t) + F_2(x + c_0 t), \quad P' = c_0^2 \rho'. \tag{7.7.42}$$

Therefore, replacing in the nonlinear term (7.7.40) ρ' with P'/c_0^2, we obtain the *Boussinesq equation:*

$$\varphi_1 \frac{\partial^4 P'}{\partial t^2 \partial x^2} + \varphi_2 \frac{\partial^3 P'}{\partial x^2 \partial t} + \frac{\varphi_3}{c_0^2} \frac{\partial^2}{\partial t^2}(P')^2 + c_0^2 \frac{\partial^2 P'}{\partial x^2} - \frac{\partial^2 P'}{\partial t^2} = 0. \tag{7.7.43}$$

Consider a wave propagating from the left to the right. Since the dispersive, viscous, and nonlinear items are small, it will weakly deform in a reference frame moving at a velocity c_0. From the Galilei transformations

$$\zeta = x - c_0 t, \quad t' = t, \tag{7.7.44}$$

the formulas follow for an arbitrary function φ:

$$\frac{\partial \varphi}{\partial t} = \frac{\partial \varphi}{\partial t'} - \frac{\partial \varphi}{\partial \zeta} c_0, \quad \frac{\partial \varphi}{\partial \zeta} = \frac{\partial \varphi}{\partial x}. \tag{7.7.45}$$

In our case $\frac{\partial \varphi}{\partial t'} \approx 0$; therefore, we obtain from (7.7.45):

$$\frac{\partial \varphi}{\partial t} \approx -c_0 \frac{\partial \varphi}{\partial x}. \tag{7.7.46}$$

Replacing the time derivatives in the first three terms of equation (7.7.43) with the aid of (7.7.46), with the derivatives with respect to the coordinate $\frac{\partial P'}{\partial t} \sim -c_0 \frac{\partial P'}{\partial x}$ and in the wave operator in (7.7.43)

$$\frac{\partial^2}{\partial t^2} - c_0^2 \frac{\partial^2}{\partial x^2} = \frac{\partial}{\partial t}\left(\frac{\partial}{\partial t} + c_0 \frac{\partial}{\partial x}\right) - c_0 \frac{\partial}{\partial x}\left(\frac{\partial}{\partial t} + c_0 \frac{\partial}{\partial x}\right) \approx 2 \frac{\partial}{\partial t}\left(\frac{\partial}{\partial t} + c_0 \frac{\partial}{\partial x}\right),$$

and integrating it once in time, we obtain the *Burgers–Korteweg–de Vries (BKdV) equation*:

$$\frac{\partial P'}{\partial t} + c_0 \frac{\partial P'}{\partial x} + \frac{\varphi_1 c_0}{2} \frac{\partial^3 P'}{\partial x^3} - \frac{\varphi_2}{2} \frac{\partial^2 P'}{\partial x^2} + \frac{\varphi_3}{2c_0} \frac{\partial (P')^2}{\partial x} = 0. \tag{7.7.47}$$

Using formulas (7.7.45), we rewrite the BKdV equation in a reference frame moving at a velocity c_0:

$$\frac{\partial P'}{\partial t} + \frac{\varphi_3}{c_0} P' \frac{\partial P'}{\partial \zeta} - \frac{\varphi_2}{2} \frac{\partial^2 P'}{\partial \zeta^2} + \frac{\varphi_1 c_0}{2} \frac{\partial^3 P'}{\partial \zeta^3} = 0, \tag{7.7.48}$$

where $\zeta = x - c_0 t$. Multiplying this equation by $\frac{c_0 L}{\varphi_3 (\delta P^0)^2}$ and introducing the nondimensional variables

$$\bar{P} = \frac{P'}{\delta P^0}, \quad \bar{t} = \frac{t}{t^*}, \bar{\zeta} = \frac{\zeta}{L}, \quad t^* = \frac{2\gamma}{(\gamma+1)} \frac{L}{c_0} \frac{P^0}{\delta P^0}, \tag{7.7.49}$$

let us rewrite the BKdV equation in the nondimensional form:

$$\frac{\partial \bar{P}}{\partial \bar{t}} + \bar{P} \frac{\partial \bar{P}}{\partial \bar{\zeta}} - \frac{1}{\mathrm{Re}} \frac{\partial^2 \bar{P}}{\partial \bar{\zeta}^2} + \frac{1}{D_\sigma^2} \frac{\partial^3 \bar{P}}{\partial \bar{\zeta}^3} = 0, \tag{7.7.50}$$

where the Reynolds number Re and the dispersion D_σ are determined by formulas

$$\mathrm{Re} = \frac{3}{4}\left(\frac{\gamma+1}{\gamma}\right) \frac{\rho^0 L c_0}{\eta} m_1^0 \frac{\delta P^0}{P^0}, \quad D_\sigma = \frac{L}{a_0} \sqrt{\frac{3(\gamma+1)}{\gamma} m_1^0 \frac{\delta P^0}{P^0}}. \tag{7.7.51}$$

The quantity δP^0 entering the definition of the nondimensional parameters (7.7.49) and (7.7.51) is a reference amplitude of the disturbance at the initial moment of time $t = 0$. It follows from equation (7.7.50) that the solution character is determined by the values of the nondimensional parameters Re and D_σ. Note that the BKdV equations were used for the first time for the study of the propagation of weak nonlinear waves in a bubbly liquid in the literature[44,45]. A more detailed bibliography on this question may be found in[5,46].

7.7.3 Progressive, Weak Nonlinear Waves in Bubbly Liquids

Let us study *progressive waves* with the aid of the BKdV equation (7.7.50), which propagate at a constant velocity D without alteration of their form. Let us rewrite equation (7.7.50) in a more convenient form:

$$\frac{\partial P}{\partial t} + P\frac{\partial P}{\partial \zeta} - \mu\frac{\partial^2 P}{\partial \zeta^2} + \beta\frac{\partial^3 P}{\partial \zeta^3} = 0, \quad \zeta = x - c_0 t, \tag{7.7.52}$$

where all quantities are nondimensional. The progressive wave type solution of (7.7.52) is given by formula

$$P = P(\zeta - Dt). \tag{7.7.53}$$

Consider at first the particular cases of equation (7.7.52).

Let the dispersion in a bubbly medium be small: $\beta \to 0$. Then we obtain the *Burgers equation* instead of (7.7.52):

$$\frac{\partial P}{\partial t} + P\frac{\partial P}{\partial \zeta} - \mu\frac{\partial^2 P}{\partial \zeta^2} = 0.$$

Introducing a new variable

$$z = \zeta - Dt, \tag{7.7.54}$$

we calculate the derivatives:

$$\frac{\partial P}{\partial t} = -D\frac{dP}{dz}, \quad \frac{\partial P}{\partial \zeta} = \frac{dP}{dz}, \quad \frac{\partial^2 P}{\partial \zeta^2} = \frac{d^2 P}{dz^2}. \tag{7.7.55}$$

Substituting these expressions into the Burgers equation, we find:

$$(P - D)\frac{dP}{dz} = \mu\frac{d^2 P}{dz^2}. \tag{7.7.56}$$

We will search for the solution of (7.7.56) in the form of a shock wave under the boundary conditions

$$\begin{aligned} P &= 0, \quad \frac{dP}{dz} = 0 \quad \text{at} \quad z \to +\infty, \\ P &= P_*, \quad \frac{dP}{dz} = 0 \quad \text{at} \quad z \to -\infty. \end{aligned} \tag{7.7.57}$$

Integrating equation (7.7.56) once, we obtain:

$$\mu\frac{dP}{dz} = \frac{P^2}{2} - DP + C. \tag{7.7.58}$$

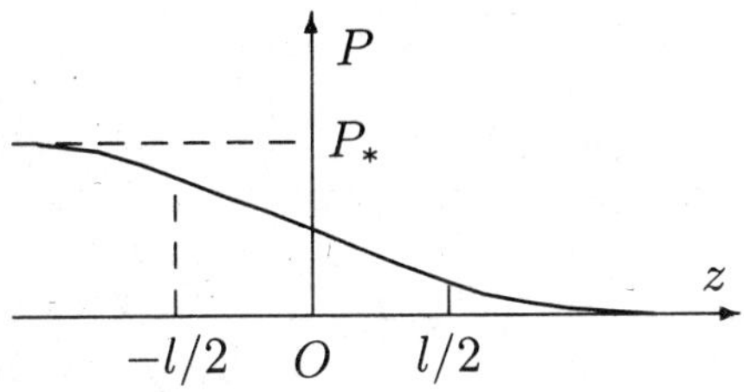

Figure 7.45: The shock wave profile corresponding to (7.7.60).

From the boundary conditions (7.7.57), we can find the velocity of the shock wave propagation D and the constant C:

$$D = P_*/2, \qquad C = 0. \tag{7.7.59}$$

Substituting (7.7.59) into (7.7.58) and integrating this equation, we obtain:

$$P = P_*\Big(1 + \exp\frac{P_* z}{2\mu}\Big)^{-1}, \quad z = \zeta - Dt. \tag{7.7.60}$$

We show in Fig. 7.45 the shock wave profile corresponding to equation (7.7.60). A characteristic thickness of the shock wave has the order $l \sim \mu/P_*$, according to (7.7.60). The velocity of the shock wave propagation with respect to a nondisturbed medium is equal to $\frac{dx}{dt} = c_0 + P_*/2$.

Consider another particular case, in which the viscosity of a bubbly medium is small: $\mu \to 0$. In this case, we obtain from (7.7.52) the *Korteweg–de Vries (KdV) equation*:

$$\frac{\partial P}{\partial t} + P\frac{\partial P}{\partial \zeta} + \beta\frac{\partial^3 P}{\partial \zeta^3} = 0, \quad \zeta = x - c_0 t. \tag{7.7.61}$$

We search for a solution of this equation in the class of the progressive waves (7.7.53). Substituting (7.7.53) into (7.7.61), we arrive with regard for (7.7.55) to the ODE

$$(P - D)\frac{dP}{dz} + \beta\frac{d^3 P}{dz^3} = 0, \quad z = \zeta - Dt. \tag{7.7.62}$$

The first integral of this equation is

$$\beta\frac{d^2 P}{dz^2} + \frac{P^2}{2} - DP = 0. \tag{7.7.63}$$

The integration constant in (7.7.63) $C = 0$ because, with the aid of the substitution $P = \tilde{P} + B$ and $D = \tilde{D} + B$, equation (7.7.63) reduces to

$$\beta\frac{d^2 \tilde{P}}{dz^2} + \frac{\tilde{P}^2}{2} - \tilde{D}\tilde{P} = B(\tilde{D} + B/2).$$

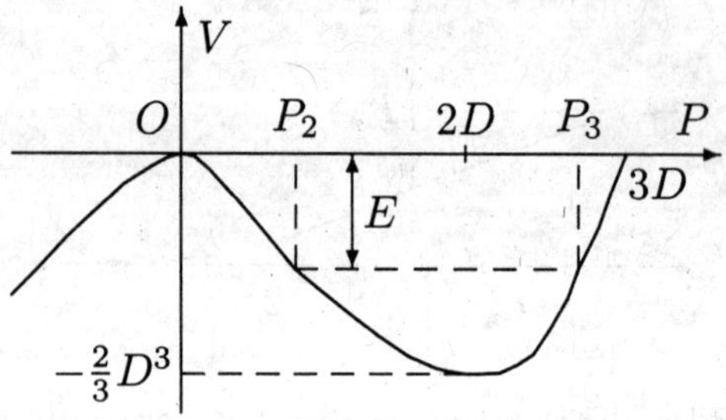

Figure 7.46: The motion of a particle with the mass β and total energy $E < 0$ in the potential hole $V(P)$.

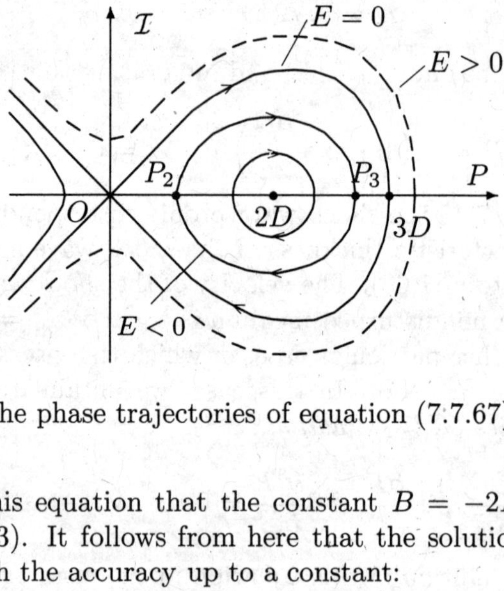

Figure 7.47: The phase trajectories of equation (7.7.67) at $E = \text{const}$.

Assuming in this equation that the constant $B = -2\tilde{D}$, we arrive at equation (7.7.63). It follows from here that the solution of (7.7.63) is determined with the accuracy up to a constant:

$$P = \tilde{P} + B, \qquad D = \tilde{D} + B. \tag{7.7.64}$$

Equation (7.7.63) is similar to the motion equation of a particle with the mass β in the potential $V(P)$:

$$\beta \frac{d^2 P}{dz^2} = -\frac{dV}{dP}, \quad V(P) = \frac{P^3}{6} - \frac{DP^2}{2}. \tag{7.7.65}$$

Multiplying the first equation by $\frac{dP}{dz}$ and integrating over z, we obtain the energy integral:

$$\frac{1}{2}\beta\left(\frac{dP}{dz}\right)^2 + V(P) = E. \tag{7.7.66}$$

The bounded (finite) solution of (7.7.65) exists only in the case in which a particle is in the potential hole $V(P)$ corresponding to (7.7.65) with the

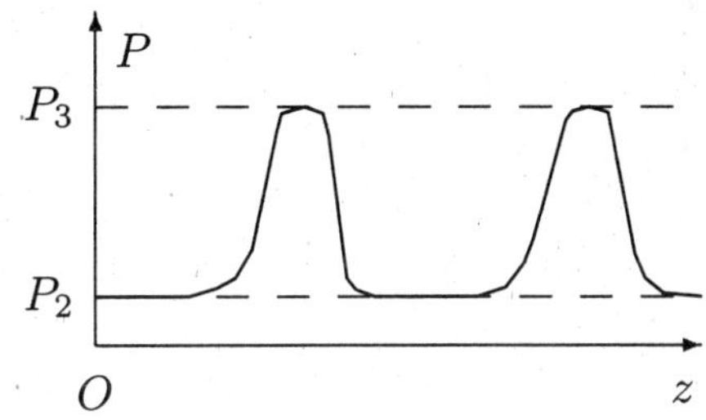

Figure 7.48: The cnoidal waves.

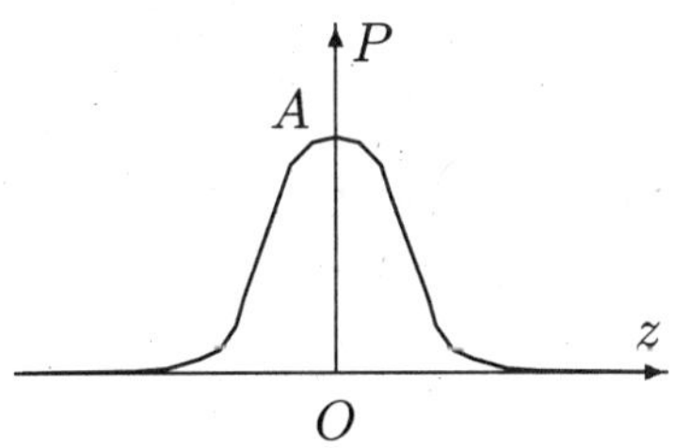

Figure 7.49: The soliton.

total energy $E < 0$ (see Fig. 7.46). In this case, the particle performs the oscillations between the turning points P_2 and P_3 at which $\frac{dP}{dz} = 0$. Determining the particle impulse by formula $\mathcal{I} = \beta \frac{dP}{dz}$, we rewrite (7.7.66) in the form

$$\frac{\mathcal{I}^2}{2\beta} + V(P) = E. \qquad (7.7.67)$$

The phase trajectories described by equation (7.7.67) at different values of $E = \text{const}$ are shown in Fig. 7.47. The finitary and infinite motions are separated from each other by a separatrix $E = 0$. In our case of physical nonlinear waves, only the bounded solutions at $E \leq 0$ have a physical meaning. If $E < 0$, then the oscillations of P between P_2 and P_3 take place, which are the roots of equation $E + \frac{1}{2}DP^2 - \frac{1}{6}P^3 = 0$. As a consequence of the asymmetry of $V(P)$, the values of P for a large interval of z are located near the point P_2. Such waves are called *cnoidal* and shown in Fig. 7.48, where $P_2 > 0$, $P_3 < 3D$. (Since an arbitrary constant B can be added to the solution, the value of P_2 can also be negative.) The increase in E leads to an increase in the period of oscillations in the wave, and in the limit as $E \to 0$, the solution tends to a sequence of solitons. (A solitary wave is called the *soliton*, as shown in Fig. 7.49.) The phase trajectory corresponding to the soliton solution $E = 0$ is a separatrix, which is shown by a thick line in Fig. 7.47. As follows from Fig. 7.47, $P_2 = 0$ and

$$P_3 = A = 3D. \tag{7.7.68}$$

That is, the soliton amplitude exceeds its velocity by a factor of three. It turns out that in this case equation (7.7.66) with potential (7.7.65) allows the exact solution.

Let a soliton be located at point $z = 0$, and we construct the solution in the region $z > 0$, where $\frac{dP}{dz} < 0$. Then we find from equation (7.7.66) at $E = 0$:

$$\frac{dP}{dz} = -\sqrt{-\frac{2}{\beta}V(P)} = -\sqrt{\frac{D}{\beta}}P\sqrt{1-\frac{P}{3D}},$$

from where

$$\int_{P_3}^{P} \frac{dP}{P\sqrt{1-P/(3D)}} = -\int_0^z \sqrt{\frac{D}{\beta}}\,dz.$$

Since $P_3 = 3D$, the integral reduces to the form

$$\int_{P/(3D)}^{1} \frac{dx}{x\sqrt{1-x}} = \sqrt{\frac{D}{\beta}}\,z.$$

This integral is easily computed with the aid of the substitution $x = \sin^2\varphi$. As a result, we obtain the soliton type solution, see Fig. 7.49:

$$P = \frac{A}{\cosh^2\left(\sqrt{\frac{A}{12\beta}}\,z\right)}, \qquad D = \frac{A}{3}. \tag{7.7.69}$$

The soliton has the width $\delta z = \sqrt{12\beta/A}$ and propagates at a velocity $\frac{dx}{dt} = c_0 + A/3$ with respect to the undisturbed medium. The soliton is stable when the nonlinear term in the KdV equation $P\frac{\partial P}{\partial \varsigma}$ is of the order of the dispersion term $\beta\frac{\partial^3 P}{\partial \varsigma^3}$. Using the expressions $P\frac{\partial P}{\partial \varsigma} \sim \frac{A^2}{\Delta z}$ and $\beta\frac{\partial^3 P}{\partial \varsigma^3} \sim \beta\frac{A}{(\Delta z)^3}$, let us estimate the stability boundary:

$$P\frac{\partial P}{\partial \varsigma}\Big/\left(\beta\frac{\partial^3 P}{\partial \varsigma^3}\right) \sim \frac{A(\Delta z)^2}{\beta} \sim 1.$$

A strict stability analysis of the soliton on the basis of the KdV equation shows that it is stable at $A(\Delta z)^2/\beta < 12$. If an inverse inequality is satisfied $A(\Delta z)^2/\beta > 12$, then the soliton solution splits into a sequence of solitons. Since the soliton velocity is proportional to its amplitude the first running soliton is the soliton with the largest amplitude, and the next solitons follow in the order of the decreasing amplitude. Note one more important property of solitons. If two solitons collide, then they will have after the collision the same shapes as before the collision. There is at present a vast bibliography on the theory of solitons. We present here the references only to some works in this domain of mathematical physics[47–49].

We finally consider a solution of the progressive wave type (7.7.53) for the BKdV equation (7.7.52) in the presence of both the nonlinearity and viscosity and dispersion. Substituting (7.7.53) into (7.7.52), we obtain with regard for (7.7.55):

$$\beta\frac{d^3P}{dz^3} - \mu\frac{d^2P}{dz^2} + (u - D)\frac{dP}{dz} = 0. \tag{7.7.70}$$

We will search for the solutions of the progressive wave type of equation (7.7.70) under the boundary conditions

$$\begin{aligned} P &= 0, \quad \frac{dP}{dz} = \frac{d^2P}{dz^2} = 0 \ \text{ at } \ z \to \infty, \\ P &= P_*, \quad \frac{dP}{dz} = \frac{d^2P}{dz^2} = 0 \ \text{ at } \ z \to -\infty. \end{aligned} \tag{7.7.71}$$

Integrating equation (7.7.70) once, we obtain with regard for the boundary conditions (7.7.71):

$$\beta\frac{d^2P}{dz^2} - \mu\frac{dP}{dz} + \frac{P^2}{2} - DP = 0, \quad D = \frac{1}{2}P_*. \tag{7.7.72}$$

It follows from here that the shock wave moves at a velocity $\frac{dx}{dt} = c_0 + P_*/2$ with respect to the undisturbed medium. This velocity coincides with the shock wave velocity (7.7.60) governed by the Burgers equation (7.7.52). The shock wave profile in this case, however, can have an oscillatory form in contrast with (7.7.60)[50]. To prove this fact, let us consider the solution behavior near the stationary point at $z \to -\infty$, $P \to P_*$. Assuming that $P = P_* + p'$ ($|p'| \ll P_*$) in equation (7.7.72), we obtain an equation for the disturbances:

$$\beta\frac{d^2p'}{dz^2} - \mu\frac{dp'}{dz} + \frac{P_*p'}{2} = 0. \tag{7.7.73}$$

Representing the solution of (7.7.73) in the form $p' = Ae^{\lambda z}$, we arrive at the equation for the eigenvalues:

$$\beta\lambda^2 - \mu\lambda + P_*/2 = 0,$$

which has the solutions

$$\lambda_{1,2} = \frac{\mu}{2\beta} \pm \sqrt{\frac{\mu^2}{4\beta^2} - \frac{P_*}{2\beta}}. \tag{7.7.74}$$

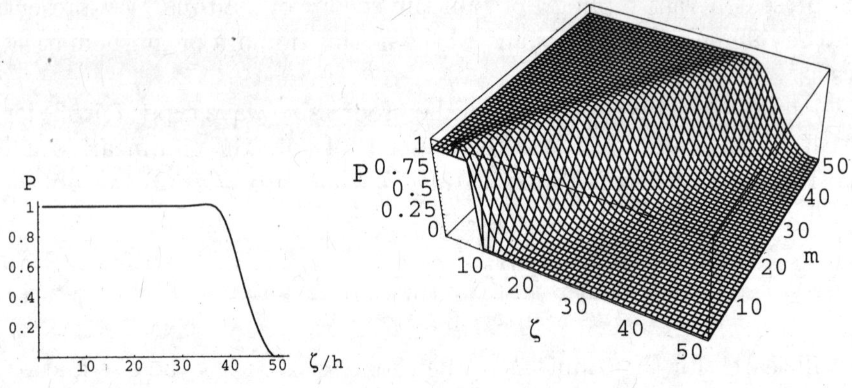

(a) (b)

Figure 7.50: The numerical solution of the BKdV equation (7.7.52) at $\mathrm{Re} = 10$, $D_\sigma = 10$: (a) the profile of the solution $P(\zeta, t)$ at $t = 6.92597$ and (b) the surface $P = P(\zeta, t)$ at $0 \leq \zeta \leq 5$, $0 \leq t \leq 6.92597$; $m = t/(8\tau)$.

It follows from here that, at $P_* > \frac{\mu^2}{2\beta}$, the shock wave will have an oscillatory form (Fig. 7.51) and at $P_* < \frac{\mu^2}{2\beta}$, it will have a monotonous form (Fig. 7.50). The decay of oscillations is related here to the viscosity effect. As was shown in[51], however, the heat exchange processes between the gas in the bubbles and the ambient liquid have an even stronger effect on the oscillations decay.

It should be noted that it is possible to obtain the closed-form solution of the BKdV equation (7.7.52) only at certain initial conditions (see above). At arbitrary initial conditions, the solution of the BKdV equation can be obtained only with the aid of the numerical methods.

Let us describe an efficient finite difference method for the numerical solution of the Burgers–Korteweg–de Vries equation (7.7.52). Let us take as the integration interval on the ζ-axis the interval $0 \leq \zeta \leq a$, where $a > 0$ is a given quantity. The time integration interval is the interval $[0, T]$, where $T > 0$ is a given quantity. Assume that the initial condition

$$P(\zeta, 0) = P_0(\zeta), \quad 0 \leq \zeta \leq a \tag{7.7.75}$$

is specified, where $P_0(\zeta)$ is an arbitrary function. Let h be the step of a uniform grid in the interval $[0, a]$, $h = a/(M-1)$ and M is the user-specified number of the grid nodes $M > 1$. Let $\tau > 0$ be the time step of a finite difference scheme. Let $P_j^n = P((j-1)h, (n-1)\tau)$, $j = 1, \ldots, M$,

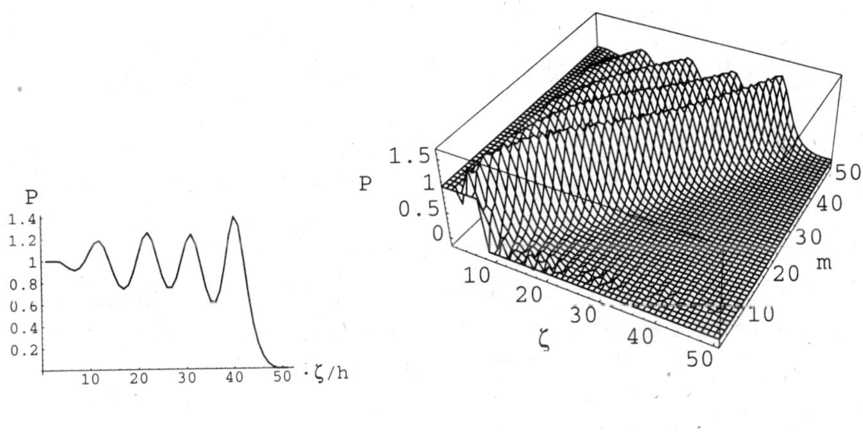

(a) (b)

Figure 7.51: The numerical solution of the BKdV equation (7.7.52) at Re $= 100$, $D_\sigma = 10$: (a) the profile of the solution $P(\zeta, t)$ at $t = 6.92597$ and (b) the surface $P = P(\zeta, t)$ at $0 \le \zeta \le 5$, $0 \le t \le 6.92597$; $m = t/(8\tau)$.

and $n = 1, 2, \ldots, [T/\tau]$, where the symbol $[T/\tau]$ denotes the integral part of the number T/τ. Let us approximate the Burgers–Korteweg–de Vries equation (7.7.52) by the following finite difference scheme[52]:

$$\begin{aligned} & \frac{P_j^{n+1} - P_j^{n-1}}{2\tau} - \frac{P_j^n}{12h}(P_{j+2}^n - 8P_{j+1}^n + 8P_{j-1}^n - P_{j-2}^n) \\ - \quad & \mu \cdot \frac{P_{j+1}^n - P_j^{n+1} - P_j^{n-1} + P_{j-1}^n}{h^2} - \frac{\beta}{8h^3}(P_{j+3}^n - 8P_{j+2}^n \\ + \quad & 13P_{j+1}^n - 13P_{j-1}^n + 8P_{j-2}^n - P_{j-3}^n) = 0, \\ & j = 4, 5, 6, \ldots, M-3; \quad n = 2, 3, 4, \ldots, [T/\tau] - 1. \end{aligned} \tag{7.7.76}$$

The du Fort–Frankel's approximation[53] is used here for the diffusive term $-\mu P_{\zeta\zeta}$ in this difference scheme. At sufficiently small steps h, the stability condition of scheme (7.7.76) has the form[52]:

$$\frac{\tau\beta}{h^3} \le \frac{108}{(43 + 7\sqrt{73})\sqrt{10\sqrt{73} - 62}} \approx 0.217. \tag{7.7.77}$$

The difference scheme (7.7.76) has the approximation error[52,53] $O(\tau^2) + O((\tau/h)^2) + O(h^4)$. By virtue of the limitation for τ (7.7.77) dictated by the stability of scheme (7.7.76), however, we have $\tau = O(h^3)$. Therefore, the scheme (7.7.76) indeed has the approximation error $O(h^6) + O(h^4)$. We have implemented this scheme in our *Mathematica* program `prog7-2.nb` (we present a more detailed description of this program in Appendix B).

Figs. 7.50 and 7.51 are obtained at the same initial condition of a step form (the step $h = 0.1$ in all of the runs presented below):

$$P(\zeta, 0) = \begin{cases} 1, & 0 \leq \zeta \leq 7h \\ 0, & \zeta > 7h. \end{cases}$$

It can be seen from Fig. 7.50 that, at Re $= 10$, $D_\sigma = 10$, the *monotonous* shock wave regime is realized. At Re $= 100$ and $D_\sigma = 10$, the smoothing effect of the viscosity becomes weaker and the predominance of the dispersion effects leads to an *oscillatory* shock wave (see Fig. 7.51).

We show in Fig. 7.52 an example of computation by scheme (7.7.76) at an initial function $P(\zeta, 0)$ containing two bell-shaped impulses of the same amplitude:

$$P(\zeta, 0) = e^{-(\zeta-2)^2} + e^{-(\zeta-6)^2}, \quad 0 \leq \zeta \leq 16.5.$$

In this computational example, several solitons form in the solution with increasing t. A soliton with a larger amplitude following the head soliton at $t = 4.2908$ [Fig. 7.52 (a)] lags yet behind the soliton of smaller amplitude following the head soliton. As the time increases, however, the above large-amplitude soliton outstrips the small-amplitude soliton and we see in Fig. 7.52 (b) two large-amplitude solitons having approximately equal amplitudes, which are followed by a train of smaller amplitude solitons. Fig. 7.52 (c) illustrates this complex picture of the solitons interaction.

Problem 7.13. Find the dispersion law $\omega(k)$ for the waves governed by equation (7.7.47) in the case $P' \to 0$ and the viscosity absence $\varphi_2 = 0$.

Solution: Neglecting the nonlinear term $\sim (P')^2$ in (7.7.47), we obtain a linear equation:

$$\frac{\partial P'}{\partial t} + c_0 \frac{\partial P'}{\partial x} + \beta \frac{\partial^3 P'}{\partial x^3} = 0, \quad \beta = \frac{\varphi_1 c_0}{2}. \qquad (7.7.78)$$

Substituting here the solution of the progressive wave form

$$P' = \delta P^0 \exp\left(i(kx - \omega t)\right),$$

we arrive at the equation:

$$(\omega - kc_0 + \beta k^3)\delta P^0 = 0.$$

The condition for the existence of a nonzero solution $\delta P^0 \neq 0$ gives the dispersion law:

$$\omega = kc_0 - \beta k^3.$$

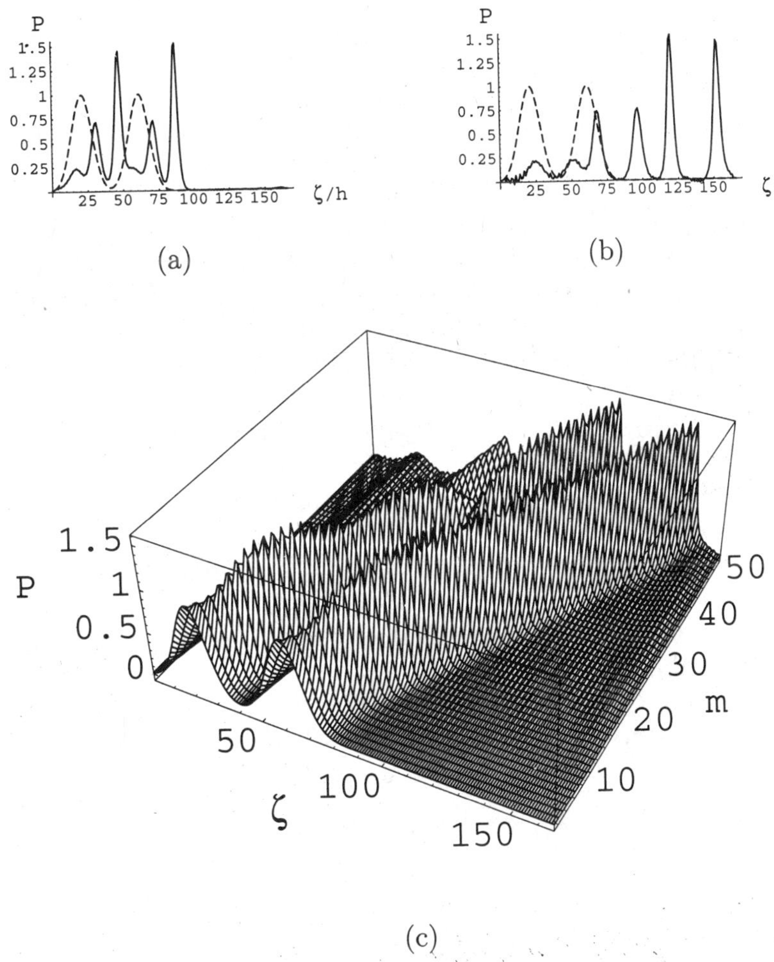

Figure 7.52: The numerical solution of the BKdV equation (7.7.52) at $\mathrm{Re} = 10^4$ and $D_\sigma = 10$: (a) the profile of the solution $P(\zeta, t)$ at $t = 200\tau = 4.2908$, (b) the solution profile at $t = 800\tau = 17.1633$ and (c) the surface $P = P(\zeta, t)$ at $0 \le \zeta \le 16.5$, $0 \le t \le 17.1633$; $m = t/(16\tau)$. The dashed line in (a) and (b) is the initial function $P(\zeta, 0)$.

The phase velocity c_ϕ and the group velocity c_g of the wave are given by formulas

$$c_\phi = \frac{\omega}{k} = c_0 - \beta k^2, \quad c_g = \frac{d\omega}{dk} = c_0 - 3\beta k^2.$$

It follows from here that the long waves $k \to \infty$ propagate faster than the short waves. The waves of an infinitely small amplitude with the wave vector $k > k^* = \sqrt{c_0/\beta}$ cannot propagate in the medium at all, which is described by equation (7.7.74).

References

1. **Rakhmatulin, Kh.A.,** The fundamentals of gas dynamics of interpenetrating motions of compressible media, *Prikladnaya Matematika i Mekhanika* (in Russian), 20(2):184, 1956.
2. **Kraiko, A.N., Nigmatulin, R.I., Starkov, V.K., and Sternin, L.E.,** Mechanics of multiphase media, *Itogi nauki i tehniki. Ser. Gidromekhanika* (in Russian), Vol. 6, Nauka, Moscow, 1972.
3. **Nigmatulin, R.I.,** *The Fundamentals of the Mechanics of Heterogeneous Media* (in Russian), Nauka, Moscow, 1978.
4. **Nigmatulin, R.I.,** *The Dynamics of the Multiphase Media* (in Russian), Part 1, Nauka, Moscow, 1987.
5. **Nigmatulin, R.I.,** *The Dynamics of the Multiphase Media* (in Russian), Part 2, Nauka, Moscow, 1987.
6. **Nikolaevskii, V.N. et al.,** *The Mechanics of Saturated Porous Media* (in Russian), Nedra, Moscow, 1970.
7. **Yanenko, N.N., Soloukhin, R.I., Papyrin, A.N., and Fomin, V.M.,** *Supersonic Two-Phase Flows under the Conditions of Velocity Nonequilibrium of Particles* (in Russian), Nauka, Novosibirsk, 1980.
8. **Kiselev, S.P., Ruev, G.A., Trunev, A.P., Fomin, V.M., and Shavaliev, M.Sh.,** *Shockwave Processes in Two-Component and Two-Phase Media* (in Russian), Nauka, Novosibirsk, 1992.
9. **Panton, P.,** Flow properties from the continuum viewpoint of a nonequilibrium gas-particle mixture, *J. Fluid Mechanics,* 31(2):273, 1968.
10. **Henderson, C.B.,** Drag coefficient of spheres in continuum and rarefied flows, *AIAA J.*, 14(6):707, 1967.
11. **Carlson, D.I. and Hogland, R.F.,** Particle drag and heat transfer in rocket nozzles, *AIAA J.*, 2(11):1980, 1964.
12. **Boiko, V.M. et al.,** Shock wave interaction with a cloud of particles, *Shock Waves,* 7:275, 1997.
13. **Kraiko, A.N. and Sternin, L.E.,** Theory of flows of a two-velocity continuous medium containing solid or liquid particles, *Prikladnaya Matematika i Mekhanika* (in Russian), 29(3):418, 1965. (English transl. in: *J. Appl. Math. Mechanics (PMM),* 29(3):482, 1965).
14. **Klebonov, L.A., Kroshilin, A.E., Nigmatulin, B.I., and Nigmatulin, R.I.,** On the hyperbolicity, stability and correctness of the Cauchy problem for the system of equations of two-velocity motion of two-phase media, *Prikladnaya Matematika i Mekhanika* (in Russian), 46(1):83, 1982.

15. **Kraiko, A.N.,** On the correctness of the Cauchy problem for a two-liquid flow model of a gas-particle mixture, *Prikladnaya Matematika i Mekhanika* (in Russian), 46(3):420, 1982.
16. **Kliegel, J.R. and Nickerson, G.R.,** Flow of gas-particle mixtures in axially symmetric nozzles, in *Detonation and Two-Phase Flow. Progress in Astronautics and Rocketry*, Vol. 6/Eds. S.S. Penner and F.A. Williams, Academic Press, New York, 1962, p. 173.
17. **Kiselev, S.P. and Fomin, V.M.,** A continual/discrete model for the gas-particle mixture at a small volume concentration of particles, *Zhurnal Prikladnoi Mekhaniki i Tehnicheskoi Fiziki* (in Russian), 2:93, 1986.
18. **Lavrentyev, M.M., Romanov, V.G., and Shishatskii, S.P.,** *Ill-Posed Problems of the Mathematical Physics and Analysis* (in Russian), Nauka, Moscow, 1980.
19. **Myasnikov, V.P.,** On the dynamic motion equations of two-component systems, *Zhurnal Prikladnoi Mekhaniki i Tehnicheskoi Fiziki* (in Russian), 2, 58, 1967.
20. **Goldshtik, M.A. and Kozlov, B.M.,** The elementary theory of the condensed disperse systems, *Zhurnal Prikladnoi Mekhaniki i Tehnicheskoi Fiziki* (in Russian), 4:89, 1973.
21. **Bird, G.A.,** *Molecular Gas Dynamics*, Clarendon Press, Oxford, 1976.
22. **Volkov, A. and Tsirkunov, Yu.,** Direct simulation Monte-Carlo modelling of two-phase gas-solid particle flows with inelastic particle-particle collisions, in *Proc. 3rd ECCOMAS Computational Fluid Dynamics Conf., 9-13 September 1996, Paris, France*, John Wiley and Sons, Inc., New York, 1996, p. 662.
23. **Kiselev, S.P. and Fomin, V.M.,** The relations at a combined discontinuity in gas with solid particles, *Zhurnal Prikladnoi Mekhaniki i Tehnicheskoi Fiziki* (in Russian), 2:112, 1984.
24. **Sternin, L.E.,** *The Fundamentals of the Gas Dynamics of Two-Phase Nozzle Flows* (in Russian), Mashinostroenie, Moscow, 1974.
25. **Kliegel, J.R.,** Gas particle nozzle flows, *Ninth Symposium (International) on Combustion*, Academic Press, New York, 1963, p. 811.
26. **Kogan, M.N.,** *Rarefied Gas Dynamics* (in Russian), Nauka, Moscow, 1967.
27. **Zeldovich, Ya.B. and Myshkis, A.D.,** *The Elements of the Mathematical Physics* (in Russian), Nauka, Moscow, 1973.
28. **Zeldovich, Ya.B.,** *Selected Works. Particles, Kernels, the Universe* (in Russian), Nauka, Moscow, 1985.
29. **Arnold, V.I.,** *Catastrophe Theory* (in Russian), Nauka, Moscow, 1990.

30. **Kiselev, S.P. and Kiselev, V.P.,** On the interaction of a shock wave with a cloud of particles with perturbed boundaries, *Zhurnal Prikladnoi Mekhaniki i Tehnicheskoi Fiziki* (in Russian), 37(4):36, 1996.
31. **Kuhl, A.L., Reichenbach, H., and Ferguson, R.E.,** Shock interaction with a dense-gas wall layer. In: *Shock Waves Proc.*, Vol. 1 /Ed. K. Takayama, Sendai, Japan, 1991.
32. **Boiko, V.M.,** *The Investigation of the Dynamics of Acceleration, Breakdown, and Ignition of Particles Behind the Shock Waves by the Methods of Laser Visualization,* a summary of Thesis (in Russian), Novosibirsk, 1984.
33. **Kiselev, S.P. and Kiselev, V.P.,** On the ignition of coal dust particles in shock waves, *Prikladnaya Mekhanika i Tehnicheskaya Fizika* (in Russian), 36(3):31, 1995.
34. **Kiselev, S.P. and Kiselev, V.P.,** On some peculiarities of gas flow arising as a result of the shock wave interaction with a cloud of particles, *Prikladnaya Mekhanika i Tehnicheskaya Fizika* (in Russian), 36(2), 8, 1995.
35. **Fomin, V.M. et al.,** On some peculiarities of gas flow arising at the shock wave interaction with a cloud of particles, *Doklady Rossiiskoi Akademii Nauk* (in Russian), 340(2):8, 1995.
36. **Landau, L.D. and Lifschitz, E.M.,** *Hydrodynamics* (in Russian), Nauka, Moscow, 1986.
37. **Landau, L.D. and Lifschitz, E.M.,** *Statistical Physics* (in Russian), Part 1, Nauka, Moscow, 1976.
38. **Larson, R.G. and Hirasaki, G.J.,** Analysis of the physical mechanisms in surfactant flooding, *Soc. Petroleum Eng. J.*, 18(1):42, 1978.
39. **Larson, R.G., Davis, H.T., and Scriven, L.E.,** Displacement of residual nonwetting fluid from porous media, *Chem. Eng. Sci.*, 18:75, 1980.
40. **Davis, S.A. and Jones, S.C.,** Displacement mechanism of micellar solution, *J. Petroleum Technol.*, 20:1415, 1968.
41. **Chernyi, I.A.,** *Soil Hydrogasdynamics* (in Russian), Gostoptehizdat, Moscow, 1963.
42. **Barenblatt, G.I., Entov, V.M., and Ryzhik, V.M.,** *The Motion of Liquids and Gases in Natural Seams* (in Russian), Nedra, Moscow, 1984.
43. **Kiselev, S.P.,** *Continuum Mechanics* (in Russian), Novosibirsk State Technical University, Novosibirsk, 1997.
44. **Nakoryakov, V.E., Sobolev, V.V., and Schreiber, I.R.,** Long wavelength disturbances in a gas-liquid mixture, *Izvestiya AN SSSR, Mekhanika Zhidkosti i Gaza* (in Russian), 5:71, 1972.

45. **Wijngaarden, L. van,** One-dimensional flow of liquids containing small gas bubbles, *Annu. Rev. Fluid Mechanics*, Vol. 4:369, 1972.
46. **Kutateladze, S.S. and Nakoryakov, V.E.,** *Heat and Mass Transfer and the Waves in Gas-Liquid Systems* (in Russian), Nauka, Novosibirsk, 1984.
47. **Karpman, V.I.,** *Nonlinear Waves in Disperse Media* (in Russian), Nauka, Moscow, 1973.
48. **Zakharov, V.E. et al.,** *The Theory of Solitons* (in Russian), Nauka, Moscow, 1980.
49. **Zabusky, N.J., and Kruškal, M.D.,** Interaction of "solitons" in a collisionless plasma and the recurrence of initial states, *Phys. Rev. Lett.*, 15:240, 1965.
50. **Sagdeev, R.Z.,** On nonlinear motions of rarefied plasma in the magnetic field, in *Physics of Plasma and the Problem of Controlled Thermonuclear Reactions* (in Russian), Vol. 4, USSR Academy of Sciences, Novosibirsk, 1958, p. 384.
51. **Nigmatulin, R.I., Ivandaev, A. I., Nigmatulin, B. I., and Milashenko, V.I.,** Nonstationary wave processes in the gas/vapor/liquid mixtures, in *Nonlinear Wave Processes in Two-Phase Media* (in Russian), Institute of Thermophysics of the USSR Academy of Sciences, Novosibirsk, 1977, p. 80.
52. **Berezin, Y.A.,** *Modeling of Nonlinear Wave Processes* (in Russian), Nauka, Novosibirsk, 1982.
53. **Richtmyer, R.D. and Morton, K.W.,** *Difference Methods for Initial-Value Problems*, Second Edition, Interscience Publishers, New York, 1967.

Appendix A: *Mathematica* Functions

In this appendix, we describe those *Mathematica* functions that we have used in this book. The use of each function is illustrated by examples. A more detailed description of these functions as well as many other functions may be found in the *Mathematica 3.0* manual:

> S. Wolfram. *The MATHEMATICA Book.* Third Edition. Wolfram Media/Cambridge University Press, Cambridge (UK), New York, Melbourne, 1996.

This version of *Mathematica* can be installed on various computer platforms, which have the following:

1) processor type Intel 80386 or higher;

2) operating system Windows 95 or Windows NT 3.51;

3) hard disk space 116 MB (needed if not running off CD-ROM); 116 MB recommended, 30 MB minimum;

4) system memory 16 MB recommended, 8 MB minimum.

The *Mathematica* kernel uses PostScript language to represent all of the graphics it produces. The *Mathematica* for Windows front end includes a PostScript interpreter to translate PostScript code from the kernel into screen images.

One of the features of *Mathematica 3.0* is that the *Mathematica* documents are called Notebooks. Each Notebook is stored as a file `name.nb`. Notebooks can contain ordinary text and graphics as well as *Mathematica* input and output. The information in a Notebook is stored in cells, whose characteristics vary with the cell's function and the kind of information it holds. The square brackets on the right-hand side of the Notebook window indicate the extent of each cell. We have presented here the texts of our Notebooks together with the results of symbolic and numeric computations. Therefore, the corresponding Notebook cells include not only the input, but also the output of a computation made within a cell.

It should be noted that a number of books have been published, in which the software system *Mathematica* is applied for the analytical or numerical solution of some continuum mechanics problems.

The main feature of the use of *Mathematica 3.0* in this book is that we have applied this powerful software system for the computer implementation of *analytical* or *approximate analytical* methods for the solution of fluid mechanics problems or for the derivation and investigation of underlying mathematical models of various continua. In this respect, this book differs from Ganzha and Vorozhtsov's book[1], where the emphasis was placed on the computer implementation of finite difference, finite volume, and finite element methods for the numerical solution of partial differential equations.

As shown in Chapter 1, many problems of tensor analysis arising at the derivation of the governing equations of fluid mechanics can be solved with the aid of *Mathematica.* Parker and Christensen[2] describe their *MathTensor* software, which provides a computer program that extends *Mathematica* capabilities to include tensor analysis.

Kythe et al. and Vvedensky[3,4] discuss the implementation of a number of analytical methods for the solution of partial differential equations: the method of characteristics, the separation of variables, etc.

We have used the bisection method for finding the roots of transcendental equations in a number of our *Mathematica* notebooks. The interested reader can find in[5] both the implementation of the bisection method and a discussion of its advantages and disadvantages.

Baumann[6] shows, in particular, how *Mathematica* can be used for the numerical solution of the Korteweg–de Vries equation (cf. Section 7.7 of the present book) with the aid of both the analytic method (the soliton solution) and the finite difference method by Zabusky and Kruskal.

We may conclude from the above survey that there is at present a huge gap in the development of big industrial programs written in the language of *Mathematica*, which would solve such complex fluid mechanics problems as the three-dimensional transonic gas flow around the complete aircraft, the hypersonic three-dimensional gas flows around the reentry vehicles, three-dimensional gas flows between the compressor blades of an aircraft engine, two-dimensional flows of gas-particle mixtures in rocket nozzles, three-dimensional incompressible gas flows around the cars, etc. This book only partly bridges this gap. This is in line with the main purpose, which may be formulated as follows: the derivation, investigation, and comparison of the mathematical models of various continua.

Before proceeding to the description of the built-in *Mathematica* functions we have used above, we present in Table A.1 the names of these functions or the options used in functions in alphabetic order together with the numbers of the sections, where these functions are used.

Table A.1
The *Mathematica* Functions

Abs	Flatten	PlotVectorField3D
AppendTo	Floor	PointSize
ArcSin	Graphics	Polygon
ArcTan	GraphicsArray	PowerExpand
AspectRatio	Hue	Print
Axes	IdentityMatrix	ReplacePart
AxesLabel	If	RGBColor
Circle	Im	ScaleFactor
ClearAll	Interpolation	Show
ColorFunction	Inverse	Sign
ComplexExpand	Join	Simplify
ContourPlot	Length	Sin
Contours	Line	Solve
ContourShading	LinearSolve	Sqrt
Cos	ListPlot	Sum
D	Log	Table
Dashing	MapThread	Text
DensityGraphics	MatrixForm	Thickness
DisplayFunction	N	TraditionalForm
DSolve	NIntegrate	Transpose
Eigenvalues	ParametricPlot3D	TrigExpand
Eigenvectors	Pi	TrigReduce
Expand	Plot3D	VectorHeads
First	PlotPoints	While

Abs[z] gives the absolute value of the real or complex number z.

Example 1

Abs[−3.5]
3.5

AppendTo[s, elem] appends *elem* to the value of *s*, and resets *s* to the result.

Example 2

$xx = \{x_1,\ x_2,\ x_3\}$; AppendTo[xx, a^2]
$\{x_1,\ x_2,\ x_3,\ a^2\}$

ArcSin[z] gives the arc sine of the complex number z.

Example 3

ArcSin [1/2]

$$\frac{\pi}{6}$$

ArcTan[z] gives the arc tangent of the complex number z. `ArcTan[x, y]` gives the arc tangent of y/x, taking into account which quadrant the point (x, y) is in.

Example 4

ArcTan [1]

$$\frac{\pi}{4}$$

AspectRatio is an option for **Show[...]** and related functions that specifies the ratio of height to width for a plot. The default value `AspectRatio -> 1/GoldenRatio` is used for two-dimensional plots. The setting `AspectRatio -> Automatic` forces *Mathematica* to take the actual ratio of height to width.

Example 5

```
a1 = 1.4779155; a2 = -0.624424; a3 = -1.727016; a4 = 1.384087;
a5 = -0.489769; t1 = 0.2;
body[z_]:= t1 (a1 Sqrt[z] + z (a2 + z (a3 + z (a4 + a5 z))));
g1 = Plot[body[x], {x, 0, 1},
          DisplayFunction - > Identity];
g2 = Plot[-body[x], {x, 0, 1},DisplayFunction - > Identity];
 Show[g1, g2, DisplayFunction - > $DisplayFunction]
```

The graphical image produced with the aid of this program is shown in Fig. A.1 (a). Compare this picture with Fig. A.2 obtained in the case where the value `AspectRatio -> Automatic` is used (see the description of the option `AxesLabel`).

Axes is an option for graphics functions that specifies whether axes should be drawn. `Axes -> True` draws all axes. `Axes -> False` removes the axes. [see also Fig. A.1 (b)]

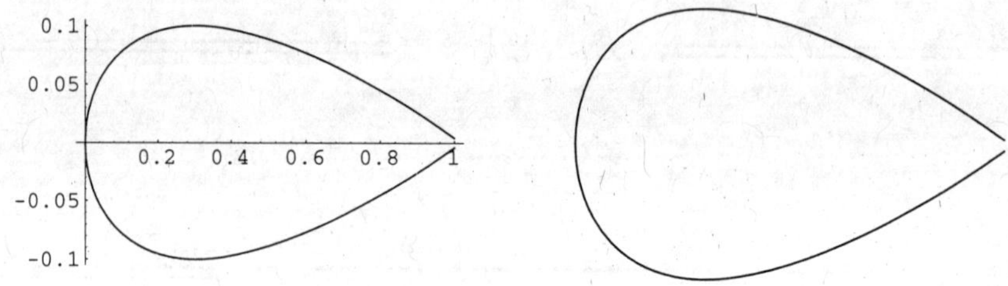

(a) (b)

Figure A.1: Illustration to the description of the option (a) `AspectRatio` and (b) `Axes -> False`.

Example 6

```
a1 = 1.4779155; a2 = -0.624424; a3 = -1.727016; a4 = 1.384087;
a5 = -0.489769; t1 = 0.2;
body[z_]:= t1 (a1 Sqrt[z] + z (a2 + z (a3 + z (a4 + a5 z))));
g1 = Plot[body[x], {x, 0, 1}, DisplayFunction -> Identity];
g2 = Plot[-body[x], {x, 0, 1}, DisplayFunction -> Identity];
Show[g1, g2, Axes -> False, DisplayFunction -> $DisplayFunction]
```

AxesLabel is an option for graphics functions that specifies labels for axes (see also Fig. A.2).

Example 7

```
a1 = 1.4779155; a2 = -0.624424; a3 = -1.727016; a4 = 1.384087;
a5 = -0.489769; t1 = 0.2;
body[z_]:= t1 (a1 Sqrt[z] + z (a2 + z (a3 + z (a4 + a5 z))));
g1 = Plot[body[x], {x, 0, 1}, DisplayFunction -> Identity];
g2 = Plot[-body[x], {x, 0, 1},DisplayFunction -> Identity];
Show[g1,g2, AxesLabel -> {"x", "y"}, AspectRatio -> Automatic,
DisplayFunction -> $DisplayFunction]
```

Circle[{x, y}, r] is a two-dimensional graphics primitive that represents a circle of radius r centered at point x, y. **Circle[{x,y}, {rx, ry}]** yields an ellipse with semi-axes rx and ry. **Circle[{x,y}, r, {theta1, theta2}]** represents a circular arc. *Example:* see the description of functions **ContourPlot[...], Text[...]**.

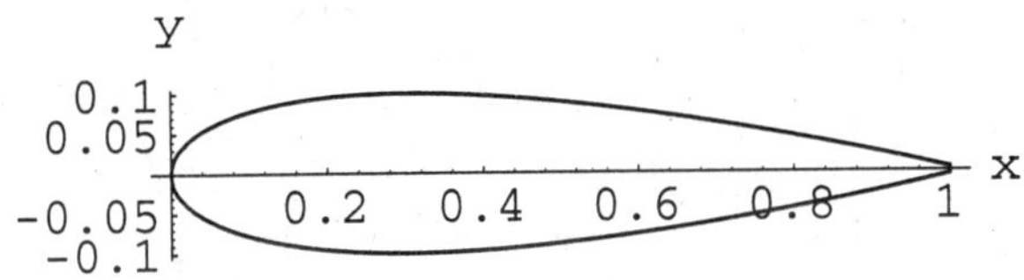

Figure A.2: Illustration to the description of the option `AxesLabel`.

ClearAll[symb1, symb2, ...] clears all values, definitions, attributes, messages and defaults associated with symbols.
ClearAll["form1", "form2", ...] clears all symbols whose names textually match any of the forms.

Example 8

```
x1 = 2Cos[t]
2 Cos [t]
ClearAll[x1]
x1
```

ColorFunction is an option for various graphics functions that specifies a function to apply to z values to determine the color to use for a particular x, y region.

ComplexExpand[expr] expands expr assuming that all variables are real. **ComplexExpand[expr, {x1, x2, ... }]** expands expr assuming that variables matching any of the xi are complex.

Example 9

ComplexExpand[Im[Cos [z^2]/. z − > x + I y]]//TraditionalForm

$$-\sin(x^2 - y^2)\sinh(2xy)$$

ContourPlot[f, {x, xmin, xmax}, {y, ymin, ymax}] generates a contour plot of f as a function of x and y. This function has the following default options:
Options[ContourPlot] = {AspectRatio − > 1,

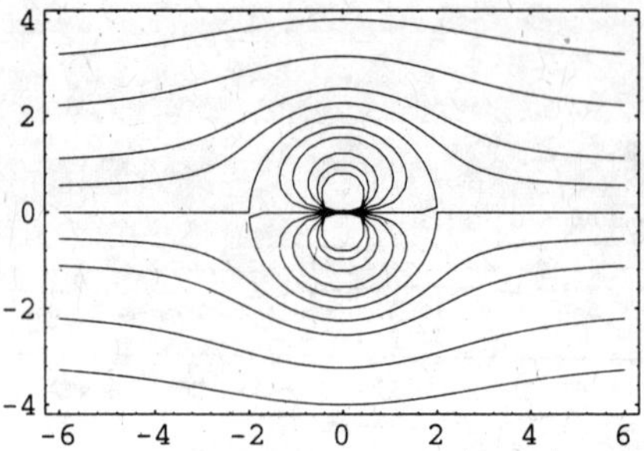

Figure A.3: Illustration to the description of the function `ContourPlot`.

Axes – > False, AxesLabel – > None,
AxesOrigin – > Automatic, AxesStyle – > Automatic,
Background – > Automatic, ColorFunction – > Automatic,
ColorOutput – > Automatic, Compiled – > True,
ContourLines – > True, Contours – > 10,
ContourShading – > True, ContourSmoothing – > True,
ContourStyle – > Automatic, DefaultColor – > Automatic,
Epilog – > {}, Frame – > True,
FrameLabel – > None, FrameStyle – > Automatic,
FrameTicks – > Automatic, PlotLabel – > None,
PlotPoints – > 15, PlotRange – > Automatic,
PlotRegion – > Automatic, Prolog – > {},
RotateLabel – > True, Ticks – > Automatic,
DefaultFont :> $DefaultFont,
DisplayFunction :> $DisplayFunction}

(see also Fig. A.3).

Example 10

$\psi[x\_, y\_] := y\left(1 - \frac{rc^2}{x^2+y^2}\right);$
rc=2;
streamlines = ContourPlot[ψ[x,y],{x,-6,6},{y,-4,4}, PlotPoints → 40,
Contours → {-4,-3,-2,-1,-0.5,0,0.5,1,2,3,4},
ContourShading → False, ContourSmoothing → Automatic,
DisplayFunction → Identity];
bound = Graphics[Circle[{0,0}, rc]];
Show[streamlines, bound, AspectRatio → Automatic,
DisplayFunction → $DisplayFunction];

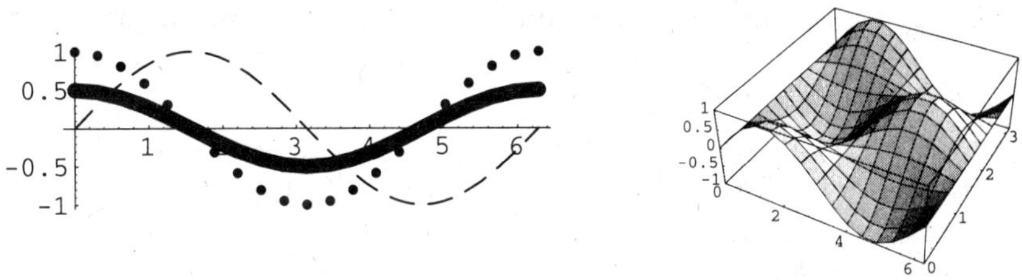

Figure A.4: Illustration to the descriptions of the functions `Dashing`, `GraphicsArray`, `Plot3D`, `Thickness`, and `PointSize`.

Contours is an option for **ContourGraphics** specifying the contours to use. The contour values are specified in the form of a list of numerical values. *Example:* see the description of the function **ContourPlot**[...].

ContourShading is an option for contour plots that specifies whether the regions between contour lines should be shaded. *Example:* see the description of the function **ContourPlot**[...].

Cos[**z**] gives the cosine of z. The argument of `Cos` is assumed to be in radians.

Example 11

TraditionalForm[Cos[Pi/10]]

$$\frac{1}{2}\sqrt{\frac{1}{2}(5+\sqrt{5})}$$

D[**f, x**] gives the partial derivative of f with respect to x. **D**[**f,** {**x, n**}] gives the nth partial derivative of f with respect to x. **D**[**f,x1,x2,...**] gives a mixed derivative.

Example 12

$x = k[1]$ Sin$[k[2]]$ Sin$[k[3]]$; D$[x,\ k[2]]$

$$\cos[k[2]]\ k[1]\ \sin[k[3]]$$

Dashing[{**r1, r2, ...**}] is a two-dimensional graphics directive that specifies that lines which follow are to be drawn dashed, with successive segments of lengths $r1, r2, \ldots$ (repeated cyclically). The ri is given as a fraction of the total width of the graph (see also Fig. A.4).

Example 13

```
g1 = Plot[Sin[x], {x,0, 2π},
PlotStyle - > {Dashing[{0.05,0.03}]}, DisplayFunction - > Identity];
 xx = Table[N[(k-1) π/10], {k, 21}]; yy = Table[N[Cos[xx[[k]]]], {k, 21}];
g2=ListPlot[MapThread[List,{xx, yy}],
              PlotStyle - > {PointSize[0.02]},
             DisplayFunction - > Identity];
g3=ListPlot[MapThread[List,{xx, 0.5yy}], PlotJoined - > True,
             PlotStyle - > {Thickness[0.03]},
             DisplayFunction - > Identity];
A2 = Plot3D[Sin[x] Cos[2y], {x, 0, 2Pi}, {y, 0, Pi},
              DisplayFunction - > Identity];
A1 = Show[g2,g1,g3, AspectRatio - > Automatic,
             DisplayFunction - > Identity];
Show[GraphicsArray[{A1, A2}]];
```

DensityGraphics[array] is a representation of a density plot.
Options[DensityGraphics] = {AspectRatio - > 1,
Axes - > False, AxesLabel - > None,
AxesOrigin - > Automatic, AxesStyle - > Automatic,
Background - > Automatic, ColorFunction - > Automatic,
ColorOutput -¿ Automatic, DefaultColor - > Automatic,
Epilog - > {}, Frame - > True,
FrameLabel - > None, FrameStyle - > Automatic,
FrameTicks - > Automatic, ImageSize - > Automatic,
Mesh - > True, MeshRange - > Automatic,
MeshStyle - > Automatic, PlotLabel - > None,
PlotRange - > Automatic, PlotRegion - > Automatic,
Prolog - > {}, RotateLabel - > True,
Ticks - > Automatic, DefaultFont:>$DefaultFont,
DisplayFunction:>$DisplayFunction,
FormatType:>$FormatType, TextStyle:>$TextStyle}.

DisplayFunction is an option for graphics and sound functions that specifies the function to apply to graphics and sound objects in order to display them. The default setting for **DisplayFunction** in graphics functions is **$DisplayFunction**. Setting **DisplayFunction - > Identity** will cause the objects to be returned, but no display to be generated. *Examples:* see the descriptions of options **AspectRatio, Axes, AxesLabel**.

DSolve[eqn, y, x] solves a differential equation for the function y, with independent variable x. **DSolve[{eqn1, eqn2, ... }, {y1, y2, ...**

}, **x**] solves a list of differential equations. **DSolve[eqn, y, {x1, x2, ... }]** solves a partial differential equation.

Example 14

DSolve[$\{x2'[x] == -x2[x]\ x,\ x2[\xi_1] == \xi_2\},\ x2[x],\ x$]

$\{\{x2[x]->E^{-\frac{x^2}{2}+\frac{\xi_1^2}{2}}\xi_2\}\}$

Eigenvalues[m] gives a list of the eigenvalues of the square matrix m.

Example 15

A = {{-1, 0, 3}, {2,1,1}, {5,1,4}}; Eigenvalues[A]

$$\{-3, \frac{1}{2}(7-\sqrt{33}), \frac{1}{2}(7+\sqrt{33})\}$$

Eigenvectors[A]

$$\{\{-3,1,2\},\{-\frac{6}{-9+\sqrt{33}},\frac{2(9+2\sqrt{33})}{-9+\sqrt{33}},1\},\{\frac{6}{9+\sqrt{33}},\frac{2(-9+2\sqrt{33})}{9+\sqrt{33}},1\}\}$$

TraditionalForm[Inverse[A]]

$$\begin{pmatrix} -\frac{1}{4} & -\frac{1}{4} & \frac{1}{4} \\ \frac{1}{4} & \frac{19}{12} & -\frac{7}{12} \\ \frac{1}{4} & -\frac{1}{12} & \frac{1}{12} \end{pmatrix}$$

Eigenvectors[m] gives a list of the eigenvectors of the square matrix m.
Options[Eigenvectors]={ZeroTest − > Automatic}.
Example: see the description of function **Eigenvalues[...]**.

Expand[expr] expands out products and positive integer powers in expr. **Expand[expr, patt]** leaves unexpanded any parts of expr that are free of the pattern patt.

Example 16

TraditionalForm[Expand[$(\alpha-3\beta)^5$]]

$$\alpha^5-15\beta\alpha^4+90\beta^2\alpha^3-270\beta^3\alpha^2+405\beta^4\alpha-243\beta^5$$

First[expr] gives the first element in expr.

Example 17

lis = {a1, Sin[t], $\pi/8$}; First[lis]
a1

Flatten[list] flattens out nested lists. **Flatten[list, n]** flattens to level n. **Flatten[list, n, h]** flattens subexpressions with head h.

Example 18

Flatten[{a1,b2, {{2,3}, {a4, s5}}}]

$$\{a1, b2, 2, 3, a4, s5\}$$

Floor[x] gives the greatest integer less than or equal to x.

Example 19

Floor[-2.4]
-3
Floor[2.4]
2

Graphics[primitives, options] represents a two-dimensional graphical image. The default options are:

Options[Graphics]={AspectRatio – > GoldenRatio(– 1),
Axes – > False, AxesLabel – > None,
AxesOrigin – > Automatic, AxesStyle – > Automatic,
Background – > Automatic, ColorOutput – > Automatic,
DefaultColor – > Automatic, Epilog – > {},
Frame – > False, FrameLabel – > None,
FrameStyle – > Automatic, FrameTicks – > Automatic,
GridLines – > None, ImageSize – > Automatic,
PlotLabel – > None, PlotRange – > Automatic,
PlotRegion – > Automatic, Prolog – >{},
RotateLabel – > True, Ticks – > Automatic,
DefaultFont:>$DefaultFont,
DisplayFunction:>$DisplayFunction,
FormatType:>$FormatType, TextStyle:>$TextStyle}.

Example: see the description of the function **Line[...]**.

GraphicsArray[{g1, g2, ...}] represents a row of graphics objects. `GraphicsArray[{{g11, g12, ... }, ... }]` produces a two-dimensional array of graphics objects.

Options[GraphicsArray] = {AspectRatio − > Automatic,
Axes − > False, AxesLabel − > None,
AxesOrigin − > Automatic, AxesStyle − > Automatic,
Background − > Automatic, ColorOutput − > Automatic,
DefaultColor − > Automatic, Epilog − > {},
Frame − > False, FrameLabel − > None,
FrameStyle − > Automatic, FrameTicks − > None,
GraphicsSpacing − > 0.1, GridLines − > None,
ImageSize − > Automatic, PlotLabel − > None,
PlotRange − > Automatic, PlotRegion − > Automatic,
Prolog − > {}, RotateLabel − > True,
Ticks − > None, DefaultFont:> $DefaultFont,
DisplayFunction:> $DisplayFunction,
FormatType:> $FormatType, TextStyle:> $TextStyle}.

See *Example 13.*

Hue[h] is a graphics directive that specifies that graphical objects which follow are to be displayed, if possible, in a color corresponding to hue h. `Hue[h, s, b]` specifies colors in terms of hue, saturation, and brightness.

IdentityMatrix[n] gives the n by n identity matrix.

Example 20

TraditionalForm[IdentityMatrix[3]]

$$\begin{pmatrix} 1 & 0 & 0 \\ 0 & 1 & 0 \\ 0 & 0 & 1 \end{pmatrix}$$

If[condition, t, f] gives t if condition evaluates to True, and f if it evaluates to False. If[condition, t, f, u] gives u if condition evaluates to neither True nor False.

Example 21

b = 5; c = Sin[t]; d = t^2; If[b < 4, c = Cos[t]; d = 5t, c = Cosh[t];
d = 4t];
Print["c = ", c, "; d = ", d];
c = Cosh[t]; d = 4 t

Im[z] gives the imaginary part of the complex number z. *Example:* see the description of the function **ComplexExpand[...]**.

Interpolation[data] constructs an InterpolatingFunction object representing an approximate function that interpolates the data. The data can have the forms {{x1, f1}, {x2, f2}, ... } or {f1, f2, ... }, where in the second case, the xi are taken to have values 1, 2, ...
Options[Interpolation] = InterpolationOrder − > 3

Inverse[m] gives the inverse of a square matrix m.
Options[Inverse] = {Method − > CofactorExpansion, Modulus − > 0, ZeroTest − > (#1 == 0 &)}.
Example: see the description of function **Eigenvalues[...]**.

Join[list1, list2, ...] concatenates lists together.

Example 22

$ls1 = \{x_1,\ x_2\};\ ls2 = \{\mathrm{Sqrt}[1 + b^2],\ \mathrm{Sin}[t]\};$
TraditionalForm[Join[$ls1$, $ls2$]]
$\{x_1,\ x_2,\ \sqrt{b^2+1},\ \sin(t)\}$

Length[expr] gives the number of elements in expr.

Example 23

Length[{a, x^3, 1998}]
3

Line[{pt1, pt2, ... }] is a graphics primitive that represents a line joining a sequence of points [see Fig. A.5 (a)].

Example 24

Show[Graphics[Line[{{4,3}, {5,4}, {2,7}, {1,6}, {4,3}, {1,3}, {1,1}, {7,1}, {7,3}, {4,3}}]], AspectRatio − > Automatic]

LinearSolve[m, b] finds an x that solves the matrix equations m.x == b.

Example 25

n = 3; b = Table[k, {k, n}]; A = Table[k/(i + k), {i, n}, {k, n}];
LinearSolve[A, b]

$$\{132, -300, 180\}$$

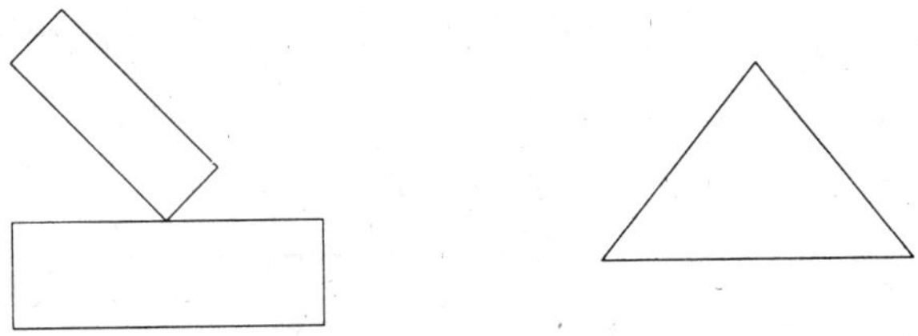

(a) (b)

Figure A.5: Illustration to the description of the functions (a) `Line` and (b) `ListPlot`.

ListPlot[{**y1, y2, ...** }] plots a list of values. The x coordinates for each point are taken to be 1, 2, ...
ListPlot[{{x1, y1}, {x2, y2}, ... }] plots a list of values with specified x and y coordinates.
Options[ListPlot] = {AspectRatio – > GoldenRatio$^{(-1)}$,
Axes – > Automatic, AxesLabel – > None,
AxesOrigin – > Automatic, AxesStyle – > Automatic,
Background – > Automatic, ColorOutput – > Automatic,
DefaultColor – > Automatic, Epilog – > {},
Frame – > False, FrameLabel – > None,
FrameStyle – > Automatic, FrameTicks – > Automatic,
GridLines – > None, ImageSize – > Automatic,
PlotJoined – > False, PlotLabel – > None,
PlotRange – > Automatic, PlotRegion – > Automatic,
PlotStyle – > Automatic, Prolog – > {},
RotateLabel – > True, Ticks – > Automatic,
DefaultFont:>$DefaultFont,
DisplayFunction:>$DisplayFunction,
FormatType:>$FormatType,
TextStyle:>$TextStyle}
[see also Fig. A.5 (b)].

Example 26

ListPlot[{{0,1},{2,1}, {1,3}, {0,1}}, Axes – > False, PlotJoined – > True]

Log[z] gives the natural logarithm of z (logarithm to base e). Log[b, z] gives the logarithm to base b.

Example 27

N[Log[10]]

2.30259

MapThread[f, {{a1, a2, ... }, {b1, b2, ... }, ... }] gives the list {f[a1, b1, ...], f[a2, b2, ...],... }. **MapThread[f, {expr1, expr2, ... }, n]}** applies f to the parts of expri at level n.

Example 28

$xx = \{x_1, x_2, x_3\}$; $yy = \{y_1, y_2, y_3\}$; MapThread[List, $\{xx, yy\}$]
$\{\{x_1, y_1\}, \{x_2, y_2\}, \{x_3, y_3\}\}$

MatrixForm[list] prints with the elements of list arranged in a regular array.

Example 29

Print["g = ", TraditionalForm[MatrixForm[IdentityMatrix[3]]]]

$$g = \begin{pmatrix} 1 & 0 & 0 \\ 0 & 1 & 0 \\ 0 & 0 & 1 \end{pmatrix}$$

N[expr] gives the numerical value of expr. **N[expr, n]** attempts to give a result with n-digit precision.

Example 30

N[Sqrt[2], 45]
1.41421356237309504880168872420969807856967188

NIntegrate[f, x, xmin, xmax] gives a numerical approximation to the integral of f with respect to x from xmin to xmax.
Options[NIntegrate] = AccuracyGoal − > ∞,
Compiled − > True, GaussPoints − > Automatic,
MaxPoints − > Automatic, MaxRecursion − > 6,
Method − > Automatic, MinRecursion − > 0,
PrecisionGoal − > Automatic, SingularityDepth − > 4,
WorkingPrecision − > 16}.

Example 31

NIntegrate[Sqrt[Sin [$\pi/2$] - Sin [x]], $\{x, 0, \pi/2\}$]

0.828427

We can check this result with the aid of the *Mathematica* function `Integrate[_]`, which enables the user to obtain the expression for the definite integral in symbolic form:

Integrate[Sqrt[Sin[$\pi/2$] – Sin[x]], $\{x, 0, \pi/2\}$]
N[%]

$$-2 + 2\sqrt{2}$$

0.828427

ParametricPlot3D[{fx, fy, fz}, {t, tmin, tmax}] produces a three-dimensional space curve parametrized by a variable t that runs from *tmin* to *tmax*. **ParametricPlot3D[{fx, fy, fz}, {t, tmin, tmax}, {u, umin, umax}]** produces a three-dimensional surface parametrized by t and u. **ParametricPlot3D[{fx, fy, fz, s}, ...]** shades the plot according to the color specification s. **ParametricPlot3D[{{fx, fy, fz}, {gx, gy, gz}, ... }, ...]** plots several objects together. This function has the following default options.

Options[ParametricPlot3D] = {AmbientLight – > GrayLevel[0.],
AspectRatio – > Automatic, Axes – > True,
AxesEdge – > Automatic, AxesLabel – > None,
AxesStyle – > Automatic, Background – > Automatic,
Boxed – > True, BoxRatios – > Automatic,
BoxStyle – > Automatic, ColorOutput – > Automatic,
Compiled – > True, DefaultColor – > Automatic,
Epilog – > {}, FaceGrids – > None,
ImageSize – > Automatic, Lighting – > True,
LightSources – > {{{1.,0.,1.}, RGBColor[1,0,0]},
{{1.,1.,1.}, RGBColor[0,1,0]}, {{0.,1.,1.},
RGBColor[0,0,1]}}, Plot3Matrix – > Automatic,
PlotLabel – > None, PlotPoints – > Automatic,
PlotRange – > Automatic, PlotRegion – > Automatic,
PolygonIntersections – > True, Prolog – > {},
RenderAll – > True, Shading – > True,
SphericalRegion – > False, Ticks – > Automatic,
ViewCenter – > Automatic,
ViewPoint – > {1.3,-2.399999999999999,2.},

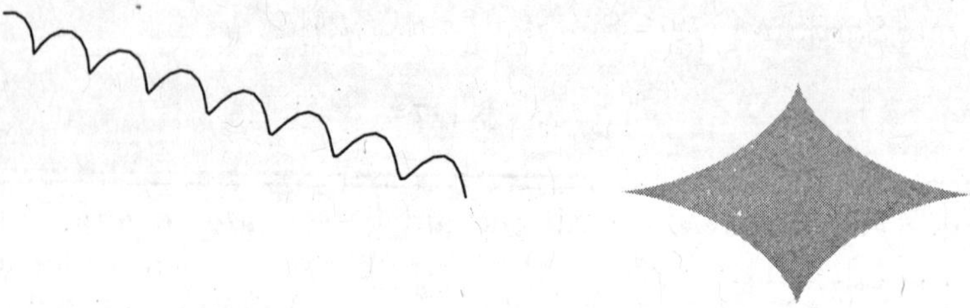

(a) (b)

Figure A.6: Illustration to the descriptions of the function (a) `ParametricPlot3D[...]` and (b) `Polygon[...]`.

ViewVertical – > {0.,0.,1.},
DefaultFont:>$DefaultFont,
DisplayFunction:>$DisplayFunction,
FormatType:>$FormatType, TextStyle:>$TextStyle}
[see also Fig. A.6 (a)].

Example 32

ParametricPlot3D[{2t, Sin[3t], Cos[3t]}, {t, 0, 5Pi},
Boxed – > False, Axes – > False]

Pi (π) is the constant pi, with numerical value approximately equal to 3.14159.

Plot3D[f, {x, xmin, xmax}, {y, ymin, ymax}] generates a three-dimensional plot of f as a function of x and y. `Plot3D[{f, s}, {x, xmin,xmax}, {y, ymin,ymax}]` generates a three-dimensional plot in which the height of the surface is specified by f, and the shading is specified by s.

Options[Plot3D] = AmbientLight – > GrayLevel[0],
AspectRatio – > Automatic, Axes – > True,
AxesEdge – > Automatic, AxesLabel – > None,
AxesStyle – > Automatic, Background – > Automatic,
Boxed – > True, BoxRatios – > {1,1,0.4},
BoxStyle – > Automatic, ClipFill – > Automatic,
ColorFunction – > Automatic, ColorOutput – > Automatic,
Compiled – > True, DefaultColor – > Automatic,
Epilog – > {}, FaceGrids – > None,

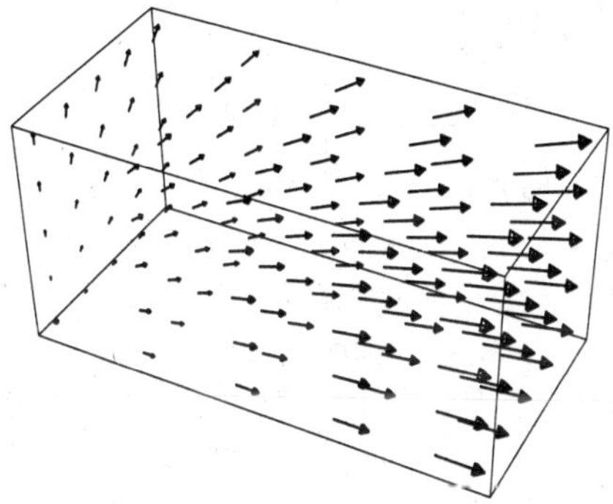

Figure A.7: Illustration to the descriptions of the function `PlotVectorField3D[...]`.

HiddenSurface − > True, ImageSize − > Automatic, Lighting − > True, LightSources − > {{{1.,0.,1.}, RGBColor[1,0,0]}, {{1.,1.,1.}, RGBColor[0,1,0]}, {{0.,1.,1.}, RGBColor[0,0,1]}}, Mesh − > True, MeshStyle − > Automatic, Plot3Matrix − > Automatic, PlotLabel − > None, PlotPoints − > 15, PlotRange − > Automatic, PlotRegion − > Automatic, Prolog − > {}, Shading − > True, SphericalRegion − > False, Ticks − > Automatic, ViewCenter − > Automatic, ViewPoint − > {1.3,-2.3999999999999999,2.}, ViewVertical − > {0.,0.,1.}, DefaultFont:>$DefaultFont, DisplayFunction:>$DisplayFunction, FormatType:>$FormatType, TextStyle:>$TextStyle}.

Example: see the description of the *Mathematica* function `Dashing[_]`.

PlotPoints is an option for plotting functions that specifies how many sample points to use. *Example:* see the description of the function **ContourPlot[...]**, where the option `PlotPoints -> 40` is used. The use of a larger number of plot points generally renders smoother curves.

PlotVectorField3D$[\{f_x, f_y, f_z\}, \{x, xmin, xmax\}, \{y, ymin, ymax\}, \{z, zmin, zmax\}]$ plots the vector field given by the vector function in the range specified. This function has the following default options:

ScaleFactor − > Automatic,
ScaleFunction − > None,

MaxArrowLength – > None,
ColorFunction – > None,
PlotPoints – > 7,
VectorHeads – > False.

The specification `VectorHeads – > True` puts heads on arrows.
The option `PlotPoints` specifies the number of evaluation points in each direction.
The option `ScaleFactor – > Automatic` ensures that the largest vector fits in the mesh. If the vectors obtained with this default option are too short, one can rescale the vectors, for example, by setting `ScaleFactor – > 1.3` (see Fig. A.7).

Example 33

```
<< Graphics'PlotField3D'
PlotVectorField3D[{2x,y,z}, {x,0,2}, {y,0,1},{z,0,1},
        PlotPoints – > 6,
        VectorHeads – > True]
```

The function `PlotVectorField3D[...]` is described in more detail in Martin's book[7].

PointSize[d] is a graphics directive that specifies that points which follow are to be shown if possible as circular regions with diameter d. The diameter d is given as a fraction of the total width of the graph. *Example:* see the description of the *Mathematica* function `Dashing[_]`.

Polygon[{pt1, pt2, ... }] is a graphics primitive that represents a filled polygon (see also Fig. A.6 (b)].

Example 34

Show[Graphics[{RGBColor[0.1, 1.0, 1.0],
Polygon[Table[{2Cos[$\frac{n\,\pi}{40}$]^3, 2Sin[$\frac{n\,\pi}{40}$]^3}, {n, 80}]]}]]

PowerExpand[expr] expands all powers of products and powers. The use of this function is efficient when one wishes to simplify expressions like $\sqrt{r^4 \sin^2 \varphi}$.

Example 35

$a = r^4$ Sin[φ]^2; TraditionalForm[Sqrt[a]]
$\sqrt{r^4\ \sin(\varphi)^2}$
TraditionalForm[PowerExpand[$a^{1/2}$]]
$r^2 \sin(\varphi)$

Print[expr1, expr2, ...] prints the expri, followed by a newline (line feed). *Example:* see the description of function **TraditionalForm[...]**.

ReplacePart[expr, new, n] yields an expression in which the nth part of *expr* is replaced by *new*. **ReplacePart[expr, new, {i, j, ... }]** replaces the part at position {i, j,... }. **ReplacePart[expr, new, {i1, j1, ... }, {i2, j2, ... }, ... }]** replaces parts at several positions by *new*.

Example 36

ReplacePart[IdentityMatrix[3], Sin[φ], {3, 1}]//TraditionalForm

$$\begin{pmatrix} 1 & 0 & 0 \\ 0 & 1 & 0 \\ \sin(\varphi) & 0 & 1 \end{pmatrix}$$

RGBColor[red, green, blue] is a graphics directive that specifies that graphical objects which follow are to be displayed, if possible, in the color given. *Example:* see the description of function **Polygon[...]**.

Show[graphics, options] displays two- and three-dimensional graphics, using the options specified. **Show[g1, g2, ...]** shows several plots combined. *Examples:* see the descriptions of option **DisplayFunction** and functions **Line[...]**, **ListPlot[...]**.

Simplify[expr] performs a sequence of algebraic transformations on expr, and returns the simplest form it finds. This function has the following default options:

Options[Simplify]={ComplexityFunction – > Automatic, TimeConstraint – > 300, Trig – > True}.

Example 37

Simplify[x^2/3 + x y/(5 - x)]

$$\frac{1}{3}x\left(x-\frac{3y}{-5+x}\right)$$

Sign[x] gives -1, 0, or 1, depending on whether x is negative, zero, or positive.

Example 38

Sign[-1.35];
-1
Sign[0.];
0
Sign[0.005]
1

Sin[z] gives the sine of z. The argument of `Sin` is assumed to be in radians.

Example 39

TraditionalForm[Sin[Pi/10]]

$$\frac{1}{4}(-1+\sqrt{5})$$

Solve[eqns, vars] attempts to solve an equation or set of equations for the variables vars. **Solve[eqns, vars, elims]** attempts to solve the equations for vars, eliminating the variables elims. This function has the following default options:

Options[Solve]= {InverseFunctions – > Automatic, MakeRules – > False, Method – > 3, Mode – > Generic, Sort – > True, VerifySolutions – > Automatic, WorkingPrecision – > ∞ }.

Example 40

ComplexExpand[Solve[x^3 - 1/3 x - 1 == 0, x]]

$$\{\{x \to \tfrac{1}{9}\left(\tfrac{729}{2}-\tfrac{135\sqrt{29}}{2}\right)^{1/3}+\tfrac{1}{3}\left(\tfrac{1}{2}(27+5\sqrt{29})\right)^{1/3}\},$$

$$\{x \to -\tfrac{1}{18}(1+i\sqrt{3})\left(\tfrac{729}{2}-\tfrac{135\sqrt{29}}{2}\right)^{1/3}-$$

$$\tfrac{1}{6}(1-i\sqrt{3})\left(\tfrac{1}{2}(27+5\sqrt{29})\right)^{1/3}\},$$

$$\{x \to -\tfrac{1}{18}(1-i\sqrt{3})\left(\tfrac{729}{2}-\tfrac{135\sqrt{29}}{2}\right)^{1/3}-$$

$$\tfrac{1}{6}(1+i\sqrt{3})\left(\tfrac{1}{2}(27+5\sqrt{29})\right)^{1/3}\}\}$$

Sqrt[**z**] or $\sqrt{z}$ gives the square root of z. *Example:* see the description of function **N[...]**.

Sum[**f, {i, imax}**] evaluates the sum of the expressions f as evaluated for each i from 1 to $imax$. **Sum[f, {i, imin, imax}]** starts with $i = imin$. **Sum[f, {i, imin, imax, di}]** uses steps di. **Sum[f, {i, imin, imax}, {j, jmin, jmax}, ...]** evaluates a sum over multiple indices.

Example 41

Simplify[Sum[$(-1)^{j+1}$ Cos[j x], {j, 4}]]//TraditionalForm
$\cos(x) - \cos(2x) + \cos(3x) - \cos(4x)$

Table[**expr, {imax}**] generates a list of imax copies of expr. **Table[expr, {i, imax}]** generates a list of the values of expr when i runs from 1 to imax. **Table[expr, {i, imin, imax}]** starts with $i = imin$. **Table[expr, {i, imin, imax, di}]** uses steps di. **Table[expr, {i, imin, imax}, {j, jmin, jmax},...]** gives a nested list. The list associated with i is outermost.

Example 42

TraditionalForm[Table[$a^i\, b^j$, $\{i,\ 3\}$, $\{j,\ 3\}$]]

$$\begin{pmatrix} ab & ab^2 & ab^3 \\ a^2b & a^2b^2 & a^2b^3 \\ a^3b & a^3b^2 & a^3b^3 \end{pmatrix}$$

Text[**expr, coords**] is a graphics primitive that represents text corresponding to the printed form of expr, centered at the point specified by coords (see also Fig. A.8).

Example 43

rc = 2; bound = Graphics[Circle[{0,0}, rc]];
arc = Graphics[Circle[{0, 0}, 0.9, {0, Pi/6}]];
lin = Graphics[Line[{{rc,0},{0,0}, {rc Cos[$\frac{\pi}{6}$], rc Sin[$\frac{\pi}{6}$]}}]];
gR = Graphics[Text["R", {0.7, 1.1}]];
gt = Graphics[Text["t", {1.3, 0.37}]];
Show[bound, arc, lin, gR, gt, AspectRatio − > Automatic]

Thickness[**r**] is a graphics directive that specifies that lines which follow are to be drawn with a thickness r. The thickness r is given as a fraction of the total width of the graph. *Example:* see the description of the *Mathematica* function `Dashing[_]`.

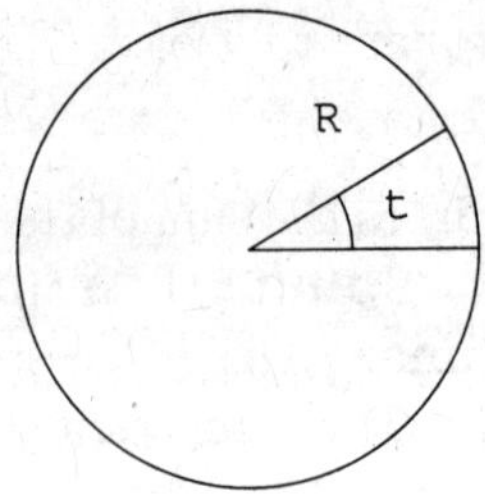

Figure A.8: Illustration to the description of the function `Text[...]`.

TraditionalForm[expr] prints as an approximation to the traditional mathematical notation for expr. *Example:* see the descriptions of the functions **TrigExpand[...]**, **Sin[...]**.

Transpose[list] transposes the first two levels in list. **Transpose[list, {n1, n2, ... }]** transposes the list so that the levels 1, 2, ... in the list correspond to levels n1, n2, ... in the result.

Example 44

A = {{a,b,c}, {1,2,3}}; TraditionalForm[A]
TraditionalForm[Transpose[A]]

$$\begin{pmatrix} a & b & c \\ 1 & 2 & 3 \end{pmatrix}$$

$$\begin{pmatrix} a & 1 \\ b & 2 \\ c & 3 \end{pmatrix}$$

TrigExpand[expr] expands out trigonometric functions in expr.

Example 45

TraditionalForm[TrigExpand[Cos[3x] Sin[4y]]]

$$4\cos^3(x)\sin(y)\cos^3(y) - 12\cos(x)\sin^2(x)\sin(y)\cos^3(y)$$
$$-4\cos^3(x)\sin^3(y)\cos(y) + 12\cos(x)\sin^2(x)\sin^3(y)\cos(y)$$

TrigReduce[expr] rewrites products and powers of trigonometric functions in expr in terms of trigonometric functions with combined arguments.

Example 46

TraditionalForm[TrigReduce[Cos[x]^5 + Sin[x]^5]]

$$\frac{1}{16}(10\cos(x) + 5\cos(3x) + \cos(5x) + 10\sin(x) - 5\sin(3x) + \sin(5x))$$

VectorHeads is an option for plotting the three-dimensional vector field with the aid of the function **PlotVectorField3D[...]**. The specification `VectorHeads` – > `True` puts heads on arrows (see Fig. A.7).

While[test, body] evaluates the test, then the body, repetitively, until the test first fails to give the `True`.

Example 47

$\xi = 0;\ y0 = 0;\ y = \{y0\};$ While[$\xi < 1,\ \xi = \xi + 0.1;\ y0 = y0 + 0.1\ \xi^2;$ AppendTo[$y,\ y0$]]; y

{0, 0.001, 0.005000000000000001, 0.014, 0.03, 0.055, 0.091, 0.14, 0.204, 0.2849999999999999, 0.3849999999999999, 0.5059999999999999}

References

1. **Ganzha, V.G. and Vorozhtsov, E.V.,** *Numerical Solutions for Partial Differential Equations: Problem Solving Using Mathematica,* CRC Press, Boca Raton, 1996.
2. **Parker, L. and Christensen, S.M.,** *MathTensor: a System for Doing Tensor Analysis by Computer,* Addison-Wesley, Reading, 1994.
3. **Kythe, P.K., Puri, P., and Schäferkotter, M.R.,** *Partial Differential Equations and Mathematica,* CRC Press, Boca Raton, 1996.
4. **Vvedensky, D.,** *Partial Differential Equations with Mathematica,* Addison-Wesley, Reading, 1993.
5. **Skeel, R.D. and Keiper, J.B.,** *Elementary Numerical Computing with Mathematica,* McGraw-Hill, Inc., New York, 1993.
6. **Baumann, G.,** *Mathematica in Theoretical Physics: Selected Examples from Classical Mechanics to Fractals,* TELOS/Springer-Verlag, New York, 1996.
7. **Martin, E. (Ed.),** *Mathematica 3.0. Standard Add-on Packages,* Wolfram Media, Cambridge University Press, Champaign, IL, 1996.

Appendix B: Glossary of Programs

As shown in Chapters 1–7, a feature of this book is the use of the advanced software system *Mathematica 3.0* to illustrate various concepts or to implement some of the analytical or numerical methods, which are presented in this book.

In this appendix, we briefly outline the purposes of all *Mathematica* notebooks, which we have made and tested on a personal computer using DOS. In order to use any specific program or the file of input data you can simply copy the needed files on your computer from the following Birkhäuser url address:

`http://www.birkhauser.com/book/isbn/0-8176-3995-0`

One of the features of *Mathematica* is that the *Mathematica* documents are called Notebooks. Notebooks can contain ordinary text and graphics as well as *Mathematica* input and output. We describe 24 *Mathematica* Notebooks in this appendix. All our Notebooks have the unified names $\texttt{prog}i - j\texttt{.nb}$, where i is the chapter number ($i = 1, 2, 3, 4, 5, 6, 7$) and j is the Notebook number within a given chapter, $j = 1, 2, \ldots$.

The information in a Notebook is stored in cells, whose characteristics vary with the cell's function and the kind of information it holds. The brackets on the right side of the Notebook window indicate the extent of each cell.

The typical structure of the Notebooks for this book is as follows. Each Notebook begins with a title, for example, **Metric Tensor Components**. The contents of a Notebook are subdivided into the following sections:

Impressum. This section contains the text as follows.
This *Mathematica* Notebook is part of the book entitled

S.P. Kiselev, E.V. Vorozhtsov, and V.M. Fomin.
Foundations of Fluid Mechanics with Applications
Problem Solving Using *Mathematica*.
Birkhäuser Boston, Basel, 1999.

General Description. This section explains the purpose of the Notebook, the method implemented in a specific program, the limitations of the program, and refers the reader to a specific section of the book, where the implemented method is described in detail.

User's Guide. This section represents an internal instruction of the Notebook for the practical use of a *Mathematica* program contained in the Notebook. This instruction can include several steps.

Program Listing. This section contains the listing of a program written in the language of *Mathematica 3.0* and usually includes the major part of the *Mathematica* program.

Parameters Used in Program *.nb. This section explains in detail the physical or mathematical meaning of each input parameter and gives the limitations on the numerical values of parameters if necessary. It should be noted that this section is absent in a number of our Notebooks, which, in particular, solve certain specific tasks of symbolic computations.

Examples of the Input Data. This section gives one or several examples of the specification of the input parameters. These examples enable the user to reproduce those analytical, numerical, or graphics examples, which were obtained with *Mathematica* in this book. At the end of this appendix, we present Tables B.13–B.14, which should help the interested reader to easily reproduce any of the figures whose numbers are indicated in the tables.

If you want to introduce some change into a file of input data, please always save our original version of the file. This will help you to find an error in your file of input data if our program fails at your data.

One of the possible errors in the process of the preparation of the file of input data is that one of the entries (divided by commas from each other) is missing, so that the total number of the entries is less than the number needed by our program. Then the *Mathematica* system shows your file of input data on the screen of the computer monitor, for example, as follows:

```
PlateFlow[1, Pi/4, -4, 4, -3, 3, 40]
```

and thus does not proceed to the computation. In this case we recommend you to compare your prepared file of input data with one of the examples of the input data in a given Notebook.

The Structure of the Output. This section explains the meaning of each entry of the data that are printed out in the process of the work of the program.

We give below the lists of our *Mathematica* Notebooks for each chapter. After the name of each Notebook file (given by bold face letters), we reproduce the meaning of each entry of the input data by using the corresponding cell of the Notebook. If the given program uses some functions, we also describe briefly the purpose of these functions (the names of the functions are printed in italics).

The limitations for the book size have not allowed us to include in this appendix the listings of our *Mathematica* Notebooks.

Chapter 1

prog1-1.nb This Notebook solves analytically Problem 1.1 (see Section 1.1.9 of this book). It enables the user to find the analytic expressions for the components of the metric tensor g_{ij} in the cylindrical coordinates r, φ, z. This program also computes the following analytic expressions:
1) the squared length element ds^2 in cylindrical coordinates;
2) the expressions for the basis vectors $\vec{e}_1, \vec{e}_2, \vec{e}_3$ of the Cartesian coordinate system in terms of the basis vectors $\vec{E}_1, \vec{E}_2, \vec{E}_3$ of the cylindrical coordinate system.

prog1-2.nb This Notebook solves analytically Problem 1.2 (see Section 1.1.9 of this book). It enables the user to find the analytic expressions for the components of the metric tensor g_{ij} in the spherical coordinates r, φ, θ. This program also finds the analytic expression for the squared length element ds^2 in spherical coordinates.

prog1-3.nb enables the user to compute the analytic expressions for the metric tensor components g_{ij}, $i, j = 1, 2$, as well as the expressions for the covariant basis vectors $\vec{e}^{\,1}$ and $\vec{e}^{\,2}$ in terms of the contravariant basis vectors $\vec{e}_1$ and $\vec{e}_2$ in the oblique-angled coordinate system (see Problem 1.3 at the end of Section 1.1.9 of this book).

prog1-4.nb enables the user to find:
1) the motion trajectories,
2) the components of velocity and acceleration,
3) the streamlines

for a given velocity field

$$v_1 = v_1(x_1), \quad v_2 = v_2(x_2), \quad v_3 = v_3(x_3),$$

where $x_1(t)$, $x_2(t)$, and $x_3(t)$ are the coordinates of a material point moving with time t in the (x_1, x_2, x_3) space. The formulation of this problem can be found in Section 1.2 (see Problem 1.4).

prog1-5.nb enables the user to find the analytic expressions for the components of the Lagrangian $\overset{o}{\varepsilon}_{ij}$ and Eulerian ε_{ij} strain tensors for a displacement field $\vec{u}(u_1, u_2, u_3)$ given in the Cartesian basis (ξ_1, ξ_2, ξ_3); that is, $u_j = u_j(\xi_1, \xi_2, \xi_3)$, $j = 1, 2, 3$. This Notebook solves Problem 1.5 (see Section 1.2 of this book).

prog1-6.nb enables the user to find the analytic expressions for the strain tensor ε and the rotation tensor ω_{ij} for the case in which the displacement vector components u_1, u_2, u_3 are given in a Cartesian basis x_1, x_2, x_3. The Notebook also computes the analytic expressions for the principal strains ε_i and the eigenvectors $\vec{n}_i$. This Notebook solves Problem 1.6 (see Section 1.2.3 of this book).

prog1-7.nb enables the user to find the analytic expressions for the vortex lines governed by the equations

$$x_1 = x_1, \quad x_2 = f_2(x_1), \quad x_3 = f_3(x_1)$$

from a velocity field given in a Cartesian basis (x_1, x_2, x_3). The coordinates $(x_1, x_2(x_1), x_3(x_1))$ of vortex lines are found by the solution of the ordinary differential equations (ODEs)

$$\frac{dx_2}{dx_1} = \frac{\omega_2}{\omega_1}, \quad \frac{dx_3}{dx_1} = \frac{\omega_3}{\omega_1},$$

where the quantities ω_1, ω_2, and ω_3 denote the components of angular velocity $\vec{\omega} = (1/2)\,\mathrm{rot}\,\vec{v}$, $\vec{v}$ is the given velocity vector. This Notebook solves Problem 1.7 (see Section 1.2.3 of this book).

prog1-8.nb is similar to the Notebook `prog1-7.nb` and differs only by another form of the functions $v_j(x_1, x_2, x_3)$, $j = 1, 2, 3$, which specify the velocity field in a Cartesian basis (x_1, x_2, x_3). This Notebook solves Problem 1.8 (see Section 1.2.3 of this book).

Chapter 2

prog2-1.nb enables the user to compute the analytic expressions for the Christoffel symbols Γ_{ij}^k in the spherical coordinates. The obtained expressions for the Γ_{ij}^k are then used in Section 2.1 of this book.

Chapter 4

prog4-1.nb implements with *Mathematica* the analytical solution (4.2.12) of the problem of a noncirculatory stationary irrotational flow of an incompressible fluid past the cylinder. The formulation of this problem and the discussion of its solution may be found in Section 4.2.1 of this book.

ψ `ψ[x_, y_]` is the function to compute the value of the stream function $\psi(x, y)$ at (x, y) point.

The main program of this Notebook is

```
NoncircFlow[R_, xl_, xr_, yb_, yt_, npp_].
```

The meaning of the input data is explained in Table B.1.

Table B.1
Parameters Used in Program prog4-1.nb

Parameter	Description
R	the cylinder radius, $R > 0$;
xl	the abscissa of the left boundary of a rectangular region D in the (x, y) plane, in which the fluid flow is to be determined; $\|xl\| \geq R$;
xr	the abscissa of the right boundary of region D, $xr > R \geq xl$;
yb	the ordinate of the lower (bottom) boundary of region D, $\|yb\| \geq R$;
yt	the ordinate of the upper boundary of region D, $\|yt\| > R$;
npp	the number of points of each streamline for the graphics function `ContourPlot`, `npp`> 10.

prog4-2.nb implements with *Mathematica* the analytical solution (4.2.14) of the problem of a circulatory stationary irrotational flow of an incompressible fluid past the cylinder for the case of one or two critical points.

The formulation of this problem and the discussion of its solution may be found in Section 4.2.1 of this book.

ψ `ψ[x_, y_]` is the function to compute the value of the stream function $\psi(x, y)$ at (x, y) point.

The main program of this Notebook is

```
CircFlow2[R_, g_, xl_, xr_, yb_, yt_, npp_]
```

The meaning of the input data is explained in Table B.2.

prog4-3.nb implements with *Mathematica* the analytical solution (4.2.14) of the problem of a circulatory stationary irrotational flow of an incompressible fluid past the cylinder for the case of a single critical point lying on the y-axis above the cylinder.

The formulation of this problem and the discussion of its solution may be found in Section 4.2.1 of this book.

ψ `ψ[x_, y_]` is the function to compute the value of the stream function $\psi(x, y)$ at (x, y) point.

The main program of this Notebook is

```
CircFlow1[R_, g_, xl_, xr_, yb_, yt_, npp_].
```

The meaning of the input data is explained in Table B.2.

prog4-4.nb implements with *Mathematica* the analytical solution (4.2.29), (4.2.30) of the problem of a circulatory planar stationary flow of an incompressible fluid past a plate.

Table B.2
Parameters Used in Program prog4-2.nb

Parameter	Description
`R`	the cylinder radius, $R > 0$;
`g`	$0 < g = \Gamma/(2\pi v_\infty) < 2R$ to ensure the existence of two critical points; $0 < g = 2R$ to ensure that the two critical points merge into one point;
`xl`	the abscissa of the left boundary of a rectangular region D in the (x, y) plane, in which the fluid flow is to be determined; $\lvert xl\rvert \geq R$;
`xr`	the abscissa of the right boundary of region D, $xr > R \geq xl$;
`yb`	the ordinate of the lower (bottom) boundary of region D, $\lvert yb\rvert \geq R$;
`yt`	the ordinate of the upper boundary of region D, $\lvert yt\rvert > R$;
`npp`	the number of points of each streamline for the graphics function `ContourPlot`, `npp`> 10.

Table B.3
Parameters Used in Program prog4-4.nb

Parameter	Description
`2a`	$2a$ is the plate length; the plate lies in the interval $-a \leq x \leq a$, $a > 0$;
`v`	the magnitude of the freestream velocity, $v > 0$;
α	the angle between the vector of freestream velocity and the positive direction of the x-axis; the value of α is specified in radians;
`xl`	the abscissa of the left boundary of a rectangular region D in the (x, y) plane, in which the fluid flow is to be determined; $\lvert xl\rvert > a$;
`xr`	the abscissa of the right boundary of D, $xr > a$;
`yb`	the ordinate of the lower (bottom) boundary of region D, $yb < 0$;
`yt`	the ordinate of the upper boundary of region D, $\lvert yt\rvert > 0$;
`npp`	the number of points of each streamline for the graphics function `ContourPlot`, `npp`> 10.

The formulation of this problem and the discussion of its solution may be found in Section 4.2.3 of this book.

w `w[z_]` is the function to compute the value of the complex potential $w(z)$ for a given value of complex variable z in accordance with formula (4.2.29).

The main program of this Notebook is

```
PlateFlow[a_, v_, α_, xl_, xr_, yb_, yt_, npp_].
```

The meaning of the input data is explained in Table B.3.

prog4-5.nb solves the problem of axially symmetric incompressible fluid flow around a body of revolution by the method of sources and sinks in the case in which the freestream velocity is directed along the body axis.

The formulation of this problem and the discussion of its solution may be found in Section 4.3.2 of this book.

body `body[z_]` is the function specifying the form of the body generatrix x = x(z) as the NACA 00t1 profile; the value t1 is specified by the program user (see below the section parameters used in program prog4-5.nb).

ψ`z_, x_]` is the approximation of the stream function with the aid of a quadrature formula.

The main program of this Notebook is

```
Axisym[v0_, tprf_, n_, npp_].
```

The meaning of the input data is explained in Table B.4.

Table B.4
Parameters Used in Program prog4-5.nb

Parameter	Description
`v0`	the magnitude of the freestream velocity, `v0`>0;
`tprf`	the relative thickness of the body profile, that is `tprf` is the ratio of the profile thickness and chord length; 0 <tprf< 1;
`n`	the number of sources and sinks; `n` is a positive integer;
`npp`	the number of points of each streamline for the graphics function `ContourPlot`, `npp`> 10.

prog4-6.nb solves the problem of three-dimensional incompressible fluid flow around a body of revolution by the method of sources

and sinks in the case in which the freestream velocity is directed along a normal to the symmetry axis of the body.

The formulation of this problem and the discussion of its solution may be found in Section 4.3.4 of this book.

body `body[z_]` is the function specifying the form of the body generatrix x = x(z) as the NACA 00t1 profile; the value t1 is specified by the program user (see below the section parameters used in program prog4-6.nb).

uxyz `uxyz[x5_, y5_, z5_]` is the function that computes the approximations by the method of sources and sinks of the velocity components in the Cartesian coordinate system in terms of the velocity components in the cylindrical coordinate system.

grlin `grlin[j_]` is the function that computes the (x, y, z) coordinates of the individual (jth) streamline and generates a graphics picture of this streamline, which is plotted subsequently on a single graphics picture along with all other streamlines. The streamline coordinates (x, y, z) are determined by numerical integration of the system of equations (4.3.19).

The main program of this Notebook is

```
ThreeDimFlow[v0_, tprf_, n_].
```

The meaning of the input data is explained in Table B.5.

Table B.5
Parameters Used in Program prog4-6.nb

Parameter	Description
`v0`	the magnitude of the freestream velocity, v0>0;
`tprf`	the relative thickness of the body profile; that is, `tprf` is the ratio of the profile thickness and chord length; 0 <tprf< 1;
`n`	the number of sources and sinks; **n** is a positive integer;
`vpts`	the number of evaluation points in each of the spatial directions x, y, z for the plot of the velocity vectors in a region around the body; **vpts** is a positive integer, **vpts**> 1.

Chapter 5

prog5-1.nb enables the user to compute the analytic expressions for the components of the velocity Laplacian $\Delta\vec{v}$ in the spherical coordinates r, φ, θ.

The discussion of underlying formulas may be found in Section 5.1 of this book.

prog5-2.nb enables the user to solve numerically the problem of a boundary layer of a flat plate. The problem under consideration is the initial-value problem (5.3.18), (5.3.19) (see Section 5.3 of this book). It is solved numerically with the aid of the standard fourth-order Runge–Kutta method.

fyz `fyz[f_, y_, z_]` is the function to compute the right-hand sides of equations (5.3.20).

The main program of this Notebook is

```
Blayer[ξ_, hmin_]
```

The meaning of the input data is explained in Table B.6.

Table B.6
Parameters Used in Program prog5-2.nb

Parameter	Description
ξmax	the abscissa of the right end of the integration interval $0 \le \xi \le \xi max$, $\xi max > 0$;
`hmin`	the start value of the integration stepsize along the ξ axis, $0 < hmin < 1$.

Chapter 6

prog6-1.nb enables the user to solve numerically the problem of a quasi-one-dimensional stationary isentropic gas flow in a variable section duct. The derivation of the underlying solution formulas may be found in Section 6.1.2 of this book.

fw `fw[x_]` computes the ordinate y of the nozzle wall in accordance with Fig. 6.5.

fm `fm[M_]` is a function constructed to solve numerically the transcendental equation (6.1.10) by the bisection method.

The main program of this Notebook is

```
nozzle[hc_, cy0_, cR0_, cR_, cx0_, cx4_, thet1_, thet2_, γ_,
        M1_, Mmax_, eps_, ix_, reg_].
```

The meaning of the input data is explained in Table B.7.

prog6-2.nb enables the user to determine the curve $V = V(x)$ in the shock transition in a viscous, non-heat-conducting compressible gas.

The formulation of this problem and the discussion of its solution may be found in Section 6.2 of this book.

The main program of this Notebook is

```
Struct[V1_, V2_, γ_, η_, xl_, xr_, ε_, np_].
```

Table B.7
Parameters Used in Program prog6-1.nb

Parameter	Description
`hc`	the nozzle wall height in the critical section of the nozzle, `hc > 0`;
`cy0`	the ordinate of the nozzle wall in the inlet section x = 0 related to `hc`;
`cR0`	the radius of the curvature of the nozzle wall in its subsonic part related to `hc`;
`cR`	the nozzle wall radius of the curvature in the critical section related to `hc`;
`cx0`	the abscissa of the critical nozzle section related to `hc`;
`cx4`	the nozzle length related to `hc`, `cx4>cx0`;
`thet1`	the inclination angle of the nozzle wall in the subsonic part of the nozzle; the angle is specified in radians;
`thet2`	the inclination angle of the nozzle wall in the supersonic part of the nozzle; the angle is specified in radians;
γ	the ratio of the gas specific heats, $\gamma > 1$;
`M1`	the numerical value of the Mach number at the duct inlet; `M1 > 0`;
`Mmax`	the solution of equation (6.1.10) is sought for in the interval `[0, Mmax], Mmax > 0]`; if `reg = 1`, then the value `Mmax > 1` should be specified;
`eps`	the user-specified accuracy of the numerical solution of equation (6.1.10) by the bisection method; `0 < eps < 1`;
`ix`	the number of uniform grid nodes in the interval $0 \leq x \leq x4$, where $x4$ is the nozzle length along the x axis; `ix` is a positive integer, `ix > 1`;
`reg`	if `reg = 0`, then the subsonic flow regime is chosen in the diverging part of the Laval nozzle; if `reg = 1`, then the supersonic flow regime is chosen in the diverging part of the Laval nozzle.

The meaning of the input data is explained in Table B.8.

prog6-3.nb enables the user to construct the solution in the centered Prandtl–Meyer rarefaction wave arising in a supersonic flow around a dihedral corner for the case in which $|\varphi_0| < |\theta_0|$.

The formulation of this problem and the discussion of its solution may be found in Section 6.4 of this book.

Table B.8
Parameters Used in Program prog6-2.nb

Parameter	Description
`V1`	the value of the gas specific volume ahead of the shock wave; $V1 > 0$;
`V2`	the value of the gas specific volume behind the shock wave; $0 < V2 < V1$;
γ	the ratio of the gas specific heats; $\gamma > 0$;
η	the nondimensionalized viscosity coefficient; $\eta > 0$;
`xl`	the abscissa of the left end of the integration interval on the x axis; $xl < 0$;
`xr`	the abscissa of the right end of the integration interval on the x axis; $xr > xl$;
ε	the accuracy of the computation of $V(x)$ by the bisection method for a given value of x;
`np`	the number of uniform grid nodes in the interval $[xl, xr]$; `np` is a positive integer;

Table B.9
Parameters Used in Program prog6-3.nb

Parameter	Description
γ	the ratio of the gas specific heats, $\gamma > 1$;
`M1r`	the right end of the interval `1, M1r]`, in which the local value $M(x, y)$ of the Mach number is computed; `M1r`> 3 for $\gamma = 1.4$;
`M10`	the freestream Mach number, `M10`≥ 1;
θ`max`	the maximum value of the angle θ, in degrees, for making the one-dimensional plots of $M(\theta), P(\theta)/P_1, \rho(\theta)/\rho_1$; θ is a positive integer, $1 < \theta < \theta_+$
`surf`	if `surf = 1` then the surface $\theta = f(M_1, M)$ is plotted; otherwise this surface is not plotted;
`xl`	the abscissa of the left end `(xl, 0)` of a dihedral corner, `xl`< 0;
`xr`	the abscissa of the right end of the right facet of the dihedral corner;
`yr`	the ordinate of the right end of the right facet of the dihedral corner, `yr`< 0;
`h0`	the number in the interval $0 <$`h0`≤ 1 to specify the color corresponding to the minimum value of the function $P(x, y)/P_1$ in the spatial region under consideration;
`ix`	the number of cells of an uniform rectangular (rough) grid in the interval `xl`$\leq x \leq$`xr`; `ix` is a positive integer, `ix`> 1;
`m1int`	the number of cells of a fine mesh in the interval `xl`$\leq x \leq$`xr`, `m1int`$>$`ix`.

θ `θ[M1_, M_]` computes the numerical value of the right-hand side of equation (6.4.42) (see Section 6.4).

cl2 `cl2[f_]` is a function to assign to a point of a color map certain color corresponding to the numerical value of the argument `f`.

fij `fij[mm_, mm1_]` is a function obtained from the system of equations (6.4.38) for the purpose of the determination of the Mach number M in the Prandtl–Meyer rarefaction wave at a given value of M_1.

The main program of this Notebook is

```
Meyer[γ_,M1r_, M10_,θmax_,surf_, xl_,xr_,yr_,h0_,ix_,m1int_].
```

The meaning of the input data is explained in Table B.9.

prog6-4.nb enables the user to construct the solution of a problem of three-dimensional supersonic inviscid compressible gas flow around a V-shaped wing.

The formulation of this problem and the discussion of its solution may be found in Section 6.5 of this book.

The main program of this Notebook is

```
Wing[β_, ω_, M1_,γ_, xl_, xr_, h0_].
```

The meaning of the input data is explained in Table B.10.

Table B.10
Parameters Used in Program prog6-4.nb

Parameter	Description
β	the angle between the shock wave and the x-axis; the angle $\beta > 0$ and is specified in radians;
ω	the angle between the upper wing facet and the z-axis, $0 < \omega < \pi/2$;
`M1`	the freestream Mach number; `M1`> 1;
γ	the ratio of the gas specific heats, $\gamma > 1$;
`xl`	the x-coordinate of the left end of the interval on the x-axis, in which the gas flow is considered; `xl`< 0 (at x$= 0$, the wing nose tip is located;
`xr`	the x-coordinate of the right end of the interval on the x-axis, in which the gas flow is considered; `xr`> 0
`h0`	the number in the interval $0 <$`h0`≤ 1 to specify the color corresponding to the minimum value of the function $P(x, y, z)$ in the spatial region under consideration.

Chapter 7

prog7-1.nb enables the user to investigate the zeroes of the characteristic equation (7.2.5) corresponding to the mathematical model of gas-particle mixture (see Section 7.2 of this book). The program finds the analytic expression for all four zeroes of equation (7.2.5) and shows the region of real roots in the (ϵ, M_{12}) plane, where $\epsilon = m_2\rho_{11}/(\rho_{22}m_1)$.

A more detailed discussion of the properties of the zeroes of equation (7.2.5) may be found in Section 7.2 of this book.

ff `f[x_, e_, m_]` computes the left-hand side of equation (7.2.5).

The main program of this Notebook is

```
gaspart[nroot_, np_, npict_, epsmax_, M12_, neps_, nM12_].
```

The meaning of the input data is explained in Table B.11.

Table B.11
Parameters Used in Program prog7-1.nb

Parameter	Description
`nroot`	the number of the root of equation (7.2.5); `nroot` is a positive integer, $1 \leq$`nroot`≤ 4;
`np`	if `np = 1`, then the analytical expression for the root λ_{nroot} is printed;
`npict`	if `npict = 1`, then the picture of the region of real roots λ_j, $j = 1, 2, 3, 4$, is plotted;
`epsmax`	the size of the region in the (ϵ, M_{12}) plane along the ϵ-axis, where $\epsilon = m_2\rho_{11}/(\rho_{22}m_1)$; `epsmax`$> 0$;
`M12`	the size of the region in the (ϵ, M_{12}) plane along the M_{12}-axis; `M12`> 0;
`neps`	the number of nodes of an uniform grid along the ϵ-axis; `neps` is a positive integer;
`nM12`	the number of nodes of an uniform grid along the M_{12}-axis; `nM12` is a positive integer.

prog7-2.nb enables the user to integrate numerically the Burgers–Korteweg–de Vries equation (BKdV equation) at different constant values of the equation coefficients specified by the user. This equation is used as a model of nonlinear wave processes in bubbly liquids.

A more detailed discussion of the Burgers–Korteweg–de Vries equation may be found in Section 7.7 of this book.

P0 `P0[ζ_]` computes the value of the function $P_0(\zeta)$ specifying the initial condition for the BKdV equation at a given point ζ.

The main program of this Notebook is

```
BKdV[Re_, dsig_, a_, M_, thet_, Nt_, cf0_].
```

The meaning of the input data is explained in Table B.12.

Table B.12
Parameters Used in Program prog7-2.nb

Parameter	Description
`Re`	the positive similarity parameter Re in the BKdV equation;
`dsig`	the positive similarity parameter D_σ in the BKdV equation;
`a`	the right end of the integration interval $0 \le \zeta \le a$ on the ζ-axis; $a > 0$;
`M`	the number of the uniform grid nodes in the interval $0 \le \zeta \le a$; `M` is a positive integer, `M`> 3;
`thet`	the safety factor in a formula for computation of time step from the stability condition; 0 ≤`thet`≤ 1;
`Nt`	the number of time steps to be computed; `Nt` is a positive integer, `Nt`> 2;
`cf0`	the value `cf0*h` determines the abscissa of the discontinuity in the initial distribution $P(\zeta, 0)$ having a form of a step; `cf0` is a positive integer in the case of a step form initial condition, 4 <`cf0`<`M`; if `cf0`≤ 0, the bell-shaped initial condition $P(\zeta, 0) = e^{-(\zeta-2)^2} + \lvert cf0 \rvert e^{-(\zeta-6)^2}$ is specified.

Appendix A

appendA.nb contains the examples of small *Mathematica* programs, which show how the various built-in *Mathematica* functions are used. The syntax of the functions is given in Appendix A.

A number of figures presented in this book were obtained with the aid of the above-listed programs. For the convenience of the reader, we present in Tables B.13 and B.14 the names of the *Mathematica* source files and the examples of input data, which were used by us to obtain a number of the figures in the book. The left column of each table shows the number of the book section, in which a specific figure is located.

By using these tables, the interested reader can easily reproduce the figures whose numbers are indicated in Tables B.13 and B.14.

Table B.13
The *Mathematica* Source Files and the Files of the Input Data Needed for Obtaining the Figures of Chapters 1, 4, and 5

Section	Figure number	Program file	Example of input data
1.2	Fig. 1.13	`prog1-4.nb`	`Example 1`
4.2	Fig. 4.9	`prog4-1.nb`	`Example 1`
4.2	Fig. 4.10	`prog4-2.nb`	`Example 1`
4.2	Fig. 4.11	`prog4-2.nb`	`Example 2`
4.2	Fig. 4.12	`prog4-3.nb`	`Example 1`
4.2	Fig. 4.16	`prog4-4.nb`	`Example 1`
4.3	Fig. 4.20	`prog4-5.nb`	`Example 1`
4.3	Fig. 4.21	`prog4-6.nb`	`Example 1`
4.3	Fig. 4.22	`prog4-6.nb`	`Example 1`
5.3	Fig. 5.35	`prog5-2.nb`	`Example 1`

Table B.14
The *Mathematica* Source Files and the Files of the Input Data Needed for Obtaining the Figures of Chapters 6–7

Section	Figure number	Program file	Example of input data
6.1	Fig. 6.6,a,b,c	`prog6-1.nb`	`Example 1`
6.1	Fig. 6.6,d,e	`prog6-1.nb`	`Example 2`
6.1	Fig. 6.13	`prog6-2.nb`	`Example 1`
6.4	Fig. 6.30	`prog6-3.nb`	`Example 1`
6.4	Fig. 6.31	`prog6-3.nb`	`Example 1`
6.5	Fig. 6.44	`prog6-4.nb`	`Example 1`
6.5	Fig. 6.45	`prog6-4.nb`	`Example 1`
7.2	Fig. 7.4	`prog7-1.nb`	`Examples 3,4`
7.7	Fig. 7.50	`prog7-2.nb`	`Example 1.1`
7.7	Fig. 7.51	`prog7-2.nb`	`Example 1.3`
7.7	Fig. 7.52	`prog7-2.nb`	`Example 3.1`

Index